This is a volume in the Arno Press collection

HISTORY OF ECOLOGY

Advisory Editor
Frank N. Egerton III

Editorial Board
John F. Lussenhop
Robert P. McIntosh

*See last pages of this volume for a
complete list of titles.*

THE LIFE FORMS OF PLANTS
AND
STATISTICAL PLANT GEOGRAPHY

LIFE FORMS OF PLANTS
AND
STATISTICAL PLANT GEOGRAPHY

Christian Raunkiaer

ARNO PRESS
A New York Times Company
New York / 1977

QK
3
.R24
1977

Editorial Supervision: LUCILLE MAIORCA

Reprint Edition 1977 by Arno Press Inc.

This reprint has been authorized by
   the Clarendon Press Oxford

Reprinted from a copy in
   The University of Michigan Library

HISTORY OF ECOLOGY   # 3071581
ISBN for complete set: 0-405-10369-7
See last pages of this volume for titles.

Manufactured in the United States of America

Library of Congress Cataloging in Publication Data

Raunkiaer, Christen, 1860-1938.
   The life forms of plants and statistical plant
geography.

   (History of ecology)
   Reprint of the 1934 ed. published by Clarendon
Press, Oxford.
   Includes bibliographies.
   1. Botany--Collected works. 2. Botany--
Ecology--Collected works. 3. Phytogeography--
Statistical methods--Collected works. I. Title.
II. Series.
QK3.R24 1977   581'.08   77-74249
ISBN 0-405-10418-9

THE
LIFE FORMS OF PLANTS
AND
STATISTICAL PLANT GEOGRAPHY

THE
LIFE FORMS OF PLANTS
AND
STATISTICAL PLANT GEOGRAPHY

BEING

THE COLLECTED PAPERS OF

C. RAUNKIAER

*WITH 189 PHOTOGRAPHS
AND FIGURES*

OXFORD
AT THE CLARENDON PRESS
1934

OXFORD UNIVERSITY PRESS
AMEN HOUSE, E.C. 4
LONDON EDINBURGH GLASGOW
LEIPZIG NEW YORK TORONTO
MELBOURNE CAPETOWN BOMBAY
CALCUTTA MADRAS SHANGHAI
HUMPHREY MILFORD
PUBLISHER TO THE
UNIVERSITY

PRINTED IN GREAT BRITAIN

PREFATORY NOTE

THE works of Professor Raunkiaer which here appear in English have all but one (the last) been published before in Danish periodicals or as books. Most of them were written in Danish; two were in French, and one in German. The English translations are here arranged chronologically in order to show the development of Raunkiaer's system of life-forms and its applications. Parts of some of the original papers have been omitted because they are repetitions of former writings.

The works in Danish have been translated by Mr. H. Gilbert-Carter of Cambridge, with the exception of the last (unpublished) paper, which was translated by Miss A. Fausbøll. The translation of the three papers written in French and German has been done by Professor A. G. Tansley of Oxford.

The Rask-Oersted Fund has contributed to the cost of publication.

Raunkiaer's works on Plant Geography, because of difficulties of language, and because they were published in different periodicals, have not awakened the attention they deserve. The undersigned hope that a complete English edition of the works will be received with interest.

    K. GRAM   JOH. GRØNTVED
    H. MOLHOLM HANSEN   C. H. OSTENFELD (*deceased*)
    OVE PAULSEN

COPENHAGEN
1932

CONTENTS

PREFATORY NOTE — v

LIST OF PLATES — viii

INTRODUCTION — xi

I. BIOLOGICAL TYPES WITH REFERENCE TO THE ADAPTATION OF PLANTS TO SURVIVE THE UNFAVOURABLE SEASON — 1
(Om biologiske Typer, med Hensyn til Planternes Tilpasning til at overleve ugunstige Aarstider, *Botanisk Tidsskrift*, xxvi, 1904)

II. THE LIFE-FORMS OF PLANTS AND THEIR BEARING ON GEOGRAPHY (Figs. 1–77) — 2
(*Planterigets Livsformer og deres Betydning for Geografien*, København, 1907)

III. THE LIFE-FORM OF *TUSSILAGO FARFARUS* — 105
(Om Livsformen hos *Tussilago farfarus*. Et lille Bidrag til Følfodens Naturhistorie, *Botanisk Tidsskrift*, xxviii, 1907)

IV. THE STATISTICS OF LIFE-FORMS AS A BASIS FOR BIOLOGICAL PLANT GEOGRAPHY — 111
(Livsformernes Statistik som Grundlag for biologisk Plantegeografi, *Botanisk Tidsskrift*, xxix, 1908)

V. THE LIFE-FORMS OF PLANTS ON NEW SOIL (Figs. 78–104) — 148
(Livsformen hos Planter paa ny Jord, *Kgl. Danske Vidensk. Selsk. Skrifter*, vii. 8, 1909)

VI. INVESTIGATIONS AND STATISTICS OF PLANT FORMATIONS (Figs. 105–24) — 201
(Formationsundersøgelse og Formationsstatistik, *Botanisk Tidsskrift*, xxx, 1909–10)

VII. THE ARCTIC AND ANTARCTIC CHAMAEPHYTE CLIMATE (Fig. 125) — 283
(Det arktiske og det antarktiske Chamæfytklima, *Biologiske Arbejder tilegnede Eug. Warming*, 1911)

VIII. STATISTICAL INVESTIGATIONS OF THE PLANT FORMATIONS OF SKAGENS ODDE (THE SKAW) (Figs. 126–31) — 303
(Formationsstatistiske Undersøgelser paa Skagens Odde, *Botanisk Tidsskrift*, xxxiii, 1913)

IX. ON THE VEGETATION OF THE FRENCH MEDITERRANEAN ALLUVIA (Figs. 132–8) — 343
(Sur la végétation des alluvions méditerranéennes françaises, *Mindeskrift for Japetus Steenstrup*, nr. xxxiii, 1914)

CONTENTS

X. THE USE OF LEAF SIZE IN BIOLOGICAL PLANT GEOGRAPHY (Fig. 139) 368
(Om Bladstørrelsens Anvendelse i den biologiske Plantegeografi, *Botanisk Tidsskrift*, xxxiv, 1916)

XI. STATISTICAL RESEARCHES ON PLANT FORMATIONS (Figs. 140–2) 379
(Recherches statistiques sur les formations végétales, *Kgl. Danske Videnskabernes Selskab. Biologiske Meddelelser*, i. 3, 1918)

XII. ON THE BIOLOGICAL NORMAL SPECTRUM 425
(Über das biologische Normalspektrum, *Kgl. Danske Videnskabernes Selskab. Biologiske Meddelelser*, i. 4, 1918)

XIII. ON THE SIGNIFICANCE OF CRYPTOGAMS FOR CHARACTERIZING PLANT CLIMATES 435
(Om Kryptogamernes Betydning for Karakteriseringen af Planteklimaterne, *Botanisk Tidsskrift*, xxxvii, 1920)

XIV. THE DIFFERENT INFLUENCE EXERCISED BY VARIOUS TYPES OF VEGETATION ON THE 'DEGREE OF ACIDITY (HYDROGEN-ION CONCENTRATION) OF THE SOIL (Fig. 143) 443
(Forskellige Vegetationstypers forskellige Indflydelse paa Jordbundens Surhedsgrad (Brintionkoncentration), *Kgl. Danske Videnskabernes Selskab. Biologiske Meddelelser*, iii, 10, 1922)

XV. THE NITRATE CONTENT OF *ANEMONE NEMOROSA* GROWING IN VARIOUS LOCALITIES 488
(Nitratindholdet hos *Anemone nemorosa* paa forskellige Standpladser, *Kgl. Danske Videnskabernes Selskab. Biologiske Meddelelser*, v. 5, 1926)

XVI. THE AREA OF DOMINANCE, SPECIES DENSITY, AND FORMATION DOMINANTS (Figs. 144–6) 517
(Dominansareal, Artstæthed og Formationsdominanter, *Kgl. Danske Videnskabernes Selskab. Biologiske Meddelelser*, vii. 1, 1928)

XVII. BOTANICAL STUDIES IN THE MEDITERRANEAN REGION (Figs. 147–89) (hitherto unpublished) 547

INDEX 621

LIST OF PLATES

(The figures omitted from this list are in the text.)

Frontispiece

Fig. 78. Coral Bay on St. Jan, West Indies	*facing page* 148
Fig. 79. The east end of St. Jan	,, 149
Fig. 80. The west coast of Løvenlund Bay, St. Thomas, West Indies	,, 150
Fig. 81. The east coast of Magens Bay, St. Thomas	,, 150
Fig. 82. Sandy beach at the head of Magens Bay, St. Thomas	,, 152
Fig. 83. Ditto	,, 153
Fig. 84. The west coast of St. Croix, West Indies	,, 153
Fig. 85. The east end of St. Jan. The Mangrove-Formation	,, 154
Fig. 86. Ditto	,, 154
Fig. 87. Great Cruz Bay at the west end of St. Jan. Mangrove	,, 155
Fig. 88. The East end of St. Thomas	,, 155
Fig. 90. Sand-bank bordering Krauses Lagoon, St. Croix	,, 160
Fig. 91. The north side of Krauses Lagoon	,, 160
Fig. 92. Destroyed *Avicennia* wood, Krauses Lagoon	,, 164
Fig. 93. Sand-bank, Krauses Lagoon	,, 164
Fig. 95. Westend Salt Pond on St. Croix	,, 172
Fig. 96. South end of Westend Salt Pond on St. Croix	,, 172
Fig. 97. Sandy Point on St. Croix: *Pescaprae*-Formation	,, 174
Fig. 98. Western corner of Sandy Point on St. Croix	,, 174
Fig. 99. South coast of Sandy Point on St. Croix	,, 175
Fig. 100. South-east coast of Sandy Point on St. Croix	,, 175
Fig. 101. Ditto	,, 176
Fig. 102. The scrub—*Coccoloba*-Formation—on Sandy Point on St. Croix	,, 176
Fig. 103. Ditto	,, 177
Fig. 104. Ditto	,, 177
Fig. 106. 'Maaløv Krat.' Annual grassfield in June	,, 226
Fig. 107. *Anemone nemorosa*-Facies in beechwood; Jonstrup Vang	,, 226
Fig. 108. *Allium ursinum*-Facies in beechwood; Bognaes Wood	,, 233
Fig. 109. *Oxalis acetosella*-Facies, sprucewood near Jonstrup Vang	,, 233
Fig. 110. A little bog in Maaløv Krat dug out and filled with water	,, 247
Fig. 111. Lyngby Bog. *Calluna–Oxycoccus*-Facies	,, 247
Fig. 112. Skrædermose in Jonstrup Vang	,, 253
Fig. 113. Ditto	,, 254
Fig. 114. Alder swamp in Jonstrup Vang	,, 255
Fig. 116. Snogeeng in Jonstrup Vang, Sedge Meadow	,, 262
Fig. 117. Ditto	,, 263
Fig. 118. Ting Heath with knolls of Ting in the background	,, 264
Fig. 119. Knude Heath: *Erica tetralix*-Formation	,, 265

LIST OF PLATES

Fig. 120. Ting Heath: *Calluna*-Formation	facing page	270
Fig. 121. Fanø. Salt marsh south of Nordby: *Salicornia*-Formation	,,	274
Fig. 122. Fanø. Salt marsh south of Nordby at low tide	,,	274
Fig. 123. Fanø. Salt marsh south of Nordby: *Glyceria*-Formation	,,	276
Fig. 124. Nymindegab. Damp saline sand in process of being overgrown	,,	276
Fig. 126. Heath-Plain between Skagen and Højen	,,	321
Fig. 127. Low dunes covered with small *Salix repens*.	,,	321
Fig. 128. North side of Engklit north of Højen station	,,	334
Fig. 129. Meadow dune north of Højen station	,,	334
Fig. 130. Ditto	,,	335
Fig. 131. *Salix repens*- and *Armeria vulgaris*-Formations south of Højen	,,	335
Fig. 132. Dunes to the west of Palavas.	,,	356
Fig. 133. Dunes to the east of Palavas.	,,	356
Fig. 134. Low dunes near Cette	,,	357
Fig. 135. Low dunes along the mouth of the Petit-Rhône	,,	357
Fig. 136. Beach to the east of Saintes-Maries.	,,	366
Fig. 137. Zone immediately interior to that represented in Fig. 136	,,	366
Fig. 138. Formation of *Salicornia glauca* west of Saintes-Maries	,,	367
Fig. 139. Diagram for determination of the size class of a leaf	,,	371
Fig. 154. The dune area south of Marina di Pisa	,,	584
Fig. 155. Ditto	,,	584
Fig. 156. Ditto	,,	585
Fig. 157. Tenuto del Tombolo, west of Tombolo railway station	,,	585
Fig. 158. Near Fig. 157.	,,	586
Fig. 159. Near Figs. 157 and 158	,,	586
Fig. 160. Tombolo della Gianella	,,	588
Fig. 161. Ditto	,,	588
Fig. 162. The north-eastern end of Tombolo della Gianella.	,,	590
Fig. 163. Tombolo di Feniglia	,,	590
Fig. 164. South coast of Tombolo di Feniglia, looking west.	,,	591
Fig. 165. South coast of Tombolo di Feniglia, looking east	,,	591
Fig. 166. Tombolo di Feniglia: partly denuded shifting dune	,,	595
Fig. 167. South side of Tombolo di Feniglia	,,	595
Fig. 168. Tombolo di Feniglia: dune clad with *Clematis*	,,	596
Fig. 169. Tombolo di Feniglia: dunes	,,	596
Fig. 170. Tombolo di Feniglia. View of Stagno di Orbetello	,,	597
Fig. 171. Tombolo di Feniglia, looking north-west	,,	597
Fig. 172. Tombolo di Feniglia: low dunes with *Euphorbia Paralias*, &c.	,,	596
Fig. 173. Tombolo di Feniglia: big dune burying the Maremma wood	,,	596
Fig. 174. Tombolo di Feniglia: maquis being choked by the shifting dune	,,	597
Fig. 175. Tombolo di Feniglia: high shifting dune burying the maquis	,,	597
Fig. 176. Tombolo di Feniglia. *Populus-Ulmus* wood partly buried in dune	,,	596
Fig. 177. Tombolo di Feniglia: northern border of a large shifting dune	,,	596
Fig. 178. Tombolo di Feniglia: border of the easternmost shifting dune	,,	597

LIST OF PLATES

FIG. 179. The coast between Ansedonia and Capalbio . . . *facing page* 597
FIG. 180. Between Ansedonia and Capalbio: seaside dune with *Juniperus* scrub „ 600
FIG. 181. North of Fiumicino: beach with wreckage, then colonizing vegetation „ 600
FIG. 182. North of Fiumicino, looking north . . . . . „ 601
FIG. 183. Ditto . . . . „ 601
FIG. 184. North of Fiumicino: Maremma maquis . . . „ 602
FIG. 185. West of Terracina, looking west . . . . „ 602
FIG. 186. South coast of the Pontine Swamps. . . . „ 604
FIG. 187. Near Fig. 186: shifting dunes burying the maquis. . . „ 604
FIG. 188. Near Figs. 186 and 187: wood and maquis cleared . . „ 605
FIG. 189. Seaside dune west of Lago di Fusaro . . . „ 605

INTRODUCTION

Though we know neither what life is, nor whence the first germs of life arose, yet there is no reasonable doubt that all the myriads of diversely fashioned plants and animals inhabiting the world to-day are the result of the process of evolution.

In the absence of certain knowledge opinions are divided about the mechanism which brings new species into being.

Some investigators think that new species arise congenitally, that is to say, that individuals arise which differ from their parents in potentialities and properties, and that some of these differences, which are demonstrably hereditary, are handed on to their offspring, making the primordia of a new species. This, known as the Darwinian Theory, has recently been amplified and placed upon an experimental basis chiefly by the Dutch botanist Hugo de Vries, who calls the forms which show hereditary deviations mutations. His theory is called the mutation theory. It can scarcely be doubted that new species can arise in this way; but the causes of the mutations are unknown.

Others think that it is the environment that makes new species; that individuals arise having identical structure and propensities, and that these individuals growing and developing in different environments come to differ from each other in form and structure. As a matter of fact these differences are almost certainly not hereditary in the ordinary sense of the word, and do not occur again in the offspring which grow in an environment differing from the one that brought about their existence, so that they cannot be looked upon as the beginnings of new species. But even if the difference brought about by the environment, i.e. the acquired characters, is not directly transmissible, yet the possibility is by no means excluded that a gradual alteration takes place in the individuals, so that in the course of many generations of growth in a different environment they gradually come to inherit the peculiarities impressed on the individuals by this environment when the environment which caused these peculiarities is no longer present. Thus one can imagine a species gradually becoming split into several species by growth of individuals during long periods in environments differing widely from the original one. There is much in nature which makes it difficult for us to doubt that this is one of the sources of new species; but exact scientific proof can only be obtained by means of cultural experiments carried out during decades or even centuries, and such experiments can scarcely be carried out by an individual investigator, but need Phylogenetic Institutions guaranteed by the State. As yet no such institutions exist.

Others believe that new species arise by hybridization, and we can scarcely doubt that this may take place. These three possible modes of recent elaboration by Dr. Braun-Blanquet is further testimony to its general acceptance and usefulness. Though the system has been so widely used it is doubtful if many of the users have read all or nearly all that Raunkiaer has written on the subject: some have certainly contented themselves with the convenient partial summaries to which I have referred.

The seventeen chapters of this book, the majority of which, though by no means all, relate to life-forms and their distribution, are arranged in the chronological order of the original publications, thus enabling the reader to follow step by step the development of the author's investigations, though a certain amount of repetition is necessarily entailed by this mode of publication. The full titles and sources of the original papers are given in the Table of Contents.

Chapter I is of great historical interest because it consists of the brief record of Raunkiaer's original communication to the Danish Botanical Society in December 1903. In this he established the various 'biological types' of higher plants, afterwards called 'life-forms', on the basis of the protection afforded to the perennating buds or shoot-apices by their position in relation to the surface of the soil during the unfavourable season. Four years later, in 1907, Raunkiaer published the book *Planterigets Livsformer og deres Betydning for Geografien* (here translated as Chapter II), which is the fundamental exposition of his System of Life-Forms. This admirable work is based upon an analysis of the yearly incidence of the season or seasons unfavourable to plant life, illustrated for different regions by 'hydrotherm charts', which show graphically the courses of the curves of precipitation and temperature during the year. The detailed description of the classes of life-forms and their subdivisions is illustrated by a series of 70 excellent drawings from nature, mostly by Mrs. Raunkiaer. Finally the widely different percentages of the various life-forms represented among all the species of the separate floras of different climatic regions are illustrated by a comparison of these percentages in Denmark and in the Danish West Indies. In the same year was published a little study (Chapter III) of the life-form of the common Coltsfoot (*Tussilago farfarus*), showing the care which has sometimes to be exercised in determining the life-form of a species, and the possibility of the same species differing in life-form from one climate to another.

In the following year (1908) Raunkiaer published a systematic treatment (Chapter IV) of the different 'plant climates' and the percentages of the different life-forms which characterize them. Here he introduces the conceptions of the 'biological spectrum' and of the 'biochore'. The former, which is simply the list of percentages of the different life-forms

INTRODUCTION

represented in a given region, has since become very familiar to phytogeographers. The latter, originally Köppen's term and analogous to the climatological terms isotherm and isohyet, is the line passing through all the areas whose floras show the same percentage number of a characteristic life-form: biochores can only be determined by the analysis of a large number of adjacent local floras. A great part of this chapter is devoted to an analysis, in terms of biological spectra and biochores, of the vegetation of arctic and subarctic regions, and of the higher altitudinal zones of mountain ranges, both of which are characterized by a preponderance of the life-form type known as the Chamaephyte.

The next year (1909) there appeared a study of the life-forms of plants on 'new' soil, i.e. of the first colonists of ground newly exposed to invasion by plant life (Chapter V). This is based on a detailed investigation of parts of the alluvial coasts of the Danish West Indian islands, illustrated by a series of photographs, and on a comparison with the vegetation of the alluvial coasts of Denmark. It is shown that the life-forms of such pioneer colonists are correlated with climate just as are those of the more mature vegetation.

Shortly afterwards (1909–10) Raunkiaer published an important contribution on a different topic, one with which his name has also become closely associated. This is the quantitative and statistical analysis of 'plant formations'. In Chapter VI, which is illustrated by a number of excellent photographs of the Danish formations investigated, he aims at establishing an objective method of determining the 'valency' or quantitative status of the different species of a given plant community, free from the uncertainty inherent in all subjective estimates. The paper of which this chapter is a translation formed the starting-point for a great deal of modern work on this subject, which is, however, still in a state of considerable confusion, partly owing to the failure of many workers to formulate a sufficiently clear and comprehensive definition of their real objectives, partly to lack of an adequate knowledge of statistical theory, and partly perhaps to inadequate development of those portions of statistical theory which are applicable to these problems.

In 1911 (Chapter VII) Raunkiaer returns to the relation of life-form to climate in a further study of the distribution of life-forms in the Arctic and Antarctic Chamaephyte climates, already dealt with in Chapter IV. Here he shows that the two polar regions correspond broadly, though they have certain differences, and that in the severest climates the formation-dominants as well as most (or in extreme cases all) of the species are Chamaephytes.

Chapter VIII (1913) contains a detailed study of the vegetation of the northern point of Jutland, known to the English as the Skaw, with its coastal vegetation on sand and mud and its wide stretches of heath. In this the author makes use of the life-form system in combination with his system of calculating 'valency' by means of 'points', as worked out in Chapter VI. A few photographs illustrate some of the Skaw vegetation.

Chapter IX (1914) deals with the vegetation of the French Mediterranean coastal alluvia, as seen on the sea-shores of the Rhone delta. Here again use is made of life-form in relation to climate, and also of the analysis of 'formations' by the valency method. Different methods of judging vegetation are compared, and the biochore separating the Mediterranean Therophyte climate from the Central European Hemicryptophyte climate is traced. This chapter again is illustrated by photographs of the coastal vegetation.

In Chapter X (1916) new ground is broken with a consideration of the use of leaf-size in studying vegetation. Leaves are grouped into six classes according to their area, the dividing-lines between the higher classes being exact multiples of the figure chosen to separate the lowest class from the one next above. Each class is given a distinctive name. By using this method the life-form groups, as well as the dominants and other species of the 'formations', can be analysed, and the results treated statistically. These results are shown to be of considerable significance, and the method has in fact been used by subsequent workers (for example by Adamson in his studies of the vegetation of the Cape Peninsula), though not so much as it might have been.

Chapter XI (1918) returns to a consideration of the statistics of the distribution of species, in and between plant-formations, and carries the analysis, initiated in Chapter VI, a good deal farther, introducing a number of new conceptions and definitions, including the 'areal percentage' or 'degree of cover' of individual species. An ingenious arrangement is described for delimiting and dividing a circular area of one-tenth of a square metre by means of radial arms attached to a stick thrust into the ground, so that the areal percentages of different species can be rapidly estimated.

In the same year (Chapter XII) the 'normal spectrum', i.e. the percentages of the different life-forms in the flora of the whole world, which can be used as a standard of comparison for the spectra of the different phytoclimatic regions, a conception introduced in 1908 (Chapter IV), is worked out on a wider basis.

In Chapter XIII (1920) the Pteridophytes are for the first time introduced into the comparison of the floras of different regions, by means of the 'Pteridophyte Quotient', which is a measure of the ratio of Pteridophytes to Phanerogams in any given flora in terms of the corresponding ratio in the whole world. In the same way the Moss, Liverwort, and Lichen quotients can be obtained and used for comparing regional floras.

Chapters XIV (1922) and XV (1926) deal with quite different topics: the effects of different types of vegetation in Denmark on the hydrogen-ion concentration of the soil, and the nitrate contents of the wood anemone growing in different localities. These two chapters contain a large amount of valuable ecological data, and in both the author is able to arrive at general conclusions bearing on the theory of plant-formations.

In Chapter XVI (1928) a number of difficult questions relating to the density and dominance of species are dealt with. The material of this chapter will have to be taken into account in any future statistical theories and practical field methods which may enable students of the subject to deal satisfactorily with the analysis and comparison of plant-communities.

The concluding chapter (XVII) deals first with various aspects of the Mediterranean Therophyte climate, and includes a very attractive account, illustrated by a large number of photographs, of the vegetation of parts of the west coast of Italy, especially of the coastal sand-dunes and the adjoining vegetation, comparing the life-forms of these Mediterranean sandy coasts with those of the corresponding regions in the Hemicryptophyte climate of Jutland.

This brief indication of the contents of the present volume will give the reader who is unacquainted with Raunkiaer's works at first hand some idea of the breadth of his interest in the problems of plant geography, of the penetrating and fundamental character of his researches, of the ingenuity and at the same time the practical nature of his conceptions and devices, and finally of the laborious patience with which the most complex phenomena are worked out. But it is necessary to follow the various investigations in detail to gain an adequate appreciation of the high degree in which all these qualities and powers are displayed. The author's deep love of nature, also, is vividly brought out by occasional passages in which he describes the aesthetic effect of the Danish heath or woodland, or of the Mediterranean maquis.

The work of translation has not been without its difficulties; but it is hoped that most of them have been satisfactorily solved. Mr. Gilbert-Carter, who translated most of the papers, has taken infinite pains to secure the best renderings of Danish expressions. The Latin names of plants have been given in the forms in which they appear in the original publications, though these are not always uniform as between the earlier and the later, and the spellings of the names are not always those most familiar to English readers. Most of the lists of literature referred to are printed at the ends of the chapters as they appear in the originals, though this involves a good deal of repetition. Only where all the references are to the author's own works which appear as chapters of the book are the bibliographical lists omitted, the references to the proper chapters appearing in the text. The Index has been constructed by Miss J. E. Salzman.

In commending the volume to a wide circle of readers, I must express my sense of the privilege it has been to have had a share in presenting to the world this remarkable monument of the life-work of one of the greatest of the minds which have concerned themselves with the deeper problems of plant geography during the first three decades of the present century.

A. G. TANSLEY

OXFORD
March, 1934

I
BIOLOGICAL TYPES WITH REFERENCE TO THE ADAPTATION OF PLANTS TO SURVIVE THE UNFAVOURABLE SEASON

December 5th, 1903

DR. RAUNKIAER gave a contribution on Biological Types and how they are adapted to survive the unfavourable season. He characterized the types by the amount and kind of protection afforded to the buds and shoot-apices. Dr. Raunkiaer established the following types:

I. *Phanerophytes.* The surviving buds or shoot-apices are borne on negatively geotropic shoots which project into the air.
    1. Evergreen Phanerophytes without bud-covering.
    2. Evergreen Phanerophytes with bud-covering.
    3. Deciduous Phanerophytes with bud-covering.
    4. Nanophanerophytes.

II. *Chamaephytes.* The surviving buds or shoot-apices are borne on shoots very close to the ground.
    5. Suffruticose Chamaephytes. The aerial shoots are erect and negatively geotropic; at the beginning of the unfavourable season they die back to the portion, of varying length, that bears the surviving buds.
    6. Passive Chamaephytes. The shoots are persistent and negatively geotropic but they are not furnished with sufficient strengthening tissue to keep them erect. They are therefore procumbent.
    7. Active Chamaephytes. The shoots are persistent and transversely geotropic in light, and for this reason they are procumbent.
    8. Cushion Plants.

III. *Hemicryptophytes.* The surviving buds or shoot-apices are situated in the soil-surface.
    9. Protohemicryptophytes. From the base upwards the aerial shoots have elongated internodes and bear foliage leaves; the lowermost leaves are less perfectly developed than the other ones.
    10. Partial Rosette Plants. The internodes near the base of the shoot are short and it is in this region that the most and the largest foliage leaves are borne. Upwards the internodes elongate, the leaves are fewer, and flowers are borne.
    11. Rosette Plants. The shoots are contracted at the base, where all the foliage leaves are borne. The elongated aerial shoot bears only flowers.

IV. *Cryptophytes.* The surviving buds or shoot-apices are buried in the ground at a distance from the surface that varies in the different species.
    12. Geocryptophytes or Geophytes.
        a. Rhizome Geophytes. *b.* Bulb Geophytes. *c.* Stem Tuber Geophytes. *d.* Root Tuber Geophytes.
    13. Limnocryptophytes or Limnophytes (Marsh Plants).[1]
    14. Hydrocryptophytes or Hydrophytes.

V. *Annuals or Therophytes.* Plants of the summer or of the favourable season.

[1] Now called Helophytes.—*Editor's note.*

II
THE LIFE-FORMS OF PLANTS AND THEIR BEARING ON GEOGRAPHY

PREFACE

In this little book I have attempted to classify and describe the life-forms of plants as a basis of biological Plant Geography by using in my definitions of the life-forms only those structural characters that reflect the essential dependence of the plants upon climate, i.e. their adaptation to survive the unfavourable season. Here then Plant Geography as botanical science gives place to Plant Geography as geographical science. We shall consider vegetation as an expression of the climate, and life-forms of plants as a means of determining the biological characteristics of the different climates.

As long ago as 1903 I made a statement of this general theme and later elaborated it in *Videnskabernes Selskabs Oversigt* for 1905. This account is written in French and is not easily accessible, and as I consider that others besides professional botanists might perhaps be interested in my views, I have here attempted to give a more popular account of them written in Danish and illustrated with figures, some of which are from my book *De danske Blomsterplanters Naturhistorie*, Vol. I, and some of which are new. Most of the illustrations have been drawn by my wife, Mrs. Ingeborg Raunkiaer.

The information about the structure of the individual plants is in part already known, especially from the works of Irmisch and Warming, and in part is the result of my own investigations; but I did not think it necessary to overburden the descriptions by mentioning which facts are new and which were already known, especially since the morphological and the physiological relationship have one and only one significance in illustrating my principal theme, which is a coherent sketch of life-forms accurately defined and the relationship they bear to the climate.

My thanks are due to the Committee of the Carlsberg fund. It is their assistance that has made it possible for this work, in spite of the probability of a small circle of readers, to appear as a book.

CONTENTS

INTRODUCTION. The Plant's Environment and the Plant's Demand. Climate and Life-form. Hydrotherm Figures 3

LIFE-FORMS (BIOLOGICAL TYPES) 16

I. PHANEROPHYTES. Bud-covering. Herbaceous Phanerophytes. Conspectus of Subtypes 1–15 19

II. CHAMAEPHYTES. 16. Suffruticose Chamaephytes. 17. Passive Chamaephytes. 18. Active Chamaephytes. 19. Cushion Plants. 34

III. HEMICRYPTOPHYTES. 20. Proto-Hemicryptophytes. 21. Partial Rosette Plants. 22. Rosette Plants. Monopodial Rosette Plants. General view of Hemicryptophytes. 39

IV. CRYPTOPHYTES. 23. Rhizome Geophytes. 24. Stem-Tuber Geophytes. 25. Root-Tuber Geophytes. 26. Bulb Geophytes. 27. Root Geophytes. 28. Helophytes. 29. Hydrophytes. 64

V. 30. THEROPHYTES 97

The Use of Life-forms in Plant Geography to characterize Equiconditional Regions . 98

INTRODUCTION

Though we know neither what life is, nor whence the first germs of life arose, yet there is no reasonable doubt that all the myriads of diversely fashioned plants and animals inhabiting the world to-day are the result of the process of evolution.

In the absence of certain knowledge opinions are divided about the mechanism which brings new species into being.

Some investigators think that new species arise congenitally, that is to say, that individuals arise which differ from their parents in potentialities and properties, and that some of these differences, which are demonstrably hereditary, are handed on to their offspring, making the primordia of a new species. This, known as the Darwinian Theory, has recently been amplified and placed upon an experimental basis chiefly by the Dutch botanist Hugo de Vries, who calls the forms which show hereditary deviations mutations. His theory is called the mutation theory. It can scarcely be doubted that new species can arise in this way; but the causes of the mutations are unknown.

Others think that it is the environment that makes new species; that individuals arise having identical structure and propensities, and that these individuals growing and developing in different environments come to differ from each other in form and structure. As a matter of fact these differences are almost certainly not hereditary in the ordinary sense of the word, and do not occur again in the offspring which grow in an environment differing from the one that brought about their existence, so that they cannot be looked upon as the beginnings of new species. But even if the difference brought about by the environment, i.e. the acquired characters, is not directly transmissible, yet the possibility is by no means excluded that a gradual alteration takes place in the individuals, so that in the course of many generations of growth in a different environment they gradually come to inherit the peculiarities impressed on the individuals by this environment when the environment which caused these peculiarities is no longer present. Thus one can imagine a species gradually becoming split into several species by growth of individuals during long periods in environments differing widely from the original one. There is much in nature which makes it difficult for us to doubt that this is one of the sources of new species; but exact scientific proof can only be obtained by means of cultural experiments carried out during decades or even centuries, and such experiments can scarcely be carried out by an individual investigator, but need Phylogenetic Institutions guaranteed by the State. As yet no such institutions exist.

Others believe that new species arise by hybridization, and we can scarcely doubt that this may take place. These three possible modes of origin of new species are not mutually exclusive, but presumably can operate at the same time. Exact experiments to illustrate this problem are only in their infancy.

But in whatever way species arise it is certainly **the environment that determines the destiny of the species**, determines whether they shall perish as unfit for life or whether they survive to be a link in the chain of species. It is the environment that determines the place in nature occupied by the individual species. Often more individuals are produced than there is room for. Competition for space and food of necessity entails that each individual species shall within limits become confined to those situations to which it is best adapted and better adapted than the other competing species. In all places where the plant community is dense and there is competition for space and food we may safely assume that there is complete harmony between the species and their environment, in other words that the plants are adapted to their environment. It was Darwin who first opened our eyes to the fact that this adaptation, this orderly arrangement of the world, was not an incomprehensible phenomenon, but the necessary consequence of the relationship between the environment and the demands of the individual organisms. The structure of any plant is not in itself effective; it is the relationship to its environment that will show whether an organism is constructed on a plan favourable to its life. A plant of *Corynephorus canescens* with its marked xerophily is not serviceably constructed in relation to itself; this plant would perish if there existed none of the peculiar dry localities on which it is able to live and win the battle with competing species. It is the environment, the competition, &c., which determine the place of the species in nature; adaptation follows as a derived phenomenon. The environment determines everything.

Most plants produce many seeds which are distributed with more or less facility, bringing the individuals gradually into places where they are best adapted to live. But since the methods of migration and the factors which limit it are manifold, and since many species migrate slowly, it often happens that plants have not yet arrived at places where under the present conditions they are best adapted to live, and have not yet arrived at all places where they are better adapted to live than the species which at present occupy those places. We see this by the behaviour of certain species introduced by man into new territories; in some localities such aliens have succeeded in partially supplanting the original vegetation. But in most parts of the world where cultivation has not interfered excessively with nature the plant world is doubtless in a kind of equilibrium, the individual species growing in environments to which they are best suited. But this equilibrium is not constant, for the environment is always altering, slowly, it may be, in some places, and faster in others.

Since the time of Darwin the study of the correspondence between the demands of plants and the environments in which the plants grow has influenced botanical science deeply. This subject is boundless; the number of species amounts to hundreds of thousands, and the demands of each species are manifold. Plants are dependent for pollination on insects and on the wind, for seed distribution they are dependent upon animals, wind, water, &c. But this is not the subject that concerns us; we are interested in the plant's dependence upon factors which bring about its vegetative welfare and in the resulting dispersion of the species on the earth's surface; in other words in how the demands of plants are expressed in various life-forms in harmony with the various environments offered by the different regions of the earth.

All over the world environments varying from place to place determine the existence of different life-forms, because the demands of the plants, which are, at any rate partially, expressed by their structure, must of necessity be in harmony with the environment, if life is to continue. The plant world therefore varies from place to place. In order to obtain a clear view of the complexity involved we must find out what in the main is common to equiconditional regions and to the vegetation of these regions. Since it is much easier to determine the environment than the demands of the plant, the best results are to be expected by beginning with the environments and trying to define those which are in the main alike. In investigating the plants within these individual environments we must first try to discover the chief peculiarities in their structure, the life-forms by means of which the harmony between the plant world and the environment has expressed itself clearly. The requirements for the life of plants are all of equal importance inasmuch as none of them can be dispensed with; but when these requirements are used as a foundation for dividing up the earth into equiconditional regions they are very far from being of equal importance. Some, for example the amount of oxygen and carbon dioxide in the air, differ so little in different places that they have no significance for the life-forms, and therefore cannot be used as characters for equiconditional regions. Others, for example the chemical and physical nature of the soil, the relationship between plants and animals, and between plants themselves, vary so widely even within the smallest districts that they cannot be used for limiting large equiconditional regions; but on the other hand they are useful in the detailed analysis of vegetation within these regions.

The same is approximately true of light. If the demand for light always expressed itself sufficiently obviously in the structure of plants, and if the plants were all of equal height and shaded each other equally, then the different intensity of sunlight in the different degrees of latitude would be an important factor for limiting large equiconditional areas.

But there is a vast difference in the size of plants, and some grow in the shade of others, so that the relationship of light even in very small areas differs so greatly that it is impossible to use it for determining what is common to the environment over extensive tracts.

The most important factors determining the environment which still have to be mentioned are moisture, water (here more closely defined as precipitation), and temperature. Temperature regulates transpiration and thus alters the significance of the amount of water present. The relationship of temperature to humidity is the factor which makes the deepest impression on vegetation at the present epoch. Heat *quâ* heat certainly plays a very important role in the distribution of plants. Each species demands its own degree of heat, and consequently each occupies a corresponding geographical position. Megatherms[1] demand much heat, and will only grow where the temperature is relatively high throughout the year; they are consequently found only in the tropics. Mesotherms can endure a considerably lower temperature during a longer or shorter period of the year; they can grow in tropical and sub-tropical regions; but it is only in the sub-tropical region that they can conquer competitors which demand a different degree of heat. Microtherms are plants of the temperate regions. They demand a still lower temperature, not needing so high a summer temperature, and enduring a much lower winter temperature. It does not necessarily follow, however, that the plants mentioned grow best under the physical and chemical conditions available. All that is meant is that under these conditions they are able to prevail against competitors which demand a different degree of heat. A fourth group, Hecistotherms, comprises plants belonging to the cold regions. They have the lowest heat demand of all plants, will grow where the summer is short, and are able to endure a long and very cold winter. In spite of the fact that heat *quâ* heat has such great importance in determining the distribution of plants, yet it is impossible to use heat as the basis for delimiting equiconditional regions and characterizing these regions by their plant life. The reason of this is that the temperature demanded by plants has scarcely any influence on their structure, or perhaps it would be better to say that our knowledge is at present insufficient to enable us to draw from a plant's structure any conclusions about its heat demand. If our knowledge were sufficient to enable us with ease and certainty to determine the heat demand of a plant from its structure, then temperature would become perhaps the most important factor in delimiting equiconditional regions. But at present the use of temperature for this purpose must be imperfect, as we have to rely on our knowledge of the temperature in which

---

[1] This and the following terms are here used to signify only the different demands for heat by plants. A. de Candolle, who first used them, included, at any rate partially, demands for moisture as well.

plants grow in a state of nature. On the other hand temperature as a factor affecting transpiration is recognized as a very important character in delimiting equiconditional regions.

Though the factors necessary for plant life are, as I have said before, of equal importance, inasmuch as plant life cannot exist in the absence of any one of them; yet it is not incorrect to say that some factors are more important than others. The same kind of relationship prevails between the plant and the ten elements which are necessary for its perfect development. All these elements are equally important in that each is necessary for the development of the plant; but the element which is present in the smallest quantities plays the most important part because it *is* present in such very small quantities; its complete absence would prevent the development of the plant altogether. This element then might be considered more important than the others. The same is true of environmental factors in general: the necessary factors which are present in the lowest degree are comparatively of most importance, because their complete absence can stop the further distribution of plants, and thus become factors defining regions of common environment.

Of all the factors necessary for plant life water is the one which over vast expanses of the earth's surface most nearly approaches the status of a limiting factor. This is true even of regions where a sufficiency of water is actually present, as it is in extensive tracts in the colder regions of the world, where at some times of the year the temperature is so low that the plants cannot absorb the water; they can actually die of drought standing in saturated soil, perishing of what Schimper calls physiological drought, which has the same effect on plants that physical drought has in warmer and drier regions.

While then, in addition to the fact that the water demand of plants compared with their other demands usually finds marked expression in their external and internal structure, the relationship of plants to water influences vegetation to such a degree that it is by far the most important factor in the 'plant climate'—we can, in fact, use the structural expression of the demands of plants as a reagent, as it were, to 'test' their environment. The series of types we are thus able to make will at the same time reveal the historical development of the plant world, of the changes that have taken place throughout the ages according to the operation of fixed laws. This we can do because there is every reason to suppose that such water conditions in the course of time have become increasingly unfavourable to plants, and that species have gradually arisen which are essentially new in that their demand for water has decreased in harmony with the curtailment of available water. The series of types then is a natural one; it is the expression of an organic development.

We may safely assume that the conditions of life were formerly more favourable than they are now. There was a time when heat and moisture were less dependent upon seasons than they are now. The luxuriance of vegetation during the Carboniferous age and the uniformity of that vegetation in all parts of the world lead us to suppose that the climate must have been warm and humid and approximately constant from the Equator to the Poles.

But gradually, as dissimilar areas became differentiated, the character of the vegetation became differentiated too. This was because the deterioration of the environment in any region destroyed those species whose demands were too exacting, and of the new species which arose only those could survive whose demands could be satisfied by the environment. Thus existing vegetation is not only an expression of conditions to-day, but it also is a link in the chain of organic evolution, a chain which has been forged by alterations of the environment.

We must then consider the life-forms belonging to constantly warm and constantly humid regions as the most primitive life-forms, and the life-forms adapted to other climates as later developments. There is, for example, no doubt that trees and shrubs with covered buds arose later than those with naked buds; that species adapted to life in an unfavourable climate by bearing their buds on underground shoots have arisen later than those life-forms which do not possess this peculiarity. There is then good reason to characterize life-forms simply by the means through which they are able to exist in progressively more unfavourable environments which have arisen because of the gradual changes in climate.

But this statement needs qualification. The important point is that environments not only differ in space but in time. Conditions differ not only from place to place, but in the same place from month to month. The seasons impose different conditions. Apart from the tropics, where the climate is always warm and humid and thus fairly uniform and favourable during the whole year, all other regions have at least two seasons, a favourable and an unfavourable, or more correctly a more favourable and less favourable season. Those structural characters which enable plants to harmonize the demands of their vegetative organs with their environment are on the whole the characters which make the most obvious impression on vegetation. From the nature of the case, however, the difference between the favourable seasons of two regions must be far less than the difference between their unfavourable seasons. This makes it exceedingly probable that those structural differences which enable plants to survive unfavourable seasons are greater than those which harmonize the same plants with the favourable seasons. If we then wish to use vegetation as a test of the plant climate, to delimit the equiconditional regions by means of the vegetation of those regions, we must, I think, do so by observing the structural peculiarities which enable the plants to survive the unfavourable seasons.

Since water in relation to temperature has the greatest significance in defining equiconditional areas, and since those two factors rise and fall during the course of the year, it becomes of great interest to make clear diagrammatic illustrations, such as are shown in Figs. 1–6, of the relationship of these factors in given regions during the course of the year.

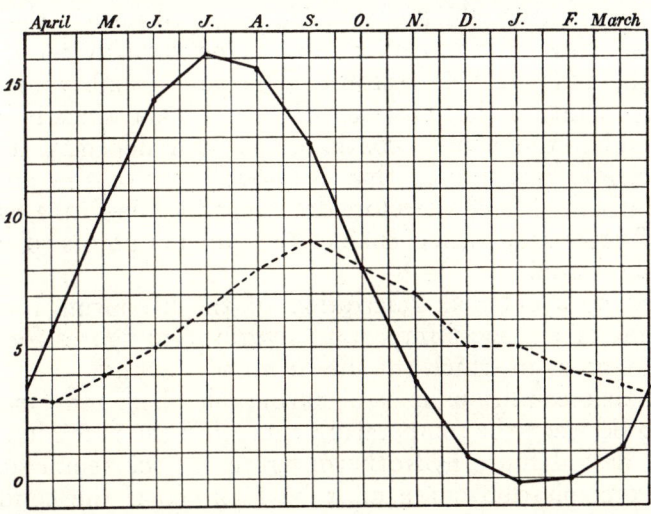

FIG. 1. Hydrotherm figure for Denmark. ——— Temperature curve; - - - - - - Precipitation curve. The numbers denote degrees Centigrade for the Temperature curve and centimetres for the Precipitation curve.

The figures on the vertical line to the left denote degrees Centigrade; the names of the months are given on the horizontal line above, beginning with April for the Northern Hemisphere and with October for the Southern Hemisphere. On the vertical lines the mean temperatures of the months are given, and the points marked with these mean temperatures are joined by a ruled line: the temperature curve thus resulting portrays the yearly variations of temperature in the locality in question. The precipitation curves are made in the same way. As far as I know these two curves have hitherto always been made separately, but here we can look at both together; indeed, both curves are combined in the same figure, the same number on the vertical line to the left representing degrees Centigrade for the temperature curve and centimetres for the precipitation curve. In the figures the ruled line always represents the temperature curve and the dotted line the precipitation curve.

The favourable and unfavourable seasons, in so far as they are dependent upon the two essential factors, heat and moisture, are shown in these figures as crests and troughs in one or both curves. The two troughs may occur at the same time of the year or at different times.

Thus each equiconditional region shows a characteristic figure which I shall here designate its hydrotherm figure, and which exhibits the most important environmental properties of the district.

The most important adaptation for plants is that which enables them to survive the seasonal trough. I have therefore used as a basis for defining life-forms, that are to characterize equiconditional regions, structural peculiarities which enable the plants to survive the unfavourable seasons. Now all parts of plants are not equally sensitive to the effects of the unfavourable seasons. The young embryonic tissue of the growing points is the most sensitive of all, and since it is this very tissue on which the plant's continued growth depends, it is of the greatest possible importance that it should survive the unfavourable season unscathed. I have therefore delimited the life-forms by means of the kind of protection which enables the growing points to survive the unfavourable season.

Before giving a detailed description of the life-forms I will give a short account of the most important hydrotherm figures for the equiconditional regions to which the life-forms correspond.

In order that conditions may be favourable there must be a certain relationship between the temperature curve and the precipitation curve, there must be a definite hydrotherm figure. If the temperature curve is high the precipitation curve must be high too if the conditions are to be favourable. A high temperature promotes transpiration, so that precipitation must be abundant in order to make good the water thus lost.

In tropical regions, apart of course from mountainous districts, the temperature curve remains high all the year round; the temperature is thus sufficiently favourable for the continuous vital activities of plants. It is the course of the precipitation curve which here demarcates the great regions. There are of course imperceptible gradations between the different types of precipitation curve, from high precipitation all the year round (Fig. 2), through different degrees of varyingly high precipitation (Figs. 3 and 4) to very low precipitation all the year round (Tropical desert regions). We must not suppose therefore that large and extensive regions have approximately the same environments. The expression 'equiconditional regions' must not be taken literally; it is only meant to imply that in certain broad features there is approximate uniformity in the vegetation, which shows in those regions the same relationship to the life-forms composing it. In spite of the imperceptible gradations encountered we are obliged, in order to obtain a clear view of the matter, to draw boundaries. It is important that these boundaries should be as natural as possible.

In certain comparatively well-demarcated tropical regions, e.g. parts of the East Indies, especially the East Indian Islands, the West coast of

Africa, parts of South America, especially the region of the Amazon, &c., the climate is not merely constantly warm, but it is also constantly wet; the precipitation is high throughout the year. This is seen in the hydrotherm figure for Sumatra (Fig. 2). The conditions are favourable throughout the year, and there is no really unfavourable season; the vegetation is composed of the least protected life-forms; even the most exacting demands for water and heat can be satisfied. From this ideal state of affairs there are two divergent lines of climates. Firstly, we have the

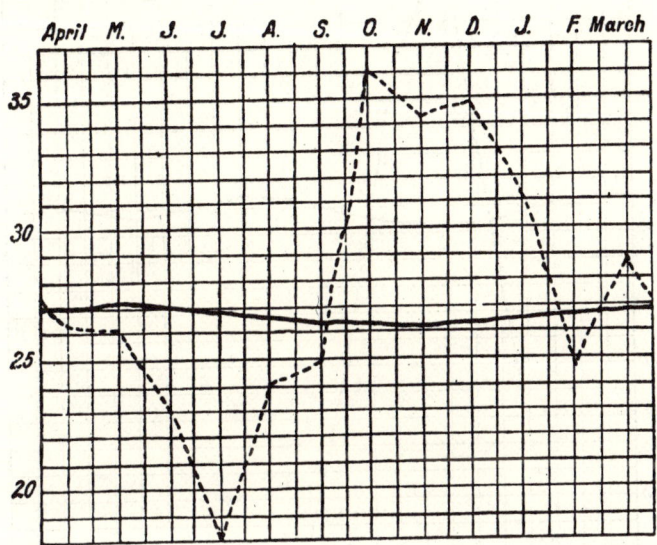

Fig. 2. Hydrotherm figure for the East coast of Sumatra near the Equator. Numbers, &c., as in Fig. 1.

tropical series with the constantly high temperature curve, but with the decreasingly favourable precipitation curve. Secondly, there is the series of climates outside the tropical regions extending through the subtropical zones of the temperate and cold parts of the world. These climates have a different kind of temperature curve, and their precipitation curves differ widely.

Within the tropical series of climates we can differentiate, apart from the constantly humid regions, at least two main areas. In the first of these the precipitation is still comparatively high, but its curve shows a very marked trough, signifying that the precipitation is unevenly distributed during the course of the year. These regions have two different seasons, a rainy season and a dry season. The boundary line between this and the foregoing region must be drawn in accordance with statistical investigations of the comparative relationship between the life-forms. Owing to the high temperature, if the precipitation amounts for a long period to less than 5 cm. a month a great change is seen in the relationship of the life-forms. But it is apparent that if the yearly precipitation is fairly high, the vegetation can well tolerate a falling of the precipitation curve for a single month to below 5 cm. without causing essential change in the composition of the flora as far as life-forms are concerned (Fig. 3).

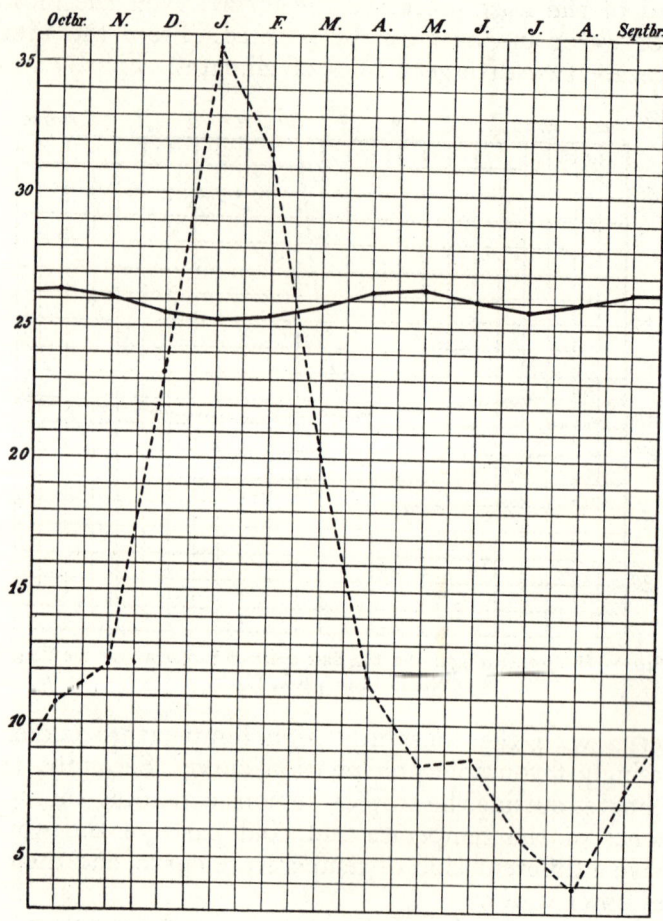

Fig. 3. Hydrotherm figure for Batavia. Numbers, &c., as in Fig. 1.

We can here scarcely speak of an actually dry season if the precipitation curve during two months or more does not fall below 5 cm. In that equiconditional region known as the savannah region, where there is a marked dry season in combination with a yearly precipitation which is not excessively low (Fig. 4), the dominant life-forms differ from those of the regions of tropical rain forests. It is probable that statistical investigation of life-forms will make it necessary for us to divide the savannah region into two or more regions.

Where the precipitation curve as a whole sinks very low because the annual precipitation is very small, we find tropical deserts, and with

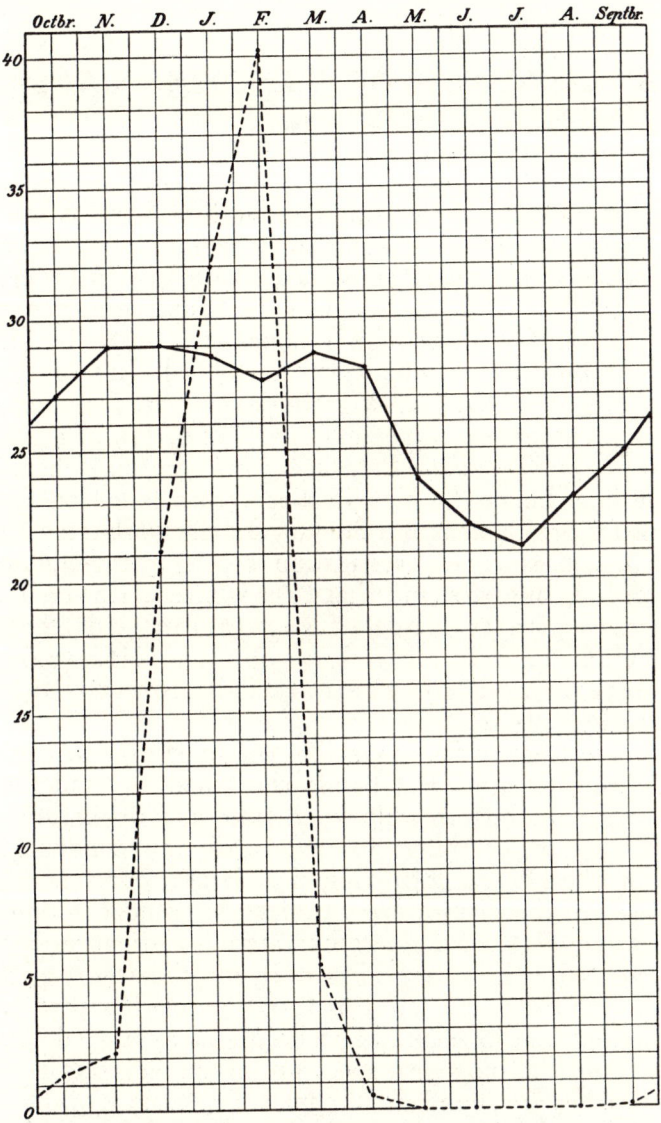

FIG. 4. Hydrotherm figure for Sweers Island, at the head of the Gulf of Carpentaria. Numbers, &c., as in Fig. 1.

these perhaps we might unite the sub-tropical deserts, calling them respectively, Megathermic and Mesothermic deserts.

In the sub-tropical zone the temperature and the precipitation curves both show a conspicuous trough. Apart from the tropical deserts and

V
THE LIFE-FORMS OF PLANTS ON NEW SOIL
INTRODUCTION

THE chief object of my journey to the West Indies in 1905-6 was the opportunity it afforded of determining the life-forms of the species of flowering plants found within a circumscribed area of tropical climate, viz. the Danish West Indies. I wished to do this in order to obtain material for the biological spectrum of a tropical country with a comparatively low rainfall. Although I hope later on to be able to give more detailed information about the biological data of some of the flowering plants of the Danish West Indies and St. Domingo, the chief result of my investigations is already given on p. 129 of my book called *Planterigets Livsformer og deres Betydning for Geografien*,[1] where it occupies but a single line, being the biological spectrum based on the statistics of life-forms for St. Thomas and St. Jan. In the work cited I have given a basis for a comparative investigation of the plant-climates of the different tropical countries, and that basis is the climate as a factor determining definite kinds of vegetation and expressed by the statistical relationship between the life-forms of the species, determined by their adaptation to survive the unfavourable season.

To determine then as far as possible the life-forms of the individual species was my chief aim. But beside this I had in mind various small tasks with which to busy myself when time and opportunity allowed. I wished among other things to investigate the conditions and the vegetation of the alluvial formations which are found here and there along the coast of the Danish West Indies. My object in doing this was especially to obtain a basis for a comparative investigation of vegetation found on essentially the same soil in different climates, e.g. the Danish West Indies and Denmark. By this investigation I wished to find out how far climate, even in such a specialized area as alluvial beach formations, determines the biological spectrum of the vegetation.

Alluvial beaches are found in all climates, and show wide differences. It is true indeed that we cannot always separate sharply the different beach formations, but in general we can differentiate those made up of deposits of coarser inorganic components (sand and gravel), and the deposits of finer components (clay, mud, &c.). The former are found, as is well known, especially on unprotected coasts, the latter on protected

[1] Included in the present volume as Chapter II, *The Life-forms of Plants and their Bearing on Geography*.

ship to the temperature curve (Fig. 1). In certain districts however the precipitation causes a covering of snow in winter, and this is of great importance in delimiting regions dominated by the same environment. I shall not here enter into the problem of delimiting equiconditional areas in the colder parts of the world, where the different regions pro-

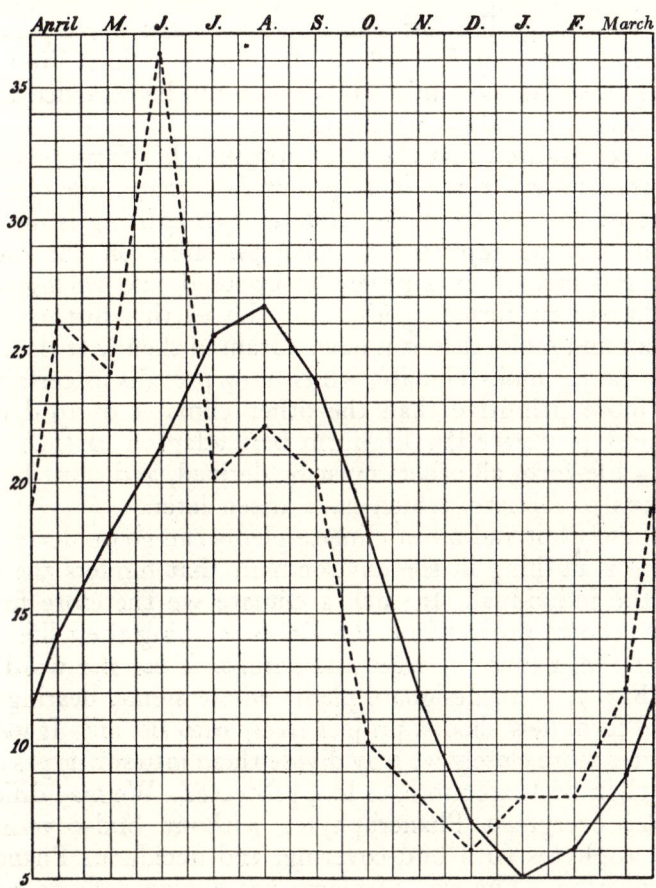

FIG. 6. Hydrotherm figure for South Japan (Nagasaki). Numbers, &c., as in Fig. 1.

bably show a wide difference in their hydrotherm figures, without, however, causing such a marked difference in life-forms as is seen in the tropical regions. The reason for this peculiarity is that the life-form specially adapted to colder regions, viz. the Hemicryptophyte, is only above the ground during the favourable season; in the unfavourable season it is protected by the soil, and therefore less influenced by the degree of severity of the unfavourable season.

Now that we have completed our short description of the great equiconditional regions let us study the life-forms which belong to them, and in order to understand the matter fundamentally let us first describe the five main types of life-form.

LIFE-FORMS (BIOLOGICAL TYPES)

In order to arrange life-forms in a natural series corresponding with their evolution, we must first ask the questions: 'which life-form must be looked upon as the most primitive, which can be regarded as standing in the relationship of parent to the others?'

If we take for granted that the most primitive life-form is the one that is best adapted to life in the most primitive climate, our question may then be altered to, 'which of the climates now existing is most like the climate that may be supposed to have prevailed on the earth when flowering plants arose?' It is generally supposed, and with good reason, that the climate in former periods of the earth's history was more uniformly hot and moist than it is now. Granted then that the constantly hot and constantly moist climate, which now prevails in certain tropical regions, is more primitive than the other climates of to-day, we can regard as most primitive the life-form best adapted to such a climate. From such a life-form all others must be derived, and must be regarded as adaptations to climates which have arisen later.

In a constantly hot and moist climate plants can go on developing new shoots; there is nothing in the environment that hinders the continual growth of the individual. In such a climate we therefore find chiefly plants whose shoots project into the air, continuing their life from year to year after the manner of trees and shrubs. I use the word **Phanerophyte** (Fig. 7, 1) to designate plants whose stems, bearing the buds which are to form new shoots, project freely into the air. If we examine Phanerophytes more closely we may divide them into sub-types according to whether their buds are more, or less, protected. We may differentiate, for example, evergreen Phanerophytes without bud-covering, evergreen Phanerophytes with bud-covering, and deciduous Phanerophytes with bud-covering. The size of plants has a very definite bearing on their relationship to humidity, so that we may further divide the Phanerophytes into Mega-phanerophytes, Meso-phanerophytes, Micro-phanerophytes, and Nano-phanerophytes.

Phanerophytes, especially the least protected among them, are the plants of those portions of the earth which are most favoured climatically. In tropical regions which are constantly warm and constantly moist, they form the bulk of the species; but as we leave the tropics and reach less and less favourable climates, climates with a long and hot dry season, or climates with a severe winter, we find that Phanerophytes decrease in proportion to the remaining life-forms. In these less favoured localities we meet only the best protected sub-types of Phanerophytes, especially those which in the unfavourable season lose their leaves and whose apical shoots are protected by bud-scales. Gradually, as we approach the boundary beyond which Phanerophytes cannot live, they become lower and lower in stature till we meet only small bushes and dwarf trees. Because of the lowness of their growth these plants are not as exposed to the dangers of excessive transpiration as the tall Phanerophytes. Indeed, in regions where snow lies in the winter they are often so low as to be protected by the covering of snow; thus forming a transition to the second main type of life-form namely the Chamaephyte. Finally in the far north, in the higher regions of lofty mountains, and in dry steppes, the Phanerophytes disappear altogether.

The Chamaephyte, the second main type, is characterized by having the surviving buds situated close to the ground. This position may come about either because the whole shoot lies flat on the ground (Fig. 7, 3), or because the distal portion of the shoot, which projects into the air, dies at the beginning of the unfavourable season, so that only the shoot's lowest portion together with the surviving buds remains behind (Fig. 7, 2). In Cushion plants the surviving shoots are densely massed together and so short that the plants may be considered to be Chamaephytes.

The third type is the Hemicryptophyte which has the surviving buds actually in the soil-surface, protected by the soil itself and by the dry dead portions of the plant (Fig. 7, 4). The aerial shoots, bearing leaves and flowers, survive for only a single period of vegetation, and then die back as far as the part of the shoot situated in the ground and bearing the surviving buds. The majority of our biennial and perennial herbs, and indeed the bulk of plants in the cold and temperate parts of the earth, belong to this life-form, which can be subdivided according to whether the shoots are more or less modified for their life in the soil.

The Cryptophyte, as the fourth type is called (Fig. 7, 5–9), is characterized by having its buds completely concealed in the ground or at the bottom of the water. The depth at which the buds are buried varies in different species. This life-form is better protected against desiccation than the Hemicryptophyte, and it is specially adapted to regions with a longer dry period; hence steppes are especially rich in Cryptophytes. Many herbs with horizontal root-stocks (Fig. 7, 5) and the majority of bulbous and tuberous plants belong to this life-form (Fig. 7, 6). In the favourable season they send up shoots bearing leaves and flowers: in the dry period the plant is lost to sight, the buds being hidden at a greater or less depth in the ground.

The four life-forms mentioned show a series of progressions in the same direction. As we travel away from the most favoured portions of the earth, which are characterized by Phanerophytes, to less clement

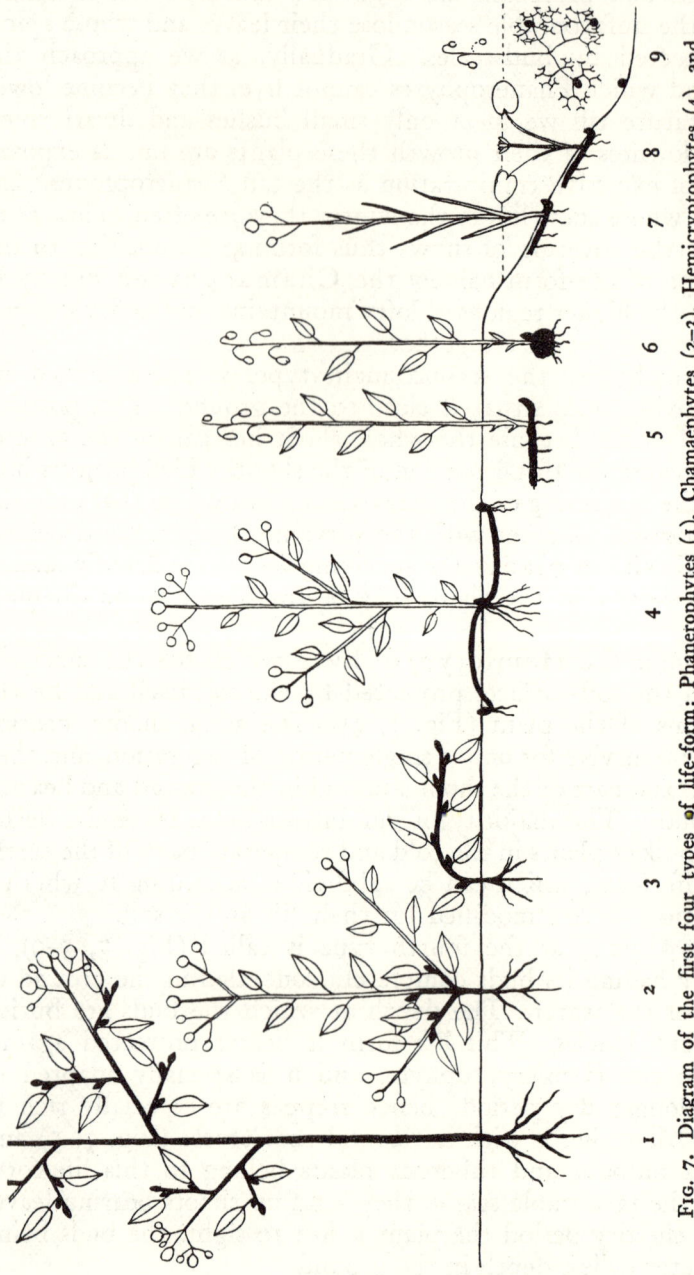

FIG. 7. Diagram of the first four types of life-form: Phanerophytes (1), Chamaephytes (2-3), Hemicryptophytes (4), and Cryptophytes (5-9). The parts of the plant which die in the unfavourable season are unshaded; the persistent axes with the surviving buds are black. Proceeding from Phanerophytes on the left and going farther and farther to the right, it is seen how the plants enjoy progressively better protection during the unfavourable season, the surviving buds being attached lower and lower. In Chamaephytes the buds are on the surface of the ground (2 and 3), in Hemicryptophytes they are in the soil-surface (4), and lastly in genuine Cryptophytes (5 and 6) the buds are actually in the soil, or at the bottom of the water in Helophytes (7) and Hydrophytes (8-9).

climates, we observe a tendency for surviving buds to be borne, firstly near the surface of the earth (Chamaephytes), secondly in the soil-surface (Hemicryptophytes), and finally entirely buried beneath the soil or water (Cryptophytes), where of course they are best protected from desiccation, which is their greatest danger.

The fifth life-form, the Therophyte, comprises plants which complete their life-cycle within a single favourable season, and remain dormant in the form of seed throughout the unfavourable periods. The shorter the time that these plants take to complete their life-cycle, the more unfavourable the climate they are able to endure. Because these plants are able to complete their life-cycle in a comparatively short period, often in a space of a few weeks, this life-form is the best protected of all. Therophytes disappear entirely during the unfavourable season of the year, existing only in the best protected state of all, that of the seed. They abound particularly in unfavoured portions of the earth, such as steppes and deserts.

I. Phanerophytes

By Phanerophytes I mean plants whose buds and apical shoots destined to survive the unfavourable period of the year project into the air on stems which live for several, often for many, years. In Phanerophytes the primary shoots, and often also certain lateral shoots, are negatively geotropic, so that the buds are bound to be carried upwards. The direction assumed by branches has been shown to be due to several causes. It may differ not only in different species, but in different individuals of the same species, and even, indeed, in different branches of the same individual. These diversities are largely responsible for the peculiar habit of individual plants.

Apart from certain weeping trees, which I shall discuss later, the shoots of all Phanerophytes are more or less negatively geotropic, but the direction of their growth is usually determined by other causes than geotropism, so that the angle they subtend to the mother axis and the direction they take up in space are exceedingly various. In fastigiate trees the shoots grow erect, forming an acute angle with the stem (e.g. *Populus italica*). There are innumerable gradations between fastigiate trees and those whose primary branches arise at an angle of 90°, an arrangement common in the *Coniferae*. The antithesis of the fastigiate trees is the weeping tree, whose branches, or at least some of them, hang downwards, especially towards their apices. In so-called 'passively weeping trees' it is the weight of the branches that causes them to arch towards the ground. In these trees most of the young shoots, as long as they are so short that their rigidity can maintain them in a definite position, ascend more or less obliquely according to the angle they subtend to the mother axis. Gradually as the branch lengthens without its rigidity increasing sufficiently to maintain it in an erect position, it sinks more and more, finally becoming pendulous. The apex is often horizontal, or it may ascend to form an S. A great many trees, when growing under conditions that make them produce long flaccid branches, show a tendency to become passively weeping trees, e.g. species of *Fagus, Tilia, Fraxinus, Salix*, and many others.

The ultimate position taken up by the branches is the result of several concomitant causes, partly external, such as weight, and in a less degree light, and partly internal. The direction taken by the branch in many Phanerophytes is determined partly by its vegetative strength, so that the more vigorous the branches are the more acute is the angle they subtend to the mother shoot, and the more they follow the direction of the mother shoot. For example on vigorous shoots of *Prunus spinosa* the lower and middle buds grow out into short thorn-tipped branches which subtend approximately a right angle with the mother axis, whether the mother axis itself is placed vertically or obliquely. Only a few of the upper buds form long shoots, and these arise at an acute angle. *Prunus Padus* and *Rhamnus catharticus* behave in the same way. In the latter the short branches, which are often thorn-tipped, are frequently curved towards the base of the mother axis, while the upper ones arise at an acute angle.

Internal causes which influence the direction of the twigs operate only when the buds develop into shoots the same year they are laid down, or the year after. The direction of branches which arise two or more years after the formation of the mother shoots seems to be determined almost exclusively by weight. Under certain circumstances light may also influence the direction. Such shoots are therefore nearly always vertical or approximately vertical, whatever the direction of the mother shoots may be. A familiar example of this is the shoot from the stool. Many Phanerophytes produce now and again from older branches new vigorous shoots which, without regard to the direction of the mother shoot, ascend vertically into the air. Weak lateral illumination will cause these branches to rise obliquely. Such shoots are frequently seen in the following genera: *Prunus, Pyrus, Crataegus, Platanus, Diervillea, Lonicera, Viburnum, Sambucus, Philadelphus, Deutzia, Fraxinus, Ligustrum, Hippophaë, Elaeagnus, Catalpa, Ribes*, and many others.

Here it may be mentioned that the direction of suckers, as is well known, is determined by gravity. Suckers are markedly negatively geotropic.

Thus we see in Phanerophytes all possible gradations in the influence exerted by geotropism on the direction of the shoot. The direction of the primary shoot, which of course has no mother shoot, apart from the effects of illumination, is usually exclusively determined by the negative geotropism of the shoot. Primary shoots are negatively geotropic, as also are suckers and late shoots. By late shoots I mean those that grow out from mother shoots two or three years old. The direction of ordinary side shoots which grow out from annual mother shoots is on the contrary influenced to a less extent by geotropism; in these internal causes play the principal part.

It is a well-known fact which has already been touched upon that the weight of the twigs of Phanerophytes plays an important role in determining their direction. Especially in individuals which stand free, or nearly so, the branch system may become entirely pendulous, though the young shoots may curve upwards so that the system becomes more or less 'S' shaped. It seems to me that the condition of affairs in markedly weeping trees is somewhat different. In these the branches hang down more vertically and the young shoots are not curved upwards at the apex, as always happens in plants whose shoots are negatively geotropic and at the same time not sufficiently rigid to remain erect for their whole length nor to maintain a definite direction. Many physiologists (Hofmeister, Frank, Voechting, Baranetzky, &c.) think that the branches of the pendulous Ash and pendulous Elm hang by their own weight. According to Baranetzky the shoots of these trees are really negatively geotropic, but as the pendulous varieties have longer and weaker internodes than the type species, the negative geotropism is held in abeyance. Of these forms I have had an opportunity of observing only the pendulous Ash, whose shoots are certainly weak, but that they are so weak that they have not even the power to show a slight negatively geotropic bend, if they are really negatively geotropic, in a short portion of their distal ends seems to me improbable. I have, however, carried out no experiments with these forms, but I have done so with a few corresponding trees such as *Betula alba* var. *pendula* and *Sambucus nigra* var. *pendula*.

I do not know how the branch system of *Betula alba* var. *pendula* initiates its weeping habit; presumably it occurs passively; but when the branches have at last taken up a pendulous position, apparently the direction of the lateral shoots is not affected by geotropism; these grow obliquely downwards, forming a rather definite angle with the mother shoot. In their growing apices we see no kind of negatively geotropic turning upwards, although they are sufficiently rigid, at any rate in a short length, to maintain an obliquely erect position. If a twig be placed so that its apical shoot points obliquely upwards, it always remains for a long time in that position; it seems, therefore, remarkable that if they are really negatively geotropic they should not show the same upward curve shown by negatively geotropic plants which have weak, and therefore procumbent pendulous twigs, and exhibited by all other negatively geotropic plants. And, even if it should turn out, in conformity with Baranetzky's experiments on Elms, that the shoot-apices bend upwards in pendulous Birches and other weeping trees when the leaves are removed from the shoots, it cannot from that fact alone be concluded that it is weight that makes shoots droop, because to deprive a shoot of leaves is to make a profound disturbance in its economy and all the phenomena which depend upon its economy. It is as reasonable to suppose that the altered direction of the shoot when the leaves are removed is brought about by the removal of the leaves causing an alteration of the shoot's geotropism, as is the supposition that a negative geotropism, which is present before the removal of the leaves, is first able to make the shoot bend upwards when it is relieved of their weight.

Sambucus nigra var. *pendula* has its twigs directed vertically downwards. Both in light and darkness, and when the mother shoots are placed vertically with the apices pointing either upwards or downwards, the lateral shoots grow downwards. At length the shoots reach the ground and are compelled then to grow horizontally. If the youngest part of such a shoot be fastened so that its apex is turned upwards it then bends down, and the curve it makes does not give the impression of being brought about solely by the weight of the shoot, for the shoot apex is sufficiently rigid to stand vertically or horizontally.

Now whether the shoots of *Sambucus nigra* var. *pendula* become drooping, and at length procumbent on the ground, because their reaction towards gravity is different from that which obtains in ordinary Phanerophytes, or whether they behave in this way because of their own weight, we have in either case, within the same species, an interesting transition between the Phanerophytes and the Chamaephytes; for while the type *Sambucus nigra* is a marked Phanerophyte, var. *pendula* is a Chamaephyte, whose shoots lie horizontally on the ground.

Such a change in the shoot's reaction towards gravity occasioning a transition from Phanerophyte to Chamaephyte is met with in many genera, so that some species within an individual genus have negatively geotropic shoots, and are therefore Phanerophytes, while other species whose shoots grow approximately at right angles to the gravitational pull are Chamaephytes. Examples may be seen in *Salix* and *Rosa*.

Phanerophytes include a long series of gradations in their adaptation. We therefore see that although Phanerophytes are predominantly plants of favourable climates, yet they are found in many other regions; only from the most unfavourable of all climates are they entirely excluded. The vast majority of Phanerophytes are indigenous in tropical and subtropical regions, where there is not an excessively long dry season.

The causes which have enabled comparatively few Phanerophytes to extend beyond the boundary of the warmer parts of the earth are presumably various; some of these causes are not yet understood. Apart from the question of water it is certain that the reduction in temperature is the most important cause. Different species of plants are able to endure during the resting season different degrees of cold, that is to say, a difference in the lowness of temperature both maximal and total, and they demand a difference in the height of the temperature in the vegetative season; but however the matter stands for individual species it is quite certain that the Phanerophyte is the type of plant which belongs properly to the warmer and comparatively moist regions of the earth, and that comparatively few species of Phanerophytes have become

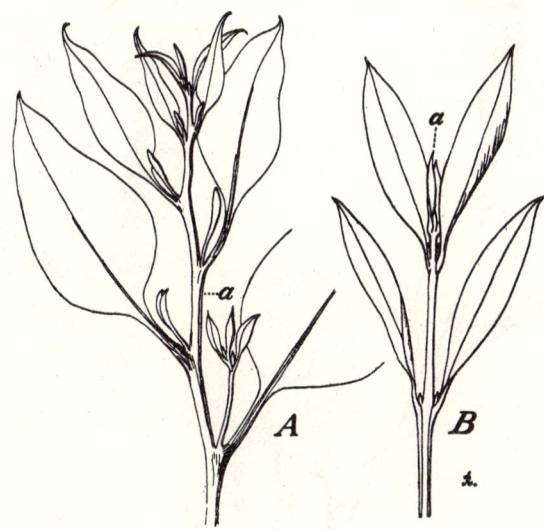

FIG. 8. Phanerophytes without bud-covering. A, *Eucalyptus orientalis*, the top part of a shoot; *a*, the line of demarcation between two periods of growth (⅔ nat. size). B, *Olea europaea*, a shooting twig; *a*, the leaves of the latest period of growth, which are about to grow out (nat. size).

adapted to life in colder climates. It is true that we have in the temperate, fairly humid regions large areas with Phanerophytic vegetation, such as the extensive woodland formation of the temperate zones; but this Phanerophytic formation is made up of comparatively few species, and that they are rich in individuals obviously does not invalidate the statement that the Phanerophyte is the plant type that belongs to warm regions; it only shows that those species which have become adapted to life in less favourable climates are specialized.

We must now try to discover and describe the characters of a series of sub-types of Phanerophytes which will be useful to us. Here we meet, as we always do in this domain of investigation, the great difficulty that the degree of adaptation cannot be determined by merely looking at the plant; while the sub-types in order to be useful to us must be recognized at sight. Our difficulty is that the protection of the growing point against the effects of the unfavourable season can take place in many ways, and we are often unable to determine the efficiency of the protection afforded.

The growing point may be protected directly by bud-scales, or the protection may be indirect. Indirect protection is of four kinds. Firstly, the foliage may be thrown off at the onset of the unfavourable season.

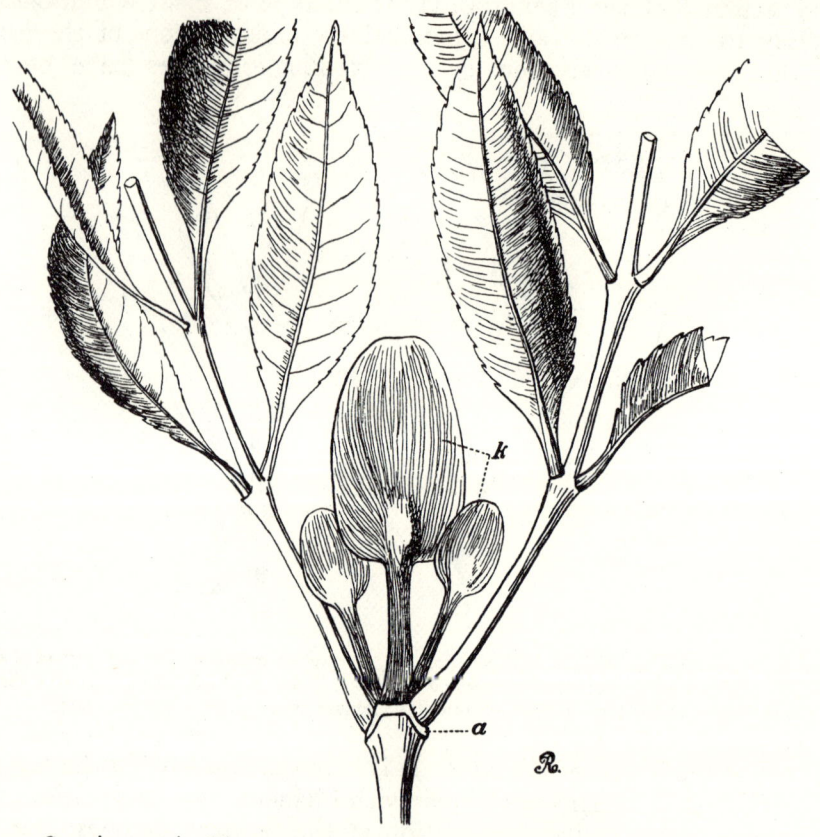

Fig. 9. *Cunonia capensis.* The unexpanded leaves are entirely enclosed by the large interpetiolar stipules (k) whose margins are firmly united ($\frac{3}{4}$).

Secondly, the plant may be reduced in size, so that the shoot apices or buds are not so exposed to desiccation. This adaptation is of course especially efficacious in Phanerophytes which grow beneath taller Phanerophytes. Thirdly, the plant may be anatomically xerophilous, so that its whole structure is protected against drought: the growing points, of course, participate in this protection. Fourthly, the actual structure of the cells may be drought-proof (intracellular xerophily). The order in which these five adaptations are mentioned correspond essentially with their usefulness in establishing sub-types.

Bud-covering. We can divide Phanerophytes into two groups;

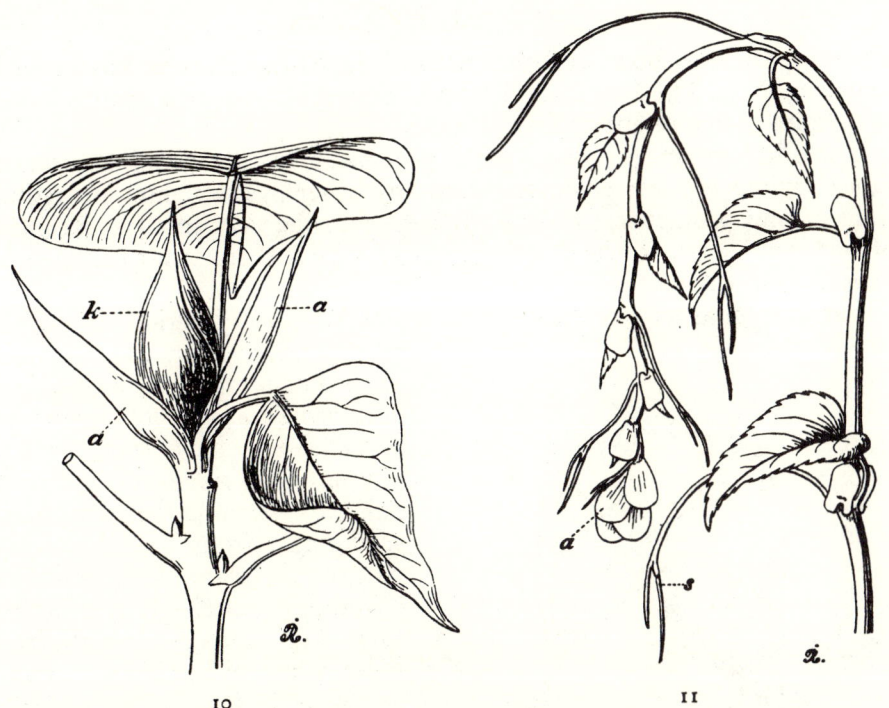

Fig. 10. *Homalanthus Leschenaultiana*. The leaves become very large before they unfold, and remain enclosed and protected by the balloon-shaped stipules of the latest developed leaves, *k*. (¼)

Fig. 11. *Cissus* sp. The young leaves are protected by stipules, *a*. (¼)

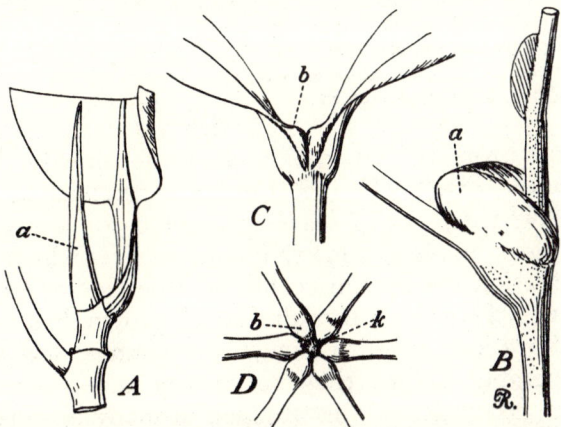

Fig. 12. Examples of evergreen Phanerophytes whose shoot apices with their young leaves are protected by the already developed foliage leaves. A, *Ficus rubiginosa*; *a*, the ligule of the last developed leaf, which fits like a cap over the shoot apex. (¾). B, *Leea sambucina*; *a*, the large stipules of the latest developed leaf surround the apex of the shoot, which is now about to sprout. (¾). C, *Fagraea zeylanica*; the shoot apex is protected by the expanded bases of the latest developed leaves. These leaf-bases are closely applied to each other, *b*. (¾). D, *Allamanda verticillata*; the shoot apex, *k*, is protected by an expansion, *b*, at the base of each leaf of the last developed whorl. (¾)

Phanerophytes without bud-protection (Fig. 8) and Phanerophytes with bud-covering. The boundary however between these two groups is not as definite as one might at first suppose.

The Phanerophytes which grow in constantly warm and constantly moist regions are less protected than any others. They show no visible signs of bud protection; their buds are, as a rule, very small and consist of but few leaves, in the development of which there is no perceptible pause.

But as soon as we come to regions which have a dry period, though it may be only a short one, we meet with Phanerophytes whose leaf primordia are protected in various ways. Some have their young, still unfolded leaves covered with hairs, or surrounded by hairs springing from neighbouring parts of the plant. Others have a covering of mucilage, and others again are protected by the stipules, which either belong to undeveloped leaves, *Cunonia capensis* (Fig. 9), *Homalanthus Leschenaultiana* (Fig. 10), species of *Cissus* (Fig. 11), *Leea* (Fig. 12, B), *Coprosma*, &c., or the stipules may belong to leaves which are already developed, e.g. *Ficus* sp. (Fig. 12, A), or the shoot apex with the leaf primordia may be surrounded and protected by the base of the already developed leaf (*Fagraea zeylanica*, Fig. 12, C, and *Allamanda verticillata*, Fig. 12, D).

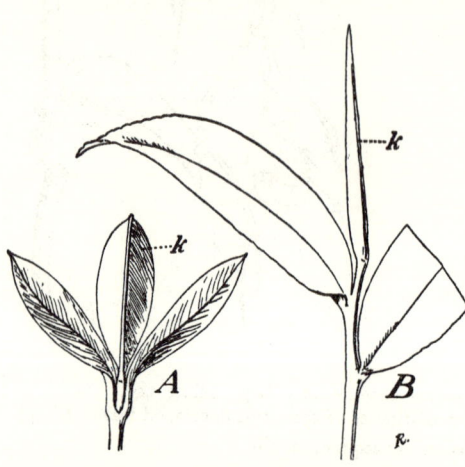

FIG. 13. Phanerophytes without bud-scales. The leaves grow nearly to their full extent before they unfold. The younger leaves are protected for a long time by the older ones. A, *Veronica elliptica*. (½). B, *Marcgravia* sp. (⅔); *k*, still unexpanded leaves.

All plants have their young leaves protected for a longer or shorter period by the older leaves. In many Phanerophytes which have no bud-scales this protection is very noticeable because the leaves attain a considerable size, they become in fact almost full grown, before they unfold, and continue for a long time to enclose the younger leaves, which thus develop considerably before they suffer any direct exposure to the desiccating air. Examples of this are seen in many evergreen species of *Veronica*, e.g. *V. elliptica* (Fig. 13, A), *V. salicifolia*, *V. Kirkii*, *V. Traversii*, *V. speciosa*, *V. parviflora*, &c. Further examples are *Marcgravia* (Fig. 13, B) and *Melaleuca*, e.g. *Melaleuca hypericifolia* and *M. violacea*. In some plants the margins of the leaves which surround the apex adhere together, or they are held together by hairs, e.g. *Veronica elliptica* and *V. salicifolia*. In some species the space enclosed by the outermost leaves is completely packed with young leaves, e.g. *Veronica parviflora* and *V. Traversii*; in others there are much fewer young leaves, and the space is partially filled with fluid, e.g. *Veronica salicifolia*.

Fig. 14. Examples of the transition between foliage leaves and bud-scales in evergreen Phanerophytes. A, *Erica multiflora*; four-year-old branch with foliage leaves separated by sets of less developed leaves, *a*, which are however proper foliage leaves, that have developed at the beginning of the unfavourable season and have served as a protection for the young unexpanded foliage leaves; these protective leaves are influenced by the unfavourable season in such a way that they are not able to continue their development further when the favourable season begins; they are smaller and more erect than the real foliage leaves and fall off sooner, a_1 and a_2. (⅓). B, *Escallonia rubra*; branch with transitional leaves, *a*, between two sets of leaves belonging to periods of growth (⅓). C and D, *Myrtus ugni*; branch with transitional leaves, *a*, between two sets of ordinary leaves; a_1 (in D), transitional leaves which indicate a temporary pause in development during the last period of growth. (⅓)

A somewhat similar protection is seen in a large number of evergreen small-leaved Phanerophytes, chiefly Nanophanerophytes, which have at the ends of their stems a great many young leaves in different stages of development. These are surrounded and protected by older leaves until they are full grown. Examples are: *Calluna vulgaris, Pentapera sicula*,

Fabiana imbricata, Calocephalus Brownii, Diosma oppositifolia, Agathosma apiculata, Phylica ericoides, Melaleuca armillaris, species of *Erica* (Fig. 14, A), *Gnidia* and many others.

Finally we have numerous species in which the protection takes place by real bud-scales, that is to say by leaves whose function is exclusively protective.

Now it is probable that buds which are protected by the stipules or

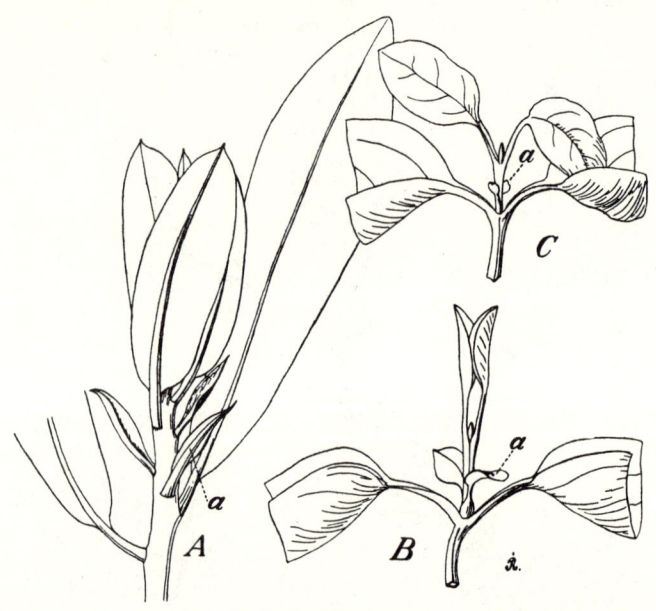

FIG. 15. Evergreen Phanerophytes which show the transition to real bud-scales. A, *Drimys Winteri*; *a*, undeveloped leaves which have served as a protection for the young foliage leaves during the unfavourable season. (⅔). B and C, *Viburnum tinus*; shoot with more or less reduced transitional leaves, *a*, between those belonging to two periods of growth. (⅔)

leaf-bases of ordinary foliage leaves are just as well protected as those furnished with bud-scales. For example, the young leaves of *Cunonia* are very efficiently protected by the unexpanded adherent stipules. *Homalanthus Leschenaultiana, Ficus, Philodendron*, &c., are other examples of efficient protection without the aid of bud-scales. In spite of this fact, however, I have deemed it fitter to draw the boundary between the two groups according to whether proper bud-scales, as defined above, are present or absent. Where there are real bud-scales (Figs. 17, 18 and 19) the climatic periodicity is expressed by a corresponding morphological periodicity, which is easily observed and which is therefore of practical use to us in our classification. Those plants which, according to this division, come under the category of those without proper bud-scales, but whose growing points are well protected, whether by stipules, leaf-bases, or any other such means, might if desired be made into a group by themselves.

The boundary between plants with bud-scales and those without is not a sharp one; there are a number of Phanerophytes without well-marked bud-scales which have, however, a morphological periodicity dependent on the climate. At the beginning of the unfavourable season they form

Fig. 16. *Jacquinia acuminata* sprouting; *a*, the bud-scales are separated by distinct internodes so that the young foliage leaves, l_1, are borne above the older ones, *l*. ($\frac{2}{3}$)

leaves which are indeed green, and on the whole resemble foliage leaves, but never attain their full size. They remain small, and the internodes which separate them elongate but little; when the favourable season comes they have so far completed their structure that they are not able to grow any more. These leaves without having the special structure of bud-scales serve to protect the younger unexpanded leaves during the unfavourable season. As examples may be mentioned *Escallonia rubra* (Fig. 14, B), *Myrtus ugni* (Fig. 14, C and D), *Drimys Winteri* (Fig. 15, A), and *Viburnum tinus* (Fig. 15, B).

We find a second more advanced transition in plants which indeed have bud-scales, in that the protecting leaves always remain scale-like, but the internodes between these leaves elongate to a greater or lesser extent when the shoot develops. By this means the assimilating part of the shoot is borne farther away from the mother shoot; examples *Jambosa australis, Carapa procera, Eriodendron anfractuosum, Jacquinia acuminata* (Fig. 16) and *J. aurantiaca*.

Making use of leaf-fall and bud-covering as characters we obtain at once three obvious and useful types: (1) Evergreen Phanerophytes without bud-covering, (2) Evergreen Phanerophytes with bud-covering, and (3) Deciduous Phanerophytes with bud-covering. The distribution of these three types is on the whole closely correlated with the changes experienced in passing from constantly warm and humid tropical climates, through tropical and sub-tropical regions with a rainy and a dry season, to the colder parts of the earth, where there are long and more or less severe winters.

If we examine more closely the behaviour of these types within their regions of distribution, it becomes obvious that gradually, as the unfavourable season becomes longer and more severe towards the boundaries of the areas of the types, the plants become lower and lower. Gradually, as one departs from the coastal regions with tropical forest to regions farther inland where the dry season becomes longer and more severe, the woods become progressively lower. Similarly, in the colder regions of the earth the nearer we approach the poles, or the higher we climb into the mountains, the stature of the Phanerophytic vegetation dwindles, till at last it becomes composed entirely of Nanophanerophytes, which, by their lowness, have the advantage of being protected during the winter by snow, and by that means alone are able to survive the unfavourable season. Now dwarf growth (nanism), as understood here, has often perhaps not arisen as an adaptation for surviving the unfavourable season, but is probably a direct result of all the unfavourable

FIG. 17. *Griselinia litoralis* Raool. Evergreen Phanerophyte with bud-covering formed of one to several bud-scales. a_1, bud-scales on the boundary between this year's and last year's growth, a, scar of the corresponding bud-scale between growth period of last year and the year before. Side shoots (s) have appeared a little above their axillant leaves. ($\frac{2}{3}$)

FIG. 18. *Photinia serrulata* sprouting; *a*, Bud-scales. (⅔)

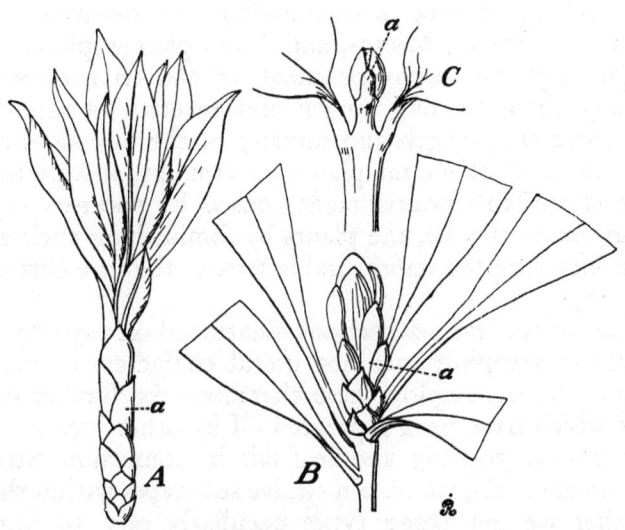

FIG. 19. Evergreen Phanerophytes with bud-coverings, *a*. A, *Pittosporum undulatum*. (¾). B, *Pittosporum crassifolium*. C, *Metrosideros tomentosa*. (⅔)

factors; but on the other hand dwarf growth is certainly an advantage to the plant that is obliged to survive unfavourable seasons, and since this character is very useful to us, it is convenient to subdivide the three above-mentioned types according to their size. Investigations in the field must decide how many divisions according to size it is useful to make, and at what height the boundary between them must be drawn. According to my experience it is useful to make four classes within each of the already mentioned types of Phanerophytes. In descriptions of Phanerophytic vegetation the specification of height is often left out; and the absence of the information makes our task a difficult one. I shall attempt it however by making use partly of what I have seen, and partly of what I have read. If the results thus obtained do not lead to the establishment of useful boundaries between heights, we must of course employ other methods. Our results must accommodate themselves to the picture of vegetation as a whole, and not to the vegetation of one particular region.

According to my own knowledge and according to the scanty information about the height of Phanerophytic vegetation in different regions which I have culled from literature, I assume that the boundaries between the four chief types can be drawn at 2, 8, and 30 metres, thus:

Nanophanerophytes	under 2 metres
Microphanerophytes	2–8 ,,
Mesophanerophytes	8–30 ,,
Megaphanerophytes	over 30 ,,

Various causes co-operate in occasioning the transition from Megaphanerophytes to Meso-, Micro-, and Nano-phanerophytes. We must consider especially the impoverishment of the environment brought about partially by increasing length and severity of the dry period, partially by increasing length and severity of the period of cold, which indeed has the same effect on plants as drought, and partially also by the decrease in available nourishment caused by poorness of soil. But whatever the causes may be, the plants by diminishing their size gain an advantage in surviving the unfavourable season; they are better protected against drought.

The two last of the five adaptations mentioned on page 24, anatomical and intracellular xerophily, are not useful characters in making types, but we may of course employ these characters for further demarcation of sub-types which have been separated off by other means.

Considering bud-covering and leaf-fall in connexion with the four above-mentioned classes, we obtain twelve sub-types within the Phanerophytes; further we get three types peculiarly easy to recognize: (1) Phanerophytic Herbs, (2) Phanerophytic epiphytes, (3) Phanerophytic stem-succulents.

Herbaceous Phanerophytes. Under this heading I include the vast number of Phanerophytes, more or less herbaceous, and found especially in regions with an almost constantly humid tropical climate. They resemble our large herbs, but their aerial shoots remain alive for several years without becoming woody; their stems are as a rule weaker than in real woody plants; strengthening tissue is not so much developed, and there is comparatively more parenchyma than there is in woody plants. Because of their lighter herbaceous structure herbaceous Phanerophytes are the least protected of all Phanerophytes; they are therefore found particularly in the most favourable climates, where they live for the most part protected by the higher Phanerophytic vegetation. As examples can be named: *Scaevola Koenigii*, *Rhytidophyllum tomentosum*, *Myriocarpa macrophylla* and other *Urticaceae*, species of *Impatiens*, many Begonias, *Acalypha hispida*; and many other *Euphorbiaceae*, species of *Piperaceae*, *Phytolaccaceae*, *Acanthaceae*, *Labiatae*, *Compositae*, *Verbenaceae*, *Araceae*, *Commelinaceae*, and other families.

As long as the concept of the herbaceous Phanerophyte is omitted by students of floras, we shall have to refer plants belonging to this category to the above-mentioned classes which are demarcated by size; most of them will be included among the Nanophanerophytes; but the importance of attempting to separate herbaceous Phanerophytes from Nanophanerophytes is apparent from the fact that while Nanophanerophytes (and Chamaephytes) characterize the tropical and the sub-tropical regions with a marked, but not an excessively severe, dry period, the herbaceous Phanerophytes, on the other hand, belong to types characteristic of constantly warm and constantly moist tropical climates.

In regions with a well-marked dry season the type of herbaceous Phanerophyte passes imperceptibly into the Chamaephyte, especially into the suffruticose Chamaephyte; because in this type the upper parts of the shoot die in the unfavourable season, leaving only the lowest part of the shoots with the buds attached to them to survive the unfavourable season. From the suffruticose Chamaephyte development, or rather reduction, proceeds as far as the Protohemicryptophyte.

The two other types, stem-succulent Phanerophytes, which have succulent stems without proper foliage leaves, and Epiphytic Phanerophytes (under this heading we must reckon also the Phanerophytic parasites), need no further description. They are characteristic of certain definite climates, and as they are easy to separate from other Phanerophytes they therefore form useful types.

Conspectus of Phanerophytic Sub-types. Within the Phanerophytes we have thus the following fifteen sub-types:

1. Herbaceous Phanerophytes
2. Evergreen Megaphanerophytes without bud-covering
3. ,, Mesophanerophytes ,, ,,

4. Evergreen Microphanerophytes without bud-covering
5.   ,,   Nanophanerophytes   ,,   ,,
6. Epiphytic Phanerophytes
7. Evergreen Megaphanerophytes with bud-covering
8.   ,,   Mesophanerophytes ,,   ,,
9.   ,,   Microphanerophytes ,,   ,,
10.   ,,   Nanophanerophytes ,,   ,,
11. Stem-succulent Phanerophytes
12. Deciduous Megaphanerophytes ,,   ,,
13.   ,,   Mesophanerophytes ,,   ,,
14.   ,,   Microphanerophytes ,,   ,,
15.   ,,   Nanophanerophytes ,,   ,,

We have here, of course, no continuous series in the sense that each member is better adapted than the one before to survive the unfavourable season. This, however, shows no defect in the types; it is, on the other hand, a natural expression of the fact that climate-forms do not constitute one, but several series. It is in harmony with this fact that there are no sharp boundaries between the types; the same species passing from one climate to another often changes its type; species which in a more favourable climate are Evergreen Phanerophytes may in a less favourable climate occur as Deciduous Phanerophytes; species which in certain climates are Microphanerophytes very often become in more unfavourable climates Nanophanerophytes. This is, as already said, not a defect in the types, but on the other hand a merit, for it shows that the types are in harmony with nature: with a change in climate there occurs a corresponding change in life-form, which is indeed an expression of the climate.

II. Chamaephytes

By Chamaephytes I mean plants whose buds or shoot-apices destined to survive the unfavourable season are situated on shoots or portions of shoots which either lie on the surface of the earth or are situated quite near to it, so that in regions where the ground in winter is covered with snow this can protect them, or in warmer regions with a dry season the buds can be partially protected by the withered remains of the plants on the surface of the ground. In either case, because the buds of these plants are near the surface of the earth, they are *ceteris paribus* much better protected than the buds of Phanerophytes, which are situated on shoots projecting a long way into the atmosphere.

The flowering shoots are, as a rule, negatively geotropic, and project freely into the air; they live only during the favourable season, and have to occupy a conspicuous position to ensure pollination, whether by animals or by the wind.

On the other hand the persistent shoots which bear the surviving buds lie along the earth, or at any rate do not project more than 20 or 30 cm. above it. If they project to a greater height the plants pass into the category of Phanerophytes. Different causes determine the proximity of the shoots to the ground in different Chamaephytes, and the greater or less degree of adaptation to life near the ground can be used for making divisions within the type.

In some Chamaephytes the shoots that bear the surviving buds are near the ground, because the shoots, which project into the air in the favourable season, die back to near the ground at the beginning of the unfavourable season. I call this type Suffruticose Chamaephytes. In others the shoots lie on the ground because of their weight; these shoots are long, slender, and comparatively flaccid, and plants possessing them I call Passive Chamaephytes. The most pronounced Chamaephytes lay their shoots flat on the ground because their shoots are transversely geotropic, i.e. they take up a horizontal position in response to the action of gravity. Plants possessing such shoots I call Active Chamaephytes; they are the most marked of all Chamaephytes because they have become Chamaephytes by an alteration of the physiological reaction of their shoots, the reaction towards gravity. Cushion Plants constitute a fourth type, forming a transition to the Hemicryptophytes. In Cushion plants the surviving buds or shoot-apices are close to the ground because the shoots are very low; they are, as a rule, so closely packed together that they both support and protect one another.

16. **Suffruticose Chamaephytes.** The shoots which develop in the favourable season are as a rule markedly negatively geotropic and bear leaves and flowers. At the end of the vegetative period the upper parts of these shoots die away so that only the lower portions survive the unfavourable season, and these surviving portions bear the buds which in the next period of growth are destined to grow out into shoots bearing leaves and flowers (Fig. 7, 2). This proximity of the buds to the ground saves them from being exposed to the desiccation to which the buds of Phanerophytes are exposed. The buds of the Suffruticose Chamaephytes are in part protected from desiccation by the wholly or partially dead plant-remains which lie on the surface of the ground.

The length of the persistent portion of the shoot varies widely, and in individual species is to a certain extent adapted to the climate. If the conditions are particularly favourable it is only the most distal portions of the shoots that die, and the plants then often very closely resemble Herbaceous Phanerophytes, from which type they have certainly been in great part derived. If the conditions are very unfavourable then the shoots die back nearly to the ground, forming imperceptible gradations towards the first type of Hemicryptophytes. Nanophanerophytes also often make transitions to Suffruticose Chamaephytes. The fact that the spaces between the life-forms are thus bridged within a single species when the unfavourable season is more severe, shows that the plan on which the life-forms are built is able to reflect with accuracy changes in climate; this indeed is its chief task.

Suffruticose Chamaephytes are doubtless especially widely distributed in the warm and warm-temperate regions of the earth with a rather long dry season, where the factors are less favourable than in the regions where Nanophanerophytes have the upper hand; but the two types often grow together. Suffruticose Chamaephytes are especially in evidence in many parts of the Mediterranean region, where they are represented by species of *Labiatae, Caryophyllaceae, Leguminosae*, and other families.

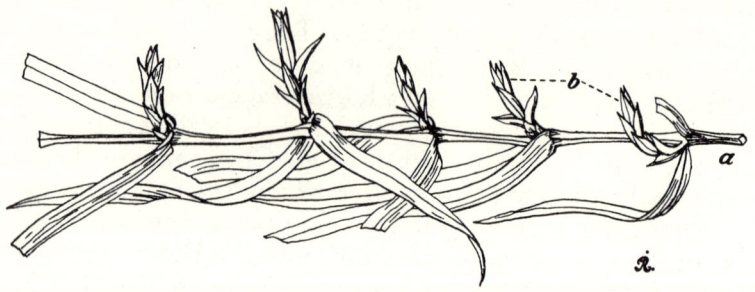

Fig. 20. *Stellaria holostea*; detached portion of an aerial procumbent shoot, *a*, from whose leaf-axils the buds, *b*, destined to survive the winter, are shooting.

In this sub-type I include also a small group of Chamaephytes whose shoots are upright and negatively geotropic but always of meagre stature. Their shoot-apices do not die like those of the real Suffruticose Chamaephytes; but these plants cannot be included among the Cushion plants because they form no cushions.

17. **Passive Chamaephytes.** The vegetative shoots are, like those of Phanerophytes, negatively geotropic and persistent, remaining intact at the beginning of the unfavourable season; but because they are weak in proportion to their length and not furnished with sufficient strengthening tissue, they are not able to stand erect, but fall over and lie along the ground. It is only the apices of these shoots that are erect or ascending, because it is only at the apex that the negative geotropism is able to influence growth; the extent of this influence depends upon the rigidity of the shoot. By observing the growing apices we can easily determine whether the plant is a passive or an active Chamaephyte; if it is active even the growing apices are horizontal. Among passive Chamaephytes we find evergreen and deciduous species, species with and species without bud-covering. Some are pronounced woody plants, others more herbaceous. These and other characters can guide us in making sub-types. They all belong especially to the Alpine region.

Examples are: *Arabis albida, A. alpina, Aubrietia* sp., *Veronica* sp.,

Sedum, Saxifraga, Polygonum brunonis, Kernera saxatilis, Stellaria holostea (Fig. 20), *Cerastium tomentosum, C. trigynum,* and several other species of *Cerastium, Campanula fragilis, Lotus peliorrhynchus,* and many others.

18. **Active Chamaephytes.** These too have persistent vegetative aerial shoots, but they differ from the Passive Chamaephytes, in that the shoots lie along the ground because they have a different reaction towards gravity, pursuing a direction at right angles to the gravitational pull (transverse geotropism) (Fig. 7, 3). The growing apices therefore do not ascend like those of the Passive Chamaephytes, though they sometimes appear to do so from the leaves being bent upwards. Sometimes the shoots lie on the ground because they are negatively heliotropic.

This type, like the last, contains both evergreen and deciduous species, some with and some without bud-covering; some are pronounced woody plants, others herbaceous. As shoots which lie on the ground are inclined to strike root when they come into contact with damp earth, this type passes by imperceptible gradations into Hemicryptophytes with aerial stolons. The difference between the two types depends upon whether or not the roots draw the surviving buds down into the soil. Unless this takes place throughout a long series of buds, I include the plant among the Hemicryptophytes. Because of the nature of the case we should not, and indeed cannot, draw a sharp boundary.

Examples of Active Chamaephytes are: *Thymus* sp. (e.g. *Thymus zygis*), *Veronica officinalis, Vinca, Acaena Novae Zealandiae, Cerastium caespitosum, Arctostaphylos uva ursi, Empetrum nigrum, Linnaea borealis, Lysimachia nummularia,* and many others.

Looked at biologically the Active and the Passive Chamaephytes are essentially the same, and their occurrence points to approximately the same conditions. They differ chiefly in the physiological method by which they have become Chamaephytes; the Active Chamaephytes have adopted the more thorough method, so this type may be looked upon as the more perfectly adapted of the two. Both types are specially constructed to live where there is a snow covering which protects the shoots and buds against loss of water; they are both comparatively abundant in alpine regions.

If the shoot of an Active Chamaephyte be fixed in the light with the apex pointing upwards, when growth continues it will bend again in a horizontal or obliquely descending direction, and go on growing until the shoot, partly or entirely of its own weight, reaches the surface of the ground, and then grows along it. On the other hand in the dark the shoots of these plants are negatively geotropic, at any rate in the species I have investigated, e.g. *Cerastium caespitosum, Veronica officinalis,* and *Lysimachia nummularia.* The behaviour of *Lysimachia* has already been discussed by Oltmanns, and *Vinca major* behaves according to Czapek in the same way. In the three species mentioned above, *Cerastium caespitosum*, *Veronica officinalis*, and *Lysimachia nummularia*, the shoots in my experiments showed themselves to be weakly positively heliotropic. In weak unilateral illumination they grew towards the light. Czapek has found that *Lysimachia nummularia* behaves in this way.

An interesting transition between Passive and Active Chamaephytes occurs in some species which seem to have entirely lost their sensitiveness to the stimulus of gravity. The shoots of these plants continue to grow in any direction in which they are placed; only the weight of the shoot forces it gradually out of that position, unless, of course, it has been placed with the apex pointing downwards. As in the Passive Chamaephyte the shoots of these plants rest on the ground because of their own weight, but they differ from the shoots of the Passive Chamaephyte in that the apices show no upward curving due to negative geotropism. On the other hand the horizontal direction of the shoot-apices is not, as in the Active Chamaephytes, determined by transverse geotropism, because the shoots are indifferent to gravity; their direction is due to the position their own weight has caused them to assume.

Rosa Wichuraiana is an example. This plant has procumbent shoots, which in the course of a season can grow several metres in length. The shoot-apices are sufficiently rigid to enable them to maintain a definite direction, at any rate for a short distance. This can be demonstrated by allowing a shoot lying on the ground to grow over a depression; the apex will grow on in its original direction without support; not until it is several centimetres long does its own weight begin to cause it to sink. Shoots fastened in a vertical direction with their apex upwards continue to grow upwards until they become so long that they lose their equilibrium and topple over by their own weight, which causes the shoots as they elongate to droop more and more until they become pendulous. Then growing vertically downwards they at last reach the earth, which compels them to grow to one side. Not even in darkness do the shoots of *Rosa Wichuraiana* appear to be negatively geotropic, as are many species with horizontal aerial shoots.

19. **Cushion Plants.** As in the Passive Chamaephytes the shoots are negatively geotropic, but arranged so close together that they prevent each other from falling over, even if they are not furnished with sufficient strengthening tissue to enable them to stand erect. The shoots, which are as a rule low, are often packed together so tightly that the cushion becomes quite solid. This solidity brings about a mutual protection to the shoots during the unfavourable season. The surviving shoot-apices are protected by withered leaves, as often happens in Hemicryptophytes. There are transitions between Cushion plants and the tussock-forming Hemicryptophytes. On the other hand Cushion plants are connected by imperceptible gradations to Passive Chamaephytes, from which they are certainly derived. A number of Passive Chamaephytes show a ten-

dency to form cushions. In favourable conditions the shoots are so lax that the branchlets are not pressed so tightly together as to support one another. Under these conditions a proper cushion is not formed. When, however, the conditions are less favourable the shoots are so short and arranged so close together that they become real cushion plants. Examples are seen in *Sedum, Saxifrage, Aubrietia,* and other genera.

Real Cushion plants belong, even more than the Passive Chamaephytes, to the alpine region. Examples are *Maja, Merope,* and *Azorella* in the Andes; *Myosotis Hookeri, Crepis glomerata, Saussurea hemisphaerica, S. gossypiphora, Acantholimon diapensioides,* and *Saxifraga hemisphaerica* in the highlands of Central Asia; *Myosotis uniflora, Veronica pulvinaris, Azorella elegans, Hectorella caespitosa, Raoulia mammillaris, R. eximia, R. australis,* and other spp. of *Raoulia, Haastia pulvinaris, H. Greenii,* and *Dracophyllum muscoides* in the mountains of New Zealand.

III. HEMICRYPTOPHYTES

This type is better adapted than the last to survive unfavourable seasons, the shoots dying back to the level of the ground at the beginning of the unfavourable period, so that only the lowest parts of the plant, protected by the soil and withered leaves, remain alive and bear buds destined in the next period of growth to form shoots bearing leaves and flowers (Fig. 7, 4). Thus in all Hemicryptophytes the shoot-apices which are to survive the unfavourable season are situated in the soil-surface, protected by the surrounding soil and by the withered remains of the plant itself.

In most Hemicryptophytes the surviving portion of the single shoot persists for several years, so that there is formed a more or less branched contracted system of shoot-bases corresponding to the shoot-bases of Suffruticose Chamaephytes, but differing from them in being situated in the ground and not above it.

Outside the warmer and the moderately humid regions of the earth, where Phanerophytes abound, the Hemicryptophytes are manifestly dominant; apart, that is to say, from the driest regions, where Cryptophytes, and under certain conditions Therophytes, form an essential component of the vegetation. Most of our herbs are Hemicryptophytes, and probably about half of the plants belonging to Central Europe belong to this type.

Morphologically great diversity prevails in these plants, a diversity connected with the manifold devices for vegetative reproduction and distribution which often characterize Hemicryptophytes. They attain no great height, but spread extensively by means of horizontal runners. In some the primary root persists as it does in most Phanerophytes, and the whole shoot system lives as long as the individual; this type is connected with the Suffruticose Chamaephytes. In most of them the primary root dies after a longer or shorter time and the individual becomes attached by adventitious roots. Not only does the primary root die, but the shoot system gradually dies back from behind, so that vegetative reproduction takes place by division, the individual parts of the shoot system becoming free by the dying of the older portions which held the younger together. The extreme in this direction is reached by those plants whose persisting shoot dies before the buds attached to it unfold. This, for example, may happen in *Epilobium montanum* (Fig. 21, B), *Samolus Valerandi*, species of *Aconitum*, &c. Persisting buds in these plants develop adventitious roots at a later stage, so that they are able to continue life independently.

The length of the persisting portion of the shoot varies greatly. If we confine ourselves first to those plants whose shoots are from the very beginning all negatively geotropic and therefore erect, then we find that the length of the persisting shoot depends upon the depth in the soil from which the shoot sprang, because the whole of the subterranean portion of the shoot, as a rule, remains alive, and only the aerial portion dies back. Normally the new shoots spring from the uppermost part of the persisting portion of the old ones, which are, of course, in the soil; but sometimes shoots arise lower down from dormant buds on portions of old stems. These buds have to break through a thicker layer of soil, so that their underground persistent portions are comparatively longer; but the buds destined to survive the unfavourable period and which in the next growing season are to form aerial shoots, are normally laid down in the portion of the shoots situated in the soil.

While the new shoots thus spring from that portion of the mother shoot which is in the soil, only the lowermost portion of these shoots remains covered by the soil; but that portion which is to bear the surviving buds must have a certain length even if its internodes are very short, and the plant's surviving shoot system will therefore ultimately grow up over the surface of the ground and die back if it has no other means of keeping itself covered. This protective covering is obtained in different ways, partly passively, partly actively. Passively it takes place, as is well known, by 'hilling', partly by means of worms and other animals, partly by the layer of dried plant-remains which every year is deposited on the ground. Actively the covering takes place by the contraction of the roots dragging the plant down into the soil; this has been shown to take place in many plants. By this means the underground stems are kept at such a depth that they are covered and the buds do not come above the surface of the ground.

It may however happen that the underground shoot system becomes buried too deeply. This may occur through the activity of moles or in other ways. Hemicryptophytes correct this state of affairs by laying down their rejuvenating buds not at the base of the shoots but on the portion of the shoot which is in the upper layer of the soil; so that the surviving buds come to occupy the position characteristic of Hemicryptophytes.

Even if it be true that all Hemicryptophytes have their surviving buds or shoot-apices situated in the soil-surface, yet the various species belonging to this category are not all equally well protected, because there are also other means of protection than that afforded by the soil.

The side shoots which are laid down in the course of the favourable season and which are destined to develop in the following period of growth, always have some of the lowest leaves modified to be mainly protective. The degree of this protection varies widely, but it is difficult to measure, and is therefore unsuited to serve as a character for the formation of sub-types.

The apices of shoots which have already developed foliage leaves are as a rule not furnished with protective scales, but are protected by the lowermost portions of the living or withered leaves. This is well known in rosette plants. Only rarely does the shoot which has not yet formed an aerial portion, though it may possess radical foliage leaves, continue to the end of its period of growth to form scale leaves for protecting the apex: examples are *Carex caespitosa* and *C. stricta*.

The protection can thus be of various kinds. One must remember that there are also anatomical and intracellular adaptations; but as their degree is difficult to determine they cannot be used in classification, however tempting it may be so to use them. The sub-types within these life-forms will therefore be demarcated according to the degree in which the plant has been externally modified for its life in the soil, taking for granted that the ordinary aerial shoot is the most primitive.

As in Phanerophytes which have covered buds, the shoots of the Hemicryptophytes start with imperfectly formed leaves. Some Hemicryptophytes continue to produce imperfect leaves as far as the lowest part of the aerial shoot, on which are therefore found the largest foliage leaves and the greatest number. The part of the aerial shoot which has elongated internodes is the only flower-bearing and at the same time the only assimilating part of the plant; there is no radical rosette of leaves. This, the least altered type of Hemicryptophytes, I propose to call the Proto-Hemicryptophyte.

In other Hemicryptophytes, though the elongated portion of the aerial shoot bears foliage leaves, the greatest number and the largest of them are found on the lowermost part of the shoot where the internodes are shorter, partly in the soil, and partly immediately above its surface; they gradually decrease in size upwards; thus we have, besides the foliage leaves on the elongated portion of the aerial stem, also a more or less marked radical rosette; I therefore call these partial rosette plants (Fig. 25).

Finally we have the **genuine rosette plants** in which all or nearly all of the foliage leaves are situated in a radical rosette, and the elongated aerial shoots produce flowers only.

20. **Proto-Hemicryptophytes** are Hemicryptophytes whose leaf- and flower-bearing aerial shoots are elongated from the base. The largest foliage leaves are as a rule borne by the middle portion of the stem or thereabouts, and from this point they decrease in size towards the base and towards the apex of the shoot. The leaves situated immediately above the soil are more or less scale-like and serve to cover the buds during the unfavourable season. The shoots are thus essentially like those of Phanerophytes, and this resemblance is even more marked if we look at seedlings and at shoots not about to flower. Even in the year in which they germinate Proto-Hemicryptophytes develop an elongated aerial shoot entirely resembling those of the Phanerophyte and unlike that of the other Hemicryptophytes. Even later in the plant-life we usually see that even the shoot which is not destined to flower develops into an elongated aerial shoot (Fig. 21, A). By this behaviour Proto-Hemicryptophytes demonstrate their position as representing the first stage in adaptation to life in the soil.

Just as within the Chamaephytes, Suffruticose Chamaephytes often pass imperceptibly into Herbaceous Phanerophytes and Nanopherophytes, so the Proto-Hemicryptophytes pass imperceptibly into the Suffruticose Chamaephytes (Fig. 22), from which most of them are certainly descended. The type is distributed both in warm parts of the world with an unfavourable season, a dry period, and also in the colder regions of the world, where the unfavourable season is a cold winter. In the latter regions tuberous structures for food storage appear to be rare (Fig. 23); in the former regions on the other hand we often see large tuberous parts of the stem situated in the soil; these organs presumably function as water storers in the dry season. Vegetative distribution over wide areas is not common, since most, at any rate of the European species, are without stolons.

A. WITHOUT STOLONS. In some the primary root persists, in others, doubtless the majority, it ultimately dies, and the older portions of the subterranean stem system gradually die too, so that reproduction by division takes place, when the connexion between the different shoot systems is severed. But since these plants have no stolons they cannot multiply copiously in a vegetative manner; their growth is usually more or less densely tufted. Examples are: *Thalictrum flavum, T. flexuosum, T. minus, Hypericum hirsutum, H. pulchrum, H. montanum, H. perforatum, H. quadrangulum, H. tetrapterum, Euphorbia salicifolia, E. dulcis, E. virgata,* &c., *Veronica longifolia, V. latifolia, V. sibirica, V. teucrium, V. austriaca, Verbena officinalis, V. littoralis, V. bonariensis, V. urticifolia, Linaria genistifolia, L. purpurea, L. italica, L. dalmatica, Scrophularia lateriflora,*

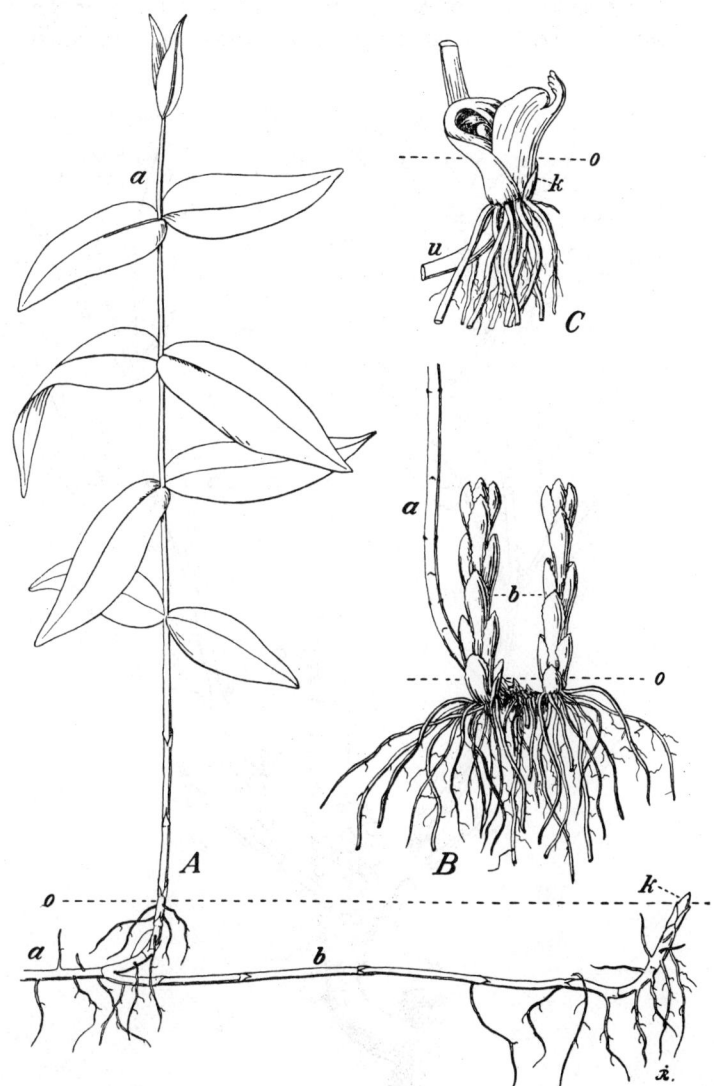

FIG. 21. A, *Lysimachia vulgaris*; a proto-hemicryptophyte; the shoot *a*, although not destined to flower, has developed an elongated aerial portion, which will die at the close of the growing period; from the underground portion of this shoot there springs a new subterranean lateral stolon, *b*, which bends up at its apex so that the winter bud, *k*, is situated in the soil, *o*. (½). B, *Epilobium montanum* in its spring condition; *a*, last year's dead shoot; the winter buds, *b*, which are now expanding, were situated in the soil, *o*. (½). C, *Aegopodium podagraria* in spring; *u*, the apex of a stolon with the bud, *k*, which is situated in the soil, *o*, and is now expanding. (¼.)

S. nodosa, S. vernalis, S. canina, S. alata, Astragalus glycyphyllus, Onobrychis viciaefolia, Melilotus albus, M. altissimus, M. officinalis, Medicago

FIG. 22. *Nepeta latifolia*: shoot system in the autumn stage; *a*, persistent portion of a shoot that flowered in 1903. *b*, lowermost persisting portion of shoot that flowered in 1904; the buds attached to this are in the soil; they will develop next year, and have already in the autumn of 1904 grown out into leafy shoots, *c*, with a short lateral shoot, *d*, situated in the soil-surface. Under favourable conditions the aerial portion of the shoot, *c*, may survive the winter, and the plant can thus grow as a Suffruticose Chamaephyte. (⅔)

falcata, M. sativa, Epilobium roseum, E. montanum, Ballota nigra, Lamium album, Marrubium vulgare, Nepeta cataria, N. latifolia and many other Labiates; further *Cynanchum vincetoxicum*, many species of *Galium*, many

Composites: species of *Artemisia, Inula, Aster, Hieracium, Solidago,* and many other genera.

B. WITH STOLONS. Some species may have both subterranean and epigeal stolons, e.g. *Stachys silvatica* and other Labiates; further *Urtica dioica*. Examples of plants with subterranean stolons are *Epilobium tetragonum, E. obscurum, E. palustre, E. parviflorum, E. hirsutum, Lysimachia vulgaris* (Fig. 21, A), *Mercurialis perennis, Saponaria officinalis,* various *Labiatae* and *Leguminosae* (for example species of *Lathyrus* and *Orobus*).

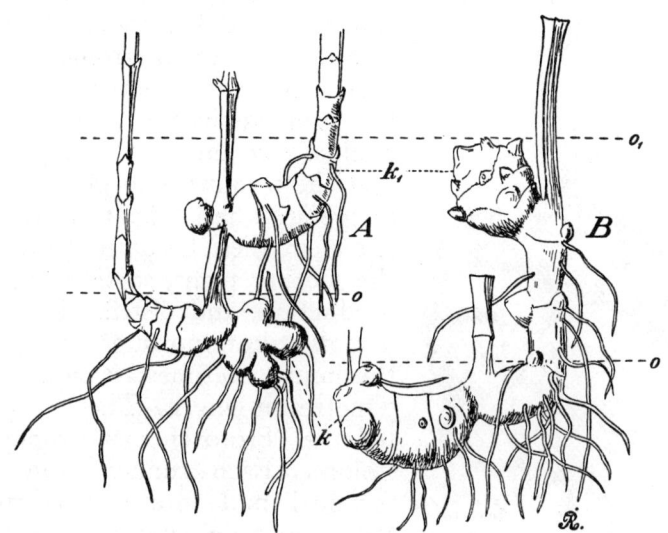

FIG. 23. *Scrophularia nodosa;* two plants, which have become covered with a layer of soil, $o-o_1$; the new tubers, k_1, with the winter buds, are not formed like the others immediately beside the old tubers, k, but higher up, i.e. in the soil-surface. (¾)

Among Proto-Hemicryptophytes must also be reckoned our species of *Rubus*, of which, however, most occupy a peculiar position within this sub-type. Our species of the sub-genus *Cylactis*, namely *Rubus chamaemorus* and *R. saxatilis*, are genuine Proto-Hemicryptophytes, the former without and the latter with stolons. All our other species, namely species of the sub-genera *Idaeobatus* and *Eubatus*, deviate in that the aerial shoots do not die at the end of the growing period, but surviving the unfavourable season, live for one more growing period, and then die; they thus have a biennial aerial life; the first year they are vegetative, bearing only leaves, in the second year they produce the flowering lateral shoots, and then die down to the subterranean portion, from which the shoots arise that are to carry on the next year's growth. Although the aerial shoots thus live for two years it is only the buds on the flowering shoots which survive the unfavourable season on shoots which project into the air, while the vegetative buds, on which the continuation of the individual life depends, are always found on the part of the shoot situated in the soil-surface (Fig. 24), as in the other Proto-Hemicryptophytes. I therefore consider that these species of *Rubus* belong to that group. It is not necessary for the continuance of the individual that the aerial shoot should survive the unfavourable season. In some species the vegetative shoots are erect or arched, but throughout their whole length they are free aerial shoots. These species belong to the Proto-Hemicryptophytes without stolons: examples are *Rubus idaeus*, *R. suberectus*, and *R. plicatus*. In most of the species they are so strongly arched that their apices with the terminal buds reach the ground, where they root and are gradually covered by 'hilling'. The next year the terminal bud grows out to form a vegetative aerial shoot, which, when the old stem dies, becomes quite free from the mother plant, so that vegetative reproduction, and distribution take place. The distribution can go on very actively, the stem sometimes attaining as much as 7 metres in length. These species belong to the Proto-Hemicryptophytes with stolons. Each runner grows from its terminal bud into a new runner the following year, and in this way there ultimately arises a monopodium which forms an undulating line; the crests of the waves in this line are formed of the arched aerial portions of the shoot, and the troughs are formed of the shoot-apices which are carried down into the soil-surface so that the terminal bud may be protected during the unfavourable season. In this way new seats are formed for the origin of several monopodia, for besides the terminal bud which continues the monopodium already laid down, there are also a varying number of lateral buds from which the beginnings of new monopodia arise.

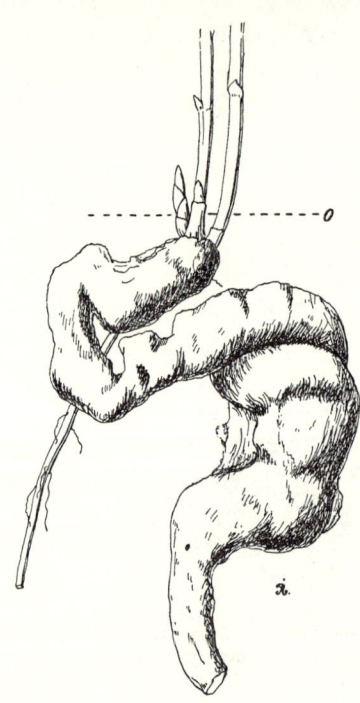

FIG. 24. *Rubus* sp. in spring. *o*, soil-surface. (⅔)

21. **Partial Rosette Plants.** In these plants too the aerial shoot bears foliage leaves as well as flowers, but the largest leaves and also often the greatest number of them are attached to the lower portion of the shoot, where the internodes are more or less contracted, so that they form a kind of rosette (Fig. 25). The shoot is usually biennial; in the first year the radical rosette is formed, in the second year an elongated aerial shoot with foliage leaves and flowers.

LIFE-FORMS OF PLANTS

As in the Proto-Hemicryptophytes the aerial portion of the shoot bears foliage leaves, but in the partial rosette plant the greatest development of leaves extends from the middle of the stem to its contracted

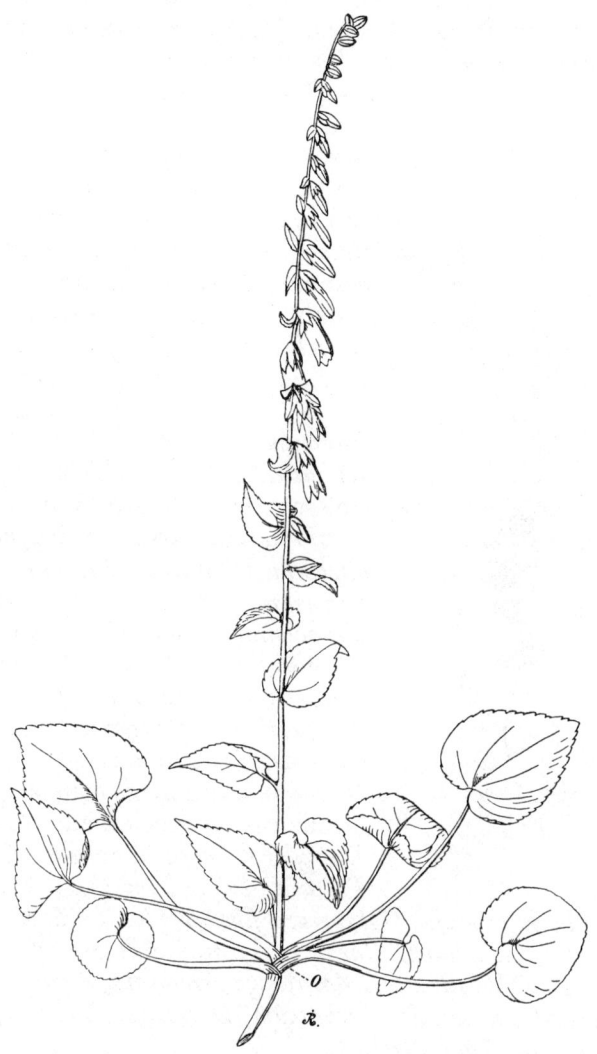

Fig. 25. *Campanula lamiifolia*; partial rosette plant. *o*, soil-surface. (¼)

lowest portion, so that a transition is formed from the rosette plants in which the elongated aerial portion of the shoot bears only flowers and the foliage leaves are all gathered into a rosette at the base. Most of these plants are without stolons. They live principally in temperate regions where the summer is not too dry, and where the ground is covered with snow for a longer or shorter period; Central Europe is the centre of their distribution. Besides a great number of perennial herbs the bulk of biennials belong here too.

A. WITHOUT STOLONS. Many species of the following families: *Caryophyllaceae*, e.g. *Viscaria viscosa*, *Melandrium rubrum* and *M. album*, *Lychnis flos cuculi*, species of *Silene*, *Dianthus*; *Ranunculaceae*, e.g. *Caltha palustris*, *Ficaria ranunculoides*, *Ranunculus flammula*, *R. auricomus*, *R. acer*, *R. lanuginosus*, *R. polyanthemus*, and other species, *Helleborus*, *Aquilegia*, *Aconitum*, *Delphinium*, and many others; *Rosaceae*, e.g. *Agrimonia*, *Poterium*, *Sanguisorba*, and *Spiraea*; *Umbelliferae*, e.g. *Anthriscus silvester*, *Oenanthe Lachenalii*, *Cnidium venosum*, *Libanotis montana*, *Angelica silvestris*, *Selinum carvifolium*, *Peucedanum oreoselinum* and *P. palustre*, *Laserpitium latifolium* and many others; *Campanulaceae*: many species of *Campanula*, and *Phyteuma*; *Dipsaceae*: species of *Dipsacus*, *Knautia*, *Scabiosa*, and *Succisa*; *Compositae*, e.g. species of *Lappa*, *Serratula*, *Carduus*, *Cirsium*, *Centaurea*, *Arnica montana*, *Solidago virga aurea*, *Erigeron acer*, *Aster tripolium*, *Aracium paludosum*, *Picris hieracioides*, *Lactuca muralis*, *L. scariola*, *Chrysanthemum leucanthemum*, *Cineraria campestris*, *Senecio erucaefolius*, *S. jacobaea*, *S. aquatica* and others, *Inula helenium*, *I. vulgaris*, *Hieracium vulgatum*, *H. murorum*, *H. caesium* and many others; *Grasses*, most of our tufted species, e.g. *Aira caespitosa*, *Dactylis glomerata*, species of *Schedonorus*, *Festuca*, *Poa*, *Cynosurus*, &c. Species of *Rumex*, *Chelidonium majus*, *Glaucium flavum*, *Nasturtium anceps*, *Cardamine pratensis*, *C. amara*, *Barbarea lyrata*, *Alliaria officinalis*, *Malva alcea*, *M. moschata*, *M. silvestris*, *Saxifraga granulata*, *Anthyllis vulneraria*, *Samolus Valerandi*, *Myosotis*, *Verbascum*, *Digitalis*, *Veronica*, *Pedicularis*, *Betonica*, *Brunella*, *Ajuga*, &c., &c.

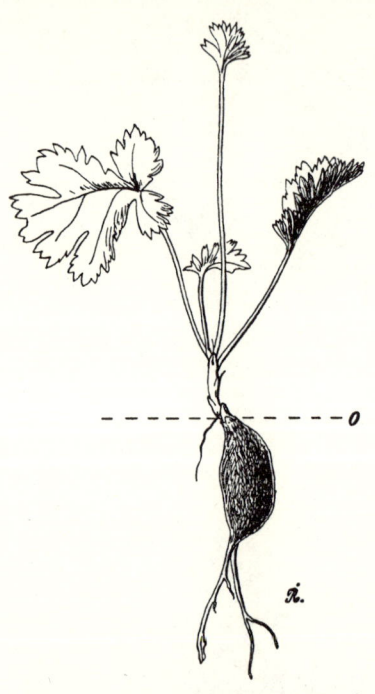

FIG. 26. *Pelargonium flavum*. Hemicryptophyte with hypocotyl (?) stem tuber; a young plant. *o*, surface of soil. (½)

B. WITH STOLONS. The following have aerial stolons: *Ajuga reptans* and *Ranunculus repens*. The following have subterranean stolons: *Aegopodium podagraria*, *Cirsium heterophyllum*, *Tanacetum vulgare*, *Achillea millefolium*, and a number of grasses and sedges.

22. Rosette Plants. The aerial elongated portion of the shoot in these plants is almost exclusively flower-bearing, while the foliage leaves are attached close together to that portion of the shoot which is situated in the soil-surface (Fig. 27). This life-form is the one that has undergone most transformation for adaptation to life in the soil-surface; during the whole of the vegetative period the shoots remain in the soil-surface; only when the plant flowers is there formed an aerial shoot, or scape.

In most of them the underground stem is a sympodium, and the development of the shoot is biennial (not counting the bud stage); in the first year a rosette of leaves is formed and the second year a flowering aerial portion, which ends the life of the shoot.

For efficient carbon assimilation it is very important that the closely packed leaves of the rosette should not shade one another more than is

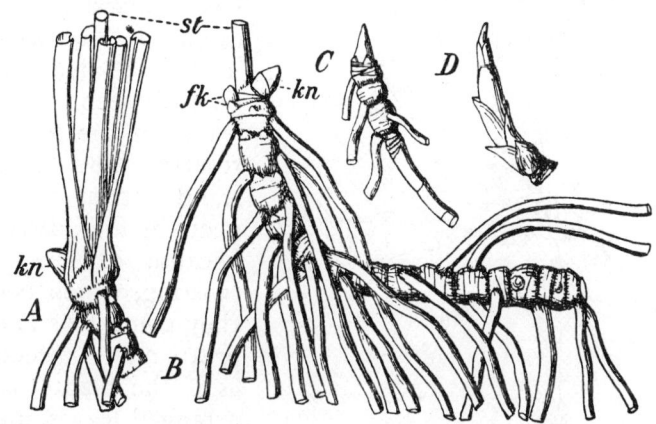

FIG. 27. Rhizomes of *Anthericus ramosus*; a Hemicryptophyte (Rosette plant). The winter buds—*kn* in A; *fk* and *kn* in B, C, and D—are situated in the soil-surface; *st*, the lowest portion of the flowering stem with the radical rosette. (c. ⅓)

absolutely necessary. This end is reached by various devices; sometimes the leaves are borne on long stalks, so that the blades are separated (examples *Drosera*, *Viola*, *Petasites*); sometimes the same result is attained by the leaves being long and narrow (examples *Luzula*, Grasses and Sedges). But the most characteristic rosette leaf has a narrow base from which it gradually broadens towards the apex, in other words it gradually broadens as the space at its disposal increases. Leaves of this kind may be obovate, spathulate, or of other forms; examples include the most characteristic rosette plants, *Sempervivum*, *Primula*, *Plantago*, *Bellis* (Fig. 28), *Hieracium pilosella*, *H. auricula*, *Taraxacum*, *Hypochaeris maculata*, and many others. But even if the form of the leaf be the one best suited for its own illumination, that would not suffice if the arrangement and direction of the leaves were not such as to prevent them standing immediately above one another. Thus the leaves of rosette plants are not merely as a rule alternate, but there are a great many leaves in each spiral, so that many leaves are developed before one arrives at a point immediately above an already developed leaf. Even when this occurs, it does not follow that the last developed leaf shades the leaf below, for it often happens that a leaf does not project in the same direction as the one immediately below it, but deviates slightly to one side, thus scarcely obscuring the lower leaf. We must consider too that the youngest leaves are still small and cover only the narrow basal portion of the older leaves, so that all the space is used in the best possible way (Fig. 28). We see in such rosettes a greater number of rows than the phyllotaxy would lead us to suppose possible: an especially good example of this is *Crassula orbicularis* (Fig. 29), which has a marked rosette with opposite decussate leaves. The leaf arrangement can be easily seen from the young leaves in the middle of the rosette. The leaves do not stand out, as one would expect, in four directions, but because of the process of twisting, which begins early, they come to stand out in a great many directions, just as if the plants had spirally arranged leaves, numerous in each turn of the spiral.

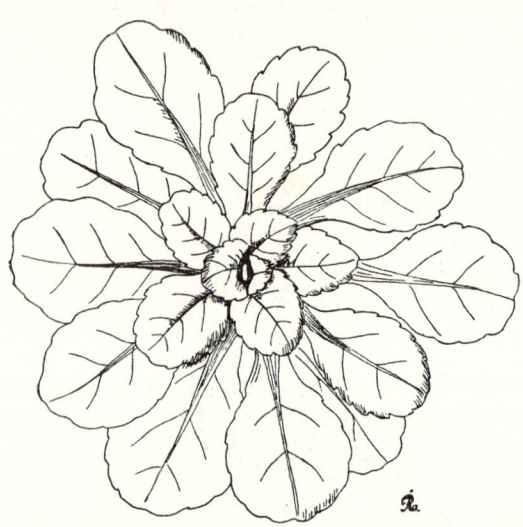

Fig. 28. Rosette of *Bellis perennis* seen from above. (⅓)

Rosette plants belong principally to regions where the ground is covered with snow in winter. Most of them have winter-green leaves, which are protected by the snow; they can begin assimilation immediately at the beginning of the favourable season.

A. WITHOUT STOLONS. Examples of Danish plants: *Drosera, Statice, Limonium, Primula, Bellis, Taraxacum, Thrincia, Leontodon, Hypochoeris, Triglochin maritimum, Liparis Loeselii* (Fig. 31), *Malaxis paludosa, Spiranthes spiralis* (Fig. 32, B), some species of *Luzula*, Grasses and Sedges, especially species of *Carex*.

B. WITH STOLONS. *Petasites, Hieracium pilosella, H. auricula*, &c., *Triglochin palustre, Scheuchzeria palustris, Goodyera repens* (Fig. 32, A); and some Grasses and Sedges. In some the stolons are epigeal, for example *Hieracium* sp., but in most they are subterranean.

MONOPODIAL ROSETTE PLANTS. In the examples of Rosette plants mentioned above, the subterranean stem system is a sympodium, but I include among Rosette plants also Hemicryptophytes whose rhizomes are monopodia. Some of the species belonging here are marked Rosette plants; but others on close examination are found to resemble partial

LIFE-FORMS OF PLANTS

Rosette plants, or even Proto-Hemicryptophytes, since the aerial shoots bear foliage leaves. If I include them among Rosette plants it is because the aerial shoots have no direct significance for the plants' ability to survive the unfavourable seasons. The aerial shoots die back to the base, so that their lowermost parts situated in the soil-surface do not bear buds destined to survive the unfavourable period of the year; the surviving buds are found exclusively on the monopodium, which is situated in the

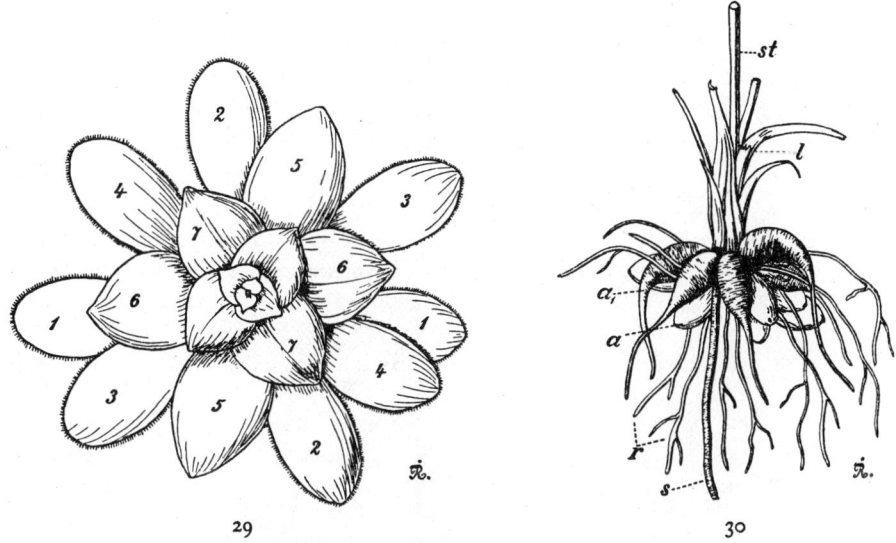

Fig. 29. *Crassula orbicularis*; a leaf rosette formed of opposite decussate leaves, which during development twist so as to produce a many-rayed rosette; the leaves belonging to the same pair are marked with the same number.

Fig. 30. *Ranunculus millefolius*. Hemicryptophyte (Rosette plant), with storage roots, a; l, the sheaths of the radical leaves; st, the lowest portion of the flowering stem. (†)

soil-surface itself, usually has short internodes, and never grows out to form an aerial shoot. Moreover this type of plant appears to be best adapted for life in the soil-surface, showing as it does, the most thorough division of labour for the tasks imposed upon the plant in adapting itself to favourable and unfavourable periods. By grouping both types together it is possible to give a complete picture of their interesting behaviour. In the following short conspectus the most primitive types are put at the beginning.

A. The monopodial axis bears only foliage leaves, and no scales. In the axils of these leaves arise partly new contracted monopodial axes, which are situated in the soil-surface, partly aerial shoots which either bear both flowers and foliage leaves, or flowers alone; in the latter case the species may or may not have stolons.

a. The aerial shoots bear foliage leaves; proper stolons usually absent. Examples of Danish plants: *Viola silvatica, V. mirabilis, Geum, Alchemilla vulgaris* and allied species, *Potentilla incana, P. verna, P. opaca, P. silvestris, Trifolium pratense. Carex strigosa* also most nearly approximates to this division.

b. The aerial shoots without foliage leaves, bearing only flowers.

o. Without stolons.

Examples: *Viola hirta, Plantago major, P. media, P. lanceolata,* and *P. maritima, Pinguicula vulgaris* and *Carex digitata.*

oo. With stolons.

Examples: *Viola odorata, Fragaria, Potentilla sterilis, P. reptans, P. procumbens, P. anserina, Trifolium repens, T. fragiferum,* and *Sagina procumbens.*

In the species of *Potentilla* it is

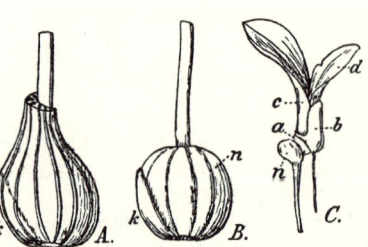

FIG. 31. *Liparis Loeselii*; Hemicryptophyte (Rosette plant) with stem tubers formed of the lowermost portion of the flowering stem enclosed by leaf-bases; the winter bud is situated at the base of the tuber in the soil-surface; it arises in the axil of the uppermost leaf, 5; I, last year's tuber with the lower portion of last year's flower-stem. (⅓.)

A, the lowest tuberous portion of the flowering stem surrounded by the sheath of the uppermost leaf; the presence of the bud, *k*, can be made out within the sheath; in B all the leaves have been removed and the bud, *k*, is seen more clearly. C, young plant with no flowering axis; *n*, the old tuber; *a–d*, the new shoot's four leaves which surround the new tuber. (*c.* ⅔.)

the flower-bearing stem that forms the runners, its reaction to gravity being changed so that instead of being negatively geotropic it grows at right angles to the direction of gravity, thus lying along the earth, to which it soon becomes rooted; it is only the flowering stalks that are negatively geotropic.

B. The monopodial axis bears both scales and foliage leaves; in each period of growth there are developed a series of each kind; after the foliage leaves have unfolded at the beginning of the period of growth there are formed a series of scales for the protection of the young foliage leaves and flowers, which unfold the following year.

a. Without stolons.

Example: *Anemone hepatica*; with solitary flowers in the axils of scales; the foliage leaves remain green through the winter.

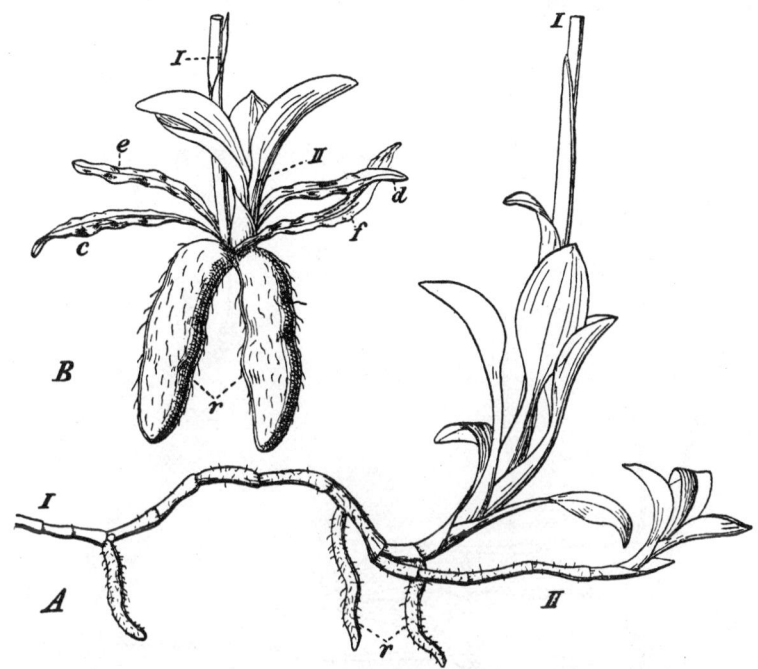

FIG. 32. A, *Goodyera repens*; Hemicryptophyte (Rosette plant) with Stolons; I–I, Flowering shoot; II, a lateral axis arisen from I, with a winter rosette. (↑). B, *Spiranthes spiralis*; Hemicryptophyte (Rosette plant) with storage roots, *r*; I, the flower-shoot whose leaves, *c–f*, are now withered; II, lateral shoot destined to spend the winter in the soil-surface. (↑)

b. With stolons.

Convallaria majalis; the vertical rhizome usually produces only two foliage leaves during each period of growth, and after that three to seven scales which surround and protect the foliage leaves and flowers of the coming year. The flowers arise in the axils of the uppermost scale. The stolons, which are subterranean and sometimes branched, arise in the axils of foliage leaves. After a longer or shorter period of growth the tips of the stolons bend up into the soil-surface, where they grow into a monopodial axis like the mother shoot.

Oxalis acetosella (Fig. 33) is best included here. It certainly does not possess a vertical rhizome with runners like the plants just mentioned, but the monopodium, itself horizontal, resembles a stolon, bearing thick scales (Fig. 33, A, *n*) and foliage leaves, whose swollen basal portion (Fig. 33, A, *b*) remains attached after the petiole and leaf are dead, and, like scale-leaves, serves for storing nourishment. The flowers are attached singly in the leaf axils. The rhizome of this plant is indeed situated in the soil-surface, but it is

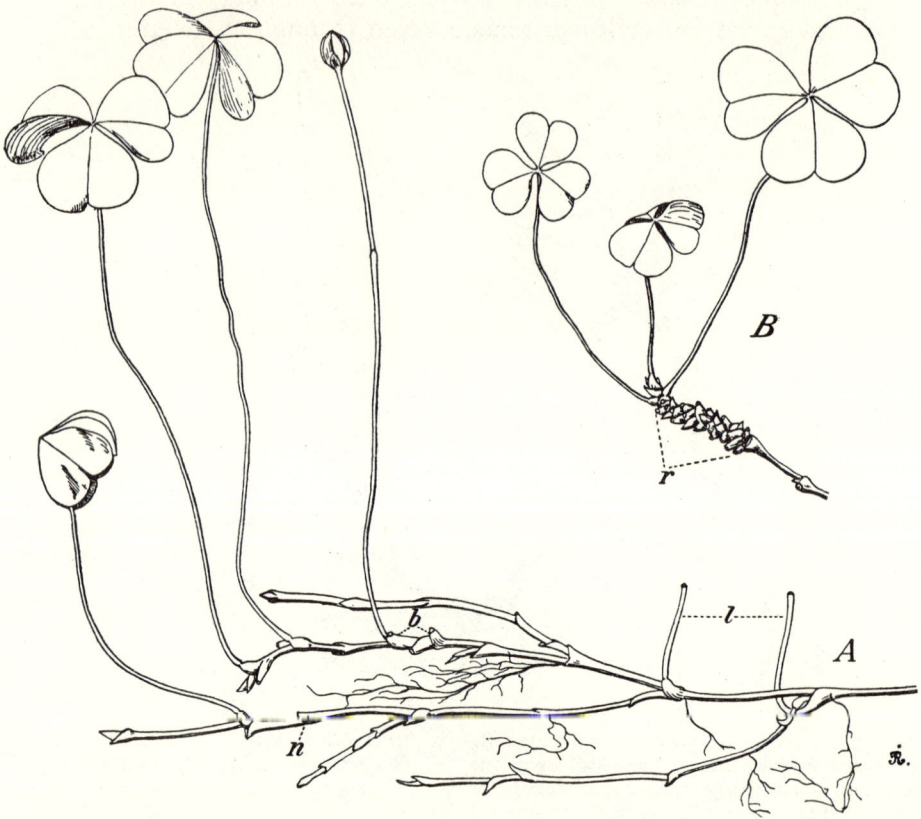

FIG. 33. *Oxalis acetosella.* A, plant with elongated rhizome creeping horizontally under and amongst fallen leaves; the single shoots are monopodial axes with flowers arising laterally in the leaf axils; *l*, the lowest portion of two leaf stalks, *n*, scale leaf; *b*, persistent fleshy basal portion of a fallen leaf; in the axil of one of these is seen a fruiting stem. B, a contracted shoot in the soil-surface on somewhat firm ground not covered with leaves; it bears a great number of closely packed scale leaves and a very great number of leaf bases of fallen leaves, *r*. ($\frac{2}{3}$)

covered by a layer of leaves thus forming a transition to the Cryptophytes; but as it can also live where there is no leaf covering, when the rhizomes remain in the soil-surface, the plant may be reckoned among the Hemicryptophytes.

C. The monopodial axis bears only scale-leaves; the aerial shoots therefore always bear foliage leaves as well as flowers. Of Danish plants the only representative is *Gentiana pneumonanthe*; *Sedum rhodiola* is an exotic example.

A General View of the Hemicryptophytes. Now that the most important types of Hemicryptophytes have been mentioned, let us take a general view of these plants collectively in order to see in what manner and in what degree they have become modified in comparison with the Phanerophytes, from which they are certainly derived. We must consider here not only those changes of external form, but also the alterations in the internal economy of the plants which are connected with their capacity to react in a definite manner to the stimuli of gravity and light, the reactions which are of special importance in determining the direction of growth.

In Proto-Hemicryptophytes there is, apart from the stolons possessed by some species, usually no alteration in the reaction of the shoots to gravity and light; the shoots are constructed like those of Phanerophytes, and they share their negatively geotropic growth. These plants have become Hemicryptophytes only because the shoots, by reason of the unfavourable climate, die back to that portion of the stem which is situated in the soil-surface, and which alone bears winter buds. The close relationship between Proto-Hemicryptophytes and Phanerophytes is seen from the fact that they form an aerial shoot in the year in which they germinate, even if the shoot is not destined to bear flowers. And, even if the individual has become capable of flowering, it forms shoots that are purely vegetative, besides the flowering shoots. In this connexion it is worth noting that when flowering ceases there is often a copious development of purely vegetative lateral shoots even from the uppermost flower-bearing parts of the plants. All these shoots die just as the mother shoot does at the advent of the unfavourable season. But these plants (e.g. *Urtica dioeca*) give the impression that, if the external conditions remained favourable during the whole year, they would behave like Phanerophytes, and it is therefore not surprising that certain plants (e.g. some *Labiatae*), which in the Mediterranean region are small shrubs, grow with us as Proto-Hemicryptophytes.

Looking next at the other kinds of Hemicryptophytes we observe that there has occurred in the more pronounced partial Rosette plants and in the real Rosette plants an alteration in their reaction; the shoots during their first purely vegetative stage of development react to light by almost entire suspension of growth in length, while a copious development of foliage leaves is taking place, thus bringing about the development of a rosette. Not till the shoot is about to flower does its reaction alter, so that a marked elongation occurs. In the few plants which have an upright monopodial axis, whose apex never develops an inflorescence, the shoot has to remain at the first stage, in which growth in length does not occur.

That the shoots in partial rosette plants and rosette plants do not grow in length when they reach the soil-surface is not due only to internal causes, but is determined by light. This is seen from the fact that the shoots which begin growing at some depth in the soil develop elongated internodes as long as they remain in the soil, but become contracted as soon as they reach the surface. If such plants are covered with a layer of earth of varying thickness they develop elongated shoots until they again reach the surface; examples: *Campanula trachelium* (Fig. 34), *Ranunculus bulbosus* (Fig. 35), *Taraxacum vulgare* (Fig. 36) and *Bellis*. The rosettes

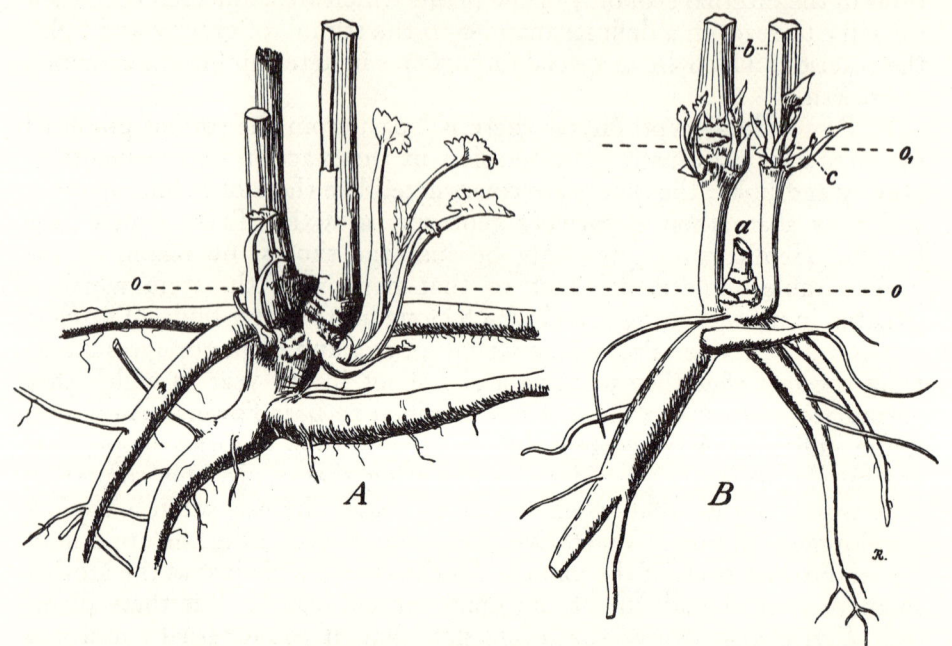

FIG. 34. *Campanula trachelium* in spring; partial rosette Hemicryptophyte. A, the plant under ordinary circumstances; o, the soil-surface. B, a plant which has become covered with a layer of soil, $o-o_1$; the shoot, b, which has developed after this covering has been laid down, has made an elongated stem in its portion covered with earth; only after reaching the light above the new soil-surface, o_1, has it formed a rosette of foliage leaves, in whose axils, c, the winter buds are laid down. (⅔)

are in this way insured both against rising too high above the ground and against being buried, and this distinguishes these plants as marked Hemicryptophytes. Both Rosette plants and partial Rosette plants are, already in the first stage of their life, i.e. in the year in which they germinate, differentiated from Proto-Hemicryptophytes by producing during that year only rosettes and not elongated aerial shoots, which are not developed until the plants are about to flower. The primary shoots of Proto-Hemicryptophytes, as already mentioned, grow out in the year of germination to form elongated aerial shoots.

We have already seen how the rhizomes of Hemicryptophytes, because of their annual growth, however small this may be, rise higher and higher up, and how they are prevented from rising too high above the earth, partly by the deposition of dry plant remains, partly by 'hilling', especially through the activity of worms, and partly because the rhizomes of many are dragged down by the contraction of the roots.

Lastly let us briefly consider the fact that many Hemicryptophytes possess shoots which, at any rate during the first stage of their life, react to gravity differently from the ordinary aerial shoots. I refer to the so-called 'stolons' which serve for vegetative distribution and multiplication.

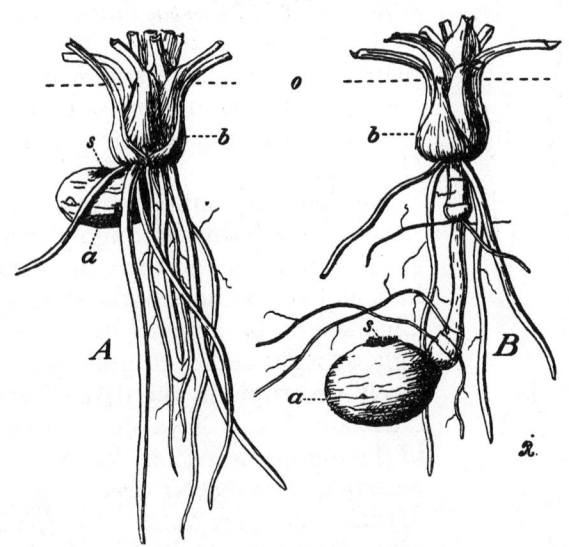

FIG. 35. *Ranunculus bulbosus*; Hemicryptophyte (Rosette plant) with tuberous underground stem. *a*, last year's tuber; *s*, scar of last year's flowering stem; the sheaths of the radical foliage leaves, *b*, are situated on and surround the new tuber. During the development of this tuber and the leaves and inflorescence connected with it, the stored-up material in the old tuber, *a*, is consumed. *o*, the present soil-surface; in B, the last year's soil-surface was a little above *s*; after that the plant was covered with earth, so that the soil-surface was raised to *o*; the new plant, which is a lateral shoot of the old tuber, has not formed the new tuber and leaf rosette, as it otherwise would, beside the old tuber, but it has first formed an elongated slender shoot, and when this has reached the soil-surface a new tuber and leaf rosette were formed.

They are, at any rate to begin with, horizontal or nearly so, but sooner or later they bend upwards to form an upright portion which strikes root and becomes a separate individual when the horizontal portion of the stolon dies. Only very rarely are the vegetative shoots of plants constantly horizontal; e.g. *Oxalis acetosella*. We often find in the axils of the leaves of the stolons buds which are also capable of forming new individuals, either by immediately making an upright shoot which becomes rooted, or by growing out into horizontal stolons, which ultimately bend their tips upwards and become rooted at the base. The buds and shoot-apices of these stolons, which are destined to survive the unfavourable period, are lodged in the soil-surface as are the buds of the ordinary shoots. This further serves to characterize these plants as marked Hemicryptophytes.

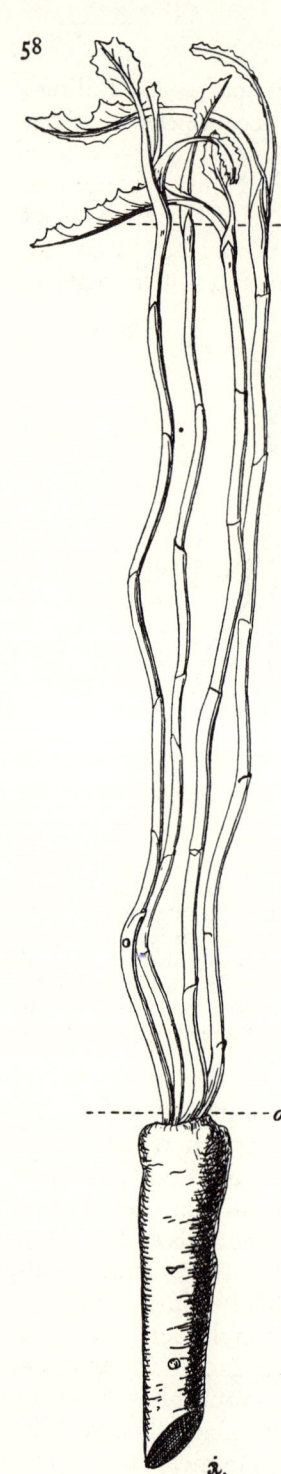

LIFE-FORMS OF PLANTS

In some species the stolons are always epigeal (*Fragaria, Potentilla anserina, P. reptans*, and other species, *Hieracium pilosella*, and many others). In other plants they are always subterranean (*Petasites*, some Grasses and Sedges, &c.). As mentioned before there are some plants which have both epigeal and subterranean stolons (*Urtica dioeca, Stachys silvaticus*, species of *Mentha*, &c.). It sometimes happens in these plants that the same stolon during part of its course is epigeal and in part of its course subterranean or at least covered by fallen leaves (Fig. 37).

Neither the subterranean nor the epigeal stolons have been thoroughly investigated; but as far as inquiries go the epigeal stolons seem to behave like the horizontal shoots of the active Chamaephytes already described, growing in light at right angles to gravity and being negatively geotropic in the dark. Czapek and more especially Maige have shown this to take place in the following species: *Rubus caesius, Potentilla anserina, Ranunculus repens, Trifolium repens, Mentha aquatica, Stachys silvaticus, Ajuga reptans, Hieracium pilosella*. I found the same behaviour in my experiments with *Veronica serpyllifolia* and *Brunella vulgaris*. When I grew the stolons of these plants in the dark, keeping them in a case impervious to light, but not otherwise interfering with them, the apex of the stolons grew upwards. When on the other hand the stolons were held in a vertical position with their apices upwards and in light, they continued their growth horizontally or obliquely downwards. Growth obliquely downwards in my experiments was marked in *Hieracium pilosella*, whose shoots by this growth became tightly pressed to the soil.

Among the plants whose runners grow

Fig. 36. *Taraxacum vulgare*; a Rosette plant which after being covered with earth is bringing its bud up to the soil-surface by the elongation of the shoots; not till the apices have reached the light are there developed short internodes with leaves and rosette. o, the original soil-surface; o_1, the later soil-surface. (⅓)

horizontally in light, but are negatively geotropic in the dark, Maige mentions *Nepeta glechoma* and *Potentilla reptans*. According to Czapek on the contrary these two species grow horizontally both in light and in darkness; that investigator finds that the shoots of *Linaria cymbalaria* behave in the same way. I found this kind of horizontal growth both in light and darkness in *Ajuga reptans* (a variety with reddish-brown leaves) and *Ranunculus repens*, thus deviating from the results of Maige. In some species the stolons are negatively heliotropic (*Nepeta glechoma* according to Wiesner and Maige, and *Potentilla reptans* according to Maige). In other species they are slightly positively heliotropic (*Hieracium pilosella* and *Mentha aquatica* according to Maige). The heliotropism

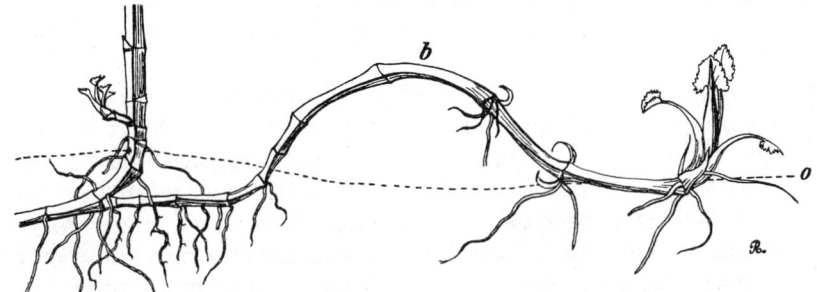

Fig. 37. *Stachys silvaticus*; the lowermost portion of a flowering shoot with a stolon whose apex has arisen above the soil-surface, *o*; it is then bent downwards, and on again reaching the soil-surface it is rooted and begun to form foliage leaves.

varies no doubt with the light's intensity, and this behaviour may explain why Wiesner and H. de Vries find the stolons of *Fragaria* negatively heliotropic, and Czapek describes them as positively heliotropic.

As already mentioned both subterranean and epigeal stolons may be found in *Stachys silvaticus* and *Urtica dioeca*; doubtless as a rule the stolons are situated actually in the soil-surface; the subterranean stolons most usually lie near the surface, so near the surface perhaps that sufficient light can reach them to counteract their geotropism and make them grow at right angles to the direction of gravity; otherwise it is difficult to understand why the subterranean stolons should grow horizontally, for in the absolute darkness of a case impervious to light the stolons of these plants grew erect, and were thus negatively geotropic. Moreover I planted some individuals of the above named species in the soil-surface and some at a depth of 15 cm.: in the former the new stolons lay above the soil or actually in the soil-surface partially exposed to light; in the latter some stolons came off at a considerable depth, but they grew at once almost vertically up through the soil until they reached the surface, when they grew horizontally. Even if it seems to be general for the epigeal stolons of Hemicryptophytes to be negatively geotropic in the dark, and, because of an altered reaction towards gravity, vertical in the light, yet all of them do not behave in this way. The slender stolons of the well-known *Saxifraga sarmentosa* appear to be entirely insensitive to gravity, at any rate in the dark. A number of stolons, connected all the time of course with the mother plant, were brought into the dark and so fixed that some were horizontal, others vertical with their apex upwards, others again vertical with their apex downwards. They all went on growing in the direction in which they had been fixed. On the other hand they were very sensitive to light, being negatively heliotropic in moderately strong light. When lighted from one side in a greenhouse which had windows only on its southern side, four stolons, which were fastened with their apices erect, first curved over in a horizontal direction, and then arched downwards, always away from the light, even if by so growing they came to have the side which had been formerly turned upwards directed downwards.

Although much remains to be investigated about the causes determining the direction of the stolons in Hemicryptophytes, we do at least know something about these causes, and we know the methods to employ in order to learn still more. On the other hand we are entirely in the dark about how stolons arose, about what determined negatively geotropic shoots to alter their reaction towards gravity and grow at right angles to its attraction. We encounter here, as at so many other points, the question of the origin of species, or what comes to the same thing, the formation of new hereditary characters.

Let us here discuss some circumstances and some recent investigations which may possibly have significance for the solution of the problem of the origin of stolons.

There are species of plants in whose flowering shoots we see imperceptible gradations between lateral flowering shoots and stolons. For example in *Stachys silvaticus* we can trace lateral branches from the top of an erect mother shoot to the portion situated in the soil-surface. The uppermost of these branches are comparatively short, erecto-patent, and bear flowers; but the farther we go down the stem the longer they become, and their proximal vegetative portion becomes longer and spreads at a wider angle. Right down on the soil, but still wholly in the light, we find long branches which either project into the air or rest on the ground, and bear only a few flowers at their apex, or are quite flowerless. If they are flowerless they can root and live through the winter: they have in fact become stolons, which, in this species either lie in the soil-surface, or partially or entirely above the ground. The stolons of this species are, like those of all plants furnished with a sympodium and stolons, lateral shoots which do not flower the same year as the mother shoot, but delay their flowering until the following year. Stolons thus differ from the branches which arise higher up on the mother shoot. In the first growth period they grow horizontally along the earth or buried in it, thus obtaining protection from the effects of the unfavourable season. It is by this means, in fact, that they are enabled to survive the winter. Not till the next growth period do they turn upwards to form a vertical flowering shoot. Thus we see in all sympodial plants with stolons that it is only in the first period of development that the stolons grow at right angles to the pull of gravity; at another period, during the next season of growth, they alter their reaction to gravity, becoming negatively geotropic, and growing up into vertical flowering shoots.

In typical stolons we do not know the circumstances in the shoot's development which determine this altered reaction to gravity. On the other hand in some plants without real stolons, but with shoots that under certain circumstances change their reaction to gravity, we do know the external factor which determines this change of reaction. Thus the German botanist Vöchting has observed in *Mimulus Tilingii* that the shoots in spring and when they begin to flower grow horizontally in a low temperature, and vertically in a high temperature. Similarly the Swede Lidforss showed that the shoots in different plants grow horizontally in a low temperature, but vertically in a high temperature (e.g. *Holosteum umbellatum, Lamium purpureum, Chrysanthemum leucanthemum, Stellaria media, Veronica hederifolia,* and *Anagallis arvensis,* &c.). By experiments with the first two of these species Lidforss showed that the changed direction of growth was dependent upon the changed sensitiveness to the stimulus of gravity, so that the shoots in a low temperature grew at right angles to the direction of gravity, while in a higher temperature they were negatively geotropic. We know nothing of how the change of sensitiveness arises, what it is in the plant that occasions the rising or falling temperature to alter its reaction; but recently some observations have been made which may perhaps help us to discover some of the links in the chain which brings about these changes.

We know that numerous roots, both primary roots and a great number of adventitious roots, so react to gravity that they grow vertically downwards; they are positively geotropic. If such a root be placed horizontally its continued growth makes it bend vertically downwards. The botanist Nemec has established the theory, which he supports by his anatomical investigations, that it is through the pressure of the starch grains on the cell protoplasm that the horizontally placed root learns, if I may use the word, the alteration of its position. Nemec has always found starch grains in the extreme tips of the roots in certain of the cells of the rootcap. Because the starch grains have a higher specific gravity than the cell-sap, they always sink down to the bottom wall of the cell and rest on the protoplasmic lining of this wall. When the root assumes a horizontal position the wall on which the starch grains have been resting becomes vertical, and the starch grains sink down on to what was formerly the lateral wall, but has now become the floor. Nemec supposes that it is by the pressure which the starch grains now exert on the protoplasmic lining of this wall that the root becomes aware of its altered position. How this actually happens it is impossible for us to understand; but if the theory proves to be true we have discovered one link in the chain of events, and that is at any rate a beginning.

In support of this theory may be mentioned the investigations of the German botanist Haberlandt on the direction of the stem in *Linum perenne*. The stems of this species always grow erect in the summer and are markedly negatively geotropic; but in the autumn when the temperature has sunk to a certain point the shoots no longer grow vertically upwards, but proceed irregularly in various directions. If such plants are brought into the higher temperature of a greenhouse the shoots after some time again become strongly negatively geotropic and grow vertically upwards. Anatomical investigations showed that the autumn shoot, which had lost its sensitiveness to gravity, had no starch grains in its cells; but starch grains were formed when the plant was moved into the warmth, and since everything takes time, an interval elapsed before the starch grains were formed, and during this interval the stem showed no tendency to grow vertically upwards in the warmth; but as soon as the warmth increased the vital activities of the plant sufficiently to make it begin to form starch grains, the stem began to grow vertically upwards. Not until the starch grains were formed did the stems regain their old sensitiveness to gravitation.

These investigations show that there is very probably a relationship between the presence of starch grains and the alteration of reaction of shoots towards gravity. This brings us back to the experiments of Vöchting and Lidforss which showed that the shoots of certain plants are negatively geotropic in warmth but grow at right angles to gravity when the temperature is lower. This reaction explains how sensitiveness to gravity may alter with changes to temperature. Besides the plants mentioned by Vöchting and Lidforss many other species have horizontal shoots in the autumn (e.g. *Anthemis arvensis* and *Arenaria serpyllifolia*). It is perhaps along this line that we may hope to investigate the causes which brought about the origin of stolons.

We encounter here, as we always do when the question of new characters arises, which is the same question as that of the origin of new species, two hypotheses which we may call that of Darwin and that of Lamarck, even if these hypotheses at the present day are altered and more sharply defined than they were originally. We must consider whether the stolons, which are shoots with a changed reaction towards gravity, have arisen because of an internal unknown force operating by means of hereditary variation (or 'mutation' as it is usually called to-day), or whether they have arisen from shoots whose reaction has been altered by the influence of such factors as their own weight, or a snow covering, affecting the shoots generation after generation, and at length gradually altering their internal economy, so that they continue to grow in the changed direction even after the factors which brought about the change no longer operate.

Several investigators are inquiring into the question of hereditary variation, into the hereditary characters, but such investigations miss the main point with which we are concerned, the origin of the characters in question, and only concern themselves with the transmission of characters which already exist. The question of heredity is being much more thoroughly investigated now than it was formerly; but of the origin of the characters, which is the main question, we know nothing; we know neither how nor when any given hereditary character came into being; we merely see that it exists. This is of course true of the much discussed 'new' characters which the Dutch botanist H. de Vries has found in part of the offspring of *Oenothera Lamarckiana*, a plant he has investigated thoroughly. Whether these characters are new, whether they arose in the course of H. de Vries's cultivation of this plant, or whether they had been present for hundreds or indeed thousands of years, we do not know for certain. For the fact that a given character has not been seen before does not necessarily mean that it has not existed before. It would be an extraordinary coincidence, one chance out of an infinity of chances, that a botanist should be present at the time when a character first came into existence. We can never be certain that a character not observed before is new. Of itself this fact is of no great importance: it would be of much greater importance if, in the course of our investigations of plants in which such sources of error as self-pollination and parthenogenesis were excluded, we could succeed in producing hereditary variations by known causes, if we could succeed, in fact, in producing hereditary variations by submitting the plant to definite influences. Whether we should ever succeed in doing this in the higher plants is a difficult question; but we are certainly justified in attempting to do so.

While in Darwin's hypothesis the inheritance of the characters, or of the variations, is the known fact and the origin of the variations is unknown, in Lamarck's hypothesis the origin of the variation is the known quantity and their inheritance the unknown quantity. It is just these 'conditioned' characters we have to deal with here, and we either know the causes of their origin, or we at least know how we must proceed to find out these causes; we also know that these characters are not inheritable in the ordinary sense of the word. The unknown, the undecided and difficult question is whether the presence of the conditioned characters throughout a long series of generations can so influence the organism that characters originally conditioned at length become hereditary, i.e. occur also when the causes necessary to their origin are no longer present.

Many facts in Ecology and Plant Geography are best explained by the help of this hypothesis. The hypothesis helps, too, in explaining facts I have used in characterizing life-forms. We do not know whether the hypothesis is true, but it has this great advantage that we know how to set to work to prove it, we know how to start the necessary experiments. We can cultivate plants experimentally throughout a long series of generations under conditions that originate certain characters which are not hereditary. Then we can find out, after the lapse of a long time, by cultivating the descendants of these plants, whether the characters will be present even if the conditions which brought them into being are no longer present. Some ten years ago I began experiments of this kind, but I gave them up again, for even granted that the hypothesis be true, such experiments would probably extend over a greater number of years than the lifetime of a single man, so that if the work is to be the affair of a private individual it will probably be done in vain. I proposed therefore eight years ago, and again later, the establishment of Phylogenetic Institutions, where experiments of this kind could be carried out independently of the lifetime of a single investigator. Such institutions would be best united to Botanic Gardens. Experiments like the one indicated above might be set on foot and I do not doubt that one day this will be done; for we can only make headway in solving the problem of the origin of characters, which is the same as the origin of species, by submitting Lamarck's hypothesis to experimental proof.

IV. Cryptophytes

Cryptophytes are plants whose buds or shoot-apices destined to survive the unfavourable season are situated under the surface of the ground, or at the bottom of water; the depth below the surface of the ground varies in different species.

The adaptation to environment, which consists in the buds destined to survive the unfavourable period being so situated that they are protected as well as they possibly can be against desiccation and sudden strong changes of temperature, is here attained in the most perfect manner, the buds being placed, not in the air as they are in Phanerophytes, nor on the surface of the ground as they are in Chamaephytes, nor in the soil-surface as they are in Hemicryptophytes, but completely buried in the soil at a distance from the surface, or at the bottom of water.

Since water plants, and albeit in a lesser degree, marsh plants, make well-defined groups, it is convenient to divide Cryptophytes into Geophytes, Helophytes, and Hydrophytes.

Geophytes include land plants whose surviving buds or shoot-apices are borne on subterranean shoots at a distance from the surface of the ground. They are particularly well adapted to live in districts with long and marked dry periods; Geophytes are actually found predominantly in such districts as dry steppes where they form a large component of the flora; but there are also a number of Geophytes that are adapted to and live in regions with a comparatively long period of vegetation, and where the unfavourable period is not a hot dry season but a more or less severe winter.

The Geophytes living in regions with a long dry period and short season of vegetation are adapted to finishing their epigeal life in a short period; as soon as the short, favourable, sufficiently humid season sets in they can at once begin and in a short time complete the development of the aerial shoots which bear foliage leaves and flowers. But in order that this can happen these organs must already have attained a comparatively high degree of development during their life under the ground, and this development must take place during the dry period. But since the plant during this season has no foliage leaves and therefore is not able to form the plastic material it needs, it must previously be furnished with sufficient stored-up nourishment. It is to meet this need that the plants are furnished with special organs for storage by means of which the plant can, during the dry season, develop leaves and flowers sufficiently far to make them ready to break forth as soon as the short favourable period begins. The rich floral display that appears at the beginning of spring is always emphasized as a characteristic feature in the flora of steppes.

The flowers often emerge before the leaves. We are familiar with this in various steppe plants cultivated in our gardens (e.g. *Crocus* spp.). The development of the flowers has to take place in absolute dependence upon the nourishment stored up during the last vegetative period; but this is usually true also of the species whose leaves and flowers develop at the same time, as may be demonstrated by removing the new leaves, which may be done without interfering with full development of flower or with seed formation.

Because of the shortness of the vegetative period, the formation of plastic materials and of the organs have taken place very early. In the short vegetative period the plastic materials are elaborated and in the long so-called 'resting period', which is really no resting period at all, but merely a time during which development is removed from our sight, the new organs are formed to the extent that only elongation in the individual cells is necessary to bring the leaves and flowers forth into the daylight. Since the nourishment formed in the vegetative period by means of the foliage leaves is not used till after the leaves are dead Geophytes must possess organs in which plastic material may be stored. That is the reason for the peculiar feature of these plants, that they are furnished with the special storage organs which are so characteristic of them.

In some the food is stored in thick fleshy leaves which are packed close together like the leaves of a bud, forming what is called a **bulb**; in others the store for the food is a short thick tuberous stem or tuberously swollen roots. We thus have three groups of Geophytes: Bulb Geophytes, Stem Tuber Geophytes, and Root Tuber Geophytes, all of which live preferably where there is a long dry period and a short period of vegetation, and even if some of them are met with in regions with a comparatively long favourable season they usually complete their aerial life in a short time.

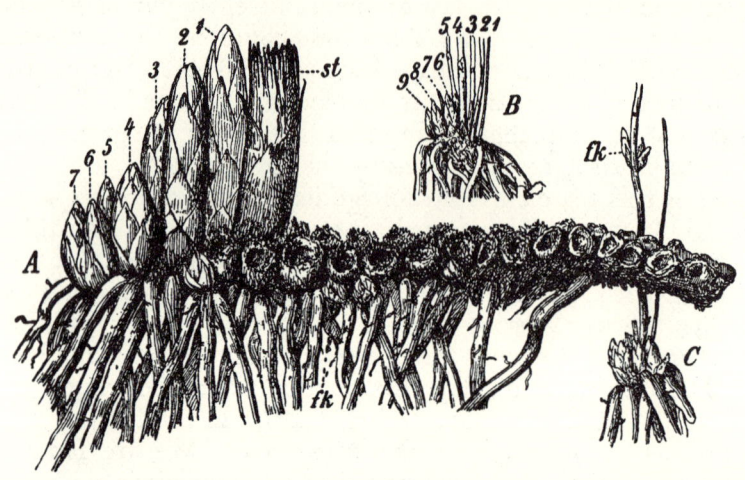

FIG. 38. *Asparagus officinalis*; Rhizome Geophyte; the subterranean stem of an older plant, A, and one of two younger plants, B and C; *st* in A is the lowest part of the last developed of last year's aerial shoots; to the right of this are seen the scars of the aerial shoots of the last three years; 1-7 in A and 6-9 in B are the buds which spend the winter in the earth; they are now about to grow into epigeal shoots. The young plant, C, has been planted comparatively deep—presumably too deep—in the earth; at some distance on the erect shoot, but still under the soil-surface, there are laid down some buds (*fk*) forming a new centre of growth higher up in the earth.

Besides these three groups of Geophytes we can recognize a fourth, namely Rhizome Geophytes, which are very likely just as well adapted to their life in the ground, but their underground stem is not nearly so extensively modified, differing indeed little from the underground stem of the Hemicryptophytes. Rhizome Geophytes are adapted to life preferably in regions which indeed have a severe unfavourable period, for example a hard winter, but have at the same time a long period of vegetation, so that they do not need to have a large amount of food stored up, and do not therefore need special organs to store it. The stems of Rhizome Geophytes are more or less elongated, and as a rule horizontal (Figs. 38 and 39).

As a fifth small group may be mentioned the Root Geophytes, which persist wholly or principally during the unfavourable season by means of buds on the roots, which are hidden in the ground and which are protected by it, the shoot dying away at the beginning of the unfavourable period.

Although Geophytes seem to be well protected by being wholly hidden in the ground during the unfavourable season, yet the influence of drought can sometimes reach them there. In such cases we find that the underground organs have the need of special protection. We recognize amongst the underground organs of Geophytes a series of structures specially designed to serve as a protection against drought. Thus in

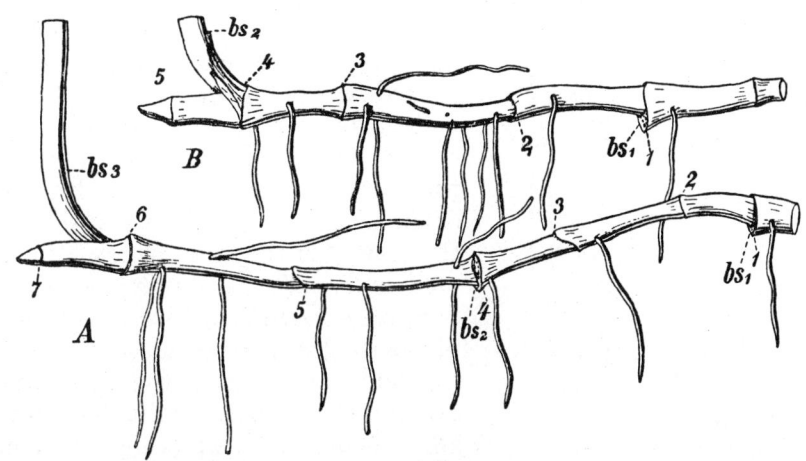

Fig. 39. *Paris quadrifolia*; Rhizome Geophyte; the front part of two flower-bearing subterranean stems; bs_3 in A and bs_2 in B is the lowermost portion of the erect flower-bearing aerial shoot; bs_1 and bs_2 in A and bs_1 in B are scars of the flowering shoots of the preceding year; the subterranean stem, which is a monopodium, is horizontal and lies several centimetres from the surface of the ground; the winter buds are also at this level (7 in A and 5 in B).

Bulb Geophytes there are either especially formed protective leaves, or the outermost leaves of the bulb already emptied of their food material serve for protection. Thus in many Stem Tuber Geophytes (*Crocus* and other *Iridaceae*) we find protective leaves enclosing the tuber. Other Stem Tuber Geophytes protect their tuber from desiccation by developing in the periphery of the tuber a special protective tissue, notably cork (*Arum, Solanum tuberosum*, &c.). Lastly other plants are protected by the composition of their cell-sap, which prevents water being easily lost, e.g. root tubers of ground Orchids, and many bulbs, &c., whose cells are rich in mucus.

Since the buds of Geophytes are situated deep down in the earth before they develop into aerial shoots, they have to break through a layer of earth of varying depth. Even if the soil at the time when the shoots sprout is soft and moist, yet the resistance it offers is usually so great that the tender young growing plants and flower buds would not be able to break through it without suffering damage were it not for

68 LIFE-FORMS OF PLANTS

some special adaptation. We also find therefore among Geophytes different structural devices which serve to facilitate the growth of the shoots through the soil.

Some species grow with their shoot-apices not directed forwards, but the stem is bent slightly behind the apex so that the growing point is turned downwards or backwards, while the older and firmer part of the stem, which forms a bow, breaks a way through the soil, if I may use the expression, by putting its back into the work (*Anemone nemorosa, Eranthis,* &c.).

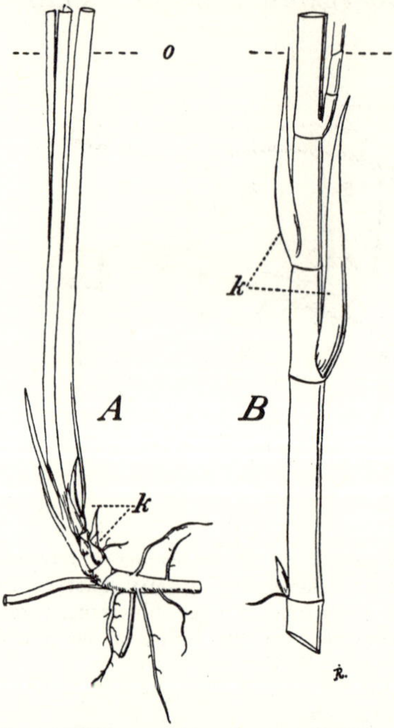

In others whose shoots are not thus bent the older and firmer leaves are packed tightly together around the shoot-apex forming a protection to the young and tender leaves, and sometimes also to the flowers, thus taking the first brunt of the shock from them; it is still however the growth of the shoot which is the force used in breaking through the soil. An example is *Fritillaria imperialis*. In other plants the leaves serve not only to protect the bud and to pave the way, but the driving force comes actually from the growth of the leaves (*Gagea* and many other Bulb Geophytes). In these plants the leaves are packed tightly together and protect the flower-bearing aerial stem, while it is growing up to the surface of the soil. The leaf-apices, which are tightly packed and as a rule have a specially dense structure, penetrate higher and higher up through the earth because the leaves grow at their base. When they reach the surface the leaves diverge from one another, and the flowering stem, which they have hitherto enclosed, now grows freely up into the air.

FIG. 40. A, *Calamagrostis epigeios*; B, land form of *Phragmites communis*; *k*, winter buds; *o*, soil-surface. (⅔)

23. **Rhizome Geophytes** have as a rule a more or less elongated, usually horizontal rhizome, which is almost invariably a sympodium composed of the lowermost horizontal portions of the individual shoots whose uppermost vertical parts project into the air and bear foliage leaves and flowers. If the rhizome be situated at a depth proper for the species the individual shoot in the first part of its life is transversely geotropic, growing at right angles to gravity, but it changes its reaction towards gravity at a certain time of its life and becomes negatively geotropic. We have already seen that this happens in Hemicryptophytes with subterranean stolons; but the peculiarity of Rhizome Geophytes, which

gives them their cryptophytic character, is that they are not only able, by change in the direction of growth, to attain the proper depth for their own species, but to place their winter buds at this depth, and not in the soil-surface like the subterranean stolons of Hemicryptophytes.

Examples of Rhizome Geophytes are: *Polygonatum multiflorum* (Figs. 45–8), *P. anceps*, *P. latifolium* and other species, *Paris quadrifolia* (Fig. 39), *Asparagus officinalis* (Fig. 38), species of *Cephalanthera, Epipactis, Listera, Neottia, Coralliorrhiza, Epipogon, Eriophorum alpinum* (Fig. 42), *E. polystachyum, E. gracile, Heleocharis palustris* (Fig. 44), *H. uniglumis, Scirpus compressus, S. rufus, Carex incurva, C. arenaria* (Fig. 43), *C. disticha, C. flacca, C. ericetorum, C. verna, C. acutiformis* and others, some Grasses (Fig. 40), *Maranta arundinacea* (Fig. 41), *Curcuma longa, Anemone nemorosa, A. ranunculoides, Dentaria bulbifera, Tussilago farfarus.*

Many Rhizome Geophytes can travel both up and down in the soil according to whether the rhizome is placed too deeply or not deeply enough, e.g. *Polygonatum multiflorum* (Figs. 45 and 48), *Paris quadrifolia, Asparagus officinalis, Cephalanthera rubra, Listera ovata, Juncus obtusiflorus,* and *J. filiformis.* When the rhizome-apices have arrived by the altered direction of growth at a depth proper for the species, which is presumably the depth at which the plant thrives best under the circumstances, the rhizome again takes up a horizontal direction. This depth is different for the

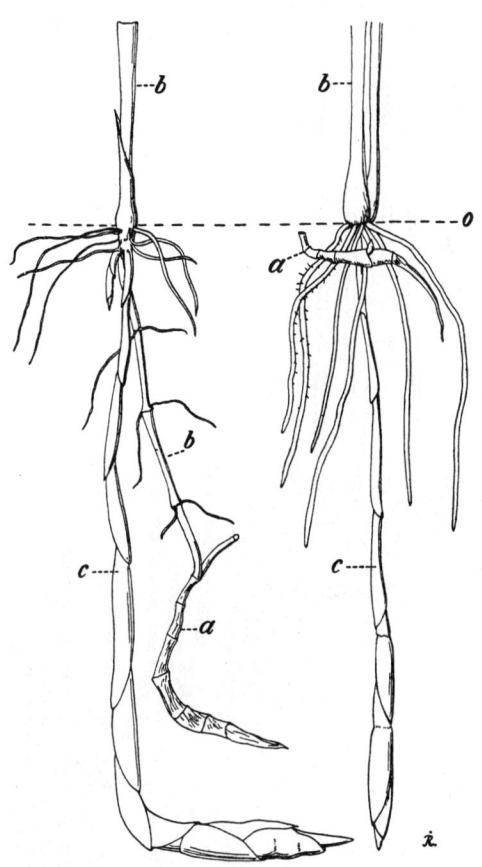

FIG. 41. *Maranta arundinacea*, a Rhizome Geophyte, the apices of whose rhizomes lie at a considerable depth; *a*, the mother rhizome of the right-hand plant was laid in the soil-surface, of the left-hand plant about 10 cm. below the soil-surface. From each of the mother rhizomes, *a*, there was given off an aerial shoot, *b*, which developed many roots in the soil-surface, *o*, from the bases of the radical leaves; from the axils of these leaves are developed several stolons, *c*, which grow vertically downwards to a depth of over 25 cm.; the stolon in the figure to the left, reaching the bottom of the box in which the plant was cultivated, was compelled to grow to one side. (⅓)

different species and seems also to differ at different stages in the development of the same species. The depth perhaps also varies slightly with the external conditions.

In some species the rhizome is able to ascend when buried too deeply in the soil, but on the other hand it seems unable to descend, however high in the soil it is situated, provided that it is covered with soil, e.g. *Anemone nemorosa*, *Heleocharis palustris*, *Carex disticha*, *C. arenaria* (Fig. 43) and *Eriophorum alpinum* (Fig. 42). If on the other hand the rhizomes are laid bare and exposed to light they grow at once obliquely downwards until the apex is buried in the earth, where it resumes its horizontal growth; it is this process that produces the curious loop formations that we often see in *Heleocharis palustris* (Fig. 44) and other plants; *Anemone* often produces much smaller loops, and similar loops are met with also in Hemicryptophytes with subterranean stolons, e.g. *Achillea millefolium*.

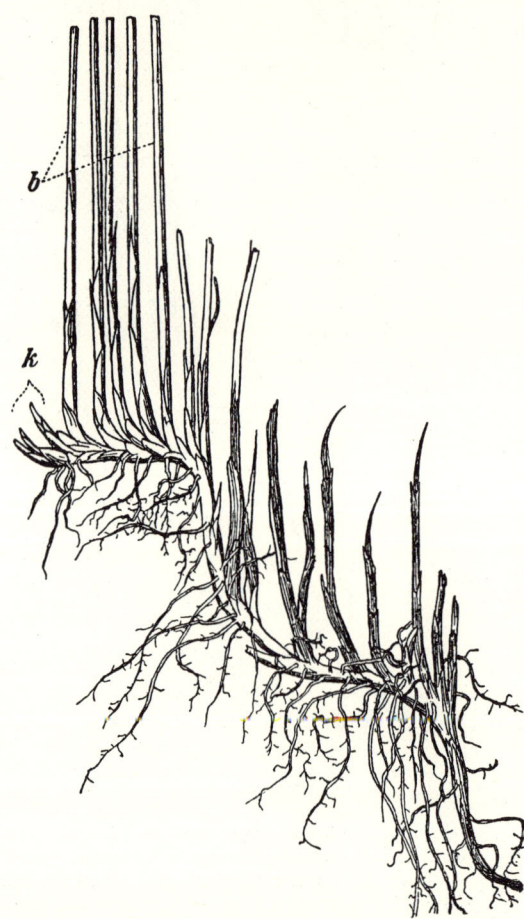

Fig. 42. Rhizome of *Eriophorum alpinum*, a Rhizome Geophyte which has grown in a rapidly increasing layer of peat, ascending to bring its winter buds, *k*, to a suitable depth; the surface of the soil is a little above *k* (*c*. ¾).

That Rhizome Geophytes are able, by changing the direction of their growth, to take up their position at a definite depth can be demonstrated by planting the rhizomes at different depths; but it is a far more difficult matter to discover how rhizomes learn whether they are at a normal depth, or whether they are too high up or too low down in the soil. Only few experiments have been made with the object of elucidating this problem. I have carried out a series of experiments with *Polygonatum multiflorum* which, as before mentioned, belongs to the plants whose rhizomes are able to attain a definite depth in the soil. At this, the normal depth, the rhizome is transversely geotropic and therefore grows horizontally; at a greater depth it becomes negatively geotropic and grows obliquely upwards (Fig. 48, B–C), while at a less depth it becomes positively geotropic and grows obliquely downwards (Fig. 45), until it again reaches its normal depth, where it becomes transversely geotropic and grows horizontally.

Of the external conditions which vary with depth, and which therefore may be able to make a change in the geotropic sensitiveness of the rhizome, the following may be considered: composition of the soil air (e.g. percentage of oxygen), water content of the soil, temperature of the soil, and the resistance which the erect portion of the shoot has to overcome in piercing the soil. Experiments showed however that none of these conditions determined the altered direction in growth of the rhizome at different depths. There thus appears to be no other course open for us than to suppose that the plant is able to become aware of its depth in the soil, not by means of a determining factor in the soil, a factor which would alter with the depth, but by means of a factor above the soil and with which therefore the plant would not come into contact before the upright portion of the shoot reached the soil-surface. Of such factors light is the first and foremost one to be discussed. The influence of the light might, I think, be due

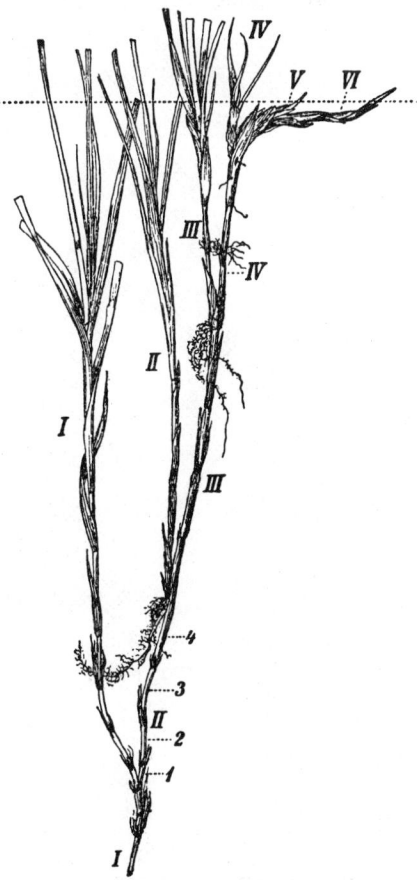

FIG. 43. *Carex arenaria*, a Rhizome Geophyte. The plant has become gradually covered with blown sand; instead of growing horizontally the rhizome portions of the shoots (II, III, and IV to the right) have grown nearly vertically upwards, thus saving the plant from being overwhelmed by the sand. The last developed generations, V and VI, which have reached the surface, bend over and grow horizontally; if the apex of the rhizome is laid bare it will then grow downwards into the sand. (*c.* ⅓)

to a kind of vital activity which enables the plant to measure the distance between the rhizome and the point where the erect portion of the shoot has reached the light. But if there is truth in this hypothesis we should be able to deceive the plant by substituting for the layer of earth a dark layer of air; in other words we should be able to make the rhizome grow obliquely upwards even when situated high up in the soil, by taking care that the upright portion of the aerial shoot was not illumined immediately it reached the soil-surface, and that it did not reach the light until it

Fig. 44. *Heleocharis palustris*, Rhizome Geophyte, showing loop formation. The dotted line is the soil-surface. (*c*. ½)

had grown through the dark space containing no soil but only air. Experiments showed that the plant did actually behave in this way.

Some rhizomes were planted in the autumn at a depth of 5 cm.; over some of them zinc cylinders were placed. These cylinders, of which one was placed over each individual, were 8 cm. wide and 10 to 25 cm. high. The rhizomes which were beneath the cylinders, when they began to sprout next spring, had to grow up through the cylinder and did not meet the light till they reached the edge of the cylinder. Each cylinder was covered with a cardboard lid, which later on, when the shoot reached it, was replaced by a cork through which a hole had been bored. The

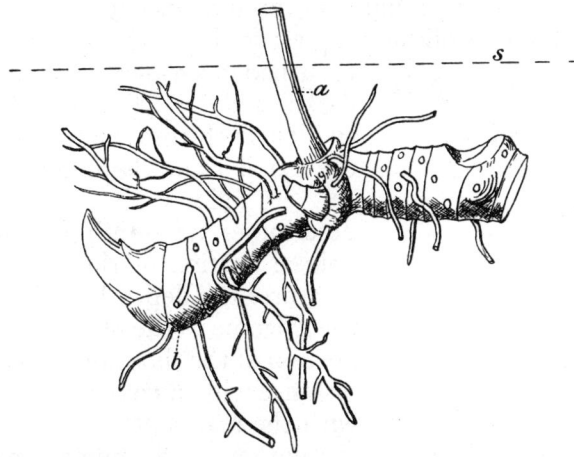

Fig. 45. *Polygonatum multiflorum*; a rhizome planted at a depth of 2 cm.; the terminal bud has formed an aerial shoot, *a*; the continuation shoot, *b*, has grown obliquely downwards; *s*, soil-surface. (⅔)

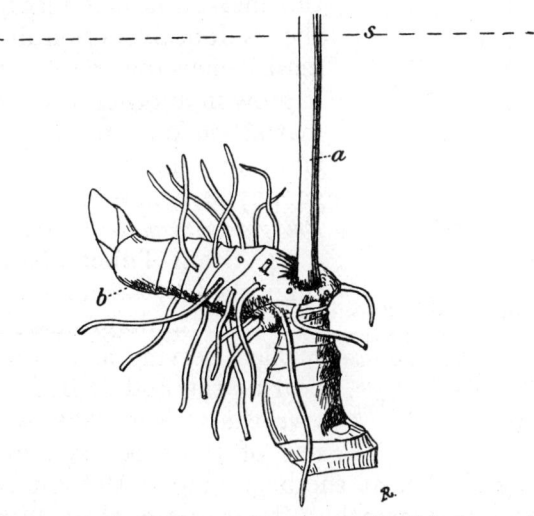

Fig. 46. *Polygonatum multiflorum*; a rhizome planted vertically with the terminal bud upwards and at a depth of 5 cm.; the terminal bud has formed an aerial shoot, *a*; the continuation shoot, *b*, has grown approximately horizontally. (⅔)

shoot grew through the hole up into the air and light; and in order to prevent the light entering the cylinder the hole in the cork was stuffed full of wadding wool. When the plants were taken up during the following August this is what was seen; in the plants whose aerial shoots had not grown up through the cylinder the new segment of rhizome had grown approximately horizontally or slightly downwards; in the individuals whose aerial shoots had grown up through a cylinder and had not come into contact with light till reaching its edge, the new portion of rhizome was directed obliquely upwards (Figs. 47, 48, A). These plants then had been deceived; because of the darkness in the cylinder they had behaved as if their rhizomes were much deeper in the earth than they really were. From this it follows that the plant really determines its depth by means of the epigeal portion of the shoot, and that it measures the depth by the distance from the rhizome to the place where the epigeal portion of the shoot reaches the light; this distance will, of course, coincide in nature with the real depth.

Another and much more difficult question is this. By means of what internal vital activity in the plant does this measuring take place, and how does the result of this action alter the rhizome's sensitiveness towards gravity so as to make it grow in another direction? About this matter we know nothing. Neither do we know in what degree the other rhizome geophytes behave in the same manner as *Polygonatum multiflorum*.

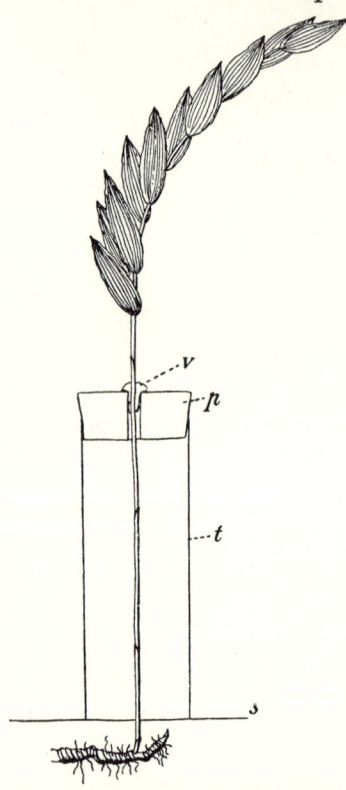

Fig. 47. An experiment with *Polygonatum multiflorum* to find how the plant comes to know at what depth its rhizome lies in the soil; *s*, soil-surface; *t*, a zinc cylinder 25 cm. high, *p*, cork; *v*, wadding wool. See text. (*c.* ⅓)

24. **Stem-Tuber Geophytes** are distinguished by having one or several portions of the underground shoots tuberously swollen, serving as a storage organ, and bearing the bud or buds destined to survive the unfavourable season, while the rest of the shoot system, whether subterranean or epigeal, dies at the beginning of the unfavourable period.

The tubers may vary greatly in their origin, their duration, and their structure. In some, which form a connecting link with Rhizome Geophytes, the tuber is more or less elongated (*Polygonum viviparum*) (Fig. 50). In a number of Stem Tuber Geophytes the tuber is formed of the portion of the stem beneath the cotyledons, in some perhaps the neighbouring portions of the roots participate. Examples are: *Eranthis* (Fig. 51), *Cyclamen*, *Corydalis cava*, *Chaerophyllum bulbosum* (Fig. 52), *Bunium bulbocastanum*, *Tropaeolum brachyceras*, and other species,

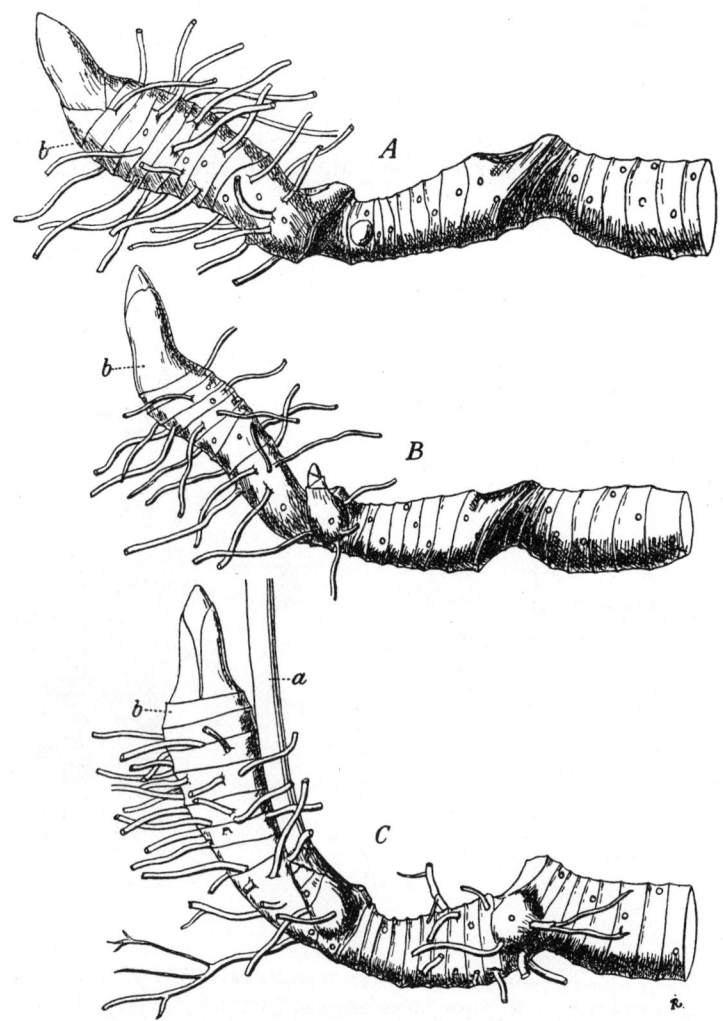

Fig. 48. An experiment with *Polygonatum multiflorum*; three rhizomes planted horizontally at different depths: A at 5 cm., B at 15 cm., C at 30 cm. In A the aerial shoot grew up through a cylinder 10 cm. high (see Fig. 47) so that it seemed to the plant that its rhizome was far deeper in the soil than it really was, and accordingly its continuation shoot, *b*, grew obliquely upwards, as did the continuation shoot, *b*, on the rhizomes B and C, which were deeply planted in the ground.

On the rhizomes that were planted at the same depth as A (5 cm.), but whose aerial shoots did not grow through a cylinder, the continuation bud grew horizontally or slightly obliquely downwards. (⅔)

Sinningia, Corytholoma, Begonia spp. (sub-genera *Huszia* and *Eupetalum*). In these plants the same tuber continues its life and serves as a starting-point for aerial shoots as long as the individual lives; all, or nearly all, of them are without means of vegetative reproduction and of determining the depth of the tuber in the soil; in some the position of the young tuber is determined merely by the position of the germinating seed, and

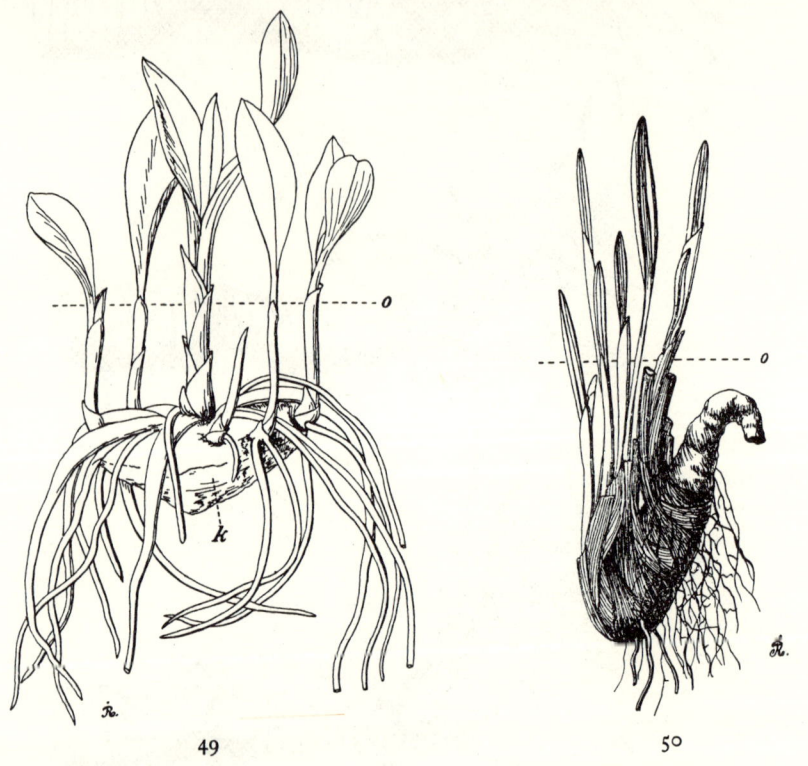

Fig. 49. *Biarum tenuifolium*; Stem Tuber Geophyte; *k*, tuber; *o*, soil-surface. (¾)

Fig. 50. *Polygonum viviparum*; Stem Tuber Geophyte, which by growing obliquely downwards comes to its proper depth in the soil. (½)

in these it is only by 'hilling' that the tubers come to occupy a deeper position in the soil, e.g. *Sinningia*, *Corytholoma*, and *Begonia* species; in others when the seed germinates the young tuber is carried a longer or shorter distance down in the soil by geotropic growth at the base of the cotyledons; e.g. *Eranthis hiemalis*, *Chaerophyllum bulbosum*, and *Bunium bulbocastanum*, or the portion of the stem immediately above the cotyledons participates in the geotropic growth, e.g. *Tropaeolum brachyceras* and other species.

In other Stem Tuber Geophytes the tuber is formed by one or more of the internodes which follow after the cotyledons, and in these plants the tuber usually lasts only for a year, a new tuber being formed every year as a starting-point for the following year's aerial shoot or shoots which develop with the help of the food stored up in the persisting tuber; the superfluous plastic material which the shoots formed during the period of growth is stored up in a new tuber formed of one or several

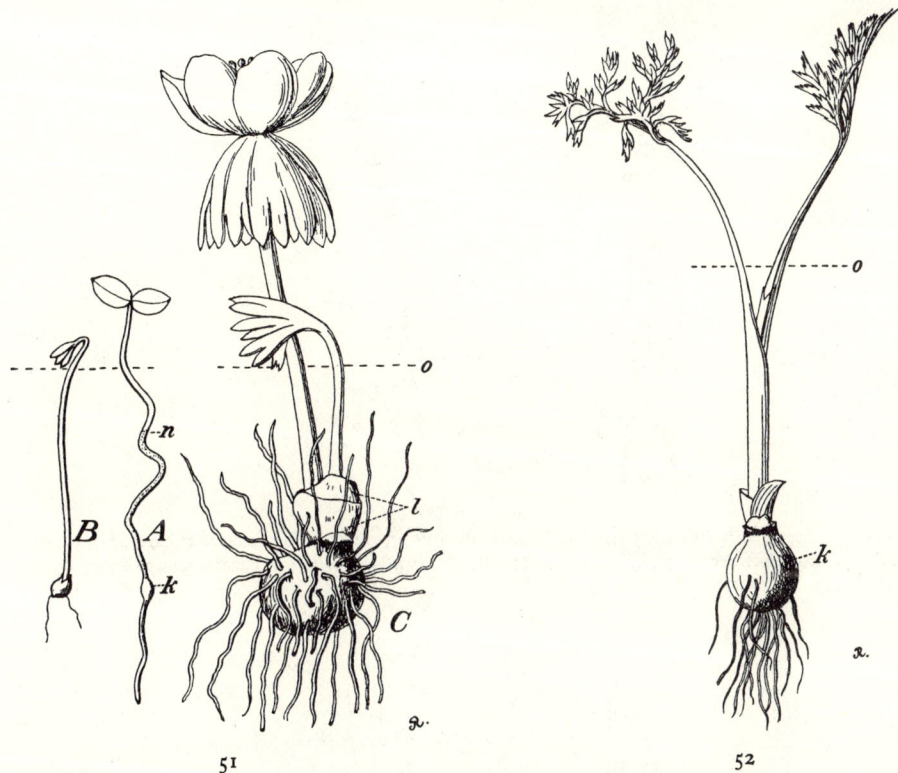

Fig. 51. *Eranthis hiemalis*, Stem Tuber Geophyte; A, seedling; *n*, tube formed by the base of the cotyledons; the bud is situated at the bottom of this tube at the apex of the young tuber (*k*). B, one-year-old tuber in the spring; C, an older plant at the flowering stage; *l*, tuber; *o*, soil-surface. (¾)

Fig. 52. *Chaerophyllum bulbosum*, Stem Tuber Geophyte; young plant; *k*, tuber; *o*, soil-surface. (⅔)

of the shoot's lowermost internodes. Examples of these are *Colchicum, Methonica, Arum,* and other *Araceae, Crocus, Gladiolus, Ixia, Freesia* and other *Iridaceae*. In many of the *Iridaceae* which behave in this way there develops from the base of the individual bud a large fleshy turnip-shaped root whose essential function is doubtless water storage, but at any rate sometimes they also serve to pull the tuber lower down into the ground. Otherwise the species of this group have a very slight capacity for altering the position of the tuber in the ground; but some do certainly possess such a capacity, e.g. *Colchicum* and *Arum*. In *Colchicum* at germination the young tuber is carried down into the soil by a special device.

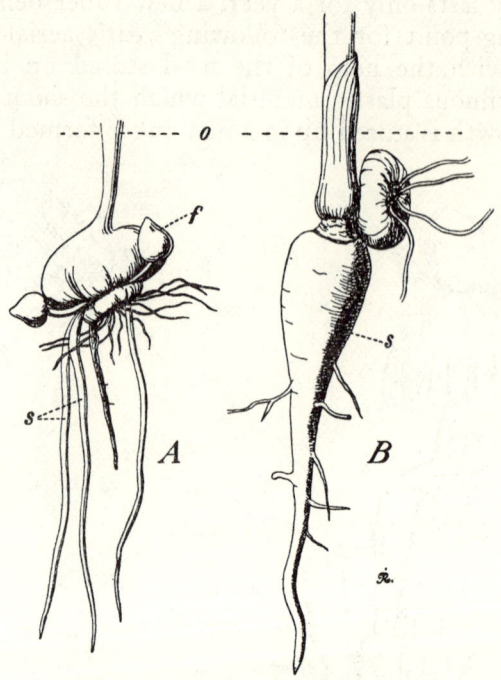

Fig. 53. Stem Tuber Geophytes. A, *Tritonia hyalina*; *f*, stalked reproductive tubers; *s*, thick roots which are not, however, turnip-shaped. (⅘). B, *Gladiolus Colvillei*; *s*, turnip-shaped root; *o*, soil-surface. (⅔)

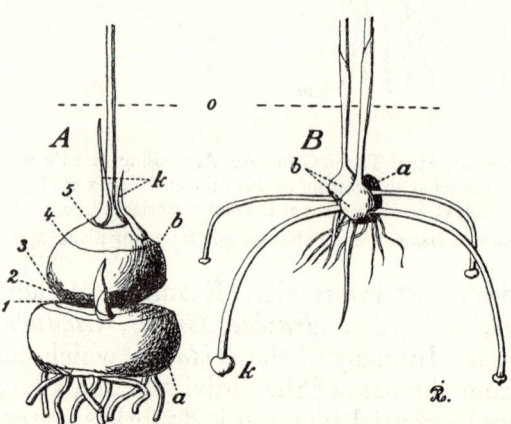

Fig. 54. Stem Tuber Geophytes. A, *Babiana plicata*; *a*, the old tuber; *b*, the young tuber, which is the lowest portion of the flowering shoot; 1–5, the five first leaves on *b*; 1, scale; 2–3, scales whose apices are like foliage leaves; 4–5, foliage leaves; above 5, there are three more foliage leaves; *k*, buds. (⅟₁) B, *Ixia conica*; *a*, old tuber; *b*, the two young tubers; from the lowermost leaf axils on *b* there arise stolons which arch downwards ending in little tubers. (¾)

Several buds are often found on the tubers, but as a rule one of them is much stronger than the others, and sometimes this is the only one to develop. On stronger tubers one or two of the other buds usually

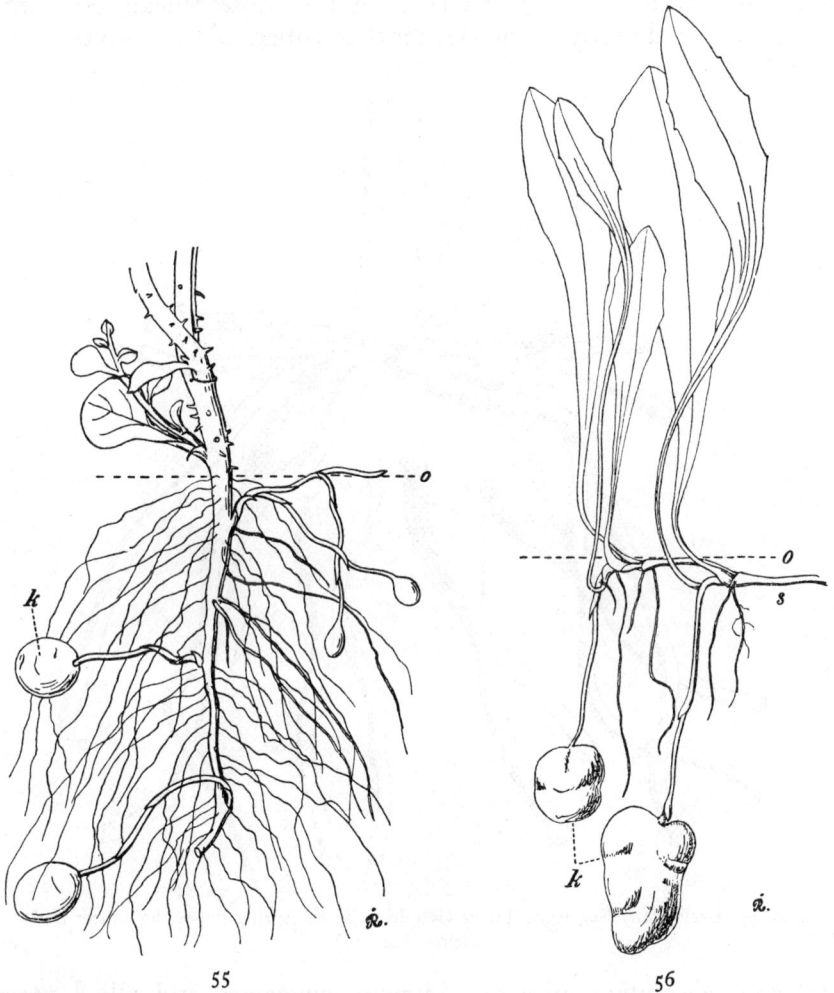

Fig. 55. *Ullucus tuberosus*; Stem Tuber Geophyte with new tubers, *k*, at the apex of the stolons; *o*, soil-surface. (¼)

Fig. 56. *Crepis bulbosa*; Stem Tuber Geophyte with stolons whose apices develop into tubers; *o*, soil-surface. (⅔)

develop into weaker shoots, each of which produces a new but smaller tuber. In this way vegetative reproduction takes place. Sometimes these new tubers are separated from the mother tuber by means of stolons of varying length (e.g. species of *Tritonia* and *Ixia*, Fig. 54, B), and by this means distribution, though not extensive, takes place.

80 LIFE-FORMS OF PLANTS

This leads us on to a third group of Stem Tuber Geophytes, in which a marked division of labour has taken place, the tubers not being formed at all on the aerial shoots, but only at the apex of subterranean stolons of varying length which arise either on the subterranean portions of aerial shoots or directly from the mother tuber. As examples may be

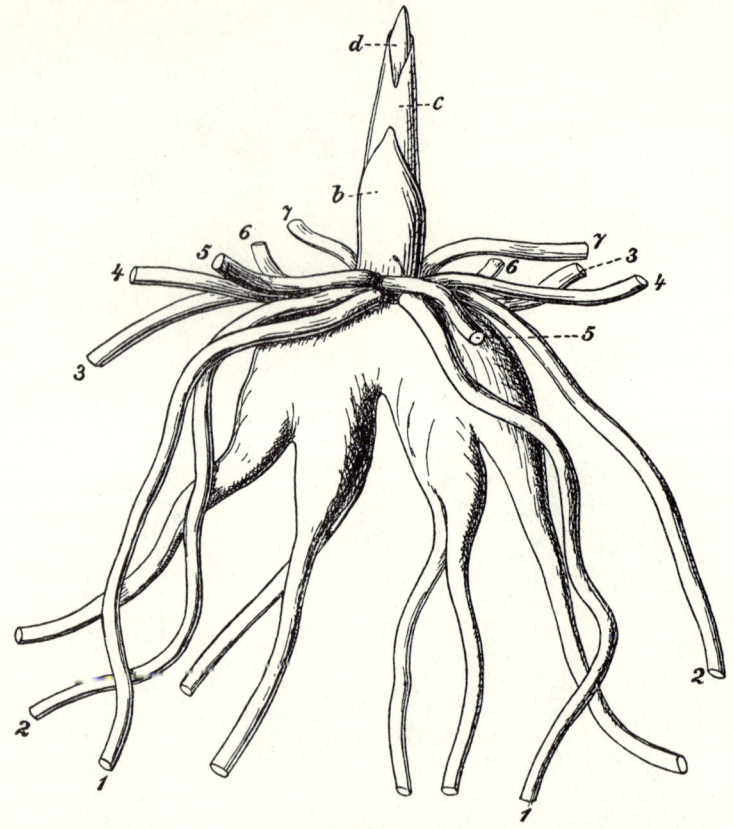

FIG. 57. *Orchis latifolius*, Root Tuber Geophyte, in the spring stage; the winter bud is sprouting. (⅓)

mentioned *Helianthus tuberosus*, *Solanum tuberosum*, and allied species, *Ullucus tuberosus* (Fig. 55), *Stachys affinis* and *S. palustris*, *Crepis bulbosa* (Fig. 56). As far as investigations have been carried out these plants seem to possess, within certain bounds, the capacity of regulating their depth in the soil by means of their stolons.

25. **Root-Tuber Geophytes.** In these the roots are swollen to form tubers or turnip-shaped bodies which serve as food stores. At the beginning of the unfavourable season the aerial shoots die right down. Only certain buds, together with the roots connected with them, survive the unfavourable season. We have been long familiar with this behaviour

Fig. 58. *Dichorisandra* sp., Root Tuber Geophyte; *a*, tuberously swollen rhizome from which the aerial shoot springs; *b*, continuation bud from whose more or less swollen axis there spring a number of tuberously swollen storage roots, *rk*; *o*, soil-surface. (⅓)

in *Orchis* (Fig. 57), and other *Ophrydeae*; *Dichorisandra* sp. (Fig. 58) behaves in the same way.

In the *Ophrydeae* the tubers live for only one year; the bud belonging to each tuber, when it shoots, uses up all the stored material, so that the tuber shrivels and dies, while the new continuation buds become furnished with their own tubers.

Tubers may get lower down into the soil by 'hilling', but the plants are also in some degree able to alter their position in the soil; if for any reason they find themselves too deep they can reach a higher level by elongating the internodes of the lowermost portion of the aerial shoot, which is situated in the soil, so that the continuation buds attached to this part of the stem are raised to a higher level in the soil (Fig. 59). Besides, at any rate in some species, sinking in the soil may be brought about by an individual bud being carried a little downward by obliquely downward growth of the part uniting the tubers with the mother axis. In *Herminium monorchis* (Fig. 60) the bud and the tuber belonging to it are carried a considerable distance from the mother plant because the part uniting the tuber with the mother axis is developed as a stolon.

26. **Bulb Geophytes** store their food in scale-like leaves or portions of leaves which are packed together like leaves in a bud; the persisting portion of the stem consists only of that part which bears leaves; the leaves surround those buds which in the next period of growth are destined to form new aerial shoots.

The numerous structural variations are well known. In some the bulb scales are complete leaves completely modified for storage, e.g. *Gagea, Lilium, Oxalis* spp. (Fig. 61), *Gloxinia, Achimenes,* and several other *Gesneriaceae* (Figs. 65–7), *Actinostemma* (Fig. 68). In others it is only the lowermost, often sheathing, portion of the foliage leaves (or scale leaves) which forms the bulb scales (e.g. *Allium, Ornithogalum, Galanthus, Leucojum, Narcissus,* &c.).

Bulbs are usually sympodial, the strong terminal bud of the bulb first developing into a flower-bearing aerial shoot when the favourable season begins and continuation taking place by means of lateral buds (Fig. 62), of which the one which arises in the axil of the uppermost of the radical leaves is usually by far the strongest. Examples occur especially in bulb-bearing *Liliaceae* and *Iridaceae*.

In others the bulb is a monopodium which every year produces a group of leaves, of which some become wholly or partially storage leaves, while the flower-bearing aerial shoots are lateral shoots (e.g. *Galanthus, Leucojum, Narcissus,* and other *Amaryllidaceae*).

In some species the bulb is made up of scales belonging to several seasons, and the food stored up in a scale belonging to one vegetative period is not used by an aerial shoot during the following vegetative period,

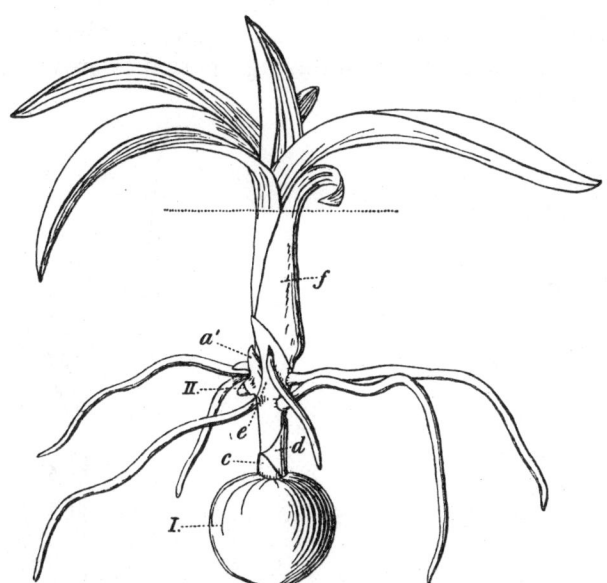

Fig. 59. *Orchis morio*, Root Tuber Geophyte; a plant in the spring stage. The continuation bud for the following year is already laid down, a', and the tuber belonging to this bud, II, is about to grow out; the long dotted line is the soil-surface.

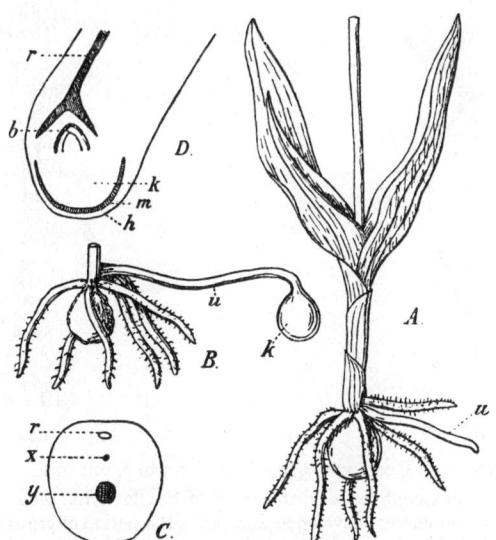

Fig. 60. *Herminium monorchis*; Root Tuber Geophyte. A, the lowermost portion of a flowering plant in June; from the base of the stem spring six adventitious roots (with root-hairs); the stolon, u, by means of which the new tuber is separated from the mother plant, is growing out. B, base of A after the new tuber, k, has been formed at the end of the stolon, u. (†). C, magnified transverse section of a stolon; $r = r$ in D; x and y, vascular bundles. D, magnified longitudinal section through the apex of a stolon at the time of formation of the new tuber k; b, winter continuation bud; r, narrow canal which runs through the stolon and opens near the mother plant; it is by means of this tube alone that the continuation bud is connected with the air in the soil.

but during the one following that, or later still (e.g. *Ornithogalum nutans*). Usually, however, the food stored during a period of growth is used by the

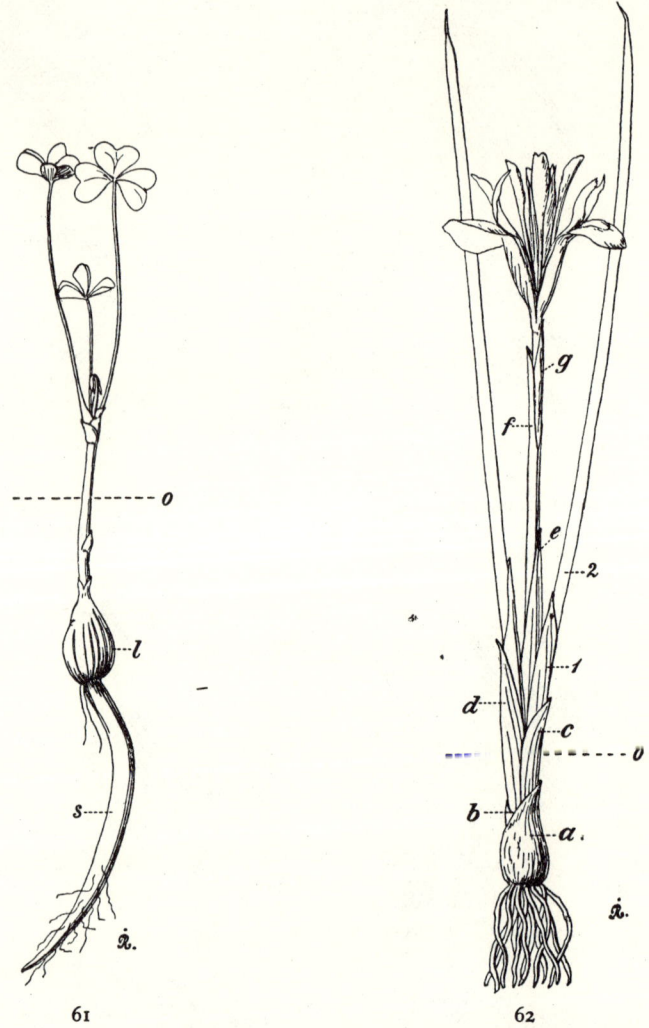

FIG. 61. *Oxalis cernua*, Bulb Geophyte; *s*, sap-root; *l*, bulb; *o*, soil-surface. (⅔)

FIG. 62. *Iris histrio*, Bulb Geophyte; *a–g*, the leaves of the flowering shoot (Nos. 3–9); the first and second leaves of the shoot, which were developed during the last period of growth, are dead; *a* is a storage leaf which also was developed during the last period of growth and is now partially empty of its food; in the axils of the fifth and sixth leaves, *c* and *d*, are lateral shoots each of which begins with a scale leaf, 1, and then makes a foliage leaf, 2, and within this a storage leaf corresponding with *a*. (½)

next year's aerial shoot, so that only one set of fresh storage leaves is present in a bud (e.g. *Ornithogalum umbellatum*). In others again the developing aerial shoot uses the food stored in its own lowermost leaves, which are

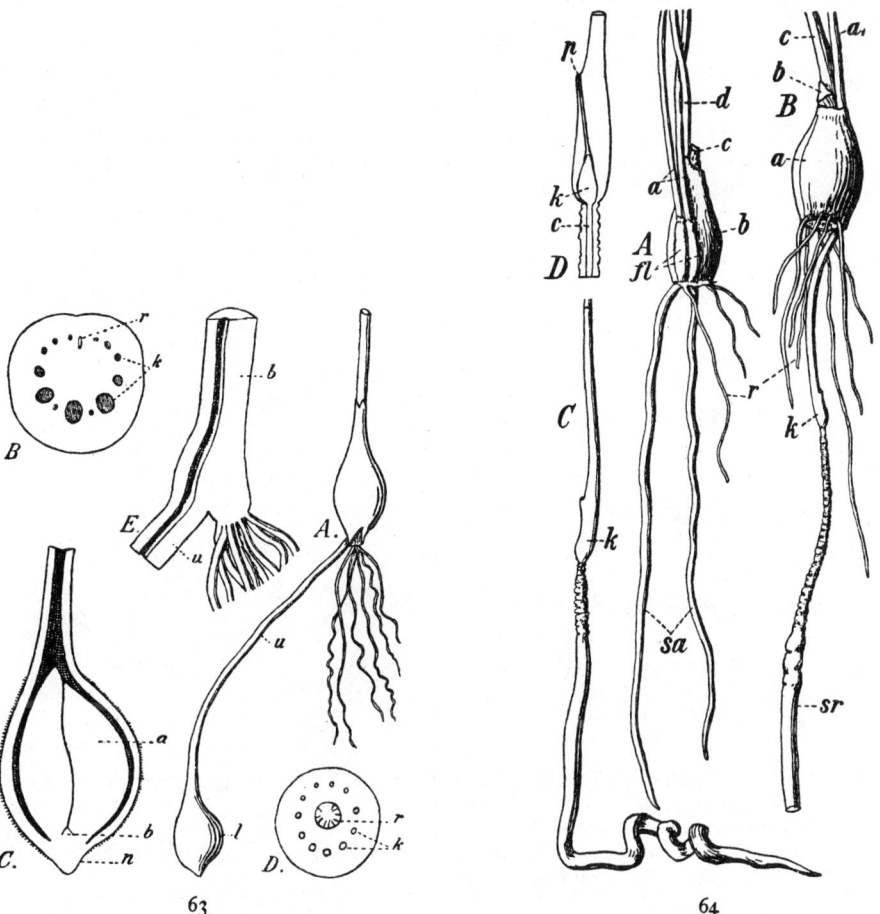

Fig. 63. *Tulipa silvestris*, Bulb Geophyte. The new bulb, *l* in A, is being carried deeper into the earth by means of a slender stolon, *u*, in whose swollen end it arises. ($\frac{1}{1}$). B, transverse section through the stalk-like sheath of the foliage leaf; *r*, air canal; *k*, vascular bundles. ($\frac{10}{1}$). C, longitudinal section through the apex of a stolon with the new bulb, which connects with the air solely by means of a narrow canal which runs through the stolon and is continued up through the sheath of the foliage leaves of the mother bulb; the opening of this sheath is above the soil-surface. ($\frac{2}{1}$). D, transverse section of the stolon; *r*, air canal; *k*, vascular bundles. E, longitudinal section through the uppermost portion of a stolon *u*, and the lowermost portion of the sheath of the foliage leaf, *b*; the continuation of the air canal is seen running completely through the section.

Fig. 64. *Ornithogalum nutans*, Bulb Geophyte; example of a plant in which the buds are dragged down in the soil by means of the roots. A, a young bulb with two small daughter bulbs, *fl*, each with a foliage leaf, *a*, and each with a long rather slender sap-root, *sa*. ($\frac{1}{1}$). B, a similar bulb with only one daughter bulb, *k*, which by the contraction of the sap-root belonging to it, *sr*, has been drawn down into the soil; a_1, foliage leaf of the daughter bulb which is wedged in between the mother bulb and its leaf, *a*, in whose axil the daughter bulb has been situated. ($\frac{1}{1}$). C, daughter bulb with its sap-root and the lowermost portion of the foliage leaf. ($\frac{1}{1}$). D, longitudinal section through a daughter bulb; *p*, the opening of the sheath of the foliage leaf, leading through a narrow canal down to the bud, *k*. ($\frac{2}{1}$)

differentiated and filled with food material before the development of the epigeal flower-bearing portion of the shoot (e.g. *Gagea*).

In some species the bulb has no other protection than that afforded by the old scales which have become mere pellicles (e.g. *Ornithogalum, Gagea, Allium,* &c.). In others each new shoot, or in a monopodium each year's increment, begins with one or more membranous but firmly constructed protective leaves (*Tulipa*).

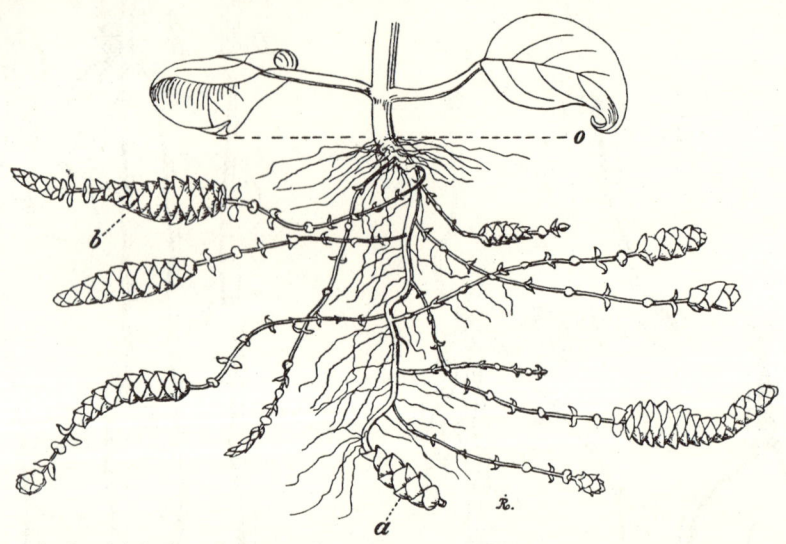

FIG. 65. *Tydaea Lindeniana,* Bulb Geophyte with catkin-like bulbs at the end of slender stolons. The terminal bud on the mother bulb, *a*, has grown out into an aerial shoot with foliage leaves and flowers; the underground portion of this shoot bears scale-leaves whose axillary buds grow out into stolons of varying length and form elongate bulbs (*b*) at their extremities, the tightly fitted leaves becoming thick and fleshy and packed full of food material. (⅔)

Most have no power, or next to none, of altering their position in the soil; growth carries them indeed gradually a little higher up, but 'hilling' increases their depth again. Some, like *Galanthus* and *Leucojum,* can, if they have sunk too deep in the soil, rise higher up again by interposing an elongated shoot-portion between the old and the new parts of the bulb. In many species of *Allium* the small daughter buds are carried on stalks to a higher level in the soil, where they presumably find the depth better suited to their size. In certain species of *Tulipa,* e.g. *T. silvestris* (Fig. 63), especially in young individuals, the young bulb is carried to a varying distance from the mother plant by a stolon, and the direction this stolon takes is within certain limits dependent on the depth of the mother bulbs, so that these plants possess a means of, at any rate partially, determining their depth in the soil. There are also species of *Oxalis* that can change their position in the soil.

In many species contraction of the roots can draw the bulbs down: in

Ornithogalum nutans (Fig. 64) the small daughter bulbs are carried by this means away from the mother bulb.

Vegetative reproduction takes place in all these species, the older plants developing, besides the new bulbs, also one or more reproductive bulbs. When these reproductive bulbs, either by the contraction of the roots, or by means of stolons, are carried away from the mother plant, vegetative distribution also takes place. Examples of this have been mentioned in *Ornithogalum*, *Allium*, and *Tulipa*; bulb-bearing *Gesneriaceae* behave in this way, and also *Actinostemma*.

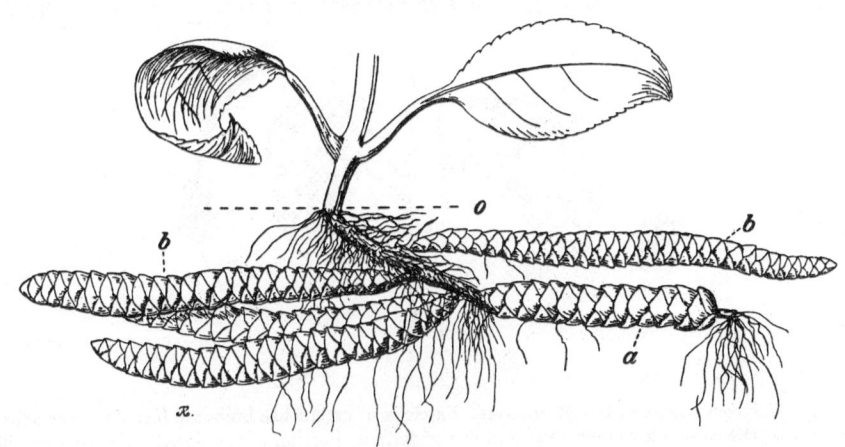

Fig. 66. *Isoloma pictum*, Bulb Geophyte; with long catkin-like bulbs. Development as in *Tydaea Lindeniana* (Fig. 65); *a*, mother bulb; *b*, new bulb; *o*, soil-surface. (⅔)

A great many *Gesneriaceae* (Figs. 65–7), viz. genera in the group *Gloxinieae*, and some genera in the groups *Kohlerieae* and *Bellonieae*, have underground stolons of varying length, ending in comparatively long cylindrical catkin-like structures which, in spite of their deviating form, can be referred to as bulbs (*bulbus articulatus, propagulum squamoso-amentaceum*): these bulbs are usually made up of a great number of small scale-like storage leaves which are packed tightly together on a slender axis. The species in question are native of Tropical and Sub-tropical America, and to judge by the manner in which they behave in our greenhouses, the epigeal portion of the plant dies in the dry season, the vegetative organs of the plant surviving this period solely by means of the bulbs which are buried in the soil. In the next period these bulbs send out new aerial shoots, and from the underground portion of these there arise new axillary stolons which bear scales and terminate their growth by the formation of a bulb. Such bulbs may also sometimes be formed on the epigeal portion of the plant, for example in the axils of prophylls (Fig. 67), afterwards descending to the ground.

In *Actinostemma* (*paniculatum* ?) (Fig. 68) belonging to the *Cucurbitaceae* the approximately spherical bulb consists of a collection of tuberous storage leaves; the terminal bud of this bulb grows out into a climbing, or under certain circumstances a prostrate, aerial shoot in whose leaf axils arise slender shoots with rudimentary leaves. The shoots grow downwards and are often branched, and the ends of the shoots often reach the ground, where the apex develops into a bulb. Some of the shoots however grow up from the ground again without forming a bulb. In these species considerable vegetative distribution can take place.

27. **Root Geophytes.** These plants, as I understand them, persist through the unfavourable season exclusively or principally by means of

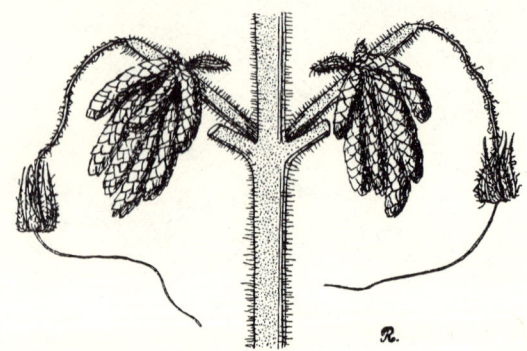

Fig. 67. *Achimenes cordata*, Bulb Geophyte. Fascicles of catkin-like bulbs are formed in the axils of the prophylls; the corolla has fallen away and the uppermost portion of the flower stalk has shrivelled up; the lowermost portion which bears the bulbs is fresh and turgid. (⅔)

buds situated on the persistent roots, while the rest of the plant dies at the beginning of the unfavourable season. *Cirsium arvense* is an example; *Chimaphila uniflora* might perhaps also be classed here; its epigeal shoot certainly does last through the winter, but as it bears no continuation bud it has no direct significance for the plant's capacity to survive the unfavourable season. In this respect it behaves like the aerial shoot in some species of *Rubus*.

There are certainly only a very few species which are typically Root Geophytes, surviving the unfavourable season solely by means of root-buds. On the other hand there are a great many species (*Linaria vulgaris*, *Rumex acetosella*, *Sonchus* spp. &c., &c.), which, according to our classification, should be classed among the Hemicryptophytes, but in which the bud formation on the roots is so common that one might suppose that those shoots which have already reached the soil-surface, or even a higher level, might die completely in the unfavourable season without the individual life being threatened, the plant surviving solely by means of the root-buds. In climates that are very unfavourable for Hemicryptophytes these species might perhaps occur as Cryptophytes, viz. Root Geophytes.

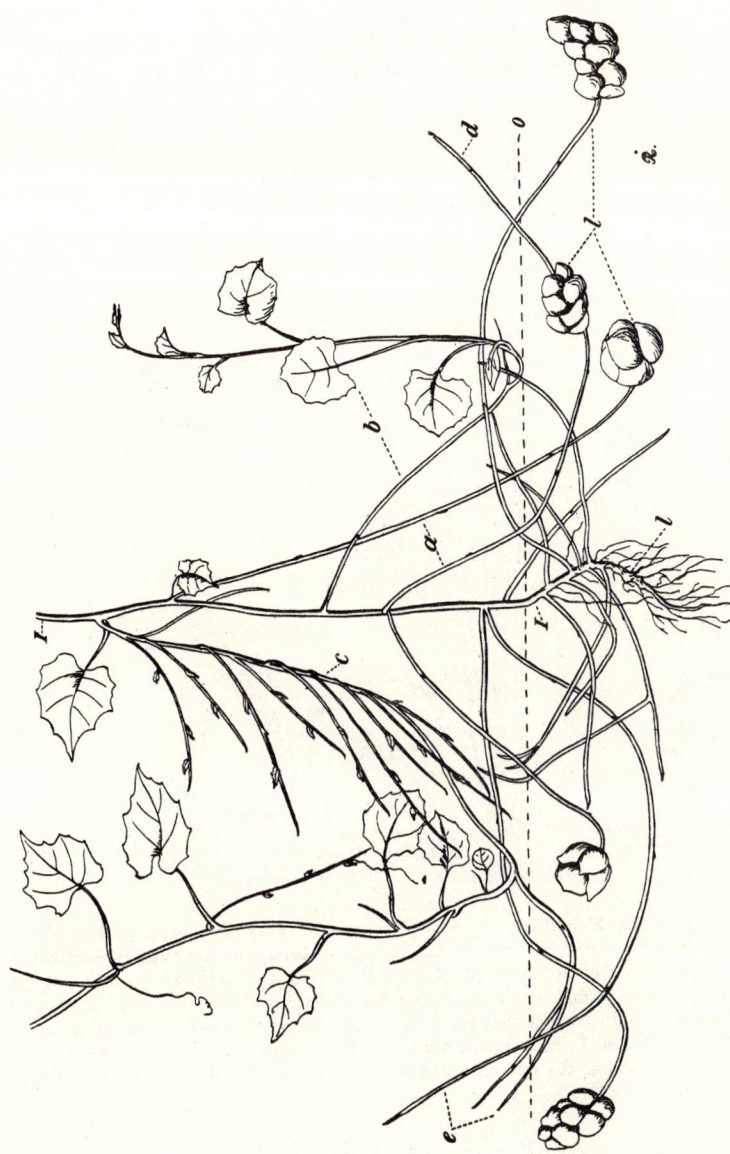

FIG. 68. *Actinostemma* sp. (*paniculatum*?), Bulb-Geophyte. The bulbs, *l*, which consist of a number of tuberously swollen leaves, are formed at the apex of elongated aerial shoots, *a*, which bear foliage leaves and grow downwards. These shoots eventually reach and penetrate the soil where the apex develops into a bulb, the internodes of the stem remaining short and the leaves swelling up and becoming full of food material. (¼)

28. **Helophytes.** By Helophytes I mean those Cryptophytes which exclusively, or at any rate chiefly, grow in soil saturated with water, or in the water itself, from which the leaf- and flower-bearing shoots emerge. Helophytes do not therefore include all the plants ordinarily

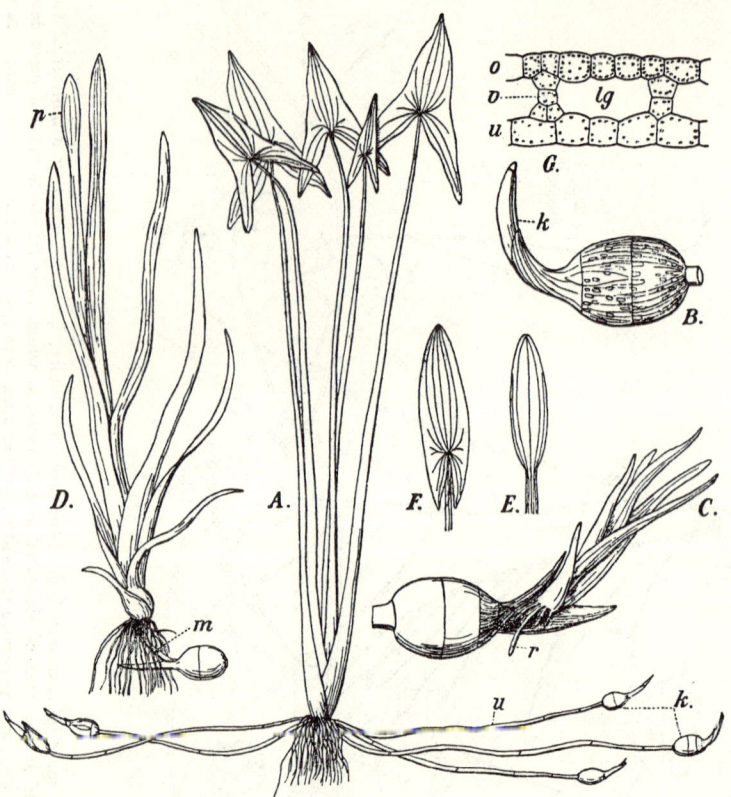

Fig. 69. *Sagittaria sagittifolia*, Helophyte. A, young plant not ready to flower, its leaf blades standing above the surface of the water, and long stolons, *u*, with tubers, *k*, at their apices, each tuber bearing a winter bud. (c. $\frac{1}{5}$.) B, tuber with winter bud, *k*. (c. $\frac{1}{1}$.) C (c. $\frac{3}{4}$) and D ($\frac{1}{4}$), tubers in the spring stage; the bud is growing out to form a new plant whose first leaves are ribbon-like and submerged; later on such long leaves are produced that they reach the surface and begin to develop laminae (*p* in D); these imperfect laminae, E and F, float on the surface of the water; after that perfect leaves are developed whose laminae stand above the water surface (A). G, part of a transverse section of a submerged leaf; *lg*, air chambers. (c. $\frac{100}{1}$.)

known as marsh plants, but only those of them which are Cryptophytes; i.e. they are able to make use of a quality of the environment for protecting their surviving buds, which are situated in the water or in the saturated mud at the bottom of the water. Examples are *Typha, Sparganium, Cyperus, Scirpus, Cladium, Sium, Acorus calamus* (Fig. 70), *Phragmites communis, Alisma plantago, Sagittaria sagittifolia* (Fig. 69), *Ranunculus lingua,* &c. Many Hemicryptophytes affect swampy places, but this does not make them Helophytes, for their persisting buds are situated in the soil-surface. Such Hemicryptophytes may perhaps under special climatic conditions occur as Helophytes because the buds in the soil-surface die without the individual perishing, regeneration taking place by buds which are situated in the soil on old parts of the shoot, and which ordinarily play no part in the plant's life. The boundary between

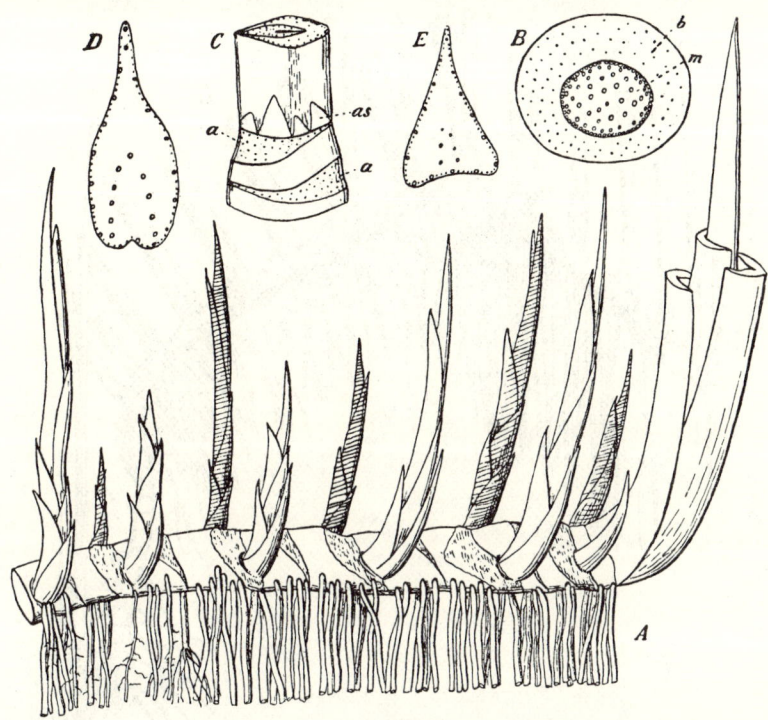

Fig. 70. *Acorus calamus*, Helophyte, A, with horizontal rhizome and two rows of winter buds or young shoots. B, D and E: transverse sections of a rhizome (B), aerial stem (D), and sheathing leaf (E); C, apex of a rhizome with terminal bud, whose leaves are cut off just above their bases; *a*, leaf scar; *as*, small scales in the leaf axils.

Helophytes and Hemicryptophytes which grow on wet soil is indistinct for another reason. It often happens that when the winter buds are laid down in the soil-surface, and when we should therefore expect the plant to be a Hemicryptophyte, the buds become submerged because of the rise in the level of the water during the autumn, and are thus protected throughout the winter like buds that are laid down in water from the very first.

Helophytes have usually elongated, horizontal, transversely geotropic rhizomes or stolons like Rhizome Geophytes: imperceptible gradations unite the two groups, many Rhizome Geophytes being able to grow as Helophytes and vice versa.

92   LIFE-FORMS OF PLANTS

The rhizomes are ordinarily buried in the mud at the bottom of the water, but where the bottom slopes steeply, as it does along banks, it often happens that rhizomes growing horizontally from the banks project into the water, becoming exposed to the light; they then grow obliquely downwards into the water and again reach the bottom. This behaviour

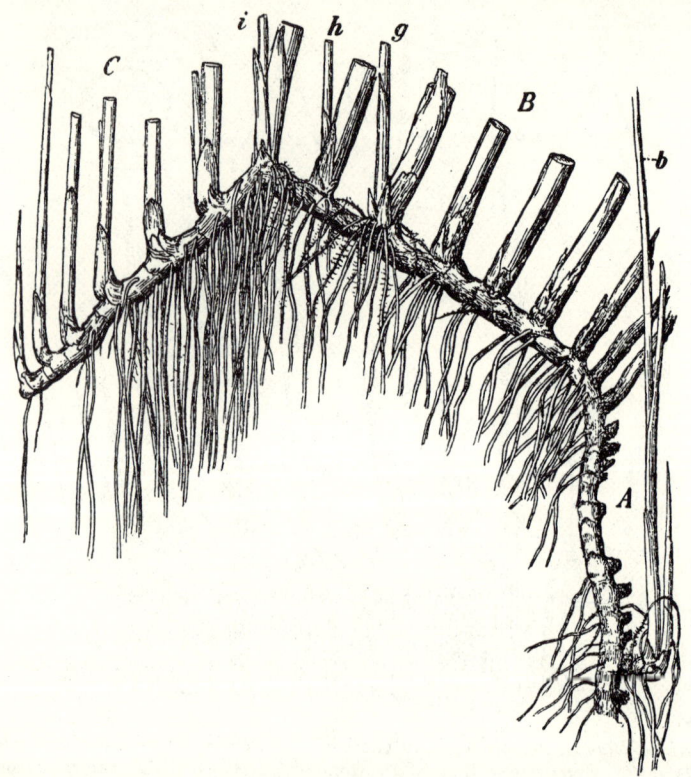

FIG. 71. *Juncus obtusiflorus*, Helophyte; a rhizome from the outermost part of a community growing in water. The vertical part, A, has originally been approximately horizontal, but at one time became loosened from the soft mud and tilted slightly upwards, whereupon part B grew out nearly horizontally, doubtless under the mud; after that the rhizome again became tilted up and came out of the mud into the water, where part C, exposed to the light, grew very obliquely downwards. Meanwhile the shoots b, g, h, and i, have grown out. (c. $\frac{1}{2}$)

can also be observed in *Juncus obtusiflorus* (Fig. 71), *Cladium mariscus* (Fig. 72), *Scirpus lacuster* and *Typha* sp. (Fig. 73). It is not certain whether this downward growth depends upon the fact that these rhizomes are negatively heliotropic, or whether it takes place because they are positively geotropic in the light. In the experiments I made to elucidate this problem the rhizomes that were placed freely in the water certainly grew obliquely downwards in the dark also. This may signify that when they are free in the water even in the dark they are

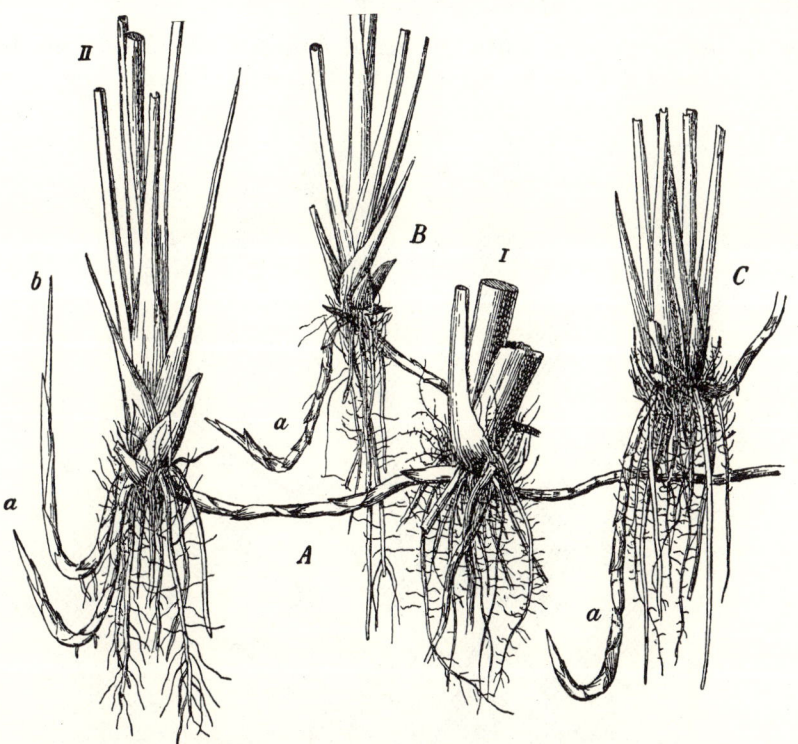

FIG. 72. *Cladium mariscus*, Helophyte; the youngest part of three rhizomes from the side of a community by open water; the rhizomes *a* exposed to the light grow almost vertically downwards. (*c.* ⅓)

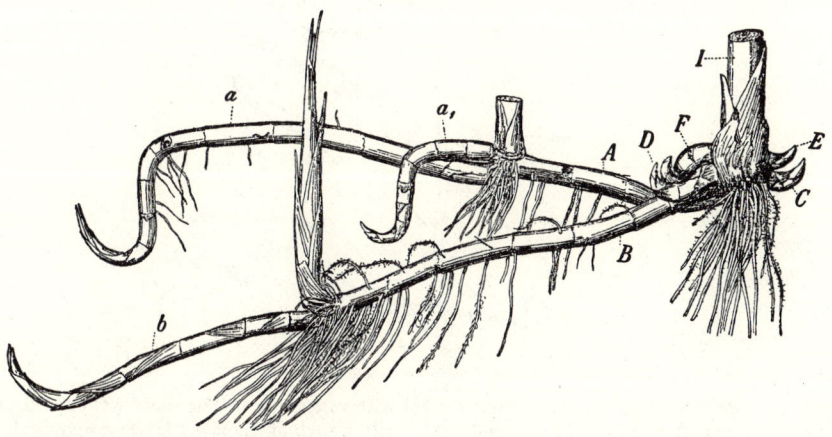

FIG. 73. *Typha latifolia*, Helophyte; the rhizome of a plant which has grown at the outer edge of a community; shoot B has grown obliquely downwards following the sloping bottom of the pond; shoot A has originally grown in the same direction, but before it became rooted suffered some kind of injury at its base; because of its lighter weight it has then been tilted higher up in the water, where it was exposed to the light, which has made the apices of the young shoots, *a* and a_1, grow vertically downwards. (*c.* ¼)

positively geotropic; but since the aerial shoots which developed during my experiments had to be directed up through the covering used for

Fig. 74. *Potamogeton alpinus*; rooted water plant with winter buds. The shoot was pressed against the bottom by the force of the stream, and in the axils of each of the lower leaves a short stolon has then developed with the winter bud at its apex. (½)

excluding the light, it was difficult completely to prevent the access of light to the rhizomes. I do not think it safe to conclude anything from my experiments, especially since the results of my observations are

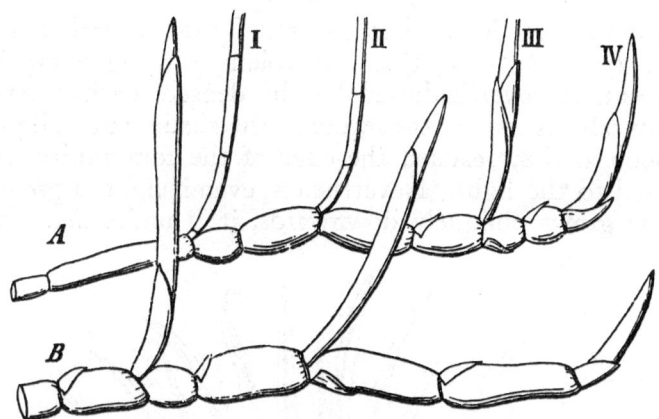

Fig. 75. *Potamogeton lucens*; rooted Hydrophyte; the apex of two rhizomes which survive the winter with some tuberously swollen joints; A with two (III and IV), B with three winter buds. (½)

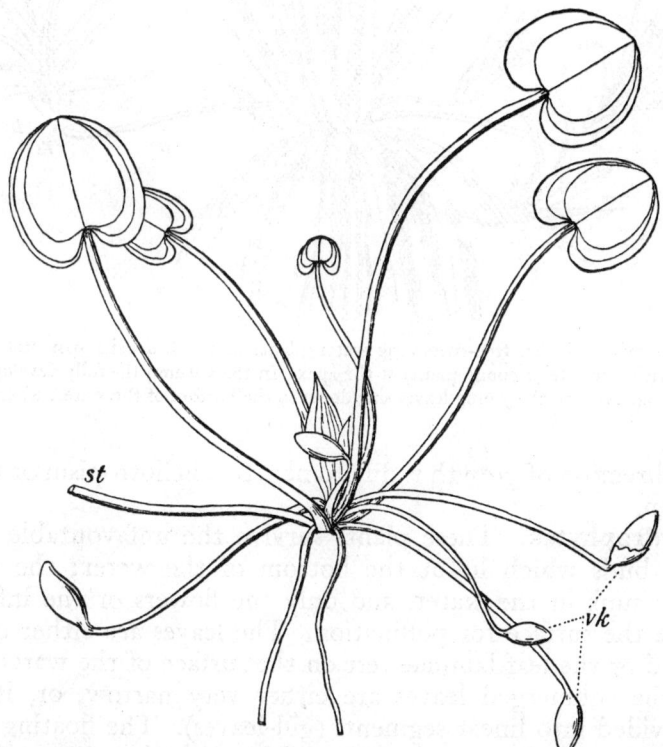

Fig. 76. *Hydrocharis morsus ranae*; a free-swimming plant in the autumn stage. From the leaf axils arise short stolons each with a winter bud, *vk*, at the apex. When the winter buds have attained their full size and are filled with food material they fall off and sink to the bottom; the mother plant then dies. Later on, in the spring, the leaves of the winter bud begin to separate, and air-bubbles are developed which are held between the leaves, making the specific gravity of the buds less than that of the water, so that the buds rise up to the surface, where the leaves unfold. (c. ⅓)

contradicted by what happens in nature. I have several times seen, for example in *Typha latifolia*, that the stolons growing freely out in the water but entirely overshadowed by the densely packed aerial shoots, grew horizontally as long as they were in the shade, but obliquely downwards as soon as they reached the edge of the community, where they were exposed to the light. Nevertheless, even if light is present so that the rhizome grows obliquely downwards, it is undecided whether the

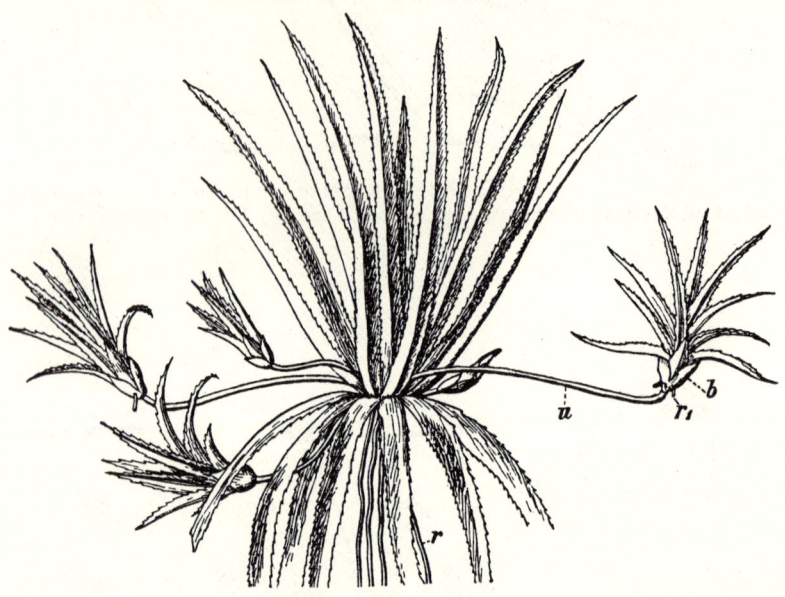

Fig. 77. *Stratiotes aloides*, free-swimming water plant; in the leaf axils arise stolons of varying length each with a rosette (a young plant) at the apex. In the autumn the fully developed leaves die, and the shoot-apices with the young leaves sink down to the bottom of the water, where they survive the winter.

changed direction of growth is due to negative heliotropism or to positive geotropism.

29. **Hydrophytes.** These plants survive the unfavourable season by means of buds which lie at the bottom of the water; the vegetative shoots are sunk in the water, and only the flowers or the inflorescence rise above the surface for pollination. The leaves are either completely submerged or the leaf laminae rest on the surface of the water (floating-leaves); the submerged leaves are either very narrow, or, if they are broad, divided into linear segments (gill-leaves). The floating leaves are undivided or orbicular, ovate or ovate oblong; the base is usually cordate.

Some Hydrophytes have rhizomes that grow at the bottom of the water, and the winter buds are found on these rhizomes, as in *Nuphar, Nymphaea, Zostera, Aponogeton, Limmanthemum, Helodea*, species of *Potamogeton* (Figs. 74 and 75). Others on the contrary do not have perennating rhizomes; but at the beginning of the unfavourable season the plant dies entirely except for certain winter buds which become detached from the plant, sink to the bottom of the water, and there survive the unfavourable season. Several of these plants are rooted, e.g. *Potamogeton zosterifolius, P. acutifolius, P. mucronatus, P. obtusifolius, P. pusillus,* and others. Others are free swimming with roots, e.g. *Hydrocharis* (Fig. 76) and *Stratiotes aloides* (Fig. 77), or without roots (species of *Utricularia*). Those Hydrophytes that survive the winter season by means of buds at the bottom of the water show the widest adaptation to periods unfavourable for growth. Between this type and the evergreen Megaphanerophyte with naked buds seen in the tropical rain forest lie the whole series of degrees of adaptation represented by the series of types that have been described.

V. THEROPHYTES

30. **Therophytes** are plants of the favourable season, or 'summer-plants'. In this category I place plants which survive the unfavourable season in the form of seed; but they complete their entire life-history from seed to seed during the favourable season, and many of them can run through the whole cycle in as short a period as a few weeks. Since seeds, because of their firm close testa, are among the parts of plants which are best protected against drought, these plants are able to live in regions with a very hot and dry climate, and only a short favourable season; steppes and deserts are comparatively rich in Therophytes. This life-form occurs also indeed in many other situations, but especially on cultivated land where there is much open ground. Many of the annual plants occurring on cultivated ground have presumably arisen as a result of cultivation by unconscious selection of variations whose life-history corresponds with that of the annual crop, and whose seeds therefore ripen and are harvested together with the crop. Besides the annual summer plants which complete their life-history during a single summer I reckon as belonging to this type at least some of the annual plants which survive the winter, and which in our climate germinate in the late summer or autumn, and pass the winter in the form of rosettes from which flowering shoots are produced during the next spring. At first sight one might suppose that these plants, like biennials, should be reckoned among the Hemicryptophytes. Those annuals which survive the winter and which can also be biennial I count as Hemicryptophytes, e.g. *Lepidium ruderale, L. campestre, Torilis anthriscus, Chaerophyllum temulum, Jasione montana, Carlina vulgaris*; and also those which besides surviving a single winter may grow as perennials, e.g. *Myosotis arvensis* and *M. silvatica.*

But those species which either invariably behave as annual plants lasting through the winter, or may also be ordinary annuals, I reckon among the Therophytes: examples of the first group are *Schedonorus tectorum, Draba muralis, Alyssum calycinum, Anthriscus vulgaris*, and many others. Examples of the second group are *Bromus arvensis, B. commutatus, B. mollis, B. secalinus, Cerastium semidecandrum, C. glomeratum, Arenaria serpyllifolia, Draba verna, Thlaspi arvense, Teesdalea nudicaulis, Arabis thaliana, Scandix pecten Veneris, Galium aparine, Veronica hederifolia, Centaurea cyanus, Anthemis arvensis, Crepis tectorum, C. virens*, &c., &c.

The fact that these plants survive our winter but not our midsummer leads us to suppose that it is the dry summer and not the cold winter which is the unfavourable period for them, and that they come from regions with a summer hotter and drier than ours. Nearly all our species are found in the Orient.

Most of the plants which grow as annuals with us are not native in this country. It is only because they grow on cultivated soil, where the interference of man protects them from competition with perennial plants, that they are able to exist. If the ground were left to itself most of the annuals would succumb in the battle with other life-forms.

THE USE OF LIFE-FORMS IN PLANT GEOGRAPHY TO CHARACTERIZE EQUICONDITIONAL REGIONS

Geography is the study of the Earth, especially in relation to its suitability for cultivation and its importance as a habitation of man. In this science the plant world plays the most important role, partly directly, by yielding cereals, &c., and partly indirectly by fulfilling the demands of animals.

The different character of vegetation in the various regions of the world results from the difference in the essential qualities of the unfavourable season. If, for example, the climate of Denmark at all seasons resembled our best summer weather, then a number of tropical and subtropical plants, including cultivated plants, would certainly thrive here. They do indeed grow remarkably well here in the summer. This is of course not true of all parts of the world; towards the poles even the most favourable period is too cold for plants belonging to hot regions. But at any rate it is true that if the climate throughout the year in every region resembled the climate of the most favourable season of the region, then a great number of plants, including cultivated plants, would have a much wider distribution than they have at present. It is especially the unfavourable season which limits their distribution.

Our task is to describe a region in terms of the plant world, but we are not concerned so much with purely botanico-biological affairs. We must also consider factors which make plant cultivation possible, and stock raising, which follows as a sequel, and also the suitability of the region as a habitation of man, which is dependent both on the plants and on the animals. This task is best accomplished by considering how plants behave during the unfavourable season, and to what extent this season hinders the existence of the life-forms to which different cultivated plants belong. It is therefore the metamorphosis of the plant body enabling it to survive unfavourable seasons which I have taken as a foundation for the life-forms I have described; they are biological types corresponding to different climates.

It is of course not true that plants in a definite climate all belong to one single life-form; but the main types of climates are characterized by the fact that one or few life-forms are, relatively or absolutely, dominant. This can be expressed numerically; the plant-climate can be characterized by a statistical survey of the life-forms. How this can be done I will show by some examples.

Here, however, we encounter a difficulty. The flora of no country has been investigated so thoroughly that the information available in books is sufficient to enable any one to decide to which life-form each individual species belongs. In some floras this can be decided for the majority of species; but for a statistical investigation all of the species, as far as possible, must be included. By investigations of the Danish flora carried out for many years I have tried to procure all the information necessary for determining the life-form of all our species, so that it is now possible to decide with some certainty the percentage of Danish species belonging to each life-form described in this book. As we shall see later, for the sake of comparative investigation it is better to reduce the number of the life-forms used.

But first we must consider how to determine the number of species. For the time being I include only flowering plants. I omit vascular cryptogams because they are not often included in lists of the higher plants of a country or a district. If they were included here a comparison between the statistics of life-forms in Denmark and other regions would encounter the difficulty that the comparison did not rest upon a uniform foundation, as of course it should.

Next we must decide which species should be included among the flowering plants. Besides the aboriginal plants, we have here, as in most other places, many species which mankind in the course of time has either voluntarily or involuntarily introduced. It is often difficult to decide whether a species has reached the country with or without the aid of man; but this is not the most important thing. It is a matter of no consequence whether a species has been introduced by the wind, by animals, or by man. Man is as much a part of nature as animals are. What seems to be decisive is that the species in question can maintain a place for themselves in our flora without the assistance of cultivation. The species which can do this I regard as belonging to the flora; and it will be apparent that these species, both here and in other regions, belong to the locally dominant life-form or life-forms, so that it does not influence the statistical result whether they are included or omitted. I exclude on the other hand all species which are dependent upon man, such as numbers of weeds in fields and gardens, especially Therophytes, which in our climate can only persist on constantly cultivated ground, where they have not to compete with the dominant life-form of this region, the Hemicryptophytes.

The flora of Denmark thus limited contains 1,084 species. We shall see later what proportion belongs to each life-form. Here it may be remarked that only 75 species (7 per cent.) are Phanerophytes, while 538 species (about 50 per cent.) are Hemicryptophytes. The numerical superiority of the Hemicryptophytes makes it probable that the Hemicryptophyte is the characteristic life-form of the plant-climate of Denmark. But at this stage we cannot be certain of this fact, for it is possible that Hemicryptophytes preponderate in every flora. We are not certain of our facts till we make a comparison between the floras of different climates.

Even if no other country or region has been investigated sufficiently to enable us to give a comparative account of the life-forms as we can for Denmark, yet there are a great many regions of which sufficient is known of the flora to enable us to say that the species are distributed among life-forms in a manner widely different from the Danish species. We know for example that in all tropical countries where the dry season is not too severe it is the trees (Phanerophytes) which dominate the other life-forms. But of the actual relative percentage of life-forms among the species we can say nothing until all the life-forms of all the species are known. The floristic handbooks of to-day resemble in this respect those written a hundred years ago. Herbaceous Chamaephytes, Hemicryptophytes, and even a number of Geophytes are included under the designation of 'Perennial Herbs'. For Phanerophytes, trees and bushes, often the height is not mentioned; and as a rule there is no information about the covering of the buds or whether the plants are deciduous or evergreen. One can indeed often find out from books whether the woods of a district are evergreen or deciduous; but as it often happens that in a region with evergreen woods deciduous trees also occur, and since in regions with deciduous woods (e.g. Denmark) there are evergreen species, it is not possible to use statistics based on the fact that the woods are evergreen or deciduous. It is obvious then that the floristic handbooks stand in need of enrichment with biological material, and that it is exceedingly desirable that those studying the flora of individual regions should strive to include in their books information about the life-forms of each species, or at any rate information sufficient to enable others to determine the life-forms. It is often considered superfluous to give life-forms of every species; but this must be done.

In the latest edition of my *Dansk Ekskursionsflora* I have added to the descriptions the main type of life-form to which each species belongs; but the information given is not quite complete; e.g. the Beech is marked as a Phanerophyte, but that it is deciduous is not mentioned. It is however mentioned when trees and shrubs are evergreen; if they are deciduous it is not so stated. I have allowed it to be taken for granted that a Phanerophyte is deciduous unless it is stated as being evergreen, because the majority of our Phanerophytes are deciduous. But this method is wrong and must eventually be altered. We must bear in mind that in Japan or some other distant country botanists might be making use of a list of plants of Denmark for comparative study. For these botanists it is very far from self-evident that a certain Danish tree is deciduous, and that all of them are not deciduous. We are placed in the same position in dealing with the floras of distant countries. I have attempted in vain to determine from books the relative numbers of evergreen and deciduous plants in different floras, e.g. New Zealand. We know indeed that most of the Phanerophytes of New Zealand are evergreen, and we also know that some of them are deciduous; but in the floras of New Zealand it is not stated which are evergreen and which are deciduous.

Since it was important for me to obtain for comparative purposes the numerical relationship of the life-forms in a climate widely different from our own, and since this information was not available in books, I was obliged to undertake a journey myself. During 1905 and 1906 I visited the West Indies with the object of finding out the life-form of all the species, as far as possible, in a sharply bounded and accessible territory with a climate very different from that of Denmark.

For comparison with Denmark I have chosen the Isles of St. Thomas and St. Jan. Besides the indigenous species I include also, as for Denmark, those plants that have been introduced and which appear to be naturalized. This brings the number of species to 904, and I shall now show how they are divided among the various life-forms.

As I said before it is best to diminish the number of life-forms used. There are two reasons for this. Firstly, it is not necessary for the portrayal of plant climatic territories to use as many life-forms as I have defined in this book. Secondly, for a long time to come the data available for determining some of the life-forms will not be procurable for all species, at any rate for the regions less frequented by botanists. This is true for example of bud-scales. Thus among the Phanerophytes those with covered buds and those with naked buds are not here distinguished; we must content ourselves with stating whether or not bud-scales appear to be dominant. Unfortunately we cannot for the time being distinguish between evergreen and deciduous plants, because our information about individual species is insufficient to enable us to do so. But so many accounts given in general terms are available that this want can be supplied by observing which condition is the commoner. Further, Megaphanerophytes and Mesophanerophytes are treated as one group: Herbaceous Phanerophytes are united to Nanophanerophytes: Hemicryptophytes are all treated as one life-form, as are also Chamaephytes; and lastly the seven groups of Geophytes are reduced to two, viz. Geophytes proper as one group and Hydrophytes and Helophytes as the other. Thus the thirty life-forms are reduced to ten.

It is needless to say that investigators should do all they can to decide to which of the thirty life-forms the individual species belong. The number of life-forms can always be reduced if this is necessary for purposes of comparison.

The table below shows the flowering plants of Denmark in one column and those of St. Thomas and St. Jan in the other column, divided into ten main life-forms.

Further explanation is unnecessary; the numbers express at once the essential relationship between the climate and vegetation, i.e. the plant climate. That of Denmark is a Hemicryptophyte climate (half of the plants belonging to this life-form). In the islands of St. Thomas and St. Jan the majority of the species (53 per cent.) are Microphanerophytes and Nanophanerophytes. In Denmark only 6 per cent. of the species belong to these groups. For the West Indies it can further be added that almost all Phanerophytes are evergreen, and that a number, especially the tall ones, have bud-scales.

	Denmark. 1,084 species. per cent.	St. Thomas and St. Jan. 904 species. per cent.
1. Megaphanerophytes and Mesophanerophytes	1	5
2. Microphanerophytes	3	23
3. Nanophanerophytes	3	30
4. Epiphytes	0	1
5. Stem succulents	0	2
6. Chamaephytes	3	12
7. Hemicryptophytes	50	9
8. Geophytes	11	3
9. Hydrophytes and Helophytes	11	1
10. Therophytes	18	14

Now if we knew the life-forms of the plants of all regions as well as we know those of Denmark and the Danish West Indies, we could then portray the climate of each individual region by means of a numerical conspectus, showing how the plants of each region were divided among the different life-forms. But for limiting the boundaries of the different plant climates there is another want to be supplied; there is no guiding principle to show where the boundaries are to be drawn. Let me illustrate this by an example.

The plant climate of North and Central Europe is characterized by a preponderance of Hemicryptophytes. If we go towards the south the Hemicryptophytes decrease, and when we reach the Mediterranean region it is other life-forms, e.g. Chamaephytes and Nanophanerophytes, which preponderate. But there is no guiding principle to tell us how far the number of Hemicryptophytes must fall before we can say that this life-form no longer characterizes the climate. We lack a common measure with which to compare all regions, and this common measure is the percentage of life-forms among all the species of plants in the world. But while such a common measure ought to be the starting-point for our demarcation of plant-climates, as a matter of fact, because of our present ignorance of the life-forms of species, it comes to be the final result of our investigations, if not their ultimate object. To overcome this difficulty I have conceived and also begun to apply the following method. In order to discover the percentage of the different species composing the various seeds in a mixture we must be content with finding this out from samples. In the same way we must determine the percentage of individual life-forms in the flora of the world by determining their relative frequency in a considerable number of species selected at random. Here we encounter at least two difficulties. Firstly, species whose life-form is to be determined must be chosen in such a way that they can be looked upon as a true picture of all the species in the world. But this may be done approximately by taking a certain number, say a thousand species, from an alphabetical list of all known species.

The second difficulty is to discover the life-forms of these thousand species; because of the method of choosing the species one is likely always to hit upon plants about which little is known. But by contenting oneself with the ten groups of life-forms used above this difficulty can be overcome.

Investigations of this kind give us a common measure by which the individual floras can be compared, and by means of which the bio-geographical character of the flora can be determined, because the plant-climate of the individual regions will be characterized by the life-form or the life-forms whose percentage rises above the percentage in the common measure. And the higher the rise of the number above that of the common measure the more strongly the life-forms in question characterize the plant-climate of the region.

It thus becomes possible to draw with certainty bio-geographical lines limiting regions which possess, taken as a whole, uniform requirements for cultivation. Such boundaries must be drawn where the percentage of given life-forms rises above their percentage in the common measure. If, for example, we are to draw a boundary line between the Hemicryptophyte climate of Central Europe and the climate of the Mediterranean region, which is characterized by Chamaephytes, Nanophanerophytes, &c., we must go southward, making biological statistics by investigations of the local floras, till at length we come to regions where the percentage of Hemicryptophytes falls below the number in the common measure, and where, therefore, it can no longer be said that Hemicryptophytes characterize the climate. Of necessity the percentage of one or more other life-forms will have risen above that of the common measure, and it will be these life-forms that now characterize the climate. Such a change always indicates a bio-geographical boundary. Gradually, as the biological study of the vegetation of individual regions advances, we shall be able in this way to divide the earth's surface into phyto-climatic regions and provinces in which are found uniform conditions for plant life and correspondingly uniform conditions for cultivation.

Such an investigation in addition to its bio-geographical interest has also a very important botanical aspect. Gradually, as the bio-geographical investigation of the plant-climates of the world proceeds by determination of the life-forms of the individual species, we procure a mass of biological and morphological facts, suitable for use in monographs dealing with the biology and morphology of the vegetative organs within the individual families. In this way it will be possible to give a comparative picture of life-forms in the individual families, showing how the behaviour of the vegetative organs and the life-forms determined by their behaviour alter in harmony with the regions of geographical distribution of the individual families. It is hoped to obtain by these means an historical explanation of the morphology of the vegetative organs, an explanation of the various morphological transformations that are encountered in the individual families, in correspondence with their distribution in different climates in the past and at the present time.

By studying adaptations to climate we shall find out whether species have arisen by mutation or whether variation brought about by environment have in the course of time become hereditary.

This chapter is no more than a small contribution to our knowledge of a small territory within Nature's far-flung domain. And even if, as it seems to me, from the point we have reached, paths can be seen that lead forward, yet the distant goals of which glimpses are caught lie so far away that one investigator can never reach them.

Many paths lead to those distant goals; and when one goal is reached, indeed often long before, fresh ones loom forth, firing the traveller with enthusiasm to reach them. As it has been the destiny of every investigator so shall it continue to be; his mind is not filled with apprehension, but rather braced for fresh efforts and the strings of his heart are tightened and tuned, in harmony, let us hope, with the unknowable. Only a short way does one investigator penetrate into Nature's land of mystery; but his efforts are not in vain, for the marvels of his observations, happiness in his work, and joy of recognition are his wages and his daily bread.

III
THE LIFE-FORM OF *TUSSILAGO FARFARUS*

WITH the object of obtaining a presentation of the flora of Denmark founded upon a statistical survey of life-forms I undertook a few years ago a provisional enumeration of the percentage of the various life-forms represented in the Danish flora.[1] With this end in view I had for several years investigated a great number of species and settled the life-forms to which they belonged. The life-form of many of the species, however, was determined from knowledge I had already acquired, partly by my own studies in the field, and partly from literature, especially the comprehensive investigations of Irmisch and Warming. From the knowledge gained by these means I thought it best to include *Tussilago farfarus* among the Hemicryptophytes, or, more narrowly, among the Rosette-hemicryptophytes. I continued however my investigations of those species I had not studied during the last few years, and in the autumn of 1905 investigated the behaviour of *Tussilago farfarus*, deciding that this species is not a Hemicryptophyte, but a Cryptophyte, at any rate in our climate. But since Irmisch's account leads us to suppose that farther south this plant can grow as a Hemicryptophyte, and since if this be confirmed it will be of great interest in relation to my bio-geographical views, I feel justified in pursuing the matter in greater detail.

I believe that the plant-climate of a region can be characterized by the numerical preponderance of one or several definite life-forms among all the species in that region. But this expression of a plant-climate is often emphasized in an interesting manner: part of the species in a region, which do not belong to the dominant life-form, change their life-form partially or entirely, thus coming to resemble the dominant life-form more closely than individuals in other climates do. In other words, at least some species have the capacity in a greater or less degree of changing their life-form in correspondence with change of climate, so that they can approach the life-form or life-forms characteristic of the climate in which they grow. Some examples illustrating this will now be given. The climate of the Danish West Indies and that of the Virgin Islands as a whole finds expression in the vegetation by a preponderance of Microphanerophytes and Nanophanerophytes over the other life-forms. This characteristic of the climate has further been emphasized by the fact that some species which usually belong to other life-forms show a tendency to approach

[1] See pp. 100-2.

these two dominant life-forms. This is true both of species which otherwise belong to life-forms less efficiently protected than the dominant ones and also of species that usually belong to better protected life-forms. Thus there are a number of species which in more favourable climates are Mesophanerophytes but which in the Virgin Islands usually occur as Microphanerophytes. Further, we come across a series of species which in less favourable climates are Therophytes, but which in the Virgin Islands occur over large expanses as Suffruticose Chamaephytes, or even as Nanophanerophytes.

In the same way there are some species which in the Chamaephyte-Nanophanerophyte climate of South Europe are Chamaephytes or perhaps even Nanophanerophytes, but which when cultivated in Denmark grow as Hemicryptophytes, thus becoming able to survive the winter in a well-marked Hemicryptophyte climate. On the other hand doubtless some of our species in more favourable climates can grow in a less protected life-form than the one they assume in Denmark. To judge by the information available in literature *Tussilago farfarus* seems to be one of these plants. It is in the hope of interesting some of the botanists of Central and South Europe in a more exact investigation of this question that I here give my observations about the life-form of this plant in Denmark.

So far as I know Irmisch was the first to give a thorough description of the morphology and biology of *Tussilago farfarus*, but as there is one feature in his description which deviates from the behaviour of the plant with us I will first describe the observations that have been made in Denmark.

In Møller-Holst's *Agricultural Dictionary*[1] P. Nielsen has given an excellent account of the morphology and biology of *Tussilago farfarus* founded upon culture experiments. Having described the young seedling P. Nielsen writes: 'After a lapse of a few months lateral shoots—stolons—arise from buds in the axils of the lowest leaves. These stolons will be subterranean if the plant is growing in soft or loose soil, where they extend horizontally in all directions at a varying depth ...', 'if the plants which have arisen from the seed grow under favourable circumstances, in the last half of September or at the latest in October, flower buds will appear in the upper leaf axils and perhaps also at the apex of the stem, but the flowers do not open until the following spring when the leaves have rotted away'. Thus the primary shoot completes its life in the course of two years, two periods of growth. The later shoots on the contrary live through three periods of growth, as described by P. Nielsen and before him by Irmisch, producing in the first year stolons which spend the winter entirely under the ground: in the second year the apex of the stolons grow up to the soil-surface where they produce a rosette of foliage leaves.

[1] Møller-Holst's *Landbrugs-Ordbog*, Anden Del (1878), 323–34.

These rosettes produce a varying number of lateral inflorescence buds and sometimes also a terminal one. In the spring of the third year these buds develop into flowering shoots, and when this occurs the system of shoots in the soil-surface dies; thus the life of the shoot endures for three years.

We may perhaps conclude from Nielsen's account of seedlings that he dealt with unfavourably situated seedling plants, which in the first year did not reach the stage at which they could develop inflorescent buds; but he does not say so definitely; neither does he say that later in the life of the plant shoots with foliage leaves arise, which by the autumn have reached a stage of formation of inflorescent buds. Such shoots nevertheless grow, both under ordinary circumstances and when the normal shoots with foliage leaves have been cut off or destroyed in some other way: thus when the stolons which were to reach the stage of assimilation the next year have grown up to the soil-surface they make rosettes of foliage leaves the same year in which the stolons are produced. It was because of the presence of these rosettes of foliage leaves devoid of inflorescence buds that I originally reckoned *Tussilago farfarus* among the Hemicryptophytes, taking for granted that the terminal buds of these rosettes survived the winter and next year, thus, after a period of rest and reinforcement, reaching the flowering stage. There are indeed a number of Hemicryptophytes in Denmark which behave in this way. Of such Rosette-hemicryptophytes we have: *Petasites officinalis* and *Triglochin palustre*, and of partial Rosette-hemicryptophytes we have: *Aegopodium podagraria, Cirsium heterophyllum, Tanacetum vulgare*, and *Achillea millefolium*. In *Achillea millefolium* one can see in the autumn stolon-like shoots in all possible stages of development; some of them have a long time ago reached the surface of the soil, where they have formed large strong rosettes destined to produce flowers next year. Many of these shoots have only just reached the surface, and have formed only weak rosettes which must live through one or more periods of growth before they become capable of flowering. Besides these two kinds there are many stolons which are still entirely subterranean and which spend the winter beneath the ground. It should here be made quite clear that the presence in *Achillea millefolium* and similar plants of completely subterranean buds which survive the winter cannot be considered a reason for including these plants among the Cryptophytes; for plants which have winter buds in two distinct positions must obviously be included in the life-form which is characterized by having the buds in the less protected position. If therefore the purely vegetative leaf rosettes of *Tussilago farfarus* survive the winter we must include this species among the Hemicryptophytes.

In supposing that the purely vegetative shoots with foliage leaves of *Tussilago farfarus* survive the winter I was in complete agreement with the account given by Warming[1] whose description of the natural history of this plant is based partly on the observations of Irmisch and Nielsen and partly on his own culture experiments. After telling us that the seedlings, according to their investigations, produce inflorescence buds already in their first year, Warming adds, 'My seedlings did not get so far; I suppose that weaker plants like these will perennate with a continuing terminal bud until they are capable of flowering.' Warming describes the three stages in the life of the shoot in older plants as follows:

(1) The scale-leaf stage.
(2) The assimilating stage (rosettes of foliage leaves).
(3) The flowering stage.

He then tells us that the shoot can remain in the assimilating stage for longer than a year, and because of this fact he includes the Coltsfoot in the same group as *Petasites officinalis*, *Achillea millefolium*, *Tanacetum vulgare*, *Aegopodium podagraria*, &c.; indeed he uses the Coltsfoot as a type of this group.

My observations of *Tussilago farfarus* during the last few years have, however, shown me that the leaf rosettes which are not yet ready to flower do not survive the winter in our climate, so that the species survive the winter solely by means of subterranean buds, and must therefore be included among the Cryptophytes, more narrowly among the Rhizome Geophytes.

Under ordinary conditions, where *Tussilago farfarus* is able to grow undisturbed, as it does in grass fields, probably a varying number of the shoots which during the course of the summer have approached the surface of the soil form purely vegetative rosettes, but most of these rosettes produce in the same growing period lateral inflorescence buds which, as is well known, survive the winter in the soil-surface. Sometimes a terminal inflorescence bud is also formed; but usually the apex of the rosette remains purely vegetative.

When I investigated *Tussilago farfarus* in the beginning of December, 1904, I found that all the purely vegetative rosettes were completely dead. Not only the leaves, which ordinarily die rather early, but also the stems were dead: some of them had died to some distance below the surface, even, indeed, down as far as the uppermost stolon, the depth of which varies considerably. Doubtless on those shoots lateral inflorescence buds had formed, but the apices of the shoots that had remained vegetative had died too. In other words all the purely vegetative shoot-apices which had reached the soil-surface and formed rosettes there had died. My observations in the autumn of 1906 yielded the same results. I examined 310 rosettes of foliage leaves, and found them as follows:

33 possessed one to several lateral inflorescence buds and at the same

---

[1] Warming, Eug., *Om Skudbygning, Overvintring og Foryngelse* (Festskrift i Anledning af den naturhistoriske Forenings Bestaaen fra 1833–83), pp. 74–5.

time one terminal inflorescence bud which was usually weaker than the lateral ones.

177 had one to several lateral inflorescence buds. The apices of the shoots were on the other hand vegetative, but dead.

100 rosettes were purely vegetative but dead.

It should be observed that these numbers do not express the true relationship between purely vegetative and floral shoots. Actually there are doubtless under ordinary circumstances far fewer purely vegetative leaf rosettes than the numbers show; but as it is my object to investigate purely vegetative shoots my attention was turned chiefly to these while I was making my collection. The chief point is that all vegetative shoot-apices that have reached the soil-surface and there formed rosettes of foliage leaves, die at the beginning of winter, so that *Tussilago farfarus* in Denmark survives the winter solely by means of subterranean buds. That inflorescence buds survive the winter in the soil-surface has obviously no significance in determining the life-form of the plant. The life-form depends entirely on the position occupied by the vegetative buds, on which alone the persistence of the individual depends. It is of great interest that the floral buds survive the winter in a less protected position than the purely vegetative buds, but this behaviour is seen in various other plants, e.g. *Rubus*.

That the results of my observations do not entirely correspond with Warming's description is, I suppose, because Warming founded that part of his account which disagrees with mine on Irmisch's description. Let us therefore see what Irmisch wrote.

Irmisch has written two small papers about *Tussilago farfarus*: in the first of these[1] he describes how the plant behaves in general, and then discusses some of the deviations from its normal behaviour, e.g. 'Sometimes the terminal inflorescence in its development lags behind the lateral ones, or we find in the place of the terminal inflorescence only a bud formed of scale leaves, ... this terminal bud later develops foliage leaves, and in the following year can produce a terminal and a lateral flowering stem' (l.c., p. 178).

In his little paper[2] which deals with the seedlings of the Coltsfoot it is apparent that Irmisch has not seen the seedling plant form inflorescence buds in the year of germination.

He ends this paper by saying, 'The main root ordinarily dies during the second year, the lateral roots persist and elongate so that the small stolons that perennate by means of a terminal bud can no longer be distinguished from those which have arisen from weak stolons and which have become independent by the perishing of the stolons.' It

[1] Irmisch, Th., 'Einige Bemerkungen über Tussilago farfara', *Flora*, 1851, pp. 177–82.
[2] Irmisch, 'Kurze botanische Mittheilungen. 1. Keimpflanze von Tussilago farfara', *Flora*, 1853, pp. 521–2.

follows from this either that Irmisch has seen the bud of the purely vegetative rosette survive the winter, or at any rate believes that it does so.

That P. Nielsen's seedlings produced inflorescence buds in the year of germination, and that Irmisch's plants did not, is almost certainly explained by the facts that Nielsen sowed fruits and watered his seedlings, while Irmisch mentions only wild seedlings which he did not find till July or August.

Hemicryptophytes with the structure and growth of *Achillea millefolium* are of great interest, because they have the faculty of growing as Cryptophytes. They need not perish even if they grow in conditions which cause the shoot-apices situated in the soil-surface to die during the winter: even under these conditions the buds on the subterranean stolons can carry the individual through the unfavourable seasons. Many Cryptophytes, especially Rhizome-geophytes, have presumably been derived by this method from Hemicryptophytes. It will therefore be of great interest, if Irmisch's description of *Tussilago farfarus* be correct, to contemplate a species which, in harmony with change of climate, forms a transition between the Hemicryptophytes and Cryptophytes. In Denmark the Coltsfoot is a Cryptophyte, but in the milder climates towards the South it occurs as a Hemicryptophyte.

It can be added here that *Tussilago farfarus*, because of its structure and physiological peculiarities, is able also to grow as a Chamaephyte. Rosette-hemicryptophytes form their rosettes by a shortening of the internodes of the shoots which is due to exposure to light. For some species, however, the light must be of considerable strength to cause a shortening of the internodes sufficient to make a rosette actually in the soil-surface. If such species are shaded by other plants the internodes of the shoot lengthen so that the rosette is situated at the apex of an erect aerial stem of varying length. This is just what happens to *Tussilago farfarus* when it grows amidst luxuriant herbaceous vegetation. The rosettes are then borne on comparatively thick aerial stems which may attain the length of several centimetres. If these shoots succeed in surviving the unfavourable season, *Tussilago farfarus* will then grow as a Chamaephyte, thus resembling *Geranium macrorhizum*.

I have written this account in the hope that it will lead to further investigations of the biological behaviour of *Tussilago farfarus* in those localities of Central and South Europe where there is a possibility of the plant occurring in a less protected life-form than that in which it occurs with us. Perhaps Irmisch's account of the behaviour of the purely vegetative rosettes is due to a mistake; but the matter certainly needs further investigation.

IV

THE STATISTICS OF LIFE-FORMS AS A BASIS FOR BIOLOGICAL PLANT GEOGRAPHY

INTRODUCTION

In my investigations of Biological Plant Geography I have attempted to find means of determining the characters of vegetation, and of measuring as accurately as possible the value of the plant-climates of different regions.

By means of physical instruments, thermometers, rain gauges, &c., we are able to measure physically the factors that determine the vegetation of a region; but the numbers obtained tell us nothing about the biological value which results from a harmonious co-operation of these factors: thus an identical value of an individual physical factor may in different combinations of factors have an entirely different biological value. There is therefore nothing to prevent different combinations of factors having essentially the same biological value. The only measure of the biological value of the factors is given by the plants themselves, that is to say their life-form, which is the sum of the organic structure by which plants are related to their environment. What we have to do is to learn to make use of this measure.

The climate determines the vegetation, so that vegetation is in a sense an expression of climate. It is well known which climatic factors are most important in determining vegetation. They are precipitation and temperature. But how the plant world is an expression of climate is much less evident, and this problem can be approached from many sides. We may investigate species, genera, families, &c. which are found in a definite climate, and which are never, or but rarely, found outside that climate, and we may thus look upon those plants as an expression of the climate in question. We may for example regard Palms and some other families as an expression of the tropical climate. This study of floristic plant geography does not, however, enable us to understand the relationship between the vegetation and the climate, because the species composition may vary greatly in its origin even though the climate be essentially uniform. The object, however, of biological plant geography is to investigate this relationship between vegetation and climate in order to obtain a phyto-biological expression for the individual climates; and since the vegetation determines all other life, this expression becomes a measure for the complete biological value of the climate.

The units of floristic plant geography are the same as those of systematic botany, viz. species, genera, families, &c., but this is not so in biological plant geography. In considering the relationship to environment, or to the climate, we do not ask to what genera or species a plant belongs, but we want to know its life-form. The life-forms are the units which here correspond to the higher units, genera, families, &c., of floristic plant geography. But what is a life-form? Purely theoretically we can say that the life-form embraces all the adaptations of the plant to the climate, climate being taken in the widest possible sense. But these adaptations are manifold; they are not merely morphological and anatomical, but also intracellular. We know many of them very well; with others we are only slightly acquainted, and there are certainly adaptations of which we are entirely ignorant. Every plant has a long series of adaptations with which we have no means of dealing as a single quantity capable of expressing the grade of adaptation, i.e. the life-form. In determining the life-form we must then content ourselves with one single essential and obvious side of the adaptations, which may be treated and presented as a unit.

I consider it necessary to make the three following demands as a basis on which to construct the life-forms:

1. The character must in the first place be essential; it must represent something fundamental in the plant's relationship to climate.
2. It must be fairly easy to use, so that we may easily see in nature to which life-form a plant belongs.
3. It must represent a single aspect of the plant, thus enabling a comparative statistical treatment of the vegetation of different regions.

It can be of no use to base one life-form on intracellular adaptations, a second on the presence of bud-scales, and a third on xerophilous leaf structure, &c. Such a foundation may, indeed, be essential, but it is heterogeneous, and renders comparative statistical treatment impossible. The necessity of statistical treatment makes it impossible for life-forms to be ideal, because they are founded upon only one side, albeit an essential side, of the adaptations. Among species in the same climate representatives therefore occur of several different life-forms; but because the statistical expression of the distribution of species among the various life-forms shows itself the same in a different region of the earth with an entirely different species composition but with essentially the same climate, we have proof that in our determination of life-forms we are treading the right path. Our starting-point must represent a single aspect so that we can obtain a coherent series of life-forms, of which each successive one is adapted to a successively more unfavourable climate, unfavourable, that is, in respect to the basis chosen.

I have therefore as a basis for my series of life-forms chosen the adaptation of the plant to survive the unfavourable season, having special regard to the protection of the surviving buds or shoot-apices.

The protection of the buds or growing-points against the effects of the unfavourable season is without any doubt a very important part of a plant's adaptation. The continued existence of the individual depends upon the preservation of the buds: plants can survive the unfavourable season by means of buds alone; and if the plant survives the unfavourable season it will manage without difficulty to get through the favourable season.

The adaptations to survive the unfavourable season make the deepest impressions upon plants, and determine to a great extent the metamorphosis of the vegetative shoot. The point of view chosen may therefore serve also as a basis for showing the metamorphosis of the vegetative shoot through the ages as an adaptation to the various climates which have prevailed during the course of the evolution of the world.

I shall here go no further into details of what I have done, nor shall I give a more exact description of the 30 life-forms I have made. This will be found in my two works *Types biologiques* and *Planterigets Livsformer*,* the latter translated in the present volume (Chapter II). In *Types biologiques* I have merely discussed in a general way how Plant Geography can be treated on a basis of the statistics of life-forms; but in *Planterigets Livsformer* I have added examples of the method of treatment, viz. the flora of Denmark and that of the Isles of St. Thomas and St. Jan in the Danish West Indies. In the latter work I have also pointed out that, at any rate in a preliminary treatment of Plant Geography on the basis of the statistics of life-forms, we must, for theoretical and practical reasons work with fewer life-forms than the 30 I have described. In *Planterigets Livsformer*, p. 129,† I have contracted these 30 into 10 life-forms or groups of life-forms, and these 10 I shall use in the following account. In only too many little investigated regions of the world it will be a long time before we can procure much of the information necessary to characterize some of the life-forms for comparative purposes. This is unfortunately true of information about bud-scales and leaf-fall in the Phanerophytes of tropical and subtropical regions. It is thus chiefly practical considerations which have determined my method of reducing the 30 life-forms to 10. In finer work and in more thorough treatment of biological plant geography it will often be necessary again to break up one or more of these 10 groups. We must, then, always strive to decide to which of the 30 life-forms each species belongs: the number can always be reduced. The facts on which the 30 life-forms are constructed are of such a kind to make it reasonable to suppose that every scientifically trained botanist or geographer will be able to determine in the field the life-forms of the species he meets, if indeed the plants are in a state to enable their life-form to be determined at all.

In the tables of the distribution of the species in life-forms, in this

* See List of Literature (p. 146), Nos. 62 and 63.   † p. 102 of the present volume.

work and in various floras, I have used abbreviations for each of the 10 groups of life-forms. These abbreviations are as follows:

S = Stem Succulents
E = Epiphytes
MM = Mega-+Meso-phanerophytes
M = Micro-phanerophytes
N = Nano-phanerophytes
Ch = Chamaephytes
H = Hemicryptophytes
G = Geophytes
HH = Helo-+Hydrophytes
Th = Therophytes

The last 8 life-forms in the above list make a continuous series, of which each successive member is on the whole better adapted to survive unfavourable seasons. The two first life-forms, however, Stem Succulents and Epiphytes, cannot be given places of this kind in the series. I have nevertheless refrained from merging them into the 3 other groups of phanerophytes especially because, apart from certain difficulties, these two life-forms are highly characteristic of certain floral regions. It is, moreover, easy to decide which species belong to these life-forms, and this information is generally treated sufficiently in the literature. Since, however, Stem Succulents and Epiphytes can find no place among the 8 other life-forms, I have put them together at the beginning of the series.

On p. 102 of the present volume there are given examples of the use of the 10 life-forms in a statistical investigation of the floras of different regions, viz. the flora of Denmark and that of the Danish West Indian islands of St. Thomas and St. Jan. In this Chapter I shall begin with another example, viz. the flora of the Seychelles, tropical islands in the Indian Ocean with a considerable rainfall.

As in that case I here deal only with flowering plants. All plants could doubtless be included in my system of life-forms; but our systematic knowledge of the lower plants is so imperfect and so motley that they have to be excluded in comparative statistical analyses. On p. 99 I have explained why I have not included even the vascular cryptogams, but have confined myself to flowering plants.

By the study of literature and herbarium material I have tried to determine the life-form of the species of flowering plants of the Seychelles with the following result:

TABLE I

	No. of species.	The percentage distribution of the species among the life-forms.									
		S	E	MM	M	N	Ch	H	G	HH	Th
Seychelles[4]*	258	1	3	10	23	24	6	12	3	2	16

* This number and the corresponding numbers in the following tables refer to the list of literature, which includes the works that have formed the floristic basis of the various spectra.

These numbers express after the manner of a spectrum the plant climate of the Seychelles, as far as it can be expressed by those adaptations to the climate which characterize the life-forms I have made. I shall, therefore, for the sake of brevity, call a statistical conspectus of the the life-forms of a flora a biological spectrum or phyto-climatic spectrum.

But what do these numbers mean? Are we to conclude from the fact that the Nanophanerophytes are the best represented life-form that it is the Nanophanerophytes that are particularly characteristic of the humid and hot tropical regions? By no means! The large number of Nano-phanerophytes might perhaps mean that this life-form is very common in the world taken as a whole. We may expect to get enlightenment by comparison. Let us for example compare the flora of the Seychelles with the flora of another tropical region, whose climate differs from that of the Seychelles. In the much drier climate of the islands of St. Thomas and St. Jan (Table 2) we see that with the increased dryness of the climate a displacement from left to right has taken place in the spectrum: the percentage of large Phanerophytes has decreased while the Nanophanerophytes and Chamaephytes have increased.

TABLE 2

	No. of species.	The percentage distribution of the species among the life-forms.									
		S	E	MM	M	N	Ch	H	G	HH	Th
Seychelles[4] . . .	258	1	3	10	23	24	6	12	3	2	16
St. Thomas and St. Jan[21]	904	2	1	5	23	30	12	9	3	1	14

By this means the various floral regions may be compared and arranged in groups according to their mutual biological affinity. We can draw, too, by this means, bio-geographical boundaries. But what we lack is a standard, a 'normal spectrum' with which to compare the spectra of the various regions, and by means of which the value of the individual numbers can be determined. It is most reasonable to suppose that a normal spectrum of this kind might be found in the spectrum of the whole world, that is to say the percentage of each life-form in the flowering plants of the world. On p. 103 I have given my reasons for supposing it possible to secure such a spectrum, and since the necessity of such a normal spectrum was first realized I have done a good deal of work on the subject, without, however, succeeding in determining the life-forms of more than four hundred out of the thousand species which my preliminary calculations lead me to think necessary for an average sample. It is true that the samples investigated are sparse, and I do not consider my result to be final; yet it may turn out not far from the truth. There are various means of controlling the relationship between the samples investigated and the accurate expression of the entire spectrum. Firstly there is the amount of change undergone in the spectrum of the first 100 species when each new 100 are investigated. Moreover, we can test the result by means of certain numbers which can be ascertained in some other way, for example the number of Stem Succulents and the number of species in the largest families. In the preliminary normal spectrum there are 1 per cent. of Stem Succulents. Taking the number of known flowering plants as 130,000, 1 per cent. corresponds to 1,300 species, and this number is not far from the number of known Stem Succulents. As an example of a large family I have investigated the *Compositae*, which is supposed to contain about $\frac{1}{10}$th of all the flowering plants, i.e. 13,000. Of the 400 species on which the preliminary normal spectrum was based 45 belonged to the *Compositae*. Supposing there to be 130,000 flowering plants this makes the number of *Compositae* 14,625 instead of 13,000, thus about $\frac{1}{8}$th instead of $\frac{1}{10}$th, a difference that cannot be considered great.

But even though 400 be too small a number in the following account, I shall use a spectrum founded on this number as a preliminary normal spectrum. I take this course because I am here dealing principally with demonstrating the principles of my technique.

TABLE 3

	No. of species.	The percentage distribution of the species among the life-forms.									
		S	E	MM	M	N	Ch	H	G	HH	Th
Seychelles[4]	258	1	3	**10**	**23**	**24**	6	12	3	**2**	**16**
St. Thomas and St. Jan[21]	904	**2**	1	5	**23**	**30**	**12**	9	3	1	**14**
Aden[46]	176	1	..	..	7	**26**	**27**	19	3	..	**17**
Normal spectrum	400	1	3	6	17	20	9	27	3	1	13

As an example of the use of the normal spectrum I have in Table 3 placed it with three tropical spectra of regions whose hydrotherm figures differ widely, especially as regards precipitation. The regions are the Seychelles, St. Thomas and St. Jan, and Aden. In this table the characteristic numbers, i.e. those exceeding the corresponding numbers in the normal spectrum, and which therefore especially characterize the plant-climate, are in heavy type. These numbers show that gradually as the climate with its decreasing precipitation becomes less friendly to plant life, the centre of gravity is moved from left to right, from the less to the more protected life-forms. But the centre of gravity is not moved out of the Phanerophytes and Chamaephytes, which are characteristic of the tropics as a whole. In the two first spectra in Table 3, those of the Seychelles and of St. Thomas and St. Jan, the centre of gravity lies in the Phanerophyte portion, in both these spectra Phanerophytes being present in a greater number than in the normal spectrum. Both spectra have 61 per cent. of Phanerophytes, while the normal spectrum shows only 47 per cent. It is characteristic of all tropical lands in which the precipitation is not too small that the centre of gravity in the biological spectrum is amongst the Phanerophytes. This is not so in any other climate. The tropical climate which is not too dry is, expressed as a plant-climate, a Phanerophyte climate.

Within the regions of Phanerophytes we can, however, draw further phyto-climatic boundaries. From Table 3 we see within the Phanerophyte region that the Seychelles and St. Thomas and St. Jan belong to different plant-climates, each characterized by their percentage of Mega- and Meso-phanerophytes. In the Seychelles this number is larger, in St. Thomas and St. Jan it is smaller than in the normal spectrum. I shall not discuss at greater length the other phyto-climatic boundary lines we can draw within the tropical regions, but shall only remark that, as far as lists of species and determinations of life-forms are available, we can draw just as many boundary lines as we have immediate occasion to use, just as we can make as many isotherms or isohyetic lines as we have occasion to use, if the meteorological data are available.

In order to draw the climatic boundaries, *Biochores*, as Köppen calls them, we must first know the biological spectra of the different regions of the earth. We should first have investigated local floras at various points scattered over the earth's surface, but principally of regions showing conspicuous differences in their hydrotherm figures, i.e. in the course of their temperature and precipitation curves. Only with these data can we reach the goal towards which I am striving, my aim being a biological plant geography based upon the statistics of life-forms.

It is evident that this preparatory work is so vast that it is impossible for any one man to carry it out. It is exceedingly difficult to determine from literature and herbarium material alone the life-forms of the plants of a flora one has not examined in the field. Spectra formed from herbarium material are likely to be too defective for use. If then in the following account I attempt, using my own investigations as a basis, to draw Biochores within the nearest floral region, i.e. the northern cold temperate and arctic region, I do this in the hope that others, seeing what can be attained by this means, may feel drawn to pursue the work farther.

THE PRINCIPAL PLANT-CLIMATES

I shall first attempt to show the position occupied by the northern plant-climate in relationship to the other plant-climates. In order to obtain a conspectus of the various plant-climates characterized statistically, which of course can still only be done in a general way, I shall begin with the climate most favourable to plant life, the humid and hot tropical climate, which amongst existing climates may be looked upon as most primitive. Since I am here only concerned with the Northern Hemisphere, I shall then proceed towards the north, ending with the arctic climate. But, as is well known, the hydrotherm figure, which is of especial importance for plant life, differs widely amongst places on the same degree of latitude. We must therefore draw several lines from the Equator to the Pole, observing what happens along more than one meridian.

Apart from the regional series of climates, whose corresponding biological spectra I shall not here discuss, we find, starting from the humid, hot tropical climate, three main series of climates:

A. A purely tropical series, in which the temperature is everywhere high but with decreasing humidity. As I have already given examples of the biological spectra of this series (Table 3) it is omitted in what follows.

B. Equator–Pole: Falling temperature with an increasing difference between summer and winter; but the precipitation in general favourable to plant life.

C. Equator–Pole: The temperature falls essentially as in B, but there is a corresponding decrease in the precipitation, at any rate during the summer. Farther towards the north the precipitation behaves as in B.

As is well known series B occurs especially in the eastern portions of the great continents, and series C in their western portions. We must first draw trial lines parallel to meridians of longitude through these series of climates, and we must base these lines on the biological spectra of a series of floras. I shall now give the chief features of three such lines, one through the western, another through the eastern part of North America, and a third through the western part of the old world.

Table 4 shows the chief points in a line drawn through the eastern part of North America, thus passing through the climatic series B. In this and the following tables heavy type is used for the numbers especially characteristic of the biological spectrum of each local flora. For ready comparison the normal spectrum is added at the bottom of each table. A glance at the table shows us clearly the relationship of the vegetation to climate expressed by means of life-forms. The gradual decrease of favourable factors towards the north affects the number of species and the species composition, as far as life-form is concerned, like a corresponding series of sieves of increasing fineness. Towards the south, where the factors are most favourable, the number of species is greatest, and a comparatively large number of them belong to the least protected life-forms. In the extreme north, where the factors are most unfavourable, the holes of the sieve are so fine that only a few specially adapted species can pass through. First of all the Phanerophytes disappear, the large ones first, then by degrees the smaller ones, until finally none remain (Ellesmereland). Even the Therophytes decrease and eventually disappear. Though it is true that the Therophyte is the best adapted life-form where decreasing humidity is the unfavourable factor while the temperature is still sufficiently high; where, on the other hand, the decreasing temperature is the limiting factor, the Therophytes are not in their element, as is seen by their

TABLE 4

	No. of species.	The percentage distribution of the species among the life-forms.									
		S	E	MM	M	N	Ch	H	G	HH	Th
Ellesmereland[72] . .	107	..	..	..	..	..	23·5	65·5	8	3	..
Baffin's Land[12,34,65,77,91].	129	..	..	..	..	1	30	51	13	3	2
Chidley Peninsula on North Coast of Labrador[92] . .	64	..	..	..	..	3	27	61	6	..	3
Coastal region of Labrador[53] . .	246	..	..	2	1	8	17	52	9	5	6
South Labrador[53]. .	334	..	..	3	3	8	9	48	12	11	6
Altamaha, Georgia[31],* .	717	(0·1)	(0·4)	5	7	11	4	55	4	6	8
St. Thomas and St. Jan[21]	904	2	1	5	23	30	12	9	3	1	14
Normal spectrum .	400	1	3	6	17	20	9	27	3	1	13

disappearance from the far north and from the snowy regions of mountains. Helophytes and Hydrophytes, which are well represented in the cold temperate regions, decrease farther towards the north. Geophytes, especially Rhizome Geophytes, are likewise very well represented in the cold temperate regions, and often also in the far north. They too must be used in characterizing certain floral regions. If, however, I do not make use of them, although they often exceed considerably in their percentage the Geophytes of the normal spectrum, this is because the Geophyte percentage given in the tables is often not quite accurate, since it is not always easy to decide from literature and herbarium material whether a species is a Geophyte or a Hemicryptophyte. Perhaps, therefore, some of the species that I have given as Geophytes are really Hemicryptophytes; but this does not affect the percentage of Hemicryptophytes, which, in the region under discussion, are so numerous that 1 or 2 per cent. more or less is of no importance. For the percentage of Geophytes, however, one or two more or less is of great importance,

* The numbers in this spectrum are less certain than in the others. But the main feature, the low Phanerophyte percentage and the high Hemicryptophyte percentage, is beyond question.

because the Geophyte percentage in the normal spectrum is small. The value of the number by which a life-form in a given flora exceeds that in the normal spectrum does not consist in the absolute number itself, but in its size in relationship to the number in the normal spectrum.

Table 4 shows two main types of climate: the tropical Phanerophyte climate, and north of it the Hemicryptophyte climate. In the southern part of the Hemicryptophyte climate there still remain a considerable minority of Phanerophytes; but gradually as these disappear towards the north the Hemicryptophyte percentage does not rise in a corresponding degree. It is, on the other hand, the Chamaephyte percentage that rises, ultimately rising enormously. It first becomes twice as high as in the normal spectrum, then three times as high or even higher. In some parts of the inhospitable north the Chamaephyte percentage rises so high that even the Hemicryptophyte percentage is lowered considerably. There are then really three chief climates: the Phanerophyte climate, the Hemicryptophyte climate, and the (Hemicryptophyte and) Chamaephyte climate. In the last climate, in correspondence with the rising Chamaephyte percentage, we can distinguish between northern, and arctic, and, if you will, nival zone.

It must here suffice to indicate the boundary lines of the most important plant-climates as shown by an experimental line drawn through eastern North America. Unfortunately my material for the corresponding line along the east coast of Asia is still insufficient; the material at hand, however, gives no reason to doubt that this line will show essential correspondence with the other.

Let us now see the result given by the experimental line drawn through the series of climates C. I shall begin with a line through western North America. Compared with the first line we have here the Californian-Mexican Desert in the place of the luxuriant vegetation of Georgia and Florida. Compared with series B we have in the sub-tropical zone a precipitation curve both low and unfavourable to plant life; the summer is a dry period making a climate unfavourable to plants, so that we expect to find the best protected life-forms well represented. Table 5 shows that the characteristic life-form here is the Therophyte; in the Death Valley there are 42 per cent. of Therophytes, over three times as many as in the normal spectrum. Beside this it should be observed that there is a high percentage of Stem Succulents and a comparatively high percentage of Nanophanerophytes, both of which life-forms are the best protected amongst Phanerophytes.

A glance at the table shows that farther towards the north this trial line corresponds with the one before, the difference merely being that in series C a Therophyte climate is interpolated between the Phanerophyte climate to the south and the Hemicryptophyte climate to the north.

TABLE 5

	No. of species.	The percentage distribution of the species among the life-forms.									
		S	E	MM	M	N	Ch	H	G	HH	Th
St. Lawrence[43,64]	126	..	..	..	..	..	23	61	11	4	1
Country of the West Esquimos[44,69]	291	..	..	(0.3)	(0.3)	5	18	61	12	2	2
Chilkat, Alaska[49]	425	..	..	3	2	6	11	57	11	4	6
Sitka[64]	222	..	..	3	3	5	7	60	10	7	5
Death Valley, California[14]	294	3	..	..	2	21	7	18	2	5	42
St. Thomas and St. Jan[21]	904	2	1	5	23	30	12	9	3	1	14
Normal spectrum	400	1	3	6	17	20	9	27	3	1	13

We shall now see that the state of affairs is exactly the same in series C of the old world. In Table 6 I have given a series of biological spectra illustrating this. I am here able to give several spectra from the sub-tropical zone. The remarkable preponderance of Therophytes characterizes all these spectra as belonging to the Therophyte climate. Chamaephytes and Geophytes also occur, usually surpassing the numbers in the normal spectrum. The Chamaephyte and Phanerophyte numbers in Table 6 indicate some of the boundary lines which can be drawn within a Therophyte climate, but I shall say no more about this at present.

The boundary between the Therophyte climate and the Hemicryptophyte climate must be drawn first of all by the percentage of Hemicryptophytes; but I have as yet no means of deciding the number at which this boundary is most naturally placed. The Therophyte percentage must of course be considered; this figure should be used with great caution. In dealing with this life-form we must use a measure with larger intervals than we used for the other life-forms. Therophytes are distributed much more easily by cultivation than the other life-forms, and even if they subsequently disappear they may be reintroduced very easily; so that if the introduced species are included the floras of cultivated land are comparatively rich in Therophytes. Although, as explained on pp. 99–100 of the present volume, I have tried to overcome this difficulty by excluding from floras species occurring only on cultivated soil, yet in all probability I have included many Therophytes that do not belong to the original flora, many that would disappear if the land were left uncultivated for a sufficiently long period. We may suppose that the boundaries between the Therophyte and Hemicryptophyte climate occurs where there are about 30 per cent. of Therophytes or somewhere near 35–40 per cent. of Hemicryptophytes. Poschiavo, for example, has only 21 per cent. of Therophytes but 55 per cent. of Hemicryptophytes. The Tuscan Islands have a low Hemicryptophyte percentage, which has not been accurately determined, but 42 per cent. of Therophytes.

TABLE 6

	No. of species.	The percentage distribution of the species among the life-forms.									
		S	E	MM	M	N	Ch	H	G	HH	Th
Hope Island[84]	7	..	..	..	..	..	43	57			
Franz Josef Land*	25	..	..	..	..	..	32	60	8		
King Karl's Land[1]	25	..	..	..	..	..	28	60	8	4	
Spitsbergen[22,54]	110	..	..	..	..	1	22	60	13	2	2
Vardø[70]	134	..	..	..	..	2	15	61	8	5	9
Iceland[75]	329	..	..	..	..	2	13	54	10	10	11
Clova[90], Scotland, under 300 metres	304	..	..	3	2	4	7	59	7	5	13
Denmark[63]	1,084	..	(0·1)	1	3	3	3	50	11	11	18
District of Stuttgart[41]	862	..	..	3	3	3	3	54	10	7	17
Poschiavo[10] under 850 metres	447	..	(0·2)	3	4	3	5	55	8	1	21
Lowland of Madeira[86]	213	..	..	..	1	14	7	24	..	3	51
Ghardaia[13]	300	(0·3)	..	..	..	3	16	20	3	..	58
El Goléa[13]	169	..	..	..	..	9	13	15	5	2	56
Tripoli[16,17,51]	369	..	..	(0·3)	..	6	13	19	9	2	51
Cyrenaica[16,17,18]	375	..	..	1	1	7	14	19	8	..	50
Samos[74]	400	..	..	1	4	4	13	32	11	2	33
Libyan Desert[3]	194	..	..	..	3	9	21	20	4	1	42
Aden[46]	176	1	..	..	7	26	27	19	3	..	17
Seychelles[4]	258	1	3	10	23	24	6	12	3	2	16
Normal spectrum	400	1	3	6	17	20	9	27	3	1	13

In Italy the boundary will thus lie between these two points, thus passing through North Italy.

In the cold temperate and cold zone the numbers in Table 6 correspond exactly with the behaviour of the experimental lines both in the eastern and western part of North America. The sequence of the characteristic numbers and the Chamaephyte percentage steadily increasing towards the north are the same. But since the climate, especially the summer temperature, is not the same at the various places on the same latitude, we cannot expect to find the same biological spectrum everywhere on the same latitude. At one place, for example, we have 20 per cent. of Chamaephytes at 65–6° north (e.g. East Greenland). In other places this percentage is not reached until 70–1° north (e.g. Novaya Zemlya). Our object then is to draw lines, by means of the biological spectra of sufficient numbers of local floras, passing through those regions whose

---

* I am not acquainted with any lists of plants of Franz Josef Land; but C. H. Ostenfeld has pointed out to me that there is in the Herbarium of the Copenhagen Botanical Museum a set of the plants collected there by H. Fischer when on the Jackson-Harmsworth Polar Expedition. The biological spectrum I have given is based on this collection, including in all twenty-five species. The species of Monocotyledons will be found in Ostenfeld's *Flora Arctica*.

spectrum shows the same characteristic numbers, in this example the same Chamaephyte percentage. These biological boundary lines or **Biochores** are analogous with the isotherms and isohyets of climatology.

The following is an attempt to draw a few such Biochores in the cold region of the Northern Hemisphere.

THE ARCTIC CHAMAEPHYTE CLIMATE

It is clear from Tables 4–6 that a very high percentage of Hemicryptophytes is characteristic of the whole of the northern temperate and cold zone, and that the cold zone is distinguished from the cold temperate zone by a high percentage of Chamaephytes; this percentage rises the farther we penetrate north, as the environment becomes progressively unfavourable. It is, then, the percentage of Chamaephytes that must here be used in demarcating the different phyto-climatic regions by means of Biochores. The number at which the boundary must be drawn depends upon the numbers in the normal spectrum and upon practical considerations. The boundary can be drawn, just like an isotherm, at any number of Chamaephytes we choose. Fractions even may be used, but for the sake of convenience we are never likely to employ them. The boundary between the Hemicryptophyte climate of the cold temperate zone and that of the Hemicryptophyte+Chamaephyte climate of the cold zone naturally occurs where the Chamaephyte percentage of the local flora exceeds that of the normal spectrum. I have chosen 10 per cent. of Chamaephytes as the boundary. The next Biochore, the boundary between the northern and the arctic zone, I have drawn at 20 per cent. of Chamaephytes, about double the number in the normal spectrum. The third Biochore, separating the arctic zone from the arctic-nival region, I draw at 30 per cent. of Chamaephytes, about treble the number in the normal spectrum.

TABLE 7

	No. of species.	The percentage distribution of the species among the life-forms.								
		S	E	MM	M	N	Ch	H	G	HH	Th
Novaya Zemlya	192	..	..	..	..	2	19	62	11	4	2

I shall begin with a 20 per cent. Chamaephyte Biochore. Table 7 shows the biological spectrum of Novaya Zemlya (with Vaygach). This shows 19 per cent. of Chamaephytes; but since Novaya Zemlya and Vaygach extend over about 8° of latitude it is reasonable to suppose that if the whole region has 19 per cent. of Chamaephytes, the 20 per cent. Chamaephytes line does not lie to the north of these islands, but must cut them at one place or another. Using Feilden's[25] list of plants as a floristic basis I have determined the biological spectrum for each degree of latitude separately, with the result shown in Table 8. Even if we

TABLE 8

Novaya Zemlya[25]	No. of species.	The percentage distribution of the species among the life-forms.									
		S	E	MM	M	N	Ch	H	G	HH	Th
76°–77° N. L.	4	..	..	..	..	..	50	50	..	..	..
75°–76° ,,	16	..	..	..	..	..	31	69	..	..	..
74°–75° ,,	49	..	..	..	..	2	25	63	10	..	..
73°–74° ,,	128	..	..	..	..	2	22	62	11	2	1
72°–73° ,,	130	..	..	..	..	2	20	60	12	3	3
71°–72° ,,	130	..	..	..	..	3	20	64	11	2	..
70°–71° ,,	141	..	..	..	..	3	20	62	12	3	..
Vaygach, 69°–70° N.	154	..	..	..	..	3	18	63	12	4	..
North coast of West Siberia (near Yugor)[70]	119	..	..	..	..	3	17	63	11	4	2

cannot emphasize the importance of the biological spectrum of the north portion of Novaya Zemlya, because its floristic investigation is very imperfect, yet the table as a whole shows in a striking manner the before-mentioned gradual increase in the Chamaephyte percentage as the environment becomes more unfavourable farther and farther towards the north. The table also shows that the 20 per cent. Chamaephyte-Biochore passes through the Straits of Vaygach between Vaygach, which has 18 per cent. of Chamaephytes, and Novaya Zemlya whose southern portion has 20 per cent.

TABLE 9

	No. of species.	The percentage distribution of the species among the life-forms.									
		S	E	MM	M	N	Ch	H	G	HH	Th
Franz Josef Land	25	..	..	..	..	..	32	60	8	..	..
King Karl's Land[1]	25	..	..	..	..	..	28	60	8	4	..
Spitsbergen[2,22,54]	110	..	..	..	..	1	22	60	13	2	2
Hope Island[84]	7	..	..	..	..	..	43	57	..	..	..
Bear Island[2,29]	38	..	..	..	..	..	32	60	5	3	..
Jan Mayen[47]	137	..	..	..	..	..	32	57	8	..	3
Kolguev[26]	137	..	..	..	..	4	14	66	9	3	4
Vardø[70]	134	..	..	..	..	2	15	61	8	5	9
Iceland[75]	329	..	..	..	..	2	13	54	10	10	11

Table 9 shows the course of the 20 per cent. Chamaephyte line towards the west to Greenland. It passes north of Kolguev (14 per cent.) and Scandinavia (Vardø 15 per cent.), but to the south of Spitsbergen (22 per cent.), Bear Island (32 per cent.), &c. Farther towards the west

STATISTICS OF LIFE-FORMS

it passes between Iceland (13 per cent.) and Jan Mayen (32 per cent.). I shall show later how the line runs in Greenland, but first I will follow its course from Novaya Zemlya eastwards.

Kjellman[42] gives a series of lists of plants from the arctic coast of Asia. I have condensed these lists to five, each representing a portion of the north coast of Asia from Yalmal to Pitlekaj, viz. (1) 68–70° 40′ E.L.: Lütkes Island; west coast of Yalmal, north coast of Yalmal and White Island; (2) 80° 58′–85° 8′: Port Dickson, Minin Island; (3) 95–103° 25′: Aktinie Harbour, Mouth of the Taimyr River, Cape Chelyuskin; (4) 113–161°: Preobrazhen Island, Mouth of the River Olenek, Mouths of the Lena and the Kolyma; (5) 177° 38′ E.–173° 24′ W.: Cape Yakan and Pitlekaj. Table 10 shows the biological spectrum for each of these five

TABLE 10

Arctic coast of Asia.	No. of species.	The percentage distribution of the species among the life-forms.									
		S	E	MM	M	N	Ch	H	G	HH	Th
68°–70° 40′ E. Long.	59	..	..	..	..	2	20	58	15	5	..
80° 58′–85° 8′.	76	..	..	..	..	..	20	66	12	2	..
95°–103° 25′	56	..	..	..	..	..	27	54	14	3	2
113°–161°	121	..	..	..	..	1	23	64	9	2	1
177° 38′ E. to 173° 24′ W.	114	..	..	..	..	2	23	61	10	3	1
Whole coast from 68° E. to 173° 24′ W.	177	..	..	..	..	2	22	62	10	3	1
Islands of New Siberia[82]	57	..	..	..	..	1·5	23	65	9	1·5	..
Chukotski[50]	272	..	..	..	..	4	21	62	9	2	2

areas and also for the whole coast taken as one area. The spectra of the Islands of New Siberia and of the Chukoten Peninsula and the northeast portion of Asia are also shown. These spectra make it clear that the arctic coast of Asia lies to the north of the 20 per cent. Chamaephyte Biochore; but the numbers indicate that this Biochore is not far from the coast. Only on the Taimyr Peninsula, which projects farthest towards the north, does the Chamaephyte percentage attain a high number (27 per cent.), while the rest of the coast has only 20–3 per cent. of Chamaephytes, and the coast taken as a whole has 22 per cent. As we shall see later the 10 per cent. Chamaephyte Biochore goes a long way north in North Asia, reaching about 68° north at the Yenisei. The 20 per cent. Chamaephyte line may therefore be drawn so that it cuts off only a comparatively narrow strip of the arctic coast of Asia. To the East towards Bering Strait the line bends south-east cutting off the Chukotski Peninsula. Between Asia and America it runs, as Table 11 shows, south of the St. Lawrence Island (23 per cent.), but north of the Commander Islands (15 per cent.), Aleutian Islands (Unalaska 15 per cent.), and the Pribilov Islands (12 per cent.).

STATISTICS OF LIFE-FORMS

TABLE 11

	No. of species.	The percentage distribution of the species among the life-forms.									
		S	E	MM	M	N	Ch	H	G	HH	Th
St. Lawrence[43,64]	126	..	..	..	..	..	23	61	11	4	1
Pribilov Islands[52]	161	..	..	..	..	0·5	12	66	13·5	4	4
Unalaska[64]	188	..	..	0·5	0·5	2	15	60·5	11·5	4	6
Commander Islands[24]	236	..	..	(0·2)	2	2	15	63	11	3	4

As far as conclusions may be drawn from the poor floristic material available for Arctic America, the 20 per cent. Chamaephyte Biochore here, as in Asia, runs in such a way that besides the Arctic Islands only a comparatively narrow strip of the Arctic coast of the mainland is to the north of this line. Among the spectra in Table 12, which are for

TABLE 12

	No. of species.	The percentage distribution of the species among the life-forms.									
		S	E	MM	M	N	Ch	H	G	HH	Th
Country of the West Esquimos[44,69]	291	..	..	(0·3)	(0·3)	5	18	61	12	2	2
Chilkat, Alaska[49]	425	..	..	3	2	6	11	57	11	4	6
Yakutat Bay[15]	122	..	..	3	4	3	6	66	12	4	2
Sitka[64]	222	..	..	3	3	5	7	60	10	7	5

four areas in north-west America, only the spectrum for the country of the West Esquimos shows a Chamaephyte percentage approaching anywhere near 20; and farther towards the east in the central part of Arctic America we have, as shown in Table 13, 19 per cent. of Chamae-

TABLE 13

	No. of species.	The percentage distribution of the species among the life-forms.									
		S	E	MM	M	N	Ch	H	G	HH	Th
Melville Island[11]	66	..	..	..	..	..	24	59	12	3	2
Banks Land[35]	52	..	..	..	..	4	25	56	11	2	2
Between Great Bear Lake and the mouth of Coppermine River	78	..	..	..	..	5	19	58	10	4	4

phytes in the district between the Great Bear Lake and the mouth of the Coppermine River. We can then scarcely doubt that in the northern part of this area along the coast of the Arctic Ocean there are over

20 per cent. of Chamaephytes. The islands lying to the north of America, Banks Land and Melville Island, have 25 and 24 per cent. respectively.

The farther course of the 20 per cent. Chamaephyte Biochore towards the east is seen from Table 14 to run south of Boothia Felix (Port Kennedy

TABLE 14

	No. of species.	The percentage distribution of the species among the life-forms.									
		S	E	MM	M	N	Ch	H	G	HH	Th
Ellesmereland[72]	107	..	..	..	..	..	23.5	65.5	8	3	..
Beechey Island[35]	26	..	..	..	..	..	31	50	19	..	..
Port Kennedy (Boothia Felix)[36]	42	..	..	..	..	..	31	53	14	2	..
Baffin Land[12,34,65,77,91]	129	..	..	..	..	1	30	51	13	3	2
Chidley Peninsula (North Labrador)[92]	69	..	..	..	..	3	27	61	6	..	3
Coast of Labrador[53]	246	..	..	2	1	8	17	52	9	5	6
South Central Labrador[53]	334	..	..	3	3	8	9	48	12	11	6
James Bay[53]	268	..	..	3	3	7	7	53	10	7	10

31 per cent.) and across the northern part of Labrador, where the Chidley Peninsula of the north coast has 27 per cent. of Chamaephytes, while the coast of Labrador taken as a whole shows 17 per cent., the southern part of Central Labrador 9 per cent., and the flora around James Bay 7 per cent.

The course of the 20 per cent. Chamaephyte Biochore in Greenland has still to be traced. The floristic material for this biochore is very rich, so that the line can be based on the biological spectra of a long series of local floras. Table 15 shows the biological spectra of the east coast

TABLE 15

East Greenland[32,48]	No. of species.	The percentage distribution of the species among the life-forms.									
		S	E	MM	M	N	Ch	H	G	HH	Th
76° 39'–77° 36' N.	28	..	..	..	..	..	36	57	7	..	..
73° 30'–75°	102	..	..	..	..	..	25	61	10	2	2
70°–73° 30'	159	..	..	..	..	1	26	55	12	4	2
69° 25'–70°	96	..	..	..	..	2	32.4	55.3	6.3	2	2
66° 20'–69° 25'	76	..	..	..	..	3	32	58	5	1	1
Entire extent from 66° 20' to 75°	170	..	..	..	..	1	25	57	11	4	2
65° 30'–66° 20'	169	..	..	..	..	1	21	60	7	7	4

of Greenland between 65½° and 77¼° N. It is clear from this table that the east coast of Greenland, at any rate north of 65° 30', lies to the north of the 20 per cent. Chamaephyte Biochore. It is true that the most northerly link in the chain of biological spectra for the east coast of Greenland is still wanting; but this will soon be available, since the material for it is present among the collections brought home by the Danish Expedition that has recently returned.

Finally Table 16* gives a conspectus of what is found on the west coast of Greenland. From this table it is clear that the 20 per cent. Chamaephyte Biochore runs from Labrador through Davis Strait only going north of 71° on the west coast of Greenland, the whole west coast south of 71° having less than 20 per cent. of Chamaephytes, and therefore lying south of the 20 per cent. Chamaephyte Biochore, which from 71° towards the south cuts the west coast and southern promontory of Greenland, then bends out into the Straits of Denmark and thence runs between Iceland and Jan Mayen.

Table 16

West Greenland[32,58,87,88]	No. of species.	The percentage distribution of the species among the life-forms.									
		S	E	MM	M	N	Ch	H	G	HH	Th
Lockwood Island . .	4	..	..	..	..	..	75	25	..	..	..
North-west Greenland, i.e. North of Meville Bay .	93	..	..	..	..	1	29	57	12	1	..
72°–74° 30′ . . .	123	..	..	..	..	1·5	31	56	9	1·5	1
69°–71° . . .	214	..	..	..	..	2	19	58	12	6	3
64°–67° . . .	239	..	..	..	..	2	19	56·5	10	7·5	5
60°–62° . . .	254	..	..	..	..	2	16	56	11	8	7

This completes the ring. We have drawn our first Biochore, the 20 per cent. Chamaephyte Biochore, as well as can be done with the floristic material available. A boundary line has been drawn on a statistical basis for a characteristic Arctic plant climate, a Chamaephyte climate, whose percentage of Chamaephytes is in some of the local floras more than twice as high as in the normal spectrum.

Within this region the Chamaephyte percentage increases gradually as the environments become less and less favourable; and, as I have already mentioned, we can take the 30 per cent. Chamaephyte line as the boundary of the region most hostile to plant life. This is an Arctic nival region or several separate nival regions. I shall, however, not follow this line in detail; it is explained in the tables already given; but shall here merely mention the regions that have over 30 per cent. of Chamaephytes, viz. Jan Mayen 32, Bear Island 32, Hope Island 43, Franz Josef Land 32, Novaya Zemlya north from 75° N., Beechey Island 31, Baffin Land 30, Port Kennedy 31. To these we must add the nival

* Besides the literature referred to I have used for the southern portion of West Greenland notes which Dr. L. Kolderup Rosenvinge has been kind enough to place at my disposal.

flora of Greenland, which occupies the lowlands on the most inhospitable stretches of the coast of Greenland. The numbers in Table 17 illustrate in a striking manner the high percentage of Chamaephytes in the nival flora of Greenland.

TABLE 17

	No. of species.	The percentage distribution of the species among the life-forms.									
		S	E	MM	M	N	Ch	H	G	HH	Th
1. Jensen's Nunataks[87]	25	..	..	..	..	..	32	68	..	..	..
2. Nausausuk Nunatak[87]	27	..	..	..	..	..	41	52	7	..	..
3. Majorarisut „ [87]	25	..	..	..	..	..	40	52	8	..	..
1–3 together	51	..	..	..	..	..	35	59	6	..	..
Mt. Schurman[65]	9	..	..	..	..	..	45	33	22	..	..
Greenland's nival Flora[87]	106	..	..	..	..	3	35	54	6	1	1

It would now be of interest to examine the biological spectrum for the whole of the extreme Arctic region. But in making this spectrum we encounter some difficulties, since we lack a systematic account of the Arctic flora. Ostenfeld's *Flora Arctica* contains unfortunately as yet only the Monocotyledons. According to a preliminary computation we have to the north of the 20 per cent. Chamaephyte Biochore, as we have drawn this line, 437 species of flowering plants, and these are shown in the spectrum in Table 18. This spectrum shows a Chamaephyte per-

TABLE 18

	No. of species.	The percentage distribution of the species among the life-forms.									
		S	E	MM	M	N	Ch	H	G	HH	Th
North of the 20% Ch-Biochore	437	..	..	..	..	3·5	19	64·5	8	2	3

centage approaching as nearly as one could wish to the number of the Biochore without actually coinciding with it. One of the reasons why the Chamaephyte percentage is not higher is, I believe, that the Chamaephyte is the life-form so specially characteristic of this region and so particularly adapted to this environment, that among Chamaephytes we find, more than amongst the other life-forms, the same species extending over the entire region and therefore occurring again and again in each local flora.

I hope on another occasion to return to this subject and to some other problems which await solution, such as tracing how the various species of Chamaephytes are distributed amongst the different types of this life-form. We should also investigate the corresponding behaviour of the other life-forms, inquiring especially into the distribution of that tendency to become Chamaephytes, to creep up from the ground, that many Hemicryptophytes apparently show. Further there remains the calculation of the statistical relationship between the life-forms in different formations, work which can scarcely be done without fresh investigations in the field.

That the Chamaephyte percentage is so high and the Chamaephytes play such a very important part in the extreme north is an expression of the fact that this life-form is in particular harmony with the Arctic climate. That it is the Chamaephyte percentage and not, for example, the Geophyte percentage that rises is connected with the fact that it is here not the severity of winter but the falling summer temperature that determines the climatic boundaries. That this is so is shown from the correspondence, as I shall show later, between the 10 per cent. and the 20 per cent. Chamaephyte lines with definite June isotherms, while there is no such correspondence between the isochamaephyte lines and the isotherms for January or the other winter months. Farther and farther towards the north the plant has not only the cold of winter to fight against, but also the cold below, from the deeply frozen ground, handicapping the plant particularly because it lasts at any rate during part of the time when the temperature of the air is high enough for vital activity. The plant has to find a suitable position between these two evils, the cold from above and from below. Chamaephytes occupy this middle position, namely the surface of the earth. Chamaephytes are protected against the cold desiccating winds of winter, just as well as Hemicryptophytes, by the covering of snow, where this is found. When the spring comes, the snow melts, and the sun imparts its heat to anything lying on the ground exposed directly to its rays. The cushions and carpets formed by the aerial shoots of the Chamaephytes then act as sun traps. When the ground beneath has thawed to a certain extent and the plants have begun to vegetate, the Chamaephytes themselves prevent the soil from freezing again beneath them, or at any rate prevent it from freezing as easily as bare ground.

The 10 per cent. Chamaephyte Biochore. The tables already given impart some information for determining the course of this line. For America it must suffice to indicate Tables 4 and 5. Table 5 shows that on the west coast of America the 10 per cent. Chamaephyte line runs between Sitka (7 per cent.) and Chilkat in Alaska (11 per cent. of Chamaephytes). I have no materials for determining the course of the line through Canada. We see from Table 4 that it runs through the middle portion of Labrador, and then turns south around Newfoundland. I have but little information about Asia. It is seen from Table 11 that the 10 per cent. Chamaephyte line runs south of the Aleutian Islands and the Commander Islands; and a preliminary investigation of the flora of Sakhalin indicates that the line passes through the northern portion of this island. In Siberia the line runs south of the land about the Kolyma River (12 per cent.), passes through the country between Chatanga and the Lena (10 per cent.), cutting the Yenisei at about 67° N. Between 65° 50' and 69° 25' the land along the Yenisei has 8 per cent. of Chamaephytes, and between 69° 25' and the Arctic Ocean it has 13 per cent. of Chamaephytes.

I have more data for the course of the 10 per cent. Chamaephyte line in Europe.

TABLE 19

Faroe Islands[56,59]	10·5% Ch.
Shetlands	about 7 ,,
Orkneys[89]	,, 6 ,,
Scilly Isles	,, 3·5 ,,

Table 19 shows that the 10 per cent. Chamaephyte line passes between the Faroe Islands (10·5 per cent.) and the Shetlands (about 7 per cent.); it then bends sharply towards the north. Table 20 gives some numbers

TABLE 20

	No. of species.	Ch %
North Norway as a whole[55]	about 622	about 8·5
Vardø[70]	,, 134	,, 15
Maasø and Magerø[28]	,, 242	,, 11·5
Nord-Reisen[27,40]	,, 323	,, 13
Tromsø[56]	,, 312	,, 10
Junkersdalen[19]	,, 345	,, 12

elucidating the course of the line in northern Norway. As a whole northern Norway (as demarcated by Norman) has about 8·5 per cent. of Chamaephytes; but the high land has over 10 per cent. (Junkersdalen and Nord-Reisen have 12 per cent. and 13 per cent. respectively) and this is true also of the most northerly portions of the lowlands. Farther towards the south the lower regions have less than 10 per cent. of Chamaephytes. The fact is that in northern Norway the 10–20 per cent. Chamaephyte zone merges into a corresponding region of the mountains of Scandinavia.

We have not, however, sufficient material to enable us to draw the boundaries of these regions. I shall now, therefore, show how the 10 per cent. Chamaephyte line again appears in the lower land to the east of the mountains. First we meet it again in northern Sweden between the region of Pajala, which has 5·5 per cent. of Chamaephytes, and *Lapponia enontekiensis* with 12 per cent. The further course of the line through northern Finland can be made out from Tables 21–2. Although Finland, as a whole, has only 5·5 per cent. of Chamaephytes, it is reasonable to suppose that, at any rate its northern part, must lie to the north of the 10 per cent. Chamaephyte line, and Tables 21–2 show that this is so. Table 21 shows the percentage of Chamaephytes in a series of local floras extending from Abo Skerries to Vardø, and Table 22 shows a corresponding series extending from the Gulf of Finland to and along the Arctic Ocean. It is clear from these tables that the 10 per cent. Chamaephyte line from its starting-point between Pajala and *Lapponia enontekiensis* first bends towards the north through *Lapponia inarensis*, which, taken as a whole, has only 9 per cent. of Chamaephytes; it then turns towards the south-east running south of the Peninsula of Kola to the White Sea.

TABLE 21

Vardø[70]	about 15 % Ch.
Lapponia inarensis[66]	„ 9 „
Ostrobottnia borealis[66]	„ 4 „
Åbo Skerries[7]	„ 3·5 „

TABLE 22

Lapponia ponojensis[66]	about 12 % Ch.
„ murmanica[66]	„ 10 „
„ tulomensis[66]	„ 10 „
„ imatrensis[66]	„ 10·5 „
Karelia keretina[66]	„ 5·3 „
The Nurmijärve-district	„ 4·5 „

The Correspondence between the 10 per cent. and 20 per cent. Chamaephyte Biochores and certain Climatic Lines. Now that we have drawn, or at any rate indicated by series of points, the 10 per cent. and 20 per cent. Chamaephyte Biochores, it will be interesting to see whether these lines correspond with certain climatic lines, especially isotherms. We are more interested in the summer isotherms, e.g. those of June or July than in the isotherms for the whole year or the winter isotherms, which have no great interest when we are dealing with plants of the extreme north.

A comparison of the 10 per cent. and 20 per cent. Chamaephyte Biochores and the June isotherms as given in *Bartholomew's Physical Atlas*, vol. iii, 1899, shows that in a general way the 20 per cent. Chamaephyte Biochore corresponds with the June isotherm of 4·44° C. (40° F.) and that the 10 per cent. Chamaephyte Biochore corresponds with the June isotherm of 10° C. (50° F.). Among failures of correspondence between the 20 per cent. Biochore and the June isotherm of 4·44° C. may be mentioned that, according to *Bartholomew's Physical Atlas*, this June isotherm runs right through Kolguev and slightly to the south of the Strait of Yugor, thus passing in this region too far to the south when compared with the 20 per cent. Chamaephyte Biochore, which runs to the north of Kolguev and through the Straits of Vaygach. Further the June isotherm 4·44° C. goes to the north of the St. Lawrence, which is too far north, because the 20 per cent. Chamaephyte Biochore passes into the south of the St. Lawrence. Finally the 20 per cent. Chamaephyte Biochore lies to the north of Disko, while the 4·44° C. June isotherm runs to the south of Disko, but in other respects cuts Greenland in the same manner as the 20 per cent. Chamaephyte Biochore. I must mention, however, that these discrepancies are not necessarily due to the 20 per cent. Chamaephyte Biochore, but may quite well arise from our ignorance of the behaviour of the isotherms in this northerly region.

The most important failure of correspondence between the 10 per cent. Chamaephyte Biochore, as indicated above, and the June 10° C. isotherm, as drawn in *Bartholomew's Physical Atlas*, is that the June 10° C. isotherm lies farther north in the Atlantic Ocean, i.e. between Iceland and the Faroe Islands, while the 10 per cent. Chamaephyte Biochore passes between the Faroe Islands and the Shetlands. But this failure of correspondence is due to the fact that this June isotherm is wrong. According to the latest information about the temperature of the Faroes, Thorshavn has a June temperature of 9·7° C., so that the June 10° C. isotherm must pass slightly to the south and thus coincide here too with the 10 per cent. Chamaephyte Biochore, which must be drawn quite close to the Faroes, which have indeed 10·5 per cent. of Chamaephytes, thus only slightly exceeding 10 per cent., while the Shetlands have only about 7 per cent.

Finally let me mention one more point. In North Scandinavia the numbers point to a correspondence between these two lines. As mentioned before the 10 per cent. Chamaephyte Biochore passes between Pajala with 5·5 per cent. of Chamaephytes and *Lapponia enontekiensis* with 12 per cent. The June temperature of Pajala is, according to Birger, 11·4° C., and Karesuanda, which lies immediately to the south of *Lapponia enontekiensis*, has a June temperature of 9·4° C. Thus the June isotherm of 10° C. runs in this region, like the 10 per cent. Chamaephyte Biochore, between Pajala and Karesuanda, and to judge by the numbers they both pass slightly to the south of the last-mentioned point.

The result of these investigations may be stated as follows: In the northern cold temperate and cold zones as we gradually go towards the north we find that the biological spectrum of the vegetation changes in a very definite manner. The Phanerophytes and Therophytes decrease and finally disappear. The Cryptophytes, too, which are well represented throughout most of the region, disappear entirely from the hostile regions of the extreme north. The percentage of Hemicryptophytes keeps fairly constant, being approximately double the percentage found in the whole world. The Chamaephyte percentage on the other hand gradually increases towards the north; in the southern parts of the region it is a long way below the Normal Spectrum, but after reaching this figure it soon doubles it. Ultimately the Chamaephyte percentage becomes three times or more that of the Normal Spectrum. All these changes follow the same series everywhere, whichever meridian we follow. But in correspondence with the fact that the climatic lines which dominate the vegetation, e.g. the June isotherms, do not run parallel with the degrees of latitude, the corresponding changes in the biological spectrum do not follow everywhere the same degree of latitude. We must then attempt to draw biological geographical lines, Biochores, i.e. lines uniting the points which have essentially the same biological spectrum. As examples I have tried to draw the 10 per cent. and 20 per cent. Chamaephyte lines and have shown that these taken as a whole coincide with the June isotherm of 4·44° and 10° C. respectively. By means of these lines 4 zones can be differentiated and demarcated. They are, from south to north:

1. A cold temperate zone. A Hemicryptophyte zone south of the 10 per cent. Chamaephyte Biochore.
2. A boreal zone. A Hemicryptophyte and Chamaephyte zone, between a 10 per cent. and 20 per cent. Chamaephyte Biochores.
3. An Arctic zone. A Chamaephyte zone between the 20 per cent. and 30 per cent. Chamaephyte Biochores.
4. An Arctic nival region with over 30 per cent. of Chamaephytes.

It is clear from this that the statistics of life-forms gives us an **exact numerically expressed basis for characterizing and demarcating plant climates**, even the most distant regions with entirely different floras but with essentially the same climatic determinants for vegetation showing essentially the same biological spectrum.

We shall now in conclusion see whether the zones and regions correspond with one another when we use the statistics of life-forms as a basis.

ALTITUDINAL PLANT CLIMATES AND THE RELATIONSHIP BETWEEN LATITUDINAL AND ALTITUDINAL ZONES IN THE NORTHERN COLD AND COLD TEMPERATE DISTRICTS

It is well known that the altitudinal zones of tropical mountains do not correspond with the horizontal climatic zones between the Equator and the Poles; for altitudinal and latitudinal zones alike in their mean annual temperatures may differ widely in their annual temperature curves. But the nearer we are to the Poles the greater becomes the correspondence between the course of the temperature curves, and in the hydrotherm figure as a whole, in the two types of zone. Within the region with which we are now dealing this correspondence is so great that one might expect to discover a similar correspondence in the biological spectra if the basis, the life-forms, upon which our spectra are constructed is correct, that is to say if it is an expression of the climate. The following tables compared with the ones already given will show the relationship between the two types of zone. The examples were chosen because they presented the floristic material in such a manner that it could be used as a starting-point for a comparative statistical biological analysis of the various regions.

I shall begin with the Alps. Table 23 shows a vertical section through the Poschiavo Valley on the south side of the Alps. The Poschiavo Valley is here divided into eight belts whose heights are given in the tables. The extent of each belt averages from 300 to 350 metres. The flora of the individual belts is taken from statements by Brockmann-Jerosch[10]. The biological spectra need no further explanation. Comparing Table 23 and Table 6 we see at a glance that in climbing from the foot of the Alps to their summits we pass through a series of biological spectra which in their characteristic features correspond completely with the spectra encountered in going from the foot of the Alps to the Polar Lands. In the Alps we find height zones quite corresponding with the latitudinal zones just described, and separated by the same biochores, viz. the 10 per cent. Chamaephyte Biochore which here lies at about 1,600 metres, the 20 per cent. Chamaephyte Biochore at about 2,500 metres, and the 39 per cent. Chamaephyte Biochore at somewhere about 2,800 metres. In this case as in the last we can draw any Biochore that we need if only the botanical material is available.

TABLE 23

Poschiavo[10].	No. of species.	The percentage distribution of the species among the life-forms.									
		S	E	MM	M	N	Ch	H	G	HH	Th
Above 2,850 metres	51	..	..	..	..	..	35	61	2	..	2
2,550–2,850 ,,	199	..	..	..	..	..	25	67	4	..	4
2,250–2,550 ,,	348	..	..	..	1	3	18	64	7	1	6
1,900–2,250 ,,	492	..	..	..	1	3	13	68	8	1	6
1,550–1,900 ,,	487	..	..	1	3	4	11	62	10	1	8
1,200–1,550 ,,	449	..	..	2·5	2·5	4	7	60	9	1	14
850–1,200 ,,	604	..	(0·2)	2	3	5	5	55	9	2	19
Under 850 ,,	447	..	(0·2)	3	4	3	5	55	8	1	21

In Table 23 the belts are 300–350 metres high. It will be interesting to see the regional spectra of floras of still narrower belts. This may be done for the nival zones of Switzerland by using the floristic material found in Heer's well-known book, *Über die nivale Flora der Schweiz*, published more than a quarter of a century ago. Using this floristic material as a basis I have given in Table 24 a vertical section through the flora of the Alps above 2,400 metres divided into eight belts, which, with the exception of the uppermost ones, are only 150 metres in height. Passing up through these belts, the number of species falls evenly, from 323 species in the lowermost belt (2,400–2,550 metres) to six species in the uppermost belt (above 3,600 metres), and at the same time the Chamaephyte percentage in the biological spectrum rises from 24 to 67.

The eight belts in Table 24 correspond closely with the two upper belts in Table 23, and by combining these two tables by inserting Table 24 instead of the two uppermost belts in Table 23 we then obtain an interesting series of biological spectra expressing a vertical section through the Alpine Flora. From belt to belt the biological spectrum changes in correspondence with the climatically determined alteration in the environment.

TABLE 24

The nival region of the Alps[33].	No. of species.	The percentage distribution of the species among the life-forms.									
		S	E	MM	M	N	Ch	H	G	HH	Th
Above 3,600 metres	6	..	..	..	..	..	67	33	..	..	..
3,300–3,600 ,,	12	..	..	..	..	..	58	42	..	..	..
3,150–3,300 ,,	19	..	..	..	..	..	58	42	..	..	..
3,000–3,150 ,,	42	..	..	..	..	..	52·5	45	..	..	2·5
2,850–3,000 ,,	117	..	..	..	..	..	33	62	2	..	3
2,700–2,850 ,,	148	..	..	..	..	..	33	61	3	..	3
2,550–2,700 ,,	229	..	..	..	..	0·5	28	65·5	2	..	4
2,400–2,550 ,,	323	..	..	..	..	1	24	67	4	(0·3)	4

The individual boundaries in Switzerland as a whole may well lie at a somewhat lower altitude than the corresponding boundaries in the Poschiavo valley, which is on the southern side of the Alps. Here we have 25 per cent. of Chamaephytes in the belt 2,500–2,850, in Switzerland as a whole there are 24 per cent. of Chamaephytes in the belt 2,400–2,550. Thus the same biological boundary is 100–200 metres lower in Switzerland as a whole than in the Poschiavo region.

It is clear that in the uppermost belts, where the flora is very poor in species, the emphasis is chiefly on a rise of the Chamaephyte percentage, while its absolute amount is of less importance. While the biological spectrum of a flora containing a few hundred or more species will alter but little if more thorough investigation adds some species to it, on the other hand the numbers in the spectrum will alter considerably if the species of a very poor flora be increased. In the four upper belts in Table 24 an increase of the flora by only a single species would alter the numbers by one or several per cent. It is all the more striking, and it is an expression of the fact that life-forms and climate are regulated by definite laws, that the Chamaephyte percentage nevertheless goes on increasing gradually as the environment becomes more and more unfavourable both in the Alpine and in the Arctic nival zones.

In order to show how a more thorough investigation can alter the numbers without altering this characteristic series, in Table 25 some biological spectra are given for the highest region of the Alps based upon the most recent lists of species.

TABLE 25

	Switzerland.	No. of species.	The percentage distribution of the species among the life-forms.									
			S	E	MM	M	N	Ch	H	G	HH	Th
1	Above 4,000 metres[68] . .	8	..	..	..	..	..	62.5	37.5	..	..	..
2	Above 3,250 metres (the extreme snow limit)[68] .	71	..	..	..	..	..	46.5	52	..	..	1.5
3	1 and 2 together with species found in the Bernina Region above 3,000 metres, i.e. the snow limit of the district[68]	108	..	..	..	..	..	39	58	1	..	2
4	The Alpine flora of the Alps above the tree limit[39] .	410	..	..	..	..	3	22	64	6	1	4

If we compare the first spectrum in Table 24, based on six species which in Heer's time were regarded as attaining the greatest altitude (over 3,600 metres) with the top spectrum in Table 25 based on eight species which are now regarded as being the most elevated (over 4,000 metres), it is true that a difference will be found of 4 per cent.; one should rather, however, say of *only* 4 per cent., for the correspondence is as great as could reasonably be expected. Of Heer's six species four are Chamaephytes and two Hemicryptophytes; of Schroeter's 8 species five are Chamaephytes and three Hemicryptophytes.

TABLE 26

	No. of species.	The percentage distribution of the species among the life-forms									
		S	E	MM	M	N	Ch	H	G	HH	Th
The arctic zone north of the 20% Chamaephyte Biochore .	437	..	..	..	..	3.5	19	64.5	8	2	3
The Alpine flora of the Alps above the tree limit . .	410	..	..	..	..	3	22	64	6	1	4

The Chamaephyte percentage in the last spectrum of Table 25, that of the region of the Alps above the tree limit is indeed somewhat higher than the Chamaephyte percentage in the Arctic Chamaephyte Zone north of the 20 per cent. Chamaephyte Biochore, but yet so close to it that it is justifiable to compare the two spectra, before even the 20 per cent. Chamaephyte Biochore of the Alps has been drawn. The numbers in Table 26 show that the correspondence between the Alpine regions of the Alps above the tree limit and the Arctic region north of the 20 per cent. Chamaephyte Biochore is almost as complete as could be wished, which further shows that a line based on the statistics of life-forms is rightly called a Biochore.

TABLE 27

Western Alps[83].	No. of species.	The percentage distribution of the species among the life-forms.									
		S	E	MM	M	N	Ch	H	G	HH	Th
Above 3,050 metres	10	..	..	..	..	..	50	50	..	..	..
2,745–3,050 ,,	84	..	..	..	..	..	30	63	2·5	1	3·5
2,440–2,745 ,,	330	..	..	..	..	1	20·5	68	6	(0·3)	4·5

There remain in the Alps two regions of which the floristic material available is sufficiently well known to make a statistical treatment possible. These places are the Western Alps and the Aosta Valley. Table 27 shows a vertical section of the Western Alps above 2,440 metres divided into three floral regions on the basis of the statements in Thompson's list. Table 28 shows a corresponding section through the Alpine region of the Aosta Valley above 2,500 metres based upon the floristic material of Vaccari. The correspondence between the biological spectra in Tables 27–8 and Table 24 is seen at once, and needs no further explanation.

TABLE 28

Alpine region of the Aosta Valley[85].	No. of species.	The percentage distribution of the species among the life-forms.									
		S	E	MM	M	N	Ch	H	G	HH	Th
3,800–4,200 metres	6	..	..	..	..	..	67	33	..	..	..
3,200–3,800 ,,	56	..	..	..	..	..	39	59	..	..	2
2,800–3,200 ,,	163	..	..	..	..	0·5	28	64·5	4	..	3
2,500–2,800 ,,	140	..	..	..	..	0·5	26·5	65	3·5	..	4·5

Unfortunately I have not been able to include the Pyrenees in the comparison, since the floristic material from that region has not been sufficiently worked out to make possible the construction of biological spectra for a series of altitudinal zones.

Though the Caucasus lies outside the region I am treating, yet it is sufficiently near to it to make a comparison desirable. In Radde's [60] work there is a list of the species occurring in the Alpine region of the Caucasus; but this list contains a large number of species many of which are quite strange to me, so that I have not succeeded in determining the life-form of all of them. But besides the species of the Alpine region Radde enumerates separately those species that are found between 3,050 and 3,660 metres, and those that ascend higher than 3,660 metres. I have determined the life-form of these species, and can give the biological spectra of the two upper belts in the Caucasus. These spectra are given in Table 29 and correspond completely with the uppermost spectra of the Alps. The Biochores of course are considerably higher up in the Caucasus than in the Alps; the 30 per cent. Chamaephyte Biochore is actually slightly above 3,050 metres.

TABLE 29

Caucasus[60].	No. of species.	The percentage distribution of the species among the life-forms.									
		S	E	MM	M	N	Ch	H	G	HH	Th
Above 3,660 metres	15	..	..	..	..	..	60	33	7	..	..
3,050–3,660 ,,	158	..	..	..	..	..	27	65	5	..	3

For the Carpathians, especially for the Tatra, Kotula's list[45] is excellently arranged for any one wanting to obtain a basis for biological spectra at different altitudes. The species in this list are given in the order they would be met descending from the summit of the Tatra; and besides the uppermost limit of the species the lowermost limit is also given when available. Above 1,400 metres I have divided the Tatra into belts of 200 metres height (with the exception of the uppermost belt) and for each of the six belts thus obtained I have determined the biological spectrum. The whole series is given in Table 30. The correspondence seen on comparing them is instructive, and needs no further explanation. It goes without saying that the individual Biochores here are much lower than the corresponding ones in the Alps.

TABLE 30

Tatra[45].	No. of species.	The percentage distribution of the species among the life-forms.									
		S	E	MM	M	N	Ch	H	G	HH	Th
Above 2,400 metres	46	..	..	..	..	..	30.5	67.5	2	..	..
2,201–2,400 ,,	86	..	..	..	..	..	26	71	2	..	1
2,001–2,200 ,,	193	..	..	..	..	2	18	75	4	..	1
1,801–2,000 ,,	295	..	..	..	1	3	16	72	5	..	3
1,601–1,800 ,,	377	..	..	1	1	3.5	14	70	6	0.5	4
1,401–1,600 ,,	469	..	..	1	1	3	10.5	70.5	7.5	0.5	6

Let me here take the opportunity of calling attention to the fact that it makes no difference to the result whether the position of the Biochore be determined by means of the spectra of small local floras, by means of the spectra of different belts, or by finding out how far we have to descend before reaching a line above which the flora, as a whole, has the percentage expressed by the Biochore. If we take the local flora as a starting-point we see from Table 30 that the 20 per cent. Chamaephyte Biochore lies between 2,000 and 2,400 metres, but probably much nearer

2,000 than 2,400 metres. In the same way we see that the 10 per cent. Chamaephyte Biochore must lie near 1,400 metres. Table 31 shows the definite boundaries for the 10, 20, and 30 per cent. Chamaephyte Biochores when their position is determined by the method of marking boundaries by lines above which the whole flora has respectively 10, 20, and 30 per cent. of Chamaephytes. For the rest the figures for the boundaries determined in these two ways do not differ widely.

TABLE 31

Tatra

The 30% Ch-Biochore is at 2,400 metres, above this Line are found 46 species of which 14 are Chamaephytes.
The 20% Ch-Biochore is at 2,030 metres, above this Line are found 157 species of which 32 are Chamaephytes.
The 10% Ch-Biochore is at 1,394 metres, above this Line are found 510 species of which 51 are Chamaephytes.

The upper boundary for *Pinus mughus* in the Tatra almost coincides with the 20 per cent. Chamaephyte Biochore, and the upper boundary of *Fagus silvatica* coincides with the 10 per cent. Chamaephyte Biochore.

For Scandinavia material has long been available in Blytt's[9] lists of flora for the regions of Valders, and in Table 32 will be found the biological spectra for the uppermost regions based on this material. The Biochores of course are lower here than in the Tatra.

TABLE 32

Valders (Norway)[9].	No. of species.	The percentage distribution of the species among the life-forms.									
		S	E	MM	M	N	Ch	H	G	HH	Th
Above the Willow limit, i.e. about 1,200 metres	93	..	..	..	..	..	30	64	3	1	2
Between the Birch and Willow limits, i.e. about 1,000–1,200 metres	266	..	..	..	..	5	15	61	10	3	6
Between the Spruce and Birch limits, i.e. about 750–1,000 metres	333	..	..	..	1	4	10	62	11	4	8

We next come to Table 33, which gives an example of a section through the Scottish Highlands (Clova) based on a list of the flora by Willis and Burkill[90]. The boundaries here are lower than in Valders, and the 10 per cent. Chamaephyte Biochore is very near that of the lowlands. In other words we approach a district where the 10–20 per cent. Chamaephyte altitudinal zone passes into the corresponding latitudinal zone, as in northern Norway. I shall now revert to this subject and treat the flora of the Faroes, completing the series of vertical sections through the mountain regions of Europe by giving the biological spectra for the regions of the Faroes.

TABLE 33

Clova (Scotland)[90].	No. of species.	The percentage distribution of the species among the life-forms.									
		S	E	MM	M	N	Ch	H	G	HH	Th
Above 1,000 metres	11	..	..	..	..	..	27	64	9	..	..
900–1,000 ,,	44	..	..	..	..	2	25	52	14	..	7
800–900 ,,	72	..	..	..	..	3	22	60	11	..	4
700–800 ,,	170	..	..	..	1	5	11	67	11	1	4
600–700 ,,	206	..	..	1·5	2	4	15	62	9	1·5	5
500–600 ,,	182	..	..	3	3	4	13	63	8	2	4
400–500 ,,	193	..	..	3	3	4	11	66	8	1	4
300–400 ,,	211	..	..	3	3	4	10	65	8	2	5
Under 300 ,,	304	..	..	3	2	4	7	59	7	5	13
Whole Region	373	..	..	2	2	5	9	58	7	4	12

Ostenfeld's work contains floristic material for the Faroes arranged so as to enable one easily to construct the six lowermost biological spectra in Table 34. The three other spectra in Table 34 are based in part on various lists in Jensen's work[38] and in part upon unpublished notes which Dr. Ostenfeld has been kind enough to place at my disposal. By these means it has been possible to construct the biological spectra for the three altitudinal belts above 500 metres.

TABLE 34

Faroes[57,59].	No. of species.	The percentage distribution of the species among the life-forms.									
		S	E	MM	M	N	Ch	H	G	HH	Th
9. Above 700 metres	37	..	..	..	..	3	27	62	8	..	..
8. ,, 600 ,,	63	..	..	..	..	3	24	65	5	..	3
7. ,, 500 ,,	78	..	..	..	..	4	23	64	5	..	4
6. Highland	122	..	..	..	..	1·5	19·5	61·5	8	3·5	6
5. Lowland	233	..	..	..	..	1·5	9	55·5	12	11·5	10·5
4. Only in Highland	21	..	..	..	..	..	28·5	62	9·5	..	..
3. Common to Lowland and Highland	101	..	..	..	..	2	18	61	8	4	7
2. Only in Lowland	132	..	..	..	..	1·5	2	51	15	17·5	13
1. Whole flora	254	..	..	..	..	1·5	10·5	56	12	10·5	9·5

If the numbers in the biological spectrum of the whole flora contained in the bottom line of Table 34 do not correspond with the numbers already given by Ostenfeld the reason is as follows. In order to obtain as floristically uniform as possible a foundation for the comparative treatment of the biological spectra of the various floras, I have been obliged to use a conservative demarcation of species, as, for example, in dealing with *Hieracium* and *Taraxacum*. In my list the number of species of flowering plants on the Faroes is only 254, while Ostenfeld makes the number 278. All the same the difference between the two series of numbers is inconsiderable.

As I have said before, and as is seen again from the lowest spectrum in Table 34, the Faroes as a whole lie within the 10–20 per cent. Chamaephyte zone, but near its southern boundary, since the percentage of Chamaephytes is only 10·5. Considering this low Chamaephyte percentage in connexion with the fact that in Scotland there are 10 per cent. of Chamaephytes at an altitude of from 300–400 metres, it seemed reasonable to suppose that on the Faroes, which extend far above 400 metres, we might find at least two of the regions just differentiated, viz. a lowland region with less than 10 per cent. of Chamaephytes and a highland region with more than 10 per cent. The biological spectra of 5 and 6 in Table 34 show that this is actually so. But since the Chamaephyte percentage in the biological spectrum of the highlands is here quite near 20 it seemed still more reasonable to suppose that the most elevated portion of the Faroes might lie above the 20 per cent. Chamaephyte Biochore, and the biological spectra 7–9 in Table 34 shows that this is true. Thus we have on the Faroes, besides the lowland with less than 10 per cent. of Chamaephytes, at least two zones, viz. a 10 per cent. to 20 per cent. Chamaephyte region and a 20 per cent. to 30 per cent. Chamaephyte region.

CONCLUSION

We are not in a position to determine the ideal life-forms of plants, which would be the sum of all their adaptations to their environment. We must content ourselves with characterizing life-forms by using one essential side of the adaptation, so that the life-forms are founded upon one point of view. The foundation must therefore not only be something essential, but it must also represent a single aspect, so that the life-forms established can come to form a coherent, continuous series, thus making possible a comparative statistical treatment. I have taken, therefore, as a foundation for my series of life-forms the adaptations of plants to survive the unfavourable season, especially in relationship to the protection afforded to their surviving buds or shoot-apices.

If we attempt to obtain an expression for the plant-climate of a district founded upon the life-forms, because of the one-sidedness of the life-forms used, we cannot content ourselves with determining the life-forms of some species only, but must investigate them all, and determine the percentages of the individual life-forms. By this means we obtain a series of numbers, a biological spectrum, as an expression for the climate, as far as this can be obtained by means of the life-forms used. Whether this biological spectrum is a true expression of the plant-climate is decided by seeing whether the same climate in different parts of the world having entirely different floras gives the same spectrum, while different climates give different spectra.

By investigating a series of local floras of different meridians from the Equator to the Pole I have shown in this paper that this is so. By this means the various plant-climates can be characterized and demarcated.

First of all, in the main, four series of climates can be differentiated:
1. A Phanerophyte climate in the tropical zone where the precipitation is not deficient.
2. A Therophyte climate in the regions of the sub-tropical zone with winter rain.
3. A Hemicryptophyte climate in the greater part of the cold temperate zone.
4. A Chamaephyte climate in the cold zone.

These main plant-climates and their subdivisions can be separated by biological boundaries, Biochores, founded on exact numbers, and analogous with climatological lines such as isotherms.

Further I have shown that in the Northern Hemisphere, travelling from the southern boundary of the Hemicryptophyte climate to the Polar lands, we pass a series of biological zones bounded by Biochores; and if in the same region we climb up from the foot of a sufficiently high mountain we pass through a corresponding series of zones characterized in exactly the same way, there being as many altitudinal zones as we find zones between the mountain in question and the Pole. Here we have the proof that the biological spectra, which express the distribution in life-forms of the individual species of the floras, can be used as an expression of the plant-climate, since they change in a definite way in correspondence with change in the climate, but remain unaltered in the same climate, even if the species composition of the flora be entirely different.

In this way a Biological Plant Geography may be built up on the basis of the statistics of life-forms.

LITERATURE

Containing, with but few exceptions, only those papers used as a floristic basis for the biological spectra

1. ANDERSSON, G., u. HESSELMAN, H., Verzeichnis der in König Karls Land während der Schwedischen Polarexpedition 1898 gefundenen Phanerogamen. Öfversigt af Kgl. Vetensk.-Akad. Förhandl., 1898, Nr. 8.
2. —— Spetsbergen och Beeren-Islands kärlväxtflora. Bihang till Kgl. Svenska Vetensk.-Akad. Handl., Bd. **26**, Afd. III, Nr. 11, 1900.
3. ASCHERSON, P., et SCHWEINFURTH, G., Illustration de la flore d'Égytpe. Mémoires de l'institut égyptien, tome ii. Le Caire 1889.

4. BAKER, J. G., Flora of Mauritius and the Seychelles. London 1877.
5. BALFOUR, J. H., and BABINGTON, CH., Account of a Botanical Excursion to Skye and the Outer Hebrides, during the month of August 1841. Transact. Bot. Soc. Edinburgh, i, 1841.
6. BARRINGTON, R. M., Notes on the Flora of St. Kilda. Journ. of Bot. **24**, 1886, p. 213.
7. BERGROTH, O., Anteckninger om Vegetationen i gränstrakterna mellan Åland och Åboområdet. 1894.
8. BIRGER, SELIM, Vegetationen och floran i Pajala socken med Muonio kapellag i arktiska Norrbotten. Arkiv för Botanik, Bd. **3**, 1904.
9. BLYTT, A., Botanisk Rejse i Valders. Christiania 1864.
10. BROCKMANN-JEROSCH, H., Die Flora des Pûschlav und ihre Pflanzengesellschaften. Leipzig 1907.
11. BROWN, R., Chloris Melvilleana, a List of Plants collected in Melville Island, in the year 1820, by the Officers of the Voyage of Discovery under the orders of Captain Parry. A supplement to the Appendix to Captain Parry's Voyage, pp. 261–300.
12. —— List of Plants collected on the Coast of Baffin's Bay, from Lat. 70° 30' to 76° 12' on the East Side; and at Possession Bay, in Lat. 73° on the West Side. I, 'A Voyage of Discovery for the purpose of exploring Baffin's Bay', by John Ross, K.S., Captain, Royal Navy. Appendix, pp. 141–4.
13. CHEVALLIER, L., Deuxième note sur la Flore du Sahara. Bull. de l'Herbier Boissier, 2me série, tome **3**, 1903.
14. COLVILLE, S. V., Botany of the Death Valley Expedition. 1893.
15. —— Botany of Jakutat Bay, Alaska. Contributions from the U.S. National Herbarium, vol. iii, No. 6, 1896.
16. COSSON, E., Plantae in Cyrenaica et Agro Tripolitano natae. Bull. Soc. Bot. Fr. **22**, 1875, p. 45.
17. —— Plantae in Cyrenaica et Agro Tripolitano, anno 1875, a. cl. J. Daveau lectae. Bull. Soc. Bot. Fr. **36**, 1889, p. 100.
18. DAVEAU, J., Excursion à Malte et en Cyrenaïque. Bull. Soc. Bot. Fr. **23**, 1876, p. 17.
19. DYRING, JOH., Junkersdalen og dens Flora. Nyt Magasin for Naturvidenskaberne, 37 Bd., Christiania 1900.
20. EASTWOOD, ALICE, A Descriptive List of the Plants collected by Dr. F. E. Blaisdell at Nome City, Alaska. Bot. Gazette, **33**, 1902, pp. 126, 199, 284.
21. EGGERS, H. F. A., The Flora of St. Croix and the Virgin Islands. Bull. of the U.S. Nat. Mus., No. 13, 1879.
22. EKSTAM, O., Beiträge zur Kenntniss der Gefässpflanzen Spitzbergens. Tromsø Museums Aarshefter, **20**, 1897.
23. —— Neue Beiträge zur Kenntniss der Gefässpflanzen Novaja Semlja's. Engl. Jahrb. **22**, 1897, pp. 184–201.
24. FEDTSCHENKO, BORIS, Flore des îles du Commandeur. Cracovie 1906.
25. FEILDEN, H. W., The Flowering Plants of Novaya Zemlya, &c. Journ. of Bot. **36**, 1898, pp. 388, 418, 468.
26. —— and GELDART, H. D., A Contribution to the Flora of Kolguev. Transact. of the Norfolk and Norwich Naturalists' Society, vol. vi, 1896.
27. FRIDTZ, R., Undersøgelser over Karplanternes Udbredelse i Nord-Reisen. Nyt Magasin for Naturvidenskaberne, 37 Bd. Christiania 1900.
28. FRIES, TH. M., En botanisk resa i Finmarken 1864. Bot. Notiser, 1865.
29. —— Om Beeren-Islands fanerogam-vegetation. Öfversigt af Kgl. Vetensk.-Akad. Förh., 1869, Nr. 2.
30. GIBSON, A. H., The Phanerogamic Flora of St. Kilda. Transact. Bot. Soc. Edinburgh, **19**, 1891–3, p. 155.
31. HARPER, R. M., A Phytogeographical Sketch of the Altamaha Grit Region of the Coastal Plain of Georgia. Annals N.Y. Acad. Sc., vol. **17**, part i, pp. 1–415, 1906.

STATISTICS OF LIFE-FORMS

32. HARTZ, N., Fanerogamer og Karkryptogamer fra Nordøst-Grønland, c. 75–70° n. Br., og Angmasalik, c. 65° 40' n. Br. 1895. Medd. om Grønland, xviii.
33. HEER, O., Ueber die nivale Flora der Schweiz. 1884.
34. HOOKER, J. D., Botanical Appendix to Captain Parry's Journal of a second Voyage for the Discovery of a North-West Passage from the Atlantic to the Pacific performed in H.M. ships 'Fury' and 'Hecla', the Years 1821–22–23, p. 381. London 1825.
35. —— On some collections of Arctic Plants chiefly made by Dr. Lyall, Dr. Anderson, Herr Miertsching, and Mr. Rae, during the Expeditions in search of Sir John Franklin, under Sir John Richardson, Sir Edward Belcher, and Sir Robert M'Clure. Journ. Linn. Soc. Botany, vol. i, 1857, pp. 114–24.
36. —— An account of the Plants collected by Dr. Walker in Greenland and Arctic America during the Expedition of Sir Francis M'Clintock, R.N., in the Yacht 'Fox'. Journ. Linn. Soc. Botany, vol. v, 1861, pp. 78–88.
37. HULT, R., Växtgeografiska anteckningar från den finska Lappmarkens skogsregioner. 1898.
38. JENSEN, C., Beretning om en Rejse til Færøerne i 1896. Bot. Tidsskr. **21**, 1897.
39. JEROSCH, MARIE CH., Geschichte und Herkunft der schweizerischen Alpenflora. Leipzig 1903.
40. JØRGENSEN, E., Om floraen i Nord-Reisen og tilstødende dele af Lyngen. Christiania Vidensk. Selsk. Forh., 1894.
41. KIRCHNER, O., Flora von Stuttgart und Umgebung. Stuttgart 1888.
42. KJELLMAN, F. R., Sibiriska nordkustens fanerogamflora. Vega-Expeditionens vetenskapeliga iakttagelser, Bd. **1**. Stockholm 1882.
43. —— Fanerogamfloran på St. Lawrence-ön. Vega-Expeditionens vetensk. iakttag., Bd. **2**, 1883, pp. 1–23.
44. —— Fanerogamer från Vest-Eskimåernes land. Vega-Expeditionens vetenskapeliga iakttagelser, Bd. 2.
45. KOTULA, B., Distributio plantarum vascularium in montibus Tatricis. Cracoviae 1889–90.
46. KRAUSE, KURT, Beiträge zur Kenntniss der Flora von Aden. Englers Jahrb. **35**, 1904–5, pp. 682–749. (Her den ældre Litteratur.)
47. KRUUSE, C., Jan Mayens Karplanter. Bot. Tiddskr. **24**, 1902, pp. 297–302.
48. —— List of Phanerogams and Vascular Cryptogams found in the Angmagsalik District on the East Coast of Greenland between 65° 30' and 66° 20' lat. N. Medd. om Grønland, **30**, 1906.
49. KURTZ, F., Die Flora des Chilcatgebietes im südöstlichen Alaska. Engl. Jabrb. **19**, 1894–5, pp. 327–431.
50. —— Die Flora der Tschuktschenhalbinsel. Engl. Jahrb. **19**, 1894–5, pp. 432–93.
51. LETOURNEUX, A., Note sur un voyage botanique à Tripoli de Barbarie. Bull. Soc. Bot. Fr. **36**, 1889, p. 91.
52. MACOUN, J. M., A List of the Plants of the Pribilof Island. (The Fur Seals and Fur-Seal Islands of the North Pacific Ocean, part iii, pp. 559–87.) Washington 1894.
53. —— List of the Plants known to occur on the Coast and in the Interior of the Labrador Peninsula. Ann. Report, Geological Survey of Canada, vol. viii, part i, Appendix VI.
54. NATHORST, A. G., Nya bidrag till kännedomen om Spetsbergens kärlväxter, och dess växtgeografiska förhållanden. Kgl. Svenska Vetensk.-Akad. Handl., Bd. **20**, Nr. 6, 1883.
55. NORMAN, J. M., Norges arktiske Flora, ii. 1895–1901.
56. NOTØ, ANDR., Florula Tromsøensis. 1904.
57. OSTENFELD, C. H., Phanerogamae and Pteridophyta, I, 'Botany of the Færøes, I', 1901.
58. —— Flora arctica, part i, 1902.
59. —— The Land-Vegetation of the Færøes. Botany of the Færøes, iii, 1908.

60. RADDE, G., Grundzüge der Pflanzenverbreitung in den Kaukasusländern, &c. Engler & Drude, Veg. d. Erde, iii.
61. RALPH, TATE, Upon the Flora of the Shetland Isles. Journ. of Bot., 1866.
62. RAUNKIÆR, C., Types biologiques pour la géographie botanique. Overs. over det Kgl. Danske Videnskabernes Selskabs Forhandlinger, 1905.
63. —— Planterigets Livsformer og deres Betydning for Geografien. Kjøbenhavn og Kristiania 1907. (Chapter II of the present work.)
64. ROTHROCK, J. T., Sketch of the Flora of Alaska. Smithsonian Reports, 1867, p. 433.
65. ROWLEE, W. W., and WIEGAND, K. M., A List of Plants collected by the Cornell Party on the Peary Voyage of 1896. Bot. Gazette, **24**, p. 417.
66. SAELAN, TH., KIHLMAN A. OSW., HJELT HJ., Herbarium Musei Fennici. Editio secunda. I, Plantae vasculares. 1889.
67. SCHEUTZ, N. J., Plantae vasculares Jeniseensis. Kgl. Svenska Vetensk.-Akad. Handl., Bd. **22**, 1888.
68. SCHROETER, C., Das Pflanzenleben der Alpen. Zürich 1908.
69. SEEMANN, B., Flora of Western Eskimoux-Land. The Botany of the Voyage of H.M.S. 'Herald'. 1852.
70. SEWELL, PHILIP, The Flora of the Coast of Lapland and of the Yugor Straits, as observed during the Voyage of the 'Labrador' in 1888; with summarized List of all the species known from the Island of Novaya Semlya and Waigats, and from the North Coast of Western Siberia. Transact. of the Bot. Soc. Edinburgh, vol. **17**, 1889, pp. 444–81.
71. SIMMONS, H. G., Preliminary Report on the Botanical Work of the Second Norwegian Polar Expedition 1898–1902. Nyt Magasin for Naturvidenskab, 1903.
72. —— The Vascular Plants in the Flora of Ellesmereland. (Rep. of the second Norwegian Arctic Expedition in the 'Fram', 1898–1902, No. 2.)
73. SOMMIER, S., La Flora dell'Arcipelago Toscano. Nuovo Giornale Botanico Italiano, Nuova Serie, vol. ix, 1902, p. 319; vol. x, 1903, p. 131.
74. STEFANI, C. DE, FORSYTH MAJOR, C. J., et WILLIAM BARBEY, Samos. Lausanne 1892.
75. STEFANSSON, S., Flora Islands. 1901.
76. STENROOS, K. E., Nurmijärven pitäjän Siemen-ja Sanais-kasvisto. 1894.
77. TAYLOR, J., Notice of Flowering Plants and Ferns collected on both sides of Davis Straits and Baffin's Bay. Transact. of the Bot. Soc. Edinburgh, vol. **7**, part 2, 1862, pp. 323–34.
78. TOWNSEND, F., Contributions to a Flora of the Scilly Isles. Journ. of Bot., 1864.
79. TRAUTVETTER, E. R. A., Plant. Sibiriae borealis ab A. Czekanowski et F. Müller annis 1874 et 1875 lect. Acta Horti Petropolitani, v, 1877, pp. 1–146.
80. —— Flora riparia Kolymensis. Sammest., pp. 495–574.
81. —— Florula taymyrensis phaenogama. I, 'Middendorff, A. Th., Reise in den äussersten Norden und Osten Sibiriens', Bd. **1**, 1887.
82. —— Syllabus plantarum Sibiriae boreali-orientalis a Dre A. a Bunge fil. lectarum. Acta Horti Petropolitani, x, 1887–9, pp. 483–540.
83. THOMPSON, S., Liste des Phanérogames et Cryptogames vasculaires recueillies au-dessus de 2440 mètres, dans les districts du Mont-Cenis, de la Savoie, du Dauphiné et des Alpes-Maritimes. Bull. de l'Académie internationale de Géographie Botanique. 17e Année, 1908, Nos. 220–1.
84. TURNBULL, R., First Record of Plants from Hope Island, Barentz Sea. Collected by W. S. Bruce. Transact. of the Bot. Soc. Edinburgh, vol. **21**, 1897–1900, pp. 166–8.
85. VACCARI, LINO, Flora cacuminale della Valle d'Aosta. Nuovo Giorn. Bot. Ital., **8**, 1901, p. 416.
86. VAHL, M., Madeiras Vegetation. 1904.
87. WARMING, EUG., Om Grønlands Vegetation. 1886–7. Medd. om Grønland, xii, 1888.

88. WARMING, EUG., Tabellarisk Oversigt over Grønlands, Islands og Færøernes Flora. Vidensk. Medd. fra d. naturh. Forening i Kjøbenhavn. 1888.
89. WATSON, H. C., Florula Orcadensis. Journ. of Bot., 1864.
90. WILLIS, J. C., and BURKILL, I. H., The Phanerogamic Flora of the Clova Mountains in special relation to Flower-Biology. Transact. Bot. Soc. Edinburgh, **22**, 1901–4, pp. 109–25.
91. Die internationale Polarforschung 1882–3. Die deutschen Expeditionen und ihre Ergebnisse, Bd. II, 1890, pp. 75–92, 97–9.
92. Plants from Labrador. Bull. of Miscell. Information, Kew, 1907, pp. 76–88.

V
THE LIFE-FORMS OF PLANTS ON NEW SOIL
INTRODUCTION

THE chief object of my journey to the West Indies in 1905–6 was the opportunity it afforded of determining the life-forms of the species of flowering plants found within a circumscribed area of tropical climate, viz. the Danish West Indies. I wished to do this in order to obtain material for the biological spectrum of a tropical country with a comparatively low rainfall. Although I hope later on to be able to give more detailed information about the biological data of some of the flowering plants of the Danish West Indies and St. Domingo, the chief result of my investigations is already given on p. 129 of my book called *Planterigets Livsformer og deres Betydning for Geografien*,[1] where it occupies but a single line, being the biological spectrum based on the statistics of life-forms for St. Thomas and St. Jan. In the work cited I have given a basis for a comparative investigation of the plant-climates of the different tropical countries, and that basis is the climate as a factor determining definite kinds of vegetation and expressed by the statistical relationship between the life-forms of the species, determined by their adaptation to survive the unfavourable season.

To determine then as far as possible the life-forms of the individual species was my chief aim. But beside this I had in mind various small tasks with which to busy myself when time and opportunity allowed. I wished among other things to investigate the conditions and the vegetation of the alluvial formations which are found here and there along the coast of the Danish West Indies. My object in doing this was especially to obtain a basis for a comparative investigation of vegetation found on essentially the same soil in different climates, e.g. the Danish West Indies and Denmark. By this investigation I wished to find out how far climate, even in such a specialized area as alluvial beach formations, determines the biological spectrum of the vegetation.

Alluvial beaches are found in all climates, and show wide differences. It is true indeed that we cannot always separate sharply the different beach formations, but in general we can differentiate those made up of deposits of coarser inorganic components (sand and gravel), and the deposits of finer components (clay, mud, &c.). The former are found, as is well known, especially on unprotected coasts, the latter on protected

[1] Included in the present volume as Chapter II, *The Life-forms of Plants and their Bearing on Geography*.

FIG. 78. Coral Bay on St. Jan (24.2.06). The North side of Popilleaus Bay looking eastward; in the background is seen 'Orkanhullet' ('the hurricane hole') and the East end of St. Jan. The picture shows the outermost prominent angle at the entrance of Popilleaus Bay, which lies behind and to the right of the spectator. From the steep foot of the cliffs a reef of large rocks projects into the sea. Loose material (stone, gravel, and sand) covers the ground at the foot of the cliff with a poor vegetation of *Croton flavens, Capparis frondosa, C. cynophallophora, Randia aculeata, Comocladia ilicifolia* (partly leafless), *Pictetia aculeata* (leafless), *Pisonia fragrans, Elaeodendron xylocarpum, Caesalpinia crista, Canavalia obtusifolia,* and *Ipomaea pescaprae*. In fissures on the face of the cliff the following plants grow: *Melocactus communis, Opuntia tuna, O. curassavica, Pilocereus Royenii, Pitcairnia angustifolia, Euphorbia petiolaris, Plumieria alba* (leafless), and *Pisonia subcordata* (leafless). The whole vegetation bears a strong impress of drought.

Fig. 79. The East end of St. Jan (3.06). The North-west side of Overhale Bay in Coral Bay, looking towards the North-east. The cliffs are not very steep and have a comparatively rich, though strongly xerophilous vegetation. Lowest down in the foreground a rosette of *Agave Morrisii*, then *Pilocereus Royenii*, and above that *Plumieria alba* (leafless). Farther back is a flowering *Agave*. The rest of the vegetation consisted of: *Citharexylum cinereum*, *Bursera simaruba* (almost leafless), *Erithalis fruticosa*, *Randia aculeata*, *Ibatia maritima*, *Ipomaea pescaprae*, *Lantana involucrata*, *Anthacanthus spinosus*, *Canavalia obtusifolia*, *Rhynchosia minima*, *Pithecolobium unguis*, *Serjania polyphylla*, *Comocladia ilicifolia*, *Elaeodendron xylocarpum*, *Melochia tomentosa*, *Euphorbia linearis*, *Croton flavens*, *C. betulinus*, *Melocactus communis*, *Opuntia tuna*, *O. curassavica*, *Cissampelos Pareira*, *Callisia repens*, and some grasses.

coasts, so that the vegetation of the two habitats is not merely determined by the different composition of the soil, but it is also partly determined by shelter. To these two kinds of beach, the sand and the clay beach, we can add a third, formed of soil made up essentially of organic material. This third kind of beach is, however, poorly represented in the Danish West Indies.

Where there is a low beach, even though it be narrow, between the foot of the higher land and the sea, we usually find along the coasts of our West Indian Islands sand formations varying in extent. These sand formations are especially in evidence in open bays. The sand consists everywhere chiefly of 'Coral Sand' which in spite of its name does not consist entirely of corals, but contains the shells of other animals, and also calcareous algae. In some places the coral sand is mixed with varying amounts of sand formed by the breaking down of rocks composing the cliffs. These sands are formed in part on the beach itself, where the coast is flanked by rocks, and partly in the beds of streams, or 'Guts' as they are called, through which the rapid mountain torrents swollen by the rains are carried out into the ocean. In some places where the beach between the sea and the foot of the cliffs is only a few metres broad, we have an opportunity of seeing all possible transitions between the final product, more or less fine sand, and the material from which the sand is formed. An example of this is Popilleaus Bay on the Island of St. Jan.

In other places the inner portion of the bay, which is more or less filled with transported material, has become separated from the sea, which has formed a bar of sand across the bay. Within the shelter of this bar we find a lagoon in which the water is calmer. The edge of the lagoon usually bears a luxuriant Mangrove vegetation. In other places the lagoon has become entirely separated from the sea, and forms a shallow lake, which in the Danish West Indies is usually called 'Salt Pond', 'Salt Panne', or simply 'Pond' or 'Panne'. A 'Pond' of this kind may ultimately become filled up, partly by material washed down into it and partly by material washed up at high tide. It then becomes dry and only exceptionally, during the rains and the spring tides, is it submerged.

The essentials of these changes have already been described by Ørsted[1] and by Eggers;[2] but since I have had the opportunity of studying them in a series of places, I shall give as an introduction to the description of Sandy Point and Krauses Lagoon a short description of the various stages of development. This I shall do by giving an account of what we now see at definite places which have names. I think that this is important as a basis for future investigations of the kinds of changes

---

[1] Ørsted, A. S., 'Dansk Vestindien i physisk-geographisk og naturhistorisk Henseende', Bergsøe, Den danske Stats Statistik, 4 Bd., Kjøbenhavn 1849.
[2] Eggers, H. F. A., 'Naturen paa de dansk-vestindiske Øer', Tiddskrift for populære Fremstillinger af Naturvidenskaben, 5te Række, 5te Bind., Kjøbenhavn 1878.

and the rapidity with which they occur, now taking place and which will continue to take place along the coast of the Danish West Indies. In my investigations of Krauses Lagoon and the recent extensive changes there I have felt the need of a more detailed description of the condition before these changes took place.

I shall attempt here no thorough description of the formations, but shall only mention those which are found in the individual localities. On the alluvial beaches of the West Indies we find four chief formations. Where the coast is exposed, we have nearest the sea the *Pes-caprae-Formation* and on the landward side of this the *Coccoloba-Formation*. On protected coasts the *Mangrove-Formation* is nearest the sea; within this and on higher ground we find the *Conocarpus-Formation*. The first three of these formations have already been mentioned by Warming[1] and Børgesen.[2] By the *Conocarpus*-Formation I understand the vegetation found on the higher and drier parts of the lagoon within the Mangrove-Formation. In such places *Conocarpus erectus* often forms an important component, especially on the lowest seaward facies nearest the Mangrove-Formation.[3]

It is only at a few promontories on the coast that the sea actually reaches the fixed rocks. But the beach in many places is formed entirely of large masses of rock that have fallen on to it. In these places it is either impossible to walk along the beach, because the waves break violently across the intervals between the rocks (Fig. 80), or else progress is made with great difficulty by jumping from one rock to another, continually embarrassed by the dense, often thorny scrub which spreads from the rocky slopes over the beach (Fig. 81). On a coast such as this, e.g. on the west side of Løvenlund Bay (Fig. 80) on the north side of St. Thomas, we already find faint traces of a sandy beach, where the spring tides have deposited masses of sand so high up that the breakers do not ordinarily reach them. In such places the hollows in and between the rocks have become partly filled with sand, on which some representatives of the *Pes-caprae*-Formation are to be found, especially *Sporobolus virginicus* and *Sesuvium portulacastrum*. Here and there too *Ipomaea pes caprae* and a few other species are to be seen. Species from the vegetation of the adjacent rocky beach are of course found; but they do not concern us here, since they do not grow on ordinary sandy beach, i.e. where the ground consists of sand to any great depth.

Popilleaus Bay. As an example of the last stage in this development

[1] Warming, E., *Plantesamfund*, Kjøbenhavn 1895.

[2] Børgesen, F. og Ove Paulsen, *Om Vegetationen paa de dansk-vestindiske Øer*, Kjøbenhavn 1898.

[3] Raunkiær, C., 'Vegetationsbilleder fra dansk Vestindien; Krauses Lagune', *Bot. Tidsskrift*, Bd. 28, 1907–8, Beretning om Foreningens Virksomhed, p. III.

FIG. 80. The West coast of Løvenlund Bay on the North coast of St. Thomas (12.12.05). The coast is rocky, and there are rocks varying in size lying along the shore. The waves break at times so violently against these rocks that the lowest belt of the vegetation on the hill-slopes is wetted by the spray. The water is often forced up between the rocks carrying with it sand, which gradually fills up the fissures and depressions, so that several of the littoral plants are to be found, e.g. *Sporobolus virginicus*, low down on the left of the picture.

FIG. 81. The East coast of Magens Bay (Great Northside Bay) on the North coast of St. Thomas (8.12.05). Rocky coast with stones and rocks along the edge of the beach.

I shall choose Popilleaus Bay on the east side of Battery Peninsula at the head of Coral Bay on the Island of St. Jan. On both sides of this bay, but especially on its south side, the waves break immediately upon the fixed rocks on the coast. The sea is actively breaking these rocks down, and they assume fantastic shapes. In some places there are still found isolated, often lofty, portions of rock at the edge of the water, usually bearing aloft blasted specimens of *Coccoloba uvifera*, a few species of *Cactus*, and other plants (*Pilocereus Royenii, Opuntia tuna, Melocactus communis, Sporobolus virginicus, Capparis cynophallophora*, and *Pictetia aculeata*). Although I shall not here deal with the vegetation of the rocky beach, yet I consider it interesting to observe which species it is that, on a definite and circumscribed locality like Popilleaus Bay, are found on the lowest portion of the steep rocks which are sometimes exposed to the full force of the sea, and on and between the fallen rock masses. Besides the species on the flat beach composed of sand and worn stones the following were found:

Sporobolus virginicus
Scleria lithosperma
Hymenocallis caribaea
Pitcairnia angustifolia
Agave Morrisii
Coccoloba uvifera
Pisonia subcordata
Melocactus communis
Opuntia curassavica
„ tuna
Pilocereus Royenii
Capparis cynophallophora
„ frondosa
Melochia tomentosa
Croton flavens

Elaeodendron xylocarpum
Pictetia aculeata
Pithecolobium unguis-cati
Conocarpus erectus
Laguncularia racemosa
Antherylium Rohrii
Bumelia cuneata
Tecoma leucoxylon
Anthacanthus spinosus
Bontia daphnoides
Plumieria alba
Erithalis fruticosa
Randia aculeata
Vernonia arborescens

As we are here dealing with but a small area, viz. the short south side of Popilleaus Bay, it is easy to understand that the list includes merely a fraction of the species that may be encountered in similar localities. It is worth noticing that many of the species mentioned are those that make up an important component of the flora of the alluvial beach formations.

Quantities of stones and rocks lie in the water near the foot of the cliff at the outermost exposed angle of the north side of the Bay (Fig. 78). Slightly to the side of this the beach consists of worn stones varying in size. These stones are fragments of coral masses, with shells of mussels, snails, and sea-urchins, which live in large numbers along the coast. When the breakers recede from the beach the stones grind against one another making a deafening uproar. Passing gradually farther towards the inner calmer angle of the bay, where the beach is flat, the materials become smaller and smaller because they are sorted out by the varying strength of the movements of the water. All sizes are found, from really large stones to stones the size of the fist, of a hen's egg, a pigeon's egg, and so on, until we come to coarser and finer gravels, and then to the innermost angle of the bay, where the beach is in general sandy.

At the place where the stones varied in size between a hazel-nut and a walnut, I determined the numerical relationship between those stones derived from the rocks of the coast and those which had arisen from the shells or calcareous skeletons of animals. I counted a thousand stones, and each hundred I divided into 4 samples at random. The result is seen below.

	Fragments of Rocks.	Corals.	Snails.	Mussels.
	79	16	4	1
	75	19	5	1
	69	30	1	0
	75	23	1	1
	73	25	0	2
	72	26	1	1
	72	20	5	3
	65	32	3	0
	81	17	2	0
	73	25	0	2
Per cent.	73·4	23·3	2·2	1·1

The coral sand and the worn fragments of coral and shells on the beach arise chiefly from the animals peculiar to the little bay. The fauna is here, as in most places along Coral Bay, very rich. At a distance of from 10 to 15 metres from the beach a dense growth of large corals is encountered; a circular species is especially in evidence, sometimes exceeding 1 metre in breadth, and closely resembling *Polyporus giganteus* in form and appearance. Each stem of this species gives off at different heights broad, flat, lobed divisions, which are greyish-yellow, while the older parts are dark brown. There are also large brown species resembling *Clavaria*, cushion- and cake-shaped Brain Corals and several others. Within and outside this belt of Coral there live shoals of other animals, Snails, Mussels, Holothurians, Sponges, and swarms of several species of Sea-urchins. That the shells of the sea-urchins are not represented among the sample of beach pebbles is because the empty shells of the sea-urchins thrown upon the beach are quickly carried so high up that the waves cannot reach them, and thus they are not broken up. As soon as a shell is washed up on to the beach and the wave recedes, the water runs out of the shell, which is then so light that the following wave, which may come in farther, carries the shell in front of it together with

FIG. 82. Sandy beach at the head of Magens Bay (Great Northside Bay) on the North side of St. Thomas, looking towards the East (12.05). Farthest to the right is the *Coccoloba*-Formation, which here consists almost entirely of *Coccoloba uvifera*, only here and there with *Caesalpinia crista* (in the foreground to the right) and a few other species. The lowest branches of the *Coccoloba* rest on the sand, and often strike root. The sandy beach outside the *Coccoloba*-Formation is narrow (4–7 metres broad), and in places the waves often reach right up to this formation, in which are found here and there some species of the *Pescaprae*-Formation (*Sporobolus virginicus*, *Ipomaea pescaprae*, *Canavalia obtusifolia*), but a real *Pescaprae*-Formation is not developed here. Only right in the foreground of the picture does the *Pescaprae*-Formation begin to show itself, represented here solely by *Sporobolus virginicus* with widely creeping subterranean runners. At definite intervals these runners throw up leafy shoots. Farther towards the West a *Pescaprae*-Formation gradually develops (see Fig. 83). To the right of the *Coccoloba*-Formation is found an extensive more or less wet area of ground covered by the *Conocarpus*-Formation. This ground is formed of the innermost portion of the bay, which was at one time cut off from the sea by the sand-bank now occupied by the *Coccoloba*-Formation, and was afterwards filled up by sand and other materials, partly washed in from the sea and partly washed down from the surrounding high ground. In the back of the picture, where the white sandy beach bends to the left and comes to an end, the rocky shore of the east coast begins, and at that point we reach Fig. 81.

FIG. 83. Sandy beach at the head of Magens Bay (Great Northside Bay) on the North coast of St. Thomas, farther towards the West than Fig. 82, but like this figure seen looking eastwards (12.05). The beach is here somewhat broader than in Fig. 82, and besides the *Coccoloba*-Formation (to the right) we have a marked *Pescaprae*-Formation consisting especially of *Sporobolus virginicus*, among which *Ipomaea pescaprae* and *Canavalia obtusifolia* are seen, and abutting on the *Coccoloba*-Formation we see *Euphorbia buxifolia*. Between the *Pescaprae*-Formation and high water-mark (the lowest corner to the left) there is a narrow belt without vegetation inhabited by tunnel-boring Staphylinid beetles.

FIG. 84. The West coast of St. Croix outside the 'William' Plantation seen looking North (1.06). Farthest to the right the *Coccoloba*-Formation consisting especially of *Coccoloba uvifera*, *Hippomane mancinilla*, and *Caesalpinia crista*; then the *Pescaprae*-Formation, in which are seen *Sporobolus virginicus* and *Ipomaea pescaprae* of which one shoot is seen almost to reach the sea. The dark line along the edge of the beach farther back is the edge of a chalky sandstone of recent origin.

other light materials, e.g. algae and leaves belonging to the 'Sea Grass' Formation. The shells of the sea-urchins are therefore found in large numbers among the strips of seaweed, where they gradually become filled with sea-borne or wind-borne sand.

On the big stones at the side of the bay we find a dense vegetation of algae, among which a number of small animals live; especially beautifully coloured limpets, and magnificent Actinidians. In the warm tropical night the water gleams with luminous animals vying with the reflection of the constellations.

On the beach in the inner portion of the bay there is found nearest the sea a weakly developed *Pes-caprae*-Formation consisting of *Ipomaea pes-caprae*, *Sesuvium portulacastrum*, *Sporobolus virginicus*, and *Canavalia obtusifolia*, with scattered *Tournefortia gnaphalodes*. To the landward of these plants there is a small *Coccoloba*-Formation mixed with species from the scrubby vegetation of the neighbouring slopes.

Essentially the same state of affairs as prevails in the inner part of Popilleaus Bay, viz. a comparatively sandy beach between the sea and the higher land, is found in many other places in the Danish West Indies; but the sandy beach is often broader, and the *Pes-caprae-* and *Coccoloba*-Formations more richly developed; as for example in John Bruce Bay and other points on the south side of St. Thomas, and at the west end of St. Croix. Fig. 84 shows the sandy beach with the *Pes-caprae-* and *Coccoloba*-Formations as they appear at St. Croix in the neighbourhood of the 'William Plantation'. The dark lines at the edge of the beach are the margin of calcareous rock recently formed of the materials of the sandy beach which have become cemented together into a firm rock. A calcareous rock of this kind is found at many places on the coast of St. Croix.

In calmer bays less exposed than Popilleaus Bay a Mangrove-Formation (Figs. 85–8) has usually arisen on the alluvial soil beside the water, and in course of time this vegetation has contributed to the formation of a broader, low beach; because mud as well as organic and inorganic matter which have been washed down are retained and become heaped up among the prop roots of the Mangroves. But the Mangrove-Formation is not the first formation to arise here. First of all we have plants of the *Coccoloba-*, *Conocarpus-*, and *Pes-caprae*-Formations, especially *Conocarpus erectus* and *Coccoloba uvifera*, which were previously components of the vegetation of the rocks at the foot of which the beach was formed. As soon as the washed-down materials and washed-up sand and stones have given rise to even quite a narrow beach, this is immediately occupied by the two species mentioned. *Conocarpus erectus* has the advantage over *Coccoloba uvifera* in being able to grow nearer the sea, indeed close to its edge, so that it can occupy the low beach which is first of all formed. Thus in the portions of the protected coast which the Mangroves have not yet reached

Conocarpus erectus as a rule forms the outermost margin of the vegetation along the coast. On soil which is higher and drier within this margin there is often some *Coccoloba uvifera*, and here and there in addition other plants of the higher sandy beach, e.g. *Caesalpinia crista, Sporobolus virginicus, Canavalia obtusifolia, Ipomaea pes-caprae*, &c.

If the conditions on the coast allow *Rhizophora* to grow, and if it be present, then things will alter; the typical species of the exposed, sunny, sandy beach will gradually be repressed. This can be seen in Otters Creek and Water Creek in Coral Bay on the Island of St. Jan. In some places where the belt of Mangroves, for a reason unknown to me, was low, and even interrupted, the beach contained some species characteristic of sand; but where the Mangroves formed a high and dense vegetation these plants had disappeared from the sandy or stony ground along the foot of the rocks, and the species of the *Conocarpus*-Formation had taken their place, while higher up on the rocks *Coccoloba* was growing. The plants of the sunny sandy beach are light-loving plants, and it is in all probability especially the shade cast by the Phanerophytes of the Mangrove- and *Conocarpus*-Formations that oust the typical representatives of the *Coccoloba*- and *Pes-caprae*-Formations.

In most places the lie of the country is such that the bays are continued inland as more or less pronounced narrow valleys or depressions running up among the surrounding mountains. The valley bottoms are occupied by stony stream beds or 'Guts' which, apart from certain streams on the Island of St. Croix, are quite dry for the greater part of the year, or contain only a few pools in the basins formed by the rocks. But during the rainy season raging mountain torrents dash down through these guts, carrying various kinds of material out into the bays. It is this material and the sand washed up from the sea that cause the beaches in the middle of the bays gradually to become broader than they are at the sides of the bays. Where this increase in breadth takes place evenly, the fresh masses of sand being laid down just outside the older masses, the vegetation essentially resembles that just described. Where the coast is exposed and the sea can bring the sand higher up, so that the newly formed ground is loose and dry, we see a *Pes-caprae-Coccoloba*-Formation. But in the calmer bays, where the newly formed ground is lower and wetter and often made up of finer materials, we see, in addition to a Mangrove-Formation, a *Laguncularia-Conocarpus*-Fringe, and a *Conocarpus*-Formation gradually merging into the scrub on the hillsides.

In the most exposed situations the new ground is formed partly or wholly of clay and fine particles of lime, the sea at times washing over the low expanses carrying with it clay and limey mud, which gradually raise the height of the ground. But because rain also often washes coarser materials out over these plains, and because during rough weather sand may be washed in from the sea, the ground may differ

FIG. 85. The East end of St. Jan (2.06). The Mangrove-Formation along the East side of Kolhale Bay in Water Creek.

FIG. 86. The East end of St. Jan (2.06). The outermost margin of the Mangrove-Formation on the North side of Otters Creek. At the surface of the water bunches of mussels are seen on the Mangrove-Roots. Bordeaux Hill in the background.

Fig. 87. Great Cruz Bay at the West end of St. Jan (2.06), illustrating the interior of a very luxurious *Rhizophora mangle* thicket growing on wet, but not submerged, ground.

Fig. 88. The East end of St. Thomas (5.06). View looking over Yersey Bay from a point near Nadir. The islands and the dark fringe along the coast are Mangrove-Formations. The high ground in the foreground of the picture is covered by a scrub of *Croton*, in which *Pilocereus Royenii* and *Agave* are seen.

widely both in height and consistence, so that it is not difficult to understand how conditions arise suited to a mixture of the species of the *Conocarpus-*, *Coccoloba-*, and *Pes-caprae*-Formations. In one of the small bays near Bovoni within Yersey Bay (Fig. 88) on the Island of St. Thomas the sequence was as follows:

1. Out in the sea a luxuriant community of Mangroves.
2. On the muddy edge of the beach a belt of *Avicennia* with *Laguncularia* 4–5 metres high; the ground almost completely covered by the breathing roots of *Avicennia*.
3. Rather wet even plain of clay, in parts completely bare, in parts with *Batis maritima*.
4. Slightly higher up the sandy ground bears *Sporobolus virginicus* and scattered individuals of *Heliotropium curassavicum*, *Acacia Farnesiana*, *Antherylium Rohrii*, and *Rhacoma crossopetalum*.
5. Xerophilous vegetation of Micro-phanerophytes on the adjacent low hills.

The three first facies belong to the *Mangrove*-Formation. No. 4 is the *Conocarpus*-Formation; but in this circumscribed area *Conocarpus* itself is absent; it is, however, plentiful on similar ground in the neighbourhood.

At other places the accretion does not always take place in immediate connexion with what is already there, the materials not always being laid down on or parallel to the coast. The sea often throws up a sand-bank across a bay, either across its mouth, e.g. Krauses Lagoon on the Island of St. Croix, or farther in, e.g. Magens Bay on the Island of St. Thomas and in Great Cruz Bay on St. Jan. Even if a sand-bank like this has at one time been coherent, completely shutting off the bay, it will later become broken through, when during the rainy season so much water comes down into the portion of the bay cut off by the sand-bank that the sand becomes breached at one or more places. In this way there arises a lagoon with calm water. The lagoon gradually becomes filled up with washed-down materials and with clay and limey mud brought in by the sea, so that the bottom, first along the shore and later everywhere, is raised so high that it is only occasionally submerged. We find such lagoons in the Danish West Indies in all possible stages of silting and drying. I do not know whether this process of silting is solely responsible for the filling up of the lagoons, or whether gradual elevation of the islands also plays a part. The relatively considerable height of an apparently old bed of Krauses Lagoon suggests elevation.

Finally in many places the portion of a bay cut off from the sea by a sand-bank is without any connexion with the sea, because, from the lie of the country, during the rainy season sufficient water is not carried down to break through the sand-bank; but the sea can at times, at any rate in many cases, wash over the sand-bank. The littoral lakes or 'Salt Ponds' thus formed gradually become filled with materials washed down into them. Some of them are now so shallow that during the dry season one can walk over the flat, plantless clay. An example of this is a salt pond in Smith's Bay on St. Thomas. Since I have investigated the conditions of only a few of the numerous salt ponds in the Danish West Indies, I do not know whether all of them are cut off from the sea merely by a sand-bank.

Krauses Lagoon and Westend Salt Pond on Sandy Point, which I shall presently describe in more detail, represent respectively the stage of a lagoon open towards the sea, not filled up, and in the salt pond stage. I shall therefore here describe only two examples of the lagoon entirely or completely filled up.

Great Cruz Bay on the Island of St. Jan. This bay is continued inland into a low region which has almost certainly itself once been a lagoon, but which has now been filled up to such an extent that the ground is higher than the sea and without a trace of water except in the lowest part of a water-course, through which the rain-streams run into the sea. Such at any rate was what I found there in February 1906. The filled-up lagoon is separated from the sea by a sand-bank 30–50 metres broad with a belt of pebbles along the beach. Only at a single place, viz. the water-course described, was there a little *Ipomaea pes-caprae*; apart from that the outermost vegetation consisted of a dense belt of *Coccoloba uvifera*. The following plants were also found abundantly on the sand-bank: *Acacia Farnesiana, Caesalpinia crista, Canella alba, Colubrina ferruginea, Pithecolobium unguis-cati, Erithalis fruticosa, Argythamnia candicans, Stigmatophyllum periplocifolium, Solanum racemosum*, and, especially down towards the water-course, *Laguncularia racemosa*.

Of the country within the sand-bank, the part lying immediately inside the bank is lowest, and the ground here is composed chiefly of organic materials, leaves, and twigs, derived partly from the luxuriant vegetation which grows here, and partly from the vegetation of the surrounding heights. Although the surface of the ground was somewhat above the water-table, yet a dense thicket of *Rhizophora mangle* grew here, forming the tallest and most luxuriant vegetation I saw anywhere in the Danish West Indies. It was 9–15 metres high with extraordinarily numerous and robust prop roots (Fig. 87) of which the uppermost were given off at a height of about 9 metres. The ground was covered by a thick layer of fallen leaves, the upper ones being dry and the lower ones more or less damp. Among these somewhat damp leaves there was a rich fauna, including a species of Amphipod, and also a fish about 4 cm. long belonging to the family *Cyprinodontidae*. Professor H. Jungersen has identified this fish as *Haplochilus Hartii* Boul. When we raked up the leaves we saw this wriggling fish moving very nimbly. It was difficult to catch, especially because of the rapidity with which it hid beneath the leaves. Since there was no trace of liquid water among the leaves we must suppose that this fish, like the Amphipod, is adapted to breathing and living in moist air.

Within the *Rhizophora* vegetation, where the ground is a little higher and less rich in organic components, there follows a belt of *Avicennia nitida* attaining 10 metres in height, with particularly numerous breathing roots 15–30 cm., and sometimes even 50 cm. long. On the higher, drier, and more sandy ground within the *Avicennia* there was open but high and luxuriant vegetation of *Bucida buceras* (up to about 18 metres high), as well as *Antherylium Rohrii, Acacia Farnesiana, Andira jamaaicensis, Melicocca bijuga, Ficus populnea,* &c., &c.

Magens Bay (Great Northside Bay) on the north coast of St. Thomas. The innermost portion of this bay resembles that of Great Cruz Bay, being cut off from the sea by a sand-bank to form a lagoon, which in the course of time has become filled in, so that now but little water is found and that only in a small area through which the rain-streams run out into the sea. The sand-bank is covered by a dense *Coccoloba*-Formation, which in some places approaches the sea so closely that there was only 1–1·5 metres between the *Coccoloba* vegetation and the line reached by the waves, which here broke up a shelving slope 3–5 metres broad. No trace of the *Pes-caprae*-Formation (Fig. 82) was found here. At most places, however, the distance between the *Coccoloba*-Formation and the sea was greater, though not exceeding a few metres, and in these situations a narrow *Pes-caprae*-Formation (Fig. 83) formed of the aerial runners of *Ipomaea pes-caprae* and of the subterranean stolons of *Sporobolus virginicus* was found. *Euphorbia buxifolia, Canavalia obtusifolia,* and *Cenchrus echinatus* also grew here. Close to the line reached by the waves there were in several places mounds of dry loose sand up to a few millimetres high formed by small Rove Beetles, which were found in the sand. The same thing is seen on Danish sandy beaches, e.g. on the west coast of Fanø, where small tunnelling beetles play a part in the formation of dunes, the little mounds of sand made by the beetles drying much more quickly than the surrounding sand which is on the comparatively firm ground. It is first and foremost these small mounds of sand carried away by the wind that contribute to the growth of new dunes.

In the *Coccoloba*-Formation were found, besides *Coccoloba uvifera, Caesalpinia crista, Dalbergia hecastophyllum,* and climbing *Canavalia obtusifolia.* On land within the sand-bank there are imperceptible gradations between submerged ground and high and dry, more or less sandy ground with luxuriant vegetation consisting of species belonging to the *Conocarpus*-Formation.

KRAUSES LAGOON

Krauses Lagoon, or Anguilla Lagoon as it is also called, lies on the south side of St. Croix, and is the largest lagoon in the Danish West

Indies. It is formed of a bay 3 kilometres broad, separated from the sea by a sand-bank that is breached in several places where the lagoon is connected with the sea. This lagoon is of particular interest, not only on account of its size, but also because it has recently undergone important changes, the luxuriant Mangrove-Formation which formerly grew here being entirely destroyed by a hurricane in 1899. In order to obtain a fixed point for our investigations of subsequent developments, I shall here give a description, as far as my investigations suffice, of the condi-

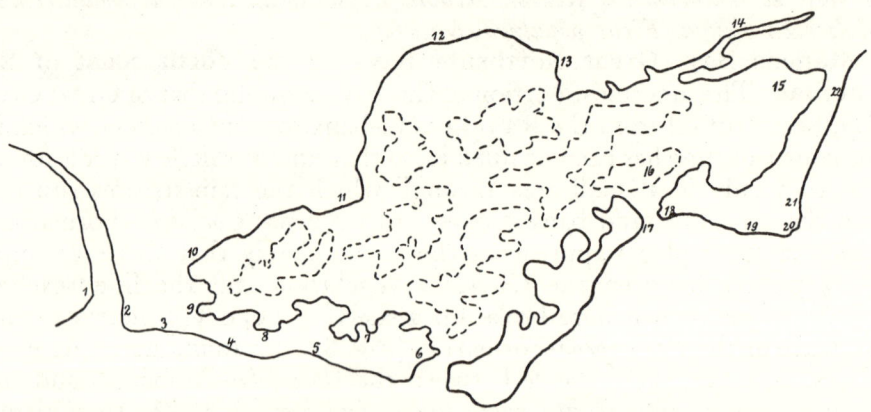

Fig. 89. Krauses Lagoon according to Oxholm's map of St. Croix made in 1828. The areas surrounded by dotted lines are in Oxholm's map designated as islands of woodland (Mangrove). These islands have all now been destroyed or removed, as seen in the western portion of the lagoon (see Fig. 91), or in its eastern portion, where dead stems are still present (see Fig. 92). For numbers see text.

tions prevailing when I visited the spot in January and February 1906. In making this description I have felt the need of a reliable map; the entries on Oxholm's map of St. Croix disagree in several points with what we now see, and these entries may never have been quite accurate, as, for example, in dealing with the extent of the Mangrove islands that were formerly scattered about in the lagoon. But in the absence of anything better I have been obliged to use this map in the following description, and for the main features it may suffice. According to this map and to the short descriptions given, first by Eggers and then by Børgesen, Krauses Lagoon was fringed by Mangrove vegetation, and there were several Mangrove islands out in the lagoon. Eggers writes (loc. cit., p. 20), that the lagoon 'is densely overgrown with Mangrove scrub and is about to become filled up'. Børgesen writes (loc. cit., p. 28): 'here too I have seen the most extensive Mangrove vegetation. Near a little brook in the western portion of the lagoon I visited a wood composed of pure *Rhizophora*... rather a tall wood casting a dense shade, under whose canopy, as far as the eye could reach, nothing but aerial roots were seen.' Børgesen gives a photograph of the lagoon taken from

LIFE-FORMS OF PLANTS ON NEW SOIL

the higher country, showing Mangrove vegetation not only along the coast, but also Mangrove islands out in the lagoon. All this is now completely changed.

Outside the opening of a narrow bay through which Kinghill Gut runs out into the sea there is on the west side (No. 1 in Fig. 89) Mangrove vegetation, and above this on the dry ground a low vegetation of Micro-phanerophytes. At the point exactly corresponding on the east side of the bay (2 in Fig. 89) the loose sandy ground is cut away to form a steep edge about $1\frac{1}{2}$ metres high. The ground to the east of this between 2, 3, and 9, and northwards, is covered with a formation of Micro- and Nano-phanerophytes whose essential components are species of *Cactaceae*, *Croton*, and *Acacia*. Many other species are found, for example *Haematoxylon campechianum*.

On the low narrow sandy peninsula separating the west portion of Krauses Lagoon from the sea the composition of the vegetation varies widely according to the height of the ground. The ground slopes evenly on both sides, especially towards the lagoon; moreover the height of the central part of the sand-bank varies in different places. In corresponding habitats along our Danish coasts we find under similar conditions a series of marked facies in the vegetation, the species arranging themselves almost entirely according to the varying humidity determined by the height of the ground. In Danish localities the illumination is essential uniform, the plants being of about the same height; at any rate none of them shade the others to an important degree. This is not so in the habitat before us in the West Indies at Krauses Lagoon. Of course the position occupied by the species is determined here too by the demand they make on soil humidity; but since a large number of the species are comparatively tall trees able to shade other species, illumination plays a much more important part than it does on Danish beaches. Where the varying height of the ground brings about many changes within short distances, the larger plants, at any rate in part, displace the smaller ones from ground they could otherwise occupy, so that the division into facies is not so conspicuous and not so finely demarcated as in Denmark, where the materials are finer, i.e. the individuals of the species are smaller and do not differ so greatly in size.

Along the seaward side of the sand-bank there is so much shelter from the trade winds that a Mangrove vegetation is able to thrive in the shallow water along the coast even a good way out in the sea. Such a Mangrove vegetation is not seen on the coast exposed to the trade winds along the sand-banks that separate the eastern part of the lagoon from the sea.

Already at 3 (see Fig. 89) we see small Mangrove islands and isolated young Mangrove plants off the coast, which is flanked by a beach of coral sand 1–2 metres broad. Here and there on this beach we find

Sporobolus virginicus and *Sesuvium portulacastrum*, and in some places on the water's edge are small individuals of *Laguncularia racemosa* and *Avicennia nitida*. These plants are the beginning of the *Laguncularia-Avicennia*-Facies, which is found farther out along the edge of the beach. Then follows on somewhat higher ground a *Sporobolus virginicus*-Facies (some metres broad) with scattered plants of *Heliotropium currassavicum*, *Batis maritima*, *Borrichia arborescens*, *Ipomaea pes-caprae*, and *Canavalia obtusifolia*, and after that a low Phanerophyte vegetation which in this locality is formed essentially of *Hippomane mancinilla*. Inside this higher sandy ground is a depression whose bottom, consisting of sand and clay mixed in varying proportions, is densely covered with *Sporobolus virginicus*. In the lowest parts of the depression there is a little water with *Ruppia rostellata*, and along the edge there is a belt of *Philoxerus vermicularis*. This depression seems to be a former continuation of the bay of the lagoon marked 9 in the map, and one is prompted to suppose that Kingshill Gut once ran out here into the lagoon. I have, however, had no opportunity of investigating the ground more thoroughly. The thorny scrub, which in places was impenetrable, offered a formidable obstacle to such an investigation.

At 4 the coast is flanked by a broad belt formed chiefly of *Conocarpus erectus*. Here and there the ground is so low that some water lies on it. In other places the ground is a little higher, and bears an open vegetation of *Batis maritima, Sesuvium portulacastrum, Salicornia ambigua, Heliotropium curassavicum*, and *Sporobolus virginicus*. On the side of the lagoon (8) there are large expanses of moist ground formed of coral sand and calcareous mud either with a vegetation of *Salicornia* and *Batis* or altogether devoid of plants.

At 5 and 7 going from the sea towards the lagoon we meet the following:

1. Outside the coast in the shallow water, *Rhizophora mangle*-Facies, formed of small Mangrove islands (Fig. 90) and many *Rhizophora* seedlings.
2. Along the edge of the beach a fringe of *Rhizophora mangle, Avicennia nitida, Laguncularia racemosa*, and *Conocarpus erectus*, the latter on the inside.
3. The somewhat higher sand-bank with *Hippomane mancinilla, Dalbergia hecastophyllum, Borrichia arborescens* (very plentiful), *Sporobolus virginicus, Euphorbia buxifolia, Scaevola Plumieri, Lantana odorata*, and *Batis maritima*.
4. In towards the lagoon the ground becomes lower and moister, bearing an *Avicennia*-Facies with *Batis*.
5. *Batis*-Facies growing on wet calcareous mud. In some places *Salicornia* is more plentiful than *Batis*. Isolated plants of *Avicennia* and *Sesuvium*.

FIG. 90. The seaward side of the sand-bank bordering Krauses Lagoon towards the South-west seen from 5 (see Fig. 89) looking towards the North-west. Along the coast *Laguncularia racemosa* and *Avicennia nitida* with breathing roots. Out in the sea an Island of partly dead Mangrove (4.2.06).

FIG. 91. The North side of Krauses Lagoon between 11 and 12 on the map (Fig. 89) seen looking East over the northern portion of the lagoon. At the edge of the lagoon are seen stumps of the Mangrove Vegetation, which was killed in the hurricane of 1899 and which has now been cut away. Young living plants of *Avicennia nitida* and *Rhizophora mangle* are also seen. Inside the lagoon there is a wide expanse of clay fissured by the drought bearing groups of *Batis maritima* and *Salicornia ambigua* (2.06).

6. Extensive tracts without vegetation. The ground, consisting chiefly of calcareous mud, is becoming fissured through drought. Isolated seedlings of *Rhizophora*.
7. Lagoon with shallow water. At one place higher ground projects into the lagoon with large bushes of *Avicennia* and *Laguncularia*.

Finally, outside the outlet at 6 (see map), the following succession occurred from the sea in towards the lagoon:
1. *Rhizophora mangle* and *Avicennia nitida* in the water.
2. Low beach with *Laguncularia* and *Batis*.
3. Slightly higher ground with *Borrichia arborescens, Conocarpus erectus*, and *Sporobolus virginicus*.
4. *Sporobolus* with *Batis maritima, Sesuvium portulacastrum, Philoxerus vermicularis, Capraria biflora*, and *Canavalia obtusifolia*.

Then inside towards the lagoon:
5. *Conocarpus erectus*-Facies.
6. *Laguncularia+Avicennia*-Facies.
7. *Rhizophora*-Facies in the water.

Apart from seedlings, and apart from the poor Mangrove vegetation on the inner side of the sand-bank bordering the lagoon on the southwest, there occur in the western portion of the lagoon neither living Mangrove vegetation nor remains of that destroyed in the hurricane of 1899, which formerly fringed the lagoon and formed islands out in its waters. It is said that the Mangroves then destroyed were cut up and used for firewood. Along the north side of the lagoon stumps were still to be seen in the water. Isolated young seedlings of *Rhizophora mangle*, the first indication of a future Mangrove vegetation, were also seen here and there.

Within 10–11–12 (see map) the hurricane sea has eaten its way into the somewhat higher ground surrounding the lagoon. The boundary here, therefore, is sharp and formed by an abrupt edge up to a metre in height (at 10) but much lower towards the east. Between this edge and the water in the lagoon there are low flat expanses up to several hundred metres in breadth covered with clay and sand in varying proportions; and these, at any rate towards the west, are entirely devoid of plants. Towards the east the edge becomes lower and lower, disappearing almost entirely at 12; at the same time a low vegetation begins to appear here and there on the expanses mentioned above.

Between 11 and 12 (see Fig. 89) the succession from the lagoon landwards was as follows (Fig. 91):
1. The lagoon with isolated tree stumps in the water and isolated young plants of *Avicennia* and *Rhizophora*: towards the land small islands with *Batis maritima*. The ground was covered with water, but it is apparently sometimes dry, for it was divided up by cracks into 5–7-angled areas, 5–20 metres in diameter.

2. Land vegetation of *Batis* about 20 metres in breadth, dense outwards and open inwards, ultimately consisting of scattered individuals. Here and there *Batis* is mixed with *Salicornia ambigua*. On the outer margin of the *Batis* vegetation there is often found a large quantity of *Sesuvium portulacastrum* and scattered young plants of *Avicennia nitida*. In many places the ground is cracked into polygonal areas.
3. A flat expanse of clay 50–80 metres broad, devoid of vegetation. The ground, which is not cracked, appears to be a little lower than at 2.
4. Somewhat higher ground, 180–250 metres broad with an open vegetation of *Sporobolus virginicus*. The ground is visible between the plants. There is also a large quantity of *Batis maritima*, which in the lowest places is absolutely dominant. There are also bushes of *Conocarpus erectus* about a metre high and *Evolvulus nummularius, Capraria biflora, Lippia nodiflora, Cynodon dactylon, Opuntia tuna, Sesuvium portulacastrum, Heliotropium curassavicum, Portulaca oleracea*, and *Salicornia ambigua*. These, as one sees, form a motley assemblage of species; but there is as yet little or no competition. Left to itself this ground would almost certainly gradually become covered with Phanerophytes of the *Conocarpus*-Formation like No. 5.
5. Vegetation of *Conocarpus* 2–4 metres high growing on ground covered with *Sporobolus virginicus*. Some species from No. 4 also present, e.g. *Evolvulus nummularius* and *Heliotropium curassavicum*.
6. Higher ground, in all probability an old raised bottom of a lagoon with xerophilous scrub of *Croton, Acacia*, &c.

Farther towards the east, from 12 and eastward, a vegetation of Phanerophytes, the innermost facies of the Mangrove vegetation, fringes the shore (*Avicennia*+*Laguncularia* Facies). Outside this all the eastern portion of the lagoon is for the most part bestrewn with dead remnants of the vegetation which previous to 1899 occupied large parts of the lagoon. Only the stems with their thick branches remain, the tallest stems scarcely exceeding 4 or 5 metres in height. All these stems belonged, as far as I could see, to *Avicennia*: I saw no stems with prop roots. Presumably *Rhizophora* was a component of this vegetation; but perhaps it was not so well able as *Avicennia* to withstand the hurricane.

At 13 the following succession occurred, from out in the lagoon passing inwards:

1. The lagoon with dead *Avicennia*: isolated plants of young *Rhizophora* and, especially towards the land, young *Avicennia*.
2. Higher ground formed chiefly of calcareous mud covered by shallow water, and in parts without vegetation: in parts an open vegetation of *Batis* with scattered groups of *Avicennia* 1–3 metres high.
3. *Avicennia*+*Laguncularia*-Facies with *Batis*. Here and there the

ground covered by water or very wet and bearing no vegetation. *Sporobolus* is seen in the inner part, where the ground is a little higher.

4. Higher but fairly wet ground, like 3, composed essentially of limey mud with a dense growth of *Sporobolus virginicus*. *Fimbristylis spadicea* forms numerous large tussocks, some exceeding 1½ metres in height. In places *Juncellus laevigatus* and *Bacopa monniera*, both abundant. This *Sporobolus* vegetation grows on ground that is covered in other places by the Phanerophytes of the *Conocarpus*-Formation.

This community of *Sporobolus*, which is also found on a similar soil in other places, is almost certainly an artifact caused by cutting down trees and bushes in order to favour grazing. That this was so here is made clear by the presence of some tree stumps and of some large trees of *Hippomane mancinilla*.

On the higher drier portions some other species were found, especially *Sida ciliaris*, *Stylosanthes hamatus*, *Evolvulus nummularius*, *Capraria biflora*, *Heliotropium curassavicum*, *Acacia Farnesiana*, and a few grasses. To the landward of this ground, a little higher up, there is cultivated land with sugar cane.

Farther to the east, towards 14, the strip of ground between the lagoon and the cultivated land narrows considerably; within the *Avicennia-Laguncularia*-Facies we find here only a very narrow *Conocarpus*-Formation with *Hippomane mancinilla*, *Pluchea odorata*, and a few other species.

In the eastern portion of the lagoon the vegetation varies greatly, because the spring tides have here cast up large quantities of sand and mud partially filling up the lagoon. We see here every transition from ground covered with shallow water to elevated land occurring especially as tongues of varying width stretching from the east into the lagoon and covered, at any rate partially, with a low Phanerophytic vegetation of *Conocarpus*, *Pluchea odorata*, *Borrichia arborescens*, *Capraria biflora*, &c. The expanses that are covered with water are either devoid of vegetation or else bear young plants of *Avicennia* and *Rhizophora*. The plants of *Avicennia* are larger and were apparently the first comers. Where the ground is a little higher and not covered with water, but yet wet and soft, there are extensive communities of *Batis* and *Salicornia*; but here also we see large flats devoid or nearly so of vegetation.

In describing what is seen along the low broad sand-bank which bounds the lagoon towards the south-east I shall begin with the outlet at 18 (see map). Here too large quantities of sand and mud are carried into the lagoon, filling it up to such an extent that extensive tracts remain dry, enabling us in some places to walk out into the dead *Avicennia* wood (Fig. 92). On the lowest submerged portions we see here scattered young plants of *Avicennia*, and on the somewhat higher ground there are groups of *Batis, Salicornia,* and *Sesuvium* varying in size. At this point, now *Batis* and now *Salicornia* extends farthest out on the low ground; *Sesuvium* prefers the slightly higher levels. The uppermost layer of sand was coloured green by blue-green algae.

I have not investigated the ground between 6 and 18 (see map), not because of any difficulty in getting there, for the channel between 17 and 18 was almost entirely blocked with sand, but because I had no time. The ground is here covered with a low Phanerophytic vegetation, which as far as I could see from 6 and 18 scarcely differs in essentials from the vegetation on the sand-bank bordering the lagoon towards the south-west. At 6 there was a luxuriant Mangrove vegetation and at 17 there were small groups of *Rhizophora* standing in the shallow water off the coast.

Along the coast eastwards from 18 no trace of Mangroves was seen. The higher portion of the sand-bank covered with low Phanerophytes is highest and broadest towards the east, becoming towards 18 lower and lower, finally passing gradually into the low flat land between 16 and 18, which is poor in vegetation. But at various places on the sand-bank there are lower portions, so that the height of the ground varies greatly. Because of these differences in level we find that in some places the species of the *Coccoloba*-Formation dominate and in others those of the *Conocarpus*-Formation. The vegetation here was not entirely killed by the hurricane. There were a good many dead individuals especially between 18 and 21; but there were some which were not killed outright and had been able to produce fresh shoots.

At 18, outside the old sand-bank, land formation is at present taking place by the heaping up along the old coast of a series of low flat sand-banks which in part are devoid of vegetation and in part have scattered *Sporobolus virginicus, Stenotaphrum americanum, Sesuvium portulacastrum,* and here and there plants of *Heliotropium curassavicum*. In places the rhizomes of the *Sporobolus* have become exposed, probably by the high tides washing the sand away from them. The bare apices of the rhizomes could here be seen growing down again into the sand, just as those of *Carex arenaria* and *Heleocharis palustris* do under similar conditions in Denmark. Towards the east the new formation becomes narrower and narrower, but it increases slightly in height. In addition to the above-mentioned plants *Borrichia arborescens* and small individuals of *Laguncularia* are found here.

At 19 (see map) where the land formation had entirely ceased, the following is what was observed from the seaward side landwards (Fig. 93):

1. Along the coast a strip bearing masses of the washed-up leaves of *Cymodocea* and *Thalassia*.
2. A strip of vegetation 3–4 metres broad consisting of *Sporobolus, Stenotaphrum,* and *Heliotropium curassavicum*, with seedlings of *Hippomane mancinilla, Laguncularia,* and *Suriana maritima*.

FIG. 92. The destroyed *Avicennia* wood in the eastern portion of Krauses Lagoon looking towards the North-west from 16 (see map Fig. 89). Mount Eagle is seen in the background and the plantation of Spanish Town shows as a white spot in front of the eastern portion of that mountain. In the foreground to the right, *Salicornia ambigua*; to the left, *Batis maritima*. In the dead wood are seen scattered young living bushes of *Avicennia* (see the dark patches farthest to the right) (6.2.06).

FIG. 93. The seaward side of the sand-bank that bounds Krauses Lagoon on the South-east, seen looking towards the North-east from 21 on the map (Fig. 89) and showing the head of Lime Tree Bay. Nearest to the sea is a belt barely 2–3 metres broad with washed-up leaves of Sea-Grasses; then a (slightly) higher strip 1–2 metres broad with scattered plants of *Sporobolus virginicus* and *Ipomaea pescaprae*; then a scrub of *Suriana maritima*. Farther back we see the outermost fringe of a *Conocarpus erectus* scrub, which here almost reaches the sea (6.2.06).

3. A fringe about 2 metres broad of *Laguncularia* about 1½ metres high.
4. A strip about 6 metres broad covered almost exclusively with *Sporobolus* and *Stenotaphrum*.
5. Low scrub of *Borrichia arborescens* with a little *Suriana maritima*, *Dalbergia hecastophyllum*, *Tournefortia gnaphalodes*, and *Ipomaea pes-caprae*.
6. Slightly higher ground about 30–35 metres in breadth bearing principally *Laguncularia*, besides which the following were also present: *Conocarpus*, a little *Coccoloba uvifera*, *Hippomane mancinilla*, *Borrichia arborescens*, and *Ipomaea pes-caprae*.
7. A depression in which *Salicornia* and *Batis* were the most important plants.
8. Higher ground: an old sand-bank covered with scrub: then the lagoon with higher tongue-like bits of land projecting from the east. Towards the west these tongues of land break up into little islands bearing low scrub.

At a single point (20 on the map) the sea is at present cutting away the coast, making an abrupt edge bearing scrub on the top. At 21 *Suriana maritima* forms the outer margin of the scrub over a long distance. This was what was seen at this point:

1. A beach 2–3 metres broad with washed-up leaves of *Cymodocea* and *Thalassia*.
2. A strip 1–2 metres broad, slightly higher, with *Sporobolus* and scattered *Ipomaea pes-caprae*.
3. *Suriana maritima* Scrub.

From 21 towards the head of Lime Tree Bay the bank of the beach becomes higher and drier and a scrub is formed, principally of *Lantana odorata*, *Corchorus hirsutus*, *Coccoloba uvifera*, *Hippomane mancinilla*, *Conocarpus erectus*, and scattered *Scaevola Plumieri*. The scrub almost reaches the sea, from which it is separated by a narrow strip of washed-up leaves of sea-grasses. Near the head of Lime Tree Bay beyond 22 the depression covered with *Salicornia* and *Batis* (No. 7 in the conspectus above) passes gradually into a grass-covered expanse with *Sporobolus* and *Stenotaphrum*, and this expanse here nearly reaches the sea. On the higher parts are found scattered bushes, and in particular low stunted scrub of *Borrichia arborescens*.

I have tried to describe what was seen at Krauses Lagoon in the year 1906. That my picture is neither complete nor clear I am fully aware. Circumstances are so diverse and complicated that it is difficult, even impossible, to write a complete and clear description. I felt therefore that the best thing I could do was to give a detailed description of what was to be seen at a series of definite places, and I hope that these descriptions together with the photographs may serve as a starting-point for investigations of changes which are bound to occur in the development of the vegetation.

Because of the extensive changes that have recently taken place there is scarcely any place on Krauses Lagoon where the original conditions remain wholly unchanged or where the various facies of the formations occurring are all present in their original form. The following conspectus of the formations on the protected coast is therefore not a description of what is seen at a single place, but an attempt to make clear the mutual relationships of the various facies, and the presumable correspondence between these facies and those of our own beach vegetation.

The Tidal Belt; Mangrove-Formation. The term tidal belt needs more accurate definition, since this belt does not, at any rate does not always, correspond exactly with the area lying between the highest and lowest water marks. This is especially so where, as in the Danish West Indies, there is but little difference between high and low tide. In those regions the Mangrove-Formation occupies in part ground that is always submerged and in part ground which the water only exceptionally covers, but which is always wet.

The term tidal belt signifies, therefore, not merely the area between the highest and lowest tides, but also the ground constantly covered by shallow water, as well as that only exceptionally submerged, but which, nevertheless, is constantly wet and saline. As I have said before, because of the coarseness of the materials the coastal divisions in the Danish West Indies are not so definite as they are in Denmark. But where the divisions are most clearly seen three facies can be distinguished, viz.

The Rhizophora-Facies. This is the outermost facies containing only *Rhizophora mangle*. Then follows:

The Avicennia-Facies in quite shallow water and on ground that is wet, but not always submerged. This facies consists of *Avicennia nitida* either alone or in company with *Rhizophora mangle*.

The Laguncularia-Facies. This is the innermost facies of the Mangrove-Formation. It usually occurs as a small strip along the coast, and most commonly consists of *Laguncularia racemosa* and *Avicennia nitida*. To landwards it abuts upon and mixes with species of the *Conocarpus*-Formation, especially with *Conocarpus erectus*, which extends farthest to the seaward. Where this Phanerophytic vegetation is not so dense as to shade the ground entirely, Chamaephytes are often seen, e.g. *Salicornia ambigua*, *Batis maritima*, and *Sesuvium portulacastrum*. The more open the Phanerophytic vegetation is, the denser the Chamaephyte vegetation becomes. Where the Phanerophytes have disappeared or nearly disappeared, for example where they have been cut down, we obtain, therefore, a more or less Chamaephytic vegetation such as is seen, as described, over large expanses along the north side of Krauses Lagoon. This vegetation closely resembles the *Salicornia* vegetation of our salt marshes; but in correspondence with the great difference in climate the life-forms differ widely from ours. The ground we are now discussing, the innermost portion of the tidal belt, corresponds in the Danish climate to the localities whose plants are particularly badly situated during the unfavourable season, the winter. Corresponding with this, the only Danish species growing in this locality, *Salicornia herbacea*, has the best protected of all life-forms, i.e. that of the Therophyte. In the West Indies the closely allied *Salicornia ambigua* is otherwise situated. Here its environment is at no season especially unfavourable to growth, and the life-form of the species present is correspondingly a less protected one, i.e. Chamaephytic. Apart from life-form there is scarcely any difference between the Danish *Salicornia herbacea* and the West Indian *S. ambigua*.

There is no reason to suppose that the boundaries between the facies of a formation in a definite climate will coincide with certain facies boundaries in the corresponding formation in an entirely different climate, whose species in general belong to other life-forms. Thus our *Salicornia*-Formation grows on approximately the same ground as the Mangrove-formation, but the outer boundaries of the two formations do not coincide. Because the typical species of the Mangrove-Formation are Phanerophytes, i.e. comparatively tall plants, they can, especially where there is no tide, extend much farther out into the water than our *Salicornia*, which under similar circumstances is almost confined to the wet ground along the coast. The outer boundary of our *Salicornia*-Formation corresponds most closely with the outer boundary of the *Laguncularia*-Facies of the Mangrove-Formation, which, as already mentioned, is also, in the West Indies, a *Salicornia*-Facies or a *Batis*-Facies, if the Phanerophytes are removed.

Judging by the ground and its humidity the innermost boundary of the Mangrove-Formation and our *Salicornia*-Formation correspond most closely. But if we compare the *Salicornia ambigua* vegetation partly covering the ground at Krauses Lagoon where the Phanerophytes of the Mangrove-Formation have been removed, with our *Salicornia*-Formation, it is seen that the *Salicornia ambigua* vegetation at Krauses Lagoon extends higher up than our *Salicornia*-Formation. This is almost certainly due to a difference in competition between the plants in these two places. With us the *Salicornia*-Formation is bounded to the landward by the *Glyceria maritima*-Formation. In the Danish West Indies the Mangrove-Formation is bounded by the *Conocarpus*-Formation; but these formations normally consist of Phanerophytes. But if the Phanerophytes be removed, and in the place of the innermost facies of the Mangrove-Formation (the *Laguncularia*-Facies) a Chamaephytic vegetation of *Salicornia ambigua*, with or without *Batis maritima*, has developed, then this vegetation abuts, as a rule, on the *Sporobolus* vegetation, which occupies at least a portion of the ground belonging to the *Conocarpus*-Formation when the Phanerophytes of the latter formation are removed. *Sporobolus virginicus* and other low plants which are found here do not, however, extend as far outwards as *Glyceria maritima* does with us. If, therefore, in the West Indies, *Salicornia ambigua* and *Batis maritima* extend higher up than *Salicornia herbacea* does with us, this is presumably because they have access, and do not encounter an overwhelming competition before they have extended so far that the moisture conditions alone put a period to their farther progress. In Denmark, on the other hand, *Glyceria maritima* extends into the territory in which *Salicornia herbacea* is still well able to grow if there is no competition, but where in competition with *Glyceria maritima* it succumbs. In places where there is no competition, because *Glyceria maritima* has not yet wholly covered the ground, we see *Salicornia herbacea* extending far into the domain of the *Glyceria*-Formation.

The ground of the *Conocarpus*-Formation corresponds most closely, as to the formations, with our salt marsh, i.e. with the *Glyceria*-Formation (excluding perhaps its outermost portion), the *Juncus Gerardi*-Formation, and the *Statice armeria*-Formation. The different species of the *Conocarpus*-Formation do not invade to the same extent the lower and moister ground. On this ground *Conocarpus erectus* usually predominates, and we see here, if the Phanerophytes are removed, as before mentioned, farthest outwards a *Salicornia*+*Batis* vegetation and farthest inwards a *Sporobolus* vegetation. The higher ground bears a motley assemblage of species, and some of the species of the xerophilous scrub begin to appear, but the conditions are so diverse that I do not think it worth while to attempt the making of different facies.

In undertaking a comparative investigation of the corresponding formations in two entirely different climates, i.e. that of Denmark and that of the Danish West Indies, in order to discover in what different ways the climates of the two regions find expression in the life-forms of the plants, we must first attempt to find out which formations correspond to one another for purposes of comparison. Since the floristic composition of the vegetation differs entirely, the species occurring will give us no help in this respect. We must, therefore, take as our starting-point a comparative investigation of the nature of the ground, and then compare the formations found on ground which corresponds in the two climates. In the case we are now considering there is correspondence between the alluvial beach formations influenced by salt water on protected coasts. As shown below I have attempted to compare those formations which are found on corresponding ground when we compare Denmark (Nordby Salt Marsh on Fanø) with the Danish West Indies (Krauses Lagoon). In dealing with Krauses Lagoon I have paid attention both

LIFE-FORMS OF PLANTS ON NEW SOIL

to the normal vegetation and also to that which is found if the Phanerophytes have been removed and have not yet again invaded the ground.

Krauses Lagoon on St. Croix		*Nordby Salt Marsh on Fanø*		
Normal vegetation.	Where the phanerophytes have been removed.			
Mangrove-Form. { Rhizophora-Facies / Avicennia-Facies / Laguncularia-Facies	Salicornia ambigua	*Salicornia herbacea-Formation*		Salt Marsh.
Conocarpus-Formation { Conocarpus erectus, &c. / Borrichia arborescens / Pluchea odorata, &c. / Acacia Farnesiana, &c.	Batis maritima / Juncellus laevigatus / Sporobolus virginicus, &c. / Stenotaphrum, &c.	Glyceria+Suaeda-Facies / Glyceria+Aster-Facies / Glyceria+Triglochin-Facies / *Juncus Gerardi-Formation* / *Statice armeria-Formation*	} *Glyceria-Form.*	

About this latter point I will add the following note:

Where Mangrove-Formation and *Conocarpus*-Formation are present as climax vegetation, as they are at the south-west margin of Krauses Lagoon, Phanerophytes of course predominate, and the species belonging to Chamaephytes and other life-forms play a subordinate part. The Chamaephytes *Salicornia ambigua* and *Batis maritima* are confined to the innermost facies of the Mangrove-Formation and the outermost portion of the *Conocarpus*-Formation, which consists chiefly of *Conocarpus erectus*. Where on the other hand the Phanerophytes have been removed, as they have been along the north side of the westerly portion of Krauses Lagoon, *Salicornia* and *Batis* are dominant and are able over wide expanses to form more or less dense vegetation, which, however, only represents an early stage of development of the innermost portion of the Mangrove-Formation and the outermost portion of the *Conocarpus*-Formation, a stage of development[1] that persists only as long as the Phanerophytes of these formations remain absent.

SANDY POINT

Sandy Point (see Fig. 94) is a peninsula formed of sand extending into the sea from the south-west corner of St. Croix for about 3 kilometres measured from the north-west extremity of Westend Salt Pond to the southern point of the peninsula. Its only connexions with the rocky coast are two small arms, a longer one on the west and a shorter one on the east. For the rest of its breadth it is separated from the coast by the above-mentioned shallow lake, Westend Salt Pond, which was formerly

---

[1] In a recent paper Dr. Børgesen describes this stage of development as an independent formation existing side by side with the Mangrove-Formation and the *Conocarpus*-Formation. (F. Børgesen, 'Notes on the Shore Vegetation of the Danish West Indian Islands', *Bot. Tidsskrift*, Bd. **29**, 1909.)

a lagoon, but which now probably, under ordinary circumstances, is entirely shut off from the sea. It is said, however, that stormy seas sometimes wash across the narrow strips of land separating the lake from the ocean on the east and west. Sandy Point thus forms a sharply bounded region especially suited to the study of the vegetation found in the West Indies on alluvial formations of this kind.

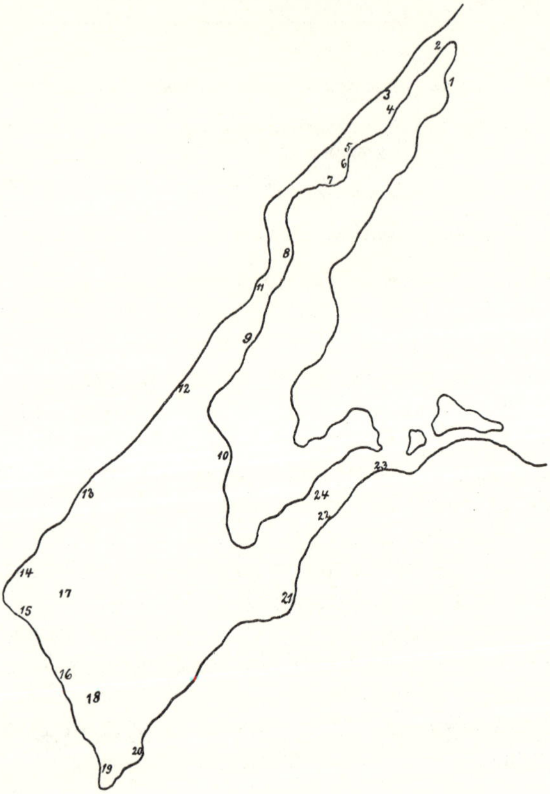

As in so many other similar localities this sandy peninsula has been subjected to changes. In some places it is increasing by the accumulation of sand along the coast, and in other places it has been washed away. It may, therefore, be of interest for future investigations to examine what now is to be seen there. Unfortunately we lack an exact map as a basis for our work. In the following account of my investigations of Sandy Point and its vegetation, as of those of Krauses Lagoon, I have had to content myself with a copy of Oxholm's map.

Sandy Point is formed exclusively of sand, the so-called 'Coral Sand', with the exception of parts by Westend Salt Pond, where some clay is mixed with the sand. But the peninsula is not dune country, resembling, for example, the west coast of Jutland. No considerable drifting of sand takes place, so that there is only a slight difference in the height of the ground above the sea at different places. The surface in all probability has the average height attained by the masses of sand thrown up at various times by the sea. The formation of the peninsula is presumably due to a southerly current along the west coast of St. Croix and a westerly current along the south coast. Where these two currents meet masses of sand, brought especially by the currents coming from the east, are

FIG. 94. Sandy Point with Westend Salt Pond on St. Croix according to Oxholm's map of St. Croix made in 1828. For figures see text.

deposited, so that the bottom becomes raised. Then during storms the sea throws up the sand along the coast forming long banks separated by more or less distinct depressions, and the whole system of banks is rounded off by the flowing back of the water and later, when the sand has dried, by the wind. If we then go from the beach inland at one of the places where the increase has most recently taken place we pass a series of broad flat banks separated by depressions a quarter to one metre deep. The surface here resembles that of the outermost beach on the west coast at the north end of Fanø,[1] where 2–3 flat sand-banks, at any rate at low tide, are exposed, separated from each other by flat depressions through which the tidal waves stream out towards the north. If we imagine this floor raised 1–2 metres then we have a state of affairs similar to what is seen on the west side of the south end of Sandy Point.

On Sandy Point, there are represented, although in a different degree, all the localities and the corresponding formations which in the Danish West Indies are found on alluvial beaches. When compared with Denmark, Sandy Point corresponds most closely with Skallingen. I shall later, therefore, compare the floras of Sandy Point and Skallingen. We shall thus be able to see how the life-forms in different regions are determined by the climate, and how even in such specialized and circumscribed tracts as the alluvial beach formations the life-forms are in entire agreement with the general biological spectrum of the climate they inhabit.

The protected coast and its formations are confined to the considerable expanse of land connecting Sandy Point with the southerly and westerly shores of Westend Salt Pond. Just as in the corresponding locality on the landward side of Skallingen, the ground here consists principally of sand, mixed here and there with more or less clay.

The Sea Grass-Formation is represented in Westend Salt Pond only by *Ruppia rostellata*.

The Tidal Belt, the Mangrove-Formation. As described in the section 'Krauses Lagoon', I understand by the term tidal belt not merely the region between the highest and lowest water-marks, but in addition to this both the perpetually wet and only exceptionally submerged ground; and also, where there is little or no tide, localities always covered by shallow water. To these latter localities belongs Westend Salt Pond (Figs. 95 and 96), at least for most of its extent, because its water is shallow, at any rate along the shores. The Mangrove-Formation that belongs to such conditions, is, however, but weakly represented; in particular the outermost characteristic facies, the *Rhizophora*-Facies, is absent; *Rhizophora mangle*, as far as I have seen, is not found in the Westend Salt Pond. The composition of the water can scarcely account

---

[1] *Vide* C. Raunkiær, *Vesterhavets Øst- og Sydkysts Vegetation*, København, 1889; Eug. Warming in *Bot. Tidsskrift*, **25**, København, 1903.

for this. I do not know exactly how much salt the water contains. Its appearance and temperature did not invite me to decide this question by tasting it. I suppose that *Rhizophora* is absent because it has never reached here. The *Avicennia*-Facies is also undeveloped, only the innermost facies of the Mangrove-Formation, the *Laguncularia-Avicennia*-Facies, is found over expanses of varying size along the southern and westerly shores, which are the only parts here considered. *Laguncularia racemosa*, *Avicennia nitida*, and usually *Conocarpus erectus*, form here a narrow fringe immediately by the water's edge. Where this fringe was not too dense *Batis maritima*, and in particular *Sesuvium portulacastrum*, grew, especially at the innermost margin of this vegetation. At other places, e.g. between 4 and 6, between 7 and 8, and at 10 (Fig. 94) the *Laguncularia-Avicennia*-Facies mentioned was not seen. Thus all the Phanerophytes of the Mangrove-Formation were absent. In general what was seen here corresponds with Fig. 96:

1. Farthest out a slightly sloping sandy beach 2–5 metres broad devoid of vegetation.
2. Then at the margin of higher ground an often quite narrow fringe of *Sesuvium*, with or without *Batis* and *Philoxerus vermicularis*. The *Sesuvium* extends farthest out. This *Sesuvium-Batis*-vegetation, which is here the only indication of a Mangrove-Formation, passes inwards into
3. A very dense *Sporobolus* vegetation reminding one vividly of the growth of *Agrostis alba* occurring on similar more or less dry sand, seen, for example, on the low ground on the inner side of Svenske Knolde on Skallingen.

In some places a few other low plants are mingled with the *Sporobolus* vegetation, especially *Sesuvium*, *Philoxerus*, *Tephrosia cinerea*, and *Canavalia obtusifolia*. At other places there are more or less densely interspersed low individuals of various Phanerophytes, especially *Conocarpus*, *Avicennia*, *Laguncularia*, *Borrichia arborescens*, *Coccoloba uvifera*, *Corchorus hirsutus*, and *Ernodea littoralis*, forming a *Conocarpus*-Formation intermixed with some species from the formations of the unprotected coast. Indeed the *Conocarpus*-Formation here marches with the *Coccoloba*-Formation on higher sandy soil between Westend Salt Pond and the sea. This corresponds entirely to what is seen at the north end of the dune region of Fanø, where the vegetation of the sandy western beach and of the low dunes meets and mixes with species from formations of the unprotected east coast.

The unprotected sandy beach; the Pes-caprae-Formation. I shall begin the following account of the sandy beach and its vegetation on the north-west coast outside the north end of Westend Salt Pond, and I shall then proceed along the coast to the south-east. The sandy beach of the north-west coast falls abruptly to the sea, and there is here a great

FIG. 95. The North end of Westend Salt Pond on St. Croix seen from 1 Fig. 94 (8.1.06). In the foreground a dense growth of *Sporobolus virginicus* with tree-stumps. Beside the water low bushes of *Laguncularia racemosa* to the right and *Avicennia nitida* to the left. The lake is, at any rate at this point, very shallow, and stones are seen above the surface of the water. On the right a tongue of dry land projects into the lake. In the background scrub on the narrow strip of land between the lake and sea.

FIG. 96. From the South end of Westend Salt Pond on St. Croix at 10 in Fig. 94 (8.1.06). From left to right: lake—line of spray—narrow sandy beach without vegetation—*Sesuvium portulacastrum*—*Sesuvium* and *Batis maritima*—*Sporobolus virginicus*, *Laguncularia racemosa*, *Conocarpus erectus*—*Coccoloba*-Formation.

difference in height between low-water mark and the line reached by the breakers. Between 2 and 11 (see map) the edge of the beach shows lime-sandstone, which is known in several other places on St. Croix. That this is a recent formation is shown by the products of civilization found in it. Identical lime-sandstone is widely distributed along the shore at the north end of Westend Salt Pond.

The sandy beach, which is bounded inwardly by a fringe of *Coccoloba uvifera*, dense, but only 1–2 metres high, is narrow between 2 and 11, attaining 10–15 metres in breadth, and almost entirely devoid of vegetation. Only here and there at the edge of the *Coccoloba*-Formation do we find scattered colonies of *Sporobolus virginicus*. At 12, however, the beach becomes wider, and from there to the south-west corner it increases steadily in breadth. At the same time the *Pes-caprae*-Formation becomes more luxuriant and spreads over the whole beach, except its outermost recent portion facing the sea. At 12 there also begins the low, broad, flat series of sand-banks mentioned above. They run parallel to the coast, and are separated from each other by depressions only $\frac{1}{4}$–1 metre deep. Usually the sand-banks slope evenly to the landward, but much more steeply towards the sea.

At 12, where the beach is about 25 metres broad, there is only one broad sand-bank running beside the sea, and within this there is a broad depression followed by slightly higher ground bearing the outermost *Coccoloba*-scrub, which is 1–2 metres high. We have here two facies in the *Pes-caprae*-Formation of the beach. Immediately outside the *Coccoloba* fringe there is a community of *Canavalia obtusifolia* about 15 metres broad mixed with varying quantities of *Ipomaea pes-caprae*. Then to the inner side of the sand-bank a belt of pure *Ipomaea pes-caprae* about 8 metres broad. Gradually as the beach broadens towards the south we find several depressions. At 13, where the beach is 50 metres broad, there are three, and at 14 where the beach is 80 metres broad, four such depressions. Conditions for the vegetation are essentially uniform at 13 and 14. I shall, therefore, confine myself to a description of what is seen at 14, where the beach is broadest.

Three facies in the *Pes-caprae*-Formation can here be distinguished. In describing these divisions I shall begin with the depressions. These occupy most of the ground and bear the most luxuriant vegetation, which extends up over the intervening flat sand-bank. I shall, therefore, delimit the four depressions by measuring from the crest of one bank to that of the following.

1. The first, outermost depression, about 12 metres broad, is entirely devoid of vegetation.
2. The second depression, which is about 16 metres broad, has an almost pure growth of *Ipomaea pes-caprae* (Fig. 97) with very long shoots lying on the surface of the sand. The older portions of these

shoots often become covered with blown sand. Of other species there was found only *Euphorbia buxifolia* and *Cakile lanceolata*.

3. The third depression, about 22 metres broad, is occupied by a very luxuriant and dense vegetation of *Canavalia obtusifolia* about ½ metre high, containing some *Ipomaea pes-caprae* and *Euphorbia buxifolia*. In some places the *Canavalia* was almost entirely covered with *Cassytha americana*, which was bright reddish-yellow and in the distance looked like *Cuscuta americana*, which, however, I did not see on Sandy Point.

4. Finally the fourth depression, which was about 30 metres broad, was covered, especially in its outermost portion, chiefly with luxuriant *Canavalia* vegetation (Fig. 98). In the innermost portion, bordering on the scrub, the *Canavalia obtusifolia* was less luxuriant; but this portion bore a dense vegetation of *Sporobolus virginicus* with *Cenchrus*, *Euphorbia buxifolia*, and a little *Ipomaea pes-caprae*.

5. Then followed a transitional formation about 30 metres broad representing facies No. 4 at a later stage of development, with a more or less dense growth $1\frac{1}{2}$–3 metres high of *Chrysobalanus icaco*, *Suriana maritima*, *Ernodea littoralis*, *Corchorus hirsutus*, and *Euphorbia linearis*.

Passing a little way inwards over the scrub-covered ground inside the sandy beach it became apparent that here also was present a system of sand-banks and depressions similar to but older than those on the beach. Here, just as amongst our Danish dunes, when we leave the young dunes and walk inland over the older ones we find that the vegetation of the older parts appears poorer and drier than that of the newly formed dunes nearest the beach.

A little way beyond the western corner, at 15, the coast forms a small bay, where the land is at present being eroded. The scrub, which here consists especially of *Coccoloba* and *Ernodea*, extends right on to the outermost angle of a slope which is undercut by the waves. Farther towards the east near 16 the dry land begins to increase again, and from here to the southern corner the beach is made up of low sand-banks and intervening depressions varying in breadth and running parallel with the coast. These sand-banks and depressions increase in number towards the south, while at the same time the beach becomes correspondingly broader.

Outside the south-west end of Sandy Point the sea is shallow. We see here in the water a large number of detached leaves of 'sea grasses', especially *Cymodocea*. This is one of the favourite fishing-places of the pelican. Numbers of these birds can be seen flying over the water, and constantly dashing down into the sea after their prey. The beach is covered by large masses of 'Tang', especially the leaves of *Cymodocea*

FIG. 97. Looking from the West corner of Sandy Point on St. Croix at 14 on the map Fig. 94 (24.1.06). The outermost facies of the *Pes-caprae*-Formation. The *Pes-caprae*-Facies is seen towards the North. In the background are the hills at the North-west corner of St. Croix. In front of these hills is the bay at Frederikssted, which lies at the innermost corner of the bay.

FIG. 98. From the Western corner of Sandy Point on St. Croix, 14 on the map Fig. 94 (24.1.06). The transition between the *Canavalia*-Facies of the *Pes-caprae*-Formation and the scrub, whose outermost portion consists here of *Chrysobalanus icaco*, *Suriana maritima*, and *Ernodea littoralis*. In the background are seen the heights at the North-west corner of St. Croix. See text.

Fig. 99. From the South coast of Sandy Point on St. Croix, at 20 on the map Fig. 94 (24.1.06). At this point the coast is being cut away.

Fig. 100. From the South-east coast of Sandy Point on St. Croix, 22 on the map Fig. 94 (26.1.06). The innermost facies of the *Pes-caprae*-Formation: the *Tournefortia*-Facies passing into the *Coccoloba*-Formation. Near the centre of the picture is a dome-shaped specimen of *Tournefortia gnaphalodes* and behind it a *Hippomane mancinilla*. The ground vegetation consists of the ordinary species of the *Pes-caprae*-Formation: *Ipomaea pes-caprae*, *Sporobolus virginicus*, *Canavalia obtusifolia*, *Euphorbia buxifolia*, and *Cenchrus echinatus*. They are mixed with an open growth of *Tournefortia*, *Hippomane*, *Scaevola Plumieri*, *Coccoloba uvifera*, and *Caesalpinia crista*.

and *Thalassia*, washed up by the sea. This is never seen on the north-west beach. Another difference between the two beaches is that on the south-west beach at 19 a few additional species are found. The beach and its vegetation form here a transition to what we see on the south-east coast, which I shall mention immediately.

At its broadest portion (19) the beach is about 75 metres broad, and shows a system of sand-banks up to eight in number and corresponding depressions. In some places the banks are fewer, two coalescing to form one. The outermost 3–5 depressions and their intervening banks are entirely devoid of vegetation; they are covered, especially nearest the sea, by large masses of cast-up weeds and shells, especially those of sea-urchins. Just as on the north-west coast the outermost vegetation is a *Pes-caprae*-Facies. Then follows a *Canavalia*-Facies, in which a number of other species are found, viz. *Ipomaea pes-caprae*, *Cenchrus echinatus*, *Sporobolus virginicus*, *Sesuvium portulacastrum*, *Euphorbia buxifolia*, *Cakile lanceolata*, and *Tournefortia gnaphalodes*. The first-mentioned species grow more luxuriously in the depressions. The three last, especially *Tournefortia gnaphalodes*, occurred chiefly on the sand-banks. *Tournefortia* is not found on the north-west coast, but is abundant on the south-east coast. Then follows the inner depression with a dense and almost pure growth of *Sporobolus virginicus*, a *Sporobolus*-Facies. From here the ground slopes gradually up towards the scrub. On this slope were found, besides some *Sporobolus*, especially *Canavalia obtusifolia* and a little *Ipomaea pes-caprae*. On the slope was a wreck of a ship partly buried in the sand. Here too was seen in the outermost portion of the scrub-bearing ground an alternation of banks and depressions. The scrub was rather open, consisting chiefly of *Coccoloba uvifera*, *Chrysobalanus icaco*, *Suriana maritima*, *Ernodea littoralis*, *Dodonaea viscosa*, and *Dalbergia hecastophyllum*. *Sporobolus virginicus* also occurs and masses of *Cassytha americana*. Of the Phanerophytes mentioned *Dalbergia* extends farthest out. It is met with in the *Canavalia* vegetation, occurring as low but vigorous individuals giving off long procumbent branches from the lower portion of the stem.

As soon as we reach the south-east coast the conditions are found to change. On the southern part of this coast (at 20 in Fig. 94) the sea is at present cutting away the land, which ends in an abrupt edge undercut by the water, bearing the ordinary scrub vegetation of low Phanerophytes, some of which have fallen down on to the beach and are partially buried in the sand (Fig. 99). Farther towards the north-east the coast is not at present being eroded, but it is not being added to. Immediately to the landward of a narrow beach the ground suddenly rises and is covered with scrub, at whose outer margin some of the ordinary beach plants were found, especially *Ipomaea*, *Canavalia*, *Sesuvium*, and *Sporobolus*. The beach is like this nearly as far as the south end of the Westend

Salt Pond. From here towards the north-east along the narrow strip of land separating Westend Salt Pond from the sea, accretion is again taking place, or has at any rate recently taken place. We see here a broad portion between the sea and the scrub covered especially by plants of the *Pes-caprae*-Formation. This new land is narrow towards the south-west, where it begins. From here it increases evenly in breadth to near its middle portion, and then becomes narrower towards the north-east. In the middle this ground is 50 metres broad. To the inside of a beach about 2 metres broad there is a strip of washed-up leaves of *Cymodocea* and *Thalassia* 1–2 metres broad. Then follows a portion about 22 metres broad formed of three low sand-banks and corresponding depressions covered with *Ipomaea pes-caprae, Sporobolus virginicus, Sesuvium portulacastrum, Cakile lanceolata,* and *Euphorbia buxifolia*. According to the changing of the dominant species from the shore landwards the following facies can usually be distinguished. Farthest out a *Sesuvium*-Facies, in which *Sesuvium portulacastrum* dominates the other species. Then a *Pes-caprae*-Facies, and a *Sporobolus*-Facies, in which *Ipomaea pes-caprae* and *Sporobolus virginicus* respectively dominate. Here and there are seen scattered individuals of *Scaevola Plumieri* and *Tournefortia gnaphalodes*.

Then follows a low flat bank up to 10 metres broad bearing the species that are found in the facies just mentioned; but here the most conspicuous objects are broad dome-shaped groups of *Tournefortia gnaphalodes* (Figs. 100–1), a *Tournefortia*-Facies.[1] This is a transition to the *Coccoloba*-Formation; scattered species of the outer margin of this formation are already to be seen, especially *Hippomane mancinilla, Coccoloba uvifera,* and *Caesalpinia crista*.

Finally inside the bank covered with the *Tournefortia*-Facies there is some low ground forming a depression up to 14 metres broad (Fig. 101), bearing chiefly *Sporobolus* and *Ipomaea*; *Euphorbia buxifolia, Canavalia obtusifolia, Cenchrus echinatus,* also occur, and here and there scattered groups of *Tournefortia* and scattered individuals of *Scaevola Plumieri*. This depression as well as the bank lying outside it becomes narrower and narrower towards the north and south, finally disappearing altogether. The depression then passes into the narrow part along the beach covered with the *Pes-caprae*-Facies.

Compared with the north-west coast the south-east coast is characterized by a richer vegetation composed of more species, by the absence of a marked *Canavalia*-Facies, and by the presence of a narrow *Sesuvium*-Facies at the outer margin of the *Pes-caprae*-Formation, as for example along the west coast of Westend Salt Pond. Besides these points the

---

[1] This is given by F. Børgesen, 'Notes on the Shore Vegetation of the Danish West Indian Islands', p. 236, as an independent formation equivalent to the *Pes-caprae-* and *Coccoloba-*Formations.

Fig. 101. From the South-east coast of Sandy Point on St. Croix, at 22 on the map, Fig. 94, looking towards the North-east (26.1.06). To the left the *Coccoloba*-Formation, to the right, the *Tournefortia*-Facies, in which is seen a specimen of *Hippomane mancinilla* (the dark bush). In the centre of the picture is a broad shallow depression bearing only the usual species of the *Pes-caprae*-Formation, i.e. *Ipomaea pes-caprae*, *Canavalia obtusifolia*, *Sporobolus virginicus*, *Cenchrus echinatus*, and *Euphorbia buxifolia*.

Fig. 102. The scrub—*Coccoloba*-Formation—on Sandy Point on St. Croix, at 24 on the map Fig. 94 (8.1.06). A growth ¼–1 metre high of *Coccoloba uvifera* and *Ernodea littoralis*. *Lantana involucrata*, *Chrysobalanus icaco*, *Erithalis fruticosa*, and *Canella alba* are also present.

FIG. 103. The Scrub—*Coccoloba*-Formation—on Sandy Point on St. Croix, 18 on the map Fig. 94 (26.1.06). To the left: prostrate shoots of *Chrysobalanus icaco* and *Ernodea littoralis*; farther back: *Coccoloba uvifera*. In the sand: *Sporobolus virginicus* and young plants of *Canavalia obtusifolia* attacked by *Cassytha americana*, whose stems grow over the sand from one host plant to another.

FIG. 104. The Scrub—*Coccoloba*-Formation—on Sandy Point on St. Croix, at 17 on Fig. 94 (24.1.06). Low scrub of *Ernodea littoralis*; open patches with *Sporobolus virginicus*, &c.; in the background scattered plants of *Coccoloba uvifera*.

vegetation transitional between the *Pes-caprae*-Formation and the *Coccoloba*-Formation is characterized by *Tournefortia gnaphalodes*.

The Coccoloba-Formation. With the exception of the comparatively narrow strip along the coast covered with *Pes-caprae*-Formation, Sandy Point is covered by a scrubby *Coccoloba*-Formation, which, besides the usual components of this formation, contains also a number of species usually absent from, or at any rate not collected in, the circumscribed bits of vegetation which compose the *Coccoloba*-Formation in other parts of the Danish West Indies. As a rule the scrub is dense, and since it often contains thorny species, especially *Caesalpinia crista*, it is often almost impenetrable. Only in places is this vegetation more open (Figs. 103 and 104). In some places *Coccoloba* and *Ernodea littoralis*, either separately or together, form a growth only ½–1 metre high. This is especially well seen on the long stretch of land between Westend Salt Pond and the south-east coast (Fig. 102). But the scrub is usually 2–4 metres high, and only here and there do individuals exceed this height. The species composition varies greatly; now one and now another species predominates. *Coccoloba uvifera* is often the dominant plant; but in other places it is *Ernodea littoralis* (Fig. 104), *Chrysobalanus icaco* (Fig. 103), *Lantana involucrata*, *Erithalis fruticosa*, or *Euphorbia linearis*, which form the bulk of the vegetation. Of species occurring abundantly, at any rate in patches, may be mentioned: *Corchorus hirsutus*, *Croton betulinus*, *C. discolor*, *Rhacoma crossopetalum*, *Elaeodendron xylocarpum*, *Colubrina ferruginea*, *Bumelia obovata*, *Jacquinia armillaris*, *Convolvulus jamaicensis*, *C. pentanthus*, *Ipomaea triloba*, *Tecoma leucoxylon*, *Anthacanthus spinosus*, *Cordia nitida*, and *Clerodendron aculeatum*. Of small plants on the floor of the scrub there are only a few individuals and species. The most important are: *Bulbostylis pauciflora*, *Fimbristylis ferruginea*, *Mariscus brunneus*, *Sporobolus virginicus*, *Kallstroemeria maxima*, and, especially along a road that has been cut through the scrub, *Pectis humifusa*, *Stylosanthes hamata*, *Dactyloctenium aegyptiacum*, *Stenotaphrum americanum*, and *Eragrostis ciliaris*.

The following list includes those species which I have found on Sandy Point as this region was demarcated above. Besides the species given I saw at least one species, a grass resembling *Psamma arenaria*, which I have also seen in a few other places in the Danish West Indies; but as it never had flowers I was not able to determine it. I seem to remember a few other plants too, but as they are represented neither in my list nor my collections I have not included them. I feel sure future investigators will add to the list; but the fact I consider of greatest importance, the preponderance of the Phanerophytes over the other life-forms, will rest unshaken. In the list I have given the life-form of every species. Some of the Phanerophytes, however, occur on Sandy Point in another, better protected life-form than that assumed in other parts of the Danish West Indies. For these plants I have added the life-form occurring at Sandy Point in brackets. The life-forms and other statistics I have dealt with in former works.[1] The flowering plants found on Sandy Point and their life-forms.

	Life-form
Epidendrum papilionaceum Vahl	E
Bulbostylis pauciflora (Liebm.) C. B. Clarke	H
Fimbristylis ferruginea (L.) Vahl	H
Mariscus brunneus (Sw.) C. B. Clarke	H
Sporobolus virginicus (L.) Kth.	G
Dactyloctenium aegyptiacum (L.) Willd.	Th
Stenotaphrum americanum Schrank	Ch
Eragrostis ciliaris (L.) Lk.	Th
Cenchrus echinatus L.	Th
Coccoloba diversifolia Jacq.	M
,, uvifera (L.) Jacq.	M
Philoxerus vermicularis (L.) R. Br.	Ch
Batis maritima L.	Ch
Pisonia subcordata Sw.	M
Sesuvium portulacastrum L.	Ch
Cassytha americana Nees.	E
Cakile lanceolata (Willd.) C. G. Schulz	Th
Canella alba Murr.	M (N)
Melochia tomentosa L.	N
Malachra capitata L.	N
Corchorus hirsutus L.	N
Croton betulinus Vahl	N
,, discolor Willd.	N
Argythamnia candicans Sw.	N
Euphorbia buxifolia Lam.	Ch
,, linearis Retz.	N
Hippomane mancinilla L.	M
Kallstroemeria maxima (L.) W. et A.	Ch
Castela recta Turp.	N
Suriana maritima L.	N
Comocladia ilicifolia Sw.	M (N)
Dodonaea viscosa L.	N
Stigmatophyllum emarginatum (Cav.) Juss.	N
,, periplocifolium (DC.) A. Juss.	M (N)
Elaeodendron xylocarpum DC.	M (N)
Rhacoma crossopetalum L.	N
Colubrina ferruginea Brongn.	M (N)

[1] Raunkiær, C., *The Life-Forms of Plants and their bearing on Geography*, *The Statistics of Life-Forms as a basis for Biological Plant Geography* (Chapters II and IV of this volume).

LIFE-FORMS OF PLANTS ON NEW SOIL

	Life-form
Chrysobalanus icaco L.	M (N)
Caesalpinia bonduc (L.) Roxb.	N
„ crista L.	N
Canavalia obtusifolia (Lam.) P. DC.	N (Ch)
Dalbergia hecastophyllum (L.) Taub.	M (N)
Stylosanthes hamata (L.) Taub.	Ch
Tephrosia cinerea (L.) Pers.	Ch
Acacia Farnesiana (L.) Willd.	M (N)
Leucaena glauca (L.) Bth.	M (N)
Pithecolobium unguis-cati (L.) Bth.	M (N)
Turnera ulmifolia L.	N
Bucida buceras L.	MM (M)
Conocarpus erectus L.	M (N)
Laguncularia racemosa Gärtn.	M (N)
Eugenia buxifolia (Sw.) Willd.	N
„ axillaris (Sw.) Willd.	M (N)
Bumelia obovata (Lam.) DC.	N
Jacquinia armillaris Jacq.	M (N)
Convolvulus jamaicensis Jacq.	N
„ pentanthus Jacq.	N
Ipomaea pes-caprae (L.) Sw.	Ch
„ triloba L.	N
Solanum racemosum L.	N
Capraria biflora L.	N
Tecoma leucoxylon (L.) Mart.	MM (M)
Anthacanthus spinosus (L.) Nees	N
Cordia nitida Vahl	M (N)
Heliotropium curassavicum L.	Ch
„ parviflorum L.	Ch
Tournefortia gnaphalodes (Jacq.) R. Br.	N
Avicennia nitida Jacq.	M (N)
Clerodendron aculeatum (L.) Griseb.	M (N)
Lantana involucrata L.	N
Stachytarpheta jamaicensis (L.) Vahl	Ch
Echites suberecta Jacq.	M (N)
Forestiera segregata (Jacq.) Kr. et Urb.	N
Erithalis fruticosa L.	N
Ernodea littoralis Sw.	N
Randia aculeata L.	M (N)
Scaevola Plumierii (L.) Vahl	N
Borrichia arborescens (L.) DC.	N
Pectis humifusa Sw.	Ch
Pluchea odorata (L.) Cass.	N

THE LIFE-FORM OF PLANTS ON NEW SOIL

A glance at the list of plants growing on Sandy Point shows us at once that Phanerophytes predominate. A closer investigation proves that this small area, containing as it does only 80 species, has a biological spectrum agreeing as closely as could be wished with the Danish West Indies taken as a whole, although Sandy Point is geologically a new piece of land consisting of soil differing widely from that prevailing in the Danish West Indies. This leads us to the question whether within the same climate there is any essential difference between the biological spectrum of the flora of geologically new ground and old, and to what extent migration of new species into a flora influences its biological spectrum.

It is true of every flora that new species enter from time to time. This is particularly so where numbers of species are introduced by cultivation both into new and into old floras. But most of these introduced species can only remain if assisted by cultivation. Such species must of course be left out of the biological spectrum, since they can scarcely be regarded as an expression of the plant climate, but are rather the result of cultivation, so that their inclusion, at any rate sometimes, will mask what is essential in the biological spectrum. This is particularly true of Therophytes, which usually form an essential component of the introduced plants attached to cultivated land, and which actually, because of their life-form, are adapted to live on ground that is tilled every year.

But amongst the migrants or introduced plants of a country some are able to maintain themselves without such help, because they are able to compete successfully with the indigenous plants. Such species may be said to have become naturalized, and must be included in the biological spectrum of the country they inhabit. The important question, however, is not how many or how few species have become naturalized in this way, but whether these plants alter essentially the original biological spectrum. If they do not do this the number of migrants does not concern my bio-geographical method. If the newly arrived species do not alter the biological spectrum this is a further proof that the biological spectrum is a true expression of the plant climate, because it shows us that the distribution amongst life-forms of even the newly arrived species follows the same law as the species of the original flora.

In cultivated countries where the original flora, or at any rate the physiognomy of the vegetation, is entirely altered, and where a large number of new species are introduced, we must not of course expect that these new species will be without effect on the biological spectrum. But we must not conclude from the alteration they produce that the biological spectrum is any the less a perfect expression of the plant climate; since the decision which of the new species must be looked upon as naturalized, and thus sharing in determining the spectrum, depends partly on individual judgement. As in similar cases, it is possible that some species will be included that perhaps ought not to be, and vice versa. As a general rule too many species will be included.

In general what occurs is that the naturalized species do not bring about any essential alteration in the biological spectrum, and especially that they never mask the characters in the biological spectrum which make it characteristic of the plant climate. At one point only in the spectrum have the species a notable influence, and that point is the Therophyte percentage. It is well known that in most places Therophytes make up a much greater part of the migrants and introduced species than of the other species of the flora.

The cause of this I shall not discuss here. In deciding which of the new species must be regarded as naturalized, in all probability a good many more Therophytes will be included than should be, so that, at any rate to some extent, the new species will somewhat increase the Therophyte percentage of the biological spectrum. But even where this is so the increased Therophyte percentage will never in any case hide what is characteristic in the spectrum of the plant climate. For example, even if the percentage of Therophytes in Denmark is for this reason computed too highly, it will never, if the flora is again left to itself, become so high as to reduce the Hemicryptophyte percentage beneath the figure characteristic of the Hemicryptophyte climate. To do this we should, in considering which Therophytes must be looked upon as naturalized in Denmark, have to reckon some 100 species more Therophytes than those found in Denmark on uncultivated soil.

We have sufficient material to enable us to decide in what degree the naturalized plants of a flora alter its biological spectrum. I shall here confine myself to one example, namely the invasion by European plants of that region of North America included in Britten and Brown's *Illustrated Flora*. This region belongs, as is clearly seen from the biological spectra in Table 1, Nos. 4–5, to the Hemicryptophyte climate. Only the most northerly portion extends into the boreal plant climate; but this fact has no bearing on our present problem. P. Klincksieck,[1] following Britten and Brown's *Illustrated Flora*, has given a list of European plants that have migrated into this region. We could now investigate which species have wandered into the smaller area of a single state and so find out how the denizens affect the spectrum of the original flora; but I have not yet made an investigation of this kind. It is possible, however, to pursue another path. It is a fact that most of the migrants into a country which have been able to maintain their life in competition with the native species, i.e. those becoming naturalized, have come from a country with the same plant climate as that of their new home. We can, therefore,

---

[1] Klincksieck, P., 'Les plantes d'Europe adventices ou naturalisées aux États-Unis d'Amérique, constatées à deux intervalles: 1832 et 1896', *Bull. Soc. Bot. Fr.*, **54**, 1907, pp. xxx–xlii.

investigate the question by determining the biological spectrum for the species which have come into the Hemicryptophyte climate of North America from the corresponding climate in Europe. If the spectrum of the migrants deviates markedly from the spectrum of the ordinary Hemicryptophyte climate, then, if the number of migrants bears a large proportion to the native species, the migration will alter the original spectrum. But if the spectrum of the migrants does not differ from that of the invaded flora, then no change will take place, however great may be the number of migrants.

Of the 370 or so European species which were found in 1896 naturalized in the eastern portion of the middle states of America, 216 belong to the Danish Flora. In general these are Central European plants. Now these species show the spectrum given in Table 1, No. 2, with 47 per cent. of Hemicryptophytes. This is thus a well-marked Hemicryptophyte climate spectrum, which, in spite of the small number of species, nevertheless shows the characteristic series of numbers for the other life-forms. The only number which deviates markedly from the normal is the percentage of Therophytes. About this point I have already remarked that in computing which species may be looked upon as naturalized it is very easy to reckon too many Therophytes, thus including a number of species that would disappear if cultivation ceased and the country was left to itself. But even if we accept this spectrum with its 32 per cent. of Therophytes, an invasion of this kind, however great it were, would never be able to alter the essential features of a Hemicryptophyte spectrum. It is the high percentage of Hemicryptophytes that is common to both spectra.

TABLE 1

	No. of species.	*Percentage distribution of the species among the life-forms.*									
		S	E	MM	M	N	Ch	H	G	HH	Th
1. European species found in Denmark which already in 1832 had invaded the eastern states of U.S.	57	..	..	..	2	2	3	58	7	..	28
2. Do. naturalized in 1896	216	..	..	0.5	4	3	4	47	6.5	3	32
3. Denmark[1]	1,084	..	..	1	3	3	3	50	11	11	18
4. Altamaha, Georgia[1]	717	..	..	5	7	11	4	55	4	6	8
5. South Labrador[1]	334	..	..	3	3	8	9	48	12	11	6

In the list of European species which were naturalized in 1896 in this region of North America P. Klincksieck has marked those species which according to Schweinitz[2] had already arrived in 1832. In the same way

---

[1] Chapter IV, pp. 119, 122.
[2] See Klincksieck, loc. cit., p. xxxi.

I have taken out the species which belong to the flora of Denmark, 57 in all, and given their spectrum in Table 1, No. 1. In spite of the small number of species this spectrum corresponds strikingly with the first spectrum. There is a surprising precision in the way that the life-form of the plants follows the climate: the biological spectrum founded on life-forms is, considering the complication of the factors, as adequate an expression of the plant climate as any one could possibly wish.

The number of introduced species able to persist in competition with the original flora is but small; and even if these species did not conform entirely with the spectrum of the original flora they would not alter the spectrum to any great extent. But we find no such lack of conformity. We find that the spectrum of the naturalized species always has its centre of gravity in the same place as the spectrum of the original flora. This further emphasizes the value of the biological spectrum as a biological expression of the climate, as a means of testing the climate biologically.

If it ever happens that species which have migrated and have become naturalized during the course of time alter the biological spectrum of a flora so as to make it indicate another plant climate, then the merest glance at the vegetation of the country where this has occurred shows us that its composition is no longer a product of nature but that it is maintained by cultivation.

I have described how the naturalized plants agree in the percentages of their life-forms with those of the flora they have invaded, so that the spectrum of this flora suffers no essential change. This too is made clear by the fact elucidated elsewhere[1] that all the investigated local floras in the same climate but in very various parts of the earth agree in their biological spectrum although both their original and adventive floras differ in their species composition.

It is not only immigration that can be regarded as altering a flora. Cultivation may banish and exterminate native species, and the biological spectrum might be supposed to be so changed by this means as to render it no longer a true expression of the plant climate. It is obvious to the meanest intellect that if all or most of the wild plants of a region are exterminated, leaving only cultivated plants and weeds, we cannot then expect such a product of cultivation to be an expression solely of natural factors. Such a flora is essentially determined by cultivation. But such a state of affairs plays no part in investigations of plant climates, where we deal not merely with isolated square kilometres, or even smaller areas, but with countries or portions of countries. In dealing even with such a small area as Denmark we find that the species which have disappeared or become exterminated during historical time have not altered a single figure in the biological spectrum of the country. In

[1] Raunkiær, C., 'The Statistics of Life-Forms', &c.

order to decrease or increase one number merely by 1 per cent., not only must 10–11 species disappear, but these species must all belong to the same life-form. In order to alter the biological spectrum of Denmark to such an extent that it no longer showed the high percentage of Hemicryptophytes characteristic of the Hemicryptophyte climate, 200 species would have to disappear; and these 200 species would all have to be Hemicryptophytes.

Even if Denmark's 6 per cent. of woodland were extirpated this would not, as Warming[1] supposes, alter the character of the biological spectrum. The extirpation of the woodland would not cause the disappearance of the species composing the woodland. My method is based on the number of species, not on the mass of individuals.

And finally even if the fantasy of the extirpation of trees and shrubs became a reality, this would entail merely a disappearance from the biological spectrum of Denmark's 7 per cent. of Phanerophytes; but their absence would not merely not decrease the characteristic feature of our biological spectrum, i.e. the high Hemicryptophyte percentage, but it would actually increase it by several per cent. And although the Phanerophytes were gone, the spectrum could be classified in no other way than that in which we at present classify it, namely as belonging to a Hemicryptophyte climate. We should have to conclude that this region must either be quite unfavourable to Phanerophytes, or rather that it must be a region from which the Phanerophytic species had been artificially removed. The low Chamaephyte percentage would at once teach us that the climate was originally neither the Arctic Chamaephyte climate nor the boreal climate.

But in dealing with plant climates we need not confine ourselves to a tiny area like that of Denmark. The cold temperate zone extends from the Atlantic Ocean to the Pacific Ocean in the old world, and the corresponding part of North America shows essentially the same biological spectrum, i.e. the Hemicryptophyte spectrum. And even if every tree and shrub were to be exterminated over thousands and thousands of square kilometres, this would not entail the disappearance of a single species from the area of this plant climate. At the moment I can remember no single species which has disappeared during historic times from the cold temperate Hemicryptophyte climate—in spite of the encroachments of cultivation. In order to alter the character of the spectrum of this Hemicryptophyte climate a number of species exceeding 2,000 would have to be exterminated, and these moreover would all have to be Hemicryptophytes.

We shall now occupy ourselves with the question whether there is any reason to suppose that there is an essential difference in the biological

[1] Warming, E., *Om Planterigets Livsformer*, 1908, p. 23.

spectrum in the floras of geologically old and new ground. I shall here refer to one of Warming's pronouncements which appears to stultify the use of my life-forms as an expression of plant climate. In mentioning the richness of the tropical rain forests in the Amazons and the poorness and uniformity of the woods of North Europe and Siberia Warming (loc. cit., pp. 23-4) writes: 'The reasons for this are in all probability two, firstly, perhaps, because the development of species, on account of the infinitely more favourable conditions of the tropics, proceeds much more quickly. This, however, I must own to be mere supposition. The other reason is that those woods are infinitely older than ours, which must have come into being after the Ice Age. Another factor here participates, namely the capacity of the species for migration, and also the struggle between species, which is still undoubtedly taking place.' And after pointing to the richness in species of certain parts of Australia and South Africa he adds: 'I see no necessity to suppose that the relationship between the life-forms is always the same in countries with the same climate but with floras of widely different age, and if this is not so, too definite conclusions should not be based on the relationship to the plant climates.'

If Warming sees 'no necessity to suppose that the relationship between the life-forms is always the same in countries with the same climate but with floras of widely different age', this is perhaps because he has not investigated this question more closely; nor does he attempt to show how the state of affairs differs from my statement of it. The problem is, however, less difficult to unravel than might be supposed, since in various climates we have old and young formations side by side, enabling us to discover whether there is any essential difference between the biological spectrum of the floras of these different formations. But even if materials for such an investigation were not available one would, in subjecting the problem to a thorough general consideration, arrive at a result corresponding with that of my investigations, which were founded on an attempt to place Biological Plant Geography on as exact a basis as possible. Before I go on to the description of a definite case I shall, therefore, subject the problem to a general survey of this kind, and for this purpose I shall choose the flora of Siberia mentioned by Warming. Warming indeed uses the expression 'woods', but I suppose that he means by this expression the flora as a whole, as it is only on this supposition that there can be any objection to my view, which does not concern itself in particular with woods, but with the flora as a whole.

The flora of Siberia is geologically young, and there are thus, according to Warming's opinion, reasons to suppose that the struggle between the species is still undecided, and that all the species able to grow in Siberia have not yet arrived there. No doubt this view is correct, for we constantly see new species in various parts of the country. But as I have already emphasized, this is not the important point. The important point is to make clear whether there is any reason at all to suppose that the migration of new species will alter what is essential in the biological spectrum of the indigenous flora, as long as the climate remains unchanged.

Let us first see whence Siberia can obtain her new species.

We know by experience that no tropical plants, or scarcely any, are hardy in the cold temperate zone, and still less will they be able to compete with plants of this region. An invasion of any importance from the Phanerophyte climate is therefore excluded and, as is well known, does not take place. If plants come into Siberia they must come from countries with essentially the same climate as Siberia: they must come from other Hemicryptophyte regions. But it is in these regions that we always find essentially the same distribution of species within the series of life-forms as is found in Siberia. If then the species which enter Siberia are to alter the biological spectrum these invading species must show a biological spectrum entirely different from that of their native land. Judging by the example given on p. 182, and by what we already know, there is no reason to suppose that this is the case. Moreover, if the spectrum of the original flora is to be altered sufficiently to make it indicate another plant climate, the Hemicryptophyte percentage of the invading species must fall from over 50 to under 30; and even then the invasion would have to be on such a scale that the number of the introduced species would far exceed those of the indigenous flora. There is nothing to show that anything of the kind has taken place either in Siberia or in any other country.

Warming (loc. cit., p. 24) thinks that the climate of Siberia is just as good a tree climate as that of the Amazon. The actual state of affairs is seen by examining, for example, the flora of the Yenisei Valley. To the north of Krasnojarsk this flora includes 926 species, which show the spectrum given in Table 2, No. 5. Of these 926 species 56 are Phanerophytes, so that the trees and shrubs in all amount to 6 per cent. If the spectrum of this flora should change in character from that of the Hemicryptophyte climate of Siberia to the Phanerophyte climate of the Amazon, over 1,200 species of trees and shrubs would have to migrate into the Yenisei Valley. It should be noted that besides the trees and shrubs plants belonging to other life-forms might come too, in which case the number of trees and shrubs would have to be even greater.

I shall now give some examples of floras geologically still younger than that of Siberia, namely the floras of some alluvial formations in different climates: that of Skallingen on the west coast of Jutland and Sandy Point on St. Croix.

Sandy Point and Skallingen. Sandy Point is not merely geologically a new formation, but in part at any rate it is new historically, since new formation of land is at present taking place there. But at the same time it is a formation whose soil differs entirely from the soil of the land of the coast beside which it has arisen; and since, looked at from the point of view of plant climate, it is a very small region, we should not be surprised to find a percentage distribution of life-forms differing from that prevailing in the Phanerophyte climate of the West Indies. But this is not so. Even such a small region as Sandy Point, which is so new and which differs so widely from the adjoining country, follows the law that the percentage distribution of life-forms among the invading species is determined by the climate and is an expression of the climate. Earlier in this book I have given the biological spectrum of St. Thomas and St. Jan. This spectrum is true for the Danish West Indies as a whole, so that we must compare with it the spectrum of Sandy Point. Table 3, Nos. 1–2, show the relationship between these two spectra as regards the classes of life-forms, mainly Phanerophytes, Chamaephytes, Hemicryptophytes, Cryptophytes, and Therophytes.

TABLE 2

	No. of species.	Percentage distribution of the species among the life-forms.								
		S	E	MM	M	N	Ch	H	G	HH	Th
1. Clova, Scotland,[1] under 300 metres	304	..	..	3	2	4	7	59	7	5	13
2. Denmark[1]	1,084	..	..	1	3	3	3	50	11	11	18
3. District of Stuttgart[1]	862	..	..	3	3	3	3	54	10	7	17
4. Obi Valley, 61°–66° 32′	265	..	..	2·5	2	7	7	50	10	10	11·5
5. The Yenisei Valley from 56° to the Arctic Ocean[2]	926	..	..	1	1	4	7	61	11	6	9
6. Amgun Bureja, north of the Amur River[3]	441	..	..	3·5	5	9	7	58	11	3	3·5
7. Southern Kuriles, south of the Straits of Keltoi[4]	231	..	..	3	6	8	8	55	12	2	6
8. N. America, Sitka[1]	222	..	..	3	3	5	7	60	10	7	5
9. „ James Bay[1]	268	..	..	3	3	7	7	53	10	7	10
10. Normal spectrum[1]	400	1	3	6	17	20	9	27	3	1	13

In order to compare this with the flora of an alluvial formation corresponding to Sandy Point, but in an entirely different climate, the Hemicryptophyte climate, I have added the corresponding numbers for Skallingen and for the flora of Denmark as a whole. I have chosen

[1] Chapter IV, Tables 5, 6 and 14.
[2] Plant list in: Scheutz, N. J., 'Plantae vasculares Jeniseenses', *Kgl. Sv. Vetensk.-Akad. Handl.*, vol. **22**, 1888.
[3] Plant list in: Schmidt, Fr., 'Florula Amguno-Burejensis', *Mém. de l'Acad. imper. des sc. de St. Petersbourg*, vii Sér., tome xii, 1868.
[4] Plant list in: Miyabe, K., 'The Flora of the Kurile Islands', *Memoirs of the Boston Society of Natural History*, vol. iv, 1886–93, pp. 203–75.

Skallingen for comparison with Sandy Point partly because Skallingen, both in its relationship to the land from which its flora came and in the nature of its soil, is one of the places in Denmark most resembling Sandy Point, and partly for the reason that of those districts which can be here considered Skallingen is amongst those whose flora I know best. Not only is there in other works, especially those of Warming[1], information about a number of species found at Skallingen, but I have myself several times, especially in 1896, investigated Skallingen in order to obtain as complete a list as possible of flowering plants. The result of these investigations is a list of 105 species showing the proportion between life-forms given in Table 3, No. 5. A glance at the table shows that in the statistics of its life-forms Skallingen agrees with Denmark just as Sandy Point agrees with the Danish West Indies as a whole. In the flora of Skallingen Hemicryptophytes predominate just as they do in Denmark, and on Sandy Point Phanerophytes are dominant just as they are in the Danish West Indies. It is worth noting that although the part of Skallingen bearing vegetation is not much larger than the corresponding part of Sandy Point, yet Skallingen has more species than Sandy Point.

TABLE 3

	No. of species.	Percentage distribution of the species among the life-forms.				
		Ph	Ch	H	Cr	Th
1. St. Thomas and St. Jan	904	61	12	9	4	14
2. Sandy Point	80	74	16	4	1	5
3. Krauses Lagoon	32	56	28	10	3	3
4. Denmark	1,084	7	3	50	22	18
5. Skallingen	105	1	7	47	19	26
6. Langlig (islet just E. of Skallingen)	108	3	9	50	18	20
7. Nordby Marsh (Fanø)	17	..	6	65	12	17
8. The Camargue (mouth of the Rhone)	233	10	8	23	11	39
9. Normal spectrum	400	47	9	27	4	13

Let us now examine more closely the distribution of the species among the classes of life-forms. In Table 4 I have given the biological spectra of Sandy Point and of the Danish West Indies. Two spectra are given for Sandy Point for this reason. Spectrum No. 2 is the spectrum shown by the species of Sandy Point when these species are counted as having the life-form they usually have in the Danish West Indies, and it is this spectrum which must first be compared with the general spectrum of the Danish West Indies if we wish to discover the relationship between the spectrum of a flora growing on new ground with that of one growing on old ground. It will be seen that the correspondence is as close as

[1] See *Botanisk Tiddskrift*, vol. **19**, pp. 73–80; vol. **25**, pp. 72–4.

could be wished in a flora consisting of only 80 species. But spectrum No. 3 is also very instructive. What actually happens is that because of the unfavourable soil conditions on Sandy Point, the same unfavourable conditions by which the dune region of Jutland differs from the neighbouring land, the Phanerophytic vegetation of Sandy Point is, in comparison with the Phanerophytic vegetation of the Danish West Indies, a dwarf vegetation. This is not, however, merely because lower Phanerophytes preponderate among the migrants, whose spectrum is given in No. 2, Table 4, but also because many of the Phanerophytes, which ordinarily belong to the tall or tallest Phanerophytes, are reduced to a lower class of life-form because of the poorness of the conditions on Sandy Point. All the species found on Sandy Point that are ordinarily Mesophanerophytes occur on Sandy Point as Microphanerophytes; and most of the species found on Sandy Point which are ordinarily Microphanerophytes occur on Sandy Point as Nanophanerophytes. Considering then the actual size of the plants, we obtain the spectrum given in Table 4, No. 3, which, besides showing the marked Phanerophyte climate, is also a striking expression, partly of the comparatively low precipitation of the climate, and partly of the unfavourable conditions on Sandy Point, due to the poorness of the environment caused by the effect of the climate on the soil.

TABLE 4

	No. of species.	Percentage distribution of the species among the life-forms.									
		S	E	MM	M	N	Ch	H	G	HH	Th
1. St. Thomas and St. Jan	904	2	1	5	23	30	12	9	3	1	14
2. Sandy Point (see text)	80	..	2·5	2·5	29	40	16	4	1	..	5
3. Sandy Point (see text)	80	..	2·5	..	7·5	62·5	17·5	4	1	..	5
4. Denmark . . .	1,084	..	..	1	3	3	3	50	11	11	18
5. Langlig . . .	108	..	..	..	..	3	9	50	14	4	20
6. Skallingen . .	105	..	..	..	..	1	7	47	16	3	26
7. Normal spectrum .	400	1	3	6	17	20	9	27	3	1	13

We must here consider the significance of the soil for the biological spectrum, a question that Warming (loc. cit., pp. 24–5) has raised as an objection to my bio-geographical view. He writes: 'There is finally still one factor which influences the relationship between the life-forms, namely the soil. This is beautifully seen in the Brazilian Campos (Savannas), where the valleys with their water-courses are covered with wood and the higher ground with Campos (Savanna), and the boundary between wood and Savanna is as sharp as if it were drawn with a line. The proportion between Phanerophytes and Hemicryptophytes is entirely different in the wood and in the Campos. In the wood there are about twice as many Phanerophytes as Hemicryptophytes, and in the

Campos there are about twice as many Hemicryptophytes as Phanerophytes, in spite of the fact that the climate of the two habitats is identical. The precipitation, the temperature, and the wind are all the same; only one factor is different, and that is the soil. The more various the number of habitats are in a country the greater will be the number of species: the more uniform the habitats are, the fewer species will there be. The flat country has not such a chance of richness in species as a mountainous country.'

The passage quoted above about the richness and poorness in species being occasioned by the variety or uniformity of habitats makes no mention of the fact that the larger or smaller number of species alters what is here of great importance, i.e. the percentage proportion of the life-forms. This, however, can be left out of consideration. As far as the nature of the soil is concerned Warming describes in support of his view what is seen on the Brazilian Campos, 'where the valleys with their water-courses are covered with wood, the higher ground with Campos (Savanna)' although 'the climate is identical', 'only one factor differs and that is the soil'. If this is an objection to my attempt to use the statistics of life-forms as a biological expression of the climate, then Warming must have overlooked the fact that he himself especially emphasizes that the climate is identical in the two localities—wood and Savanna. For as the climate is the same, then it is neither the flora of the wood alone nor of the Savanna alone which must be used in making the spectrum of that district, but the flora of both localities and the flora of the whole surrounding district with the same temperature and precipitation, i.e. the same Hydrotherm figure. In this way we obtain a biological spectrum giving an excellent picture of the plant climate of Lagoa Santa. In the wood the Phanerophytes are as numerous as in all tropical forests; but on the Savannas a number of species are found that are not Phanerophytes. The result of this is that the percentage of Phanerophytes is comparatively low for a tropical region, just as one might expect from the Hydrotherm figure of the region if my view is right.

The question will perhaps be asked: but how about the fact that only the valleys with their water-courses are covered with wood? One will perhaps say with Warming that this must be due to the different composition of the soil. Entirely apart from the fact that this question only concerns the science of formations, the grouping of species within a given climate, and does not concern the plant climate, which is not based on the species of a single formation, but, as just described, on those of all the formations, there is another reason which makes it worth while to examine more closely this question of the soil.

The soil can be looked upon in different ways. In the case before us, for example, it is not the kind of soil that is of importance, but the water content of the soil resulting from precipitation. Precipitation is one of the most important factors determining the plant climate of a country. If then there is such a great difference between the vegetation of the valleys and that of the higher Savanna this is not due to the soil in a narrow sense, but to the degree of humidity in the soil due to the amount of precipitation, that is, due to the climate. Now since the precipitation is comparatively slight, the higher parts of the country, from which the water runs off, become so dry that a rich Phanerophyte vegetation cannot thrive. On the other hand, the valleys into which the water runs become sufficiently moist to harbour a vegetation of Phanerophytes. That this is so is apparent from the fact that as soon as we encounter more copious precipitation, as we do in many other parts of Brazil, we find also the higher parts of the land covered with wood. If the precipitation of Lagoa Santa were to rise considerably there is no doubt that the Savanna would be changed into wood, although the soil remained the same and only one factor were altered, i.e. precipitation.

Warming seems here to have embarked on the same course that led Schimper into his fundamental error of separating edaphic and climatic formations. About the part of Brazil now under discussion Schimper says that woodland occurs as an edaphic formation, i.e. as a formation determined by the soil, and that Savanna, on the contrary, is a climatic formation.[1] But as it would be equally just to say the contrary, it is better to say neither. The fact is that every formation is determined first and foremost by temperature and by the moisture resulting from precipitation. The precipitation is distributed differently in the soil according to the kind and the surface of the soil. This is what determines the different formations. It cannot, therefore, be said that one formation is edaphic and another is not. It might, however, perhaps be said that they are all edaphic, that they are all determined by the humidity of the soil. But since the humidity of the soil is determined by precipitation it is most natural to say that they are all climatic. Everything that can be referred to climate should in my opinion be expressed as climatic.

While the surface and the physical properties of the soil are of great importance in deciding the distribution of the moisture derived from precipitation and thus in deciding the distribution of the species in the region of which a biological spectrum is formed, the chemical composition of the soil on the other hand seems to be of little or no influence in deciding the biological spectrum of the different localities. Sandy Point, whose soil is formed of coral sand, has been invaded by a flora which shows essentially the same biological spectrum as that of the flora covering the rocky ground of the Danish West Indies. On Sandy Point the rain-water sinks quickly into the earth: from the rocky land it runs

---

[1] Schimper, A. F. W., *Pflanzengeographie auf physiologischer Grundlage*, Jena 1898, Karte 3: 'Grasfluren als klimatische Formationen, Gehölze als edaphische Formationen'.

quickly away. Whether the comparatively dwarf vegetation on Sandy Point is due to the fact that the plants are more quickly deprived of rainwater than are those on the rocky ground, or whether it is due to the fact that the soil on Sandy Point is poorer in food material I shall not attempt to decide. This has no significance for the problem confronting us, since the biological spectrum is in both localities essentially the same.

The biological spectrum of the plants that have invaded the alluvial formations on the protected coast shows also that the spectrum of the new ground is that characteristic of the flora covering the old ground in the same climate. As an example of this I have added in Table 3 the numbers concerned for Krauses Lagoon on St. Croix and for the salt marsh south of Nordby on Fanø. Since the number of species is so small, there is no reason to give a more detailed spectrum. What is of importance here is that the flora around Krauses Lagoon shows the spectrum of the Phanerophyte climate, while the flora of the salt marsh south of Nordby shows the spectrum of the Hemicryptophyte climate; and the high percentage of Phanerophytes and Hemicryptophytes respectively corresponds as closely as could be wished to the percentages of the same life-forms in the adjacent floras.

Since it might be of interest to see the same relationship illustrated by examples from other plant climates than those named, I have given in Table 3, No. 8, an example from a region in a Therophyte climate, i.e. from the alluvial formations at the mouth of the Rhone on the Île de la Camargue. I have used as a floristic basis the list of plants of Flahault and Combres.[1] The flora that has invaded the alluvial soil of the Île de la Camargue shows the same Therophyte spectrum that is found in the adjacent part of the south of France and in the Mediterranean region as a whole.

Sandy Point, Krauses Lagoon, the Camargue, the salt-marsh at Fanø, and Skallingen thus prove that the geological age of the land has no noticeable influence on the biological spectrum of its flora.

It is a matter of course that the life-form or life-forms whose percentages in the biological spectrum rise particularly above the corresponding numbers in the normal spectrum characterize the plant climate of the region. They form the positive side of the characteristics of the spectrum. But it is not without significance to investigate the negative side of these characteristics. By this I understand those life-forms whose percentages fall below that of the normal spectrum. These recessive life-forms do not all become suppressed in the same degree as we gradually pass from one climate to another. Going from southern Central Europe towards the north it is true that the Phanerophytes gradually disappear, but the various kinds of Phanerophytes do not disappear equally rapidly. Firstly

[1] Flahault, Ch., et Combres, P., 'Sur la flore de la Camargue et des alluvions du Rhône', *Bull. Soc. Bot. Fr.*, tome **41**, 1894, p. 37.

the large Mega- and Mesophanerophytes and lastly the small Nanophanerophytes disappear.

The result of the change of climate observed when going from the south to the north is the same for all the recessive life-forms; their numbers all diminish; but the secondary causes, the combination of factors causing this diminution, may be different for the different life-forms as well as for the different species. As an example I shall choose the Phanerophytes and Therophytes. A large number of Therophytes well able to grow on cultivated land in Denmark are quickly suppressed if the land is left to itself, because they are not able to maintain themselves in competition with other life-forms, especially Hemicryptophytes, which quickly take possession of the uncultivated ground. In such special localities as these, where the conditions are such that the Hemicryptophytes only slowly take possession of land that has been previously cleared, and especially where on new formations with poorer soil, as in our dune regions, habitats remain for a long time in which Therophytes do not encounter overwhelming competition, a comparatively large number of Therophytes persist. If, therefore, a biological spectrum is made for the flora of these special localities, this spectrum will as a rule show a comparatively high percentage of Therophytes. Compared with the spectrum of the adjoining land this high percentage of Therophytes reflects in a manner the conditions prevailing. The numbers in Table 4, Nos. 4–6, illustrate this instructively.

Many Phanerophytes behave differently. There are also numbers of exotic Phanerophytes able to grow with us on cultivated ground, but which cannot maintain themselves in competition with Hemicryptophytes. But while the Therophytes disappear partly because there is no room left for them, the Phanerophytes disappear in all probability because they cannot compete with the invading Hemicryptophyte vegetation. Since the competition is here for food material, and perhaps especially for water, such Phanerophytes will as a rule not be able to grow where there is no competition if the soil is dry and poor, and evaporation is rapid, as it is in our dune region. More or less the same thing is true of our indigenous Phanerophytes, some of which are certainly able to maintain themselves in competition with Hemicryptophytes and other life-forms, and, where the conditions are more favourable, eventually to carry the day. Where the environment is particularly unfavourable they may not be able to grow at all, even in the absence of competition, as on our dunes. It follows then that in travelling from the southern part of the Hemicrypyophyte region towards the north we do not merely find that the Phanerophytes gradually decrease, but that those remaining become more and more suppressed in the unfavourable localities, eventually being found only as insignificant remnants where the conditions are most favourable.

In the Phanerophyte climate of the West Indies we find at Sandy Point 74 per cent. of Phanerophytes. At the northern boundary of the Therophyte climate in the old world we find on the Camargue still 10 per cent. of Phanerophytes. From here they decrease towards the north, receding more rapidly from the unfavourable than the favourable localities. In the climate of Holland we still find a not inconsiderable Phanerophyte vegetation in the dune valleys. In the dunes of Jutland the Phanerophyte vegetation is very sparse. At Skallingen the percentage of Phanerophytes falls to 1 per cent., the only Phanerophyte there being a Nanophanerophyte, *Salix repens*, which is so stunted that it is almost a Chamaephyte.

We shall finally consider by means of a few examples what occurs on ground not merely geologically but also historically new. This may lead us to ask whether perhaps we do not find a different relationship between the life-forms, the newly invaded flora showing a spectrum different from that of the original flora. For this purpose we can either investigate the flora that has occupied ground cleared of vegetation artificially, or we can investigate ground that has come into being by natural processes. I shall take the latter first, choosing one example from the Hemicryptophyte climate, i.e. the newly formed islands in Lake Hjelmar, and another from the Phanerophyte climate, namely, Krakatoa. These localities also give us examples from widely different kinds of soil.

The newly formed islands in Lake Hjelmar. As a result of making a canal in 1882 the level of the water in Lake Hjelmar sank 1·2 metres, and in 1886 a further 0·7 metre. The laying bare of parts of the bottom of the lake caused a series of islands to appear. Furthermore considerable extents of new land appeared along the old coast. This gives us large areas of new ground only 27 years old which has been gradually invaded by a not inconsiderable number of plants. The new flora has in the course of years been investigated several times, in 1886 by Callmé,[1] in 1892 by Grevillius,[2] and finally in 1903 and 1904 by Birger.[3] We are thus able to follow the development from the beginning. In the last-mentioned work,[3] Birger considered the work already done and brought together the results of all the investigations. I have used his floristic material, determining the life-forms of the several species, and constructing the biological spectrum of the floras found respectively in 1886, 1892, 1903-4, and also of all the species observed taken together. In delimiting the flora I

[1] Callmé, Alfr., 'Om de nybildade Hjälmaröarnas vegetation', *Bih. t. K. Vet. Akad. Handl.*, **12** (1887), iii, No. 7.

[2] Grevillius, A. Y., 'Om vegetationsförhållandena på de genom sänkningarna 1882 og 1886 nybildade skären i Hjälmaren', *Bot. Not.* 1893; 'Om vegetationens utveckling på de nybildade Hjälmaröarna', *Bih. t. K. Vet. Akad. Handl.*, **18** (1893), iii, No. 6.

[3] Birger, Selim, 'De 1882-6 nybildade Hjälmaröarnas vegetation', *Arkiv för Botanik*, vol. **5**, 1905-6.

have followed the same rules as in making the biological spectrum of the flora of Denmark. Table 5 shows the distribution of the species into the five series of life-forms. Nos. 1–3 show in this manner the biological spectrum of the newly formed islands in 1886, 1892, and 1903–4 respectively. No. 4 shows the biological spectrum of all the species that have been found on the newly formed islands. That the total number of species is considerably greater than that of the last investigation is because a number of the species observed in the first and second investigations were not found later, either because some of the species have occasionally disappeared from the islands, or because they have been overlooked. In No. 5 I have given the biological spectrum for all the species found, partly on the newly formed islands and partly on the newly formed land along the coast of the islands already present. Finally for comparison I have added in No. 6 the biological spectrum of the flora of Denmark.

TABLE 5

	No. of species.	The percentage distribution of the species among the life-forms.				
		Ph	Ch	H	Cr	Th
1. The new islands in Lake Hjelmar 1886	91	9	5	46	14	26
2. ,, ,, ,, 1892	140	12	3	52	20	13
3. ,, ,, ,, 1903–4	148	17	3	53	18	9
4. ,, ,, ,, 1886–1904	192	14	3	51	16	16
5. The new islands in Lake Hjelmar together with the new land along the coast of the old islands 1886–1904	228	13	4	52	16	15
6. Denmark	1,084	7	3	50	22	18

All these spectra agree in showing the high Hemicryptophyte percentage of a Hemicryptophyte climate. From this it is seen that the biological spectrum of the flora of new land is governed by the same laws that make the biological spectrum as a whole an appropriate expression of the plant climate. As early as four years after the new islands came into existence, although only 91 species had arrived, this new flora showed in its biological spectrum 46 per cent. of Hemicryptophytes. In 1892 and 1903–4 the percentage had risen to 52 and 53 respectively. Concerning the Hemicryptophytes there is thus nothing particularly noteworthy, and the same is true of the percentage of Chamaephytes and Cryptophytes. On the other hand, during the eighteen years from 1886 to 1904 the percentages of Phanerophytes and Therophytes have shown a somewhat wider oscillation. At the moment I cannot give the biological spectrum for the region of Lake Hjelmar as a whole, so that I have not a definite number with which to compare the spectrum of the new islands; but it may be regarded as certain that the spectrum of the new islands in 1886 had a higher percentage of Therophytes than that of the surrounding district. Annual plants indeed appear always to be comparatively richly represented among the species, at any rate in a Hemicryptophyte climate, which are found on new ground and on ground cleared for cultivation. We see also that the comparatively high percentage of Therophytes falls to 13 in 1892, and finally to 9 in 1904. It should here be noticed that even if it be true that Therophytes are more richly represented in the flora of the new soil during its first years, yet the preponderance of Therophytes usually appears greater than it really is. For example if we investigate a field that has been cultivated for a number of years and then left to itself, we shall probably find that the Therophytes are comparatively richly represented, but not so richly as would appear at first sight. Closer investigation reveals a number of species at first overlooked and belonging to other life-forms, especially to Hemicryptophytes, although these plants may be represented only by seedlings. After the lapse of a few years these other life-forms will have repressed the Therophytes both in number of individuals and of species, so that they occupy the humble position they hold in a Hemicryptophyte climate when nature is left alone for a long period.

TABLE 6

	No. of species.	The percentage distribution of the species among the life-forms.									
		S	E	MM	M	N	Ch	H	G	HH	Th
1. The new islands in Lake Hjelmar 1886	91	..	..	3	3	3	5	46	7	7	26
2. The new islands in Lake Hjelmar 1892	140	..	..	6	3	3	3	52	7	13	13
3. The new islands in Lake Hjelmar 1903–4	148	..	..	8	4	5	3	53	5	13	9
4. The new islands in Lake Hjelmar 1886–1904	192	..	..	6	3	5	3	51	5	11	16
5. The new islands in Lake Hjelmar with the new land along the coast of the old islands 1886–1904	228	..	..	5	3	5	4	52	6	10	15
6. Denmark	1,084	..	(0·1)	1	3	3	3	50	11	11	18

The Phanerophytes began in 1886 with 9 per cent., but rose during the following years, reaching 12 per cent. in 1892 and 17 per cent. in 1903–4. Although, as I have said, I cannot give the biological spectrum for the flora of the region of Lake Hjelmar as a whole, I suppose, however, that the Phanerophyte percentage shown by the flora of the new islands in 1903–4 was higher than the corresponding percentage in the flora of the surrounding district. Taking into consideration not only the new islands but also the new land on the coast of old ones the spectrum shows only

13 per cent. of Phanerophytes. This number may not be far from the normal for the woodland region of Central Sweden. But however this may be it is certain that the fact of greatest importance is established, namely that the flora that has invaded the new islands in Lake Hjelmar shows, just like the flora of the surrounding country, the spectrum of a Hemicryptophyte climate. Table 6 shows further that the distribution of the species among the classes of life-forms corresponds with the spectrum of the ordinary Hemicryptophyte climate as closely as could be expected in a flora of this size.

TABLE 7

	No. of species.	The percentage distribution of the species among the life-forms.				
		Ph	Ch	H	Cr	Th
1. Krakatoa 1886	15	59	7	20	7	7
2. „ 1897	49	70	6	14	6	4
3. „ 1906	73	78	6	8	3	5
4. „ 1886–1906	90	77	4	10	3	6
5. Edam (Java)	74	55	15	7	4	19

Krakatoa. The volcanic island of Krakatoa in the Sunda Straits is very well suited for comparison with new islands of Lake Hjelmar if we wish to investigate the biological spectrum of a flora that has invaded historically new ground. The present soil of Krakatoa came into existence about the same time as that of the islands in Hjelmar, i.e. 1883, and a new flora of Krakatoa, just as that of the new islands in Hjelmar, has been investigated and described three times, in 1886 by Treub,[1] in 1897 by Penzig,[2] and in 1906 by Ernst.[3] Before 1883 Krakatoa was covered by an impenetrable primeval forest, a vegetation of Phanerophytes. This vegetation was destroyed in the summer of 1883 by the terrific eruption of August 26–7. Two-thirds of the island disappeared into the ocean and what was left was covered to an average depth of 30 metres by red-hot scoria and ashes. The flora that invaded Krakatoa after 1883 is much poorer than that which invaded the islands in Hjelmar; in 1886 only 15 species were found, in 1897 only 49, and in 1906 only 73. Altogether 90 species were found. Table 7 shows how these species are divided among the 5 groups of life-forms, and Table 8 gives the more detailed biological spectra, i.e. the distribution of the species into classes of life-forms. It will be seen that the flora from the very beginning shows the

[1] Treub, M., 'Notice sur la nouvelle Flore de Krakatau', *Annales du Jardin bot. de Buitenzorg*, 7, 1888.
[2] Penzig, O., 'Die Fortschritte der Flora des Krakatau', ibid., 1902.
[3] Ernst, A., 'Die neue Flora der Vulkaninsel Krakatau', *Vierteljahrschrift d. naturf. Gesellsch. in Zürich*, 1907.

spectrum of a Phanerophyte climate, agreeing with the marked Phanerophyte climate of the part of the world in which Krakatoa is situated. I cannot indeed give the exact numbers of the biological spectrum of the climate; this can be done at present neither for Java nor Sumatra; but it is quite certain that we are dealing here with the spectrum of a Phanerophyte climate. In order, however, to obtain the flora of even a very small area for comparison, I have determined the life-forms of the plants growing on the little coral-island of Edam lying off Batavia. This flora contains 74 species, and its biological spectrum is given in Table 7, No. 5, and Table 8, No. 5. The correspondence between the biological spectrum of this island and that of Krakatoa is as close as could be expected in dealing with such a small number of species, in spite of the fact that the soils of the two islands differ widely in age and composition. Just as in the examples given before, this shows the significance of the biological spectrum as a biological expression for the climate, as a means of giving a proper picture of the plant climate.

TABLE 8

	No. of species.	The percentage distribution of the species among the life-forms.									
		S	E	MM	M	N	Ch	H	G	HH	Th
1. Krakatoa 1886	15	..	..	20	13	26	7	20	7	..	7
2. ,, 1897	49	..	4	23	16	27	6	14	6	..	4
3. ,, 1906	73	..	3	16	30	29	6	8	3	..	5
4. ,, 1886–1906	90	..	3	18	27	29	4	10	3	..	6
5. Edam	74	..	1	14	20	20	15	7	4	..	19

Besides new ground in the strictest sense, i.e. ground that has never before borne plants, another kind of new ground arises as the result of cultivation, i.e. ground on which space becomes available for an invasion of new plants. Neither of these two kinds of new ground, however, has any influence at all on the biological spectrum. I have already shown this to be true for new ground that has arisen naturally. Finally I shall spend a little time discussing new ground arising from cultivation.

All kinds of cultivation disturb to a greater or less extent the equilibrium of the original flora. Grazing, silviculture, mowing, agriculture, and horticulture are the most important encroachments of cultivation, disturbing the original equilibrium, and each producing a different degree of disturbance. Thus the use of the ground for grazing by itself produces a considerable interference with the original equilibrium, but this interference is insignificant compared with that produced by agriculture and horticulture, which destroy entirely or almost entirely the original plants of the land employed. Wherever cultivation encroaches, disturbance of equilibrium takes place, a disturbance which may be great or small, and a new ground, as defined above, arises. As soon as cultivation is left off, conditions are created for the invasion of new species and for a new struggle for space among the species. It is of course obvious that the biological spectrum alters when man destroys the original flora and cultivates definite plants in its place. This cannot be used as an objection to a scientific system. If, however, the biological spectrum of a flora which gradually occupies land deserted by cultivation became different from that of the original flora, this would indeed be an objection to my view. But to show that this does not occur it may not be necessary to give special examples. For the kind of observation necessary there are opportunities almost everywhere. As soon as a piece of ground goes out of cultivation a change in its vegetation occurs in the direction of gradually re-establishing, as time goes on, the vegetation that prevailed before the encroachment of cultivation. The original biological spectrum is again restored to the land deserted by cultivation. If in a woodland district of Denmark, for example North Zealand, cultivation ceases in a field, the same plants as were there before cultivation began, and amongst these Phanerophytes such as oak, beech, alder, &c., gradually reappear. These plants are the ones that at present make up the vegetation of habitats, such as woodland, common, &c., which have been less encroached upon by cultivation. In the same way in the heath region in West Jutland if cultivation ceases in a field the plants of the heath again invade that field. We see the same thing in other localities and other climates. In the West Indies if land goes out of cultivation it becomes in a few years densely covered with a Phanerophytic vegetation resembling that of the places which have never been cultivated.

I am surprised, therefore, that this process presents to Warming an objection to my system, or that it can be any kind of difficulty to him. On the contrary it confirms very beautifully the correctness of my view.

I have already mentioned the objection which Warming finds in the possibility that the woods of such a country as Denmark might become exterminated. After discussing this possibility, namely the extirpation of the 6 per cent. of wood in Denmark, Warming writes (loc. cit., p. 23): 'The same applies to other places, even to the tropics; if human beings were to desert Blumenau in South Brazil, says A. Møller, the country would in the course of ten years be completely covered with trees, and H. Cotta prophesies the same fate for Germany in the course of 100 years, if mankind were to disappear.' It should be added, however, to Warming's example, that around Blumenau the country would not merely become covered with trees, but that this Phanerophytic vegetation would also gradually become composed of a comparatively large number of species, thus resembling the woods in that district; and that the flora which occupied the land no longer cultivated would show and does show a Phanerophytic spectrum resembling that of the surrounding district.

And in discussing the example from Germany it should be added that the vegetation of trees, with which, according to Cotta, Germany would become covered 100 years after the disappearance of human beings, would include, just as the present German woods, only comparatively few species of Phanerophytes. But it would include on the other hand a great many species of Hemicryptophytes and other life-forms. It would, taken as a whole, show clearly the spectrum of a Hemicryptophyte climate, just like the flora of the large expanses in Germany which always have been and still are covered with wood. These examples then contain nothing against the use of my system, but they actually illustrate in a striking manner its importance, showing that mankind is unable even by means of the most violent encroachments to bring about a change in nature so permanent that it can persist without the help of cultivation. We see everywhere on the other hand that if the encroachments cease nature returns again in the distribution of her life-forms to her original proportions, so that a country left to itself after the lapse of time again shows the biological spectrum which corresponds to its plant climate and which it showed before the country was cultivated. In fact it will be difficult enough for man to alter essentially the biological spectrum of a country, to say nothing of that of a complete climatic zone. For example as long as there exist in Denmark roadsides, ditches, steep slopes, and other localities that cannot be treated by the plough or the spade, the flora, looking after itself, will continue to show the spectrum of the Hemicryptophyte climate, agreeing with the actually existing climate; and this is true of all countries and all climates.

While cultivation can easily radically disturb the results of floristic plant geography, which are built upon the species of the systematists, it will on the other hand be difficult for cultivation to make an essential change in the biological spectrum founded upon life-forms. And even if such a fantastic thing occurred as the extirpation of all wild plants, so that only cultivated plants remained, even these plants alone would in the main reflect the existing kinds of plant climates. It only has to be borne in mind that in the tropical Phanerophyte climate a far greater number of Phanerophytic economic plants grow than in a Hemicryptophyte climate, where many more hemicryptophytic economic plants are grown than in the tropics. As long, therefore, as there are plants in the world we shall possess in the life-forms of these plants and in the biological spectra founded on them an expression of the prevailing plant climates.

VI
INVESTIGATIONS AND STATISTICS OF PLANT FORMATIONS

Numbers are the Metrical Units of Science[1]

ONE of the first aims of biological plant geography is the solution of the difficult problem presented by the fact that the floristic composition of vegetation may differ widely in different places even if the conditions are entirely uniform and the adaptation of the plants to their environments may therefore be considered essentially the same. Our task is to find a method of explaining the uniformity of vegetation in spite of its varying floristic composition.

We have to make use of a characteristic feature which can be used for determining and demarcating the phyto-climatic regions. I think that the life-forms I have proposed, which are based upon the adaptation plants show for surviving the unfavourable seasons, form a practical characteristic feature that can be used until we succeed in finding something better. I have shown that we have in the **biological spectrum** founded on these life-forms a means of deciding how far different regions whose vegetation is floristically entirely different belong to the same or to different plant climates. I have shown that those regions whose climate is the same as far as vegetation is concerned show essentially the same biological spectrum, and that this applies to zones as well as to regions. On the other hand I have shown that regions where the climate is essentially different, as far as vegetation is concerned, show essentially different biological spectra. Finally I have made it clear that the biological spectrum has proved itself such a delicate instrument that even floras that are very poor in species have biological spectra which follow these rules accurately.

The future must decide to what extent these life-forms can be used in the theory of plant formations. But it seems reasonable in the meantime to make use of them in characterizing series of formations in such a way that these formations are called after the life-form to which its one or more dominant species belong. For example we have series of formations of Phanerophytes, Chamaephytes, Hemicryptophytes, Helophytes, and Hydrophytes. Thus the Beechwood belongs to the series of formations of deciduous Mega- and Mesophanerophytes, Heath to that of the Chamaephytes, and Meadow to that of the Hemicryptophytes.

[1] The original Danish 'Tal er Videnskabens Versefødder' is not altogether easy to translate. It might perhaps be rendered 'Numbers are to Science what Feet are to Metrical Verse.'

The theory of plant climates, as I have attempted to base it on the biological spectrum of life-forms, must not be confounded with the theory of plant formations. **Plant climatology**, if I may use this term for the sake of conciseness, seeks to delimit and characterize those great areas of the earth's surface whose vegetation is determined by climate. It seeks to discover a biological expression, independent of the floristic composition of the vegetation, an expression which is common to all districts possessing essentially the same plant climate but is different in regions where the plant climate is essentially different. **The theory of formations**, on the other hand, concerns itself with the grouping of the individual species of the flora on a given substratum. Finally **Ecology** tries to investigate the causes of this grouping, investigating the environment in individual localities and the general structure of the species that inhabit these localities. Ecology by these means seeks to demonstrate the relationship between the demands of the plants and their environment.

We may say that our ultimate aim must be to delimit and characterize formations in an ecological manner; but before doing so there must be a preliminary delimiting and a characterization of formations to serve as a starting-point for our ecological investigations. In making this preliminary sketch we may begin either with the environment, i.e. the localities, or with the physiognomic-floristic composition of the communities.

From the point of view of plant geography, we should aim at being able to undertake comparative investigations, at being able to demonstrate affinities between formations with different floristic composition but with the same ecology. Now since it is this very difficulty that we have set out to avoid, the difficulty arising from the fact that regions whose formations have widely different floristic composition may quite probably be equi-conditional, so that their formations may be ecologically the same, it is wrong to use floristic composition as a basis for the preliminary demarcation and characterizing of formations. From the very nature of physiologico-ecological investigations much time must elapse before it is possible to draw up an accurate, comparative, comprehensive ecology of formations. Our only course therefore is to make use of the environment, the nature of the habitat, as a preliminary basis in defining formations, taking for granted that by reason of competition the same environment will determine essentially the same ecology of formation, the same total of adaptations. But by no means does this hinder us from making what use we can of the results of our floristic investigations. Within the same floristic region there will always be found far-reaching floristic conformity between localities having the same environment, and since the floristic physiognomy is the most readily accessible character, this will always be the character that will be our first and foremost guide; not leading us merely to a demonstration of the unity of the formations but also of the unity of environment, to which one can add a probable unity in ecology, which amounts to a unity in formation.

When the formations are thus established in a preliminary way we can undertake a more thorough floristic analysis, and a floristic characterization based upon it: to this I shall return later. Then finally we can set to work with the ecological investigation. Although there is already available a very comprehensive ecological literature, yet an accurate ecological investigation applicable to comparative plant geography is still in its infancy. For example, it is not sufficient to content ourselves with the investigation of leaf structure, the amount of transpiration, and similar isolated factors. We must determine the true life-form, that is to say we must measure all the factors which act upon the life of a plant in a given environment. When we see two species, one with xerophilous and the other with mesophilous leaves, growing side by side in the same kind of soil we cannot conclude that they are not equally well adapted to their environment. Leaf structure is only one factor that we have to consider. The investigation of leaf structure is of course very important, but the investigation of the morphology of the root and of its physiological capacity is of equal importance if we wish to understand the ecology of the plant. *Festuca rubra* with markedly xerophilous leaf structure and *Ononis repens* whose leaves and whose whole habit are mesophilous, grow side by side, but yet both hold their own equally well during dry periods. This is because the roots of *Ononis repens*, as is well known, penetrate deeply into the soil, and obtain access to deeper and wetter layers than *Festuca rubra* is able to reach. If then ecology is to be something more than a series of superficial illustrations of certain ordinary propositions, if it is to enable us to build a foundation for an exact comprehensive science of formations, such profound and comprehensive investigations will be necessary that it seems to me long periods of time will elapse before we can obtain a firm foundation.

Until then in our studies of plant geography we must content ourselves with bringing together what seems from our studies of localities and analyses of floristic composition to belong together; and especially must we make as much use of life-forms in demarcating plant climates as we practically can.

It is of foremost importance to find a method of improving upon the uncertain picture we obtain by subjective estimates of plant communities, so that it may become possible to treat the floristic composition of a formation in such a way that we can not only recognize the same formation in various places, but that other people may be certain that various investigations of the same formation will give the same result.

In plant climatology, in demarcating and characterizing the great regions, the plant climates, the individual species are all of equal importance; the rare ones count for as much as the common ones. The adaptations of the individual to survive the unfavourable seasons is independent of the extent of the capacity possessed by the species for multiplication and distribution. In the science of formations, in which, if for no other reason, at any rate for practical considerations, the floristic physiognomy looms very large, the size and abundance of a species, as every one realizes, must be taken into consideration. Formations are in general characterized by their dominant species, which are very often used in naming an individual formation. I have therefore first and foremost striven to find a method of determining the significance of the species in the composition of the vegetation, the valency of the individual species, which is entirely independent of subjective estimate.

Physiognomically a species dominates by reason of its bulk, i.e. by reason of the number of individuals or their size, or by a combination of both. It is important then to find a practical method of determining bulk; but I consider it more important that the method be such that various investigations of the same formation will give the same result when expressed numerically, than that it should lead one to an accurate expression of the actual bulk of the species, as such a figure is in practice unobtainable.

One might at first suppose it to be comparatively easy to determine the mass of a species by the process of weighing, which is a method used in agriculture in investigating hay-fields in order to determine the percentage of the different species in a harvested crop. But it soon becomes apparent that the matter is not so easy, and that this method in practice is unfeasible in the investigation of formations. An essential difficulty is suggested by the question: at what period in the growth of the plant is the weighing to take place? In agriculture this difficulty does not arise; the time of weighing is determined by the object in view, as for example in the determination of the mass of the individual species in a harvested crop. But in investigating formations the problem is different. Either the bulk must be determined when the individual species are at their perfection, or one might attempt to determine the bulk produced by an individual species during the whole period of its growth. The latter is at the outset in practice impossible, and the former is really impossible too, for since different species reach perfection at different times of the year the method would need nearly as many investigations as there are species in the formation, amounting to a long series of investigations spread over a whole growing period. And since the composition of the vegetation is not uniform throughout the same formation each investigation would have to include a series of samples taken at random. I believe therefore that when in investigating a formation we wish to obtain a true expression of bulk the method of weighing is unfeasible.

The other method of determining bulk, by counting, is not much better. If individuals are counted one meets with the difficulty that many of the species have stolons which make it difficult to determine what constitutes an individual; and in most of the areas which serve as a basis for the determination of the bulk of the individual species fractions of individuals will occur whose size cannot be exactly determined. Besides this a justifiable criticism of the method is that it gives no expression of what it tries to express, i.e. relationship of bulk, since individuals are not merely of different size in different species but also within the same species. By substituting the counting of shoots for the counting of individuals we fare little better, for, as every botanist knows, the shoots counted differ as widely as possible in bulk. Further, in both methods, weighing and counting, there is this difficulty, that either one must count the annual crop of shoots, which in practice cannot be done, or one must determine the number of shoots of the different species at different periods of their growth according to the period at which the shoots reach their perfection. This will entail such an enormous amount of work that the value of the results obtained stands in no reasonable relationship to the labour.

Although we might at first suppose that the methods of weighing and counting would give definite results useful in comparison, which is just what is required, closer consideration shows us that, even if it be possible to procure definite numbers by these methods, in the method of weighing we are confronted with a procedure we cannot carry out in practice with accurate results, while the method of counting, apart from the fact that the procedure is exceedingly difficult, hardly, if ever, gives the measure of bulk for which we are striving.

In comparison with these methods an estimate of bulk based on a thorough floristic analysis would be preferable. This is the method actually in general use, a procedure being employed similar to that used for a long time in estimating the degree of frequency of the individual species. Here we encounter the great drawback that this method does not give exact figures. It does, it is true, give some figure, the degree of bulk being expressed by a number in a scale of numbers; but this number is founded merely upon personal judgement, which is too uncertain a basis. Not only are different persons often led by personal judgement to different results, but even the same person often judges the same object differently at different times.

I have therefore striven, as before mentioned, to find a method of getting beyond the uncertain result of personal judgement, a method of investigating formations which will express numerically the frequency of the individual species independently of personal judgement. It must be a method which can be used partly for comparative investigations of formations within the narrower confines of the same flora, and at the same time can be made the means of transforming systematic units, i.e. species, into biological units, i.e. life-forms. In doing this we must take into consideration the frequency grades which the investigation of the formations imposes upon us. This procedure opens the way to a comparative investigation of formations which correspond with one another, although they belong to different floral regions and have entirely different floristic composition.

As an example I will choose the ground flora of the beechwood on humus. In Jonstrup Vang, where my observations were principally made, this ground flora, named after the dominant species, is an *Anemone nemorosa*-Facies. I avoided all transitional regions abutting on formations of another species. A floristic investigation of this *Anemone nemorosa*-Facies yields a series of species, but *Anemone nemorosa* is overwhelmingly commoner than any of the others.

In determining the degree of frequency (valency) of the individual species I now adopt the method of making a floristic analysis of a number (e.g. 50) of areas of a definite size taken at random in the formation to be investigated. In determining the degree of frequency each species is given a number of points. This figure represents the number of areas in which it has been found. For example if one has investigated 50 areas and found 5 species in 50, 18, 14, 8, and 3 of these areas respectively, and if we use a short scale of frequency, which in practice I have found preferable, e.g. from 1 to 5, 5 representing the highest degree of frequency, then the 5 species under consideration are given the following valencies: 5, 1·8, 1·4, 0·8, and 0·3. We may round off the figures to the nearest half making them 5, 2, 1·5, 1, and 0·5. In my later investigation, however, I have used as degree of valency the number of points from 1 to 50. In making the investigations a square frame is used, which is thrown down at random at different places of the formation to be investigated.

The next question to be considered is the size and the number of the areas to be counted. This question can best be answered by trying to find out what number and size of area are at the same time feasible and lead to a result as nearly as possible in accordance with the facts.

We are not concerned here with making a list of all the species, including the rarest ones. Indeed in a general floristic investigation, however thorough it may be, we are never certain whether all the species have been observed. If the units investigated are neither too few nor too small, the species not included will all be rare or very rare. Where there is reason to suppose that it will be of special interest to make the list as complete as possible the species which have been seen but which did not occur within the quadrats may be added afterwards. Of course the method must be such as to include all species which are not particularly rare. The larger the areas are the fewer need one investigate if it is wished to include all the more important species; the smaller the areas are the more of them must be investigated. Since however we are not dependent first and foremost upon the completeness of our list, but upon an agreement as exact as possible between the number of points and the frequency of the species, we must rivet our attention particularly on this agreement.

In a formation where there is no very wide difference between the frequency of the species comparatively few areas have to be investigated in order to obtain a fairly true expression of the valency of the species. But in the investigation of a formation with one or several markedly dominant species and a number of species which are sparsely scattered, e.g. the *Anemone nemorosa*-Facies of the beechwood (see Fig. 107, p. 226), the one or several dominant species will obtain too small a number of points in proportion, unless a considerable number of areas be investigated.

Let us confine ourselves for the present to the *Anemone nemorosa*-Facies in Jonstrup Vang, where I began to make use of an Ar (100 sq. metres) for my area. It soon however became apparent, as I had expected it would, that this was too large a unit. For this reason I did not continue the experiment on this scale and will here give no example of the results obtained.

The areas afterwards adopted were a $\frac{1}{10}$, $\frac{1}{100}$, $\frac{1}{1000}$, and $\frac{1}{10000}$ of an Ar or 10 sq. metres, 1 sq. metre, $\frac{1}{10}$ sq. metre, and $\frac{1}{100}$ sq. metre. The result of the investigation of the *Anemone nemorosa*-Facies in the beechwood in Jonstrup Vang is seen in Table 1, where the degree of frequency of the individual species of the different areas is given according to a scale of 1–5, 5 representing the greatest frequency.

I have already said that I find it easiest to use a short scale, e.g. 1–5; if it seems desirable to use more than 5 degrees of frequency each degree can be divided into two or even into ten. The reason I have not begun with a longer scale, e.g. 1–10, is that I have found it much easier in surveying to determine the degree of frequency within 5 degrees and then if necessary to divide the degrees into halves or tens. This is easier than using a scale of 1–10, but if it seems more natural to use such a scale it is very easy to transform the one into the other.

The following investigations depend not so much upon the differences or correspondences between species found by using areas of different size, but rather on the relationship between the valency of the dominant species, which is here *Anemone nemorosa*, and the sum of the other species found. I have given this latter figure at the bottom of Table 1. In this table column 1 contains a figure based on the supposition that each species was counted only once, so that the individual species in this column have an equal valency whether they are common or rare. I have included only those species which were found in the different areas. The relationship between the dominant species and all the others taken together by this means comes to 6:94. This relationship would of course be still more misleading if the species observed outside the quadrats were included.

Now if we use the method I propose for determining the valency of species we shall of course obtain an entirely different result, which, however, will vary with the size of the areas used. In Table I, columns 2–5 show the result of using areas of 10, 1, $\frac{1}{10}$, and $\frac{1}{100}$ sq. metres respectively, the number of the areas for determining the valency being 10, 20, 100, and 400 respectively

Table I

	Life-form.	1	10 10 sq. metres. 2	20 1 sq. metre. 3	100 $\frac{1}{10}$ sq. metre. 4	400 $\frac{1}{100}$ sq. metre. 5	No. of shoots (1 sq. metre.) 6
Anemone nemorosa	G	1	5	5	5	5	5
Gagea lutea	G	1	3	1·5	0·85	0·26	0·03
Ficaria verna	H	1	2	0·25	0·05	0·06	0·09
Oxalis acetosella	H	1	1·5	0·75	0·8	0·25	0·2
Melica uniflora	H	1	1·5	0·25	0·25	0·01	0·008
Milium effusum	H	1	1·5	..	..	..	..
Anemone ranunculoides	G	1	0·5	0·25	0·15	0·11	..
Dactylis glomerata	H	1	0·5	0·25	..	..	..
Anemone hepatica	H	1	0·5	0·25	0·05	..	..
Mercurialis perennis	H	1	0·5	..	..	..	..
Corydalis intermedia	G	1	0·5	1	0·2	0·05	0·03
Ranunculus auricomus	H	1	..	..	..	..	..
Poa nemoralis	H	1	..	..	0·1	..	..
Convallaria majalis	G	1	..	0·25	..	..	..
Urtica dioeca	H	1	..	..	0·05	..	..
Asperula odorata	G	1	..	..	0·05	0·01	..
Stellaria holostea	Ch	1	..	..	0·05	..	..
Polygonatum multiflorum	G	1	..	..	..	0·01	..
Percentage proportion between *Anemone nemorosa* and all the other species taken together .		$\frac{6}{94}$	$\frac{29}{71}$	$\frac{51}{49}$	$\frac{66}{34}$	$\frac{85}{15}$	$\frac{94}{6}$

The valency for each species according to the areas of different size is given according to the scale 1–5. In the investigation of ten 10 sq. metre plots the relationship between *Anemone nemorosa* and the other species was 29:71. This is of course too low a figure for *Anemone*. The smaller the unit area employed the greater will be the number of plants of the dominant species in proportion to the other species; and the greater the number of areas the more closely will the result express the true relationship.

By using 1, $\frac{1}{10}$, and $\frac{1}{100}$ of a sq. metre the relations became respectively

51 : 49, 66 : 34, and 85 : 15. The last of these proportions is not far from the true relationship, but the proportional representation of the *Anemone* is still almost certainly too small. In formations such as this, therefore, in which one or few species are overwhelmingly dominant, we must use a unit area even smaller than $\frac{1}{100}$ sq. metre, if we wish to get as near as we can to the truth. We usually find, however, gradual transitions between the valency of the species, so that it is not a matter of one or few species overwhelmingly dominating the others. Now in my opinion we are not merely concerned with determining valency as exactly as possible; our chief care should be to obtain constant comparative numbers, which are of course as exact as any practical method allows. I have therefore in my later investigations both of the ground flora of the wood and of other formations always used not the smallest of the unit areas, but the next smallest, viz. $\frac{1}{10}$ sq. metre. This unit for several reasons proves the most convenient in practice.

We must also consider how we are in practice to obtain a reasonable relationship between the work done and the result obtained. If large unit areas are employed we certainly do not need such a large number of them in order to obtain a final result, that is to say in order that the investigation of still more unit areas do not alter essentially the proportions shown by the first number of unit areas. The exact investigation of even the smallest number of random areas is in most formations a great labour, and at the same time the agreement between the results of investigation of valency compared with reality is usually very unsatisfactory.

But if we take as a starting-point a very small unit area, e.g. $\frac{1}{100}$ sq. metre, the investigation will require a large number of these areas in order to give us a final stable result. Sometimes, as, for example, in formations poor in species the investigation of the single unit, the quadrat, is easy; but usually, and in all formations rich in species, the investigation is not so easy as one is at first inclined to suppose. Among other difficulties is the length of time taken to decide how many of the individuals whose shoots are encountered inside the frame are actually rooted inside the frame. Of those rooted outside the frame the herbs should not be counted. Since, as already said, in using a small unit area far more areas must be investigated to reach a stable result than if a larger unit area be used, I have chosen to work with $\frac{1}{10}$ sq. metre (about 1 sq. ft.) as this is the unit area which I have found best; it appears to give the result most in harmony with reality, and at the same time the floristic analysis of the individual areas is not particularly difficult, and the number of investigated areas does not need to be so very large to attain a stable result.

Let us here take the opportunity of making a further inquiry into the question of the number of unit areas necessary for different investigations. The greater the number of unit areas the greater will be the number of the species included, and hence the more complete will the list of the flora become. But it is of no consequence whatever to find out whether individual rare species have a valency of 0·2 or 0·5, and a complete list of the flora can be more easily procured by making a general floristic investigation, if this seems desirable. Thus the number of unit areas should depend solely upon the extent to which a stable result is approached, i.e. it will depend upon the number of unit areas by means of which the stable valency for the commoner species is attained, or where one single species in particular is dominant it will depend upon the number of unit areas which give a stable percentage relationship between the points counted for this species and the sum of the points for all the other species. This latter figure can be illustrated by my investigations of the *Anemone nemorosa*-Facies in the beechwood at Jonstrup Vang.

Table 2

The result of the investigation of the *Anemone nemorosa*-Facies in the beechwood at Jonstrup Vang. Unit areas of 10 sq. metre are used

	Life-form.	The number of times the individual species occurred in:	
		the first 5 unit areas.	the first 10 unit areas.
Anemone nemorosa.	G	5	10
Ficaria verna.	H	2	4
Gagea lutea.	G	4	6
Milium effusum.	H	2	3
Oxalis acetosella.	H	2	3
Melica uniflora.	H	1	3
Anemone ranunculoides.	G	1	1
Dactylis glomerata.	H	..	1
Anemone hepatica.	H	..	1
Mercurialis perennis.	H	..	1
Corydalis intermedia.	G	..	1
Percentage proportion between *Anemone nemorosa* and all the other species taken together . .		$\frac{29}{71}$	$\frac{29}{71}$

It will be seen from Table 2 that using a unit area of 10 sq. metres the 10 unit areas seem to be sufficient; the first 5 areas investigated show the same result as the first 10, and the next 5 unit areas would scarcely have made much difference to the result. That I did not investigate them is because when I had investigated ten 10 sq. metre plots, I had already given up ten 10 sq. metres as a unit area.

Table 3 shows that, using 1 sq. metre as a unit area, at any rate in the example given, 20 areas are sufficient. The percentage proportion between *Anemone nemorosa* and the remaining species together is in the first 10 areas 50 : 50, and this number does not alter appreciably if more areas are investigated. For example 15 areas gave 48 : 52 and 20 areas gave 51 : 49.

TABLE 3

The result of the investigation of the *Anemone nemorosa*-Facies in the beechwood of Jonstrup Vang using 1 sq. metre as a unit area

	The number of times the individual species occurred in:		
	the first 10 areas.	the first 15 areas.	the first 20 areas.
Anemone nemorosa.	10	15	20
Gagea lutea	2	3	6
Corydalis intermedia	2	4	4
Oxalis acetosella	2	3	3
Anemone ranunculoides	1	1	1
„ hepatica	1	1	1
Dactylis glomerata	1	1	1
Ficaria verna	..	1	1
Melica uniflora	..	1	1
Convallaria majalis	1	1	1
Percentage proportion between *Anemone nemorosa* and all the other species taken together	$\frac{50}{50}$	$\frac{48}{52}$	$\frac{51}{49}$

Taking a $\frac{1}{10}$ sq. metre as a unit area it will be seen from Table 4 that 50 areas suffice. The first 50 gave a proportion of 68 : 32 between the valency of *Anemone nemorosa* and all the other species together, and for 60, 70, 80, 90, and 100 areas the proportions respectively were 64 : 36, 61 : 39, 62 : 38, 64 : 36, and 65 : 35. In my later investigation therefore in which I have used $\frac{1}{10}$ sq. metre as a unit area I have usually made use of only 50 areas.

In some cases fewer areas might perhaps be necessary and in others more; the investigations themselves must be the guide. For determining the valency of species the number of unit areas is indifferent if sufficient of them be used to give a stable result; but as it depends upon the particular species composition whether few or many areas have to be investigated in order to obtain that stable result, the number of unit areas used must always be stated.

If $\frac{1}{100}$ sq. metre be chosen the number of units must, as above stated, be large. From Table 5 it is seen that in my investigations of the *Anemone nemorosa*-Facies in the beechwood of Jonstrup Vang I had to investigate 200 units before reaching a stable relationship between *Anemone nemorosa* and the other species. The first 100 gave a relationship of 93 : 7, the second 100 gave 79 : 21. Taken together these make 86 : 14, a result

TABLE 4

The result of the investigation of the *Anemone nemorosa*-Facies in the beechwood of Jonstrup Vang with $\frac{1}{10}$ sq. metre as unit area

	The number of times the individual species occurred in:					
	50 unit areas.	60 unit areas.	70 unit areas.	80 unit areas.	90 unit areas.	100 unit areas.
Anemone nemorosa	50	60	70	80	90	100
Gagea lutea	8	10	12	15	16	17
Oxalis acetosella	7	11	15	15	16	16
Melica uniflora	3	4	5	5	5	5
Corydalis intermedia	1	3	4	4	4	4
Anemone ranunculoides	3	3	3	3	3	3
Poa nemoralis	..	..	..	2	2	2
Asperula odorata	1	1	1	1	1	1
Anemone hepatica	..	..	1	1	1	1
Ficaria verna	..	..	1	1	1	1
Stellaria holostea	..	..	1	1	1	1
Urtica dioeca	1	1	1	1	1	1
Percentage proportion between *Anemone nemorosa* and all the other species taken together	$\frac{68}{32}$	$\frac{64}{36}$	$\frac{61}{39}$	$\frac{62}{38}$	$\frac{64}{36}$	$\frac{65}{35}$

TABLE 5

The result of the investigation of the *Anemone nemorosa*-Facies in the beechwood of Jonstrup Vang with $\frac{1}{100}$ sq. metre as unit area

	The number of times the individual species occurred in:			
	100 unit areas.	200 unit areas.	300 unit areas.	400 unit areas.
Anemone nemorosa	100	200	300	400
Gagea lutea	4	8	13	21
Oxalis acetosella	3	13	14	20
Anemone ranunculoides	..	3	4	9
Melica uniflora	..	3	7	8
Ficaria verna	..	2	2	5
Corydalis intermedia	..	4	4	4
Asperula odorata	..	..	1	1
Polygonatum multiflorum	..	..	1	1
Percentage proportion between *Anemone nemorosa* and all the other species taken together	$\frac{93}{7}$	$\frac{86}{14}$	$\frac{87}{13}$	$\frac{85}{15}$

which does not alter very much when 200 more areas are investigated. Adding the second, third, and fourth 100 areas the results were 86 : 14, 87 : 13, and 85 : 15 respectively.

Tables 1 and 6 give a conspectus of the results of using different unit areas. In Table 1 the whole of the investigated areas are considered; but in Table 6 only the number of areas are considered which seems to be necessary for investigations with different sizes of areas.

TABLE 6

		Percentage proportion between:	
Size of area.	Number of quadrats necessary	Anemone nemorosa	and all the other species taken together.
10 sq. metres (1 Deciar)	10	29	71
1 sq. metre (1 Centiar)	20	51	49
$\frac{1}{10}$,, (1 Milliar)	50	68	32
$\frac{1}{100}$,, ($\frac{1}{10}$ Milliar)	200	86	14
Percentage determined by counting the shoots on 1 sq. metre	..	94	6

In all the investigations the quadrats were taken at random at different places distributed over Jonstrup Vang.

Gradually as the size of the unit area is decreased and the number of quadrats necessary is increased accordingly, the numerical expression of the valency of the dominant species rises while that of the subordinate species sinks correspondingly. Using $\frac{1}{100}$ sq. metre the result, as one sees, is that *Anemone nemorosa* makes up 86 per cent. of the vegetation. I consider this number, if we wish to aim at the most accurate result possible, is probably still too small.

In order to test this question by another method capable of giving an exact figure which can be compared with the figures we have already obtained, there is no alternative but to make use of weighing or counting. Since the plants here dealt with do not differ widely in the size of their shoots I have used the method of counting the shoots. At the beginning of June I counted the radical shoots of each species on a 1 sq. metre plot, chosen because it appeared at a distance to be typical of the *Anemone nemorosa*-Facies of the beechwood in Jonstrup Vang. A frame with sides 1 metre long and with cross wires dividing it into 100 square decimetres was applied to the *Anemone nemorosa*-Facies, and the number of shoots of all the species present were counted. In Fig. 105 I have mapped in the square metre investigated all the shoots which bore foliage leaves when the investigations were made. There were in all 5 species besides *Anemone nemorosa*, namely: *Gagea lutea, Ficaria verna, Oxalis acetosella, Melica uniflora,* and *Corydalis intermedia.* This is a comparatively large

214 STATISTICS OF PLANT FORMATIONS

number, for in the 20 sq. metres which I investigated by my own method (see Table 3) 5 species was the highest number found in a single square metre. This does not mean, however, that the number of shoots of plants occurring with *Anemone nemorosa* must be supposed to be higher in the square metre investigated than generally in the *Anemone nemorosa*-

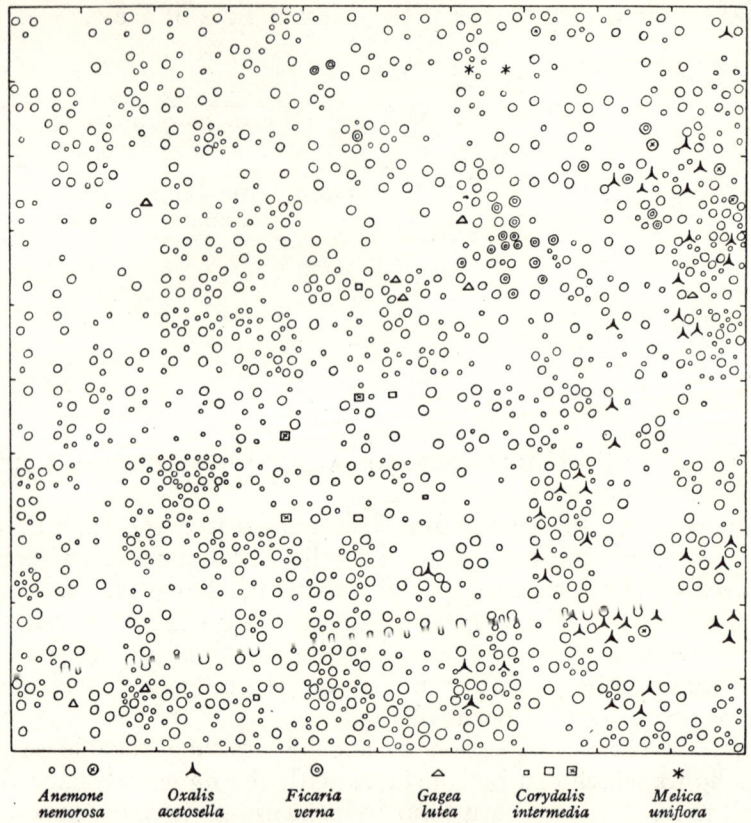

Fig. 105. *Anemone nemorosa*-Facies in the beechwood of Jonstrup Vang, 1 sq. metre (reproduced as $\frac{1}{100}$ sq. metre), showing the shoots of all the species present in this area. For *Anemone nemorosa* and *Corydalis intermedia* the size of the shoots is designated by a different size of the symbol used; and for both these species a × is included in the symbol for the shoots which bore flowers.

Facies in Jonstrup Vang. Of course this method can be tested more exactly by investigating more samples. That I have not done this is partly because, considering the innate deficiency of the method, I see no reason to regard it as certain that the investigation of more square metres would give a truer result, and partly because investigations of this kind of the number of portions of shoots inside a square metre take a great deal of time and are very wearisome, so that one is not tempted to deal with a larger area than is absolutely necessary.

Table 7 gives a conspectus of the number of shoots of the 6 species and of the valency reckoned from the number of shoots, regarding the valency of *Anemone* as 5. Of the 1,281 shoots of the *Anemone* only 5 were flowering. This is because the frame happened to be applied to a part where the *Anemone* vegetation was luxuriant and only contained a few flowering individuals. I did not see these flowering plants before making these investigations, which were carried out just after *Anemone nemorosa* had stopped flowering.

TABLE 7

The result of an investigation by the method of counting shoots of 1 sq. metre of the *Anemone nemorosa*-Facies in the beechwood at Jonstrup Vang.

	Number of shoots.	%	Valency.
Anemone nemorosa.	1,281	93·8	5
Oxalis acetosella	44	3·2	0·2
Ficaria verna .	23	1·7	0·09
Gagea lutea .	8	0·6	0·03
Corydalis intermedia	8	0·6	0·03
Melica uniflora	2	0·1	0·008

If we now calculate the proportion between *Anemone nemorosa* and the other scattered species, the *Anemone* amounts to about 94 per cent. and the other species to about 6 per cent. of the vegetation. This, I believe, will represent the true mass-relationship of the plants in question. It is seen from this that in using my method $200 \times \frac{1}{100}$ sq. metre areas give a figure that is somewhat too low for the dominant species, viz. 86 per cent. instead of 94 per cent. Using $50 \times \frac{1}{10}$ sq. metre areas the error becomes even greater, viz. 68 instead of 94 per cent. For practical reasons already described it is, however, best to use not too small a unit area; and considering the nature and aim of the method I believe a sufficiently accurate result can be obtained even with a $\frac{1}{10}$ sq. metre as a unit area. As can be seen from Table 1 not only by using a $\frac{1}{100}$ sq. metre but also with a $\frac{1}{10}$ sq. metre all the species obtain the same valency as that obtained by counting the shoots, when the valency is expressed in whole numbers from 1 to 5. If these degrees are halved or divided into tenths in determining the valency then it is true that *Gagea lutea* and *Oxalis acetosella* obtain half a degree too much when $\frac{1}{10}$ sq. metre is used. But from the nature of the case one must not draw important conclusions from the difference between two formations consisting merely in a disparity of half a degree between the valency of one or several species.

Before giving the result of my investigations of different formations carried out by means of the method described above I wish to explain shortly how this method, together with the life-forms of the plants, can be used to determine to which series of formations each individual formation belongs when using the life-forms which I have employed in characterizing and delimiting the plant climates.

Supposing we wish to characterize a formation other than by a floristic composition, for example by the life-form of the component species, it follows of itself that we must employ a scale of frequency of the same extent in biological investigations as one does in floristic investigation. If we wish to investigate the proportion between the different life-forms in a formation using the same method as that employed for investigation of plant climate, the individual species each counting as one unit, the results may differ widely, varying according to the completeness of the list made. For one thing the flora with which we are dealing is very limited and poor in species when compared with the vast phyto-climatic regions. Again, the formations have no sharp boundaries, so that in one place more and different species may occur than in another.

TABLE 8

Biological spectra of the *Anemone nemorosa*-Facies in the beechwood at Jonstrup Vang obtained by using the same method and the same unit areas that were used for determining of the valency of the species.

	Biological spectra.		
	Ch	H	G
1. When all the species count as equivalent	5·5	55·5	39
2. When the valency of the species is counted by means of 10 sq. metres as a unit area	..	50	50
3. When the valency of the species is counted by means of 1 sq. metre as a unit area	..	18	82
4. When the valency of the species is counted by means of $\frac{1}{10}$ sq. metre as a unit area	0·5	17	82·5
5. When the valency of the species is counted by means of $\frac{1}{100}$ sq. metre as a unit area.	..	7	93
6. When the valency of the species is counted by means of the method of counting shoots	..	5	95

If we still confine ourselves to the *Anemone nemorosa*-Facies in the beechwood at Jonstrup Vang, taking as our starting-point the list of flora given in Table 1, p. 208, which resulted from the investigations made with different unit areas, and if we count all the species as equivalent, then we obtain the biological spectrum given in Table 8, No. 1. This shows that the *Anemone nemorosa* community must be designated a Geophyte formation if we call it after the life-form of the majority of the species compared with the percentage of life-forms in the flora of the country as a whole, which contains only 11 per cent. of Geophytes. The *Anemone nemorosa*-Facies with which we are here dealing contains 39 per cent. of Geophytes.

But the geophytic character of the *Anemone nemorosa* vegetation becomes still more marked by the use of the method employed for determining the valency of the species. The proportion between the different life-forms may be computed either by means of the valency numbers in Table 1, which were obtained by investigations with different-sized areas, or, what comes to the same thing, direct use can be made of the numbers found by investigating the different areas, the valency being determined from these numbers. In Table 8 I have given a conspectus of the various biological spectra obtained by this means, using areas of different size in investigating the *Anemone nemorosa*-Facies in the beechwood of Jonstrup Vang. Thus if we take 10 sq. metres as a unit area it can be seen from the last column in Table 2 that *Anemone nemorosa* gives 10 Geophytes, *Ficaria verna* gives 4 Hemicryptophytes, &c. If we now transform by this means the valency of the species into the valency of the life-forms, Geophytes and Hemicryptophytes, which are the only life-forms occurring here, each obtains 16 points, i.e. 50 per cent. for each life-form. If the valency of the life-form and the biological spectrum derived from that valency be computed in the same manner, using unit areas of 1, $\frac{1}{10}$, and $\frac{1}{100}$ sq. metre, we obtain, as is seen in Table 8, respectively, 82, 82·5, and 93 per cent. of Geophytes. It will be noted that the last number approaches closely to the number obtained by counting the shoots, i.e. 95. Gradually as we use a method by which the valency agrees more and more closely with the true state of affairs, the valency of the life-form that characterizes the formation increases.

Using the smallest unit area, in the investigation concerned $\frac{1}{100}$ sq. metre, we approach most closely to the truth; but, as I shall emphasize later on, the truth cannot be determined exactly; one can only estimate it approximately. And since it is not our first care to determine the true mass-relationship, but rather to find a practical method capable of giving stable numbers which can be used in comparison, and which in some measure correspond with, and are fit to be used instead of the true mass-relationship, in my later investigations, as I have said several times, I have not used the smallest but the next smallest of the unit areas, namely $\frac{1}{10}$ sq. metre. Other unit areas, for example, $\frac{1}{25}$ sq. metre would perhaps do just as well; but for purposes of comparison it is better always to use the same size of area.

I shall now go on to give the result of some investigations of Danish plant communities made in the summer of 1909 by means of the method described, using $\frac{1}{10}$ sq. metre as a unit area. Even though the examples be few and scattered—time and opportunity allowed no more—and can in no way be looked upon as a concise treatise of the plant communities of Denmark taken as a whole, yet I have tried as far as possible to obtain examples from the different classes of communities. I shall therefore arrange them in the series I should employ if I were giving a complete account of the plant communities of Denmark. I shall begin then with a conspectus of these communities.

I have already given an account of the reasons which prompt me to believe that, at any rate for the time being, we must use the habitat, the environment, as the starting-point of our investigations of the formations. Now the factor which first and foremost determines the distribution of species in the country is indisputably soil humidity. If the soil were everywhere uniform and water everywhere contained the same substances, then the distribution of species would be determined almost entirely by soil humidity, which could be determined by analysis during different times of the period of growth or by the height of the ground above the water table at different times. This state of affairs is seen in the dune region on the island of Fanø. Here the communities follow one another in the same way in different valleys and hollows, where the height above the ground water is approximately the same. But if we consider the country as a whole we find that both water and soil differ in different places; and since I used the ground as a preliminary starting-point when treating the communities, I am obliged to arrange these in as many sections as there are combinations of different kinds of soil and water. I differentiate first between salt and fresh soil. Within the latter there are probably great differences in the composition of the water. This needs special investigation, which I have not carried out, so that I cannot found further divisions on the composition of the water. The kinds of soil I differentiate are clay, sand, and humus. From the nature of our country, however, imperceptible gradations between these soils are frequent, bringing about corresponding transitions between the different plant communities.

Confining ourselves to the main types we obtain:

   I. Formations of Fresh Soil
      A. Water Formations
      B. Land Formations
         a. Formations on clay
         b. Formations on sand
            1. Formations on drift deposits
            2. Formations on alluvial deposits
         c. Formations on humus
  II. Formations of Salt Soil
      A. Water Formations
      B. Land Formations
         a. Formations on clay
         b. Formations on sand
         c. Formations on humus

In the following account, however, I shall not keep quite to this order;

among other deviations I shall place the formations of humus after those of clays and sands upon which they abut. I shall omit the formations of water altogether, partly because they are with difficulty treated by my method, and partly because they do not present such difficulties as land formations when investigated by personal judgement.

Nearly everywhere in our country the formations are more or less affected by cultivation, and the greater part of the country consists of land actually under cultivation. Yet large expanses still remain which have been either little altered or where the formations may be supposed to differ but little from those present before the activity of man began to alter the flora. It is safe to say that enough of the primitive vegetation remains to enable us still to form an idea, at any rate of the main features, of the flora of the country before the days of agriculture. I am surprised that an inquiry of this kind into the original aspect of individual regions was not undertaken long ago. The reason such an inquiry was not made may perhaps be that botanists thought that there was so much of the original vegetation remaining that every one could see what it used to be like. But this point of view is scarcely the correct one. When we consider to what extent large uncultivated expanses of bogs, heaths, commons, &c., are more and more brought under cultivation, vanquishing the original plant communities, it is easy to understand that the time may not be far distant when it will be very difficult to form any idea of the original plant covering in the various districts and on the different kinds of soil. For example, if the time is at hand when the bogs of North Zealand are to become fields and meadows, it will then scarcely be possible by means of the available descriptions expressed in current terms to obtain any idea of how the country used to appear and of the former composition of the vegetation. Here then is a task which should be accomplished as soon as possible. If we do not set to work upon it now I believe that our descendants will have good reason to reproach us because we did not do the work when it was possible to do it, when sufficient remnants of the original vegetation were left to enable us to form a reasonable estimate of the former composition of the vegetation of different expanses of land. Our descendants will feel it a privation that they have no means of obtaining an idea of the appearance of a field or meadow when their ancestors set to work as farmers to make use of the land. In the few investigations therefore which I have undertaken by means of the method here described my attention has been turned to this question too.

I. GLACIAL CLAY DISTRICTS

A. Formations on Clay

The clay that is mixed with varying quantities of sand is for the most part under cultivation; only a small portion bears woodland which consists chiefly of beechwood. There is no reason to doubt that the present agricultural land on clay was originally covered with woodland, even if the original woodland was not as dense as the present woods which are under forest management. There must have existed, scattered here and there, open places, varying in extent, dominated by our rich Hemicryptophyte flora.

The upper and topmost layers of the wood consist entirely of Phanerophytes. I have not investigated this vegetation by means of my method. There is no reason to do so, because the beechwoods and coniferous woods usually consist of a single species. I have had no opportunity of investigating the oakwood. In other places, especially those within the Phanerophyte climate, it will be necessary to determine the degree of frequency of the species making up the Phanerophyte formations. Because of the size of the individual plants plots of large area must be used, e.g. areas of 100 sq. metres. It is the varying facies of the ground floras of the woods that I have investigated, including: (1) Oakwood, (2) Beechwood, and (3) Coniferous wood.

1. Ground Flora of the Oakwood

The forest of Jonstrup Vang, which lies at my door and which has therefore been the starting-point of my investigations, is now a beechwood, but before the beech trees stood where they now are the ground was occupied by oakwood. Of this oakwood H. Mortensen[1] wrote in 1872 'it will however in half a score of years be turned into beechwood, because the old oaks are constantly being cut down and the planting of beeches under them continues'. Jonstrup Vang is now a beechwood, apart from the alder swamps and a few small areas covered with conifers. There are still standing a few hundred scattered oaks, remnants of the ancient forest; but these are gradually being cut down. Only in small isolated areas is there any remnant of the ground flora of the original oakwood, and these remnants are becoming further and further supplanted by woods of beech and spruce. As communities they will soon disappear entirely, though here and there some of the species may persist for some time. For example: *Anthericum ramosum*, *Thesium ebracteatum*, and *Pulmonaria angustifolia* are each found in a single locality, and the last mentioned of the three is only represented by a single plant. Many others are on the brink of extermination, and must perish if the woodland meadows, whose edges in particular have afforded them a last refuge, become covered with forest, an aim that has now been pursued for a number of years. Finally the oakwood species will all disappear except those few that are able to grow beneath the deep shade of the beech trees, species which are already included in the ground flora of the beechwood.

[1] H. Mortensen, 'Nordostsjaellands Flora', *Bot. Tidsskr.* **2**, Raekke I, Bind 1872, p. 32.

'Maaløv Krat.' About 150 years ago the fields lying to the south of Jonstrup Vang between that forest and Ballerup-Maaløv were covered with wood, or at least were not farmed.[1] Presumably the ground was then covered with portions of scrub separated by open places of varying extent. The district is now called 'Maaløv Krat', and apart from individual properties the ground consists of fields belonging to different farms in Ballerup and Maaløv. The owners find these fields difficult of access, and they are indifferently cultivated. The boundaries of the fields are either stone walls partially fallen down, or else flat, in some places comparatively broad, strips of ground. These boundaries contain communities of plants which are the more or less altered remnants of the original vegetation that grew here when the land was first cultivated. Of course the numerical relationships between the species had already been altered by indiscriminate cutting and grazing; yet the species now found in these boundaries certainly belong to the original flora, except of course the few cultivated plants which have migrated from the fields. In some places the boundaries are formed of comparatively broad strips of scrub—oak, hazel, blackthorn, &c. I give below a list of the species making up this relatively original flora. Those species that have wandered in from the fields are omitted, and the list contains only plants which are found on mineral soil. The ground is beset with small bogs, some of which maintain their primitive condition, containing sphagnum, cotton grass, &c.; some of them have been wholly or partially converted into meadow. Some again have been cut by peat diggers and turned into ponds. The number of species growing only on humus is considerable; they are not included in the list.

List of the flowering plants found on mineral soil in the field boundaries of 'Maaløv Krat' (remains of the oakwood flora)

A.

Found principally in and near the parts covered with Scrub

a. Phanerophytes.
Salix capraea.
Corylus avellana.
Quercus pedunculata.
Rosa canina.
„ tomentosa.
Pyrus malus.
Crataegus oxyacantha.
Prunus spinosa.
Sambucus nigra.
Sorbus aucuparia.
Viburnum opulus.
Lonicera periclymenum.
„ xylosteum.

b. Chamaephytes.
Stellaria holostea.

c. Hemicryptophytes.
Avena elatior.
Poa nemoralis.
Urtica dioeca.
Arenaria trinervia.
Ficaria verna.
Viola hirta.
Geum urbanum.
Spiraea filipendula.
Rubus idaeus.
„ plicatus.

[1] Lutken, Ch., *Den Langenske Skovordning*, Kjöbenhavn 1899. See p. 48.

Astragalus glycyphyllos.
Epilobium montanum.
Anthriscus silvestris.
Chaerophyllum temulum.
Heracleum sphondylium.
Torilis anthriscus.
Primula officinalis.
Scrophularia nodosa.
Nepeta glechoma.
Campanula trachelium.
Tanacetum vulgare.
Lappa tomentosa.
Carduus crispus.
Taraxacum Gelertii.

d. Geophytes.
Allium oleraceum.
Gagea lutea.
Anemone nemorosa.

e. Therophytes.
Poa annua.
Galium aparine.

B

On parts devoid of Scrub

a. Chamaephytes.
Cerastium vulgatum.
Helianthemum chamaecistus.
Calluna vulgaris.
Veronica chamaedrys.
   ,,   officinalis.
Thymus serpyllum.

b. Hemicryptophytes.
Luzula campestris.
Agrostis alba.
   ,,   vulgaris.
Aira caespitosa.
   ,,   flexuosa.
Anthoxanthum odoratum.
Avena pubescens.
Dactylis glomerata.
Festuca ovina.
   ,,   pratensis.
   ,,   rubra.
Holcus lanatus.
Molinia caerulea.
Phleum pratense.
Polygonum amphibium.
Rumex acetosa.
   ,,   acetosella.
   ,,   auriculatus.
   ,,   crispus.
Silene inflata.
Stellaria graminea.
Viscaria viscosa.
Ranunculus acer.
   ,,   bulbosus.
   ,,   repens.
Viola canina.
Hypericum perforatum.
Polygala vulgare.
Saxifraga granulata.
Alchemilla vestita.
Agrimonia eupatoria.
Fragaria vesca.
Potentilla anserina.
   ,,   argentea.
   ,,   erecta.
Anthyllis vulneraria.
Lathyrus pratensis.
Lotus corniculatus.
Ononis repens.
Orobus tuberosus.
Trifolium medium.
   ,,   pratense.
   ,,   repens.
Vicia cracca.
Daucus carota.
Pastinaca sativa.
Pimpinella saxifraga.
Lysimachia vulgaris.
Plantago lanceolata.
   ,,   major.
   ,,   media.
Brunella vulgaris.
Origanum vulgare.
Galium verum.
Campanula rapunculoides.
   ,,   rotundifolia.
Jasione montana.
Knautia arvensis.
Succisa praemorsa.
Achillea millefolium.
   ,,   ptarmica.
Artemisia campestris.
   ,,   vulgaris.
Centaurea jacea.
   ,,   scabiosa.
Chrysanthemum leucanthemum.
Hieracium auricula.

Hieracium pilosella.
„ umbellatum.
Hypochaeris radicata.
Lappa minor.
Leontodon autumnalis.
„ hispidus.
Scorzonera humilis.
Senecio Jacobaea.
Solidago virga-aurea.
Taraxacum intermedium.
„ Ostenfeldii.
„ planum.
„ purpureum.
„ speciosum.
Tragopogon pratensis.

c. Geophytes.
Carex arenaria.

Carex hirta.
„ verna.
Agropyrum repens.
Holcus mollis.
Poa pratensis.
Linaria vulgaris.
Cirsium arvense.

d. Therophytes.
Medicago lupulina.
Trifolium minus.
„ procumbens.
Myosotis arenaria.
„ hispida.
Alectorolophus majus.
„ minor.
Galeopsis tetrahit.

Even if some of the species of the original oakwood flora have disappeared, and even if all the existing species are not included, there is no doubt that the list contains essentially the original oakwood ground flora of this locality, and of the corresponding portion of Jonstrup Vang. It is seen clearly from Table 9, that we have a marked Hemicryptophyte flora; even if we consider only the portions covered with scrub we find over 50 per cent. of Hemicryptophytes.

TABLE 9

'Maaløv Krat', the ordinary biological spectrum (giving each species one point) of the Flora on:

	Number of species.	Ph	Ch	H	G	Th
1. The parts with scrub	43	30	2	56	7	5
2. With no scrub	104	..	6	79	7·5	7·5
3. The whole district	147	9	5	72	7	7

In order to obtain some idea of the degree of frequency of the species and of the biological spectrum based upon that frequency I have investigated $50 \times \frac{1}{10}$ sq. metre plots in different localities within the area with no scrub. Similarly I have investigated two fields with the same soil, one of them the first year after it was left to itself (Fig. 106), the other after it had lain completely idle for eight years, being used neither for hay nor grazing. On this latter field oak and alder had long ago begun their invasion from the neighbouring wood. Table 10 gives the results of these investigations. In the list I have mentioned first those species which were found in over ⅗ths of the sample plots investigated, and which thus have a degree of frequency 4 and 5. The other species are given in systematic order, arranged alphabetically within the families. Here, as well as in the following lists of the same kind, the degree of frequency is indicated by the number obtained by investigating 50 samples. If it is required to reduce this number to the scale of 1 to 5 this can be done by moving the decimal point.

TABLE 10

'Maaløv Krat'

	Life-form.	Valency of the species ascertained by 50 $\frac{1}{10}$ sq. metre plots.		
		On annual field. 1	On octennial field. 2	In field boundaries. 3
Rumex acetosella	H	47	8	1
Anthemis arvensis	Th	41	..	..
Lolium perenne	H	44	..	..
Poa annua	Th	40	..	..
Alchemilla arvensis	Th	40	..	..
Scleranthus annuus	Th	39	..	..
Ranunculus repens	H	9	46	3
Poa pratensis	G	..	41	17
Leontodon autumnalis	H	..	33	10
Festuca rubra	H	..	45	50
Agrostis vulgaris	H	8	39	38
Achillea millefolium	H	1	48	37
Galium verum	H	..	..	37
Campanula rotundifolia	H	..	2	30
Luzula campestris	H	..	1	7
Carex hirta	G	..	..	3
„ verna	G	..	1	3
Agropyrum repens	G	18	28	11
Agrostis alba	H	..	10	12
Aira flexuosa	H	..	..	1
Airopsis caryophyllea	Th	7	3	..
Anthoxanthum odoratum	H	..	1	18
Avena elatior	H	..	4	11
„ pratensis	H	..	..	6
„ pubescens	H	..	..	4
Bromus mollis	Th	18	..	1
Cynosurus cristatus	H	..	..	5
Dactylis glomerata	H	6	7	12
Festuca ovina	H	..	..	6
„ pratensis	H	..	..	1
Holcus lanatus	H	..	25	6
„ mollis	G	..	..	4
Phleum pratense	H	4	22	16
Poa nemoralis	H	..	2	..
Secale cereale	Th	2	..	..
Allium oleraceum	G	..	..	1

Table 10—continued

	Life-form.	Valency of the species ascertained by 50 $\frac{1}{10}$ sq. metre plots.		
		On annual field. 1	On octennial field. 2	In field boundaries. 3
Rumex acetosa	H	..	6	16
Polygonum aviculare	Th	6	..	..
Arenaria trinervia	H	..	..	1
Cerastium vulgatum	Ch	21	..	1
Sagina procumbens	Ch	1	9	..
Spergula arvensis	Th	2	..	..
Spergularia rubra	Th	6	..	..
Stellaria graminea	H	..	3	11
,, holostea	Ch	..	..	5
Viscaria viscosa	H	..	..	1
Ranunculus acer	H	..	1	1
,, bulbosus	H	..	..	1
Draba verna	Th	1	..	..
Viola canina	H	..	..	8
,, hirta	H	..	..	4
,, tricolor	Th	16	..	..
Helianthemum chamaecistus	Ch	..	..	2
Hypericum humifusum	H	1	..	..
,, perforatum	H	..	..	17
Geranium molle	Th	1	..	..
Alchemilla vestita	H	..	..	1
Fragaria vesca	H	..	..	7
Potentilla argentea	H	7	..	1
,, erecta	H	..	..	4
Lathyrus pratensis	H	..	..	1
Lotus corniculatus	H	3	..	5
Ononis repens	H	..	..	5
Orobus tuberosus	H	..	..	2
Trifolium medium	H	..	3	26
,, minus	Th	..	1	1
,, pratense	H	1	1	..
,, repens	H	26	2	..
Vicia cracca	H	..	..	6
Anthriscus silvester	H	..	2	5
Pimpinella saxifraga	H	..	..	7
Epilobium montanum	H	..	1	1
Calluna vulgaris	Ch	..	..	1
Anagallis arvensis	Th	..	1	..
Primula officinalis	H	..	..	1
Echium vulgare	H	..	1	..
Myosotis versicolor	Th	..	1	..
Alectorolophus minor	Th	..	..	3
Linaria vulgaris	G	..	..	2
Veronica chamaedrys	Ch	..	2	20

TABLE 10—continued

	Life-form.	Valency of the species ascertained by 50 $\frac{1}{10}$ sq. metre plots.		
		On annual field. 1	On octennial field. 2	In field boundaries. 3
Veronica officinalis	Ch	..	..	3
„ serpyllifolia	H	4	1	..
„ verna	Th	1	..	..
Plantago lanceolata	H	..	..	26
„ major	H	13	..	..
„ media	H	..	..	1
Brunella vulgaris	H	29	12	2
Campanula rapunculoides	H	..	5	..
Knautia arvensis	H	9	8	17
Succisa praemorsa	H	..	..	4
Achillea ptarmica	H	..	..	1
Artemisia vulgaris	H	2	..	1
Bellis perennis	H	..	20	..
Centaurea jacea	H	..	..	1
„ scabiosa	H	..	..	2
Chrysanthemum leucanthemum	H	..	1	2
„ segetum	Th	28	..	..
Cirsium arvense	G	13	1	1
Filago minima	Th	1	..	..
Gnaphalium silvaticum	H	3	18	..
Hieracium auricula	H	..	..	1
„ pilosella	H	..	..	22
„ umbellatum	H	..	..	3
Hypochaeris radicata	H	..	9	..
Lampsana communis	Th	..	..	1
Senecio jacobaea	H	..	1	5
Taraxacum intermedium	H	..	1	1
„ planum	H	1	1	..
„ purpureum	H	1	9	7
Tussilago farfarus	G	2	2	..

In all three cases there are found on an average 10 to 12 species on each $\frac{1}{10}$ sq. metre, and in all three cases there are 5 to 6 species occurring in more than 66 per cent. of the samples. In the annual grass fields (Fig. 106) these 6 species are: *Rumex acetosella, Anthemis arvensis, Lolium perenne* (sown), *Poa annua, Alchemilla arvensis,* and *Scleranthus annuus*—5 of them are not found at all in the sample plots either of the 8-year-old field or of the field boundaries. Of these 6 species 4 are Therophytes.

The 8-year-old grassfield too has 6 dominant species, 1 Geophyte (*Poa pratensis*)[1] and 5 Hemicryptophytes (*Ranunculus repens, Leontodon*

[1] In spite of what I said before *Poa pratensis* is probably best regarded as a Geophyte and *Festuca rubra* as a Hemicryptophyte.

FIG. 106. 'Maaløv Krat.' Annual grassfield in June, characterized physiognomically by *Rumex acetosella* and *Anthemis arvensis*; the latter is very conspicuous in the picture. In the background to the right is seen a small area of scrub on one of the field-boundaries.

FIG. 107. Jonstrup Vang. *Anemone nemorosa*-Facies in beechwood.

autumnalis, Festuca rubra,[1] *Agrostis vulgaris,* and *Achillea millefolium*), of which 3 occur also in the field boundaries. The field boundaries themselves have 5 dominant species (*Festuca rubra, Agrostis vulgaris, Achillea millefolium, Galium verum,* and *Campanula rotundifolia*), which are all Hemicryptophytes.

The dominant species show us that the annual grassfields are characterized by Therophytes and the 8-year-old fields and the field-boundaries by Hemicryptophytes. The same is seen from the biological spectra made by taking the frequency grade of the species as a basis for determining the valency of the individual species. Table 11 shows that in the annual grassfield there are 42 per cent. of Hemicryptophytes and 48 per cent. of Therophytes, while the corresponding numbers for the 8-year-old grassfield are 82 and 1 respectively, and in the field-boundaries 87 and 1.

TABLE 11

'Maaløv Krat'; the biological spectrum of the flora made on a basis of the frequency grade of the species

	Number of species.	Points in 50 $\frac{1}{10}$ sq. metre.	Number of species per $\frac{1}{10}$ sq. metre.	Life-form.			
				Ch	H	G	Th
Annual grassfield	41	523	10·5	4	42	6	48
8-year-old grassfield	47	489	10	2	82	15	1
The field boundaries	78	613	12	5	87	7	1

Some more must be said about the various facies concerned:

The annual grassfield (Fig. 106). This is characterized physiognomically by *Rumex acetosella* and *Anthemis arvensis;* as are also many other localities within the region investigated (and outside it as well), where either no grass seed has been sown or where the fields are so badly farmed that cultivated plants are not in evidence. Besides the dominant species already mentioned, there were found also in 25 to 30 of the investigated plots the following 3 species: *Trifolium repens* (sown), *Brunella vulgaris,* and *Chrysanthemum segetum.* The last species reaches its climax as a weed among crops; it was dominant in the oats that occupied the ground before it was annual grassfield. It was very common in the annual grassfield, occurring in over 50 per cent. of the plots, but it led a miserable existence; most of the individuals were small and usually had unbranched slender stems each bearing one small head; some had the ray-florets small, but of the usual number, and in most the number of the ray-florets was reduced to below the normal. In some of these the ray-florets were large, in others they were much reduced in size and number. Last year I counted the ray-florets of *Chrysanthemum segetum* in

[1] See footnote on p. 226.

an oatfield which stood on ground now occupied by the annual grassfield. I found the maxima to be (as usual) 13 and 21. The same state of affairs prevails this year on the neighbouring fields. I cannot however give the exact number because I have lost the list I made. In order to compare the behaviour of *Chrysanthemum segetum* in the annual grassfield I must therefore use the result of an earlier investigation of the ray-florets of this species which I made on waste ground near Charlottenlund (see Table 12, A). In the same Table (in B) the result is given of an investigation of the slender individuals in the annual grassfield. There is no doubt that those individuals having a maximum of ray-florets (8) are descendants of individuals with a maximum of about 13 and 21. This is an example of the shifting of the apex of a frequency curve owing to depreciation of the environment. This behaviour is also seen in other Composites.

TABLE 12

The number of ray-florets in *Chrysanthemum segetum* growing on a waste ground near Charlottenlund (A) and in an annual grassfield in 'Maaløv Krat' (B)

Number of heads.	Number of ray-florets.	2	3	4	5	6	7	8	9	10	11
200	A	..	..	..	..	..	..	..	..	0·5	0·5
300	B	0·3	1·3	0·6	11·3	13·6	20·3	43·3	4·6	2·6	0·6
Number of heads.	Number of ray-florets.	12	13	14	15	16	17	18	19	20	21
200	A	1	16·5	16	12	8·5	9	10·5	3·5	7·5	14·5
300	B	0·6	0·3	..	..	..	..	..	..	..	..

The 8-year-old grassfield. Apart from the carpet of grass *Ranunculus repens* and *Achillea millefolium* are the two plants which characterize this community physiognomically. *Achillea millefolium* usually does not flower; *Ranunculus repens* flowers, it is true, everywhere, but is certainly on the down grade. *Trifolium medium*, *Campanula rotundifolia*, *C. rapunculoides*, and *Stellaria graminea* have begun to spread out from isolated centres forming dense colonies.

The Field Boundaries. The grassy covering is formed principally of *Festuca rubra* and *Agrostis vulgaris*. Apart from these two species the vegetation in early summer is characterized by *Hieracium pilosella* and *Veronica chamaedrys*. In the middle of summer *Achillea millefolium*, *Campanula rotundifolia*, and *Galium verum* are characteristic, as they very commonly are throughout Denmark. Here and there other plants may dominate, but looked at as a whole these three species in the middle of summer characterize the uncultivated high lying ground not only in Maaløv Krat but over a great part of Denmark wherever heath and dense wood are not prevalent.

Jonstrup Vang (the wood). On the higher parts of Jonstrup Vang, which corresponds to the portion of Maaløv Krat just described, there are no remains of the oakwood ground flora that can be looked upon as a formation, but the ground flora of the beechwood, which, though poor in species, makes such a wonderful display in April and May with its *Anemone nemorosa*-Facies, is encountered everywhere. However, on the boundary between the higher mineral ground, which is covered with beechwood, and the lower humus soil, which is covered by other formations, small remnants of the oakwood ground flora are still to be seen, partly on narrow strips of dry land on the boundary between the beechwood and meadows, partly on the corresponding soil separating the beechwood from some of the alder swamps, and also in one locality between the beechwood on the higher ground and the sprucewood on the lower, where the soil contains much humus. In this locality there is a short narrow strip devoid of trees, and on this strip there is a relic of the oakwood ground flora rich in species. This flora (A), and the flora in another somewhat similar locality (B) I have investigated thoroughly; the results of my investigations are given in Table 13.

A. This locality lies immediately to the north of the so-called 'Maaløv Mose' (bog) which is covered with alderwood and partly surrounded with sprucewood. The higher lying ground is sandy, and slopes gradually towards the south-west, ultimately becoming rich in humus. The following formations are found from north to south.

1. On the higher mineral soil beechwood with the usual ground flora, the *Anemone nemorosa*-Facies, of which something must be said later on.

2. At the margin of the beechwood a border about 5 metres broad of oak trees. Some of these trees are 12 metres high, some are quite small. Under these trees throughout about half their extent there is a ground flora characterized especially by *Carex montana* and *Anthericum ramosum*. This flora is however very recessive because of the shade, especially that cast by the progressive beeches. In the other half of the strip of oak trees a *Convallaria majalis*-Facies is seen. The result of my investigations of this facies is given in Table 13, No. 1.

3. A strip 5 to 8 metres broad devoid of trees. This is one of the few localities where some of the rarer plants of Jonstrup Vang still remain. I suppose, but this is a pure supposition, that it is due to the late training-college teacher, H. Mortensen, that this strip has been allowed to remain unplanted. Besides *Anthericum ramosum* and *Carex montana* several rare plants are found here, notably *Thesium ebracteatum* and *Trifolium alpestre*. Formerly *Pulmonaria angustifolia* grew here; but it has now disappeared. The growth of the surrounding trees continually reduces habitats suited to light-loving plants. I have investigated this vegetation partly beside the oak plantation and partly beside the spruce plantation forming the southerly and south-westerly boundary of the district.

Table 13

The result of the investigations of a few relics of oakwood ground flora of the low-lying ground in Jonstrup Vang. For Nos. 1–4 see text. In 1–3, because of the smallness of the area in question, only 25 samples were examined and the results obtained were multiplied by two for the purpose of comparison

	Life-form.	1	2	3	4
Convallaria majalis	G	50	2	..	48
Potentilla erecta	H	4	42	20	16
Orobus tuberosus	H	4	42	4	13
Aira flexuosa	H	6	14	48	39
Molinia caerulea	H	..	2	44	4
Luzula campestris	H	..	10	2	2
„ pilosa	H	..	..	..	1
Carex flacca	G	..	4	..	..
„ hirta	G	..	2	..	..
„ montana	H	24	14	2	2
„ pallescens	H	..	..	..	2
„ pilulifera	H	..	..	..	4
„ silvatica	H	..	..	..	10
Agropyrum caninum	H	..	4	..	..
Agrostis alba	H	..	2	2	..
„ vulgaris	H	..	18	2	14
Aira caespitosa	H	..	2	..	..
Anthoxanthum odoratum	H	..	2	..	..
Avena pratensis	H	..	22	4	12
„ pubescens	H	..	6	2	..
Briza media	H	..	4	2	..
Calamagrostis epigejos	G	12	2	..	7
„ lanceolata	H	..	22	..	3
Dactylis glomerata	H	..	..	..	11
Festuca ovina	H	..	6	2	1
„ pratensis	H	..	8	8	..
„ rubra	H	..	8	..	..
Holcus lanatus	H	2	8	..	..
„ mollis	G	..	4	..	3
Melica nutans	H	..	..	..	2
„ uniflora	H	..	..	..	2
Nardus strictus	H	..	2	..	..
Poa nemoralis	H	2	2	..	5
„ pratensis	G	..	4	..	1
Allium oleraceum	G	..	2	..	..
Anthericum ramosum	H	..	2	..	..
Majanthemum bifolium	G	..	..	..	25
Rumex acetosa	H	..	4	..	..
Anemone hepatica	H	2	..	..	..
„ nemorosa	G	24	16	8	27
Viola canina	H	..	24	8	..
„ silvatica	H	8	..	..	23
Hypericum perforatum	H	..	2	..	5
Oxalis acetosella	H	..	..	..	5

Table 13—continued

	Life-form.	1	2	3	4
Alchemilla vestita	H		4		
Fragaria vesca	H	8	12		2
Rubus idaeus	H				1
Spiraea filipendula	H		4		
Lotus corniculatus	H		2		
Trifolium alpestre	H		8		
Calluna vulgaris	Ch			12	
Lysimachia vulgaris	H				1
Melampyrum vulgatum	Th		2		
Veronica chamaedrys	Ch		26		2
„ officinalis	Ch			2	
Plantago lanceolata	H		6		
Clinopodium vulgare	H		2		
Stachys silvatica	H		2		
Campanula rotundifolia	H		8	8	5
Galium boreale	H		10		
„ verum	H		6		
Knautia arvensis	H		10		
Succisa praemorsa	H		4	8	
Achillea millefolium	H		22		
Hieracium pilosella	H		8	4	1
Leontodon autumnalis	H		4		
Scorzonera humilis	H		6	8	3
Number of species		12	51	21	34

The result of my investigation of the vegetation of a strip (*a*) about 2 metres broad close to the oak trees is given in Table 13, No. 2, and the result of a corresponding investigation of a strip 2 metres broad (*b*) nearest the spruce trees is given in Table 13, No. 3. The vegetation between *a* and *b* forms a transition between the two facies investigated.

B. The second locality lies along the east side of a bog immediately to the west of the Maaløv Mose described in A. The bog in question has a very uneven surface due to old peat digging; it is beset with willow, alder, bird-cherry, &c., and has a rich herbaceous flora. Between the bog and the beechwood towards the east there is a small strip of old moor with scattered oaks and a vegetation of oakwood plants very rich for such a soil. Where the shade is deepest there occurs, as in A, a *Convallaria majalis*-Facies, which, however, passes imperceptibly into vegetation on better illuminated ground, so that I have investigated the vegetation collectively and given the result in Table 13, No. 4.

It becomes apparent that three facies can be discriminated in the lower parts of the oakwood: firstly a *Convallaria majalis*-Facies, which prefers shade and is poor in species, secondly a *Potentilla-Orobus*-Facies (*Potentilla erecta, Orobus tuberosus*) which likes more light and is richer in species.

Both these facies are found on a soil more or less rich in humus. Thirdly we have an *Aira-Molinia*-Facies (*Aira flexuosa, Molinia caerulea*), which is found on the edge of the humus soil, and should probably be regarded as a formation belonging to this soil. The biological spectra (Table 14) made by determining the valency of the species in sample plots show that the *Convallaria majalis*-Facies is geophytic, while the *Potentilla-Orobus*-Facies and the *Aira-Molinia*-Facies are markedly hemicryptophytic. The vegetation described under B (Table 13, No. 4) answers most closely to the combination of *Potentilla-Orobus*-Facies and *Aira-Molinia*-Facies, and Table 14, Nos. 4 and 5, show that the biological spectra for the two localities are identical. The formations described here together with the vegetation in the field-boundaries of 'Maaløv Krat' are remnants, as already described, of the ground flora of the oakwood. But it does not, of course, include all the ground flora facies of the oakwood. Among these facies we have in the oakwood, just as in the beechwood, an *Anemone nemorosa*-Facies, even if that of the oakwood is richer in species than that of the beechwood, and not so markedly geophytic.

Table 14

The biological spectra of the various facies of the ground flora of the oakwood made on a basis of the results of the investigation in Table 13. The numbers 1–4 correspond with the columns so numbered in Table B.

	Points.	No. of species per $\frac{1}{10}$ sq. metre.	Life-form.			
			Ch	H	G	Th
1. Convallaria majalis-Facies	143	3	..	41	59	..
2. Potentilla erecta—Orobus tuberosus-F.	452	9	6	82	11·5	0·5
3. Aira flexuosa—Molinia caerulea-F.	200	4	7	89	4	..
4. Convallaria majalis—Aira flexuosa-F.	302	6	0·5	62·5	37	..
5. 1 and 2 together	..	6	3	62	35	..

If there is any original oakwood left in Denmark it is very important that it should be investigated as soon as possible and its various facies fully described.

2. The Ground Flora of the Beechwood

The facies of the ground flora that I have had the opportunity of investigating by means of the method of sample plots may be divided into three groups, in each of which the environment determines a different kind of vegetation: A. The ground flora on humus soil thickly covered with leaves and deeply shaded. B. The ground flora on more or less wind-swept, firmer, better illuminated soil with leaf-covering scanty or absent. Such situations occur especially on the outskirts of woods. C. The ground

Fig. 108. Bognaes Wood. *Allium ursinum*-Facies in beechwood.

Fig. 109. *Oxalis acetosella*-Facies in a sprucewood at Egebjergene near Jonstrup Vang.

STATISTICS OF PLANT FORMATIONS

flora on comparatively well-illuminated soil exposed to the wind, with few leaves and with more or less marked formation of raw humus.[1]

A. I have investigated the ground flora of deeply shaded soil with a thick leaf-covering in the following woods: Jonstrup Vang, Lille Hareskov, Nørreskov, Farum Lillevang, woods of Vallø, and in Bognaes. The results are given in Tables 15 and 16. I can only roughly estimate the age of these woods, which I suppose to be between 70 and 100 years, except the portion I investigated of Farum Lillevang, which is certainly considerably older, and certain parts of Jonstrup Vang, which are younger. Unfortunately I had neither time nor opportunity to determine the illumination of the woods investigated, but my impression was that Jonstrup Vang and perhaps Bognaes were the darkest. In these was found the poorest flora as far as the dominant species are concerned, containing as they did each only a single dominant species, *Anemone nemorosa* and *Allium ursinum*, respectively. In Jonstrup Vang, where I investigated 100 sample plots, the ground flora consisted of a well-marked *Anemone nemorosa*-Facies (Fig. 107) and in the portion of Bognaes Wood I investigated, where I examined 200 sample plots, there is a still more marked *Allium ursinum*-Facies (Fig. 108). It would be very interesting to know the cause of the marked difference between the ground flora of these two woods.

Farum Lillevang was the best illuminated of the woods investigated. The part of it I examined lies to the south of the Farum–Slangerup Road, and consists of enormous beeches whose tall stems stand 9–12 metres apart. The samples, 50 in number, were taken in the middle of the wood on ground thickly covered with leaves. The flora was found to be richer in dominant species than that of the other woods, there being on an average 6 species to every $\frac{1}{10}$ sq. metre. Four species occurred in over 66 per cent. of the samples; these were *Anemone nemorosa* (in all 50 samples), *Oxalis acetosella* (49), *Asperula odorata* (41), and *Milium effusum* (50). The next species in order of frequency were *Viola silvatica* (30) and *Melica uniflora* (29).

In Nørreskov, Lille Hareskov, and the two localities investigated in the woods near Vallø there were, as seen in Table 15, 2, 3, 3, and 4 dominant species respectively. *Anemone nemorosa* and *Oxalis acetosella* were dominant in them all; *Melica uniflora* was dominant in Lille Hareskov, *Asperula odorata* in the woods of Vallø, and with the latter *Stellaria holostea* (in another part of the same wood). In all of these localities three or four species per $\frac{1}{10}$ sq. metre were found.

It would be a difficult task to name briefly the facies where several species are dominant, but I take it they are somewhat as follows:

Anemone nemorosa-Facies (Table 15, No. 1).
Allium ursinum-Facies (Table 15, No. 7).

[1] Danish *Morbund*.

TABLE 15

The result of the investigation of the ground flora facies in beechwoods on leaf-covered humus in: 1. Jonstrup Vang, 2. Nørreskov, 3. Lille Hareskov, 4. Vallø Wood I, 5. Vallø Wood II, 6. Farum Lillevang, 7. Bognaes. For the rest see text.

	Life-form.	1	2	3	4	5	6	7
Anemone nemorosa	G	50	50	50	45*	46*	50	1
Oxalis acetosella	H	8	46	32	50	50	49	..
Melica uniflora	H	3	3	40	..	3	29	..
Asperula odorata	G	1	..	24	46	49	41	..
Stellaria holostea	Ch	1	..	4	..	45	13	..
Milium effusum	H	..	16	..	2	2	50	..
Viola silvatica	H	..	4	2	15	12	30	..
Allium ursinum	G	..	..	..	..	..	..	50
Juncus effusus	H	..	3	..	..	..	..	..
Luzula pilosa	H	..	..	..	..	2	1	..
Carex muricata	H	..	1	..	..	..	..	..
Agrostis alba	H	..	13	..	..	..	..	..
,, vulgaris	H	..	3	..	..	..	..	..
Aira caespitosa	H	..	9	..	..	3	..	..
,, flexuosa	H	..	..	..	..	1	3	..
Dactylis glomerata	H	..	..	..	..	..	8	..
Holcus lanatus	H	..	1	..	..	..	..	..
Poa nemoralis	H	1	4	..	..	2	5	..
,, pratensis	G	..	1	..	..	..	..	..
Schedonorus sp.	H	..	..	..	1	..	..	..
Gagea lutea	G	9	..	1	..	..	8	..
Majanthemum bifolium	G	..	..	..	..	..	9	..
Urtica dioeca	H	1	..	..	..	..	..	1
Stellaria nemorum	H	..	..	..	21	..	..	..
Anemone hepatica	H	1	..	..	..	..	4	..
,, ranunculoides	G	2	..	..	..	..	4	..
Ficaria verna	H	1	..	..	2	..	..	2
Corydalis intermedia	G	2	..	..	3	..	1	7
Mercurialis perennis	H	..	..	..	..	..	1	..
Rubus idaeus	H	..	12	..	..	..	..	..
Circaea lutetiana	G	..	..	1	..	..	..	..
Sanicula europaea	H	..	..	..	..	..	1	..
Veronica officinalis	Ch	..	3	..	..	..	..	..
Hieracium vulgatum	H	..	..	..	..	..	1	..
Lactuca muralis	H	..	..	..	..	..	3	..
No. of species.	..	12	15	8	9	11	20	5

Anemone nemorosa–Oxalis acetosella-Facies (Table 15, Nos. 2 to 6). This contained three sub-facies characterized by either *Melica uniflora* or *Asperula odorata*, or both, *Asperula odorata* and *Stellaria holostea* being dominant in addition to *Anemone nemorosa* and *Oxalis acetosella*.

Table 16 gives the biological spectra of the formations made on a basis

* Together with *Anemone ranunculoides*.

of the valency of the individual species found by means of sample plots. They show a geophytic character in varying degrees. This character is especially marked in the dark woods with a poor flora, Jonstrup Vang and Bognaes showing respectively 80 and 95 per cent. of Geophytes. In the other counts the Geophyte percentage varied between 30 and 51. The ground flora on humus with a thick covering of leaves is geophytic and poor in species.

TABLE 16

Biological spectra of the ground flora in various shaded portions of the beechwood on a basis of the results of the investigations given in Table 15. The numbers 1–7 correspond with the numbers of the columns in Table 15.

	No. of species.	Points.	Species per $\frac{1}{10}$ sq. metre.	Life-form.		
				Ch	H	G
1	12	80	1·3	1	19	80
2	15	170	3·2	2	68	30
3	8	154	3	3	48	49
4	9	185	4	..	49	51
5	11	215	4	21	35	44
6	20	311	6	4	60	36
7	5	61	1	..	5	95

B. **The ground flora of the beechwood on more or less wind-swept soil, which is firmer in texture and better illuminated, with leaf-covering scanty or absent (seen especially on the outskirts of woods).**

Whenever I have had the opportunity of studying the flora of these localities I have found that it consists physiognomically of a *Poa nemoralis*-Facies. But in different localities this facies varies very widely, the differences being due partly to differences in illumination, and partly to exposure to the wind, both factors being due to variation in density of the edge of the woods and the variation in the wind depending also on exposure. Thus the *Poa nemoralis*-Facies varies widely in different localities.

I have investigated this facies on the western side of Jonstrup Vang in a slightly elevated district bounded on the west by the bank, or dike, surrounding the wood[1] and on the north by a sprucewood. I have also tried by means of sample plots to obtain a numerical expression of this vegetation as it penetrates more and more deeply into the wood, where the *Poa nemoralis*-Facies passes imperceptibly into the ordinary *Anemone nemorosa*-Facies on the humus soil that has an ever-increasing covering of leaves. The method adopted consisted of taking 50 sample plots along lines parallel with the dike. The plots were as usual $\frac{1}{10}$ sq. metre in extent, and were taken 20, 30, 40, 50, and 60 metres from the dike.

[1] In what follows the Danish *Skovdige* is translated by the English 'dike'.

The following are the noteworthy points of what was seen along the individual lines:

20 metres from the dike: The ground fairly firm, partly with and partly without a thin covering of leaves; the vegetation comparatively dense, luxuriant *Poa nemoralis* and depauperate *Anemone nemorosa*; some *Milium effusum* and *Oxalis acetosella*. (See Table 17.)

Table 17

The result of the investigation of the vegetation at the margin of the wood on the west side of Jonstrup Vang, 20, 30, 40, 50, and 60 metres respectively from the dike bounding the wood.

	Life-form.	No. of metres from the wall of the wood.				
		20	30	40	50	60
Poa nemoralis	H	50	36	24	7	1
Milium effusum	H	9	30	9	2	..
Anemone nemorosa	G	38	41	38	47	49
Oxalis acetosella	H	10	37	23	21	..
Carex pilulifera	H	..	..	1	..	..
Aira caespitosa	H	..	..	1	..	..
Dactylis glomerata	H	3	..	..	..	..
Melica uniflora	H	..	..	..	1	..
Convallaria majalis	G	1	..	..	..	..
Gagea lutea	G	2	7	4	3	3
Stellaria holostea	Ch	..	1	..	..	..
Ficaria verna	H	..	2	..	..	..
Ranunculus auricomus	H	3	1	..	..	..
Viola silvatica	H	18	9	9	3	..
Epilobium montanum	H	..	..	..	1	..
Aegopodium podagraria	H	..	..	2	..	1
Anthriscus silvestris	H	..	4	..	..	..
Lactuca muralis	H	..	..	2	..	..
No. of species	..	9	10	10	8	4

30 metres from the dike: Darker; the soil essentially the same as in the last-mentioned place; *Poa nemoralis* less luxuriant; more *Milium* and *Oxalis*. (For details see Table 17.)

40 metres from the dike: Looser soil with more humus; fairly thick covering of leaves, but bare patches here and there. *Poa* failing; dense patches of luxuriant *Anemone nemorosa* and *Oxalis acetosella*. (See Table 17.)

50 metres from the dike: Humus; thick layer of leaves, but scattered, bare, wind-swept patches. *Poa* still more recessive; vegetation in patches consisting essentially of *Anemone nemorosa*; but in some patches *Oxalis* is dominant. (See Table 17.)

60 metres from the dike: Humus; very thick covering of leaves down

the east side of an evenly sloping hillock; large patches entirely devoid of vegetation because of the thick leaf-covering. (See Table 17.)

The results of these investigations are seen in Tables 17 and 18, they can be understood at a glance. The vegetation is open and poor in species; in the 250 samples investigated (5 ×50 $\frac{1}{10}$ sq. metres) there were found only 18 species. Along the individual lines reckoned from without inwards there were only 3, 3, 2, 2, and 1 per $\frac{1}{10}$ sq. metre. Table 18 also shows that going in the same direction the precentage of Geophytes rises from 30 to 96.

TABLE 18

The west side of Jonstrup Vang; the biological formation spectra on a basis of frequency grades given in Table 17

No. of metres from the wood-dike.	No. of species.	Points.	No. of species per $\frac{1}{10}$ sq. metre.	Ch	H	G
20	9	134	c. 3	..	69·5	30·5
30	10	168	c. 3	0·5	71	28·5
40	10	113	c. 2	..	63	37
50	8	85	c. 1	..	41	59
60	4	54	1	..	4	96

C. **The ground flora of the beechwood on comparatively light soil exposed to the wind, with scanty leaf-covering and without marked raw humus.** I have investigated this kind of soil at one place on the west side of Nørreskov with the result shown in Table 19.

TABLE 19

The west side of Nørreskov: the results of observations on the ground flora of the beechwood on rather well-illuminated more or less marked raw humus exposed to the wind and with poor leaf-covering.

	Life-form.	Points.
Aira flexuosa	H	49
Majanthemum bifolium . .	G	49
Anemone nemorosa . . .	G	37
Luzula pilosa	H	19
Milium effusum . . .	H	10
Hieracium vulgatum . . .	H	2

The ground is partially covered with mosses (*Hypnum, Polytrichum,* &c.); *Anemone nemorosa* is found in 37 of the 50 samples, but its vigour is fast failing, and it is usually flowerless. The vegetation may be called an *Aira flexuosa–Majanthemum*-Facies, which here consists of 6 species; 3 species per $\frac{1}{10}$ sq. metre; 48 per cent. Hemicryptophytes and 52 per cent. of Geophytes, the latter waning.

The flora of the felled beechwood. Before we leave the beechwood I want to say something about a single example of the flora occurring on the soil where the trees have been cut down. The great change that a flora undergoes as soon as the wood is felled is well known, but nevertheless interesting to follow. In the course of a few years the poor flora of the original beechwood is supplanted by one rich in species, whose composition may vary considerably, partly because of the altered nature of the ground (a factor in the dark beechwoods of little importance compared with light intensity) and partly owing to the relationship of the locality to the formations, especially those other than woodland, from which migration takes place.

As an example I have chosen a small area of beechwood in the middle of Jonstrup Vang, where felling has been going on during the last fifteen years or thereabouts. The felling has been so extensive that only a few scattered trees remain. The area has during the last few years been sown with beech, and the plants in the area investigated are about ·25 cm. high. The area is surrounded on all sides by dense wood, and nowhere does it abut upon formations not belonging to woodland. It is all the more strange therefore that the vast majority of the species that have migrated into the area on account of the altered illumination have come especially from formations other than woodland.

On an area scarcely $\frac{1}{2}$ hectare in extent I have investigated the flora of 50 $\frac{1}{10}$ sq. metre plots between the drills in which the beeches have been sown. These 50 plots contain 41 species, which are given in Table 20, together with their degree of frequency obtained by the ordinary method. The following species were, however, found besides those observed on the samples in the same area (the plants growing in the drills are included):

Carex pallescens	H		Veronica officinalis	Ch
Brachypodium silvaticum	H		Clinopodium vulgare	H
Poa trivialis	H		Campanula rotundifolia	H
Convallaria majalis	G		,, trachelium	H
Polygonatum multiflorum	G		Achillea millefolium	H
Urtica dioeca	H		Cirsium arvense	G
Ficaria verna	H		,, heterophyllum	H
Ranunculus acer	H		,, lanceolatum	H
,, auricomus	H		,, oleraceum	H
,, repens	H		,, palustre	H
Geum rivale	H		Hieracium auricula	H
Orobus tuberosus	H		,, vulgatum	H
Trifolium medium	H		Hypochaeris radicata	H
Vicia sepium	H		Lappa tomentosa	H
Chamaenerium angustifolium	Ch		Sonchus asper	Th
Pulmonaria officinalis	H		Tussilago farfarus	G
Scrophularia nodosa	H			

Together with the 41 species given in Table 20 there are altogether

74 species. The vegetation is thus rich in species, but it is not uniform; the species are scattered.

TABLE 20

Jonstrup Vang: species found in 50 $\frac{1}{10}$ sq. metre plots in felled beechwood, their degree of frequency and life-form

Anemone nemorosa . . . G 50	Anemone hepatica . . . H 1
Oxalis acetosella . . . . H 35	„ ranunculoides . . G 1
Rubus idaeus . . . . H 33	Viola silvatica . . . . H 13
	Hypericum perforatum . . H 1
Juncus effusus . . . . H 3	Mercurialis perennis . . . H 2
Luzula pilosa . . . . H 2	Fragaria vesca . . . . H 2
Carex muricata . . . . H 11	Epilobium montanum . . . H 5
Agrostis alba . . . . H 10	Veronica chamaedrys . . . Ch 5
„ vulgaris . . . . H 3	Stachys silvatica . . . . H 5
Aira caespitosa . . . . H 2	Asperula odorata . . . G 12
Anthoxanthum odoratum . . H 2	Gnaphalium silvaticum . . H 3
Dactylis glomerata . . . H 9	Lactuca muralis . . . . H 1
Festuca rubra . . . . H 1	Lampsana communis . . . Th 1
Melica uniflora . . . . H 11	Leontodon autumnalis . . . H 1
Milium effusum . . . . H 21	Senecio silvatica . . . . Th 3
Poa nemoralis . . . . H 17	Sonchus arvensis . . . G 1
„ pratensis . . . . G 2	Taraxacum Gelertii . . . H 17
Schedonorus sp. . . . . H 2	„ hamatum . . . H 2
Gagea lutea . . . . G 5	„ intermedium . . H 2
Majanthemum bifolium . . G 2	„ planum . . . H 3
Stellaria holostea . . . Ch 7	„ purpureum . . . H 13

The newcomers are for the most part Hemicryptophytes, and although the species density is inconsiderable, and although the original geophytic beechwood flora is present (*Anemone nemorosa* was found in all the samples) yet it is found by examining the biological spectrum based on the frequency numbers in the plots that the vegetation is essentially hemicryptophytic, with 72 per cent. of Hemicryptophytes and only 23 per cent. of Geophytes; the ordinary biological spectrum shows approximately the same numbers (see Table 21).

TABLE 21

Jonstrup Vang: Biological spectra of the ground flora of felled beechwood

		Ch	H	G	Th
1. Biological Formation spectrum . .	{ 41 Species = 322 Points }	4	72	23	1
2. The ordinary biological spectrum (i.e. each species reckoned one point) based upon species in the plots only . .	41 Species	5	73	17	5
3. The same, but all species found are included	74 Species	4	77	15	4

This is a rare case of there being only a slight difference between the ordinary biological spectrum and that founded on the basis of the degree of frequency (see Table 21).

The vegetation which thus springs up almost suddenly in the middle of a wood when the trees are felled, disappears equally abruptly when the young beech trees overshadow the ground. The conditions then become so unfavourable to ground vegetation that even the species belonging to the original ground flora of the beechwood lead a miserable existence, chiefly because the leaves of the young beech trees remain on until the new ones appear, and shade the ground even in spring, the season when the ground flora of the beechwood otherwise enjoys its short spell of sunshine. Only when the young beech trees grow taller and approach the age when they begin to cast their leaves in autumn does the ground flora begin to luxuriate. The period of transition is then passed, and a new *Anemone*-Facies appears in all its charm. In the white carpet of anemones the botanist may perhaps meet with a depauperate specimen of *Aira caespitosa*, *Juncus effusus*, or other species not belonging to the *Anemone* vegetation. He may wonder whence the plant has come, and how it has been able to live here. It is undoubtedly a relic, reminiscent of the transitional period in the wood, when for a short time the sun-loving plants, helped by man's interference, attempted the conquest of the floor of the forest.

3. The Ground Flora of the Coniferous Wood

The young beechwood with its long persisting withered leaves is a sore trial to the ground flora; even those species able to endure the deepest shade languish and droop. But even though the trial be severe and last several years, yet it is both mild and brief compared with the conditions prevalent on the floor of a dense sprucewood. The sprucewood not only gives denser shade than the beechwood, but the shade lasts as long as the wood remains closed. As soon as the young trees cover the ground, allowing scarcely a ray of light to reach the ground throughout the year, every trace of flowering plants disappears, and this condition persists until a breach is made by tempest or axe. Only then do species begin to colonize the wood, which however is still very dark, and only in those places where the light can reach the ground at some time of the day are the conditions possible for the seedlings which have sprung up from introduced seeds. The plants stand scattered, at first separated by wide intervals; here we see a *Lactuca muralis*, there a single *Taraxacum Gelertii*, an *Arenaria trinervia*, a *Veronica officinalis*, or other species able to exist in these conditions.

It is only when illumination has been improved by felling or by tempest, or where there is lateral illumination between the slender unbranched stems that colonization by gregarious species begins. These plants continue to press in serried ranks farther and farther into the wood as long as they find sufficient light.

In Egebjergene Wood, which is a continuation towards the east of

Jonstrup Vang, I have followed this colonization closely by means of sample plots. Here *Oxalis acetosella* is the first of the gregarious species to enter the sprucewood; then follows, where the ground is better illumined, *Aira flexuosa*, which as far as it spreads in a continuous carpet entirely obliterates the *Oxalis acetosella*-Facies. In other places other things may happen, depending principally on the species available near the margin of the wood. On the north side of Jonstrup Vang there is a sprucewood abutting towards the south on a woodland meadow, from which colonization has now begun. This I must say more about later on. Towards the north the same sprucewood extends to the dike of a wood, and from here has begun a colonization of plants belonging to the flora which from time immemorial has grown along the inner side of the dike. The farther one goes into the wood, naturally, the poorer and sparser the new-comers are found to be. On the outside a *Rubus idaeus*-Facies with *Urtica dioeca, Lampsana communis, Lactuca muralis, Cirsium oleraceum*, &c., then an *Urtica dioeca*-Facies with *Lampsana, Lactuca, Galium aparine, Epilobium montanum*. Finally, farther in still, we find a scattered vegetation consisting especially of *Lactuca muralis, Arenaria trinervia, Oxalis acetosella*, and a few others.

Egebjergene. The sprucewood I investigated most accurately lies eastward in Egebjergene, abutting towards the north and the south on beechwoods. In the middle of this vegetation hardly any light reaches the ground except that which comes between the dense crowns of the trees: here a ground flora is absent or consists of but few individuals. Towards the north this condition prevails right up to the beechwood. On the other hand at the southern edge of the sprucewood lateral illumination penetrates the forest, especially before the beech trees unfold their leaves, and even after that the illumination is better here than it is in the middle of the sprucewood. In this region colonization by a ground flora has begun with scattered individuals, differing widely, however, from the ground flora on the edge of the beechwood, in showing a well-marked *Oxalis acetosella*-Facies (Fig. 109, p. 233).

Although the sprucewood, at any rate its margin, is on the same soil as the beechwood, and the illumination at the base of the outer spruce trees is essentially the same as that in the neighbouring beechwood, yet the boundary between the ground floras of the two woods is peculiarly sharp. The reason for this difference certainly lies in the different properties of the fallen leaves. The ground flora of the beechwood with its *Anemone nemorosa*-Facies stops abruptly where the layer of needle leaves begins, that is to say just outside the outermost spruce trees, where an almost pure *Oxalis acetosella*-Facies begins. In the outermost portions of this facies one sees scattered individuals of *Anemone*. As these anemones appear to thrive, we may assume that it is not the chemical composition of the soil, but rather the physical properties of the needles that stop this plant from spreading. Fig. 109 and Table 22, No. 4, show the vegetation in this *Oxalis acetosella*-Facies.

Towards the north-west the sprucewood abuts upon a small sphagnum bog which has recently been planted with spruce. The sprucewood nearest to the bog is open and well illuminated, but towards the south-west it becomes darker and darker. From the bog to the south-east the following formations occur, a series seen in many other localities:

I. **Sphagnum Moor** with *Eriophorum vaginatum, E. polystachyum, Oxycoccus palustris,* &c.

II. **Grass Moor**—*Molinia caerulea*-Facies along the edge of the sphagnum moor.

III. **Spruce wood.** Ground flora:
  1. *Aira flexuosa*-Facies in open sprucewood (see Table 22, No. 1).
  2. *Aira flexuosa–Oxalis acetosella*-Facies (see Table 22, No. 2).
  3. *Oxalis acetosella*-Facies (see Table 22, No. 3). Denser and darker sprucewood.
  4. No phanerogamic ground vegetation. Dense and dark sprucewood.
  5. *Oxalis acetosella*-Facies in the southern part of the sprucewood abutting on the beechwood (see Table 22, No. 4).

IV. **Beechwood.** Ground flora: *Anemone nemorosa*-Facies.

Table 22

Egebjergene: Conspectus of the results of investigations in different facies of the flora of the sprucewood. $50 \tfrac{1}{10}$ sq. metre. 1. *Aira flexuosa*-Facies; 2. *Aira flexuosa–Oxalis acetosella*-Facies; 3 and 4. *Oxalis acetosella*-Facies.

	Life-form.	1	2	3	4
Aira flexuosa	H	**50**	**50**	..	..
Oxalis acetosella	H	1	**50**	**50**	**50**
Aira caespitosa	H	..	1	..	..
Dactylis glomerata	H	1	1	..	..
Milium effusum	H	2	2	..	..
Poa pratensis	G	1	..	..	..
Arenaria trinervia	H	..	..	1	3
Stellaria holostea	Ch	2	..	..	..
„ media	Th	..	..	..	1
Ranunculus auricomus	H	..	..	..	1
Mercurialis perennis	H	..	..	1	3
Rubus idaeus	H	2	5	1	1
Calluna vulgaris	Ch	..	1	..	..
Asperula odorata	G	..	..	4	1
Galium aparine	Th	..	..	..	1
Lactuca muralis	H	4	6	2	4
Taraxacum Gelertii	H	..	2	..	2

Table 22 gives a conspectus of the results of an investigation carried out with sample plots in the different facies of the ground flora of the sprucewood in the localities mentioned above. Table 23 shows that the ground flora of the sprucewood is markedly hemicryptophytic, the biological formation spectra showing from 93 to 99 per cent. of Hemicryptophytes.

TABLE 23

Sprucewood: 1–4. Egebjergene: biological formation spectra on the basis of the degrees of frequency given in Table 22. The numbers correspond with those in Table 22.
5. Jonstrup Vang: biological formation spectrum on the basis of the degrees of frequency given in Table 24, Nos. 1–6.

	No. of species.	Points.	Ch	H	G	Th
1. Aira flexuosa-Facies	8	63	3	95	2	..
2. Aira flexuosa–Oxalis acetosella-Facies	9	118	1	99	..	..
3. Oxalis acetosella-Facies	6	59	..	93	7	..
4. ,, ,,	10	67	..	96	1	3
5. Festuca ovina–Dactylis-Facies. 300 $\frac{1}{50}$ sq. metre	30	293	..	92·5	7·5	..

Jonstrup Vang. On the northern edge of Jonstrup Vang there is a woodland meadow (No. 11), which abuts upon spruce trees 15 to 17 metres high growing on rather sandy soil, which, nearest the meadow, is but little higher than the humus-containing soil along the edge of the meadow. On the narrow transitional slope immediately on the outside of the outermost row of trees there is found a luxuriant ground vegetation rich in species, characterized by *Galium boreale, Centaurea jacea,* and *Cirsium acaule,* &c. Up to 2 metres from the ground even the outermost spruce trees are entirely devoid of branches either living or dead. The next 6 metres of stem have numerous dead branches but only few living ones. Some light therefore enters the wood from the south side, allowing some of the plants from the woodland meadow to migrate into the sprucewood; but they cannot get far in, as the light decreases rapidly and soon becomes too weak for them.

I have used the method of sample plots for determining this migration over a distance of 70 metres, investigating the vegetation of 50 plots in the successive intervals between the rows of spruce trees. The result is given in Table 24. The intervals between the rows of trees are about 1¼ metres broad; in Table 24 they are numbered from without inwards, that is to say from the meadow into the wood. The three series of numbers at the end of Table 24, taken in conjunction with the biological formation spectrum in Table 23, No. 5, show sufficiently clearly the character of the vegetation and its invasion. The invasion decreases, by degrees but rapidly, both in the number of species and individuals.

From the intervals 1–6 the number of species sinks from 26 to 2; in interval No. 7 only a solitary individual was seen, a seedling of *Dactylis glomerata*.

As in Egebjergene we have here a very marked Hemicryptophyte vegetation invading the floor of the sprucewood.

B. Formations on Humus

I have investigated these formations particularly in Jonstrup Vang and 'Maaløv Krat'. They demonstrate most of the principal types of formations found on humus in Denmark.

The surface of the ground in Jonstrup Vang is very diversified, ridges and elevations continually separating valleys and depressions varying in depth and in their relationship to water. Some of the depressions are considerable in extent, but most of them are small, some extending only a few metres. All have a layer of humus varying in thickness, and often exceeding a metre in depth. Most probably these layers of humus have all at one time been sphagnum bogs, which have been so altered by cultivation that it would be difficult or impossible to obtain any idea of their formation and their original flora were it not for the existence of a few places where transitional states are still to be seen.

In the bogs that have undergone the greatest change not even microscopic examination of the humus will elucidate its origin and method of formation, because the influence of digging and of the new vegetation has so transformed the turf that in dry summer weather it assumes the appearance of a loose peat in which the structure of the plants cannot be determined apart from twigs and stems, some of them of oak.

Drainage makes it impossible for us to infer from the present level of the water-table anything about the wetness of the bogs when man first encroached upon them. Most of the bogs have been presumably covered, at any rate in parts, with *Calluna*; but it is difficult to decide whether some of them were not at one time too moist to allow *Calluna* to grow. It is true that there are now wet sphagnum bogs of this kind both inside the wood and outside it; but we do not know whether this was their original condition or whether it is rather that these bogs have once been higher and drier and covered with *Calluna*, but have been wholly or partially dug out for peat and then been overgrown, and are now at the stage of sphagnum bog. If an accurate investigation were made of the ground and of the depth of the drainage trenches, comparing these with the unaltered bogs, we might perhaps be able to decide with some certainty the stages the various bogs had reached when man first encroached upon them. But I have not yet made an investigation of this kind.

Of the bogs in Jonstrup Vang there are two, Dommermose and Skrædermose, which still have something of their original character,

Table 24

Jonstrup Vang. Invasion from the margin of the field into the sprucewood determined by means of sample plots. 1–6: the 6 outermost intervals between the rows of spruce trees numbered from the edge of the meadow into the wood.

	Life-form.	1	2	3	4	5	6
Festuca ovina	H	31	32	14	4	..	..
Dactylis glomerata	H	18	24	28	17	7	1
Carex flacca	G	2	..	..	..	..	..
„ hirta	G	6	3	1	..	..	..
„ acutiformis	G	1	..	..	..	..	..
Agrostis vulgaris	H	1	..	..	..	..	..
Aira caespitosa	H	..	1	..	..	..	..
„ flexuosa	H	7	7	2	1	..	..
Anthoxanthum odoratum	H	3	2	1	1	..	..
Avena elatior	H	2	4	1	2	1	..
„ pratensis	H	6	2	..	..	..	..
Festuca arundinacea	H	2	2	..	..	..	..
„ rubra	H	4	1	1	..	..	..
Holcus lanatus	H	..	1	..	..	..	..
Poa nemoralis	H	1	2	..	..	..	..
„ pratensis	G	3	4	..	..	..	..
Viola silvatica	H	1	..	..	..	..	..
Rubus idaeus	H	2	..	..	..	..	..
Lathyrus pratensis	H	2	..	..	..	..	..
Trifolium pratense	H	1	..	..	..	..	..
Vicia cracca	H	1	..	..	..	..	..
Galium boreale	H	2	..	..	..	..	..
Campanula rotundifolia	H	3	1	..	..	..	..
Achillea millefolium	H	1	..	..	..	..	..
Centaurea jacea	H	1	..	..	..	..	..
Cirsium arvense	G	2	..	..	..	..	..
Hieracium auricula	H	..	..	1	..	..	..
„ pilosella	H	11	1	..	..	..	..
Lactuca muralis	H	1	..	..	..	..	..
Taraxacum Gelertii	H	..	1	3	3	..	1
Points	..	115	88	52	28	8	2
No. of species	..	26	16	9	6	2	2
No of species per $\frac{1}{10}$ sq. metre	..	c. 2·4	c. 1·8	c. 1	c. 0·6	c. 0·2	c. 0·04

the former consisting almost entirely and the latter partially of sphagnum bog. The rest of them, over 30 in number if one includes the smallest, have no trace left of their original character. A few have been dug out for peat and so transformed into lakes. Most of them are covered with wood, chiefly alder, among which are found on the higher grounds a little birch, ash, and spruce. Some, especially the larger ones, have become poor woodland meadows, which of late years have been partially planted with alder, birch, and spruce. Under these three headings:

(1) Sphagnum bog, (2) Alder swamp, (3) Woodland meadow, I shall now describe the results of the investigations which I have carried out to elucidate the use of my method of sample plots.

1. Sphagnum Bogs

A. **Maaløv Krat**, &c. Looking at that portion of the ordnance survey map which includes the country south of Jonstrup Vang and Hareskov one meets with a veritable 'Land of a Thousand Lakes'. The teeming little blue patches on the map represent ponds or lakes which probably all owe their origin to peat-digging. In addition to these there are numerous other bogs which have been only partially dug and then overgrown with vegetation consisting partly of phanerogamic bog plants and partly of *Sphagnum*, but devoid of clear water so that they are not marked on the map as lakes.

Many of these lakes and bogs are very interesting; firstly because they show a series of stages in the transformation of lake into dry land, and secondly because they exhibit within a remarkably small compass the succession of zones of vegetation that are dependent partly upon the height of the ground above the water-table and partly upon the difference in the composition of the water in the middle of the bogs and beside the cultivated fields. They show typical processes so conveniently on a small scale that they may be looked upon as 'samples' or 'museum specimens'. Their surroundings vary greatly; some of them are mere depressions in the higher, comparatively dry, surrounding fields.

It goes without saying that all these bogs are doomed to extinction. It is only because they are in sequestered fields far from villages that they have been allowed to remain so long undisturbed by cultivation. But in recent years cottars and gardeners have begun to settle in the district, sealing the fate of the bogs. *Andromeda polifolia* and *Scheuchzeria palustris* still grow, and the white heads of *Eriophorum* still glisten in the June sunshine above sundews and cranberries. But intensive cultivation is above all things destructive to romantic relics of this kind. First the low wet portions of the bog are filled up with all kinds of refuse from house and garden, as shards, pots, pans, bottles, jam-jars, and even old cycles. Then earth is spread on the surface and in a few years the useless bog is transformed into a prosperous plot in a well-cultivated field, so that one would never suspect that a few years ago the space was occupied by a remnant of our oldest and most characteristic plant community.

Would that the Society for Preserving Nature would start a campaign to protect and preserve some of the bogs in North Zealand for our descendants, who will perhaps value them more highly than the present generation appears to do. No time should be lost.

I have used for my investigations two of the bogs (Nos. 3 and 4; see below) which are doomed. They lie on a high ridge of land to the south

FIG. 110. A little bog in Maaløv Krat dug out and filled with water: *Calamagrostis lanceolata*-Facies above, *Carex stricta*-Facies beside the water, and *Potamogeton natans*-Facies in the water. In the background is seen a portion of the field-boundary with scrub.

FIG. 111. Lyngby Bog. *Calluna–Oxycoccus*-Facies in early summer; physiognomically characterized by *Eriophorum vaginatum*.

STATISTICS OF PLANT FORMATIONS

of the west end of Jonstrup Vang. But first let me make a few observations about two other bogs (Nos. 1 and 2; see below) in the same neighbourhood. These four bogs taken together give a picture of what is happening in this region and at the same time show the course of succession.

Bog No. 1 (Fig. 110) lies close beside the road which runs between Blide (Sheet 23*b*, Maaløv village and parish) and the Ballerup–Jonstrup Road. It is an oval bog about 30 metres in length which has been dug and has filled up with water, and shows very clearly how small lakes become overgrown when they are entirely surrounded by cultivated fields. At least 30 species grow in this bog which are not found in the surrounding land. From the edge outwards three facies can be distinguished.

1. *Calamagrostis lanceolata*-Facies along the edge. Here, especially in the outer portion, is found the bulk of the species (*Iris, Alisma plantago, Sparganium simplex, Carex rostrata, C. vesicaria, Baldingera, Lysimachia thyrsiflora, L. vulgaris*).
2. *Carex stricta*-Facies: A wide belt of tussocks over a metre high standing out of the water like pillars, each surmounted by a crown of leaves projecting in all directions. In autumn, winter, and spring the entire plants are usually covered with water; in summer, when the water-level is much lower, only the bases of the pillars are submerged.
3. *Potamogeton natans*-Facies, the leaves covering the surface of the water completely.

Bog No. 2 lies about 130 metres to the east of No. 1, which it resembles in form, but it is slightly smaller. This bog represents the 'High Moor' in miniature. Outside a marginal facies, which is comparatively rich in species, there is a narrow swamp covered by water even during the summer, with scattered tufts of *Carex stricta* and with *Carex rostrata*. Outside that comes the sphagnum bog.

1. Marginal Facies.
2. Low swamp covered even in summer with water; scattered tufts of *Carex stricta; Carex rostrata*.
3. Swamp with closed vegetation; *Menyanthes–Potentilla palustris*-Facies with *Lysimachia thyrsiflora, L. vulgaris, Calamagrostis lanceolata*.
4. *Polytrichum*-Facies, with but little *Sphagnum* hidden among *Polytrichum*; scattered *Eriophorum polystachyum, Oxycoccus palustris, Peucedanum palustre, Viola palustris, Calamagrostis lanceolata,* and *Carex stricta*.

The north-western and northern portion of the bog consists of a swampy willow scrub.

Bog No. 3 lies about 150 metres south-east of No. 2, which it surpasses considerably in size. Where conditions are at their best the following facies from the margin outwards can be distinguished:

1. Higher Marginal Facies, with *Aira caespitosa*, *Molinia caerulea*, *Potentilla erecta*, *Hypericum perforatum*, &c.
2. Lower Marginal Facies, most closely allied to *Juncus effusus*-Facies, with *Agrostis canina*, *Hydrocotyle*, *Potentilla erecta*, *Peucedanum palustre*, *Lysimachia thyrsiflora*, and *Galium palustre*.
3. Water with *Menyanthes*-Facies; *Potentilla palustris*, *Carex rostrata*, and tussocks of *Carex stricta*.
4. Quaking Bog and *Sphagnum*, not bearing one's weight, with *Menyanthes*, *Lysimachia thyrsiflora*, *Carex canescens*, *Peucedanum palustre*, and *Potentilla palustris*.
5. Quaking Bog and *Sphagnum*, firmer than No. 4, bearing one's weight, but when you stand on it the water may cover it. As seen in Table 25, No. 1, the sphagnum bog at this level is here characterized by *Carex rostrata*, *C. canescens*, *Agrostis canina*, and *Oxycoccus palustris*, all of which were found in over ⅔ths of the investigated samples. In other bogs other species may be dominant. On a corresponding level in Lyngby Bog, *Carex lasiocarpa* and *Eriophorum alpinum* play an important part.
6. The central slightly higher portion with *Eriophorum vaginatum–Oxycoccus palustris*-Facies (see Table 25, No. 2).

Bog No. 4 lies immediately to the west of the garden of the house marked on Sheet No. 8*b*, Maaløv village and parish. It is an oval sphagnum bog barely 50 metres long. The edge of the bog beside the cultivated field is occupied by a narrow belt without *Eriophorum* or *Sphagnum* but carpeted with *Hypnaceae*. In this belt are found *Equisetum limosum*, *Carex canescens*, *C. stricta*, *Potentilla palustris*, *Menyanthes trifoliata*, *Galium palustre*, and a few other species. The central and larger part is a sphagnum bog of the *Eriophorum–Oxycoccus*-Facies (Table 25, No. 3), with a little *Polytrichum* and *Calluna* on its highest parts.

By way of supplement to observations of the bogs discussed above which are situated in Maaløv Krat I will add a few observations about Lyngby Bog and Sækkedammen in Ruder Hegn. It would be interesting to give a full account of Lyngby Bog made on a basis of the method of sample plots, and then to compare the results with those of corresponding investigations of a series of other bogs in different parts of the country. I have, however, investigated by this means only three of the many facies present in Lyngby Bog. The result of one of them, a *Calluna–Oxycoccus*-Facies (Fig. 111) is given in Table 25, No. 4. It will be observed that 4 species are dominant here, *Calluna*, *Empetrum*, *Eriophorum vaginatum* (Fig. 111), and *Oxycoccus*. In order to make the name of this facies conveniently short I will use only the two dominant species, which represent respectively the highest and lowest levels. On a slightly lower level we have a facies corresponding with that in Table 25, Nos. 2 and 3, where *Calluna* and *Empetrum* are absent, but where *Oxycoccus* dominates in the *Sphagnum* and where, at any rate over large stretches, *Carex lasiocarpa* is also dominant.

TABLE 25

The result of using the method of sample plots in the investigation of formations in Sphagnum bogs. Nos. 1-2 in Maaløv Krat; 3. Another bog in Maaløv Krat; 4. Lyngby Bog; 5. Sækkedammen, a bog in Ruder Hegn. (See text.)

	Life-form.	1	2	3	4	5
Carex rostrata	(HH) G	50	36	23	7	..
Agrostis canina	H	39	1	16	..	..
Eriophorum polystachyum	G	11	37	32	..	..
Carex canescens	H	36	23	16	..	..
Oxycoccus palustris	Ch	49	49	41	48	21
Eriophorum vaginatum	H	2	46	48	48	38
Empetrum nigrum	Ch	..	..	..	47	50
Calluna vulgaris	Ch	..	4	3	30	27
Scheuchzeria palustris	H	1	15	..	2	..
Juncus effusus	H	1	..	..	..	..
Carex Goodenoughii	G	4	6	..	..	..
„ lasiocarpa	(HH) G	..	..	..	1	..
„ limosa	G	..	..	..	1	..
„ stricta	H	..	..	26	..	..
Molinia caerulea	H	2	1	16	..	..
Viola palustris	H	4	2	3	..	..
Drosera rotundifolia	H	7	24	..	26	..
Potentilla erecta	H	..	1	..	..	..
„ palustris	(HH) G	11	..	6	..	..
Epilobium palustre	H	..	..	1	..	..
Peucedanum palustre	H	25	5	18	..	..
Andromeda polifolia	Ch	..	14	18	..	..
Vaccinium uliginosum	N-Ch	..	2	..	..	2
Lysimachia thyrsiflora	(HH) G?	10	4	..	..	..
Menyanthes trifoliata	(HH) G?	2	10	5	11	..
Galium palustre	H	..	..	1	..	..
Points		254	280	273	221	138
No. of species		16	18	16	10	5
No of species per $\frac{1}{10}$ sq. metre		c. 5	5·6	5·5	4·4	2·7

Where the ground becomes higher and drier than in the facies represented in Table 25, No. 4, the *Sphagnum* gradually dies, *Oxycoccus* and *Eriophorum* begin to fail, and *Cladonia* to come in. This is seen in a bog called Sækkedammen in Ruder Hegn. Years ago I knew this bog, which was then covered with *Calluna*. But when I looked for it again, meaning to apply my method of sample plots to obtain a picture of this stage of development, I found that the bog had been thoroughly drained and planted with spruce. In one part of it, however, the spruces were so young that the original flora of the bog could be seen unaltered between the rows of trees. On the 6th of July 1909, the day on which I made the investigation, the surface of the bog was 60–70 cm. higher than the level of the water in the ditches. The result of the investigation of this *Empetrum–Calluna*-Facies is seen in Table 25, No. 5. It will be observed that *Eriophorum vaginatum* is indeed still present in 38 of the samples, but it is on the down grade, nearly everywhere flowerless, and overgrown with *Calluna, Empetrum,* and Reindeer Moss.

Taken together, Nos. 1–5 in Table 25 illustrate the composition of the vegetation in five of the facies of the bogs, proceeding from a lower to a higher level, from the wet to drier ground. Table 26 shows the corresponding formation-spectra, which demonstrate that we are here dealing with Chamaephyte formations. The Chamaephyte percentage rises with the increases in height above the ground water, from 19 per cent. of Chamaephytes in the *Carex rostrata–Oxycoccus*-Facies to 71 per cent. in the *Empetrum–Calluna*-Facies. At the same time the Cryptophyte percentage decreases in this example from 35 to 0. The formations are poor in species, at an average there are 3–6 species to $\frac{1}{10}$ sq. metre. The poorness in species increases with the percentage of Chamaephytes.

Table 26

Biological formation-spectra of the facies given in Table 25 of the vegetation of sphagnum bogs. The numbers are the same as in Table 25

	Points.	No. of species.	No. of species per $\frac{1}{10}$ sq. metre.	N	Ch	H	Cr
1	254	16	c. 5	..	19	46	35
2	280	18	5·6	1	24	42	33
3	273	16	5·5	..	23	53	24
4	221	10	4·4	..	57	34	9
5	138	5	2·7	1·5	71	27·5	..

B. Jonstrup Vang. Outside the wood the bogs are transformed into meadow or elevated by the addition of soil and converted into fields. Inside the wood after draining they are either converted into meadow or planted as woods, especially alderwood.

Of the numerous original bogs in Jonstrup Vang there are, as I have already mentioned, two, or rather two groups of several closely connected bogs, which still retain some of their original character; these are Dommermose and Skrædermose.

Dommermose lies to the south on the side of the wood and is surrounded by lofty spruce forest. It really consists of two bogs united by a narrower and higher portion of peaty ground. In both are found, as a result of peat-cutting, a lower portion which is covered with water and bears a swamp vegetation, and surrounding this lower portion there is sphagnum bog. On the sides the sphagnum bog merges almost abruptly into the higher ground which bears spruce trees. At the ends the transition is more gradual, and shows a series of stages of development corresponding with the different levels.

The southern portion of Dommermose, apart from a more elevated and narrow prolongation towards the south, is 70 to 80 metres long and scarcely half as broad. At the northern end, where the vegetation is best developed, the following facies are found, and are here numbered from the swampy part in the middle to the higher ground.

1. *Menyanthes–Carex rostrata*-Swamp with *Eriophorum polystachyum*.
2. *Sphagnum: Carex rostrata*-Facies with *Menyanthes, Potentilla palustris, Juncus effusus, Carex canescens,* and *Peucedanum palustre.* (See Table 27, No. 1.)
3. *Sphagnum: Carex rostrata–Oxycoccus*-Facies invaded by *Eriophorum vaginatum.* (See Table 27, No. 2.)
4. *Polytrichum: Eriophorum vaginatum*-Facies with a conspicuous invasion of *Molinia caerulea* and *Aira flexuosa.* (See Table 27, No. 3.)
5. *Molinia–Aira flexuosa*-Facies. (See Table 27, No. 4.)
6. *Pteridium*-Facies with recessive *Molinia* and *Aira*; a little *Calluna* is seen here.
7. Sprucewood.

At a single place in the bog there is the beginning of a willow scrub; but apart from this there are no Phanerophytes on the portion of the bog described above. There is, however, a community of alders on the higher narrow part of peaty ground which stretches from the bog about 25 metres towards the south. As this portion is only 4 to 7 metres broad, is densely surrounded by high spruces, and has some alder growing on it, the ground is so deeply shaded that the original vegetation is destined to disappear. Various *Hypnaceae* are found on the ground interspersed with bare patches; for the rest the vegetation is very open, showing recessive *Molinia–Aira flexuosa*-Facies with scattered individuals of *Juncus effusus, Agrostis alba, Melica uniflora, Succisa praemorsa, Lysimachia vulgaris,* and *Anemone nemorosa.* The uppermost layer of the peat has already been transformed into a loose dust; but deeper down there is still a well-preserved sphagnum peat. The ground is not at present at the level of the sphagnum stage, but it is characterized by *Molinia–Aira flexuosa*-Facies, which is presumably due solely to long-continued draining. This is an interesting and instructive stage in the conversion of the bog into soil characteristic of alderwood.

A similar transitional stage is seen at the higher portion about 40 metres in length which unites the southern portion of Dommermose with the northern portion. We see here scattered alder and birch;

because of its greater breadth it is not so deeply shaded by the surrounding sprucewood as the southern prolongation described above. The *Molinia–Aira flexuosa*-Facies in this part is therefore dense and robust.

TABLE 27

Jonstrup Vang. Different Facies in the vegetation near the north end of the south portion of Dommermose. The area is so restricted that I have investigated only $25 \times \frac{1}{10}$ sq. metre plots in each facies, and for the sake of comparison have multiplied the frequency grade of single species by 2. (See text.)

	Life-form.	1	2	3	4
Menyanthes trifoliata	HH	22	8	..	..
Carex rostrata	HH	48	42	..	..
Eriophorum polystachyum	G	6	20	..	..
Oxycoccus palustris	Ch	2	50	28	..
Eriophorum vaginatum	H	..	20	34	2
Carex stricta	H	..	..	26	2
Molinia caerulea	H	..	..	38	50
Aira flexuosa	H	..	..	40	48
Lastraea cristata	H	..	..	..	8
„ thelypteris	H	..	..	..	2
Pteridium aquilinum	G	..	..	..	6
Juncus effusus	H	8	..	..	..
Luzula multiflora	H	..	..	2	..
Carex canescens	H	10	14	6	..
Agrostis canina	H	4	8	..	2
Potentilla erecta	H	..	..	..	4
„ palustris	HH	14	..	..	4
Rubus idaeus	H	..	..	..	2
Peucedanum palustre	H	4	2	2	4
Calluna vulgaris	Ch	..	..	4	..
Lysimachia thyrsiflora	HH	4	..	..	2
„ vulgaris	H	2	..	..	..
Cirsium palustre	..	..	..	..	4

Part of this region, however, has been invaded by *Rubus idaeus*, and in this part the *Molinia–Aira flexuosa*-Facies is very recessive and the soil has undergone great alteration.

The northern portion of Dommermose is approximately of the same form and size as the southern portion. Its northern and largest part has been dug, and is occupied by a pond surrounded by swamp vegetation formed principally of *Equisetum limosum*. To the south of this pond the conditions are essentially the same as those in the southern portion of Dommermose, but the communities are more congested. Here we have: (1) Pond with *Equisetum limosum*-Facies; (2) *Sphagnum* + *Agrostis canina*; (3) *Polytrichum* with invasion from the next facies; (4) *Molinia–Aira flexuosa*-Facies with *Potentilla erecta*, *Juncus effusus*, and *Cirsium palustre*; (5) *Rubus idaeus*-Facies; *Molinia* and *Aira* recessive.

FIG. 112. Skrædermose in Jonstrup Vang; *Lysimachia thyrsiflora*-Facies with *Phragmites*.

The vegetation at the north end of the lake differs slightly; the reason for this is probably that whereas the other parts of the bog are surrounded by sprucewood, in this portion it abuts upon deciduous forest—alder and then beech. We encounter here a transition to what is seen in Skrædermose and Lyngby Bog. The following facies occur:
1. *Equisetum limosum*-Facies farthest out in the water.
2. *Equisetum limosum–Carex rostrata*-Facies. Swamp.
3. *Carex rostrata–Eriophorum polystachyum*-Facies. Swamp.
4. *Sphagnum: Potentilla palustris*-Facies. The ground can bear one's weight but is very wet. See Table 28, No. 1.
5. *Sphagnum: Potentilla palustris–Calamagrostis lanceolata*-Facies. See Table 28, No. 2.
6. *Molinia–Aira flexuosa*-Facies.

Equisetum limosum is dominant not only in 1 and 2, but occurs also very plentifully in 3, 4, and 5. The degree of frequency of this species in 4 and 5 is shown in Table 28.

Skrædermose, which lies to the south-east of Vangehuset in Jonstrup Vang, really consists of 3 bogs very close to each other. They are situated at a fair elevation, and as they lie near a small valley which is a branch of the Værebroaa-Dal it has been comparatively easy to drain them by making ditches. The most westerly bog is already covered with alder-wood, but in the other two there are still parts devoid of trees. The original vegetation is probably nowhere unaltered, but there are still portions with sphagnum vegetation, recessive though it may be. These two bogs are surrounded by sprucewood; but on the edges we find in most places a fringe, varying in breadth, of alder or willow. This is probably one of the reasons why the original vegetation of this bog is more changed than that of Dommermose. Where sprucewood abuts upon the bog an effect is produced only by the shadow cast upon that part of the bog which lies very close to the wood. The fallen needles are not carried on to the bog. Deciduous wood on the other hand operates also by the dead leaves which fall on to the bog.

In each of these bogs there is a large, low and rather flat portion with *Sphagnum*, where water can everywhere be seen. These portions certainly owe their origin to peat-cutting, of which traces are still seen. They pass rather abruptly into the higher parts in which there are also seen traces of peat-cutting.

1. On the lowest tracts, which occupy a large proportion of each bog, the ground is still covered with *Sphagnum*; which, however, in the summer is quite overshadowed and covered by a dense *Lysimachia thyrsiflora*-Facies, in which *Agrostis canina* occurs everywhere and also *Phragmites* and *Lysimachia vulgaris*. The species of *Lysimachia* flower comparatively seldom. See Fig. 112 and Table 28, No. 3.

2. On the higher ground *Lysimachia thyrsiflora* disappears, or at any

Table 28

Various Facies in Dommermose and Skrædermose in Jonstrup Vang and in Lyngby Bog. 1 and 2, the north end of the northern Dommermose in Jonstrup Vang; 3 and 6, Skrædermose in Jonstrup Vang (the middle of the 3 bogs); 4 and 5, Lyngby Bog. N.B. In No. 6 only 25 samples were taken. (See text.)

	Life-form.	1	2	3	4	5	6
Equisetum limosum	HH	44	36	..	..	1	..
Carex rostrata	HH	32	6	20	..	..	4
Potentilla palustris	HH	42	38	25	1	..	12
Calamagrostis lanceolata	H	2	50	5	20	..	..
Agrostis canina	H	12	24	50	28	5	14
Phragmites communis	HH	..	..	38	39	..	16
Lysimachia thyrsiflora	HH	..	4	50	39	7	2
Eriophorum vaginatum	H	..	..	..	..	40	..
Lysimachia vulgaris	H	..	4	42	28	8	48
Aira flexuosa	H	..	..	..	..	..	50
Lastraea cristata	H	..	..	..	1	3	..
„ spinulosa	H	..	..	7	..	..	..
„ thelypteris	H	..	..	..	16	1	..
Juncus effusus	H	4	4	..	..	..	2
Luzula multiflora	H	..	..	..	..	2	..
Carex canescens	H	16	16	5	..	5	..
„ lasiocarpa	HH	..	..	14	..	10	14
„ muricata	H	..	..	..	..	4	..
„ paradoxa	H	..	..	..	2	..	..
„ stricta	H	..	..	..	1	1	10
Eriophorum polystachyum	G	8	..	12	..	..	2
Molinia caerulea	H	..	..	..	..	..	18
Calla palustris	HH	..	..	12	..	..	..
Viola palustris	H	..	..	17	10	..	..
Potentilla erecta	H	..	..	..	..	..	10
Rubus idaeus	H	..	..	..	4	..	..
Spiraea ulmaria	H	..	..	..	2	..	..
Epilobium palustre	H	6	4	18	..	..	4
Peucedanum palustre	H	18	22	8	..	1	10
Calluna vulgaris	Ch	..	..	..	..	5	..
Oxycoccus palustris	Ch	..	..	..	..	7	..
Trientalis europaeus	H?	..	..	..	6	..	..
Solanum dulcamara	N	..	..	..	2	..	..
Menyanthes trifoliata	HH	4	12	..	25	10	..
Galium palustre	H	4	..	2	4	..	..
Valeriana sambucifolia	H	..	..	..	8	..	..
Cirsium palustre	H	..	..	..	..	..	6
Eupatorium cannabinum	H	..	..	..	3	..	..

rate decreases very considerably, whilst *Lysimachia vulgaris* becomes dominant on two types of ground: (*a*) not wholly overshadowed, originally with *Polytrichum*-Facies, *Polytrichum* still present but recessive, *Aira flexuosa* and *Lysimachia vulgaris* dominant (Table 28, No. 6); (*b*) over-

Fig. 113. Skrædermose in Jonstrup Vang: *Athyrium filix femina*-Facies.

FIG. 114. Alder swamps in Jonstrup Vang: *Mercurialis perennis*-Facies.

shadowed by an open community of alders, with an open but luxuriant vegetation consisting of various species. On this ground *Athyrium filix femina* and *Lysimachia vulgaris* are often dominant, at any rate physiognomically (Fig. 113).

3. Older and denser alderwood on slightly more elevated ground than 2 (*b*). The luxuriant ground vegetation in 2 (*b*) is here recessive, and there are open places where the peat has been transformed into a loose dust. The ordinary ground flora of the alder swamp has not yet invaded this part; but in a separate little bog with still older alderwood the ground flora is an almost pure *Mercurialis perennis*-Facies (Fig. 114).

Because of the smallness of the area, and probably also because of the abrupt transition between the lower and higher portions, some of the stages intermediate between bog and wood are absent in Skrædermose. Among others the stage between *Eriophorum vaginatum* and wood is not seen. It is true that there are scattered individuals of *E. vaginatum*, but there is no *Eriophorum vaginatum*-Facies. This link in the chain of succession is seen in Lyngby Bog, where I have investigated the ground flora by my method of sample plots in two portions of the wood differing in the elevation of the ground.

1. Alderwood on wet ground, which is not, however, the wettest ground, but is situated near the lowest level of the ground occupied by alderwood. The wood is formed of *Alnus glutinosa*; *Betula* and *Rhamnus* are also present. The character of the ground flora is seen in Table 28, No. 4, which demonstrates its strong resemblance to the *Lysimachia thyrsiflora*-Facies in Skrædermose (Table 28, No. 3), which is not yet covered with alderwood, but merely fringed with it.

2. Birchwood on considerably higher ground than the alderwood mentioned above. The ground flora is made up of a varied collection of species suited to the varying levels of the soil, which also accounts for the large tussocks of *Eriophorum vaginatum*. In Table 28, No. 5, it is seen that the wood in this part has invaded the *Eriophorum vaginatum* bog; the *Eriophorum* is recessive because of the shade cast by the wood, but the tussocks remain even if some of them are dead or dying.

2. Alder Swamps

The humus that is covered with wood bears in Jonstrup Vang principally alderwood; but scattered portions are overgrown with birch and spruce, and there are other parts with ash, especially on the higher portions, showing a transition to the mineral soil that bears beechwood. The alderwood has taken the place of the sphagnum bog, the soil of which it has altered considerably. As far as light is concerned the conditions are comparatively favourable for the ground flora of the alderwood, which is, as a whole, very rich in species; but because of the differences in the soil brought about by the different stages in the transformation of the peat in the various bogs, and especially because of the varying humidity of the bogs the composition of the ground flora often varies considerably in different places. The light too varies, partly with the age of the wood and partly with its density.

The surface of the soil in the middle of April was between 0 and 60 cm. above the level of the ground water. In those bogs in which the surface at that time of the year was 40–50 cm. above the ground water-level and which had a thick layer of peat, both alder and birch did very badly. Again and again various large bogs of this kind have been planted, but most of the trees died when they were young, or else lead a miserable existence. An example of such a situation is Stenholms Bog and 'No. 11', where the planting, however, now seems to be more successful. During the dry months of summer the ground of these bogs consists of loose peat dust of a considerable depth.

By means of the method of sample plots I have investigated the ground flora of a series of localities in the alder swamps of Jonstrup Vang. The commonest state of affairs is given in Table 29. From the higher ground at the margin of the alder swamp up to the beginning of the *Anemone nemorosa*-Facies of the beechwood we have either an *Anemone nemorosa*-Facies mixed with a number of other species, or, if the ground be sufficiently illuminated, *Anemone nemorosa–Geum rivale*-Facies as in Table 29, No. 2, and also in Nos. 1 and 3, even if the degree of frequency of *Geum rivale* does not reach 30. No. 3, in which *Geum rivale* is recessive and *Mercurialis perennis* is dominant, represents a stage transitional to what is seen on slightly lower loose humus. Here we have in general an *Anemone nemorosa–Mercurialis perennis*-Facies (Table 29, Nos. 4 and 5) or a pure *Mercurialis perennis*-Facies (Table 29, Nos. 6 and 7), where the ground is often completely covered with a pure community of *Mercurialis perennis* (Fig. 114).

TABLE 29

Jonstrup Vang. The different Facies in the ground flora of the alder swamps. (See text.)

	Life-form.	1	2	3	4	5	6	7
Anemone nemorosa	G	32	45	44	41	49	15	..
Geum rivale	H	26	31	20	2	3	2	..
Mercurialis perennis	H	16	4	43	44	50	50	50
Athyrium filix femina	H	..	..	..	..	..	1	..
Carex acutiformis	G	..	9	..	..	..	3	..
„ stricta	H	3	2	..	3	..	..	..
Agropyrum caninum	H	..	..	1	..	..	..	..
Agrostis alba	H	3	1	5	..	..	..	..
Aira caespitosa	H	4	11	25	..	..	..	..
Anthoxanthum odoratum	H	3	..	..	..	..	..	..
Baldingera arundinacea	(HH) H)	4	..	..	2	..	1	3

STATISTICS OF PLANT FORMATIONS

Table 29—continued

	Life-form.	1	2	3	4	5	6	7
Brachypodium silvaticum	H	..	..	1	..	3	1	..
Calamagrostis lanceolata	H	12	..	5	4	3	1	..
Dactylis glomerata	H	1	3	6	..	1	..	..
Melica nutans	H	2	..	..	..	..	..	..
Milium effusum	H	..	2	..	..	..	..	..
Poa nemoralis	H	3	2	..	..	1	..	..
Allium oleraceum	G	..	1	4	..	..	..	..
Convallaria majalis	G	..	..	..	..	2	..	..
Paris quadrifolia	G	..	..	..	..	..	1	..
Polygonatum multiflorum	G	1	..	..	..	..	..	..
Urtica dioeca	H	..	..	1	2	..	..	..
Ficaria verna	H	..	..	..	1	1	..	..
Ranunculus auricomus	H	8	6	..	..	..	..	..
„ repens	H	..	1	..	..	..	..	1
Arenaria trinervia	H	..	..	..	..	1	..	..
Stellaria holostea	Ch	..	26	17	..	..	..	..
Alliaria officinalis	Th	..	1	1	..	..	..	..
Viola silvatica	H	15	..	..	..	1	..	..
Geranium Robertianum	Th	..	2	2	..	..	..	..
Rubus idaeus	H	8	4	5	8	13	1	1
„ saxatilis	H	2	..	..	..	..	..	..
Spiraea ulmaria	H	3	1	4	..	6	..	..
Circaea lutetiana	G	..	2	..	..	..	4	..
Anthriscus silvestris	H	..	1	5	..	..	..	..
Lysimachia vulgaris	H	3	..	..	..	..	..	..
Primula officinalis	H	..	..	..	..	1	..	..
Galeopsis tetrahit	Th	..	..	..	1	..	..	..
Stachys silvatica	H	..	1	4	..	7	..	..
Asperula odorata	G	..	..	..	..	2	..	..
Galium aparine	Th	..	..	..	..	5	..	..
Campanula trachelium	H	..	..	1	..	..	..	..
Aracium paludosum	H	12	..	..	2	..	..	2
Cirsium oleraceum	H	..	..	2	..	..	..	..
Lampsana communis	Th	1	3	13	..	..	..	..
Lappa sp.	H	..	..	1	..	..	..	..
Taraxacum Gelertii	H	6	..	8	..	3	..	..
„ purpureum	H	1	..	..	..	..	..	..
Points		169	159	218	110	152	80	57
No. of species		23	22	23	11	18	11	5
No. of species per $\frac{1}{10}$ sq. metre		3·4	3·2	4·4	2·2	3	1·6	1·1

In some localities conditions differ, as in a westerly prolongation of Stenholms Bog and in Rørengmose. The westerly portion of Stenholms Bog bears alderwood about 8–10 metres high growing on loose humus. Between the low ground covered with alderwood and the beechwood towards the south there is found a mixed wood of Norway maple, ash, elm, beech, and alder growing on a dark soil which contains much humus.

On the transitional ground between the beech and alder the ground flora is an *Anemone nemorosa–Aegopodium podagraria*-Facies (Table 30, No. 1). The *Aegopodium* extends from here into the neighbouring alderwood, where *Anemone nemorosa* is recessive and *Mercurialis perennis* is becoming the dominant species: *Aegopodium podagraria–Mercurialis perennis*-Facies (Table 30, No. 2).

Table 30

Jonstrup Vang. The different Facies in the ground flora of the alder swamps. (See text.)

	Life-form.	1	2	3	4
Anemone nemorosa	G	**50**	23	11	7
Aegopodium podagraria	H	**50**	**44**	..	..
Mercurialis perennis	H	5	**31**	..	1
Geum rivale	H	2	6	**50**	5
Ranunculus repens	H	..	..	5	**32**
Poa trivialis	H	..	..	7	**30**
Urtica dioeca	H	1	..	1	**29**
Carex acutiformis	G	..	..	..	1
„ stricta	H	..	..	3	1
Aira caespitosa	H	..	..	11	1
Baldingera arundinacea	(HH) H	..	..	..	1
Calamagrostis lanceolata	H	..	..	..	2
Paris quadrifolia	G	..	3	..	..
Caltha palustris	H	..	..	..	1
Ficaria verna	H	..	..	1	5
Ranunculus acer	H	..	..	1	..
„ auricomus	H	..	5	2	2
Trollius europaeus	H	..	..	1	..
Melandrium rubrum	H	..	..	1	..
Stellaria holostea	Ch	..	..	4	2
Geranium Robertianum	Th	2	..	..	1
Rubus idaeus	H	..	1	..	..
Spiraea ulmaria	H	..	16	11	5
Vicia sepium	H	..	1	..	..
Anthriscus silvestris	H	..	..	1	..
Veronica chamaedrys	Ch	..	..	2	2
Scrophularia nodosa	H	..	..	..	1
Galeopsis tetrahit	Th	..	..	..	1
Stachys silvatica	H	..	2	..	..
Galium aparine	Th	14	7	..	..
„ palustre	H	..	..	..	1
Aracium paludosum	H	..	..	10	2
Cirsium oleraceum	H	..	..	2	2
Points		124	139	124	135
No. of species		7	11	18	23
No. of species per $\frac{1}{10}$ sq. metre		2·3	2·8	2·5	2·7

A similar area of transition between the floor of the beechwood and the bog is seen in Rørengmose, which is covered with an alderwood

10–15 metres high. In a more elevated portion the soil contains much sand and humus. A small part nearest the beechwood is dominated by *Equisetum silvaticum*, an *Equisetum silvaticum*-Facies. Next to that comes a *Geum rivale*-Facies (Table 30, No. 3). The bases of some of the alders are surrounded by *Mercurialis*, and in another part of the bog which has a similar soil *Mercurialis perennis* grew very plentifully among the species mentioned in Table 30, No. 3. On slightly lower ground with rather loose humus containing much sand there was found a more luxuriant vegetation rich in species, among which especially *Urtica dioeca*, *Ranunculus repens*, and *Poa trivialis* dominated. This was physiognomically an *Urtica dioeca*-Facies (Table 30, No. 4).

On the basis of the frequencies given in Tables 29 and 30 I have given in Table 31 a series of biological formation spectra. From these it will be seen that the ground flora of the alder swamp is composed essentially of Hemicryptophytes and Geophytes.

TABLE 31

Jonstrup Vang. Biological formation spectra on a basis of the degrees of frequency given in Tables 29 and 30. (See text.)

	Points.	No. of species.	No. of species per $\frac{1}{10}$ sq. metre.	Life-form.			
				Ch	H	G	Th
Table 29 No. 1	169	23	3·4	..	80	19·5	0·5
,, ,, 2	159	22	3·2	16	44	36	4
,, ,, 3	218	23	4·4	8	63	22	7
,, ,, 4	110	11	2·2	..	62	37	1
,, ,, 5	152	18	3	..	62	35	3
,, ,, 6	80	11	1·6	..	71	29	..
,, ,, 7	57	5	1·1	..	100	..	..
Table 30 No. 1	124	7	2·3	..	47	40	13
,, ,, 2	139	11	2·8	..	76	19	5
,, ,, 3	124	18	2·5	5	86	9	..
,, ,, 4	135	23	2·7	3	89·5	6	1·5

Apart from the transitional region between beech- and alder-wood the percentage of Hemicryptophytes is high everywhere; and as far as the serried ranks of *Anemone nemorosa* penetrate the alder swamp the Geophyte percentage is also high; this is specially due to the fact that the species in general are scattered, there being only 1–4·4 species on every $\frac{1}{10}$ sq. metre, though the total number of species is fairly high. Where *Anemone nemorosa* does not grow the Hemicryptophyte percentage rises to 100.

The ground vegetation of the alder swamp is composed essentially of large leafy species. These plants, growing in the strange illumination, give the alder swamp a peculiar character, whether seen on a cloudy day with the grey, misty half-light between the stems, or when the sky is clear and the sunshine, penetrating the crowns of the trees, reaches the ground vegetation in numberless patches, producing a strange, uncertain, and flickering light. Partly because the patches of light change place rapidly and continuously with the movements of the tree-tops, and partly because the high leafy herbs of the ground vegetation are easily moved by the wind, allowing the black peaty soil or the dark shadows under and among the plants to appear by glimpses, as it were in movement, the eye, confusing the patches of light, perceives not a series of constantly shifting shadows, but a large snake-like animal moving among the luxuriant vegetation.

One cannot doubt that it is in these alder swamps under the conditions just described that the 'Lindorm'[1] has been seen by those who believe that they have seen it.

3. Woodland Meadow

In Jonstrup Vang there are six woodland meadows, of which however two have now been entirely planted with alder, birch, and spruce. On the parts that have been planted longest the transformation of the meadow flora is in full progress, and meadow plants that cannot tolerate the shade of the alder and birchwood are disappearing, whilst the other members of the alder swamp's ground flora remain in company with the newly arrived shade-enduring species.

The woodland meadows came into existence a very long time ago, and I have no means of deciding whether they sprang directly from sphagnum bogs which were gradually converted by drainage and hay-cutting, &c., into the existing meadow, or whether they arose directly from alder swamps.

The composition and character of the vegetation differ somewhat with differences in the humidity determined by the varying height of the ground water. In twelve different places, representing the different height of the ground above the water-table, I determined the height of the surface above the water-table twice every month in 1905, from 16 April to 1 October.

In all twelve places the rising and falling of the ground water during the vegetative period was essentially the same. In Fig. 115 I have given a graphic representation of the movement of the ground water during the summer in three places, representing the lowest, medium, and highest portions of the meadows, i.e. from the wettest to the driest ground. Where the level of the ground water on 16 April was less than 30 cm. below the surface of the soil (Fig. 115 A), we find sedge meadow rich in species. Where the level of the ground water is lower (Fig. 115, B and C) there is grass meadow. I have investigated by means of random plots a sample of each of these two kinds of meadow.

[1] The Lindorm is a kind of Dragon or Griffin.

The sedge meadow I have investigated on the meadow lying to the south of the west end of Jonstrup Vang called Snogeeng or Stæreng. These names are doubtless of recent origin. The older names for meadows contain the word 'Mose' (bog); examples are Krudtmose, Stenholmsmose, Bøgebakkemose, and Dyssemose, which are all meadows.

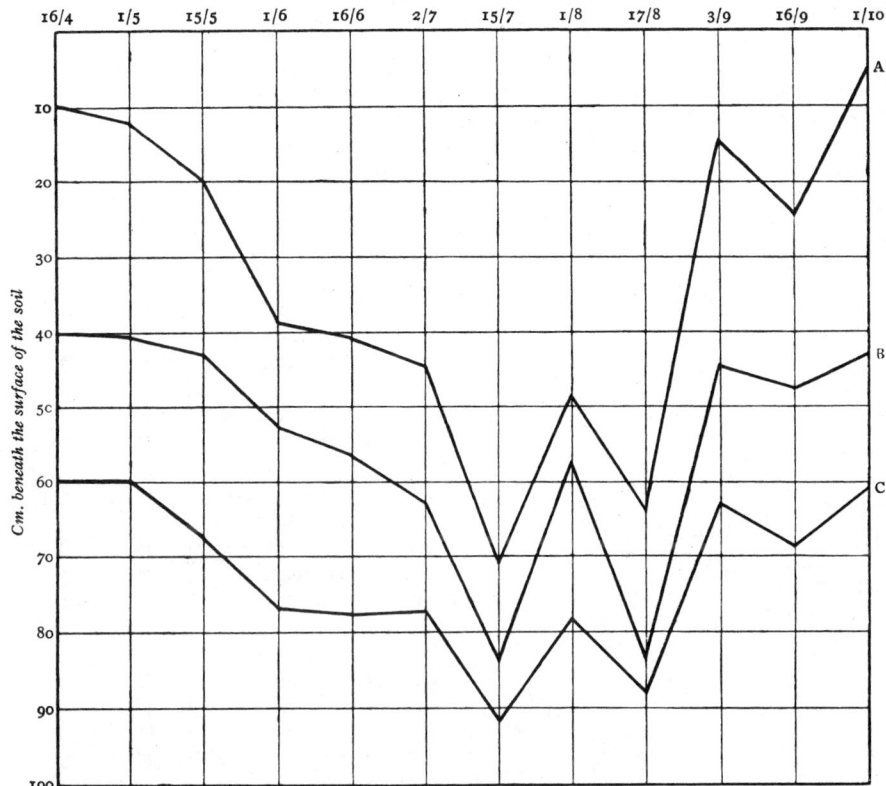

FIG. 115. Jonstrup Vang: The varying height of the ground water from the 16th of April to the 1st of October 1905 in three different parts of the woodland meadows: A, sedge meadow with high ground water; B and C, grass meadow with low ground water.

The behaviour of the ground water in the summer is shown in Fig. 115, curve A. The aspect of the vegetation differs greatly at different seasons. In May, June, and July *Carex panicea, Valeriana dioeca*, and *Cirsium palustre* may be looked upon as physiognomically dominant (Figs. 116 and 117). The result of the investigation of 50 $\frac{1}{10}$ sq. metre plots is seen in Table 32, No. 1. It will be seen that the flora is very rich in species; indeed it is the richest flora I have dealt with in my investigations with sample plots. In all no less than 78 species occurred in only 5 sq. metres; 10 of these occurred in between 60 and 100 per cent. of the samples. The species density is considerable, there being on an average 17 to 18

TABLE 32.—Jonstrup Vang: various Facies in the vegetation of the woodland meadows; 1. Sedge Meadow; 2. Grass Meadow. (See text.)

	Life-form.	1	2		Life-form.	1	2
Valeriana dioeca	H	45	..	Lychnis flos cuculi	H	3	..
Galium uliginosum	H	45	5	Stellaria uliginosum	H	1	..
Molinia caerulea	H	38	3	Anemone nemorosa	G	3	2
Carex panicea	G	37	3	Caltha palustris	H	3	..
Mentha aquatica	H	37	3	Ranunculus auricomus	H	7	21
Cirsium palustre	H	33	5	,, flammula	H	1	..
Briza media	H	33	5	,, repens	H	1	..
Ranunculus acer	H	41	37	Arabis hirsuta	H	..	6
Geum rivale	H	35	43	Cardamine pratensis	H	1	..
Festuca rubra	H	35	50	Viola canina	H	1	..
Anthoxanthum odoratum	H	26	49	,, palustris	H	20	..
Hieracium pilosella	H	1	41	Linum catharticum	Th	..	5
Campanula rotundifolia	H	..	40	Mercurialis perennis	H	..	2
Avena pubescens	H	17	39	Polygala vulgaris	H	..	9
Leontodon autumnalis	H	4	38	Potentilla erecta	H	27	..
Avena elatior	H	1	35	Spiraea ulmaria	H	28	..
Taraxacum intermedium	H	4	32	Lathyrus pratensis	H	5	..
				Trifolium pratense	H	1	1
Equisetum arvense	G	1	..	,, repens	H	17	..
,, palustre	G	7	7	Vicia cracca	H	2	3
Triglochin palustre	H	1	..	Epilobium palustre	H	2	..
Juncus conglomeratus	H	3	..	Selinum carvifolium	H	6	11
,, lamprocarpus	H	3	..	Lysimachia vulgaris	H	4	..
Luzula multiflora	H	11	23	Myosotis palustris	H	1	..
Carex acutiformis	G	11	13	Veronica chamædrys	Ch	4	1
,, flacca	G	1	1	,, officinalis	Ch	..	3
,, flava	H	24	..	Plantago lanceolata	H	13	6
,, Goodenoughii	G	16	..	Brunella vulgaris	H	8	..
,, pallescens	H	3	..	Lycopus europaeus	(HH) H	6	..
,, pulicaris	H	1	..	Scutellaria galericulata	H	2	..
,, stricta	H	1	..	Galium boreale	H	14	6
Agrostis alba	H	9	3	,, palustre	H	1	..
,, canina	H	23	..	Succisa praemorsa	H	2	1
,, vulgaris	H	1	1	Valeriana sambucifolia	H	4	..
Aira caespitosa	H	9	18	Achillea millefolium	H	4	9
Calamagrostis lanceolata	H	1	..	Aracium paludosum	H	23	..
Dactylis glomerata	H	3	..	Centaurea jacea	H	..	3
Festuca ovina	H	15	4	Cirsium acaule	H	..	2
,, pratensis	H	11	..	,, oleraceum	H	15	3
Holcus lanatus	H	17	20	Crepis praemorsa	H	..	1
Phragmites communis	(HH) G	..	5	Hieracium auricula	H	5	13
Poa pratensis	G	3	17	Sonchus arvensis	G	..	3
,, trivialis	H	2	..	Taraxacum Gelertii	H	1	26
Sieglingia decumbens	H	1	..	,, hamatum	H	..	1
Majanthemum bifolium	G	1	..	,, planum	H	..	4
Rumex acetosa	H	24	15	,, purpureum	H	..	2
Cerastium vulgatum	Ch	12	2				

FIG. 116. Snogeeng in Jonstrup Vang (looking north-west). Sedge Meadow: June aspect, here characterized especially by *Valeriana dioeca*; *Cirsium palustre* is growing vigorously, beechwood in the background, and at the very back one of the old oaks of Jonstrup Vang.

FIG. 117. Snogeeng in Jonstrup Vang (the same place as Fig. 116 but looking towards the south-east). Sedge Meadow: July aspect, here physiognomically characterized especially by *Cirsium palustre*. One of the old Oaks is seen in the centre of the background. The Phanerophytes in front of and beside this tree are *Rosa, Crataegus, Prunus spinosa, Cerasus padus, Rhamnus cartharticus, R. frangula, Lonicera xylosteum, L. periclymenum, Sambucus nigra, Viburnum opulus, Corylus avellana,* and *Populus tremula,* i.e. Phanerophytes belonging to the edge of the wood.

species per $\frac{1}{10}$ sq. metre (see Table 33). Lastly it should be observed that it is a well-marked hemicryptophytic vegetation, the formation-spectrum showing 89 per cent. Hemicryptophytes (Table 33), almost equalling the general biological spectrum of the list of species in Table 32, No. 1, which shows 87 per cent.

TABLE 33

Biological formation-spectra of Sedge and Grass Meadow made on the basis of the degree of frequency given in Table 32

	Points.	No. of species.	No. of species per $\frac{1}{10}$ sq. metre.	Life-form.			
				Ch	H	G	Th
1. Sedge meadow	883	78	17·6	2	89	9	..
2. Grass meadow	702	54	14	1	91	7	1

The grass meadow I have investigated in a large meadow to the north side of Jonstrup Vang, called No. 11. This meadow is probably now for the most part planted, except its south-east corner, which is the part I investigated. The behaviour of the ground water during the vegetative season is represented in Fig. 115, curve c. In spring the uppermost layer of the ground is still quite moist; but it soon dries, and in the summer the soil for a considerable depth becomes a dry loose peat. We encounter in this meadow a peculiar medley of mesophilous and xerophilous species.

The result of the investigation is given in Table 32, No. 2, and Table 33. The flora is scarcely so rich in species as in the sedge meadow investigated, but there were found 54 species of Vascular Plants on the 50 $\frac{1}{10}$ sq. metre plots investigated, making an average of 14 species per $\frac{1}{10}$ sq. metre. Just as in the sedge meadow investigated 10 of the species occurred in 60 per cent. of the plots. Of these 10 species only 3 belong to the sedge meadow as dominant species, viz. *Festuca rubra*, *Ranunculus acer*, and *Geum rivale*. The grass meadow resembles the sedge meadow in this well-marked hemicryptophytic vegetation, the formation spectrum showing 91 per cent. of Hemicryptophytes (Table 33). Both belong to the Danish formations richest and densest in species.

II. THE REGION OF GLACIAL SAND

In this region the mineral soil bears heath, a relatively primitive formation; the cultivated parts are arable land. The humus bears as an original formation the various facies of 'Mose'; low fields and meadow are also present.

It was only in West Jutland, on Aadum-Varde Bakkeø,[1] that I had the opportunity of trying my method in the investigation of this kind of

[1] Bakkeø is a low hill situated on a plain, from which it stands out like an island.

formation. I had only a few days at my disposal, and this did not allow of a thorough study of the rich facies of these meadows. It was also too late in the year, August had already begun, and this season is not favourable for such investigations, which should be carried out in July, before the hay harvest. I made therefore no attempt to investigate the meadows of West Jutland by means of the method of sample plots, but used the available time exclusively for the study of the facies of heath and 'Mose'. Since these investigations are so limited, and since bog ('Mose') and heath merge gradually into one another, I shall treat the formations belonging here as one, and include also some investigations of heath in the dune region near Nymindegab and on Fanø.

A. The Bog ('Mose') and Heath on Aadum-Varde Bakkeø

From the west boundary of the fields in Ølgod and Strellev parishes, immense heaths and bogs extend almost to Ringkjøbing Fjord. Towards the north they are bounded by the fields of Egvad and Lønborg, and towards the south by Lyhne, Sønder Vium, and Hemmet parishes. Here and there, especially on the outskirts of the heath, we come across cultivated patches, though the areas of heath are still large and continuous, especially to the west of the main road between Varde and Ringkjøbing, i.e. Tinghede (Ting Heath) and Østerhede.

The surface is smoothly undulating with depressions and series of valleys, which have flat bottoms and evenly sloping sides, except the deeper and more sharply cut valleys in which the larger streams run, e.g. the series of valleys between Østergaard, Brosbøl, and Varisbøl. All these depressions and valleys are covered by a layer of peat varying in thickness and in some places very thick indeed. It is in the main this layer of peat that contributes to the rounding and smoothing of the landscape by filling up its larger depressions.

The varying character of the peaty and mineral soil doubtless influences the composition of the vegetation; but it is principally the degree of humidity which determines the formations.

Going from the lowest, wettest parts of the moor upwards towards the highest and driest parts of the heath, we pass through a series of different facies, some of which can be recognized at a distance by the different colours of the dominant species of the vegetation. In contrast to the sombre brown heath the vegetation on the lower and wetter parts of the moor shows a more or less pronounced green colour, due to the presence of a considerable amount of grass and especially sedge amongst the *Calluna* and *Erica tetralix*. In these places *Vaccinium uliginosum* and *Myrica* are often present; the latter because of its stature often dominates the vegetation, at least when seen at a distance. We might almost call the *Myrica* moor a *Myrica*-Formation (Fig. 118). It is such moors as these which were transformed long ago over wide areas into meadows

FIG. 118. Ting Heath with knolls of Ting in the background. In the foreground *Myrica*-bog with *Erica*; behind that *Erica-tetralix*-Formation and around the knolls *Calluna*-Formation.

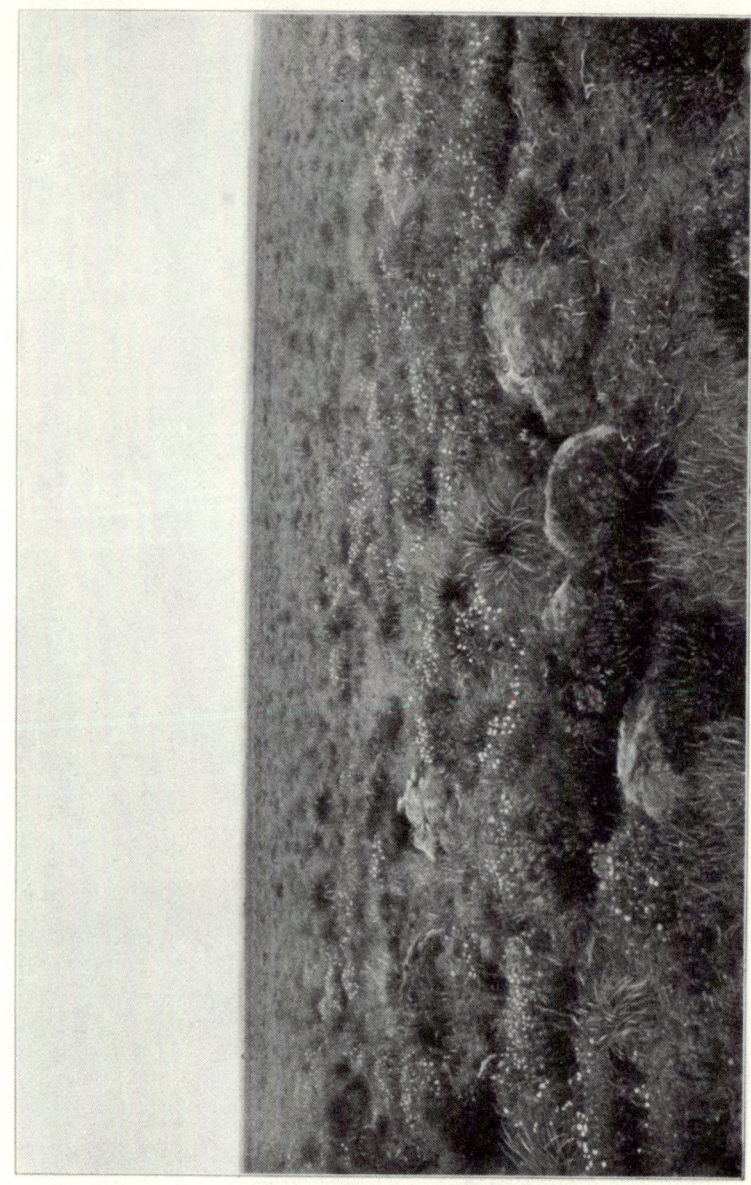

FIG. 119. Knude Heath: *Erica tetralix*-Formation with *Scirpus caespitosus* and tussocks of *Leucobryum* (in the foreground).

rich in species and in flowers, adorning with their gayness the heath-land, which is otherwise a poor region.

Higher up still the *Myrica* disappears, and the vegetation becomes darker, greyish-green, greenish-brown, or greyish-brown, according to the species and number of the herbaceous plants which grow among the Chamaephytes of the moor. Now we are at the level of *Erica tetralix*, the *Erica tetralix*-Formation (Figs. 118 and 119), whose lowest and dampest portion is lightest in colour because of the presence of *Eriophorum, Carex,* and *Molinia*. Higher up the colour darkens, though it is often enlivened by the green brush-like tussocks of *Scirpus caespitosus* (Fig. 119). Higher up still *Empetrum* often forms an essential constituent of the vegetation. Various mosses are found on the ground in many places, especially where it is drier, together with Reindeer Moss. This *Erica tetralix*-Facies is the nesting-place of the Golden Plover. But the Golden Plover is disappearing. The soil of the *Erica tetralix* moor is rich, and as time goes on it is being converted over extensive areas into fertile fields, the consolation and hope of those who farm in the district, where during the dry summer the sun scorches the sandy fields of the higher ground. In these fields the lark nests; but the Golden Plover has disappeared.

Higher up still the *Erica tetralix* disappears, and *Calluna* reigns in sole glory, at any rate as seen from a distance. The mosses decrease, and Reindeer Moss dominates alone between the clumps of *Calluna*. On closer inspection we find, however, that *Empetrum* still continues to accompany the *Calluna* as the dominant species, though it becomes sparser and sparser. But here *Arctostaphylos uva-ursi* (Fig. 120) also occurs, accompanying the *Calluna* and Reindeer Moss. Here we meet too *Lycopodium clavatum* and *Antennaria dioica*, with scattered individuals of *Scorzonera humilis* and *Solidago virga aurea*. Special mention too must be made of the glory of the heath, *Arnica montana*.

Finally, on the driest and most barren parts of the heath, *Calluna*, apart from Reindeer Moss and some other lichens, is the only plant remaining. It forms a low mantle of vegetation which does not even conceal the soil. It is not in reality the only flowering plant; closer examination reveals almost everywhere the small, conspicuously glaucous shoots of *Carex panicea*.

The warm breeze during its leisurely passage over moor and heath has become imbued with their delicious aroma. Gently and steadily it sweeps across the immense expanses, which offer it no resistance. Peace and quiet reign supreme; their dominance is strengthened by the lonely call of a Golden Plover up in the *Erica tetralix* moor. In order to enjoy fully the view of the heath you love, and to feel fully the peace of its vast expanses you should stand on an ancient barrow. Far away on the horizon there is a white church tower, or the thatched roofs of the nearest village. The eye wanders round and round the vast horizon; the hot air trembles; but the feeling of joy is abruptly supplanted by a sensation of being harshly awakened. This painful awakening is caused by one of the inevitable plantations, whose sharp lines gash, as it were, the harmony of the heath. The plantations bear to the heath approximately the same relationship as the soul-killing small settlements[1] do to the cultivated land.

But it is insisted that usefulness and life's needs must be put before everything else. We may plant to our heart's content around the homesteads and the fields, and plant to form boundaries between properties. The heath must be planted wherever it is certain that planting will pay well. But surely these invasions of the heath might stop where the ground is so poor that trees are planted merely for some aesthetic reason. A contrary opinion, also aesthetic in its origin, can be even more strongly justified. May the transformation of the heath into sprucewood be long delayed. For even if the heath is botanically poor, it is a botanical paradise compared with the desolation of the sprucewood, altogether apart from the emotion raised by the heath in the breast of one whose cradle was rocked amidst the endless brown expanses which stretch from one village to another.

The facies just described in general terms must now be elucidated by means of a series of numbers obtained by the method of sample plots. By this means the method will be shown to be applicable also to formations of Chamaephytes. It should here be observed that in counting the Chamaephytes I make use of a method slightly different from that employed in counting species belonging exclusively to herbaceous lifeforms. While the latter are included in the count only when their shoots emerge from the ground or are rooted within the frame of the quadrat, the Chamaephytes (and Phanerophytes) on the other hand are included whenever their winter-surviving shoots come within the frame, even though they are not rooted within it. The same principle is of course followed in all the investigations described above. The rule for the inclusion and exclusion can be briefly stated thus: All the species are included which have within the frame either shoots arising from the ground or perennial shoots or portions of shoots, irrespective of the height above the ground of these shoots or portions of shoots. By this means one is able to consider all the life-forms at once. If we wish in a formation of Phanerophytes to treat the upper phanerophytic layer by itself then we can make use of a larger area, e.g. an Ar, and include only those phanerophytic species which are rooted within the bounds of this area. In phanerophytic formations rich in species this method will probably be the only practical one.

Investigations were undertaken in the following localities of Aadum-Varde Bakkeø: Knude Hede between Knude and Lyhne; Østergaard-

[1] *Stationsbyer*, settlements sprung up around railway stations.

Forsom Hede between Østergaard and Forsom; Tinghede and Østerhede between Lyhne and Lønborg; and a heath between Malle and Hallum, south of Starbæk Mølle; Frøstrup Heath south of Kvong and a heath south-west of Lunde. Tables 34–6 give a conspectus of what was found in the localities investigated, which are arranged in a series from the lowest, wettest moors to the highest and driest, most barren heaths. The numbers used are consecutive and correspond with those in the text.

Table 34 includes the moor and the lower damper portion of the heath extending as far up as *Erica tetralix* occurs as a co-dominant. From this lower portion, which often has a thick layer of peat (Nos. 1–3), we pass to the higher ground with a humus pan of varying thickness (Nos. 4–6), finally reaching soil with no actual humus (No. 7).

TABLE 34

Aadum-Varde Bakkeø. *Myrica*-Formation (Nos. 1–3) and *Erica*-Formations (Nos 4–7). (See text.)

	Life-form.	1	2	3	4	5	6	7
Molinia caerulea	H	**43**	13	19	..	..	10	..
Myrica gale	N	**29**	**35**	**35**	..	..	..	..
Eriophorum polystachyum	G	14	17	8	**31**	..	..	..
„ vaginatum	H	2	**35**	..	..	..	..	..
Scirpus caespitosus	H	**27**	2	**39**	**29**	**21**	7	1
Erica tetralix	Ch	**49**	**41**	**50**	**48**	**46**	**46**	**40**
Empetrum nigrum	Ch	..	26	2	..	2	15	11
Arctostaphylos uva ursi	Ch	..	..	..	..	..	2	..
Calluna vulgaris	Ch	**30**	**33**	**35**	**40**	**43**	**44**	**50**
Narthecium ossifragum	H	19	..	..	..	..	..	..
Juncus squarrosus	H	..	..	1	3	..	..	..
Carex Goodenoughii	G	1	5	14	8	10	..	..
„ panicea	G	5	..	12	13	14	10	13
Salix repens	N	2	..	5	..	..	2	..
Potentilla erecta	H	2	..	..	..	..	..	..
Andromeda polifolia	Ch	..	2	..	1	..	..	..
Vaccinium uliginosum	N	..	26	..	..	..	3	..
„ vitis idaea	Ch	..	6	..	..	..	..	..
Oxycoccus palustris	Ch	..	26	..	1	..	..	..
Succisa praemorsa	H	1	..	..	..	..	..	..
Points		224	267	220	174	136	139	115
No. of species		13	13	11	9	6	9	5
No. of species per 1/10 sq. metre		4·5	5·3	4·4	3·5	2·7	2·8	2·3

No. 1. Wet bog in a valley north-west of Tinghøjene on Ting Heath; deep layer of peat. The green colour is conspicuous, partly owing to *Myrica gale* (degree of frequency (Hg.) 29), partly on account of sedges (particularly *Scirpus caespitosus* (Hg.) 27) and *Molinia caerulea*, which occurs here as a dominant species (Hg. 43). *Calluna vulgaris* was found

indeed in only 30 of the samples, but it too belongs among the numerically dominant species with *Erica* and *Molinia*. *Myrica gale* in its frequency forms a transition to the numerically dominant species, and as it is physiognomically very conspicuous I treat this locality together with the following as pure *Myrica*-Formation (Gale bog[1]).

No. 2. Gale bog to the east of Klangshøj on Østergaard-Forsom Heath. Less moist than No. 1. *Eriophorum vaginatum* replaces *Molinia* as the dominant species (*Myrica–Eriophorum*-Facies). *Oxycoccus palustris*, *Vaccinium uliginosum*, and *Empetrum nigrum* general (all three with Hg. 26). In other places both inside and outside the Gale bog *Vaccinium uliginosum* and *Oxycoccus palustris* occur in patches as dominant species, but at different levels. I must however add that I have not investigated such localities by means of the method of sample plots.

No. 3. Gale bog south-east of Tinghøjene on Tinghede. Slightly drier than No. 2. Instead of *Eriophorum vaginatum*, *Scirpus caespitosus* is here added as a dominant species (*Myrica–Scirpus caespitosus*-Facies).

Nos. 4–7. *Myrica* disappears, *Erica tetralix* (and usually also *Calluna vulgaris*) still dominant (*Erica tetralix*-Formation). Here and there *Empetrum nigrum* may also be dominant, but I have not investigated such places. On lower and wetter ground with a thick layer of peat much *Eriophorum polystachyum* and *E. vaginatum* occur, though the latter is not present in the samples investigated. On damp soil *Scirpus caespitosus* occurs in abundance (Fig. 119) and also *Carex Goodenoughii*. *C. panicea* is also scattered over the higher as well as the lower ground, with or without a thick layer of peat.

No. 4. Knude Heath, north-west of Stabelhøj. Rather thick layer of peat; no Reindeer Moss. Here and there huge tufts of *Leucobryum* (Fig. 119). The oldest of these tufts are generally alive only on their north side; on the side turned to the south they are dead; and other plants, especially *Calluna* and *Erica* are growing on them.

No. 5. South-east of Tinghøjene on Tinghede. Less moist than No. 4. Reindeer Moss on the ground.

No. 6. Frøstrup Heath south of Kvong. Rather dry, thin layer of peat.

No. 7. Rather low, flat heath south-west of Lunde. Peat layer poorly developed, like that on the poor *Calluna* Heath.

Nos. 8–17 (Tables 35 and 36) include the *Calluna vulgaris*-Formations in which *Erica* has completely disappeared, or at any rate is never a dominant species (Hg. less than 30). The boundary between the *Calluna* and *Erica* Formation is often very sharp, and is particularly conspicuous at the beginning of August, when the difference in colour between the profuse flowers of *Erica* and the dull *Calluna*, which flowers later, is very noticeable. Where the ground is not too dry, and particularly where it is not too poor, the *Calluna* is often not the only dominant

[1] Danish *Porsmose*.

flowering plant in the *Calluna* Heath; it is accompanied either by *Empetrum nigrum* or by *Arctostaphylos uva ursi* or by both together. *Empetrum nigrum* affects the lower parts, *Arctostaphylos* the higher levels, and the two plants together occupy an intermediate position. Reindeer Moss is seen almost everywhere.

TABLE 35

Aadum-Varde Bakkeø. The lower Facies of the *Calluna*-Formation. (See text.)

	Life-form.	8	9	10	11	12
Molinia caerulea	H	..	..	3	3	..
Scirpus caespitosus	H	..	..	2	..	..
Erica tetralix	Ch	..	..	..	..	..
Empetrum nigrum	Ch	41	49	38	31	22
Arctostaphylos uva ursi	Ch	4	7	39	41	43
Calluna vulgaris	Ch	48	50	48	50	49
Carex Goodenoughii	G	1	..	..	..	..
„ panicea	G	..	1	7	2	1
Agrostis canina	H	..	..	..	4	1
Genista anglica	Ch	..	..	..	..	1
„ pilosa	Ch	..	..	..	..	2
Vaccinium uliginosum	N	..	1	..	..	..
„ vitis idaea	Ch	3	..	3	..	..
Antennaria dioeca	Ch	..	..	..	..	1
Arnica montana	H	..	..	..	..	2
Scorzonera humilis	H	..	..	1	..	1
Solidago virga aurea	H	..	..	..	..	1
Points		97	108	141	131	124
No. of species		5	5	8	6	11
No. of species per $\frac{1}{10}$ sq. metre		2	2·2	2·8	2·6	2·5

Nos. 8 and 9. Besides *Calluna vulgaris*, *Empetrum nigrum* is also a dominant (*Calluna–Empetrum*-Facies).

No. 8. High ground east of Klangshøj on Østergaard–Forsom Heath. A thin layer of heath peat, low *Calluna*, a thick carpet of Reindeer Moss. Besides the species found in the sample plots *Carex panicea*, *Scirpus caespitosus*, and *Erica tetralix* are also present.

No. 9. Low ridge to the east of Klangshøj; in all probability formed by blown sand. Thin sandy layer of heath peat. Reindeer Moss everywhere.

Nos. 10 and 11. Besides *Calluna*, both *Empetrum nigrum* and *Arctostaphylos uva ursi* are also dominant. (*Calluna–Empetrum–Arctostaphylos*-Facies.)

No. 10. High heath south of Starbæk Mølle and Hallum. Besides the species found in the sample plots *Solidago virga aurea* and *Arnica montana* are also present.

No. 11. High ground north-north-east of Tinghøjene on Ting Heath.

Besides *Calluna* and Reindeer Moss *Arctostaphylos uva ursi* is the only dominant. (*Calluna–Arctostaphylos*-Facies.)

No. 12. High ground west of Tinghøjene on Ting Heath. Besides *Calluna* and Reindeer Moss only *Arctostaphylos uva ursi* dominant (*Calluna–Arctostaphylos*-Facies) (Fig. 120).

Nos. 13–17 (Table 36). Besides Reindeer Moss *Calluna vulgaris* is the only dominant (*Calluna*-Facies).

TABLE 36

Aadum-Vadum Bakkeø. Upper Facies of **Calluna**-Formation. (See text.)

	Life-form.	13	14	15	16	17
Molinia caerulea	H	2	6	7	..	..
Scirpus caespitosus	H	2	..	..	..	..
Erica tetralix	Ch	1	..	..	..	..
Empetrum nigrum	Ch	20	23	21	28	6
Arctostaphylos uva ursi	Ch	..	7	17	28	..
Calluna vulgaris	Ch	**50**	**48**	**49**	**50**	**50**
Carex panicea	G	3	5	..	2	18
Agrostis canina	H	..	..	..	1	..
Orchis maculata	G	..	1	..	..	..
Genista anglica	Ch	..	1	..	..	..
Arnica montana	H	..	1	1	..	..
Points		78	92	95	109	74
No. of species		6	8	5	5	3
No. of species per $\frac{1}{10}$ sq. metre		1·6	1·8	1·9	2·2	1·5

No. 13. Heath south-west of Lunde. Vegetation consisting of *Calluna*, very low and open.

No. 14. High flat Heath south-west of Klangshøj between Østergaard and Forsom. Low and open *Calluna* vegetation.

No. 15. High ground on Frøstrup Heath south of Kvong.

No. 16. High, poor Heath west of Stabelhøj between Lyhne and Knude.

No. 17. High, very poor, flat ground on Østerhede between Lønborg and Lyhne. Very low and poor *Calluna*.

In Tables 34–6, at the foot of each column the number of species found in all the sample plots in the locality is stated, and also how many species on an average were found on each plot. From these numbers it is apparent that Moor and Heath are both formations poor in species, whose species are widely scattered. All the more noticeable then is the difference which prevails between the damper more luxuriant moor and the elevated, dry, hungry heath. A glance at Table 37 shows this at once. The last two lines of this table give the number of species and the species density of the individual facies, computed on a basis of all the observations. From the wet moor with thick peat to the highest and

FIG. 120. Ting Heath: *Calluna*-Formation with Reindeer Moss and *Arctostaphylos uva ursi*.

Table 37

Aadum-Varde Bakkeø. Average number for the individual facies computed on a basis of the individual investigations given in Tables 34–6. 1. *Myrica*-Formation; 2. *Erica*-Formation; 3. *Calluna–Empetrum*-Facies; 4. *Calluna–Arctostaphylos–Empetrum*-Facies; 5. *Calluna–Arctostaphylos*-Facies; 6. *Calluna*-Facies.

	Life-form.	1	2	3	4	5	6
Myrica gale	N	33	..	..	..	..	..
Erica tetralix	Ch	43	45	..	..	..	..
Calluna vulgaris	Ch	33	44	49	49	49	49
Empetrum nigrum	Ch	9	7	45	30	22	20
Arctostaphylos uva ursi	Ch	..	1	6	40	43	10
Narthecium ossifragum	H	6	..	..	..	..	..
Juncus squarrosus	H	..	1	..	..	..	..
Carex Goodenoughii	G	7	5	1	..	..	..
„ panicea	G	6	13	1	5	1	6
Eriophorum polystachyum	G	13	8	..	..	..	..
„ vaginatum	H	12	..	..	..	..	..
Scirpus caespitosus	H	23	15	..	1	..	..
Agrostis canina	H	..	..	..	2	1	..
Molinia caerulea	H	25	3	..	3	..	3
Salix repens	N	2	1	..	..	..	..
Potentilla erecta	H	1	..	..	..	..	..
Genista anglica	Ch	..	..	..	..	1	..
„ pilosa	Ch	..	..	..	..	2	..
Andromeda polifolia	Ch	1	..	..	..	..	..
Vaccinium uliginosum	N	9	1	1	..	..	..
„ vitis idaea	Ch	2	..	2	2	..	..
Oxycoccus palustris	Ch	9	..	..	..	..	..
Antennaria dioeca	Ch	..	..	..	..	1	..
Arnica montana	H	..	..	..	..	2	..
Scorzonera humilis	H	..	..	..	1	1	..
Solidago virga aurea	H	..	..	..	..	1	..
Points		234	144	105	133	124	88
No. of species		17	12	7	9	11	5
No. of species per $\frac{1}{10}$ sq. metre		4·7	2·9	2·1	2·7	2·5	1·7

poorest heath the average number of species per 50 $\frac{1}{10}$ sq. metre falls from 17 to 5, and the species density from 4·7 to 1·7. Table 38 shows the various formation spectra based on all the observations. It will be seen from this table that the Gale bog is a Nano-phanerophyte and Chamaephyte formation; the examples investigated show 19 per cent. of Nano-phanerophytes and 41·5 per cent. of Chamaephytes. In the *Erica*-Formation the percentage of Chamaephytes rises, in the examples under discussion, to 67·5, so that this formation forms a transition to *Calluna*-Heath, which is a particularly well-marked Chamaephyte formation whose various facies show 90–97 per cent. of Chamaephytes.

The Island of Fanø. The belt of dunes on the west coast of Jutland

Table 38

Aadum-Varde Bakkeø. Biological formation-spectra of moor and heath on a basis of the numbers given in Tables 34–6

	Points.	No. of species.	No. of species per $\tfrac{1}{10}$ sq. metre.	Life-form.			
				N	Ch	H	G
1. Myrica-Formation	234	17	4·7	19	41·5	28·5	11
2. Erica-Formation	144	12	2·9	1·5	67·5	13	18
3. Calluna-Formation: Calluna–Empetrum-Facies	105	7	2·1	1	97	..	2
4. Calluna-F.: Calluna–Empetrum–Arctostaphylos-Facies	133	9	2·7	..	91	5	4
5. Calluna-F.: Calluna–Arctostaphylos-Facies	124	11	2·5	..	95	4	1
6. Calluna-F.: Calluna-Facies	88	5	1·7	..	90	3	7

and particularly the dunes of the island of Fanø, are, as it were, a long series of sand culture experiments, which nature has made for us. They show very beautifully the significance of the varying humidity of the soil for the grouping of the different species into formations. Apart from the hollows, where humus has collected, the soil is everywhere essentially uniform, consisting of sand. But the moisture contained by the sand varies, and especially on the island of Fanø where the sand is exceedingly deep the sequence of individual localities in relationship to the amount of moisture available can be, to a certain extent, determined by the height of the surface above the ground water. Owing to the uniformity of the sub-soil the water-table may be considered to be fairly uniform all over the island. In the hollows and on the evenly sloping sides of the dune-valleys, from the lowest places, which at any rate in winter are covered with water, right up to the crest of the dunes we find in this region a long series of zones of vegetation (facies) varying in breadth and sometimes not more than 1 metre broad. Each of these is characterized by one or more dominant species. These species may be regarded as competitors, each best adapted by its peculiar structure to grow in the given soil humidity. Hence it is seen that the various facies not only invariably follow one another in the same order in different localities, but that they begin at nearly the same height above the ground water. In the summer of 1896, in order to investigate the height of the ground above the water-table, I made a series of determinations of the height of the formation-boundaries above the surface of the water in holes dug at the lowest points of adjacent depressions and dune valleys. In Table 39 I have given as examples the result of my determination of the level of the lowermost and uppermost border of the *Erica*-Formation. It may seem perhaps that the height found is very different at the different points, the lowermost border of the *Erica*-Formation at the ten places investigated varying between 30 and 85 cm. above the ground water, or, perhaps one should say, above the surface of the water in holes dug in the nearest depression. But this is due partly to the fact that the level of the ground water, as already stated, was not directly determined, and partly to the fact that the levelling apparatus used was very simple, consisting merely of a diopter with a water-level which had been made by a local joiner. Lastly and finally the heights of the boundaries were determined without consideration of the exposure and slope of the ground; factors which exercise a considerable influence on the drying of the soil and which should therefore be taken into consideration. If, in determining the height of the ground above the soil water in the individual formations, localities with the same exposure and the same slope are compared, and if a more accurate estimate of the levels is made than I have had the opportunity of making, then doubtless the figures would correspond better.

TABLE 39.

Fanø. The height above ground water of the lowermost and uppermost boundaries of the *Erica*-Formation. (See text.)

	Time of investigation.	Erica-Formation.		
		Lowermost boundary.	Uppermost boundary.	Vertical extent.
		cm.	cm.	cm.
1. Sandflod Heath	6/7 1896	51	96	45
2. ,, ,,	..	53	83	30
3. Bydal.	8/7 1896	30	74	44
4. ,,	..	40	71	31
5. ,,	..	67	98	31
6. ,,	..	70	125	55
7. Near the northern decoy	..	40	70	30
8. ,, ,, ,,	..	53	103	50
9. ,, ,, ,,	..	59	76	17
10. ,, ,, ,,	..	85	120	35

The correspondence of dampness in different parts of the same formation boundaries can also be determined by an investigation of the amount of moisture in the sand at the same depth and at the same time. I shall here give no further details about the investigations of this problem that I made in 1896; I hope to say more about them in another work.

There is no place in Denmark, and I dare say there is no other place in the whole world, where the series of facies dependent upon humidity are so copious, so beautiful, and so clearly developed, as on the island of Fanø. I shall here give one example of such a series of facies extending from the lowest portions in a hollow in Sandflod Heath up to the crest of the low dunes which surround the hollow.

1. *Littorella*-Facies.
2. *Heleocharis multicaulis*-Facies.
3. *Aira setacea*-Facies.
4. Low sedge-meadow, approximately *Rhynchospora*-Facies.
5. Higher sedge-meadow, approximately *Carex Goodenoughii*-Facies.
6. *Erica*-Formation.
7. *Erica–Empetrum*-Facies.
8. *Calluna–Agrostis*-Facies.
9. *Thymus–Salix–Psamma*-Facies.
10. *Psamma*-Facies.

Of these facies I have investigated Nos. 6–9 by the method of sample plots. The result is seen in Table 40, Nos. 1–4. No. 5 in the same Table shows the result of the investigation of a dune plain covered with *Calluna* to the west of Lønne near Nymindegab. The state of affairs on this dune plain is seen to be approximately intermediate between No. 1 and No. 2 on Fanø.

In comparison with the heaths on the 'Bakkeøes'[1] the small portions of dune heath are here rich in species and especially their species density is greater. Their chamaephytic character is not nearly so marked.

III. SALINE FORMATION

On Clay. In order to test the applicability of the method of sample plots to the investigation of saline formations I examined a series of facies in the salt marsh south of Nordby on the island of Fanø, and also on the 'Grønning' at the north end of that island. The result is given in Table 41. Nos. 1–6 are from the salt marsh south of Nordby:

No. 1. *Salicornia*-Formation; *Salicornia herbacea* is the only dominant (Figs. 121 and 122).

Nos. 2–6. *Glyceria*-Formation (Fig. 122), which extends as far as *Glyceria maritima* is dominant. In this formation we can, at any rate in some places, distinguish four facies extending from the sea landwards.

No. 2. *Glyceria–Salicornia*-Facies (Fig. 122), in which, besides *Glyceria maritima*, *Salicornia herbacea* is also dominant, though not nearly so vigorous as in the *Salicornia*-Formation. As a rule scattered plants of *Suaeda maritima*, *Triglochin maritimum*, and *Aster tripolium* are also seen.

No. 3. *Glyceria–Suaeda*-Facies, in which, besides *Glyceria* and *Salicornia*, *Suaeda maritima* is also a dominant. *Triglochin* and *Aster* are increasing.

No. 4. *Glyceria–Triglochin*-Facies, in which, besides the dominants of No. 3, *Triglochin maritimum* is added as a dominant species. *Salicornia* weakly developed.

[1] See footnote, p. 263.

Fig. 121. Fanø; Salt marsh south of Nordby: the outer edge of the *Salicornia*-Formation. The tide is coming in.

Fig. 122. Fanø. Salt marsh south of Nordby at low tide. In the background the bare mud with trenches for reclamation of the land. In the middle the ground is higher with dense *Salicornia*-Formation (dark), in the foreground *Glyceria*-Formation (light).

TABLE 40

Nos. 1–4. Fanø: 1. *Erica*-Formation; 2. *Calluna–Empetrum*-Facies; 3. *Agrostis*-Facies; 4. *Thymus–Salix–Psamma*–Facies. 5. Dune plain to the west of Lønne near Nymindegab: *Erica*-Formation. (See text.)

	Life-form.	1	2	3	4	5
Carex Goodenoughii	G	42	..	..	..	36
,, panicea	G	46	..	..	..	14
Erica tetralix	Ch	50	48	..	..	50
Calluna vulgaris	Ch	50	50	48	10	50
Salix repens	N	20	36	30	44	48
Nardus strictus	H	..	40	4	..	5
Empetrum nigrum	Ch	6	48	32	..	30
Agrostis vulgaris	H	..	10	38	10	..
Hieracium umbellatum	H	..	..	40	28	..
Thymus serpyllum	Ch	..	..	34	42	..
Psamma arenaria	G	..	..	22	46	..
Juncus anceps	G	..	..	..	..	2
,, squarrosus	H	2	2	..	..	3
Luzula campestris	H	..	22	2	..	3
Carex arenaria	G	..	18	24	8	..
Festuca rubra	H	..	4	26	12	1
Molinia coerulea	H	2	..	..	..	2
Weingärtneria canescens	H	..	..	..	2	..
Drosera rotundifolia	H	6	..	..	..	..
Viola canina	H	..	..	20	16	..
,, tricolor	Th-H	..	..	..	2	..
Potentilla erecta	H	4	..	..	..	3
Genista anglica	Ch	12	18	2	..	4
Lotus corniculatus	H	..	4	2	..	..
Oxycoccus palustris	Ch	14	..	..	..	6
Vaccinium uliginosum	N-Ch	18	..	..	..	14
Pedicularis silvatica	H	2	..	..	..	..
Veronica officinalis	Ch	..	..	..	2	..
Galium silvestre	H	..	..	..	2	..
,, verum	H	..	..	4	14	..
Jasione montana	H	..	..	2	20	..
Hypochaeris radicata	H	2	..	14	12	..
Points		276	300	344	270	271
No. of species		15	12	17	16	16
No. of species per $\frac{1}{10}$ sq. metre		5·5	6	6·7	5·4	5·4

No. 5. *Glyceria–Aster*-Facies; besides *Glyceria*, *Aster tripolium* is always dominant, *Salicornia* strongly recessive. *Suaeda* and *Triglochin* are also often recessive. *Plantago maritima* and *Spergularia media* have appeared.

No. 6. *Glyceria–Plantago*-Facies: *Plantago maritima* now approaches the state of a dominant species. This facies forms a transition to the higher ground on which the next formation is situated.

I should make the following point clear about the *Glyceria–Aster*-Facies (No. 5). The ground occupied by the facies I investigated in 1909 lay to the landward of the *Glyceria–Triglochin*-Facies, but I am not prepared to say that it stood at a greater elevation than the *Glyceria–Triglochin*-Facies. In 1896 when I investigated the same marsh I differentiated at sight the same facies which I observed in 1909; but the *Glyceria–Aster*-Facies investigated in 1896 lay to the seaward of the *Glyceria–Triglochin*-Facies (see Chapter V, *The Life Forms of Plants on New Soil*, pp. 167–8). The *Glyceria–Triglochin*-Facies I investigated in 1896 (Fig. 123) was on the same ground as that occupied in 1909 by the *Glyceria–Aster*-Facies given in No. 5. Considerable remnants of formerly much richer *Triglochin* vegetation were seen in the shape of numerous remains of dead *Triglochin* tussocks. In the *Glyceria–Triglochin*-Facies found in 1909 and given in Table 41, No. 4, the *Triglochin maritimum* is not markedly dominant, its degree of frequency being only 37. It was perhaps already present in 1896, but not to the same extent as in 1909, and it was especially unimportant in comparison with the conspicuous *Glyceria–Triglochin*-Facies which at that time was found within the *Glyceria–Aster*-Facies. I cannot enter into an explanation of the difference between the former and the latter conditions before an exact determination is made of the height of the ground occupied by the various facies.

Juncus Gerardi-Formation. As the meadows had been mown when I visited Fanø in 1909 I was not able to investigate this formation in the salt marsh to the south of Nordby: but No. 7 in Table 41 represents this formation on the 'Grønning', a piece of land used for grazing at the north end of the island. Besides *Juncus Gerardi, Glaux maritima, Plantago maritima*, and *Festuca rubra* were also dominant. *Statice armeria* occurred plentifully also.

Artemisia-Formation (Table 41, No. 8, from the north coast of Fanø). *Artemisia maritima* is here a dominant species.

On the higher ground which is reached in going from the landward facies of the *Glyceria*-Formation to the *Juncus Gerardi*-Formation one often comes across a great deal of *Artemisia maritima*, but this plant is not statistically dominant, even though, on account of its size, it is a physiognomical dominant. Where, however, the sea has washed on to the higher ground of the *Juncus Gerardi*-Formation, or where the water has made a course through this ground near the sea, there is often found an *Artemisia maritima*-Formation in which *Artemisia maritima* is statistically dominant. If I do not include this formation as a facies within the *Juncus Gerardi*-Formation, one reason for this is that *Juncus Gerardi* is not always a dominant species, as is seen in Table 41, No. 8. In the example given, besides *Artemisia, Glaux maritima* and *Plantago maritima* were also dominant.

FIG. 123. Fanø. Salt marsh south of Nordby. The *Glyceria–Triglochin*-Facies of the *Glyceria*-Formation (1896)

FIG. 124. Nymindegab. Damp saline sand in the process of being overgrown. The small clumps are *Juncus bufonius* and *Spergularia salina*. The broad patches are *Glyceria maritima* and *Agrostis alba*. (See Text.)

Table 41

Fanø. 1. *Salicornia*-Formation; 2. *Glyceria–Salicornia*-Facies; 3. *Glyceria–Suaeda*-Facies; 4. *Glyceria–Triglochin*-Facies; 5. *Glyceria–Aster*-Facies; 6. *Glyceria–Plantago*-Facies; 7. *Juncus Gerardi*-Formation; 8. *Artemisia* Formation. (See Text.)

	Life-form.	1	2	3	4	5	6	7	8
Salicornia herbacea	Th	50	50	50	49	22	27	..	..
Glyceria maritima	H	..	50	50	50	50	50	..	..
Suaeda maritima	Th	..	9	37	43	30	50	..	..
Triglochin maritimum	H	..	2	12	37	24	5	5	1
Aster tripolium	H	..	4	20	9	50	48	..	..
Plantago maritima	H	..	..	..	..	23	50	50	50
Glaux maritima	H	..	..	..	..	1	..	50	32
Juncus Gerardi	G	..	..	..	..	..	..	50	18
Festuca rubra	H	..	..	..	..	..	..	50	10
Artemisia maritima	Ch	..	..	..	..	..	1	..	50
Statice armeria	H	..	..	..	..	..	..	27	..
Agrostis alba	H	..	..	..	..	1	..	5	..
Lepturus filiformis	Th	..	..	..	..	..	..	2	..
Spergularia media	H	..	..	..	..	13	19	..	2
Trifolium repens	H	..	..	..	..	..	..	1	..
Erythraea pulchella	Th	..	..	..	..	..	..	14	..
Points		50	115	169	188	214	250	254	163
No. of species		1	5	5	5	9	8	10	7
No. of species per $\frac{1}{10}$ sq. metre		1	2·3	3·4	3·8	4·3	5·2	5·1	3·3

Let us consider the question of life-forms. The *Salicornia*-Formation is a pure Therophyte Formation (Table 42, No. 1); but also in the 2–3 (Table 42) outermost facies of the *Glyceria*-Formation the Therophytes not merely statistically but also physiognomically bulk very large. In the 2–3 innermost facies of the *Glyceria*-Formation Therophytes as well as Hemicryptophytes are indeed statistically dominant, but they are here small and repressed, and some of them are completely concealed among the hemicryptophytic vegetation, though their presence is of great practical importance. The *Glyceria*-Formation resembles the gill-rakers of a fish. It retains particles of clay and other small bodies which increase the bulk of the soil when the high water ebbs back into the ocean. The innumerable small Therophytes growing between and among the shoots of the *Glyceria* increase considerably the closeness of this 'gill-raker'.

The farther one goes inland, and the higher the ground becomes, the fewer are the Therophytes encountered. In the *Juncus Gerardi*-Formation they disappear almost completely (Table 42, No. 7). They can here find no place on the soil, which is densely overgrown with Hemicryptophytes and Geophytes.

Table 42

Nos. 1–8. Fanø; No. 9. Nymindegab. The biological spectra are based on the numbers given in Tables 41 and 43. (See Text.)

	Points.	No. of species.	No. of species per $\frac{1}{10}$ sq. metre.	Life-form.			
				Ch	H	G	Th
1. Salicornia-Formation	50	1	1	..	..	..	100
2. Glyceria-F.: Glyceria–Salicornia Facies	115	5	2·3	..	49	..	51
3. Glyceria-F.: Glyceria–Suaeda-Facies	169	5	3·4	..	48·5	..	51·5
4. Glyceria-F.: Glyceria–Triglochin-Facies	188	5	3·8	..	51	..	49
5. Glyceria-F.: Glyceria–Aster-Facies	214	9	4·3	..	76	..	24
6. Glyceria-F.: Glyceria–Plantago-Facies	250	8	5·2	..	69	..	31
7. Juncus Gerardi-Facies	254	10	5·1	..	74·5	19·5	6
8. Artemisia maritima-F.	163	7	3·3	31	58	11	..
9. Sheltered sands near Nymindegab	115	8	2·3	..	41	..	59

Table 43

Nymindegab. An example of the outermost facies of sheltered sand. (See Text.)

	Life-form.	
Juncus bufonius	Th	41
Spergularia salina	Th	23
Glyceria maritima	H	18
Agrostis alba	H	17
Glaux maritima	H	5
Salicornia herbacea	Th	4
Sagina nodosa	H	4
Triglochin palustre	H	3
Points		115
No. of species		8
No. of species per $\frac{1}{10}$ sq. metre		2·3

On sand. I have made only one analysis by means of sample plots of Saline Formations on sand. This analysis was made in a creek temporarily shut in by a sand-bank on the west side of the old estuary near Nymindegab. The damp level sand which sloped gradually from the water up to the higher ground contained scattered plants, principally of *Juncus bufonius* and *Spergularia salina*. There were also scattered clumps of *Glyceria maritima* and *Agrostis alba* (Fig. 124). As seen in Tables 43 and 42, No. 9, the Therophytes are still dominant, but unless the external conditions alter, the clumps of *Glyceria* and *Agrostis* are bound to close gradually over the ground, so that the present vegetation of Therophytes will be supplanted by a vegetation of Hemicryptophytes. It is interesting to observe the correspondence between the formation-spectrum of the outer facies of this sandy marsh (Table 42, No. 9) and the spectrum of the outermost facies of the *Glyceria*-Formation growing on pure clay (Table 42, No. 2).

The sequence of the formations investigated is that determined essentially by the kind of soil. Making use of the biological spectra as a basis the plant associations of Denmark can be arranged in the following series:

1. Evergreen Meso-Megaphanerophyte Formation: Coniferous wood.
2. Deciduous Meso-Megaphanerophyte Formation: Beechwood, Oakwood, &c.
3. Deciduous Microphanerophyte Formation: Scrub.
4. Nano-phanerophyte Formation: e.g. Gale bog.
5. Chamaephyte Formation: *Calluna*-Heath, *Erica tetralix*-Heath, &c.
6. Hemicryptophyte Formation: Meadows, permanent pastures, ground flora of well-lighted woods, &c.
7. Geophyte Formation: The ground flora of the beechwood.
8. Helophyte Formation.
9. Hydrophyte Formation.
10. Therophyte Formation: e.g. *Salicornia*-Formation, the flora of agricultural fields which are tilled annually, &c.

Of all these the Hemicryptophyte Formations are both richest in species and have the greatest species density of all the plant communities of Denmark.

FINAL REMARKS

This brings me to an end of the material I have collected to solve the problems confronting us. It will be seen that the method I have developed for investigating formations is one by which it has become possible to get beyond the uncertainty of subjective estimation. By its means formations can be demarcated and characterized by actual figures independently of subjective estimation. By investigating a series of Danish plant associations I have shown how the method works in practice. Lastly I have shown how the frequency numbers for individual species found by this method may be used in investigating the biology of formations. This is a necessity when the study of formations demands a statement of the significance of bulk. The method enables us to transform floristic systematic units into biological units, or in general to transform them into units of another kind than floristic systematic units. As an example I have made use of the frequency number found out by this method as a link enabling one to transform floristic systematic units into life-form units. By this means I have been able to show how the plant communities of Denmark fall into series of formations characterized numerically by the dominant life-forms.

I believe that by this means I have succeeded in laying a proper foundation for a comparative science of formations, thereby overcoming the difficulty that the flora of different regions have different floristic systematic composition. This difficulty has up to now made an exact comparison impossible, because we had no means of converting floristic systematic units into units of another kind, e.g. into biological units.

In another Chapter[1] I have developed the theme of delimiting and characterizing the great phyto-climatic regions by means of a biological phyto-climatic spectrum founded on a statistical enumeration of life-forms. In the present chapter it is shown that within these great regions the individual formations can be determined and characterized by biological formation-spectra, and how, in this way, the demands of the science of formations with respect to the valency of species may be fulfilled.

The method and the investigations are the results of studies undertaken in preparation for a proposed biological and phyto-geographical journey to the Mediterranean region in the winter of 1909–10. A subjective estimate of the valency of the individual species in the different formations is quite uncertain even in a flora with which we are familiar; but it is far more difficult, not to say impossible, to obtain a practical result by this method when dealing with a flora with which we are not familiar. If we wish to make an approximately correct subjective estimate of the valency of the individual species in a flora containing many species that are unknown to us, we must first learn to recognize these species. The consequence of this is that in the comparatively short time one can usually devote to each place, no useful result can be obtained. By means of the method of sample plots possibilities at any rate are opened up of investigating a flora with whose species we are not hitherto familiar, even if such an investigation will occupy, as it must do, much more time than an investigation of a flora we know.

The use of the method in an imperfectly known flora is made possible by the fact that one does not need to recognize the species on the spot before proceeding to the determination of their valency. It is actually only necessary to determine the species occurring in the necessary number of samples; but their identity need not be determined on the spot. Lack of flowers, fruit, &c., may make this impossible. The determination of the species can well await the investigator's return to a museum or

[1] II. The Life-Forms of Plants and their bearing on Geography, pp. 3–104.

library. It is however necessary to be able to discriminate the species found in the sample plots, even if one cannot name them. And this demand is incumbent upon every scientifically trained and experienced botanist who wishes to investigate the formations with more accuracy than a mere tourist. Mistakes will indeed often be made of the kind which botanists have often made even when they had access to various sources of assistance. But if trouble be taken to differentiate as many systematic units as the practised eye of the botanist is able to differentiate, the mistakes will practically always be in the direction of differentiating more species than are usually recognized in the flora in question, because one looks upon varieties and forms as species. These mistakes are easily corrected when the material is worked out at home. If the investigations are made by a trained botanist we may even say that the mistakes will all be in the direction of multiplying species.

In the first sample taken one differentiates by this means as many units as the practised eye of the botanist can see. Those already known, or which can be determined on the spot, can be included at once in the list, and specimens of them need not be taken. Those that cannot be determined on the spot are included in the list, each under a number, and the same number is tied on to an individual 'species' which is taken to the next sample to be investigated. In this manner one goes on from sample to sample until the necessary numbers have been investigated. The undetermined plants bearing numbers are preserved either by drying or some other method and can be determined at home by the usual means. Names can then be substituted for numbers in the list. By the determination of these plants one learns how far the forms one has differentiated belong to the same species, and corrections can be made in the list in which each entry has its definite place. Mistakes in the other direction of including several species under one, cannot be corrected later; in order to avoid this error it is necessary to discriminate as many forms as possible.

It goes without saying that if one stays only a short time in each place where investigations of this kind are being made, it may easily happen that one does not include all the species which each sample would contain if all the plants were included that grow within the investigated area at all times of the year. Even if withered, still recognizable portions of the plants remain, others, as for example bulbous plants, will have completely disappeared from the surface of the soil. The picture given by the investigation is therefore not one of the formation as a whole, but represents a seasonal aspect. In a country or in formations where a great difference prevails between the appearance of the hypogeal vegetation and the species composition at different times in the vegetative period the representations of the formations may actually form a series of seasonal aspects.

It is not even always necessary to include the species that cannot be determined on the spot. Thus if our task is not to determine the floristic composition of the formation and the valency of its individual species, but merely to obtain a biological formation-spectrum; then instead of counting one point for every species met in each sample, we count one point for the life-form to which the species is seen to belong. This procedure can always be used in formations poor in species and in formations with well-differentiated species. Even if an otherwise trained botanist did not know a single one of the species composing the various facies of the heaths in Jutland, nevertheless he could by this method determine the formation-spectrum of this community with ease and certainty.

'Numbers are the metrical units of science.' Verses may halt, and so also may the numbers of science. I hope that the numbers given in this work will be found to halt no more than their human origin inevitably entails.

VII

THE ARCTIC AND ANTARCTIC CHAMAEPHYTE CLIMATE

The study of the life-forms of plants plays an important role in modern botany. In the science of ecology founded by E. Warming the life-form may be regarded both as the End and the Means. It is the End because ecology aims at studying the adaptation of plants to their environment, or in other words their life-forms. It is the Means inasmuch as plant communities must be defined by the life-forms of the component species.

Our present knowledge of the adaptation of plants to their environment is fragmentary, and it is easy to understand why the systems of different investigators differ. The oldest known system is the division of plants into trees, shrubs, and herbs, an arrangement used by Aristotle. Modern botany has witnessed the beginnings of three arrangements of life-forms. The first is in A. P. de Candolle's *Regni vegetabilis systema naturale*, i, p. 12; the system in that book may be considered an elaboration of Aristotle's classification. Secondly, we have Humboldt's physiognomical types in his *Ideen zu einer Physiognomik der Gewächse* [1806]. Thirdly, there is the thorough morphologico-biological investigation of the structure of shoots and of their rejuvenescence begun by A. Braun, Th. Irmisch, and others. Their studies were given new life by Darwin, and were copiously enriched by the 'physiological plant anatomy' of which Schwendener was the founder. The subject was actively pursued during the latter half of the last century, Warming, Reiter, Drude, and others making use of all the materials mentioned in founding their systems of life-forms.

The life-form is the basis of Biological Plant Geography. But since the word life-form can have different meanings it is necessary to state what we understand by the term. The word as it is used in this book means one of a series of life-forms which I have based upon the adaptation of plants to survive the unfavourable season. The percentages of the various life-forms among the species of a country I call the Biological Spectrum of that country. It has been shown that all regions whose climate is essentially the same possess essentially the same biological spectrum, while essentially different climates have essentially different biological spectra. The biological spectrum can be used as a biological expression of the climate, as an expression of the plant climate.

Just as individual climates merge imperceptibly into one another, so also there are no sharp boundaries between the types of biological spectra. Merely by comparing biological spectra we cannot therefore immediately see how the plant climates are to be characterized, and still less can we ascertain where the boundaries between them are to be drawn. For this we need a constant measure, a 'normal spectrum', with which the individual biological spectra can be compared, and in relation to which their positions can be determined. The obvious normal spectrum to use is the spectrum of the flora of the entire world. I shall not here enter into details about how such a normal spectrum is produced; but when we have obtained it, at any rate provisionally, it is easy to see how the individual plant climates can be defined and limited by means of the biological spectrum. The plant climate of a given region may be characterized by the life-form or life-forms which in the biological spectrum of that region exceeds the percentage of the same life-form or life-forms in the normal spectrum.

I shall give examples of how and where the boundaries between the individual plant climates are to be drawn by determining the southern boundary of the Arctic Chamaephyte climate and the northern boundary of the same climate in the Antarctic region. I shall also elucidate individual features which are useful in defining the Chamaephyte climate.

I. THE ARCTIC CHAMAEPHYTE CLIMATE

1. **The Southern boundary of the Arctic Chamaephyte climate.** The temperate regions of the earth are in general characterized by an absolute preponderance of Hemicryptophytes in the biological spectrum of every local flora. I therefore call the climate of these regions the Hemicryptophyte climate. In the Northern Hemisphere Therophytes are present in the southern portion of this region in greater number than in the normal spectrum; but towards the north the percentage of Therophytes gradually decreases, ultimately falling far below the percentage of Therophytes in the normal spectrum. Throughout the region the percentage of Cryptophytes is considerably higher than in the normal spectrum, and can therefore be used as a positive character in defining the Hemicryptophyte climate. Phanerophytes on the other hand form an insignificant minority, becoming fewer and fewer towards the north, while the stature of the plants themselves becomes lower and lower. In the south the percentage of Chamaephytes is considerably below that of the normal spectrum, but towards the north it increases; and when we reach the line that corresponds approximately with the June isotherm of 10° C., it exceeds that of the normal spectrum. Here there are about 10 per cent. of Chamaephytes, so that Chamaephytes become a positive character of the plant climate. I therefore look upon the line where Chamaephytes attain 10 per cent. as the Northern boundary of the Hemicryptophyte climate.

To the north of this line Therophytes and Phanerophytes become fewer and fewer. Hemicryptophytes and Cryptophytes however continue to be present in comparatively large numbers. The percentage of Chamaephytes rises rapidly, and at a line corresponding approximately to the June isotherm of 4·5° C. the Chamaephyte percentage has risen to more than double that of the normal spectrum, becoming comparatively as high as the normal Hemicryptophyte percentage. That boreal transitional climate of which the biological spectra of the local floras shows between 10 and 20 per cent. of Chamaephytes, and where, therefore, besides Hemicryptophytes and Cryptophytes, Chamaephytes have also become a positive character in defining the plant climate, I have called the Hemicryptophyte and Chamaephyte climate.

To the north of this plant climate every local flora shows a higher percentage of Chamaephytes than of Hemicryptophytes. The climate is specially characterized by Chamaephytes, so that I have called this the Chamaephyte climate. Eventually Phanerophytes and Therophytes disappear altogether; Cryptophytes may indeed be present in comparatively large numbers; but at any rate over large areas they too disappear, finally leaving the Chamaephytes and Hemicryptophytes as the only remaining life-forms. In the most unfavourable environments the percentage of Chamaephytes does not rise merely relatively, but even absolutely, above the percentage of Hemicryptophytes.

If we thus limit the Arctic Chamaephyte climate to regions where the Chamaephytes in the spectrum are dominant to a greater degree than the Hemicryptophytes, we get at the same time a southern boundary for the Chamaephyte climate, which must be drawn where the Chamaephyte and Hemicryptophyte percentages in the spectra of the local floras are equivalent in relation to the corresponding numbers in the normal spectrum. By studying the biological spectra of a series of local floras from the regions of the boundary between the Boreal and the Arctic climate we see that the southern boundary of the Chamaephyte climate may be placed where there are 20 per cent. of Chamaephytes. In Chapter IV, *The Statistics of Life-forms as a basis for Biological Plant Geography*,[1] in which I wished particularly to show where such a boundary could be drawn, I therefore made the line of 20 per cent. Chamaephytes the southern boundary of the Chamaephyte climate. We shall now see how far I was justified in doing so.

In Table 1 I have compared with the normal spectrum a number of biological spectra having a Chamaephyte percentage of about 20; the actual range is from 18 to 22. As will be seen from the Table, Vaygach has 18 per cent. of Chamaephytes, thus lying south of the 20 per cent. Chamaephyte line, so that it is therefore not included in the Arctic Chamaephyte climate. If this is correct the percentage of Chamaephytes in the local floras should not have risen in the same degree as the Hemicryptophytes. A simple calculation will soon show how matters stand.

[1] See p. 123 ff.

In the local floras under discussion the Phanerophytes, Cryptophytes, and Therophytes together amount only to 17 per cent., which is 47 per cent. less than in the normal spectrum. The Chamaephytes and Hemicryptophytes are therefore together 47 per cent. higher than in the normal spectrum. Now let x be the sum of the numbers by which the Chamaephytes and Hemicryptophytes of the local flora exceed those of the normal spectrum, and let $l\text{Ch}$ and $l\text{H}$ be the percentages of Chamaephytes and Hemicryptophytes in the local flora, and let $n\text{Ch}$ and $n\text{H}$ be the percentages of Chamaephytes and Hemicryptophytes in the normal spectrum; then if the Chamaephyte percentage of the local spectrum has increased in the same degree as the Hemicryptophyte percentage we have:

$$l\text{Ch} = n\text{Ch} + \frac{x}{n\text{Ch}+n\text{H}} \cdot n\text{Ch};$$

and thus in the flora discussed above

$$l\text{Ch} = 9 + \left(\frac{47}{36} \cdot 9\right) = 20\cdot75.$$

In the same way we get $l\text{H} = 62\cdot25$. As is seen from Table 1 the Chamaephyte percentage in the local flora is 18, thus about 3 per cent. lower than it should be had the Chamaephyte percentage risen in proportion to the Hemicryptophyte percentage. If we now work out by the same method the Chamaephyte percentage which should occur in the other local floras, if the percentage of Chamaephytes and Hemicryptophytes have risen uniformly, we get the percentage given in the last column (the numbers are rounded off to the nearest half). From this it will be seen that with 18 per cent. of Chamaephytes the boundary in question is not yet reached, with 22 per cent. it is overstepped and at 21 per cent. the boundary is either just reached or just overstepped by ½ per cent.; with 19 per cent. it is not reached in two cases but reached in one; with 20 per cent. it is just reached in one case, in three cases it is not reached, and in one it is overstepped. The boundary then is near 20. If we use whole numbers, and our material does not allow greater accuracy, the 20 per cent. Chamaephyte line then stands as the southern boundary of the Chamaephyte climate if the boundary is to be drawn along a line with a uniform percentage. If on the other hand it is to be drawn where the biological spectra show that $\frac{n\text{Ch}}{l\text{Ch}} = \frac{n\text{H}}{l\text{H}}$, then the boundary, as Table 1 shows, will vary between 19 and 21.

2. The Arctic Chamaephyte climate is not merely characterized by its high Chamaephyte percentage, but the species of Chamaephytes have at the same time a comparatively wide distribution within the area. This follows from the fact that the biological spectrum for the region of the Arctic Chamaephyte climate has as a

whole a lower Chamaephyte percentage than the biological spectrum of each local flora. The average of the 76 local floras investigated shows 36 per cent. of Chamaephytes (Table 2, No. 1). This high Chamaephyte percentage is due to the fact that many of the local floras come from Arctic nival regions, having a particular high Chamaephyte percentage, which may, indeed, taken absolutely, surpass the Hemicryptophyte percentage.

TABLE 1

	No. of species.	Ph	Ch	H	Cr	Th	
Normal spectrum	..	47	9	27	4	13	..
Land of the West Eskimos	291	5	18	61	14	2	20
Vaygach, 69°–70° N.	154	3	18	63	16	..	21
Between Great Bear Lake and the mouth of the Coppermine River	78	5	19	58	14	4	19·5
West Greenland 69°–71° N.	214	2	19	58	18	3	19·5
„ 64°–70° „	239	2	19	56·5	17·5	5	19
Novaya Zemlya 72°–73° N.	130	2	20	60	15	3	20
„ 71°–72° „	130	3	20	64	13	..	21
„ 70°–71° „	141	3	20	62	15	..	20·5
Arctic coast of Asia 68°–70° 40′ E. . .	54	2	20	58	20	..	19·5
„ „ 80° 58′–85° 8′ E. . .	76	..	20	66	14	..	21·5
Chukotski	272	4	21	62	11	2	21
East Greenland 65° 30′–66° 20′ N. . . .	169	1	21	60	14	4	20·5
Novaya Zemlya 73°–74° N.	128	2	22	62	13	1	21
Spitsbergen	110	1	22	60	15	2	20·5

TABLE 2

Region of the Arctic Chamaephyte climate.	Ph	Ch	H	Cr	Th
1. Average of the biological spectra of 76 local floras .	1	36	54	8	1
2. Average of 25 local floras whose Chamaephyte percentage is between 20 and 30	1·5	23·5	61	13	1
3. Average of 17 local floras in which all 5 series of life-forms are represented	2	26	58	12	2
4. Biological spectrum of the whole area . . .	3·5	19	64·5	10	3

The twenty-five local floras which have a Chamaephyte percentage of under 30 taken by themselves show a Chamaephyte percentage of 23·5 (Table 2, No. 2). The seventeen local floras in which all five series of life-forms are represented show the spectrum given in Table 2, No. 3, with 26 per cent. of Chamaephytes. If we compare this spectrum with the spectrum of the Arctic Chamaephyte region taken as a whole (Table 2, No. 4) we see that in the latter the Chamaephyte and Cryptophyte percentages are lower than in the average spectrum; but the percentages of Phanerophytes, Hemicryptophytes, and Therophytes are higher. Thus the Cryptophytes as well as the Chamaephytes taken as a whole must have a wider distribution than the other life-forms. The Arctic Cryptophytes are partially Helophytes and Rhizome Geophytes which grow in wet places. The Cryptophytes have a comparatively wide distribution, so that they merely conform to the general rule that marsh and water plants are widely distributed. The extensive distribution of the Chamaephytes within the Arctic Chamaephyte climate is on the contrary a peculiarity of that climate, emphasizing the character of the climate as a Chamaephyte climate. The Arctic flora in the course of time has invaded the Arctic regions along different meridians; and the biological spectrum of the flora which has thus entered the region at various places, in all probability originally had a comparatively low percentage of Chamaephytes, about the same percentage as the entire region now has. But Chamaephytic species, because they are Chamaephytes, are in general particularly adapted to the Arctic climate, so that they have been better able than other life-forms, for example the Hemicryptophytes, to spread into regions in which the conditions deviate considerably from those prevailing at the places where they first entered the Arctic region. They have thus become much more widely distributed in that region. Among Chamaephytes more often than among the other life-forms it is the same species that occurs repeatedly in the various local floras. This is why the percentage of Chamaephytes is higher in the local floras than it is in the whole region.

3. The Chamaephytic character of the Chamaephyte climate is further emphasized by the fact that the Chamaephytes are dominant (relatively) in the biological spectrum of the species that are common to the Arctic Chamaephyte Zone and to the corresponding zones in the mountains of the temperate parts of the world. We have seen that within districts with a Chamaephyte climate the species of Chamaephytes are as a whole more widely distributed than the species with other life-forms. It seems then reasonable to suppose that we shall meet with a similar state of affairs on comparing this latitudinal zone with the corresponding altitudinal zone of the mountains lying within the Hemicryptophyte climate; in other words, that the Chamaephytes will be dominant particularly among the species common to the Chamaephyte latitudinal and altitudinal zones. In order to prove this I have compared the Alpine regions of the Alps above the tree limit with the districts of the Chamaephyte climate in the North of Europe between about 8° W. and about 56° E., Jan Mayen, Bear and Hope Islands, Spitsbergen, Dolgoi, Novaya Zemlya, and Franz Josef's Land. Of the 208 species in this area 56 also occur in the Chamaephyte region of the Alps, and the biological spectrum of these common species is given in Table 3, No. 3. In the same table will be found the biological spectrum not only for the Arctic area mentioned (Table 3, No. 1), but

also for the species in the Chamaephyte zone of the Alps (Table 3, No. 2). From this table it is evident that while the Chamaephyte percentages in the two last-mentioned spectra are only 21 and 22 respectively, the biological spectrum of the species common to both contains 30 per cent. of Chamaephytes.

TABLE 3

	No. of species.	Ph	Ch	H	Cr	Th
1. Arctic Islands with a Chamaephyte climate between about 8° W. and 56° E.	208	2	21	63	12	2
2. The Alpine Zone of the Alps above the tree limit	410	3	22	64	7	4
3. Common to 1 and 2	56	4	30	55	9	2

4. **The Chamaephyte species have not only wider geographical distribution but also wider distribution within the individual areas than the other species.**

Where the available floristic material allows us to divide the flora of a large area into a series of small local floras we shall find that the chamaephytic species as a whole have the widest distribution. This was shown in Section 2 (pp. 286–8) to be true of the Chamaephyte area as a whole. I shall go into this more exactly for certain areas, including North Greenland. At the same time we shall have an opportunity of widening our knowledge of the biological spectra of North Greenland, which I could not give completely when I published 'Statistics of Life-forms', since the result of the 'Danmarks-Expedition's' investigations had not then been published.

In 'A revised list of the Flowering Plants and Ferns of North-western Greenland'[1] Simmons has given a critical account of what we know of the flora of North-west Greenland to the north of Melville Bay. Starting with this floristic material I have given in Table 4 the biological spectrum of the flora as a whole, of the flora south of the Humboldt Glacier, and of the flora north of it.

In Table 4 will also be found the biological spectra of North-east Greenland,[2] King William's Land,[3] Hare Island, and Nugsuak Peninsula.[4] These spectra are based upon floristic material recently available in the works of Ostenfeld, Lundager, and Porsild.

[1] *Report of the second Norwegian Arctic Expedition in the 'Fram'*, 1898–1902, No. 16, 1909.

[2] Ostenfeld, C. H., and Lundager, Andr., 'List of Vascular Plants from North-east Greenland', *Danmark-Ekspeditionen til Grønlands Nordøstkyst*, 1906–8, vol. iii, No. 1 (Meddelelser om Grønland, xliii), Copenhagen 1910.

[3] Ostenfeld, C. H., 'Vascular Plants collected in Arctic North America by the Gjöa Expedition under Captain Roald Amundsen, 1904–5', *Videnskabs-Selskabets Skrifter*, i. Mathem.-Naturv. Klasse, 1909, No. 8, Christiania 1910.

[4] Porsild, Morten P., 'The Plant-Life of Hare Island off the coast of West Greenland',

Table 4

Biological spectrum of the flora.	No. of species.	S	E	MM	M	N	Ch	H	G	HH	Th
1. Between Cape York and the Humboldt Glacier	102	..	..	..	..	1	25	62	9	2	1
2. North of the Humboldt Glacier	26	..	..	..	..	..	27	61·5	11·5	..	..
3. 1 and 2 together	103	..	..	..	..	1	25	61	10	2	1
4. NE. Greenland north of 76° N.	87	..	..	..	..	..	26	56	14	4	..
5. Hare Island	108	..	..	..	..	1	26	60	10	2	1
6. Nugsuak Peninsula	149	..	..	..	..	1	24	59	10	5	1
7. King William's Land	62	..	..	..	..	..	24	55	18	3	..

Simmons's work contains a tabular conspectus showing which species are found in each of the following nine districts: (1) Cape York, (2) Wolstensholme Sound, (3) Carey Island, (4) Inglefield Gulf, (5) Foulke Fjord, (6) Renselear Bay, (7) Bessels Bay, and other places in Washington Land, (8) Polaris Bay, (9) The northernmost part of the area. The biological spectra for each of these nine localities, whose species numbers lie between 5 and 78, are given in Table 5. It is clear from these tables that each individual district has a higher Chamaephyte percentage than the region as a whole, or in other words that the chamaephytic species as a whole have a wider distribution than species having other life-forms.

Table 5

NW. Greenland.	No. of species.	Ph	Ch	H	Cr	Th
1. Cape York	73	1	32	55	12	..
2. Wolstensholme Sound	34	..	35	59	6	..
3. Carey Island	5	..	60	40	..	..
4. Inglefield Gulf, &c.	69	..	30	57	12	1
5. Foulke Fjord	78	1	26	64	9	..
6. Renselear Bay	31	..	36	55	9	..
7. Bessels Bay, &c.	14	..	43	50	7	..
8. Polaris Bay	23	..	31	52	17	..
9. The northern portion	5	..	80	20	..	..
10. The whole region (1–9 together)	103	1	25	61	12	1

This is particularly well seen if, instead of considering the biological spectra of the individual districts, we look at the spectra of groups of species which differ in the number of districts in which they occur.

Arbejder fra den danske arktiske Station paa Disko (Meddelelser om Grønland, xlvii), Copenhagen 1910; id., 'List of Vascular Plants from the south coast of the Nugsuak Peninsula in West Greenland', ibid.

Table 6 shows the number of such biological spectra for North-west Greenland. At the bottom of the table is placed the biological spectrum for the species that occur in at least one of the districts given by Simmons. Thus all the species in North-west Greenland are included. From the bottom line upwards will be found the biological spectra for the species that are found in at least 2, 3, 4, 5, 6, 7, or 8 districts. The top line gives the biological spectra for three species which are found in all nine districts; this spectrum shows 100 per cent. of Chamaephytes, the three species, *Saxifraga oppositifolia*, *Cerastium alpinum*, and *Salix arctica* all having that life-form. The table shows that the Chamaephyte percentage increases with increasing distributions of the groups of species: in this flora the increase is from 25 to 100 per cent.

TABLE 6

NW. Greenland (see Table 5).	No. of species.	Ph	Ch	H	Cr	Th
In all 9 Districts	3	..	100	..	..	..
In at least 8 Districts	5	..	80	20	..	..
,, 7 ,,	9	..	67	22	11	..
,, 6 ,,	15	..	46	46	7	..
,, 5 ,,	27	..	37	52	11	..
,, 4 ,,	37	..	35	54	11	..
,, 3 ,,	57	..	31	58	11	..
,, 2 ,,	74	1	26	64	9	..
,, 1 ,,	103	1	25	61	12	1

The rise is here uninterrupted; but we must not expect groups of few widely distributed species always to follow the general rule; for example in cases where only one species is found in all the districts we must not expect this species always to be a Chamaephyte. The influence exerted by an imperfect floristic investigation may, moreover, play too great a part. We should not expect the rule that the Chamaephyte percentage rises uninterruptedly with the increasing distribution of groups of species always to hold if the number of species sinks below a certain minimum. In the example already given there are no exceptions to the rule; but exceptions are found, as we shall presently see, in other cases.

In the 'List of Vascular Plants from North-east Greenland' quoted on p. 289, Ostenfeld and Lundager have given a flora of North-east Greenland on the basis of the material collected on the 'Danmarks-Expedition'. The biological class-spectrum for North-east Greenland north of 76° N. will be seen in Table 4, No. 4. In Table 7 I have given the biological series-spectra for eleven districts (and localities), of which all the known species are mentioned in the work quoted. The number of species in the individual districts (or localities) varies, as is seen in the table, between 1 (Cape Amélie) and 87 (Germania Land). The numbers show that in the latter district all the species are found that have ever been found in North-east Greenland north of 76° N. Apart from Germania Land, whose list of species and Chamaephyte percentage coincide with that of the whole area, the table shows the usual rule that the Chamaephyte percentage in the flora of each individual district is much larger than that of the flora of the whole area. An exception to this is furnished by the nine species in Lambert's Land, only 22 per cent. of these species being Chamaephytes.

TABLE 7

NE. Greenland.	No. of species.	Ph	Ch	H	Cr	Th
1. Peary Land	9	..	44	56	..	..
2. Mallemukfjeld	8	..	50	50	..	..
3. Lambert's Land	9	..	22	67	11	..
4. Bjørneskær	11	..	36	55	9	..
5. Cape St. Jacques	24	..	33	63	4	..
6. Cape Amélie	1	..	100	..	..	..
7. Cape Marie Valdemar	13	..	54	46	..	..
8. Ymers Nunatak	17	..	41	59	..	..
9. Germania Land	87	..	26	56	18	..
10. Maroussia Island	11	..	36	55	9	..
11. St. Koldewey Island	13	..	38	62	..	..
Whole region	87	..	26	56	18	..

TABLE 8

NE. Greenland.	No. of species.	Ph	Ch	H	Cr	Th
In 9 Districts	1	..	..	100	..	..
In at least 8 Districts	3	..	33	67	..	..
,, 7 ,,	5	..	60	40	..	..
,, 6 ,,	8	..	62.5	37.5	..	..
,, 5 ,,	12	..	50	50	..	..
,, 4 ,,	19	..	47.5	47.5	5	..
,, 3 ,,	27	..	33	63	4	..
,, 2 ,,	40	..	32.5	62.5	5	..
,, 1 ,,	87	..	26	56	18	..

If we investigate in the same way the Chamaephyte percentage of North-east Greenland in relationship to the area of distribution of the species we obtain the results given in Table 8, which confirms the rule given for the Chamaephyte climate, that the percentage of Chamaephytes rises with the distribution of the species in such a way that the more the districts in which species occur the higher is the Chamaephyte percentage of the group of species. In the example given the rule only holds as long as the number of species does not fall below eight. Of the eight species found in at least six districts five are Chamaephytes, i.e. 62·5 per cent. But for this case we here reach the culminating point. Of the five species which are found in at least seven districts only three (60 per cent.) are Chamaephytes, and of the three species which are found in at least eight districts only one (33 per cent.) is a Chamaephyte. Finally the one species, *Papaver radicatum*, which is found in all nine districts, is a Hemicryptophyte, making a Chamaephyte percentage of 0. But as already mentioned this exception must not actually be looked upon as an exception to the rule given, but merely as a result of the smallness of the numbers.

As a third example I will use the flora of the islands west of North Greenland.

TABLE 9

Local floras of the Islands west of N. Greenland.	No. of species.	Ph	Ch	H	Cr	Th
9. Buckingham Island	2	..	50	50	..	..
8. Graham Island	6	..	67	33	..	..
12. Ringnes Land	6	..	67	33	..	..
1. North Devon: Boat Cape	7	..	43	57	..	..
11. Big Island	10	..	80	20	..	..
6. Devil's Island	10	..	30	60	10	..
5. Castle Island	11	..	45·5	45·5	9	..
4. North Devon: Mount Belcher	17	..	59	35	6	..
2. „ „ Lowness in the outer part of West Fjord	21	..	48	52	..	..
3. „ „ Cape Vera	23	..	39	57	4	..
7. North Kent	32	..	31	63	6	..
10. Heiberg Land	33	..	39	55	6	..
Whole Region	54	..	28	67	5	..

I have taken from Simmons's 'Stray Contributions to the Botany of North Devon and some other Islands, visited in 1900-1902'[1] twelve districts (or localities) of which Simmons gives lists of plants. Even though these lists may in many cases be imperfect, they contain all the plants hitherto found. In Table 9 I have given the biological spectrum for the flora in each of these districts, arranged not according to their position on the map, but according to the number of species they contain. I have added the biological spectrum for all twelve districts taken as a whole. In this table, as usual, we see that the Chamaephyte percentage in each separate district is higher than it is for the biological spectrum for the whole flora, showing that the Chamaephytes have a wider distribution than the species belonging to other life-forms. The same thing is seen beautifully in Table 10, which, like Tables 6 and 8, shows that the greater

[1] *Report of the Second Norwegian Arctic Expedition in the 'Fram'*, 1898-1902, No. 19, 1909.

the number of districts in which the species are found, i.e. the greater the distribution of the group of species, the higher will be the percentage of Chamaephytes in that group of species—at any rate as long as the number of species in the group does not sink below a certain minimum. Apart from the insignificant difference between the Chamaephyte percentage of the fourteen species found in at least six districts and the ten species found in at least seven districts, the percentage of Chamaephytes continues to rise until the number of species falls below six. Of the two solitary species found in at least nine districts one is a Chamaephyte (*Saxifraga oppositifolia*) and the other a Hemicryptophyte (*Papaver radicatum*); and the one species found in ten districts is a Hemicryptophyte. This one species was *Papaver radicatum*, which in the example given before was the species occurring in most districts.

TABLE 10

Local floras of the Islands west of N. Greenland.	No. of species.	Ph	Ch	H	Cr	Th
In 10 districts or local floras	1	..	..	100	..	..
In at least 9 Districts	2	..	50	50	..	..
,, 8 ,,	6	..	83	17	..	..
,, 7 ,,	10	..	70	30	..	..
,, 6 ,,	14	..	71	29	..	..
,, 5 ,,	17	..	59	35	6	..
,, 4 ,,	19	..	53	42	5	..
,, 3 ,,	24	..	42	54	4	..
,, 2 ,,	32	..	38	56	6	..
,, 1 ,,	54	..	28	67	5	..

As a last example I shall take the area whose flora I used in Section 3 (p. 288) for comparison with the Chamaephyte region of the Alps. If we arrange the species in groups according to the number of local floras in which they occur, and if we determine the biological spectrum in each of these groups, we obtain the result given in Table 11, which shows us the same as the previous examples, i.e. that the Chamaephyte percentage of the groups of species increases with the distribution of those species. With the exception of the third spectrum from the top, which shows a Chamaephyte percentage slightly too low, the percentage of Chamaephytes continues to increase, as in the example from Northwest Greenland, until we meet the smallest group of species, consisting of the two species occurring in all the local floras, *Saxifraga oppositifolia* and *Saxifraga decipiens*.

The wider the distribution of a species is within the area of the Chamaephyte climate the greater is the probability of the species being a Chamaephyte.

5. The more hostile to life the conditions are within the

Chamaephyte climate the higher is the percentage of Chamaephytes./Evidence for this statement is furnished, amongst other ways, by the fact that the Chamaephyte percentage rises as we travel gradually in the same country or along the same coast towards the north. Where exceptions to this rule occur explanations will often be found in local conditions which in part determine the environment.

TABLE 11

Jan Mayen, Bear Island, Hope Island, Spitsbergen, Dolgoi, Novaya Zemlya, Franz Josef's Land.	No. of species.	Ph	Ch	H	Cr	Th
In all 7 local floras	2	..	100	..	..	..
In at least 6 local floras	7	..	43	57	..	..
,, 5 ,,	18	..	28	67	5	..
,, 4 ,,	38	..	29	63	8	..
,, 3 ,,	70	1	29	54	13	3
,, 2 ,,	129	1·5	22	60	15	1·5
,, 1 ,,	208	2	21	63	12	2

In favour of this proposition is the circumstance that the poorer in species a flora is the higher as a rule is its Chamaephyte percentage. I shall demonstrate this by a few examples; but since factors other than the environment may be the cause of the different sizes of the plant lists, such factors for example as lack of uniformity in the floristic investigation, too wide a difference in the size of the areas, &c., we should not lay too much emphasis on the individual numbers.

In Table 5 we see that in three lists of plants (local floras) from Greenland north of the Humboldt Glacier the Chamaephyte percentage rises from 31 to 80 with a fall in the number of species from 23 to 5, and in the six local floras south of the Humboldt Glacier the Chamaephyte percentage rises from 26 to 60 with a fall in the number of species from 78 to 5. A slight irregularity is presented by Cape York, whose 73 species show a Chamaephyte percentage 2 degrees higher than Inglefield Gulf, which has only 69 species.

If we arrange the localities in Table 7 according to the number of species, which extends from 87 to 1, it is true that the Chamaephyte percentage rises from 26 to 100; but the rise is not uninterrupted. It is only when we arrange the local floras in groups and take the average of these numbers that we obtain an uninterrupted increase in Chamaephytes. If we divide up the material before us so that the local floras which have 1 to 10 species form one group, those with 11–20 species a second group, those with 21 to 30 species a third group, and so on, and if we then take the average of the number of species and also of the Chamaephyte percentages in each group, the result is that while the number of species decreases from 87 to 24, 13, and 7, the Chamaephyte percentage rises from 26 to 33, 41, and 54.

Treating the local floras in Table 9 in the same way the result is found to correspond entirely. While the number of species falls from 33 to 22, 14, and 7, the percentage of Chamaephytes rises from 35 to 44, 52, and 56.

Thus the Chamaephyte climate may be characterized, apart from its high percentage of Chamaephytes, by the fact that the Chamaephytic species have wide geographical as well as local distribution, and the more unfavourable the environment the greater is the part played by the Chamaephytes in the composition of the flora.

6. The thesis expounded above makes it reasonable to suppose that the Chamaephytes do not merely dominate the spectrum of the plant climate, but that they also belong to the life-form dominant in the formation spectrum of the dominating formations, in other words that the dominating formations are Chamaephyte formations.

In order to avoid misunderstanding I should here state that in determining the formations I intend it to be understood that the valencies of the species and those of the life-forms have been arrived at by the method described in Chapter VI, 'Investigations and Statistics of Formations'.

The formations are called after the numerically dominant life-form. There is one exception to this rule: closed formations which have more than one layer are called after the life-form that makes the uppermost layer during the unfavourable season. Phanerophytes take precedence of Chamaephytes, and Chamaephytes take precedence of Hemicryptophytes, Cryptophytes, and Therophytes. Thus among the Arctic formations willow scrubs will of course be Phanerophyte formations even if Hemicryptophytes or any other life-forms are numerically dominant in them. In the same way the Arctic heath has a Chamaephyte formation even though it may happen that Hemicryptophytes dominate in the formation-spectrum, and during the period of growth perhaps even overtop the chamaephytic vegetation.

We shall now examine more closely the nature and the geographical significance of those Arctic plant formations which are treated by Warming in his *Om Grønlands Vegetation*.[1]

The willow scrubs and the heath have already been mentioned. Of the remaining communities the 'herb field' is presumably a Hemicryptophyte formation; but we cannot be quite certain of this; doubtless Chamaephytes too play an important part here. The vegetation of the seashore and the Helophyte and Hydrophyte communities of fresh water have, just like the communities mentioned, no extensive distribution as dominant formations within the Chamaephyte climate. There is no doubt that 'Fjeldmark' and 'tundra' have the widest distribution. In tundras that are wet, even in the summer, so that bogs occur in them,

[1] *Meddelelser om Grønland*, xii, 1888.

there are found a considerable number of species, probably very often rich in individuals, of rhizome Geophytes (and Helophytes). These plants cause a high Cryptophyte percentage in the southern part of the area of the Chamaephyte climate. But in tundras that are dry in the summer, at any rate over large expanses, we see a heath-like formation of Chamaephytes. Including mosses and lichens the tundra is as a whole always a Chamaephyte formation.

There now remains only the 'Fjeldmark', which is the most widely distributed formation within the Chamaephyte climate. The open vegetation of the Fjeldmark is composed especially of Chamaephytes and Hemicryptophytes, but the literature does not tell us which of these two life-forms dominates statistically. The few species and the open vegetation will make it easy for any one having the opportunity to undertake a thorough statistical investigation of the Fjeldmark. Until this has been done we must content ourselves with a more general consideration.

Let us for a moment look at the numerous small local floras whose biological spectra are given in Tables 5, 7, and 9, and let us treat these local floras like so many sample plots of the vegetation of North Greenland, which is composed chiefly of Fjeldmark.

First let us take North-east Greenland, where we have eleven local floras which we shall call eleven 'sample plots' containing in all 87 species. If we give to each species, as usual, one point for every sample in which it is found we obtain in all 203 points; and counting each species as a life-form and then working out their percentages we obtain the spectrum in Table 12, No. 2. This spectrum compared with the general spectrum of the whole area (Table 12, No. 1) shows a distinct rise in the Chamaephyte percentage, viz. from 26 to 34. North-west Greenland (Table 12, Nos. 3 and 4) and the islands to the west of North Greenland (Table 12, Nos. 5 and 6) when treated in the same manner show a corresponding result, viz. a rise in the Chamaephyte percentage from 25 and 28 to 32 and 44 respectively.

TABLE 12

	No. of species.	Points.	Ph	Ch	H	Cr	Th
1. NE. Greenland	87	..	..	26	56	18	..
2. ,, ,,	..	203	..	34	57	9	..
3. NW. Greenland	103	..	1	25	61	12	1
4. ,, ,,	..	332	1	32	57	10	..
5. The islands to the west of N. Greenland	54	..	..	28	67	5	..
6. ,, ,, ,,	..	178	..	44	51	5	..

Now these 'sample plots' are actually areas of land of varying extent, and not the $\frac{1}{10}$ sq. metre plot, which is the unit I have usually employed in my statistical investigations of formations; but since we have found that by using unit areas varying in size from 10 sq. metres to 100 sq. metres the percentage of the dominating life-form rises as the area of the sample plots decrease, there is no reason to take for granted that the Chamaephyte percentage should fall if the area of the sample plot instead of being that of a whole district be reduced to $\frac{1}{10}$ sq. metre. I think therefore that there is reason to suppose that Chamaephytes characterize the Arctic climate even when considered from the point of view of statistics of formations.

II. THE ANTARCTIC CHAMAEPHYTE CLIMATE

It is true, at any rate as far as the phanerogamic vegetation is concerned, that there is considerable difference between the Arctic and Antarctic climates. But on the other hand it is the Antarctic climate, of all climatic zones, that most nearly approaches the Arctic. It will therefore be interesting to examine the relationship between the spectra of the Arctic and Antarctic plant climates, especially with a view to seeing whether the Antarctic climate, like the Arctic, is characterized by Chamaephytes, and if so where we are to draw the northern boundary line of the Antarctic climate when this is determined by the same means as the southern boundary line of the Arctic climate.

Within the area of the Antarctic climate the extents of land covered with a phanerogamic vegetation are so inconsiderable that we have but few local floras as a basis for our work; and since these local floras are all of more or less isolated islands they are as a rule very poor in species. This decreases the value of the biological spectra, because, amongst other reasons, defective floristic investigation in such cases can easily play an important part. But we must take things as we find them.

In Table 13 I have given the biological spectra for the individual local floras that must be considered here; they are arranged in five groups according to the position of the islands: (1) islands to the south of New Zealand: Macquarie, Campbell, Auckland Island, Antipodes Island, and Snares, of which only the first has a Chamaephyte percentage over 20. If we proceed westward we find in the southern part of the Indian Ocean: (2) Heard and Kerguelen with over 20 per cent. and Amsterdam and St. Paul with under 20 per cent. of Chamaephytes. After that we come to (3) Crozet and, farther towards the west, Prince Edward (with Marion), both with more than 20 per cent. of Chamaephytes. In the southern part of the Atlantic Ocean we have: (4) South Georgia with more than 20 per cent. and north-eastwards, Tristan da Cunha with under 20 per cent of Chamaephytes. Finally (5) includes the islands around the south point of America: of these the Falkland and Staten Islands both have over 20 per cent. of Chamaephytes, and, as seen in Table 14, the same is true of the islands south of Tierra del Fuego with the exception of Navarin Island, which lies close to the south coast of Tierra del Fuego. As far, then, as we can see from a preliminary treatment of the life-forms of the flowering plants of Tierra del Fuego that country has less than 20 per cent. of Chamaephytes.

TABLE 13

	No. of species.	Ph	Ch	H	Cr	Th
1. Snares	22	13·5	9	50	4·5	23
2. Antipodes	42	5	12	64	9	10
3. Aucklands	99	12	10	63	7	8
4. Campbell	84	12·5	14	61	4·5	6
5. Macquarie	27	..	26	59	4	11
6. Amsterdam+St. Paul	17	6	6	70	12	6
7. Kerguelen	21	..	24	52	10	14
8. Heard	7	..	43	43	14	..
9. Crozet	13	..	23	54	8	15
10. Prince Edward+Marion	8	..	37·5	25	12·5	25
11. Tristan da Cunha	29	3	17	59	..	21
12. South Georgia	15	..	33	53	7	7
13. The local floras 1–5 and 7–11 in Tab. 14	232	8	23	57	4	8
14. Staten Island	115	12	32	48	5	3
15. Falkland	117	3	28	50	8	9
16. The whole of the Antarctic Chamaephyte region	318	7	23	56	5	9
17. The same according to the points of the local floras (877 points)	..	8	31	49	6	6
18. The region of the Arctic Chamaephyte climate	437	3·5	19	64·5	10	3
19. Sierra Nevada (Spain) above 1,500 metres	478	7	21	53	2	17
20. Djurdjura (Algeria) above 1,800 metres	100	6	32	48	5	9

If we now look at these islands on the map we shall find that the northern boundary of the Chamaephyte climate, determined by the 20 per cent. Chamaephyte line, passes between Macquarie and Campbell, whence it runs westward north of Heard, Kerguelen, Crozet, and Prince Edward, but south of Amsterdam and St. Paul. It then goes north of South Georgia and south of Tristan da Cunha; finally passing north of Falkland and Staten Islands, then coming, as far as my observations extend, southwards around Navarin or perhaps through this island and towards the north-west to the Straits of Magellan, which it follows on its course towards the north-west.

According to a preliminary computation the number of species within the Chamaephyte area thus demarcated is 318. The spectrum of these species is given in Table 13, No. 16. This spectrum, though not so pronounced, shows the same phenomenon as the Arctic Chamaephyte climate spectrum (Table 13, No. 18), viz. a Chamaephyte percentage much lower than that of the local floras; in the local floras no Chamaephyte percentage is lower than that of the whole area.

If the local floras are treated in the same way, as though they were sample plots used in investigating a formation, we obtain the same result as in the corresponding treatment of the Arctic local floras; viz. that a spectrum formed on the basis of such sample plots has a much higher Chamaephyte percentage than the ordinary spectrum of the plant climate, namely 31 per cent. against 23 (see Table 13, Nos. 17 and 16).

It is thus fairly clear that there is a considerable resemblance between the Arctic and the Antarctic plant climates. When indeed we consider the differences between the physical climates of the two regions, especially with respect to the course of the temperature curve, we must expect to find corresponding differences in the biological spectra. A comparison between the biological spectra of the two regions immediately shows such differences. Compared with the biological spectrum of the Arctic Chamaephyte region (Table 13, No. 18) that of the Antarctic Chamaephyte region (Table 13, No. 16) shows a comparatively high percentage of Therophytes and Phanerophytes, viz. 9 and 7 respectively. The corresponding numbers for the Arctic region being 3 and 3·5.

TABLE 14

	No. of species.	Ph	E	MM	M	N	Ch	H	G	HH	Th
1. Desolation	69	..	..	3	4	7	32	48	2	..	4
2. Elisabeth	66	..	..	..	..	3	23	56	..	..	18
3. Clarence	52	..	..	4	6	6	34	40	2	2	6
4. Basket	65	..	1·5	3	1·5	5	31	49	3	3	3
5. Chair	40	..	2·5	2·5	7·5	5	24	45	7·5	2·5	2·5
6. Navarin	89	..	1	2	2	7	22	50	4	1	11
7. Picton	34	..	3	3	6	6	26	44	6	..	6
8. Hoste	90	..	1	2	2	5	29	52	3	1	5
9. Burne & Smoke	75	..	1	1	3	5	37	45	4	1	1
10. Hermite & Horn.	65	..	..	..	1·5	1·5	35·5	51	6	1·5	3
11. Staten Island	115	..	3	2	1	6	32	48	3	2	3
12. Falkland	117	..	..	..	..	3	28	50	7	3	9
1–5 and 7–11 together	232	..	1	1	2	4	23	57	3	1	8

Apart from the small isolated islands exposed to violent storms and either devoid of Phanerophytes or else possessing a few low ones, the most favourably situated regions of the Antarctic area are not characterized merely by a comparatively high Phanerophyte percentage: some of the Phanerophytes even attain a considerable height. As seen clearly in Table 14 the local floras from the southern point of America show as much as 18 per cent. of Phanerophytes, and these are not exclusively Nanophanerophytes, but many are Microphanerophytes, some being even Mesophanerophytes, viz. *Nothofagus antarctica* and *N. betuloides*. Whether these species actually attain in the Antarctic area sufficient height to enable us to call them Mesophanerophytes I do not exactly know; but at any rate they occur as Mesophanerophytes within the main area of their distribution. Finally, a fact of considerable importance is that some of the Phanerophytes are epiphytic, viz. three species of *Myzodendron*.

The differences between the biological spectra of the two regions correspond with the differences in their hydrotherm figures, especially

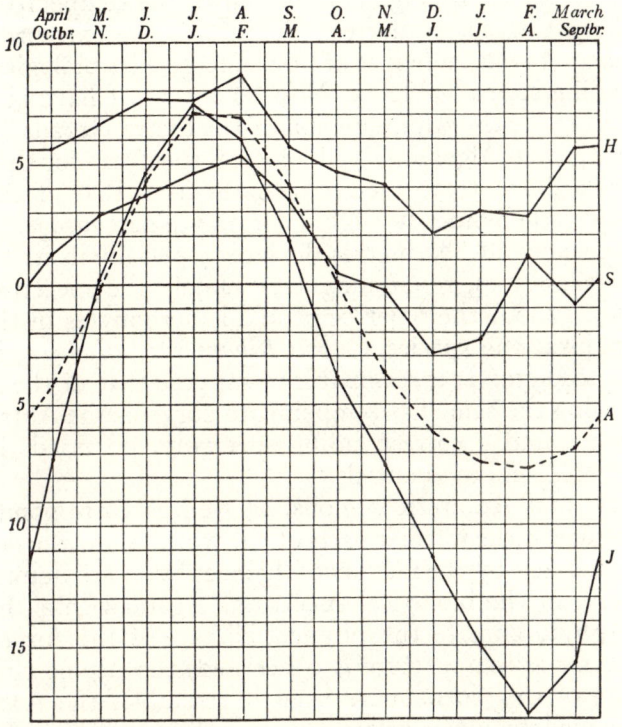

FIG. 125. Temperature curve for: H, Cape Horn; S, South Georgia; A, Etna at a height of 2,950 metres; J, Jakobshavn.

the difference in the course of the temperature curves. For example, if we compare the temperature curve of Jakobshavn (Fig. 125 J) with that of South Georgia (Fig. 125 S), and Cape Horn (Fig. 125 H), we find that the two latter have the flat curve characteristic of the tropics, while the curves of the Arctic region show a wide difference between winter and summer. If we suppose for example that the temperature curve for Tamatave, situated on the east coast of Madagascar about 18° S., has been lowered 20°, we obtain a curve intermediate between those for South Georgia and Cape Horn, having essentially the same course as these curves. Such a temperature curve at the latitude of Tamatave corresponds to an altitude of about 3,500 metres. What kind of biological spectrum is to be found at this altitude on a mountain near the boundary of the tropics I have unfortunately not been able to determine, so that I cannot compare it with the biological spectrum for the Antarctic region. I must here content myself with comparing the latter spectrum with the Alpine flora of the sub-tropical zone.

The farther we travel from the Arctic region towards the Equator the flatter as a rule the temperature curve will be for the region corresponding with the Arctic zone. In Fig. 125 A is shown the temperature curve for Etna at a height of 2,950 metres. Because of the extraordinarily poor flora of Etna at this altitude it is of no use to give its biological spectrum. Instead of doing so, in Table 13, Nos. 19 and 20, I have given the biological spectrum for the Alpine region of the Sierra Nevada in Spain and for the Djurdjura in Algeria. The biological spectrum for the flora of the Sierra Nevada above 1,500 metres deviates from that of the Antarctic region in very little besides its high Therophyte percentage. The flora of the Djurdjura above 1,800 metres deviates especially in its high percentage of Chamaephytes, resembling very closely the spectrum shown in Table 13, No. 17. Of the 6 per cent. of Phanerophytes in the spectrum of the Djurdjura, one species, *Cedrus Libani* var. *atlantica*, is ordinarily a Mesophanerophyte. It is reasonable to suppose that in the Alpine flora of a tract having the same flat temperature curve as the Antarctic region we shall always find a corresponding spectrum, provided the moisture conditions correspond in essentials.

The Arctic and Antarctic regions, as well as corresponding regions in the mountains of the warmer parts of the world, show the same climatic series, characterized by the fact that Chamaephytes are dominant, at any rate relatively, in the biological spectrum. Again within the region of the Chamaephyte climate the physical climate of the Antarctic region resembles, as far as temperature is concerned, the Alpine climate of the warmer parts of the world more than it resembles the Arctic climate. This is expressed biologically by corresponding relatively high Therophyte and Phanerophyte percentages in the Antarctic spectrum, and by the fact that the Phanerophytes belong partly to classes which grow to a greater height, and partly to the epiphytic Phanerophytes.

VIII

STATISTICAL INVESTIGATIONS OF THE PLANT FORMATIONS OF SKAGENS ODDE

THE object of this Chapter is twofold. Firstly it is a contribution to the knowledge of the composition of the vegetation of the area investigated, obtained by a method which makes it possible to use the results as a basis for studying comparatively the statistical composition of the vegetation of our country when other regions are from time to time investigated by the same method. Secondly the method used makes it possible to reproduce the plant communities investigated in such a place as a Botanic Garden, a procedure which will perhaps be of importance if the communities in question (e.g. the heath formation, &c.) disappear from our country. This disaster may actually take place. Many people make their living, at any rate in part, out of the dogma 'Where plough can't go, there trees shall grow'.[1] The State assists them to make their living out of this dogma, and year by year the areas of heath in Jutland diminish. Heaths are afforested where afforestation will be profitable; but planting of trees takes place on some heaths where there is no prospect whatever of profit.

When the Heath Society began its activities West Jutland was almost without woodland. There were, on the other hand, vast expanses of heath. It is easy to understand how the peasant of West Jutland could be persuaded to agitate for the planting at any rate of those heaths that could not be used for remunerative agriculture, if he were assured that their planting was a paying proposition. Neither do I doubt that even if each and all of the existing heaths had been obviously unfit for profitable cultivation or planting, yet many of the peasants would have favoured at any rate a partial planting of these heaths. Where there are no woods people long for them, and will give anything to get them. Of heath, on the other hand, there was such a superabundance that it was no difficult task to persuade people, on the spur of the moment, to put their strength into its entire extirpation. But circumstances have changed. Innumerable remunerative plantations are now scattered over West Jutland, and there are also large areas of heath not fit for cultivation, but on which, it is affirmed, planting could be made to pay. When all these areas are covered with trees I presume that the craving for woods, which is experienced by most of the West Jutlanders (and by myself), will be satisfied. But if these peasants, now that plantations are distributed over the whole of West Jutland, keep on planting areas which will not pay, they must be acting from lack of knowledge; they believe, in fact, that these plantations too can be made to pay.

[1] 'Hvor Ploven ej kan gaa, der bør et Træ at staa.'

Conversations I have had with country people during my journeys through the heath areas of Jutland have confirmed me in this opinion. When I have tried to put in a good word for the preservation of at least part of the heath they have glibly told me that it is unreasonable to let the heath remain to no purpose when it can be transformed into useful woodland, while at the same time the country is adorned with delightful woods in the place of barren heathy wastes. And when I have told them that many of the plantations in the arid heathland will certainly never pay they would not believe me. When I have tried to appeal to the peasant's feelings and aesthetic sense by asking him whether he cannot see something beautiful in the 'desolate' heath, whether he cherishes no tender memories fondly intertwined with the barren heath, and whether he would really be willing to witness the entire disappearance of the heath, then I have learnt that the people of West Jutland do not really wish these heaths swept entirely away, but that they console themselves with the thought that the disappearance of the heaths lies in the remote future, and that some small areas of heath, which for one reason or another no one will trouble to cultivate or plant, will always be left intact. This belief is, however, incorrect, and even if it were right but little would be gained by the retention of such small areas of heath lying amidst plantations and fields. In the course of time they would become covered by a vegetation dominated by trees, or at any rate they would gradually change their character completely. Even if such changes did not take place, mere patches of heathland would from their very lack of size give an utterly inadequate idea of the limitless expanse of primeval heath.

Even if the vaster expanses of heath have vanished, there still persist coherent areas large enough to demonstrate the original character of the heath. But if the present activities continue, and there is no prospect of them ceasing, the time will soon come when our great continuous tracts of heath must cease to exist. The efforts of the Heath Society to extirpate the heath show no signs of abating till they have far exceeded their previous attempts to justify their title-deeds. The name 'Heath Society' is misleading; as their object is to extirpate the heath, a better designation would be 'Anti-Heath Society'. The name however is in use and must be retained. How I wish there might exist an 'Anti-Heath Society' with the objects of a real Heath Society, i.e. to preserve considerable areas of heath. For this purpose we must obtain money to buy large areas of land. This might be done by such fashionable means as a 'Flower Day': flowers of *Calluna* might be sold, and a 'Heath Day' instituted all over the country. I do not doubt that if such a venture were led by capable and practical men and women who could advertise the matter well in the papers we should then be able to secure without difficulty a sum of about 100,000 Kroners. This amount would be sufficient to purchase for permanent protection an area of heath large enough to enable our descendants to form a just idea of the appearance of a great part of Jutland during a long and rich period of our country's history, and also to be a living memorial enabling them to understand the loving descriptions and songs that their forefathers wrote in praise of the heaths of Jutland. If, on the other hand, we are not careful to preserve a considerable area of heath as a memorial, then I do not doubt that our descendants will censure us for our short-sightedness and lack of feeling. I have however neither the influence nor the ability to take the matter up with prospect of success; but one thing at least I can do—I can investigate and describe the different facies or plant communities of the heath so that they can be reproduced in a Botanic Garden or elsewhere, on a larger or smaller scale according to our wishes or our means. This can be done by the methods of investigation at our disposal, and in this paper I have tried to depict a number of plant communities, especially those of the heath in the neighbourhood of Skagen. Even if, in the course of transformation of heath into plantations, these natural communities were to disappear entirely from our country, we shall be able to reproduce them essentially in their original form, if not in their original extent. Though it is my heartfelt wish that this will be unnecessary, yet I feel satisfied that it could be done. The method used is entirely adequate for my second, and scientifically my chief aim: to investigate and describe the plant communities in such a way that the results can be used for exact comparative treatment of plant communities in different regions and with different floristic compositions. This follows because if the portrayal and description of a plant community be sufficiently exact to enable it to be reproduced, then on the basis of such a description that community can be compared minutely with other communities which have been investigated by the same method.

The main points of the method in question will be found described in Chapter VI, 'Investigations and Statistics of Formations'; it will here suffice to give a general outline of it, and to add some supplementary remarks.

Just like other sciences, for example Systematic Botany, the study of plant communities may be said to have its dominating and subordinate concepts and a classification or system founded upon them. But it will certainly be much more difficult to reach a natural system in dealing with communities than it has been with species.

As an immediate result of observation and of an unconscious process of induction men like Ray and Linnaeus came to recognize the unity of certain groups of plants, e.g. *Umbelliferae, Compositae*, and Palms; but that there was no actual certainty that these groups were natural is seen by the fact that certain genera then placed in those families were found, when more closely studied, to belong to widely different circles of affinity.

It was only when a thorough, methodical, and comparative study of genera and species was undertaken that a stable foundation for the arrangement of species was reached, and a system that was more and more natural was approached. There is no reason to doubt that we must pursue a similar course if we wish to reach a natural classification of plant communities. Here too on the basis of immediate observations we can form certain groups which at least appear to be natural, e.g. Woodland, Heath, Meadow, &c., and these may again be subdivided. I prefer to call these groups by such names as: Phanerophyte community, Chamaephyte community, Hemicryptophyte community, Cryptophyte community, and Therophyte community, each with their subdivisions. But it does not follow that this classification even approaches finality. This is especially apparent if we start, as we should do, by defining the plant community as an assemblage of plants with the same life-form; 'life-form' being understood as adaptation to the same environment. For since the alder, willow, *Parnassia, Poa pratensis, Orchis maculata*, &c., grow in the same meadow with exactly the same environment they should belong to the same life-form. But since we are for the time being unable to perceive unity of life-form in species with diverse structure and habit, we are compelled for the present to take for granted that plants with different structure belong to different life-forms, when we can see that the difference is an expression of adaptation to environment. It follows from this that we must delimit life-forms not finally but provisionally, as I have attempted to do in my system of life-forms. This then cannot be the 'natural system', but an artificial system, which however does not mean that closely related forms are always separated. This stage in the history of the classification of plant communities corresponds to the stage Systematic Botany reached when the system was recognized to be artificial, as Linnaeus recognized his system to be.

My opinion, which I have already indicated, is that the endeavour to attain to a natural classification of plant communities must start from a thorough study of the ultimate units of the communities. These units must be defined, and the attempt must be made to show how they are contingent upon their environment. It is neither necessary nor desirable to establish a final system of nomenclature, since this will gradually emerge as a result of our investigations.

The ultimate unit in the Science of Plant Formations, and indeed in Plant Geography as a whole, is the qualitatively as well as the quantitatively homogeneous Plant Community.

The qualitative composition of the Plant Community. By this I mean the species-composition and the adaptation to environment of those species; the latter is of the greatest possible significance in Biological Plant Geography, whose very essence it expresses. While the individuals of a species, whatever their external appearance may be, still belong to the same species, it is not necessary for them to have the same life-form, the same adaptation to environment. Even if two plant communities are made up of the same species which occur in the same relative frequency, and even if we are dealing with micro-species or 'pure lines', these two communities need not be identical. In Systematic Botany the impression made by the environment on the individual is a matter of no consequence if the individuals are true to type. In Biological Plant Geography, on the other hand, the life-form is the essential concept; individuals belonging to the same species, micro-species, or 'pure line' may react to the locality by being xeromorphic or mesomorphic, and their difference in this character may be so strong that two plant communities with the same species composition and with the same relative frequency of species may belong to different formations. The Nanophanerophytic or Microphanerophytic oak scrub of West Jutland is an expression of definite environmental conditions, and although its dominant species are the same as those of an oak wood, yet the oak scrub as a plant community differs widely from tall oakwood, which is an expression of entirely different environment.

The quantity of the plant community means the total and proportional amounts of its constituent species.

The total amount or density of the vegetation can be determined by the eye as closed, open, scattered, &c.; but it can also be determined by the valency method, a means which should always be used in dealing with open vegetation, where it is difficult to estimate by the eye with anything like precision.

The relative amounts of the species are determined only by the statistical method, viz. by the valency method alone, or this may be combined with ocular estimate.

By the statistical method alone the relative amounts of the species of any given area are obtained, i.e. by comparing the valency numbers obtained by the valency method. If there be reason to suppose that this method does not give a true picture then the valency method combined with ocular estimate must be employed. By means both of the random plots by which the valency of the species is determined and of the relative amounts in the plots determined by ocular estimate the relative amounts of the species according to a definite scale (1–5) are determined. It is difficult to judge correctly the relative amounts within a formation, i.e. in a large area, but easy to judge sufficiently accurately within an area of a tenth of a square metre.

By comparison it is decided whether the vegetation of two or more localities belong to the same formation. It is obvious that a comparison and decision can only be made if we take for granted that the vegetation in the given environments is in equilibrium, that it is an end result. If this be taken for granted those types of vegetation which are alike both qualitatively and quantitatively belong presumably to the same formation in the strictest sense. But between similarity and the widest differences we find an endless series. How are we to classify these intermediate states? For the present I will satisfy myself by investigating those plant communities which are floristically and biologically essentially the same. These I shall gather together into groups characterized by their dominant life-form. These units will perhaps sometimes correspond to micro-species and sometimes to collective species, but I shall call both of them formations, just as the word species is used by systematists.

The word 'formation' must be used to express a floristic individuality, and I see no reason for using the word 'association' for a concept subordinate to that of formation. Such a nomenclature takes for granted the uncertain and usually wrong supposition that a floristic difference between two plant communities is not at the same time the expression of a difference in environment.

Since, however, a thorough investigation of plant communities is still only in its early stages, we need not hasten to formulate a system of concepts. Gradually the advance of an exact investigation of plant communities in various climates will show us the way to a system of concepts and to a natural classification of the communities. In the meanwhile I attempt to bring together in groups founded upon life-forms the smallest units I have investigated, the formations.

The districts within which I have carried out the investigation described in this chapter is that part of the North Cape of Jutland which lies to the north of a line joining Pælebakkeklit and Skagen. I shall make no detailed description of the character of the vegetation as a whole, but refer instead to Eug. Warming's 'Exkursionen til Skagen i Juli 1896' (*Bot. Tidsskrift*, vol. 21) and to a number of sections in the same author's great work, *Dansk Plantevækst: 2. Klitterne*.

Floristically the district is included in the region of which F. Kölpin Ravn has given an account in his 'Fortegnelse over Karplanter, fundne paa Jyllands Nordspids' (*Bot. Tidsskrift*, vol. 21), a work which treats of Skagens Odde to the north of a line drawn from Aalbæk through Gaardbogaard to Tværsted. In includes 395 species of flowering plants (with a few additions). Of these species, according to Ravn's statements and my own lists, 216 are found in the district that I have investigated. In Table 1, Nos. 4 and 5, I have given the biological spectra for each of the two local floras, and for comparison I have added the biological spectrum of the flora of Sæby (E. Rostrup, 'Sæbys Flora', *Bot. Tidsskrift*, vol. 21); I have also given the spectrum of the flora of Sæby and that of Skagens Odde added together, and that of the whole of Denmark. The correspondence of these spectra is striking, and the slight differences are instructive.

TABLE I

	No. of species.	S	E	MM	M	N	Ch	H	G	HH	Th
1. Denmark	1,084	..	..	1	3	3	3	50	11	11	18
2. Flora of Sæby	377	..	..	2	2	3	4	53	9	7	20
3. Skagens Odde+Sæby	502	..	..	2	2	2	4	49	10	12	19
4. Skagens Odde	395	..	..	1	2	3	5	44	10	14	21
5. North of the line Pælebakkeklit–Skagen	216	..	..	..	1	3	5	45	11	11	24

The soil of Denmark in general varies greatly both in chemical composition and physical properties, so that it is difficult to decide what determines the differences in the vegetation observable in separated but apparently identical localities. The relationship between cause and effect is rather clearer in the North Cape of Jutland and in similar districts, as, for example, Fanø, Skallingen, Holmslands Klit, &c., where the substratum, although not everywhere identical, is uniformly composed of sand, not only superficially, but to such a depth that the differences in the underlying strata do not affect the distribution of the species. Although the elevation of the surface varies continually, and although the composition of the vegetation alters correspondingly, it is comparatively easy to say that the essential cause of the distribution of the plant communities is their degree of elevation above the ground water. We have here, as it were, a series of great sand-culture experiments, which nature has made for us, demonstrating the relationship between the presence of a plant community and the demands of that community on the humidity of the soil. The consecutive order in which the communities occur in relationship to the varying heights of the surface above the ground water is invariably the same in all the innumerable situations in which those communities are found.

Here and there factors other than the elevation of the surface above the ground water play an important role. These factors are: (1) the relationship of the ground to the sea, (2) its age, (3) its exposure.

It is well known that there is a striking difference both in floristic composition and luxuriance of growth between the vegetation in the newer and more prosperous dunes fringing the coast and the older dunes farther inland.

Even without entering into an investigation and valuation of the individual factors determining this difference, it is natural to treat the Coastal Region and the Inland Region separately, though of course no sharp boundary separates them. As far as sand from the beach penetrates into each of the regions the vegetation is perceptibly coastal in character: where land is in active formation, and where we therefore find a broad belt of comparatively recent dunes, the vegetation, even far inland, tells us that these dunes not long ago have been 'sea dunes' even if they no longer receive sand from the beach. I include this transitional zone in the coastal region, taking as the boundary the line where we reach plant communities characteristic of the inland region.

Within that portion of the north cape of Jutland which has been investigated the coastal region falls naturally into two divisions. (1) A southern part from Pælebakkeklit in the south to Højen Fyr, Butteren in the north, where no new dune formation has recently taken place on the beach. (2) A northern part from Butteren to Grenen, where active dune formation has occurred and is still constantly taking place. The inland region consists for the most part of flat, but not horizontal, plains with series of low dunes, which at a distance are not a conspicuous feature of the landscape. The only conspicuous chain of dunes in this district is one which stretches from Højen to Skagen, and the dunes and dune-plains can be treated collectively in so far as the series of formations are determined essentially by the humidity of the ground. But while the humidity of the ground on the plains is determined essentially by the level of the ground above the ground water, on the dunes it is largely determined by exposure. And besides this, as there is a striking difference in the physiognomy of the vegetation of the dunes and plains it is of course expedient to treat the two separately; the boundary between the two can be made where the heath, which is the vegetation dominating the plains, ceases. A fifth and very limited region is formed by the stony plains, where the sand has blown away, leaving a horizontal layer consisting chiefly of shingle. I shall now proceed to give a series of examples of the composition of the vegetation within these five districts.

THE COASTAL REGION BETWEEN PÆLEBAKKEKLIT AND BUTTEREN

From Butteren in the north to Pælebakkeklit in the south, and for the matter of that for a long way farther south, no new formation of dunes has recently taken place; on the contrary the dunes are being broken down, as is easily seen from the abrupt edge which has been formed by the spring tides at the base of the sea dunes. The beach is rather narrow: in some places it is formed of sand alone, in others of sand and stones, stony beaches being especially prevalent in the north. The beaches are practically devoid of vegetation. Along the abrupt base of the sea-dunes a low rampart of sand has formed which covers the lowermost part of the dune. This rampart bears no vegetation besides the *Agropyrum junceum*-Formation which is always wont to occur on such soils. Apparently the growth of this formation is interrupted often, perhaps yearly. But the *Agropyrum junceum*-Formation is represented elsewhere. To the seaward of the dune-plain which lies between Engklit and Pælebakkeklit portions of the sea-dune have recently been cut away in several places by the sea or the wind, or perhaps by both. The openings thus formed have been filled up with sand-drifts in which the sea has subsequently cut a low vertical edge flush with that of the old sea-dunes mentioned above. On one of these new dunes the vegetation was found to consist of a well-marked *Agropyrum junceum*-Formation; in 25 ×0·1 sq. metre there were found only *Agropyrum junceum* and *Psamma arenaria* with a relative frequency of 100 and 36 (see Table 2, No. 1). In another place the vegetation was richer and contained much *Lathyrus maritimus* (Table 2, No. 2).

TABLE 2

Vegetation on recent sand-drifts in openings in the sea-dune south of Højen (25×0·1 sq. metre)

	Life-form.	1	2
Agropyrum junceum	G	100	80
Psamma arenaria	G	36	68
Lathyrus maritimus	H?	..	88
Sonchus arvensis	G	..	28
Hieracium umbellatum	H	..	28
Festuca rubra	H	..	4
Elymus arenarius	G	..	4
Points	..	136	300

The sea-dune, which is narrow, bears the open but luxuriant *Psamma*-Formation usually seen on the dunes of West Jutland. In addition to the dominant *Psamma arenaria* there are found in this district scattered plants of *Elymus arenarius, Agropyrum junceum, Sonchus arvensis, Festuca rubra, Eryngium maritimum, Weingaertneria canescens, Lathyrus maritimus, Cakile maritima, Hieracium umbellatum, Viola tricolor,* and *Senecio vulgaris* var. *radiatus*. Most of these species are hardly ever found on the inland dunes. The *Agropyrum junceum*-Formation on the very young dunes and the *Psamma arenaria*-Formation on the crest of the sea-dunes are both characterized by the Geophyte being the dominant life-form. But we find another state of affairs on the land side of the sea-dune. Here *Psamma arenaria* is probably always a dominant species, but partly through the waning of its luxuriance and partly through the increasing frequency of other species, especially Hemicryptophytes, we gradually reach what Warming called the 'Grey Dune' which is characterized by the dominance of Hemicryptophytes. First of all *Weingaertneria canescens* and *Festuca rubra* occur as dominant species either singly or together, so that we must distinguish between a *Psamma-Weingaertneria*-Formation, a *Psamma-Festuca*-Formation, and a *Psamma-Weingaertneria-Festuca*-Formation. At a lower level *Agrostis vulgaris* occurs as a dominant species, so that we get a *Psamma-Festuca-Agrostis*-Formation, and last of all a *Psamma-Agrostis*-Formation. Here moss begins to appear on the ground and a number of inland phanerogams are added, especially *Galium verum*, *Thymus serpyllum*, *Jasione montana*, *Campanula rotundifolia*, *Veronica officinalis*, and *Viola canina*; here and there *Thalictrum minus* is seen. Besides these plants *Carex arenaria* and *Sedum acre* occur in abundance. As the *Psamma* becomes more and more recessive the Hemicryptophytic vegetation gradually gets denser, until finally the *Psamma* yields its dominance in places to some of the above-mentioned species, especially *Galium verum*, *Sedum acre*, *Campanula rotundifolia*, *Thymus serpyllum*, and *Carex arenaria*. Thus the composition of the vegetation varies greatly; but it is not yet possible to assign causes for the composition of individual communities; many factors are at work together: the age of the soil, the position in relation to the points of the compass, the angle of slope, the height above ground water, &c. (see Warming's *Klitterne*, p. 102). The relationship to migration must also be considered. We are dealing with a more or less open vegetation; a given species dominates in one place not necessarily because it is specially adapted to grow under the conditions prevailing there. It may be dominant because it was the first comer, and it may be destined to be gradually supplanted when species make their appearance that are on the whole better adapted to grow there.

In general it may be said that the vegetation that first appears on the older dunes, and which succeeds the *Psamma*-Formation is composed of more or less xeromorphic Hemicryptophytes; unless the conditions are too unfavourable this vegetation is supplanted later by the heath vegetation, consisting chiefly of evergreen xeromorphic Chamaephytes.

In two places north of Højen where the sea-dunes gradually merge through a regular slope on the east side into the plains which lie inland, and where it therefore is easier to obtain a bird's eye view of the vegetation, this is what is seen. In the places nearer to Højen the plain is cultivated. In the other locality, south of Butteren, the plain is still covered with heath vegetation, and the following shows the observed transition between sea-dune and plain vegetation. To the species mentioned above as dominating the east side of the sea-dune *Salix repens* must be added. It is the plant that gives the vegetation its character. From the crest of the sea-dune downwards to the plain the vegetation was as follows:

1. *Psamma*-Formation.
2. *Psamma–Salix repens*-Formation with various herbs, and here and there low *Rosa pimpinellifolia* as the dominant species.
3. *Salix repens + Thymus serpyllum + Sedum acre*-Formation especially with *Weingaertneria canescens*, *Lotus corniculatus*, *Viola canina*, *Festuca rubra*, *Galium verum*, *Agrostis vulgaris*, *Armeria vulgaris*, *Hieracium pilosella*, *Hypochaeris radicata*, *Campanula rotundifolia*. The sand is seen here and there.

4. *Calluna vulgaris+Empetrum nigrum+Salix repens+Thymus serpyllum*-Formation; clumps and cushions of Chamaephytes partially confluent; scattered among the patches are *Sedum acre, Agrostis vulgaris, Festuca rubra, Lotus corniculatus, Carex arenaria*, and *Galium verum*.
5. *Calluna+Erica*-Formation covering the plain in various configurations.

TABLE 3

Transitional formations between the sea-dune and cultivated plains north of Højen
(25×0·1 sq. metre)

	Life-form.	1	2	3
Festuca rubra	H	100	100	100
Carex arenaria	G	100	..	12
Galium verum	H	4	..	100
Sedum acre	Ch	4	..	100
Agropyrum repens	G	..	..	48
Agrostis vulgaris	H	..	4	..
Taraxacum erythrospermum (sp. coll.)	H	8	4	24
Points	..	216	108	384
No. of species	..	5	3	6

Seawards from the cultivated portion of the plain the eastern slopes of the sea-dunes are monotonous and level, and the vegetation too is very uniform:

1. *Psamma arenaria*-Formation on the crest of the sea-dune and a long way down its eastern side; very dense; planted in some places.
2. *Festuca rubra+Carex arenaria*-Formation (see Table 3, No. 1) the sand is seen everywhere between the plants.
3. *Festuca rubra*-Formation (Table 3, No. 2) denser than before, but the sand is still visible between the plants.
4. *Festuca rubra+Galium verum+Sedum acre*-Formation (Table 3, No. 3); at another place than 3, and at a somewhat lower level.
5. *Festuca rubra+Agrostis vulgaris*-Formation on the dry meagre soil of the cultivated heath.

Besides the communities composed of Geophytes and Hemicryptophytes, which form the bulk of the plants covering the coastal regions, there is found here and there on the inner side of the sea-dune a low scrub of the Nanophanerophytes *Hippophaë rhamnoides, Salix repens*, and *Rosa pimpinellifolia*. Of these the *Hippophaë* scrubs extend farthest, often reaching the crest of the sea-dunes; where they are still open enough to admit sufficient light, the soil is covered with a copious vegetation of the herbs mentioned as belonging to the land side of the sea-dune;[1] but

[1] See p. 312.

where the scrub is closed the shade is dense and the ground vegetation entirely or almost entirely disappears. Table 4 shows the result of the investigations of vegetation of this kind covering the inner side of a dune south of Højen. The ground was covered with dead *Hippophaë* leaves and the herbs that occurred stood singly and were very etiolated; individuals of *Sonchus arvensis* especially had unusually large flaccid leaves; only of *Senecio vulgaris* var. *radiatus* a few individuals were flowering.

TABLE 4

Hippophaë-Scrub on the inner side of a sea-dune south of Højen; 0·7–1·5 metres high (25×0·1 sq. metre)

	Life-form.	
Hippophaë rhamnoides	N	**100**
Sonchus arvensis	G	28
Carex arenaria	G	36
Festuca rubra	H	16
Psamma arenaria	G	8
Senecio vulgaris var. radiata	Th	4
Points	..	192
No. of species	..	6

Salix repens too can form a small, dense, scrub, about 1 metre high, with a leaf-covered floor either devoid of vegetation or else with single etiolated shoots of *Carex arenaria*, *Festuca rubra*, and other species. But the scrub of *Salix repens* is usually low, 10–25 cm. high, in some places because the soil is dry and poor, in others because the plant covering is cut down with a scythe to be used as hay; here the soil bears a dense herbaceous vegetation consisting of *Agrostis vulgaris*, *Festuca rubra*, *Poa pratensis*, *Carex arenaria*, *Galium verum*, *Sedum acre* (etiolated), *Campanula rotundifolia*, *Viola canina*, *Psamma arenaria*, *Pimpinella saxifraga*, *Linaria vulgaris*, *Thymus serpyllum*, *Armeria vulgaris*, *Veronica officinalis*, and *Lotus corniculatus*. The same ground vegetation is found in the usually open and quite low scrub of *Rosa pimpinellifolia* or in the scrub composed of that species and *Salix repens*.

THE COASTAL REGION BETWEEN BUTTEREN AND GRENEN

Along this stretch of the coast much accretion has recently taken place: on the broad beach new series of dunes are constantly being formed outside the older ones; the ground therefore consists of alternating rows of dunes and valleys running parallel with the coast. Walking inland from the beach we pass many rows of dunes and valleys of varying breadth before the heath vegetation of the inland district begins to appear.

I have undertaken no statistical investigation of the plant communities within this continually varying region; but in order to visualize what is seen I have made notes in two places of the different vegetation encountered on passing from the beach across this coastal region.

A. Half-way between Butteren and Batteribakke:
1. Sandy Beach (23 metres broad) with scanty shingle.
2. Higher beach with much shingle (23 metres broad).
3. Still higher shingle banks (23 metres broad): scattered vegetation of *Agropyrum junceum*, *Psamma arenaria*, and *Ammodenia peploides* begins to appear.
4. New dunes up to a man's height and about 12 metres broad. *Psamma arenaria* and *Agropyrum junceum*.
5. Flat shingle (about 15 metres) and small dunes with *Psamma* and *Agropyrum*.
6. *Psamma* dunes up to 1·5 metres high (about 18 metres broad) with *Agropyrum junceum*, *Ammodenia peploides*, *Sonchus arvensis*, *Eryngium maritimum*, and *Hieracium umbellatum*.
7. Flat shingle (about 7 metres) and small dunes with *Festuca rubra*, *Armeria vulgaris*, *Psamma*, *Ammodenia*, and *Eryngium maritimum*.
8. Older *Psamma* dunes up to a man's height (8 metres broad) with single plants of *Hippophaë rhamnoides* on the south side.
9. Flat shingle (about 16 metres).
10. *Psamma-Hippophaë* dunes (about 16 metres) up to 3·5 metres high.
11. Valley (20 metres broad) with *Hippophaë*.

Then comes a confused region of dunes with *Psamma*, *Weingaertneria*, and *Hippophaë*, and valleys with *Hippophaë* scrub. To the landward of this region we meet the outermost of the valleys bearing heath vegetation.

B. Immediately to the west of the most westerly house by 'Grenens Badested':
1. Sand-bank and a flat valley to the inner side of it (23 metres).
2. Higher and shingly beach with an indication of a valley in its inner part (23 metres).
3. Higher stony beach (about 50 metres) with a little *Psamma* and *Agropyrum* on the inner part which slopes to the south.
4. Valley.
   4a. 22 metres of stony ground with small *Psamma* dunes, with *Agropyrum*, *Ammodenia*, *Cakile*, and *Glaux*.
   4b. About 20 metres of sandy hollow without vegetation.
5. Low *Psamma* dune (about 14 metres) with *Agropyrum junceum* and *Agrostis alba*.
6. Sandy hollow (15 metres) with scattered clumps of *Agrostis alba* and with *Psamma* and *Agrostis* on the margin against the low dunes.
7. Luxuriant *Psamma* dune, the height of a man (11 metres broad).

8. Valley about 23 metres broad, partly stony, partly with drifted sand bearing especially *Psamma, Ammodenia,* and *Agropyrum.*
9. *Psamma* dune as high as a man (about 20 metres).
10. Valley (about 20 metres) with *Agrostis alba*-Formation with *Sagina nodosa, Juncus alpinus, Epilobium parviflorum, Juncus lamprocarpus, Erythraea littoralis, Carex flava, Potentilla anserina.*
11. Older *Psamma* dune about 3 metres high (about 20 metres).
12. Broad moist valley with dense vegetation (about 50 metres). From the lower to the higher ground:
    a. *Heleocharis palustris*-Formation. Ground very moist, spots with *Scirpus maritimus* and *Typha.*
    b. *Agrostis alba*-Formation with patches of *Phragmites communis.*
    c. *Juncus alpinus + J. lamprocarpus*-Formation with *Salix repens, Carex flava, Agrostis alba, Euphrasia, Erythraea littoralis, E. pulchella, Linum catharticum, Potentilla anserina.*
13. Old *Psamma* dune as high as a man (about 13 metres broad).
14. Valley (about 25 metres) with *Juncus-Carex*-Formation with various species; on the inner side *Hippophaë.*
15. *Psamma* dune (about 20 metres).
16. Valley (about 25 metres) with, from lower to higher ground:
    a. *Carex-Juncus*-Formation.
    b. *Juncus-Salix*-Formation.
    c. *Hippophaë*-Scrub.
17. Irregular dune region (about 55 metres) with remains of drifted up valleys; locally with *Hippophaë* scrub.
18. Valley, very broad (over 100 metres) with low *Hippophaë* scrub, much *Juncus, Salix repens,* and a little *Empetrum nigrum*: also many herbs, among them *Epipactis palustris* in abundance.

    The lowest portions are *Carex Goodenoughii*-Meadow with *Salix repens* and *Juncus* (especially *J. alpinus* and *J. lamprocarpus*). On the lowest ground, occupied by *Hippophaë* abutting on the *Carex Goodenoughii*-Meadow, large patches of the *Hippophaë* have died. In the southern portion of the valley *Calluna vulgaris* is added.
19. Low *Festuca rubra* dunes with *Hippophaë* Scrub.
20. Hollow with *Carex Goodenoughii*; then slightly higher ground with heath and scattered low dunes. It is here that we see the beginning of the landscape and vegetation characteristic of the northern part of Jutland.

THE INLAND PLAINS

The plant communities of the inland plains and their habitats are determined first and foremost by the humidity of the ground. Since the soil in this region is essentially uniform, the communities here are determined by the elevation of the surface above the ground water. In the course of time a layer of humus (peat) up to several cm. thick has indeed formed in the lower situations, and doubtless this alteration in the nature of the soil has influenced the species composition and the relative frequency of the species; but a comparative investigation of what has happened on the older soils with peat formation and on the newer soils without peat shows that the peat has altered neither the kind of the formations nor their succession, though a shifting of the boundaries of the formations naturally occurs as the height of the ground increases.

Within the entire inland region taken collectively the following kinds of landscape are determined by the varying character of the vegetation. They succeed one another from the lowest to the highest ground; and each of them is characterized by its own type of life-form. They are:

1. Water . . . Hydrophyte community.
2. Marsh . . . Helophyte community.
3. Meadow . . . Community of mesomorphic Geophytes and Hemicryptophytes.
4. Scrub . . . Community of deciduous Phanerophytes.
5. Heath . . . Chamaephyte community.
[6. Scrub . . . Community of deciduous Phanerophytes.]
7. Steppe . . . Community of xeromorphic Hemicryptophytes.

The communities of Hydro- and Helophytes. While in many places large expanses of ground are submerged for a varying length of time during the winter months, areas covered with water during the whole of the year are few and limited. Apart from diggings the only really submerged areas are the lowest portions of the hollows within the coastal region between Butteren and Grenen, e.g. the lakes Brovande south-east of Butteren, Nedermose between Butteren and Grenen, and a third lake nearer to Grenen.

Only few Hydrophytes grow in these lakes, and *Chara* alone makes formations over large expanses. Less frequent plants are *Echinodorus ranunculoides*, *Littorella uniflora*, *Juncus supinus*, and *Potamogeton gramineus*.

The zone of Helophytes, which here is even poorer in species, consists essentially of the *Heleocharis palustris*-Formation. Towards the land this formation often passes imperceptibly into the plant community of the meadow.

The Meadow is a closed community usually consisting of mesomorphic Hemicryptophytes and Geophytes; both within the region investigated and in the heath regions it is in general easy to demarcate, being bounded below by the marsh and above by the characteristic heath association. The life-forms of both these communities differ from those of the meadow. In the woodland region, too, it is easy to demarcate the meadow, the wood here taking the place of the heath. On the other hand in certain parts of the littoral region, in particular where neither wood nor heath can thrive, it is difficult to demarcate the upper boundary of the meadow, which here passes imperceptibly into steppe. Herbaceous steppe is characterized by the same life-forms as the meadows, i.e. Hemicryptophytes and Geophytes. Steppe differs from meadow in being a more open association and in the plants being more xerophytic. But this last character is variable, altering as the ground grades imperceptibly from lower and damper situations to localities which are higher and drier.

Within the area investigated the meadow community occupies only limited expanses. Meadow forms a zone around the above-mentioned lakes, and it occupies parts of other low-lying places covered by water in winter but not in summer. Hollows of this kind are found especially in the same districts as the lakes and also north of Højen station on the plain south of Engklit. The factor which determines the upper boundary of the meadow where it abuts on heath is obviously the length of time the ground is submerged in winter. A short period of submersion does not cause the heath to be supplanted by meadow. The height of the water in winter can often be seen clearly in the summer as a distinct 'high water mark' formed by bits of loose moss and portions of plants which in winter are washed ashore and are left when the water gradually recedes. Such a 'tide line' was always found to occupy the boundary between the heath and meadow communities.

In general there can be distinguished a *Carex Goodenoughii*-meadow on the lower moister ground and a *Carex panicea*-meadow in the higher localities bordering on the lowest plant-community of the heath; naturally where the ground rises gradually the two types of meadows pass imperceptibly into one another.

Carex Goodenoughii-meadow. Besides *Carex Goodenoughii*, which is dominant over the widest areas, the following species are also dominant in varying degrees: *Equisetum limosum, Agrostis canina, Carex panicea, Heleocharis palustris, Glyceria fluitans, Aira setacea, Comarum palustre, Ranunculus flammula, Lysimachia vulgaris, Lythrum salicaria, Hydrocotyle vulgaris, Pedicularis palustris, Galium palustre, Juncus supinus* (in the lowest situations), *Carex flava, Juncus lamprocarpus,* and *J. alpinus.* In some places *Comarum palustre* and *Agrostis canina* occur in such abundance that one can speak of *Comarum palustre–Carex Goodenoughii*-meadow and *Agrostis canina–Carex Goodenoughii*-meadow. *Lythrum salicaria* also occurs in such abundance that it has a frequency grade of from 80 to 100. In some places *Carex Goodenoughii* diminishes considerably. This happens particularly in a few rather damp hollows whose bottoms are covered by *Aira setacea*-Formation with a small quantity of low *Sphagnum*. In those places the vegetation consists, apart from *Aira setacea*, especially of *Hydrocotyle vulgaris*, *Carex Goodenoughii*, *Ranunculus flammula*, *Juncus filiformis*, *Carex panicea*, *Drosera intermedia*, and *D. rotundifolia*.

The *Carex panicea*-meadow consists in general of a narrow belt between the *Carex Goodenoughii*-meadow and the lower margin of the heath.

Even in the *Carex panicea*-meadow *C. Goodenoughii* usually has a high degree of frequency, but because of the dryness of the ground *Carex Goodenoughii* develops but scantily, so that the glaucous *Carex panicea* is able to dominate the vegetation. At the lower margin of the *Carex panicea*-meadow, besides the two sedges, particularly *Hydrocotyle vulgaris* and *Agrostis canina* are prominent, and in places *Potentilla anserina*, *Mentha aquatica*, *Juncus balticus*, and others. On the higher ground *Potentilla erecta*, *Oxycoccus palustris*, *Sieglingia decumbens*, and *Nardus strictus* become prominent species. Here and there *Epipactis palustris* is found in abundance, actually on the boundary between the *Carex panicea*-meadow and the heath.

Where the ground rises rather abruptly the boundary between the *Carex panicea*-meadow and the heath is often quite sharp; for even though many plants of the meadow occur with a high degree of frequency in the lower part of the heath, yet Chamaephytes dominate in these places, characterizing the formation as heath. But where the ground rises gradually the boundary between meadow and heath is often very ill defined. In such places one sees a transitional zone with a dense ground formation of meadow species, among which are interspersed individuals of *Salix repens*, *Myrica gale*, and *Erica tetralix*. These plants become more densely aggregated as the ground rises.

The meadows are more or less influenced by cultivation, being partly used for grazing and hay, and partly for peat-cutting. In the course of time layers of peat many centimetres thick are formed in some of the hollows. This peat is cut for burning. A new vegetation quickly appears on the parts that have been dug, but this vegetation does not reach an equilibrium for a long time; here and there differences between it and the ordinary *Carex Goodenoughii*-meadow can be observed, e.g. where *Agrostis canina* dominates this is perhaps merely because the ground has been dug.

Haymaking causes, among other changes, an approach to meadow in the zone transitional between meadow and heath. This is because *Myrica Gale* and *Salix repens* are cut down at each hay harvest, so that they come to exist only as shoots which are used as hay like the herbaceous vegetation. In Table 5, No. 2, I have given the results of a statistical investigation carried out in an altered transitional zone of this kind. Table 5, No. 1, gives the results of a corresponding investigation on the lower typical meadow in the same place.

Table 5

1. *Cyperaceae*-meadow; 2. Transitional zone between meadow and heath. 3. *Myrica-gale*-Formation; 4. *Calluna-Erica*-Formation with *Myrica* and *Salix* (1 and 2: 10×0·1 sq. metre; 3 and 4: 25×0·1 sq. metre)

		1	2	3	4
Myrica gale	N	..	100	100	76
Calluna vulgaris	Ch	..	..	24	100
Erica tetralix	Ch	..	90	40	100
Salix repens	(N) Ch	20	100	24	72
Agrostis canina	H	100	100	60	..
Comarum palustre	(HH) H	100	60	4	..
Galium palustre	H	90	..	8	..
Hydrocotyle vulgaris	H	100	40	..	..
Mentha (aquatica)	H	90	..	..	..
Nardus strictus	H	10	90	..	..
Ranunculus flammula	H	100	10	..	..
Carex Goodenoughii	G	100	100	44	16
„ panicea	G	100	100	4	8
Heleocharis palustris	G	70	..	..	..
Empetrum nigrum	Ch	..	..	36	20
Vaccinium uliginosum	(N) Ch	..	..	..	4
Oxycoccus palustris	Ch	20	20	4	..
Drosera rotundifolia	H	..	10	..	..
Epilobium palustre	H	..	..	8	..
Festuca rubra	H	..	..	8	4
Hieracium pilosella	H	..	..	..	4
Juncus lamprocarpus	H	10	..	..	..
„ squarrosus	H	..	..	..	4
„ supinus	H	40	..	..	..
Leontodon autumnalis	H	10	..	..	..
Lotus corniculatus	H	..	..	..	20
Luzula campestris	H	..	..	..	4
Lythrum salicaria	H	30	20	..	..
Pedicularis palustris	H	30	..	..	..
Potentilla anserina	H	40	10	..	..
„ erecta	H	..	60	52	..
Sieglingia decumbens	H	..	60	12	..
Viola palustris	H	20	60	16	..
Carex arenaria	G	..	..	..	8
Eriophorum polystachyum	G	30	60	..	..
Juncus (alpinus ?)	G	10	40	..	..
„ balticus	G	..	..	8	..

Both investigations are very imperfect, as only ten samples were examined in each locality, but since the composition of the vegetation was very uniform, what was found must give a fairly satisfactory picture both of the species composition and the degree of frequency.

Deciduous Scrub. Above the level of the meadow in our woodland districts are found communities of deciduous Phanerophytes consisting

FIG. 126. Heath-Plain between Skagen and Højen; *Myrica Gale* in the foreground; in the background on the right the high road between Højen and Skagen. The three little girls are collecting the berries of *Vaccinium uliginosum*. Looking towards the East.

FIG. 127. Low dunes covered with small *Salix repens* on the dune-plain between Højen and Paelebakkeklit.

especially of species of *Alnus, Salix, Betula,* and *Rhamnus*, which affect, or at least tolerate, great soil humidity, and which appear at once, often indeed with the meadow, when the level of the ground rises above that suited to the Helophytes. This greater height of the ground level above the water results either from the sinking of the water level or from the elevation of the ground. All transitions between meadow and wood are encountered; there is meadow with scattered Phanerophytes, open wood or scrub with the ground vegetation consisting essentially of meadow species, and denser wood or scrub, where the meadow species diminish or disappear almost entirely.

These more or less hygrophilous communities are succeeded by mesophilous woods of oak and beech.

In the region of Skagen, as in the heath regions of Jutland taken as a whole, the heath in general takes the place of the wood; yet neither the hygrophilous nor mesophilous communities of deciduous Phanerophytes are entirely suppressed; the latter are represented in the region of Skagen, especially on the uppermost edge of the heath towards the dunes, by low nanophanerophytic-chamaephytic scrub of *Salix repens*, and here and there of *Rosa pimpinellifolia*. Farther towards the south oak scrub also occurs at the corresponding level (E. Warming, *Bot. Tidsskr.*, vol. 21).

On the lowermost boundary of the heath the formation of Phanerophytes affecting humid soils is represented by a nanophanerophytic scrub of *Myrica gale, Vaccinium uliginosum*, and damp-soil forms of *Salix repens* (Figs 126, 127). These three plants are all very common in the region of Skagen, but *Myrica* alone forms dense communities of any extent. Table 5, No. 3, shows the result of a statistical investigation of such a piece of *Myrica* scrub, a densely closed formation about 40–70 cm. high growing on the boundary between the meadow and the heath to the east of Butteren. The ground here was covered with dead leaves of *Myrica*, and with the exception of *Myrica* and *Salix* all the species present were more or less etiolated and starved. In general all three of the small deciduous bushes grow scattered; thus the *Myrica* and *Salix* usually grow both on the transitional region between meadow and heath, where the ground vegetation is still formed of meadow herbs (Table 5, No. 2), and also on the lowermost portion of the heath (Table 5, No. 4). Where the conditions are uniformly favourable all three are Nanophanerophytes, and where they form thickets in the region of Skagen they are in general low Nanophanerophytes. But *Vaccinium uliginosum* and *Salix repens* are usually Chamaephytes, and *Myrica gale* is often only a little taller; we can therefore treat them among the small deciduous bushes, Chamaephytes or low Nanophanerophytes. If we examine the four formations on the boundary between meadow and heath (Table 5, No. 2) with the view of discovering the percentage of the life-forms (evergreen Chamaephytes and Nanophanerophytes and deciduous

Chamaephytes and Nanophanerophytes) we obtain the result shown in Table 6, columns 5 and 6.

TABLE 6

Biological spectra of the formations analysed in Table 5

	No. of species.	Points.	Formations-spectrum.		N & Ch	
			Ch & N	H & G	deciduous.	evergreen.
1. Sedge meadow	21	1,120	3	97	..	..
2. Transitional zone	19	1,130	27	73	69	31
3. *Myrica gale*-Formation.	17	452	50	50	66	34
4. *Calluna-Erica*-Formation with *Myrica* and *Salix*	14	440	85	15	41	59

Heath. In the survey of vegetation given on p. 315 bog ('Mose') is left out. We usually find this kind of vegetation mentioned in descriptions of the heaths of Jutland as occurring between meadow and heath. The omission of bog as an independent type of formation is due to the fact that descriptions of heath founded upon life-forms will always include the heath-covered bog, the dominant life-forms being the same. Similarly, meadow-covered bog will be included in the meadow.

In choosing designations for plant communities we should always strive to make use of the names, often primeval, which exist in the popular language. These names often crystallize the results of the experience and observation of nature, e.g. Lake, Marsh, Meadow, Bog ('Mose'), Heath, &c. The words are used to express the ordinary typical vegetation, and they have no accurate definitions; but the scientist has to define the boundaries between the different formations in order to obtain a basis for comparison. We must therefore delimit and define the concepts of the communities existing in popular speech using them as frames surrounding the results of scientific investigations. Usually this can be done readily, but the concept bog ('Mose') presents difficulties. I shall not enter here into the different fine shades of meaning possessed by this word, but shall accept the usage prevailing in districts where heath and 'bog' occur, i.e. by 'bog' is understood an area with peat covered by such plants as *Erica tetralix*, *Myrica gale*, *Vaccinium uliginosum*, *Eriophorum*, and other species which grow in company with them. Now in trying to use the word bog to express a definite scientific concept we encounter the difficulty that the characteristic plants of this community are not found exclusively on peat, and that the differences in the composition of the vegetation is caused principally by soil humidity, which is much more important than the kind and thickness of the peat. In the popular concept of the word bog the presence of peat is certainly the most important factor, even if it is not the only factor. I am well able therefore to see eye to eye with those authors such as Mentz[1] and others who wish to keep the word bog to express a kind of soil, a geological formation. Mentz maintains however that a bog must possess peat 30 cm. deep yielding not more than 30 per cent. of ash; but my disagreement with him on this point is of no importance here, where we are concerned with limiting and investigating plant communities and not soils. It is easier to characterize the bog purely geologically, since it is not possible to maintain that the vegetation of the bog belongs to a different community from that of the heath. This is clearly seen in that part of Jutland with which we are now dealing, and in corresponding districts of the dune region of West Jutland. These districts have a sandy soil either devoid of peat or at any rate with the layer of peat no thicker than that of the *Calluna* heath. Yet as soon as we pass the uppermost edge of the meadow we encounter a community composed of the species which grow in areas customarily called bog. In the region of Skagen, beginning with species that belong to the dominant life-form, we have *Erica tetralix* and *Calluna vulgaris*. *Empetrum nigrum* is also present but is not ubiquitous (see Table 7). All three species are xeromorphic evergreen Chamaephytes; they belong therefore to the life-form dominant in both the bog and the heath. Less widely distributed, and dominant in patches, are *Salix repens*, *Vaccinium uliginosum*, and *Myrica gale*, which are mesomorphic deciduous small bushes, either low Nanophanerophytes or, where the conditions are poor, Chamaephytes. This is especially seen in *Salix repens* and *Vaccinium uliginosum*. Next come a series of species of herbs, Hemicryptophytes and Geophytes of which some, e.g. *Carex Goodenoughii*, &c., may occur in a high degree of frequency giving the ground in summer a green tint, though of course never altering the type of vegetation, which is a formation of Chamaephytes.

But even if the heath and bog communities belong to the same class of formation and also have in part the same dominant species, it would perhaps be possible to keep the word bog as a designation for a division of the heath characterized by subordinate communities of the vegetation. But this is not so, for none of the species in question (*Myrica gale*, *Vaccinium uliginosum*, and *Eriophorum vaginatum*) have a distribution within the region that could be used to characterize those soils which are usually called bog. I deplore this fact especially, because even if the name bog be kept as a designation for a kind of soil, we must face the fact that the word will continue in popular speech to denote the areas covered with bog plants without reference to the amount of peat present. Out of respect for popular usage I am inclined to use the word 'heath-bog'[2] for a bog covered with heath. Bog can then be used as a designation for a kind of soil, and in discussing the vegetation, such a soil covered with

---

[1] A. Mentz, *Studier over Danske Mosers recente Vegetation*, Copenhagen, 1912.
[2] Danish *Hedemose*.

heath plants can be called heath-bog. It must however be remembered that heath-bog is not a group of equal importance with heath but is subordinate to it; indeed I am not even able to delimit and characterize heath-bog as a part of heath; for the division of heath that has been proposed into *Calluna*-heath and *Erica*-heath[1] does not coincide with a division into heath-bog on the higher portions of land and heath on the other parts, and even if heath-bog and *Erica*-heath really should coincide, this division would only be of subordinate significance, because the dominant and characteristic heath species (*Calluna vulgaris*) is at the same time one of the dominant species of the *Erica*-heath and has the same life-form as the *Erica* itself. The difference is only a floristic one, and does not correspond with difference in life-form, so that the boundary cannot be drawn accurately where we have both *Calluna* and *Erica* dominating.

In the region of Skagen, as in the whole of the heath regions of Jutland, *Calluna* and *Erica* are dominant species over wide areas, and as the development of the vegetation has suffered no interference for long periods we may take it for granted that the vegetation of the older heath areas is in a state of equilibrium. We are justified in supposing that the dominant species occupy the soil in a manner dictated by the climate, by the humidity, and by competition. We may therefore use the occurrence of the two species in the heath as a means of subdividing the heath into two regions separated by the line up to which *Erica tetralix* occurs as a dominant species. This line is generally well defined, for as soon as the humidity of the soil decreases sufficiently to allow the degree of frequency of *Erica tetralix* to fall below 60, a very slight elevation of level and a corresponding decrease in soil humidity bring about the disappearance of *Erica tetralix*.

The composition of the *Erica*-heath is seen in Table 7, which shows the results of statistical investigations in seven places. The ground was everywhere more or less rich in Cryptogams, especially *Hypnaceae*, *Dicranum scoparium*, *Peltigera canina*, and *Cladonia rangiferina*. The difference in species composition and in the degree of frequency of the species was certainly due to the varying humidity of the soil and to digging and cutting. The high degree of frequency of *Carex Goodenoughii*, *Carex panicea*, *Sieglingia decumbens*, *Potentilla erecta*, and *Nardus strictus* in No. 1 is due to the situation of this locality in the lowest portion of the *Erica*-heath near the meadow.

On the other hand the low degree of frequency of *Empetrum nigrum* in most of the localities is almost certainly not everywhere determined by soil conditions, but is often a result of digging and cutting. *Empetrum nigrum* does not recolonize places from which it has been driven as readily as *Calluna* and *Erica*. In such situations, from which plants have been banished by digging, cutting, or burning, *Erica tetralix* is always the first

[1] *Erica tetralix* is the only species of *Erica* in Denmark.—[Editor.]

species to re-establish itself, and for a certain period of time dominates the vegetation. The thick flowering carpet of *Erica* which is conspicuous in summer in many parts of Skagen must be looked upon as a stage in the development of normal *Erica*-heath on disturbed ground rather than as vegetation in equilibrium.

Table 7

The results of statistical investigations on different parts of *Erica*-heath (25×0·1 sq. metre)

		1	2	3	4	5	6	7
Calluna vulgaris	Ch	**100**	**100**	**100**	**100**	**100**	**100**	**100**
Erica tetralix	Ch	**100**	**100**	**100**	**100**	**100**	**100**	**96**
Empetrum nigrum	Ch	32	60	4	..	8	20	**100**
Salix repens	(N) Ch	**100**	**100**	76	**100**	84	96	72
Lotus corniculatus	H	40	88	92	..	76	72	88
Antennaria dioeca	Ch	..	..	4	..	..	..	..
Oxycoccus palustris	Ch	40	..	..	..	..	..	..
Vaccinium uliginosum	(N) Ch	20	8	..	16	12	..	8
Veronica officinalis	Ch	..	8	..	..	..	..	..
Agrostis vulgaris	H	..	..	..	..	4	..	..
Festuca rubra	H	..	..	..	..	4	4	..
Hieracium pilosella	H	..	4	..	..	..	..	..
„ umbellatum	H	..	..	..	..	4	..	..
Luzula campestris	H	8	12	24	12	8	20	28
Nardus strictus	H	80	32	32	16	..	..	..
Plantago maritima	H	..	4	..	..	..	..	4
Potentilla erecta	H	84	32	12	8	..	8	..
Sieglingia decumbens	H	76	12	..	4	..	12	..
Trifolium pratense	H	..	8	..	..	..	..	..
„ repens	H	..	4	..	..	..	..	..
Carex arenaria	G	..	..	..	4	..	4	..
„ Goodenoughii	G	**100**	40	28	40	4	8	8
„ panicea	G	56	32	8	4	4	..	8
Juncus anceps	G	..	..	8	..	..	..	..
Points		836	644	488	404	408	444	512
No. of species		13	17	12	11	12	11	10

The composition of *Calluna*-heath is shown in Table 8, which gives the results of statistical investigations in seven different places. The ground is here poorer in Cryptogams than the *Erica*-heath. Mosses and *Peltigera canina* are almost absent; but *Cladonia rangiferina* is seen nearly everywhere. Locality No. 7 differs from the others in the large number of species and in their density, arising from the proximity of the north foot of the 'grey dune', which is fairly rich in species. This locality may be looked upon as a transition to the grey dune, or as being enriched by migration from it.

As is seen from Tables 7 and 8 there are therefore only six species found dominating in one or more of the fourteen heath localities

TABLE 8

The result of statistical investigations on different parts of *Calluna*-heath (25×0·1 sq. metre)

		1	2	3	4	5	6	7
Calluna vulgaris	Ch	100	100	100	100	100	100	100
Empetrum nigrum	Ch	..	..	96	28	100	44	4
Salix repens	Ch	12	100	4	76	56	64	100
Lotus corniculatus	H	40	44	4	96	84	12	40
Festuca ovina	H	..	..	..	..	..	..	100
Erica tetralix	Ch	..	..	4	..	12	44	..
Thymus serpyllum	Ch	..	4	..	..	..	..	28
Veronica officinalis	Ch	..	..	..	..	..	..	32
Achillea millefolium	H	..	..	..	..	..	..	8
Agrostis vulgaris	H	4	4	..	..	..	..	44
Anthoxanthum odoratum	H	..	12	..	..	..	..	28
Anthyllis vulneraria	H	..	..	..	4	..	..	4
Campanula rotundifolia	H	..	..	..	..	..	..	16
Festuca rubra	H	4	12	..	16	..	4	28
Galium verum	H	..	..	..	..	..	..	24
Hieracium pilosella	H	4	..	..	..	..	..	..
„ umbellatum	H	..	..	..	4	..	4	..
Hypochoeris radicata	H	8	..	..	4	..	..	..
Jasione montana	H	..	..	..	..	..	..	8
Koeleria glauca	H	..	..	..	..	..	8	20
Luzula campestris	H	8	16	..	4	..	..	48
Plantago maritima	H	4	..	..	..	..	..	..
Polygala vulgaris	H	..	..	..	..	..	..	8
Sieglingia decumbens	H	20	..	..	..	..	..	12
Viola canina	H	..	..	..	..	..	..	12
Carex arenaria	G	4	4	36	..	24	28	40
„ Goodenoughii	G	4	..	..	..	..	..	..
Linaria vulgaris	G	..	..	..	..	..	..	4
Trifolium minus	Th	..	..	..	..	..	..	4
Vicia lathyroides	Th	..	..	..	..	..	..	4
Points	..	212	296	244	332	376	308	716
No. of species	..	12	9	6	9	6	9	24

investigated. Four of these six species *Calluna vulgaris*, *Empetrum nigrum*, *Salix repens*, and *Lotus corniculatus* are common to the *Erica*-heath and the *Calluna*-heath. The fifth species, *Erica tetralix*, is naturally dominant only on the *Erica*-heath, since the line between the *Calluna* and *Erica*-heath is drawn by means of this plant. The sixth species, *Festuca ovina*, dominates only in regions merging into the grey dune, so that it need not be taken into consideration. It is seen from Table 9 that there is no difference in the numerical relationship of life-forms between the *Erica*-heath and *Calluna*-heath.

Setting aside the highest part of the *Calluna*-heath (Table 8, No. 7, and Table 9, No. 14), which has only 37 per cent. of Chamaephytes, thus

TABLE 9

Biological spectrum for each of the heath localities given in Tables 7 and 8. 1–7. *Erica*-heath; 8–14. *Calluna*-heath; 15. average of seven localities in the *Erica*-heath; 16. average of seven localities in the *Calluna*-heath.

	No. of species.	Points.	Ch	H	K (G)	Th
1. Table 7, No. 1	13	836	47	34	19	..
2. ,, ,, 2	17	644	58	31	11	..
3. ,, ,, 3	12	488	57	34	9	..
4. ,, ,, 4	11	404	78	10	12	..
5. ,, ,, 5	12	408	74	24	2	..
6. ,, ,, 6	11	444	71	26	3	..
7. ,, ,, 7	10	512	74	23	3	..
8. ,, 8, ,, 1	12	212	53	43	4	..
9. ,, ,, 2	9	296	69	29	2	..
10. ,, ,, 3	6	244	83	2	15	..
11. ,, ,, 4	9	332	61	39	..	..
12. ,, ,, 5	6	376	71	22	7	..
13. ,, ,, 6	9	308	82	9	9	..
14. ,, ,, 7	24	716	37	56	6	1
15. Average of 7 localities in the Erica-heath	24	3,696	66	26	8	..
16. Average of 7 localities in the Calluna-heath	30	2,484	65	29	6	..

showing itself to be a transition to the steppe vegetation of the grey dune, the percentage of Chamaephytes varies between 47 and 83. But taking the average of the seven localities in the *Calluna*-heath and also of the seven localities in the *Erica*-heath we see that the two spectra almost coincide (Table 9, Nos. 15 and 16). That the number of species in the *Calluna*-heath is larger than that of the *Erica*-heath is due to the species from the grey dune enriching the highest locality of the *Calluna*-heath (Table 8, No. 7). Of the twenty-four species in this locality eleven are entirely absent from the other parts of the *Calluna*-heath which were investigated (Table 8, Nos. 1–6). Considering finally the relationship between the deciduous and evergreen Chamaephytes (and Nano-phanerophytes) no difference is seen between the *Erica*-heath and *Calluna*-heath; comparing Tables 7–8 we find that the *Erica*-heath has 29 per cent. of deciduous Chamaephytes and the *Calluna*-heath 27 per cent. On the other hand the percentage of deciduous Chamaephytes on the heath in the region of Skagen is large in comparison with the heaths on glacial formations. In general there is considerable difference in the composition of the vegetation of the heaths in the region of Skagen (indeed in all heaths in the neighbourhood of dunes) compared with those heaths on the glacial formation. This is seen by the comparison between the results of investigations in the region of Skagen with heaths on Aadum-Varde Bakkeø and on Fanø, which will be found in Chapter VI, pp. 266–74.

Also apart from the localities in the region of Skagen where *Myrica gale* dominates (localities Nos. 1–3 in Table 34, l.c., p. 267), we have on the *Erica*-heath four localities (Nos. 4–7 in Table 34) which contain in all 14 species. But in the *Calluna*-heath there are ten localities (Table 35, Nos. 8–12, Table 36, Nos. 13–17) with 16 species. The corresponding numbers in the heath of Skagen were 24 and 30. The four dune-heath localities given in Chapter VI (p. 275, Table 40, Nos. 1–3 and 5), three from the *Erica*-heath (l.c., p. 275, Table 40, Nos. 1–2 and No. 5) and one from the *Calluna*-heath (Table 40, No. 3) show, in spite of the fewness of species in the localities, a greater number of species than the heath on Aadum-Varde Bakkeø (i.e. 19 and 17 species for *Erica*-heath and *Calluna*-heath respectively).

A corresponding difference prevails in the density of the species and in the biological spectrum. The heaths of Skagen have a greater species-density than those of Aadum-Varde Bakkeø, and have a much higher Hemicryptophyte percentage in the *Calluna*-heath. This difference is almost certainly due to the fact that the heaths in the region of Skagen are so recent that they still contain a part of the herbaceous vegetation which occupied the dune plains and the grey dune before the heath invaded them. The invasion of the dunes by the heath is twofold. On the one hand the heath migrates on to the older dunes inland. In Table 8, No. 7, I have given an example of the composition of the vegetation on such soil. On the other hand the heath presses farther and farther over the new dunes along the coast.

Among the species belonging to the dominant life-form of heath *Empetrum* is the first to migrate, and the one which reaches nearest to the beach; at any rate in some places *Erica* precedes *Calluna*. The following is a sketch of the vegetation in the outermost of the heath-clad dune plains to the east of Butteren, extending from the lowest portion of the plain to the outermost dune:

1. Lowest region: *Erica-Empetrum-Salix*-Formation (see Table 10, No. 1).
2. Higher part of the plain: *Empetrum*-Formation or *Empetrum-Salix*-Formation (Table 10, No. 2).
3. The foot of the dune with *Salix*-Formation (this zone may be absent).
4. *Weingaertneria-Psamma*-dune.

In the *Erica-Empetrum-Salix*-Formation *Calluna* was certainly present, but so sparsely that it did not occur in the samples investigated. Before leaving the heath I would like to give a short description of those areas that were once cultivated but which have been neglected for a long time, and form poor grazing land. I have made a statistical investigation of the composition of the vegetation in three of these localities, and the result will be found in Table 11.

PLANT FORMATIONS OF SKAGENS ODDE

Table 10

Migration of the heath in coastal regions (see Text) (25×0·1 sq. metre)

		1	2
Salix repens	(N) Ch	100	60
Empetrum nigrum	Ch	100	100
Erica tetralix	Ch	100	..
Juncus squarrosus	H	4	..
Lotus corniculatus	H	..	4
Carex arenaria	G	..	12
„ flacca	G	92	..
„ panicea	G	4	4
Juncus alpinus	G	8	..
„ balticus	G	20	..
Euphrasia sp.	Th	8	..
Points	..	436	180
No. of species	..	9	5

No. 1 (Table 11) is from an old grass field in the highest and driest part of the cultivated plain north-east of Højen. The phanerogams formed an open vegetation with low *Cladonia furcata* between the plants; and here and there small quantities of *Dicranum*. *Agrostis vulgaris*, *Armeria vulgaris*, and *Sedum acre* dominated numerically. Physiognomically the formation may be described as an *Armeria vulgaris*-Formation 10–15 cm. high.

On slightly higher and drier ground evenly covered by drifted sand the *Armeria* decreased and *Agrostis* and *Sedum* alone dominated. Some places were dominated by *Agrostis vulgaris* and others by *Festuca ovina*.

On newer fields there was a *Weingaertneria-Agrostis*-Formation.

Outwards towards the sea-dune, where the ground was evenly covered by drifted sand, there was an almost pure *Festuca rubra*-Formation (Table 3, No. 2), which, towards the dune, merged into a *Festuca-Psamma*-Formation followed on the sea-dunes by a *Psamma*-Formation.

No. 2 is from a cultivated dune-plain south of Højen between Engklit and Paelebakkeklit; here we have an old field with an open and very poor *Agrostis-Festuca-Sedum*-Formation (Table 11, No. 2); the ground is covered by low *Stereodon cupressiforme* which is seen everywhere between the flowering plants. *Cladonia furcata* and *Dicranum scoparium* are also present.

No. 3 (Table 11) is from flat ground along the north side of the high dunes south of Højen. This land has certainly been cultivated, but now bears a low *Salix repens*-Formation with a thick carpet of herbs and with *Peltigera canina* and *Hypnaceae* on the ground. The vegetation is cut down annually, for the *Salix repens* consisted only of the shoots of the current year. In the samples investigated 33 species were found with an

Table 11

The vegetation of pasture on cultivated Heath soil (25 × 0·1 sq. metre)

	Life-form.	1	2	3
Salix repens	(N) Ch	..	..	100
Sedum acre	Ch	100	72	36
Agrostis vulgaris	H	100	100	88
Armeria vulgaris	H	96	16	36
Festuca rubra	H	20	96	92
„ ovina	H	..	..	76
Achillea millefolium	H	..	..	80
Galium verum	H	4	..	84
Stellaria graminea	H	..	..	100
Carex arenaria	G	16	20	96
Rosa pimpinellifolia	(N) Ch	..	..	8
Thymus serpyllum	Ch	..	..	4
Veronica officinalis	Ch	..	..	52
Anthoxanthum odoratum	H	..	..	16
Anthyllis vulneraria	H	..	..	8
Campanula rotundifolia	H	..	..	56
Hieracium pilosella	H	..	..	40
Hypochoeris radicata	H	32	4	..
Jasione montana	H	..	..	8
Koeleria glauca	H	..	..	52
Lotus corniculatus	H	..	..	28
Luzula campestris	H	..	..	24
Pimpinella saxifraga	H	..	..	32
Plantago lanceolata	H	..	..	4
Potentilla argentea	H	64	20	..
Ranunculus sp.	H	8	..	..
Scleranthus perennis	H	..	..	8
Sieglingia decumbens	H	..	..	12
Taraxacum erythrospermum (sp. coll.)	H	44	..	..
Trifolium repens	H	..	..	12
Viola canina	H	..	..	40
Agropyrum repens	G	28	..	..
Linaria vulgaris	G	..	..	16
Poa pratensis	G	20	..	52
Arropsis praecox	Th	..	4	..
Arenaria serpyllifolia	Th	48	4	8
Bromus mollis	Th	44	4	8
Cerastium semidecandrum	Th	16	16	..
Erodium cicutarium	Th	44	44	..
Myosotis stricta	Th	..	24	..
Teesdalea nudicaulis	Th	4	..	..
Trifolium arvense	Th	12	..	4
Veronica verna?	Th	4	..	..
Viola tricolor	Th	..	..	8
Points	..	704	424	1,288
No. of species	..	19	13	33

average of 13 species in each sample; thus the formation had a high species number as well as a high species-density.

In all three examples mentioned the ground belongs to the lower parts of the grey dune and was certainly originally covered with the steppe vegetation of the grey dune, which has later been supplanted by heath. The heath in turn has been destroyed by cultivation, and a steppe vegetation from the surrounding grey dune has again invaded the area. As seen from the biological spectrum in Table 12, the Hemicryptophytes dominate numerically. The percentage of Chamaephytes is rather high, as seen in Table 11; but the Chamaephytes in the two poorer formations (Nos. 1 and 2) are represented only by one low leaf-succulent, *Sedum acre*. In the same two formations the Therophyte percentage is high too. The great richness and density of species in locality No. 3 is determined by the position of the land, which lies along the north side of the foot of the dune and agrees with the comparatively rich vegetation in the heath in corresponding localities. This is shown by comparing Table 11, No. 3, with Table 8, No. 7.

TABLE 12

Biological spectrum for the formations represented in Table 11

	No. of species.	Points.	Ch	H	G	Th
1	19	704	14	**52**	9	25
2	13	424	17	**56**	5	22
3	33	1,288	15	**70**	13	2

That the heath will again invade and supplant the steppe vegetation if cultivation ceases is apparent to the north-east of Højen, where the cultivated plain abuts on the heath, which has already begun to migrate on to ground once tilled.

THE INLAND DUNES

Apart from the heath vegetation, which covers the lowest and flattest dunes, and may rise some distance on to the bases of the other dunes, the vegetation of the inland dunes is essentially herbaceous. Here and there are small areas of deciduous scrub of *Salix repens* or *Rosa pimpinellifolia*.

Partly on account of differences in the age of the soil, but more especially because of factors varying from place to place, which are due to differences in exposure, slope, and elevation, the character of the plant covering differs greatly both in luxuriance and composition. To obtain an approximately complete picture of the changing character of this vegetation I have undertaken a series of statistical investigations in thirty-three places situated for the most part on the chain of dunes stretching from Bavneklitten south of Højen towards Skagen; the middle portion of this chain is called Engklit.

Apart from the foot of the dune, where the conditions are more favourable the south, south-east, and south-west slopes, if they are steep, have an open Phanerogamic vegetation devoid or nearly so of Cryptogams. On the less steep south, east, and west slopes, and on the upper portions of the north slopes the ground is usually covered with a carpet, varying in density, of *Cladonia furcata* or *C. rangiferina* or both these species together. Among the Lichens there is scattered a more or less open and poor Phanerogamic vegetation. On the most favourable localities a little moss grows, especially *Dicranum scoparium*. Steep north, north-west, and north-east slopes have in general a rather dense carpet of moss with more luxuriant Phanerogamic vegetation. Especially on the rather steep north sides and down towards the foot of the dunes the vegetation is luxuriant. Even if there are, as we should expect, gradual transitions between the different types of vegetation, yet, for the sake of perspicuity, I will arrange the localities investigated in the three groups above mentioned: (A) *Weingaertneria-Psamma*-dunes, free of Cryptogams and with an open Phanerogamic vegetation; (B) Lichen dune, with usually an open Phanerogamic vegetation; (C) Moss dune, with open to dense Phanerogamic vegetation. To these may be added: (D) Deciduous Nanophanerophyte-scrub.

A. *Weingaertneria-Psamma*-dune. On the dune-sides which are most strongly exposed to the scorching of the sun and which are often very steep the conditions are so unfavourable and the vegetation remains so open that the localities can be looked upon as desert. If it has not rained for some time the surface of the sand in summer is dry and loose and so easily moved by the wind that lichens and mosses can only exceptionally get a footing. *Cladonia furcata* is seen here and there. The sand is therefore visible everywhere between the scattered Phanerogams, whose dominant species are those which dominate the older parts of the sea-dune, i.e. *Psamma arenaria*, *Carex arenaria*, and *Weingaertneria canescens* (Table 13). The only other species occurring with a considerable degree of frequency is *Koeleria glauca*, a plant which is dominant in the investigated portions of the following formation, the lichen dune. In harmony with the environment (loose soil, &c.) the Geophytes are conspicuous in the biological spectrum of the *Weingaertneria-Psamma*-dune (Table 14), amounting on the average to 51 per cent. in the localities investigated. This is an open formation poor in species and consisting essentially of xeromorphic Geophytes and Hemicryptophytes.

B. Lichen-dune. The less steep south-eastern and western slopes and the upper parts of the northern slopes are in general covered with an open to quite dense carpet of *Cladonia furcata* or *C. rangiferina*, or both species together. It is a very dry and poor soil, and the intermixed

TABLE 13

Vegetation in three localities on the steep south side of high dunes south of Højen; *Weingaertneria-Psamma*-dune (25×0·1 sq. metre)

	Life-form.	1	2	3
Psamma arenaria	G	36	80	72
Carex arenaria	G	88	92	84
Weingaertneria canescens	H	68	44	92
Artemisia campestris	H–Ch	4	..	..
Festuca rubra	H	8	..	16
Galium verum	H	..	..	8
Hieracium umbellatum	H	..	8	16
Hypochoeris radicata	H	4	..	20
Jasione montana	H	..	4	..
Koeleria glauca	H	40	56	44
Viola tricolor	Th	..	..	8
Points.	..	248	284	360
No. of species	..	7	6	9

TABLE 14

The biological spectrum of the localities investigated in Table 13. *Weingaertneria-Psamma*-dune south of Højen. The numbers 1–3 correspond with the numbers in Table 13

	No. of species.	Points.	Ch	H	G	Th
1	7	248	..	50	50	..
2	6	284	..	39	61	..
3	9	360	..	55	43	2
1–3 together	11	892	..	48	51	1

Phanerogamic vegetation is usually very open and poor. Of the two lichens *Cladonia furcata* gives the impression of being the more modest in its demands and therefore often the only one or the most frequent one on the driest ground; it is usually accompanied by some *Cornicularia aculeata*. But *Cladonia furcata* is not always the first to appear, and this is very probably due to the encroachment of man into these localities. The slopes of the lichen dunes are caused by the wind: they may arise either from the removal of sand or by its deposition. In order to fix the sand such areas have been covered with a layer of cut *Calluna*. Large quantities of *Cladonia rangiferina*, which is always abundant in the *Calluna*-heath, are thus introduced, and this lichen now begins to occupy the ground in the shelter of the cut *Calluna*. In the course of time the *Calluna* crumbles away, and the *Cladonia rangiferina* gradually forms a continuous carpet covering the remains of the *Calluna*, so that it is not at once obvious how the development of this plant began.

On the moving sand that has been covered with cut *Calluna* isolated Phanerogams soon appear, which at first are scattered but vigorous. Table 15, No. 1, shows vegetation in such a locality. The remains of the *Calluna* were seen under the carpet of lichens, which consisted of *Cladonia rangiferina* with a little *Cladonia furcata*.

TABLE 15

Very poor lichen dune (25 × 0·1 sq. metre)

	Life-form.	1	2	3	4	5
Koeleria glauca	H	32	76	96	52	100
Weingaertneria canescens	H	32	52	100	64	92
Carex arenaria	G	12	28	12	100	52
Salix repens	(N) Ch	..	..	4	4	..
Festuca rubra	H	28	20	4	..	16
Hieracium umbellatum	H	..	4	12	12	..
Hypochoeris radicata	H	8	4	..	..	..
Lotus corniculatus	H	8	4	..	..	..
Viola canina	H	20	44	..	..	..
Psamma arenaria	G	..	..	..	36	16
Viola tricolor	Th	..	..	24	..	..
Points	..	140	232	252	268	280
No. of species	..	7	8	7	6	6

The Phanerogamic vegetation was so scattered that not one of the seven species had a degree of frequency amounting to 60; indeed the two commonest species, *Weingaertneria canescens* and *Koeleria glauca*, had a frequency degree of only 32.

Tables 15, 16, 18, and 19 give the results of statistical investigations of twenty-three localities within the region of the lichen dune. The localities are arranged approximately in a series, beginning with those whose vegetation is poor in species and open, and ending with those in which it is richer in species and denser. Tables 15 and 16 represent the poorest localities. The grey carpet of lichen is spotted with white bare patches of sand; here and there a little *Dicranum scoparium* is seen. The Phanerogamic vegetation is open and poor in species, there being only between 6 and 11 species in each locality and 16 species in all. The species-density is very low, less than 3. One to three dominant species were found in each of the localities except the first (Table 15, No. 1). These species were *Koeleria glauca*, *Weingaertneria canescens*, and *Carex arenaria*. On the single locality represented in Table 16, No. 3, the degree of frequency of *Festuca rubra* exceeded 60. It will be seen from Table 17 that the vegetation is composed essentially of Hemicryptophytes and Geophytes, more rarely of Hemicryptophytes alone.

Table 18 includes localities essentially similar to those of Tables 15 and

FIG. 128. The north side of Engklit north of Højen station; *Empetrum nigrum* migrating on to the *Psamma-Koeleria*-Dune; *Psamma arenaria* scattered among the *Empetrum*; isolated plants of *Lotus corniculatus, Galium verum, Carex arenaria, Salix repens*; at the edge of the *Empetrum* some *Koeleria glauca* still survives. This species forms an essential component of the original dune vegetation.

FIG. 129. Meadow dune north of Højen station: small dune plain with *Calluna-Erica-Empetrum*-Formation; here and there *Myrica Gale* is seen (as in the foreground) and *Salix repens*, which forms a fringe around the foot of the dunes. The surrounding dune was originally smothered by the application of cut *Calluna* (see page 333), and is now coverered with a poor *Koeleria glauca*-Formation and with *Cladonia*.

FIG. 130. Meadow dune north of Højen station. A small dune plain with *Calluna-Erica-Myrica-Salix*-Formation. The surrounding dune was originally smothered with cut *Calluna*, and is now covered with a poor *Weingaertneria*-Formation.

FIG. 131. *Salix repens*-Formation (to the left) and *Armeria vulgaris*-Formation (to the right) along the south side of a series of dunes south of Højen.

16, but they are richer and their species-density greater (between 3 and 4). The biological spectrum (Table 20, No. 16) is essentially the same.

TABLE 16

Poor lichen dune (25×0·1 sq. metre)

	Life-form.	1	2	3	4	5	6
Koeleria glauca	H	72	56	84	64	64	68
Weingaertneria canescens	H	64	12	32	68	32	60
Carex arenaria	G	20	100	72	84	64	48
Empetrum nigrum	Ch	..	..	4	..	..	..
Thymus serpyllum	Ch	12	..	..	4	..	..
Artemisia campestris	H–Ch	..	..	..	..	4	..
Festuca rubra	H	32	40	64	8	40	52
Galium verum	H	8	4	..	..	20	8
Hieracium umbellatum	H	16	8	24	8	..	24
Hypochoeris radicata	H	..	4	8	8	4	4
Jasione montana	H	4	16	4	..	8	..
Lotus corniculatus	H	..	..	..	..	..	12
Viola canina	H	4	..	..	44	16	8
Psamma arenaria	G	4	52	4	8	32	4
Points	..	236	292	296	296	284	288
No. of species	..	10	9	9	9	10	10

The vegetation of the localities given in Table 19 is still richer. Here the species-density is between 4 and 7 and the number of species varies between 11 and 17. The most striking feature of these localities is the invasion of Chamaephytes, the forerunners of the heath. In one locality, which abutted on the heath, *Empetrum nigrum* (Table 19, No. 1) occurred; in another locality near the sea dune *Sedum acre* (Table 19, No. 6); but the commonest Chamaephyte was either *Salix repens* or *Thymus serpyllum*. As is seen from Table 20, Nos. 7–12, the Chamaephyte percentage is high; the vegetation consisted essentially of Hemicryptophytes and Chamaephytes.

C. Moss-dune. On the slopes of the dunes, especially the north slopes, which are not exposed to full sunshine, the ground in course of time becomes covered with a vegetation of mosses more or less mixed with lichen from the lichen dunes. In the least favourable places in the moss dune the Cryptogam covering is still very open and consists, as far as mosses are concerned, essentially of *Dicranum scoparium*, which indeed occurs here and there in the region of the lichen dune. In more favourable localities other species are added, especially *Hypnaceae* (*Stereodon cupressiforme, Hylocomium parietinum,* and *H. triquetrum*). In such places there is often a thick and rather luxuriant carpet of mosses. The Phanerogamic vegetation is open, but in the most favourable localities sufficiently dense when seen from a distance to appear to cover the ground.

TABLE 17

Biological spectrum for each of the localities given in Tables 15 and 16

	No. of species.	Points.	Ch	H	G	Th
1. Table 15, No. 1	7	140	..	91	9	..
2. ,, ,, 2	8	232	..	88	12	..
3. ,, ,, 3	7	252	2	84	5	9
4. ,, ,, 4	6	268	1	48	51	..
5. ,, ,, 5	6	280	..	76	24	..
6. ,, 16, ,, 1	10	236	5	85	10	..
7. ,, ,, 2	9	292	..	48	52	..
8. ,, ,, 3	9	296	1	73	26	..
9. ,, ,, 4	9	296	1	68	31	..
10. ,, ,, 5	10	284	..	66	34	..
11. ,, ,, 6	10	288	..	82	18	..
12. 1–11 together	..	2,864	1	73	25	1
13. 1–5 ,,	..	..	1	77	20	2
14. 6–11 ,,	..	..	1	70	29	..

TABLE 18

Lichen dune a little richer than the localities in Table 16 (25×0·1 sq. metre)

	Life-form.	1	2	3	4	5	6
Koeleria glauca	H	100	60	100	100	92	32
Weingaertneria canescens	H	64	72	52	76	88	80
Carex arenaria	G	60	32	24	20	68	84
Thymus serpyllum	Ch	..	..	..	..	16	..
Veronica officinalis	Ch	..	..	..	..	..	4
Anthyllis vulneraria	H	..	..	..	..	..	4
Artemisia campestris	H–Ch	..	..	4	..	4	..
Campanula rotundifolia	H	..	..	4	..	..	4
Festuca rubra	H	40	64	88	44	..	56
Galium verum	H	8	20	..	8	12	..
Hieracium umbellatum	H	24	20	12	8	20	20
Hypochoeris radicata	H	..	..	4	4	..	4
Jasione montana	H	..	..	8	12	8	4
Lotus corniculatus	H	..	12	..	4	..	4
Viola canina	H	12	16	..	36	12	12
Psamma arenaria	G	48	4	..	16	28	8
Teesdalea nudicaulis	Th	..	..	4	..	..	..
Viola tricolor	Th	4	4	4	..	4	24
Points	..	360	304	304	328	352	340
No. of species	..	9	10	11	11	11	14

PLANT FORMATIONS OF SKAGENS ODDE

Where the Phanerogams appear to cover the ground, even when looked at from close quarters, we have what Warming calls 'Greensward dune'.[1]

TABLE 19

Comparatively rich lichen dune (25×0·1 sq. metre)

	Life-form.	1	2	3	4	5	6
Koeleria glauca	H	100	72	100	92	96	64
Weingaertneria canescens	H	8	64	52	40	92	60
Carex arenaria	G	12	60	4	..	44	12
Empetrum nigrum	Ch	100	..	..	..	..	..
Salix repens	(N-) Ch	100	64	100	72	..	..
Sedum acre	Ch	..	..	..	..	..	84
Thymus serpyllum	Ch	..	72	8	100	72	100
Agrostis vulgaris	H	4	..	..	..	..	..
Anthyllis vulneraria	H	..	..	..	8	..	..
Artemisia campestris	H-Ch	..	..	..	..	4	..
Campanula rotundifolia	H	..	..	..	4	..	20
Festuca rubra	H	8	16	32	68	20	80
Galium verum	H	..	4	..	72	4	88
Hieracium umbellatum	H	16	4	20	36	12	16
Hypochoeris radicata	H	4	4	8	4	8	4
Jasione montana	H	36	..	40	28	16	8
Lotus corniculatus	H	24	20	64	16	..	16
Viola canina	H	..	12	8	52	8	16
Poa pratensis	G	..	..	..	..	..	8
Psamma arenaria	G	..	12	4	..	20	48
Cerastium semidecandrum	Th	..	..	..	..	..	24
Viola tricolor	Th	..	..	..	4	4	12
Points	..	412	404	440	596	400	660
No. of species	..	11	12	12	14	13	17

Table 21 gives a conspectus of the composition of the Phanerogamic vegetation and the degree of frequency of the species in four different places in the moss dune; the total number of species is here 24, the number for each sample varying between 12 and 15. The species are the same as those found in the most favourable localities of the lichen dune, and the density of the species, which varies between 4 and 7, is also the same. The biological spectrum (Table 22) shows that the vegetation consists essentially of Hemicryptophytes and Geophytes; in two of the localities (Tables 21 and 22, Nos. 3 and 4) there is also a high percentage of Chamaephytes.

D. Deciduous Nanophanerophyte scrub. This kind of vegetation covers only limited areas, including but two formations: the *Rosa pimpinellifolia*-Formation and *Salix repens*-Formation.

[1] Danish *Grønsværklit*.

TABLE 20

Biological spectrum for each of the localities given in Tables 18 and 19

	No. of species.	Points.	Ch	H	G	Th
1. Table 18, No. 1	9	260	..	69	30	1
2. „ „ 2	10	304	..	87	12	1
3. „ „ 3	11	304	..	89	8	3
4. „ „ 4	11	328	..	89	11	..
5. „ „ 5	12	352	5	67	27	1
6. „ „ 6	14	340	1	65	27	7
7. „ 19, „ 1	11	412	48.5	48.5	3	..
8. „ „ 2	12	404	34	48	18	..
9. „ „ 3	12	440	24	74	2	..
10. „ „ 4	14	596	29	70	..	1
11. „ „ 5	13	400	18	65	16	1
12. „ „ 6	17	660	28	56	10	6
13. 1–12 in all	..	..	15	69	14	2
14. 1–6 „	..	..	1	78	19	2
15. 7–12 „	..	..	30	60	8	2

Rosa pimpinellifolia-Formation. In the inland dune region I have seen only a few small patches covered with this formation, which, as before mentioned, is found chiefly on the inner slopes of the sea-dune. Only rarely is the community thick enough to cast sufficient shade to obliterate the ground vegetation entirely or nearly entirely. Usually the formation is sufficiently open to allow the development of a rich ground vegetation of mosses and Phanerogams. Table 23, No. 3, contains a conspectus of a more or less open *Rosa pimpinellifolia*-scrub 20–50 cm. high, situated on the north side of the foot of the dunes south-east of Højen. Except where the vegetation was densest, the ground was covered by a thick carpet of *Hypnaceae* (including *Hylocomium triquetrum*, *H. proliferum*, and *H. parietinum*) and *Dicranum scoparium*.

The herbaceous vegetation consisted of Hemicryptophytes and Geophytes, the commonest species being *Festuca rubra*, *Carex arenaria*, and *Poa pratensis*. On corresponding ground in the neighbourhood the herbaceous vegetation consisted essentially of *Festuca rubra*, *Poa pratensis*, *Agrostis vulgaris*, *Galium verum*, *Koeleria glauca*, *Psamma arenaria*, and *Thymus serpyllum*, forming a vegetation corresponding with that of Table 19.

Salix repens-Formation. This also, as before mentioned, is common on the inner side of the sea-dune; but unlike the *Rosa pimpinellifolia*-Formation it is also common in the inland dune region, especially along the foot of the dunes. Table 23, Nos. 1 and 2, show the composition of the vegetation on comparatively low flat land along the inner side of

Table 21

Localities in the moss dune (25×0·1 sq. metre)

	Life-form.	1	2	3	4
Koeleria glauca	H	100	92	72	96
Weingaertneria canescens	H	92	52	..	60
Carex arenaria	G	88	100	100	92
Sedum acre	Ch	8	..	..	..
Thymus serpyllum	Ch	4	..	..	84
Veronica officinalis	Ch	..	8	64	56
Agrostis vulgaris	H	..	..	40	..
Anthyllis vulneraria	H	..	..	4	4
Artemisia campestris	H–Ch	..	8	..	..
Campanula rotundifolia	H	8	..	60	24
Festuca rubra	H	64	100	100	96
Galium verum	H	16	4	4	72
Hieracium umbellatum	H	..	4	..	..
Hypochoeris radicata	H	..	4	..	..
Jasione montana	H	8	60	40	12
Lotus corniculatus	H	24	24	4	20
Armeria vulgaris	H	4	..	..	20
Viola canina	H	..	..	..	16
Linaria vulgaris	G	..	..	36	..
Poa pratensis	G	..	8	28	..
Psamma arenaria	G	4	40	24	40
Cerastium semidecandrum	Th	..	..	..	12
Teesdalea nudicaulis	Th	..	16	4	..
Viola tricolor	Th	..	32	4	..
Points	..	420	552	584	704
No. of species	..	12	15	15	15

Table 22

Biological spectrum of the localities given in Table 21

	No. of species.	Points.	Ch	H	G	Th
1. Table 21, No. 1	12	420	3	75	22	..
2. ,, ,, 2	15	552	1	63	27	9
3. ,, ,, 3	15	584	11	56	32	1
4. ,, ,, 4	15	704	20	59	19	2
5. 1–4 in all	24	..	9	63	25	3

Table 23
Deciduous Nanophanerophyte scrub (25×0·1 sq. metre)

	Life-form.	1	2	3
Rosa pimpinellifolia	N–Ch	..	..	100
Salix repens	(N–) Ch	..	100	..
Sedum acre	Ch	100	92	..
Festuca rubra	H	100	100	96
Galium verum	H	100	84	24
Hieracium umbellatum	H	84	60	..
Lotus corniculatus	H	96	68	4
Weingaertneria canescens	H	72	20	..
Carex arenaria	G	12	44	72
Veronica chamaedrys	Ch	..	4	..
,, officinalis	Ch	..	..	4
Achillea millefolium	H	..	4	..
Agrostis vulgaris	H	..	24	16
Anthyllis vulneraria	H	4	24	..
Campanula rotundifolia	H	..	16	24
Erigeron acer	H	..	4	..
Hieracium pilosella	H	..	4	..
Hypochoeris radicata	H	..	8	..
Jasione montana	H	..	..	4
Leontodon autumnalis	H	..	36	..
Koeleria glauca	H	4	36	8
Plantago maritima	H	..	28	..
Viola canina	H	40	72	24
Linaria vulgaris	G	..	16	8
Poa pratensis	G	..	60	44
Psamma arenaria	G	24	..	24
Arabis thaliana	Th	..	..	4
Bromus mollis	Th	..	4	..
Viola tricolor	Th	24	8	..
Points	..	660	916	456
No. of species	..	12	24	15

the sea-dune. This locality forms a transition to that seen inland. No. 1 shows the composition of the vegetation on parts devoid of *Salix repens*; mosses begin to appear (*Polytrichum piliferum*, and a few *Hypnaceae*, &c.). The Phanerogamic vegetation is rather dense, but sand is seen here and there between the plants. No. 2 shows the state of affairs on corresponding soil, but the ground is here covered with a very dense growth of *Salix repens* only 10–30 cm. high. This was so low and poor in shoots that a rich herbaceous vegetation was able to grow and thrive in it. Its diminutive stature is perhaps due to cutting. The same thing was seen in a corresponding chamaephytic *Salix repens*-Formation farther inland. Nos. 1–4, in Table 19, are best placed here; but in the last locality the *Salix repens* thickets were very open, and the herbaceous vegetation was essentially the same as that of adjacent ground devoid of *Salix repens*. But in individual places along the foot of the inland dunes the thickets of *Salix repens* grew high and often so dense that the ground vegetation was partially or entirely suppressed, so that we have here a Nanophanerophyte scrub without or almost without ground vegetation. This corresponds to the *Hippophaë* scrub mentioned before.

THE STONY PLAINS

Although the stony plains are horizontal or nearly so, the environments they offer to plants are as diversified as those of the dunes, and the vegetation is correspondingly various. Their elevation above the ground water and the composition of their soil vary from place to place. The ground in some places consists almost entirely of stones varying in size, and these localities are almost devoid of plants. Where there is sand between the stones or the stones are almost covered with sand the vegetation may be more or less luxuriant. Table 24, No. 1, shows the

TABLE 24

Vegetation on the Stony Plains

	Life-form.	1	2
Festuca ovina	H	..	100
„ rubra	H	68	4
Viola canina	H	100	..
Weingaertneria canescens	H	88	..
Sedum acre	Ch	..	32
Agrostis vulgaris	H	..	28
Campanula rotundifolia	H	..	4
Galium verum	H	..	32
Hieracium umbellatum	H	12	..
Hypochoeris radicata	H	..	24
Jasione montana	H	4	..
Koeleria glauca	H	12	..
Lotus corniculatus	H	24	4
Plantago maritima	H	40	44
Sagina nodosa	H	4	..
Scleranthus perennis	H	..	4
Armeria vulgaris	H	4	36
Carex arenaria	G	16	4
Psamma arenaria	G	4	..
Points	..	376	316
No. of species	..	12	12

vegetation on a small stony plain immediately within the sea-dune southwest of Højen. There was sand everywhere between the stones, and the number and density of species was considerable, though the vegetation was very open. As a rule such places are almost certainly destined to be covered with blown sand, and the composition of the vegetation is altered as it becomes denser; but even if sand does not drift in, the stony plains will nevertheless in time certainly develop a denser vegetation. Table 24, No. 2, shows the vegetation of an old stony plain north-west of Skagen. The ground was almost covered with a thin carpet of *Cladonia furcata* and a little *C. rangiferina*, accompanied by an open phanerogamic vegetation. While the first-mentioned open plain might be called desert as it at present exists, the other locality on the contrary corresponds to lichen dune, and may therefore be classed as steppe.

TABLE 25

Biological spectrum of each of the localities in Tables 23 and 24

	No. of species.	Points.	N	Ch	H	H	Th
1. Table 23, No. 1	12	660	..	15	77	4	4
2. ,, ,, 2	24	916	..	21	70	8	1
3. ,, ,, 3	15	456	22	1	59	17	1
4. ,, 24, ,, 1	12	376	..	..	95	5	..
5. ,, ,, 2	12	316	..	10	89	1	..

IX
ON THE VEGETATION OF THE FRENCH MEDITERRANEAN ALLUVIA[1]

In the course of a journey which I undertook in 1909-10 in the countries bordering on the Western Mediterranean, I made, at the beginning of June 1910, a short sojourn in the south of France, and there, as elsewhere, I took occasion to study the vegetation of the marine alluvial deposits. My researches were carried out at three different points of the coast: the neighbourhood of Cette, of Palavas, and of the Saintes-Maries.

From the point of view of the biological geography of plants, the region which comprises the three localities considered belongs to the Therophyte climate, that is to say the biological type of Therophytes clearly predominates over all the other types which are represented in the 'biological spectrum' of its flora. That this is so will be seen by comparing one by one the biological spectra of the local floras with the 'normal spectrum', which represents the flora of the entire world. In default of a better, I shall use, in that which follows, the normal spectrum that I have already given in my work 'Livsformernes Statistik som Grundlag for biologisk Plantegeografi' (Chapter IV, The Statistics of Life-Forms as a basis for Biological Plant Geography, pp. 115-17), while regretting that I have not had the leisure to establish a new basis of comparison on more ample data.

The immediate impression that one receives in traversing the south of France, after having visited still more southerly regions such as the lowlands of eastern and southern Spain, is that of an impoverishment in Therophytes, and also an increase in the proportion of Hemicryptophytes. In passing northwards we are approaching the countries where the Therophyte climate, which is that of the Mediterranean countries, is becoming replaced by the Hemicryptophyte climate of Central Europe, and our comparisons of the biological spectra of the various local floras with the normal spectrum should especially consider their relative proportions of Hemicryptophytes and Therophytes.

My starting-point was the flora of the department of Hérault, which occupies a central position among the Mediterranean departments of France, and in the interior of which most of my excursions were made. As a floristic basis I used the well-known work of H. Loret and A. Barrandon, *La Flore de Montpellier ou analyse descriptive des plantes vasculaires de l'Hérault* (2nd edition, 1886). According to the definitions I have adopted, the 2,044 species of phanerogams cited in this work furnish

[1] *Mindeskrift for Japetus Steenstrup*, Copenhagen, 1914.

at least 651 Therophytes, or a proportion of at least 32 per cent. Comparing this figure with the 13 per cent. of the normal spectrum, one can see that the normal proportion is more than doubled. If the department of Hérault be situated in its entirety on the frontier separating the Therophyte from the Hemicryptophyte climate, it ought to show, in relation to the normal spectrum, an increase of Hemicryptophytes equal to that of the Therophytes; and as the normal percentage of Hemicryptophytes is 27 per cent., this life-form ought here to exceed 60 per cent. We see, however, that the Hemicryptophyte figure is very much lower, scarcely exceeding 40 per cent. By reason of the great difficulties which occur in the establishment of the life-forms of all the species belonging to such a varied flora, I have not attempted to determine the proportion of Hemicryptophytes in the biological spectrum of the flora of Hérault; but a comparison with the related floras of less extensive domains less rich in species seems to confirm the hypothesis according to which the proportion of Hemicryptophytes in the flora of Hérault does not exceed 40 per cent. Hence it follows that this department, or at least its greater part, belongs to the Therophyte climate.

The localities which have furnished floras for comparison are, first, the Ligurian Islands situated along the coast of the Gulf of Genoa, and, secondly, the Euganean Hills in the neighbourhood of Padua; both these domains, like the south of France, abutting on the northern frontier of the Therophyte climate. The lists of the two floras comprise respectively 436 and 966 species of phanerogams.

TABLE I

Biological spectra of the Ligurian Islands and the Euganean Hills

	No. of Species	S	E	MM	M	N	Ch	H	G	HH	Th
Ligurian Islands	436	..	..	1	3	9	6	36	11	1	33
Euganean Hills	966	..	..	1	3	3	3	43	8	6	33
Normal Spectrum	400	1	3	6	17	20	9	27	3	1	13

In Table 1 the biological spectra of these two domains are recorded, and below, for the sake of comparison, our provisional normal spectrum. From this Table it can be seen that in the two floras the proportion of Therophytes is 33 per cent.; and thus sensibly the same as in the flora of Hérault, where the Therophytes constitute 32 per cent. of the total flora—perhaps even a little greater, since we probably ought to class among the Therophytes some of the species placed in the category of biennials: the annual plants alone amount to 32 per cent. of the whole flora. Contrasted with the 33 per cent. of Therophytes in the floras of the Ligurian Islands and the Euganean Hills, there are respectively 36 and 43 per cent. of Hemicryptophytes in the biological spectra of these localities. These figures indicate well, I think, the approximate limits between which the proportion of Hemicryptophytes in the flora can vary, when the proportion of Therophytes is 33 per cent. In floras with a higher proportion of Therophytes the percentage of Hemicryptophytes will diminish, and reciprocally; we may cite, for example, the neighbourhood of the Island of Argentario on the west coast of Italy with 42 per cent. of Therophytes and only 29 per cent. of Hemicryptophytes, and, contrasted with this, the flora of Stuttgart which gives only 17 per cent. of Therophytes and 54 per cent. of Hemicryptophytes.

But while thus presenting the characteristic spectrum of the Therophyte climate, the flora of Hérault doubtless marks a transitional zone: it is even extremely probable that the northern frontier of the said climate passes through the northern part of the department. The precise tracing of the frontier would require an examination of the biological spectra of a series of local floras which characterize the areas considered, and these spectra should, in their turn, be based on the one hand on a series of floristic lists of suitable domains, and on the other hand on the determination of the life-forms of all the diverse species represented. Of these two conditions the first could be strictly realized with the aid of already existing literature, completed, at need, by new investigations, while the last demands, on the other hand, for a country as rich in species as the eastern part of southern France, a considerable work which I have not yet had time to undertake. I must therefore content myself with indicating certain significant features well adapted to show the rapid diminution in the number of Therophytes as one advances towards the north, proceeding from the lowlands bordering the Mediterranean to the mountain complexes which make up the northern part of the department of Hérault.

In the *Flore de Montpellier* cited above, Loret (pp. xiv–xxiii) gives a list of characteristic species of each of the three regions that are ordinarily distinguished: first, the littoral; secondly, the region of olives; thirdly, the mountain region. Since we have to distinguish between the low-lying southern region and the mountains of the north, we ought to include in a single group of species Loret's two first groups and thus to assemble on the one side the species which are peculiar to the country less than 350 metres above the sea, and, on the other, those which characterize the heights above 350 metres. It goes without saying that the two lists thus obtained ought not to be considered as two local floras: in the first place some of the species recorded as peculiar to one of the two domains are represented in the other, although in small proportion, and, secondly, all the very widely distributed species in both domains have been omitted. Thus the biological spectra that can be constructed with these species particularly characteristic of the two domains will not coincide in any way with the true biological spectra of the lowlands and highlands of Hérault. Nevertheless one can find in such partial spectra useful information on the divergences of the two floras, notably in the proportion of Therophytes.

For the aim which we have in view—namely the establishment of the direction of these divergences—it will not even be necessary to consider the totality of the species recorded as characteristic; it will suffice to select the proportion of Therophytes in some of the families which are richest in species, for example the *Gramineae*, the *Papilionaceae*, and the *Compositae*.

TABLE 2

The proportion of Therophytes in certain families of plants in the department of Hérault

	No. of species.	Species of Therophytes.	
		Total.	Per cent.
Gramineae	185	86	46
Papilionaceae	186	104	56
Compositae	238	65	27
Total	609	255	42
The whole flora	2,044	651[1]	32

[1] Minimum (exclusively annual).

In Table 2 is given the proportion of Therophytes in each of these three families in the flora of Hérault, the proportion of Therophytes in the three families taken *en bloc*, and finally the total number of Therophytes in the entire flora of the department; and it appears that the three families together show a larger proportion of Therophytes than the flora as a whole.

To facilitate the comparison with a flora belonging to the Hemicryptophyte climate I give in Table 3 the corresponding figures for the Danish flora. The two tables show the difference which exists between these two climates.

TABLE 3

The proportion of Therophytes in the same families in Denmark

	No. of species.	Species of Therophytes.	
		Total.	Per cent.
Gramineae	90	16	18
Papilionaceae	55	15	27
Compositae	112	19	17
Total	257	50	19
The whole flora	1,084	196	18

If we consider the proportion of Therophytes in the three families taken together, first the species which Loret gives as characteristic of the lowlands (less than 350 metres alt.), and then the species enumerated in the same work as characteristic of the mountain region (more than 350 metres alt.) (Tables 4 and 5), the difference is striking.

TABLE 4

The proportion of Therophyte species in the *Gramineae, Papilionaceae,* and *Compositae* in the flora of Hérault below 350 metres

	No. of species.	Species of Therophytes.	
		Total.	Per cent.
Gramineae	91	53	58
Papilionaceae	71	52	73
Compositae	62	22	35
Total	224	127	56

TABLE 5

The proportion of Therophyte species in the *Gramineae, Papilionaceae,* and *Compositae* in the flora of Hérault above 350 metres

	No. of species.	Species of Therophytes.	
		Total.	Per cent.
Gramineae	10	1	10
Papilionaceae	19	4	21
Compositae	42	5	12
Total	71	10	14

The first total of species belonging to the three families furnishes 56 per cent. of Therophytes, the second only 14 per cent. It is true that these figures do not represent the proportion of Therophytes in the two floras taken in their entirety, but they do indicate the direction of movement, and we may believe that the flora of the northern highlands of the department of Hérault would give the characteristic spectrum of the Hemicryptophyte climate—for the impoverishment in Therophytes on the northern frontier of the Therophyte climate is always accompanied by an increase in the proportion of Hemicryptophytes.

But while admitting that the northern region of the department of Hérault belongs to the Hemicryptophyte climate, we ought to recognize that the southern (Mediterranean) region definitely extends, as shown by its spectrum, into the Therophyte climate: the whole flora of Hérault has 32 per cent. of Therophytes, and this figure, too high to find a place in the biological spectrum of a flora belonging to the Hemicryptophyte climate, finds a very plausible position in the biological spectrum of a flora inhabiting areas with a Therophyte climate. We may also cite other figures which tend to place the Mediterranean regions of France in the Therophyte climate: the 39 per cent. of Therophytes which characterize, according to the list given by Flahault and Combres,[1] the flora of the Camargue at the mouth of the Rhone, and the 45 per cent. (at least) of Therophytes presented by the spectrum of the 508 species of Phanerogams in the Island of Porquerolles, near Toulon.[2] No one will dispute that floras which show such a very high proportion of Therophytes belong to the corresponding climate, for even if all the non-Therophytes were Hemicryptophytes the relative increase of the proportion of Hemicryptophytes, related to the normal spectrum, would be less great than that of the Therophytes. Let us consider, for example, the flora of the Camargue: if all the plants other than Therophytes were Hemicryptophytes it could not contain more than 61 per cent. of Hemicryptophytes; but in order that the relative increase of the proportion of Hemicryptophytes should be equal to that of the Therophytes, the figure should be 81 per cent. The actual state of things can be seen from Table 6, which gives the biological spectrum of the flora of the Camargue, assigning 32 per cent. to the Hemicryptophytes; and this spectrum agrees well on the whole with those of other regions belonging to the Therophyte (Mediterranean) climate.

TABLE 6

Biological Spectrum of the Camargue

	No. of Species.	Ph	Ch	H	Cr	Th
Camargue . .	233	10	8	32	11	39

But while distinctly belonging to the Therophyte climate the Mediterranean coasts of France are far from having a physiognomy primarily characterized by the existence of communities of Therophytes. One might even go so far as to say that Therophyte communities play quite an inconsiderable role in the vegetation of these countries; it is only certain regions of quite special nature, such as the sandy maritime alluvia, &c., where the impression made is that of a predominantly Therophyte community. This is because in considering plant communities we ordinarily take up a purely physiognomic point of view. To a certain extent this point of view is unavoidable, and besides, it is generally adopted by geographers when they attempt to characterize diverse regions with the aid of their vegetation. When we consider vegetation physio-

[1] Flahault, Ch., and Combres, P., 'Sur la flore de la Camargue et des alluvions du Rhône', *Bull. Soc. Bot. Fr.*, t. xli. [2] For the list of species, see Jahandiez, E., *Les îles d'Hyères*, 1905.

gnomically, we attend especially to the great dominant masses, the features which leap directly to the eye, rather to the number and height of the individuals than to the number of species: whereas in determining plant climates based on the life-forms of plants we seek, first of all, in the flora of a given country the life-form which is represented by the greatest number of species, and relate this to the characteristic conditions of the environment, that is to say to the climate in the widest sense of the word.

Let us consider a formation in which Phanerophytes, Chamaephytes, Hemicryptophytes, and Therophytes are represented with the same degree of frequency. Here one would be naturally led to characterize the formation as essentially phanerophytic, because the Phanerophyte life-form is the most conspicuous; and if we ignore the Phanerophytes the plants that remain will give us the impression of a formation of Chamaephytes, for these, though masked a little during the growing season by the taller Hemicryptophytes and Therophytes, will predominate in the physiognomy during the resting season. Now if we suppress the Chamaephytes the formation will take a hemicryptophytic character; and finally, if we suppress these too we shall have a formation of which the therophytic character will be disputed by no one! In other words, life-forms of equal and considerable frequency will predominate in the order of succession in which I have enumerated them in the biological spectrum, and this gives the explanation of the fact of experience that even in the Therophyte climate the cases are rare in which the Therophytes sufficiently dominate a formation to make it natural for us to call the formation by their name on simple inspection.

In considering the actual state of things and the more or less natural communities of plants, we can, I believe, say that it is the maquis, a community composed for the greater part of Nanophanerophytes and Chamaephytes, which is especially widespread and predominant in the Therophyte climate of the Mediterranean region. But we must not forget that the existing vegetation is to a great extent a product of culture and that the plant covering of the soil differs very much in appearance from that which existed before man had exerted his modifying influence. In this far-distant epoch, the extension of phanerophytic communities was much greater than it is to-day. If we may judge by what remains of the communities of Meso- and Microphanerophytes, and also by the species of Phanerophytes cultivated or tolerated at present in areas long since devoted to the cultivation of useful plants, I cannot doubt that at one time almost all the low-lying fertile lands were covered with forest, that is to say with communities of Mesophanerophytes, or—in the drier parts, those less favourable to vegetation—of Microphanerophytes. The forest was the first to succumb, and to yield its place, which was the best place, to useful plants; and then it became the turn of the communities consisting of lower-growing Phanerophytes belonging to the poorer soils to be replaced by cultivated plants.

But it is far from true that the suppression of primitive vegetation on the soils actually cultivated is the only change which has occurred in the course of the ages: the soils which have remained uncultivated and of which the majority are little suited for culture, have also undergone great modifications in the aspect of their vegetation. These modifications are due in the first place to the utilization of the Phanerophytes of the maquis as firewood; the maquis has suffered much, and the frequency of cutting and the inhospitable soil, which permits only of slow growth, have hindered any return to a primitive condition. Where the exploitation was ruthless the taller-growing Phanerophytes have suffered most, because they furnished the greatest quantity of wood which could be worked or burned; after them the nanophanerophytic shrubs of the maquis were attacked, apart from those whose spines rendered them disagreeable to touch, with the result that, notably in the neighbourhood of inhabited regions, the taller and unarmed bushes have disappeared because they have been sought out, and there remain only spiny shrubs and Chamaephytes.

Any one who has once remarked this state of things cannot travel in any Mediterranean country without being struck almost everywhere by this disappearance of phanerophytic communities. Sometimes one meets with great stretches of wooded country where the abomination of desolation prevails: thus in my travels I saw on the west coast of Italy, to the south-west of the mouth of the Albegno, great areas of forest and maquis of the Maremma felled and converted into charcoal, of which heaps as large as houses, along the whole length of the quay, were awaiting the arrival of ships to take them away. Sometimes one sees the bushes and shrubs cut in the maquis taken in great wagons to the stores of the firewood merchants in the neighbouring towns; but most often on donkeys and mules, so loaded with small wood resulting from the devastation of the maquis that one can scarcely see the animals under their burdens. In all the tracks and foot-paths leading from the maquis to the town one meets these walking faggots; in a path crossing a narrow valley to the south of Beniajan (to the east of Murcia) I met in less than two hours fifteen donkeys and mules loaded with *Spartium, Cistus*, and other shrubs and under-shrubs of the maquis, which had apparently been brought from a considerable distance, for all the hill-slopes that one could see around were almost entirely denuded of Phanerophytes.

In the steppes and deserts of North Africa, where shrubs are often not to be found, recourse is had to Chamaephytes, and particularly to Salicornias, carried from a distance on the backs of camels, as one can see, for example, at Kairouan and at Biskra. In the pastures of the Spanish Estremadura I observed to the east of the town of Caceres a locality where not only the Phanerophytes had been removed but also most of the Chamaephytes, with the exception of a spiny *Genista* which still formed an almost continuous vegetation; and it even seemed that this spiny brushwood bore traces of recent cutting.

As with our own meadows, the maquis of the Mediterranean region is probably most often a semi-natural formation, although it is not easy to reconstruct the primitive one. If we ignore the cultivated areas and the influence exerted by man in the course of centuries on the development of Phanerophytes, we must recognize that at present it is the maquis, with its communities of xeromorphic and evergreen Nanophanerophytes and Microphanerophytes, which characterizes physiognomically the Mediterranean countries. But it does not follow that the maquis can serve as the principal characteristic of the climate of Mediterranean vegetation and help us to establish a limit of demarcation between this climate and others. It is easy to understand that many, notably geographers, are tempted to adopt the physiognomic point of view and to class regions according to the dominant formation: to a certain degree this manner of seeing the facts has justification, but it cannot be made the basis of a scientific classification. It is only when we have established the main boundaries by other means that we can usefully take up the physiognomic standpoint in order to differentiate the diverse vegetation in the interior of a climatic region already delimited. To tell the truth, the physiognomic principle has no very great scientific value. The features which jump to the eye are not necessarily the best suited to estimate the scientific importance either of species or of biological types in an investigation of the vegetation as a response to climate. Such a research ought not to content itself with considering a relatively small number of species distinguished by their height or by other conspicuous features; as far as possible it ought to take into consideration all the species represented in the vegetation, since all will show themselves adapted, each in its own way, to the climate in question. We shall thus have a system of biological types arranged in a hierarchy according to the relative proportions of the species they contain.

An attempt to fix, with the aid of the maquis formation, the boundary which separates the climate of Mediterranean vegetation from that of Central Europe at once meets with the difficulty that there is no visible boundary; the maquis continues into certain plant communities of Central Europe, whose biological definition coincides with that of some forms of maquis, and which are composed partly of species which may constitute an essential element of maquis. In passing northwards from northern Italy we meet in the Alpine valleys a whole series of Phanerophytes and Chamaephytes of the maquis, though the biological spectrum of these countries clearly belongs with those of Central Europe. The same observation can be made in France in ascending the valley of the

Rhone; and the vegetation of the heaths of Central and North Europe again offer strong resemblances to the formations of Mediterranean maquis—like these they are composed of evergreen xeromorphic Nanophanerophytes and Chamaephytes, but with this difference, that ordinarily the number of phanerophytic and chamaephytic species is smaller. This last difference, however, is by no means absolute: forms of maquis exist which are poor in species, and, on the other hand, heath formations, for example the Irish heaths, are composed of as great a number of species of Phanerophytes and Chamaephytes as many maquis possess. The transitions occur by imperceptible degrees, and the limits that one is in the habit of distinguishing are neither clearly distinct nor are they scientifically well founded.

It is clear that we cannot complain that the physiognomic method does not establish sharply marked limits in vegetation where the climate shows gradual transitions; but we can, nevertheless, justly say that it is radically incapable of tracing any limits of demarcation in regions of this nature. It has no means of justifying the choice of a line nor of deciding whether a line adopted has the same biological value throughout its course. It is otherwise with the method which delimits vegetational climates by the aid of life-form statistics, for here the course of the boundary is determined by the relation between the biological spectrum of the local flora and the normal spectrum, that is to say by the relation of two quantities which can be expressed numerically. According to this method the line of demarcation between the Therophyte climate of the Mediterranean region and the Hemicryptophyte climate of Central Europe ought to be drawn through the points where the Therophytes are replaced by Hemicryptophytes as the predominant biological type in the spectra of the local floras compared with the normal spectrum. Such a line would start from a point on the Atlantic coast and go as far towards the east as the Therophyte climate extends: it would have everywhere the same value, the same relation of life-forms in the biological spectra.

When the climate and also the vegetation change gradually and there is nothing striking from the physiognomic point of view, this line can be traced in absolutely the same manner as the analogous lines of climatology. In passing from one region to the other the temperature and precipitation alter imperceptibly, but that does not prevent the June isotherm of 15° C. being a definite line that we can follow from point to point in spite of all the variations which it may show in various relations. The same thing is true with lines constructed by the help of biological spectra. The Mediterranean therophytic climate, which extends from the southern limit of the Sahara to the foot of the Alps, and from the Atlantic to the mountain masses and plateaux of Central Asia, is characterized by the predominance of Therophytes in the biological spectra of the various local floras. Towards the north it is replaced by the Hemicryptophyte climate at the precise point where the Hemicryptophytes become the predominant life-form in the biological spectrum compared with the normal spectrum; and in the south of France, with which we are particularly concerned here, this limit probably passes across the northern portion of the department of Hérault.

VEGETATION OF THE MARINE ALLUVIA

Along the Mediterranean littoral of France, and notably behind the coast-line of the department of Hérault, there stretches a series of maritime lakes or lagoons separated from the sea by a tongue of sandy alluvium ordinarily quite narrow and bordered on the other side by alluvia of variable origin and extent. These maritime lagoons constitute a phenomenon quite common in the Mediterranean region as well as in many other parts of the world. The coasts of northern countries offer many examples of analogous formations; and a comparative study of the communities of plants inhabiting the immediate neighbourhood of the lagoons of the south and of those of the north should help us to determine the response of the vegetation of this type of soil to different climates.

The Haff of Ringkjøbing (Ringkjøbing fjord), behind the 'Lido' of Holmsland; the Haff of Nissum behind the Husby line of dunes, as well as the islands of Manø and Fanø, the peninsula of Skalling, and the dune country which extends from Blaavand as far as Blaabjerg in front of the Lake of Fil and of other smaller lakes; the sandy point of Agger; the northern point of Jutland, &c.—these are all localities offering perfect geological analogies with the lagoons of southern France and the alluvia which surround them. The coasts of the Baltic Sea present similar formations, such as the Kurisches Haff and Frisches Haff with the sandy alluvia called 'Nehrung' which separate them from the sea; and they occur again on certain parts of the coasts of Holland and of France and also elsewhere.

In the Mediterranean region I was able to study the vegetation of this sort of formation not only on the French littoral, but also along the Italian coasts, in Tunis, and on the southern and eastern coasts of Spain. On the east coast of Italy I visited the Venetian lagoon with the Lido which forms a rampart on the Adriatic side, just as the dunes of Holmsland protect the fjord of Ringkjøbing against the North Sea; and on the west coast I explored, between Pisa and Leghorn, the maritime dunes and the Maremma which they bound; the sandy alluvia called Tombolo della Gianella and Tombolo di Feniglia which border, towards the northwest and towards the south, the Stagno di Orbitello; the range of dunes which extend from Capalbio to Ansedonia; the sandy formations of the mouth of the Tiber; the line of dunes which form, between Terracino and the promontory of Circe, the southern limit of the Pontine marshes; and, finally, the dunes between the lake of Fusaro and the sea. I have also been able to study the analogous formations in Tunis: the neighbourhood of the lake of Tunis, El Bahira; the sandy alluvia between Sebkret Er Riana and Cape Kamart; and the country covered with dunes to the north of Susa. In Spain I explored the alluvia near Huelva and those of the mouths of the Rio Tinto and of the Odiel; the Arenas Gordas, the great system of dunes which border, towards the south, on the side of the sea, Las Marismas, the marshes extending along the lower course of the Guadalquivir; the dunes of the neighbourhood of Sanlucar; those which surround the Bay of Cadiz; and, finally, the alluvia of the Mar Menor and of the delta of the Ebro.

In the regions abutting on the Mediterranean lagoons, just as in the analogous formations that we meet with on the coasts of the North Sea, one can distinguish the dunes from the lower-lying alluvia, and again subdivide these last into clayey and sandy alluvia.

1. The soil of the dunes is made up of sandy aeolian formations with an irregular surface of variable level.

2. The lower alluvia are marine deposits with the surface little disturbed and almost horizontal. They are essentially composed (*a*) of clay (salty mud), or (*b*) of sand, ordinarily more or less mixed with clay. These sandy alluvia often form a transition between the dunes and the clayey tracts: in many cases they are composed of alternating layers, greater or less in thickness, of sand and pure clay, and in places their surface is undulating, the dry periods giving rise to little heaps of moving sand which form embryonic dunes. In the south of France I was able to study the vegetation of two of these formations: that of the dunes (Palavas, Cette, Saintes-Maries, mouth of the Petit-Rhône), and that of the lower alluvia of sandy nature (Saintes-Maries).

1. Vegetation of the dunes

PALAVAS.—In the narrow tongue of alluvial formation which separates the Lake of Prévost from the sea to the west of Palavas one can distinguish, in passing from the sea inland, the following zones of soil and vegetation:

1. The sea.
2. The tidal zone (included between the ordinary limits of ebb and flow).
3. 17–20 metres of beach with a gentle slope, composed of sand and gravel mixed with many shells: no vegetation.
4. About 20 metres of quite low dunes rising here and there to a higher level and occupied by an *Agropyrum*-Formation. The difference between the inner and outer regions of this zone is quite considerable:
    a. The outer region, a little wider than the other and characterized

by a formation of comparatively pure *Agropyrum* (*A. junceum*) mixed with *Psamma arenaria, Malcolmia littorea, Medicago marina, Cakile maritima, Euphorbia paralias, Eryngium maritimum, Echinophora spinosa, Crucianella maritima*.

 b. The inner region forming a transition to the following formation; *Agropyrum* predominant but strongly mixed with species cited under 5.

5. A higher-lying zone characterized by an *Anthemis maritima*-Formation (Fig. 132). The dunes are of no great height, scarcely exceeding the stature of a man, and the vegetation is essentially of the same character as in the lower regions, consisting especially of the following species: *Anthemis maritima, Crucianella maritima, Malcolmia littorea*, and, besides, *Medicago marina, Vulpia uniglumis, Agropyrum junceum*, and *Psamma arenaria*. See also Table 7. Besides these species, *Convolvulus soldanella* and *Medicago littoralis* were also represented.
6. A lower-lying zone partly devoted to vine culture.
7. A marsh vegetation along the Lake of Prévost.

Table 7, Nos. 1–3, gives the results of a more detailed analysis of the composition of the plant formations occupying the dune area and represented by the zones 4*a*, 4*b*, and 5 above. The figures opposite the names of the species are proportional: they indicate the percentage of sample plots in which each species occurs. To determine this proportion, which measures the degree of frequency of the species, I have used a method explained in detail in an earlier chapter.[1] I have examined fifty samples of each formation, each sample occupying a surface of 0·1 sq. metre; the shape of these samples was not square as in my earlier researches, but circular, and to trace it I used a rod attached to my stick, the point of which described in turning a figure of 0·1 sq. metre area.[2] The formations considered have quite a loose structure, the sand appearing everywhere between the individual plants. Furthermore, they are poor in species and, as the Tables show, each species taken separately is very sparsely distributed: a comparison between the number of samples and the total of the frequency numbers shows that on the average the samples each contained only from 1–3 species. In the outer zone of the *Agropyrum*-Formation the discontinuity of the species was especially striking, most of the samples only containing a single species: in the inner region the distribution of the species was less sparse, from 2 to 3 in each sample, and that in spite of the fact that the total number of species represented was almost the same in the two regions (10 against 9). The number of the species and their continuity of growth attained their maximum in the

---

[1] Raunkiaer, C.: 'Formationsundersøgelse og Formationsstatistik', *Bot. Tidsskrift*, t. xxx, 1909. (Chapter VI of the present volume.)

[2] Raunkiaer, C.: 'Measuring-Apparatus for Statistical Investigations of Plant-Formations', *Bot. Tidsskrift*, t. xxxiii, 1912.

Anthemis maritima-Formation, where each sample contained on an average 3 species.

TABLE 7

Degrees of frequency of the species contained in the plant formations of the dunes in the neighbourhood of Palavas, 50×0·1 sq. metre.

	Life-form.	1	2	3	4
Agropyrum (junceum?)	G	98	98	18	..
Anthemis maritima	Ch	..	40	96	8
Helichrysum stoechas	Ch	..	..	..	80
Malcolmia littorea	Ch	..	36	48	42
Crucianella maritima	Ch	2	24	46	34
Medicago marina	Ch	2	16	24	18
Artemisia glutinosa	Ch	..	..	10	4
Euphorbia paralias	Ch	6	2	..	..
Coris monspeliensis	Ch	..	..	2	..
Centaurea aspera	H?	..	..	2	..
Echinophora spinosa	H	2	4	..	..
Sporobolus pungens	H?	..	..	16	..
Psamma arenaria	G	2	32	16	44
Galilaea mucronata	G	..	2	..	8
Scleropoa maritima	Th	4	..	2	2
Vulpia uniglumis	Th	..	..	20	2
Lolium rigidum	Th	2	..	..	..
Silene conica	Th	..	..	2	2
Cakile maritima	Th	2	2	..	..
Points		120	256	302	244
No. of species		9	10	13	11

To the east of Palavas, between the lake of Pérols and the sea, the state of things was somewhat different. Here the zone of dunes was more irregular, more exposed to the wind, and had a more varied and irregularly distributed vegetation: at the same time the plants looked more flourishing, as is usually the case in dunes that are not too sheltered. In the outer formation, the one next the sea, *Psamma arenaria* predominated in frequency and in luxuriance; but it was also common enough in the interior zone of lower dunes, where the most abundant species was *Helichrysum stoechas*, which occurred in forty samples out of 50, that is to say its degree of frequency was 80. Meanwhile *Psamma arenaria*, whose degree of frequency was here only 44, was conspicuous by the extraordinary development of the individuals, so much so that from the physiognomic point of view one would not hesitate to characterize the formation by the name of *Helichrysum stoechas–Psamma arenaria* (Fig. 133). The composition of this formation is clear from Table 7, column 4, which gives the results obtained from fifty samples. The little valleys separating the summits of the dunes were specially peopled by *Juncus acutus* and *Scirpus holoschoenus*. On the inner side of the dunes, here

Fig. 132. Dunes to the west of Palavas: *Anthemis maritima*-Formation, with much *Crucianella maritima* and *Malcolmia littorea* associated with the *Anthemis*. See Table 7, column 3.

Fig. 133. Dunes to the east of Palavas: *Helichrysum stoechas-Psamma arenaria*-Formation. See Table 7, col. 4.

FIG. 134. Neighbourhood of Cette: along the sea, at the base of low dunes covered with *Psamma arenaria*, *Agropyrum*, *Scleropoa maritima*, &c. Cf. p. 358, No. 4. In the foreground, a depression containing a formation of *Inula crithmoides*, whose constitution is indicated on p. 358, No. 5.

FIG. 135. Low dunes along the left bank of the mouth of the Petit-Rhône; the vegetation consists essentially of *Salicornia glauca* in the depressions and *Psamma arenaria* on the summits.

VEGETATION OF THE FRENCH MEDITERRANEAN ALLUVIA

as in the locality spoken of above, a lower-lying zone was partly devoted to vine culture, and at the base, along the side of the lagoon, occupied by formations of *Salicornia*.

The list of species shown in Table 7, completed by their figures of frequency in the different formations, gives us information on the floristic composition of these formations and also, to a certain extent, on the quantitative relations of the species represented; but the comparison which we wish to make from the point of view of Biology and Plant Geography between these formations and those of analogous localities situated in other climates and of very different floristic composition has necessitated the conversion of floristic units such as species into biological units or life-forms. I have therefore added in Table 7 the life-forms represented by the different species, using the system adopted in earlier chapters,[1] to which I refer the reader, adding here only the abbreviations used:

> S = Stem Succulents.
> E = Epiphytes and phanerophytic parasites.
> MM = Mega- and Meso-phanerophytes.
> M = Micro-phanerophytes.
> N = Nano-phanerophytes.
> Ch = Chamaephytes.
> H = Hemicryptophytes.
> G = Geophytes.
> HH = Helo- and Hydro-phytes.
> Th = Therophytes.

If we take, for example, the outer formation of the dunes in question (given in column 1 of Table 7) and add together the figures of frequency of the species belonging to each of the life-forms, we obtain, for the different types, the totals (points) as follows: Ch = 10, H = 2, G = 100, Th = 8, and expressed in percentages these form the biological spectrum given in Table 8, line 1. Considered alone, and especially in relation to the normal spectrum, this spectrum shows that the formation represented is characterized by the predominance of Geophytes: that it is essentially geophytic. If we now do the same for the other formations we find other types predominating; thus the formation of *Anthemis*

---

[1] Raunkiaer, C.: 'Types biologiques pour la géographie botanique', *Bull. de l'Académie Royale des Sciences et des Lettres de Danemark*. 1905.
——— 'Planterigets Livsformer og deres Betydning for Geografien'. Copenhagen, 1907. ('The Life-Forms of Plants and their bearing on Geography', Chapter II of the present work.)
——— 'Livsformernes Statistik som Grundlag for biologisk Plantegeografi', *Bot. Tidsskrift*, t. xxix, 1908. ('The Statistics of Life-forms as a basis for Biological Plant Geography', Chapter IV of the present work.)
——— 'Livsformen hos Planter paa ny Jord', *Mémoires de l'Académie Royale des Sciences et des Lettres de Danemark*, 7° série, section des Sciences, t. viii, Copenhagen 1909. ('The Life-forms of Plants on New Soil', Chapter V of the present work.)

maritima is characterized by the predominance of Chamaephytes (Table 7, column 2; and Table 8, line 3): as is also the formation of *Helichrysum stoechas* to the east of Palavas (Table 7, column 4; and Table 8, line 4). While showing quite different floristic compositions and different predominant species, the two formations show essentially identical biological spectra. According to the spectra, the dunes of Palavas bear, then, an outer geophytic formation and an inner chamaephytic formation occupying an older soil, and we see that these two formations are connected by an intermediate zone which is transitional not only from the floristic point of view (see Table 7, column 2), but also from the biological, the Chamaephytes and Geophytes having almost equal value in the biological spectrum (Table 8, line 2).

TABLE 8

Biological spectrum of the formations in Table 7

	Points.				*Percentage.*			
	Ch	H	G	Th	Ch	H	G	Th
1..	10	2	100	8	8	2	83	7
2..	118	4	132	2	46	1·5	51·5	1
3..	226	18	34	24	75	6	11	8
4..	186	..	52	6	76	..	21	3

CETTE.—Immediately to the east of Cette, in passing from the sea inland, one has the following zones:

1. The sea.
2. The tidal zone, about 15 metres broad, with many small shells.
3. About 25 metres of beach with a gentle slope, composed of sand, gravel, and shells: no vegetation; bounded towards the inside by the line of the highest tides marked by sea-weed drift and all sorts of materials thrown up by the sea.
4. A sandy zone at a somewhat higher level, but quite low-lying and covered by very sparse vegetation: *Agropyrum (junceum?)*, *Psamma arenaria*, *Scleropoa maritima*, *Cakile maritima*, *Eryngium maritimum*, *Lepturus incurvatus*, *Polygonum maritimum*, *Medicago (littoralis?)*, *M. marina* (Fig. 134).
5. A lower damp zone occupied by an *Inula crithmoides*-Formation (Fig. 134), and the following associated species: *Glyceria distans*, *Salicornia herbacea*, *Juncus acutus*, *Obione portulacoides*, *Spergularia marginata*, *Suaeda (maritima?)*, *Atriplex rosea*.
6. Pools of water and a small lagoon with *Scirpus maritimus*.

The fourth zone shown in this flora corresponds with the *Agropyrum*-Formation in the neighbourhood of Palavas; but this Cette locality has no equivalent of the *Anthemis*-Formation, since suitable ground is lacking;

the zone of *Agropyrum* is followed immediately by a humid depression occupied by the *Inula crithmoides*-Formation.

THE CAMARGUE.—At Saintes-Maries on the south coast of the Camargue, which forms a great island in the Rhône delta, the area of dunes is still more reduced; here the soil is formed by a sandy (and partly clayey) beach where the *Salicornia*-Formation begins at the inner limit of the shore properly so-called; here and there, and especially between Saintes-Maries and the mouth of the Petit-Rhône, we find, along the sea, little hillocks formed by sand caught in and around the tufts of *Salicornia*; on the slight elevations of the ground thus produced representatives of the *Agropyrum-Psamma*-Formation have established themselves: for example, *Agropyrum, Psamma arenaria, Euphorbia paralias, Sporobolus pungens, Eryngium maritimum, Echinophora spinosa*, &c. As the level rises the Salicornias disappear. At one isolated point a dune reaches a height of 2 metres; but its formation was doubtless partly artificial owing to the presence of an embankment of earth against the end of which the dune rested. Along the left bank of the Petit-Rhône, not far from its mouth, is a line of lower dunes from 1·5 to 2 metres in height and covered with a vegetation like that which has already been described (see Fig. 135). On the right of the mouth of the Petit-Rhône one could see in the distance a much higher dune which appeared to be bare of vegetation but was bordered on the landward side by wind-beaten Phanerophytes.

Judging by the map, higher and more extensive dunes occur at other points of the littoral, such as the 'lidos' which fringe the outer edge of the Lake of Pouent and the eastern extremity of the Lake of Mauguio; but generally one can say that the dune area of the Mediterranean littoral of France is poorly developed in comparison, for example, with the dunes of the North Sea coasts.

Flahault and Combres[1] tell us that in the interior of the Camargue there is an ancient dune area now covered by forest and maquis, that is to say by phanerophytic formations; I have not encountered these on the maritime dunes, for the vegetation of the low hills partly covered by Phanerophytes between Saintes-Maries and the sea I regard as a product of cultivation. However that may be, the state of the dunes of the south of France, at least in the localities I have studied, differs very much from that of the west coast of Italy and certain regions of Spain, where the geophytic and chamaephytic vegetation of the maritime dunes gives place somewhat quickly to a phanerophytic vegetation, an evergreen xeromorphic formation of Nano- and Micro-phanerophytes—in other words a maquis, of which the outer portion is generally a *Juniperus*-Formation. Between Pisa and Leghorn this Juniper formation comes fairly near the sea, being separated from the shore-line only by 50–60

[1] See *Bull. Soc. Bot. Fr.*, t. xli, 1894, pp. 37–58.

metres of dunes mainly occupied by geophytic, chamaephytic, and therophytic vegetation. Farther towards the south the *Juniperus*-Formation approaches still closer to the sea; such is the case between Ansedonia and Capalbio, to the west of the Lake of Fusaro, where it even invades the little curtain of maritime dunes, of which only the foot next the sea is occupied by the geophytic and chamaephytic communities which constitute, on the French Mediterranean littoral, the whole of the dune vegetation. In southern Spain the Arenas Gordas show a very similar vegetation to that of the Italian localities quoted.

A comparison between the dunes of the south of France and those of the west coast of Italy shows, then, that the geophytic and chamaephytic communities which occupy the main part of the low dunes of the French littoral are pushed farther and farther towards the sea as one advances southwards; so that the invading phanerophytic communities may even cover the dunes up to the beach limit. If we compare the dunes of Mediterranean France with those of more northern countries, and particularly with those of the coasts of the North Sea, which belong to the climate of Hemicryptophytes, we find the contrary state of things.

As a basis of comparison I shall choose the area of dunes stretching along the west coast of Jutland, since this is the only area which has been the subject of methodical researches that can furnish a series of data complete enough to be compared with my results from the Mediterranean regions. Table 9, columns 1–4, contains the results of statistical researches on the series of formations examined in order from the shoreline inland, passing from the more recent to the older zones. The four populations represent the four stages distinguished in the dry area of the dunes described in the classical work of Warming:[1] they are (1) flattened, recently formed elevations of the beach (Table 9, column 1); (2) maritime dune of *Psamma arenaria* (column 2); (3) 'grey' dune (column 3); (4) dune-heaths (columns 4–6). The floristic composition being here very different from that of the dunes of Mediterranean France, it is necessary, in order to render the floras comparable, to convert the specific frequencies of Table 9 into life-form frequencies which will enable us to establish the biological spectra of the different formations. These are given in Table 10 for the formations analysed in Table 9. And if we compare the spectra recorded in lines 1–6 of Table 10 with the spectra of the corresponding localities of Mediterranean France, we shall see that the outer formations of the dunes are in both cases geophytic formations (Table 8, line 1; Table 10, lines 1–2), with which are associated, in Jutland a small number of Hemicryptophytes, and in the French dunes especially Chamaephytes, of which the proportion quickly increases as one passes inland (Table 8, line 4), the chamaephytic formation replacing the geophytic with considerable suddenness (Table 8, lines 1–2).

[1] Warming, Eug.: *Dansk Plantevækst.* 2. Klitterne, 1909.

TABLE 9

List showing the degree of frequency of the species in a series of dune formations at various points on the west coast of Jutland.

1. *Agropyrum junceum*-Formation. Fanø.
2. *Psamma arenaria*-Formation. Fanø.
3. *Weingärtneria canescens*-Formation. Skagen.
4. *Calluna-Empetrum*-Formation. Bjergegaard, on the 'lido' of Holmsland.
5. *Calluna-Empetrum*-Formation. Fanø.
6. *Calluna-Salix*-Formation. Skagen.
7. *Hippophaë*-Formation. Skagen.
8. *Rosa pimpinellifolia*-Formation. Skagen.

	Life-form.	1	2	3	4	5	6	7	8
Hippophaë rhamnoides	N	..	..	..	..	..	..	100	..
Rosa pimpinellifolia	N	..	..	..	..	..	..	..	100
Salix repens	N (Ch)	..	..	..	28	40	100	..	..
Empetrum nigrum	Ch	..	..	..	100	64	..	..	..
Calluna vulgaris	Ch	..	..	..	100	96	100	..	..
Thymus Serpyllum	Ch	..	..	..	..	68	4	..	..
Festuca rubra	H	14	16	44	..	52	12	16	96
Weingärtneria canescens	H	..	..	96	..	..	..	..	..
Koeleria glauca	H	..	..	100	..	..	..	..	8
Agrostis vulgaris	H	..	..	..	..	76	4	..	16
Hieracium umbellatum	H	..	8	8	..	80	..	..	..
Carex arenaria	G	..	..	20	..	48	4	36	72
Psamma arenaria	G	..	96	16	..	44	..	8	24
Agropyrum junceum	G	96	12	..	..	..	..	..	..
Sonchus arvensis	G	..	72	..	..	..	..	28	..
Sedum acre	Ch	..	4	..	..	..	..	..	..
Genista anglica	Ch	..	..	..	..	4	..	..	..
Erica tetralix	Ch	..	..	..	4	..	..	..	..
Veronica officinalis	Ch	..	..	..	..	..	..	..	4
Luzula campestris	H	..	..	..	12	4	16	..	..
Anthoxanthum odoratum	H	..	..	..	..	..	12	..	..
Nardus strictus	H	..	..	..	..	8	..	..	..
Viola canina	H	..	..	36	..	40	..	..	24
Anthyllis vulneraria	H	..	2	..	..	..	..	..	..
Vicia cracca	H	..	10	..	..	..	..	..	..
Lotus corniculatus	H	..	..	4	12	4	44	..	4
Galium verum	H	..	2	8	..	8	..	..	24
Campanula rotundifolia	H	..	..	..	..	..	..	..	24
Jasione montana	H	..	..	12	..	4	..	..	4
Hypochaeris radicata	H	..	2	4	..	28	..	..	..
Hieracium pilosella	H	..	..	..	4	..	..	..	..
Poa pratensis	G	..	..	..	..	..	..	..	44
Elymus arenarius	G	..	14	..	..	..	..	..	..
Linaria vulgaris	G	..	..	..	..	..	..	..	8
Salsola Kali	Th	2	..	..	..	..	..	..	..
Cerastium semidecandrum	Th	..	4	..	..	..	..	..	..
Arabis thaliana	Th	..	..	..	..	..	..	..	4
Cakile maritima	Th	4	..	..	..	..	..	..	..
Viola tricolor	Th	..	20	..	..	..	..	..	..
Senecio vulgaris	Th	..	8	..	..	..	..	4	..
Points		116	270	348	260	668	296	192	456
No. of species		4	14	11	7	17	9	6	15

This chamaephytic formation of the French dunes corresponds to the hemicryptophytic formation of the 'grey dunes' of Jutland, as is seen not only by the place which it occupies but also by its composition; this contradicts the hypothesis that the evergreen xeromorphic chamaephytic formation which with us gradually invades the grey dune, establishing dune-heath, corresponds with the French chamaephytic formation. The Chamaephytes of the Mediterranean dunes have to me quite the appearance of transformed Hemicryptophytes and Therophytes, while the Chamaephytes of our dune-heaths correspond to the formation of Nano- and Microphanerophytes, which in the Mediterranean region often invades the dune area quite rapidly, extending as far as the maritime dune, so that the preceding sparse communities of Geophytes and Chamaephytes are reduced to a narrow fringe on the outer limit of the dune.

TABLE 10

Biological spectrum of the formations represented in Table 9

	No. of Species.	Points.	N	Ch	H	G	Th
1	4	116	..	..	12	83	5
2	14	270	..	1	15	72	12
3	11	348	..	..	90	10	..
4	7	260	..	89	11	..	..
5	17	668	..	41	45	14	..
6	9	296	..	69	30	1	..
7	6	192	52	..	8	38	2
8	15	456	22	1	44	32	1

Behind the Nano-microphanerophytic maquis of the dunes one meets here and there, notably in the old dunes of the Tuscan Maremma between Pisa and Leghorn, mesophanerophytic societies composed of evergreen xeromorphic, with some deciduous mesomorphic, species. The Jutland dunes present also phanerophytic formations of but slight extent, composed of Nanophanerophytes: formation of *Salix repens* and formation of *Hippophaë rhamnoides* (Table 9, column 7, and Table 10, line 7), formation of *Rosa pimpinellifolia* (Table 9, column 8, and Table 10, line 8); or of Nano-microphanerophytic oak scrub. As one might expect, all these formations are formed by deciduous Phanerophytes of mesomorphic, or at the most only slightly xeromorphic, structure.

2. Vegetation of the low-lying, moist, marine alluvia

The saline marine alluvia, of clayey or sandy nature, were examined at several points of the Mediterranean littoral; among the localities visited were: in Tunis, the neighbourhood of El Bahira and the delta of the Medjerda; in Spain, the mouths of the Odiel and the Rio Tinto; that of the Guadalquivir; the coast of the Bay of Cadiz; the neighbourhood of Mar Menor; the region extending to the south of Alicante; the delta of the Ebro. In southern France, which particularly interests us here, I have unfortunately only seen the alluvia near Saintes-Maries, on the southern coast of the Camargue.

THE CAMARGUE.—The Rhône delta, which is the most extensive area of alluvium in the south of France, is mainly formed by the Camargue, an island of 750 sq. kilometres bordered by the Mediterranean and the two arms of the Rhône, the Grand-Rhône to the east and the Petit-Rhône to the west. To the east of the Camargue stretch the alluvia of the Grand-Plan-du-Bourg; and, towards the west, the right bank of the Petit-Rhône is also formed by an alluvial soil of which the southern part is called the Petite-Camargue; altogether the delta comprises about 1,400 sq. kilometres of land regained from the sea. Apart from the left bank of the mouth of the Petit-Rhône, I can only notice what can be seen from the railway in the course of the journey of about 40 kilometres from Arles, at the northern point of the island, to Saintes-Maries on the Mediterranean.

The surface of the soil, which has absolutely no relief, is only from a few centimetres up to 1 metre above the neighbouring marshes and lakes which cut up the country, notably along the southern side. The soil is partly sand and partly clay; here and there it is composed of pure sand, as in the little maritime dunes and the regions of the older dunes scattered in the interior. In other places the soil is formed, at least in its upper layers, of clay or clayey mud: for example forming the substratum of the *Salicornia* vegetation along the shore to the west of Saintes-Maries. But in general the soil is composed of sand mixed with clay, or of sandy clay, or sometimes of alternating layers of clay and sand, such as one generally sees in recent formations on the coast: in calm weather during flooding mud is deposited, covering the sandy layers which have been laid down by the wind or during violent incursions of the sea.

The Camargue is a silent country: the only sounds are the song of the birds and other voices of nature. This is not to say that it is a desert: the upper parts are largely cultivated, especially those which occupy the northern region of the island. Around Arles the areas devoted to cereal culture (wheat, barley, oats) are interspersed by quite a number of pastures and vineyards, and even by areas used for horticulture (orchards and vegetables). A great abundance of deciduous trees are planted round houses and along the roads and hedges, giving the country a smiling aspect recalling that of the Danish islands. Meanwhile, in travelling from Arles southwards by either of the two lines of railway, even in the higher areas most suited for cultivation on the two sides of the lakes of Vaccarès, uncultivated areas will be met with, and these become more and more numerous as the Mediterranean is approached. By the line from Arles to Saintes-Maries one reaches after 8 kilometres, to the south of Benchouet, areas covered with a nano-phanerophytic formation of *Salicornia* much mixed with herbs (Therophytes and Hemicryptophytes); of these last *Juncus acutus* forms in places an essential element of the vegetation. The farther one travels south, the more these infertile areas increase in number and extent; and those at a higher level are often separated by considerable depressions covered with shallow water or with marshes communicating with the lakes, and occupied by hydrophytic, and especially helophytic, formations. The houses become farther and farther apart and end by being quite isolated, like the planted trees; finally the cultivated areas appear only as spots of various sizes in a landscape which is generally wild. The transition is not regular; after passing long stretches of low-lying land with here and there some raised island of fertile soil, one may come to quite extensive regions of cultivated fields, for example to the north of Balarin-Dufoure. To the south of Pioch-Badet one can see a little Pinetum at some distance from the railway.

In the older areas the formation of *Salicornia* is ordinarily mixed; apart from some isolated depressions inhabited by a very characteristic formation of *Salicornia glauca* which strongly recalls that occupying the immediate neighbourhood of the shore, the *Salicornia*-Formation of the older areas is often considerably modified by the introduction of therophytic and hemicryptophytic elements, which invade the intervals between the tufts of *Salicornia* to the point of covering them completely during the summer. After suffering this régime for some time the Salicornias disappear and the formation becomes exclusively therophytic-hemicryptophytic, serving for the pasturage of great flocks of sheep.

Tamarix gallica is quite frequent in the *Salicornia*-Formation, although it is too sparsely distributed to be dominant, in spite of its height; but where it is commonest one ought to speak of a *Tamarix-Salicornia*-Formation.

According to Flahault and Combres, the lower-lying areas particularly exposed to inundations are generally covered with *Salicornia sarmentosa*, while *Salicornia fruticosa* is characteristic of the higher levels which are not constantly submerged in winter, and *Salicornia glauca* belongs to the depressions along the edge of the sea. I myself have not had occasion to study the first two formations, which probably occur in the neighbourhood of the lakes, but at Saintes-Maries I found the formation of *Salicornia glauca*, which I will shortly describe.

The soil, of recent origin, is here composed of low-lying sandy alluvia, quite level and humid. The sands which make it up are more or less mixed with clay; layers of almost pure sand alternate with others which are clayey or even composed of almost pure clay.

At about 1 kilometre to the east of Saintes-Maries, in passing from the sea towards the interior, one can distinguish the following zones:

1. The sea.
2. The tidal zone.
3. A bank of sand about 40 metres broad, very slightly elevated, and with little surface relief, not exceeding a height of 0·25 metre above the level of the high tides; the surface of the sand dry, and absolutely without vegetation.
4. About 75 metres, slightly depressed; surface almost horizontal; of greyish tone on account of the humidity; of firm consistence; vegetation absent or presenting here and there a few individuals of *Salicornia glauca* of considerable height, sometimes very degenerate or even dead as a result of the shifting sand (Fig. 136); sparse vestiges of 'dunes' only a few centimetres high and composed of loose sand of bright tone.
5. An area covered by a formation of *Salicornia glauca* composed almost exclusively of this species (Fig. 137).

In its outer portion the formation of *Salicornia glauca* is very irregularly distributed: in three groups of 50 samples, each 0·1 sq. metre, the degree of frequency of *Salicornia glauca* was 28, 26, and 30 respectively, out of 100. In the inner and notably in the lower portions, with clayey surface layer and protected on the side of the sea by long low flattened banks of sand, the formation becomes denser and the degree of frequency of *Salicornia glauca* is here 80–100 per cent. A small number of other species are associated: *Statice bellidifolia, Statice limonium, Obione portulacoides, Glyceria convoluta, Inula crithmoides, Spergularia* sp. One may often observe that *Statice bellidifolia*, when it is covered by a layer of sand, escapes smothering by the great elongation of its internodes, thus pushing up into the light. Here and there are shallow depressions where *Salicornia perennis* replaces *S. glauca*.

As has been said above, and as Fig. 137 shows, the phenomenon of moving sand is here exhibited, though in very restricted measure, just as we see it in analogous situations in the north, for example the 'Tippersande' of the Ringkjøbing Fjord and the sandy alluvia of the shore of the north coast of the island of Fanø). The sand, caught in the tufts of *Salicornia*, gives rise to sandy hillocks: when these hillocks have reached a height of about 0·5 metre various dune plants can settle, for example, *Psamma arenaria, Euphorbia paralias, Sporobolus pungens, Eryngium maritimum, Echinophora spinosa*. A dune area is thus formed, and *Salicornia* disappears.

To the west of Saintes-Maries one has the following zones:
1. The sea.
2. The tidal zone.
3. Flat shore without vegetation.
4. A sandy zone occupied by a formation of *Salicornia glauca* and presenting, at some metres from the outer limit of the Salicornias, little hillocks covered with *Psamma arenaria* and *Agropyrum junceum* with

Table 11

Statistical analysis of the vegetation of the alluvia of the south coast of the Camargue, to the west of Saintes-Maries. 1–2, on a low-lying, wet, clayey soil; 3, on a sandy soil mixed with clayey mud; 4, on a clayey soil fissured as a result of desiccation.

	Life-form.	Degree of frequency.			
		1	2	3	4
Salicornia glauca	N	100	100	98	78
Obione portulacoides	Ch	74	74	70	..
Statice bellidifolia	H	32	..	72	..
Aeluropus littoralis	G-H-Ch	8	40	..	..
Juncus acutus	H	..	38	..	..
Glyceria convoluta	H	10	..	34	..
Statice Limonium	H	10	6	2	..
Lepturus cylindricus	Th	..	..	4	..
Points		234	258	280	78
No. of species		6	5	6	1
No. of species per 0·1 sq. metre		2·3	2·6	2·8	0·8

Table 12

Statistical analysis of the vegetation of a series of formations on the clayey alluvia of the island of Fanø, and of an isolated formation on sandy alluvium of this island (7).

	Life-form.	Degree of frequency.						
		1	2	3	4	5	6	7
Salicornia herbacea	Th	100	100	98	44	54	..	..
Glyceria maritima	H	100	100	100	100	100	..	100
Suaeda maritima	Th	18	74	86	60	100	..	..
Triglochin maritimum	H	4	24	74	48	10	10	..
Aster tripolium	H	8	40	18	100	96	..	..
Plantago maritima	H	..	..	..	46	100	100	32
Glaux maritima	H	..	..	..	2	..	100	6
Juncus Gerardi	G	..	..	..	..	..	100	..
Festuca rubra	H	..	..	..	..	..	100	6
Artemisia maritima	Ch	..	..	..	..	2	..	..
Agropyrum junceum	G	..	..	..	..	..	..	4
Agrostis alba	H	..	..	..	2	..	10	8
Erythraea pulchella	Th	..	..	..	..	..	28	..
Lepturus filiformis	Th	..	..	..	..	..	4	4
Spergularia media	H	..	..	..	26	38	..	..
Trifolium repens	H	..	..	..	..	..	2	..
Armeria vulgaris	H	..	..	..	..	..	54	..
Points	..	230	338	376	428	500	508	160
No. of species	..	5	5	5	9	8	10	7
No. of species per per 0·1 sq. metre	..	2·3	3·4	3·8	4·3	5	5·1	1·6

Fig. 136. Beach to the east of Saintes-Maries, looking east; in the background to the right the sea; to the left the edge of a sparse formation of *Salicornia glauca*; many individuals dead or very degenerate.

Fig. 137. Zone immediately interior to that represented in Fig. 136 (beach to the east of Saintes-Maries). Here the *Salicornia glauca*-Formation is less sparse and more vigorous; nevertheless, there are many individuals more or less dead and covered with sand blown by the wind.

FIG. 138. Formation of *Salicornia glauca* on a soil of clay mixed with sand, to the west of Saintes-Maries; here and there, areas entirely devoid of vegetation and cracked by the dryness.

some *Euphorbia paralias, Eryngium maritimum,* and other dune plants. On the inner side of this sandy border we have:

5. A slightly depressed area, of which the surface is sometimes clayey and sometimes sandy, covered with a formation of *Salicornia* of low stature and in character nano-phanerophytic-chamaephytic.

Table 11 contains the results of a statistical study of the vegetation of this area, represented by samples taken at four different places: Nos. 1 and 2 were taken on a rather wet clayey soil, No. 3 on a sandy soil mixed with a little clay, the surface layer often clayey: in the three cases the predominating species were the same, *Salicornia glauca* and *Obione portulacoides*: No. 4 represents a rather sparse vegetation inhabiting a clayey fissured soil (Fig. 138). For the rest, this formation, composed solely of *Salicornia glauca*, occupied great stretches where the clayey soil was decidedly wet.

TABLE 13

Biological spectra of the formations which characterize respectively the low-lying and saline alluvia of the Camargue and of the island of Fanø.

	No. of Species.	No. of Species per 0·1 sq. metre.	N	Ch	H	G	Th
Alluvia of the Camargue	8	2·1	44	26	24	6	..
Alluvia of Fanø . .	17	3·1	..	..	78	5	17

From an examination of all these studies of the clayey and sandy alluvia of the Camargue we thus conclude that the vegetation is a nano-phanerophytic formation of very small stature, assuming under specially unfavourable conditions a chamaephytic character. The same thing is true of all the Mediterranean localities in which I have had occasion to study the vegetation of low-lying marine alluvia. The predominating species of these vegetations are almost everywhere Salicornias. The hemicryptophytic climate shows, on the other hand, a very different state of things. In Denmark, for instance, the low-lying marine alluvia, clayey as well as sandy, are populated by hemicryptophytic formations (see Table 12). Only the tidal zone is an exception; it is characterized, here as in France, by a therophytic formation of *Salicornia herbacea*. In Table 13 I give the biological spectra of the formations which characterize respectively the low-lying alluvia of the Camargue and of the Island of Fanø, based on the statistical results of Tables 11 and 12. From a comparison of the two spectra the differences between these two alluvial vegetations belonging to the chamaephytic and hemicryptophytic climates respectively are clearly shown by their life-forms.

X

THE USE OF LEAF SIZE IN BIOLOGICAL PLANT GEOGRAPHY

Science consists in the recognition of similarity and dissimilarity, so that the beginning and end of science is comparison. We must therefore reduce the materials yielded by our investigations to a state capable of comparison. In other words we must measure and count phenomena so as to be able to express them in numbers. I do not mean that we must always try to express ourselves in numerical formulae—far from it—but where scientific comparison has to be made, we should attain to numerical expression of phenomena, or to expression in terms of definite concepts, which are as stable and unequivocal as numbers. This is, however, exceedingly difficult when dealing with the biological sciences, where everything is in a state of perpetual change, and where, moreover, the phenomena are so complicated and made up of so many different components, conditions, or characters, or whatever we like to call them, which we cannot analyse perfectly, that it is often impossible, or at any rate very difficult, to count or measure the individual elements, and to determine their significance.

One of the most difficult branches of botany in this respect is ecology. This is not surprising if we consider that ecology embraces the teachings and difficulties of the whole of botany. All botanical teaching can find use in ecology; ecology indeed embraces the relationship of plants to all the environments of the world.

Now it has been emphasized again and again that under conditions prevailing at the present time it is the water problem which presents plants with their greatest dangers, and which characterizes them most sharply. During the last forty years, and especially during the last quarter of the nineteenth century, the relationship of plants to water has been often investigated and described in more or less detail, and the main features of the methods of adaptation have been made clear. We have for a long time been aware of a series of different adaptations in the structure of plants enabling them to endure excessive evaporation, and thus allowing them to live in places where the environment determines intense evaporation, or where the conditions of water absorption of the ground are unfavourable either physically or physiologically.

Examples of such structure are: (1) covering of wax, (2) thick cuticle, (3) sub-epidermal protective tissue, (4) water tissue, (5) covering of hairs, (6) covering of the stomata, (7) sinking of the stomata, (8) inclusion of the stomata in a space protected from air currents, (9) diminution of the evaporating surface, &c.

The matter however is so complicated that it is very difficult to reach an exact appraisal of these adaptations in characterizing the individual plant communities biologically. The fact is that in a community which survives dry periods, some species are adapted to their environment in one way, and others in other ways, and we are still unable to determine quantitatively the value of the individual adaptations or the different combinations of different adaptations.

In general we must content ourselves with showing the most frequently occurring adaptations, without going farther into the statistical investigation. A statistical investigation in connexion with an attempt to determine the degree of adaptation should be aimed at as far as possible if we are to succeed in an exact comparative treatment of different plant communities with respect to the xerophily of these communities.

A preliminary direct consideration of a series of evergreen phanerophytic communities, for example the marked tropical rain forest, the drier tropical evergreen forest (as for instance that of the West Indies), and the maquis of the Mediterranean region, show that amongst the adaptations named, diminution of the transpiring surface, diminution in leaf size, is one of the adaptations generally in evidence; and since this adaptation is easy to observe and comparatively easy to measure, it is convenient to begin with it if we wish to use the statistical method in this domain. In trying to use the method we encounter at once difficulties of various kinds.

What are we to consider as the unit to be measured? It need not necessarily follow that the physiologico-biological units here dealt with correspond with morphological units. Are we, for example, to compare the simple leaf with the compound leaf, or must we compare the simple leaf with the individual leaflet of the compound leaf?

If we examine and compare the simple and the compound leaves of plants with the same life-forms in the same formation, or more properly in the same layer of the formation, for example the upper layer of the deciduous phanerophytic vegetation of our woods, or the upper layer of the West Indian evergreen phanerophytic vegetation, we find that the compound leaves are on an average much larger than the simple leaves, which belong to a smaller class of size. On the other hand the leaflets approach more closely the size of undivided leaves, even if they are on an average somewhat smaller than the undivided leaves. It seems then, if I may so express myself, that nature does not regard the whole compound leaf as a biological unit, but the individual leaflet of the compound leaf. It is, therefore, the leaflet of the compound leaf which must be compared with the individual leaf.

Leaves which show imperceptible transitions between simple and compound leaves present great difficulties.

First of all leaves that are lobed may be regarded as belonging to the group of undivided leaves.

It will not be easy to arrange the more deeply divided leaves. Nature in some species appears to regard the whole leaf as a unit, in other species the single division, or again amongst other species groups of divisions are the units which must be compared with the undivided leaf.

Because of these difficulties I consider that before making use of leaf size in characterizing a given formation, we must at first differentiate and determine the numerical relationship between: (1) species with undivided or lobed leaves, and (2) species with: (*a*) deeply divided leaves, and (*b*) compound leaves.

It is quite probable that the mere relationship between these three groups of species will form an interesting contribution to our knowledge of the characters of the formation in question.

It goes without saying that each life-form must be treated separately; for example, evergreen Mesophanerophytes must not be treated in the same group as deciduous Mesophanerophytes, and so on.

When the material is thus divided into natural groups the leaf size is determined for every individual species, either in all the groups, or only in some of them which have some special interest.

If the uppermost layer of a tropical Phanerophyte formation consists of 1 per cent. of deciduous Phanerophytes and 99 per cent. of evergreen Phanerophytes, of which (*a*) 9 per cent. have compound or deeply divided leaves, and (*b*) 90 per cent. have simple or lobed leaves, it will be of most immediate interest to determine the leaf size among the species in the last and largest group.

As a rule the group or groups with undivided or lobed leaves should be treated first, because this is generally the most numerous group, at any rate in formations of Phanerophytes and Chamaephytes, and also because the material is easy to treat. For the time being I shall confine myself to this group: what we have to do is to determine the leaf size for every species.

An exact determination of the size of the leaf surface must not be attempted, partly because leaves on the same plant vary more or less in leaf size, and partly because such a determination would take time out of all proportion to the use that could be made of the determination of the leaf size.

In using leaf size for a comparative investigation it will be most convenient to divide leaves into a number of sizes. In order to make our task of classification easier there must not be too many classes. In general we should be able at a glance to refer a given plant to its proper class.

We must first decide how many classes there are to be, and then where the boundaries between them are to be drawn.

For the sake of facilitating our survey I prefer few classes, which can be

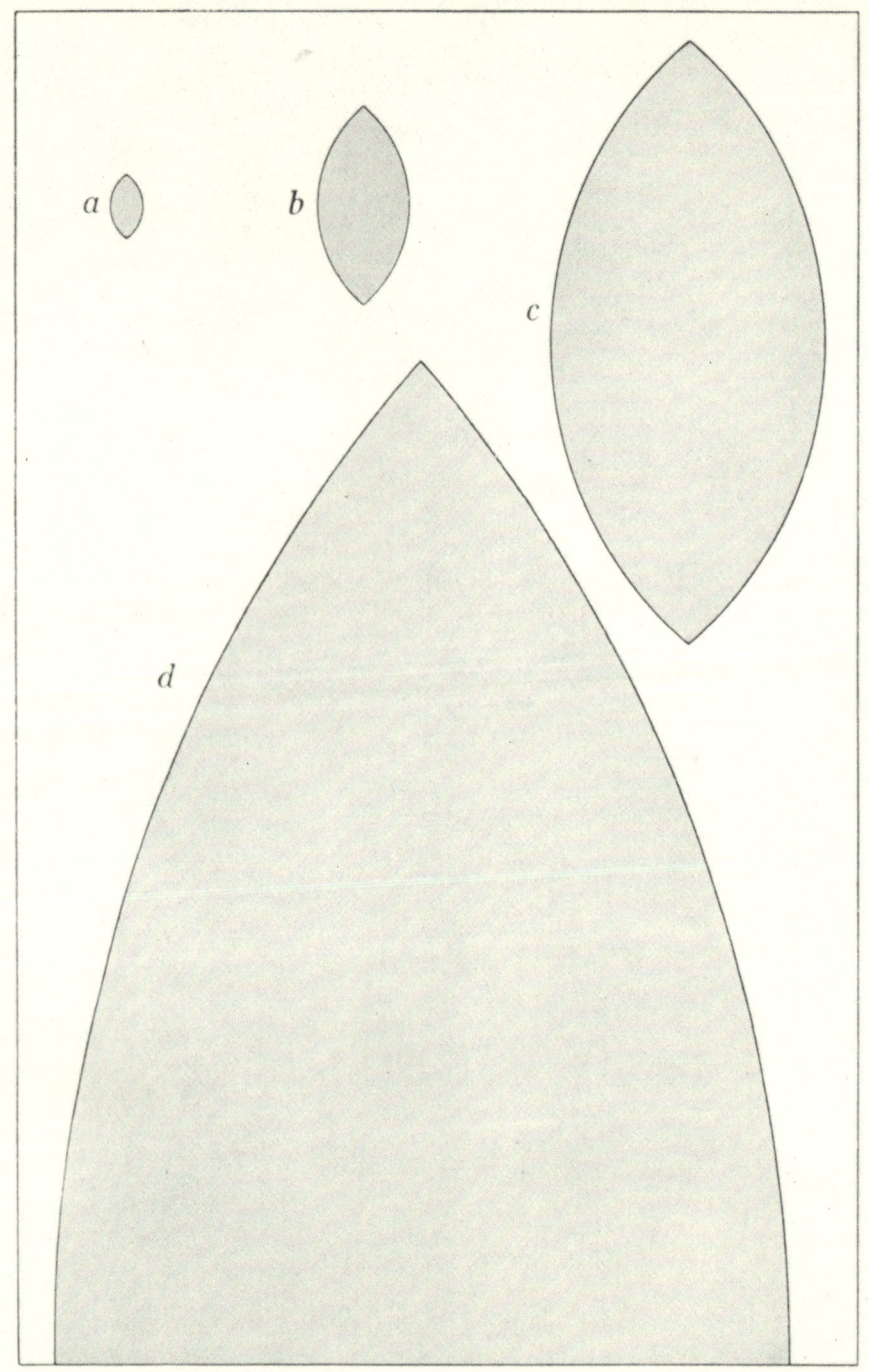

Fig. 139. Diagram for use in rapid determination of the class size of a leaf. The figures show the boundaries between the individual classes, thus:

less than a = leptophyll.
between a and b = nanophyll.
between b and c = microphyll.
between c and $2 \times d$ = mesophyll.
between $2 \times d$ and $8 \times$ the size of the diagram as bounded by the black line = macrophyll.
more than $8 \times$ the size of the diagram as bounded by the black line = megaphyll.

LEAF SIZE IN PLANT GEOGRAPHY

further divided if necessary. After making various trials I have decided on six classes, and given them the following names:

Leptophyll
Nanophyll
Microphyll
Mesophyll
Macrophyll
Megaphyll

I also made experiments in establishing the boundaries between the individual classes. First I prepared a continuous series of dried leaves from the smallest to the largest, and attempted by rough estimate to distinguish six classes. Then I made a series of quadrats on millimetre paper evenly increasing in size from very small to very large, corresponding with a series of sizes from the smallest to the largest leaves. I then considered where the boundaries between the six classes could most conveniently be drawn. Then I persuaded a number of my botanical colleagues to study the same problem. All this resulted in establishing as the uppermost boundary of the lowest class, the Leptophylls at 25 sq. mm. (0·000025 sq. metre). By multiplying this number by 9, and the result again by 9, and so on four times, we obtain all we want, i.e. all the leaf sizes divided pretty evenly into six classes, which correspond fairly with the preliminary rough estimate. The boundaries between the single classes will thus be drawn as follows:

Leptophylls
  25 sq. mm. or 0·000025 sq. metre.
Nanophylls
  $9 \times 25 = 225$ sq. mm. or 0·000225 sq. metre.
Microphylls
  $9^2 \times 25 = 2,025$ sq. mm. or 0·002025 sq. metre.
Mesophylls
  $9^3 \times 25 = 18,225$ sq. mm. or 0·018225 sq. metre.
Macrophylls
  $9^4 \times 25 = 164,025$ sq. mm. or 0·164025 sq. metre.
Megaphylls

Originally I multiplied by 10, but I found that the boundaries obtained in this way were not so natural as those obtained by multiplying by 9. Besides this, when I multiplied by 9 it was easier to subdivide the single classes into three groups: small, medium, and big, as for example in dealing with Mesophylls.

The figures in the diagram (Fig. 139) represent the sizes corresponding to this scheme, and also show the boundaries between the individual classes. Leaves smaller than a (25 sq. mm.) are Leptophylls. Those intermediate in size between a and b (9×25 sq. mm.) are Nanophylls. Those between

b and c ($9^2 \times 25$ sq. mm.) are Microphylls, and those between c and $2 \times d$ Mesophylls. The size representing the boundary between Macrophylls and Megaphylls corresponds approximately to eight times the size of the whole plate (within the black lines).

Immediate comparison with this figure will usually make it easy to decide to what class any leaf belongs, and if we have used the method for some time, we shall usually be able to determine the leaf size without the help of the figure. It is obvious of course, that plants will sometimes be met with leaves on the boundary between two classes. We can then designate the size accordingly. For example, if the leaves of a plant are on the boundary between the Nanophylls and Microphylls, they can be called Nano-Microphylls. If we wish in a given case to decide accurately whether the leaf belongs to one of two classes, for example, whether it is a Microphyll or a Mesophyll, we can use paper whose weight per sq. cm. is known, cutting out the size of the leaf on this paper. We can then compare the weight of this bit of paper with an area of the same paper corresponding to the size of Micro-Mesophylls, viz. 2,025 sq. mm.

Now if we ask what is to be gained by treating leaf size in this way, I conceive the answer to be that it enables us to prepare definitely and numerically the biological expression for the climate and environment as far as these factors express themselves in leaf size. That the environment affects leaf size is often immediately obvious; but it can be demonstrated clearly by the method described, since we can, by this means, compare the conditions of two climates which differ essentially only in one respect, e.g. humidity, and then observe how this difference is expressed in the leaf size. We can thus find out whether the corresponding difference is apparent when we investigate a country with a corresponding climate, but with a different flora.

We can also by this means investigate and compare formations in climates differing in several respects, but with biological values corresponding sufficiently to give the vegetation something of a common character. We have here a means of expressing definitely to what extent this character is different, so far as it expresses itself in leaf size.

I shall attempt to illustrate this by means of examples, comparing some European evergreen scrub formations. The choice of examples is determined by the opportunities I have had of making investigations by means of my valency[1] method; for it is only by first determining the degree of frequency of species in a formation, and by using this as a foundation

---

[1] Chapter VI, Investigations and Statistics of Plant Formations.
C. Raunkiaer, 'Measuring-Apparatus for Statistical Investigations of Plant-Formations', *Bot. Tidsskr.* vol. xxxiii, Kjøbenhavn 1912.
Chapter VIII, Statistical Investigations of the Plant Formations of Skagens Odde.
Chapter IX, On the Vegetation of the French Mediterranean Alluvia.

for comparison, that it is possible to obtain a definite result. As examples of evergreen scrub formations I have chosen (1) A west Jutland *Calluna*-heath: *Calluna vulgaris*-Formation. (2) The maquis (garigue) in the south of France: *Erica multiflora*-Formation. (3) The mountain maquis in north-east Spain: *Arbutus unedo* + *Quercus coccifera*-Formation. (4) A Thyme-heath on dry inland hills in the north of Spain: *Thymus hiemalis*-Formation.

These four formations have been investigated by the valency method, and the result is given in tables in 1, 2, 3, and 4. As far as the Phanerophytes and Chamaephytes of the formation are concerned, each of these tables can be looked upon as a sample of the formation in question. For the west Jutland *Calluna*-heath and for the maquis of the south of France all of the species in the selected samples are included, so that we have here a sample of the phanerogamic flora of the formation taken as a whole. But this is not so with the two formations of the north of Spain. Here only the Phanerophytes and the Chamaephytes are included. We cannot therefore compare the numerical proportion between the small bushes (Nanophanerophytes and Chamaephytes) of the four formations, and the other life-forms. But this is less important here as these formations taken as a whole are formations of small shrubs, i.e. they are dominated by Nanophanerophytes and Chamaephytes.

In Tables 1 to 4 I have first given one or more species dominating the formation in question. The other species are alphabetically arranged within the classes of life-forms. In the first column, after the species named, the class of life-form of the species is given. The second column gives the valency of the species expressed as the percentage of sample plots in which the species is found. Finally in the third column there is given, for evergreen Phanerophytes and Chamaephytes with entire or lobed leaves, the class of leaf size in which they are included. The abbreviations used are: l. = leptophyll, n. = nanophyll; mic. = microphyll.

As seen by comparing the tables, the four formations differ in different ways. In the number of species of Phanerophytes and Chamaephytes the *Calluna vulgaris*-Formation is the poorest (5 species). Then come the *Thymus hiemalis*-Formation (seven species), the *Arbutus unedo*+*Quercus coccifera*-Formation (12 species), and the *Erica multiflora*-Formation (22 species). But in the sum of the valency numbers of Phanerophytes and Chamaephytes the *Thymus hiemalis*-Formation is the richest. If we arrange them physiognomically, especially according to their luxurance, the *Thymus hiemalis*-Formation is the least luxuriant. Then follow the *Calluna vulgaris*-Formation and the most luxuriant *Arbutus unedo*+ *Quercus coccifera*-Formation.

We must next consider the number of evergreen Phanerophytes and Chamaephytes with entire or lobed leaves that are here to be dealt with, and ask whether the relationship between the four formations in this respect is such to make it sensible to undertake a comparison of their leaf size. If we use the information in Tables 1 to 4, and reckon for each of the four formations the percentage relationship between, on the one hand, the evergreen Phanerophytes and Chamaephytes with entire or lobed leaves, and on the other hand, the remaining Phanerophytes and Chamaephytes, both in respect of the number of species and valency, we obtain the result given in Table 5. From this Table we first see that the groups in question (evergreen Phanerophytes and Chamaephytes with entire or lobed leaves) dominate everywhere in the number of their species, the percentages in the four formations lying between 58 and 80. We further see that in considering the valency of the species, which is the most important point here, the dominance of this group is even more marked, the percentage lying between 82 and 99·5. From this it follows that the difference between the formations is slightly less than could be seen by the number of species. The four formations thus correspond essentially in the dominance of the evergreen Phanerophytes and Chamaephytes with entire or lobed leaves. We must therefore undertake a comparative investigation of the leaf sizes of this group of plants in the four formations, in order to try to characterize them.

In Tables 1 to 4 the last column gives the leaf size to which the individual species belong; and by means of the valency numbers of the species in column 2, the percentage relationship between the classes of leaf size in the formations can be determined.

TABLE I

Statistical-biological analysis of the *Calluna-Empetrum*-Heath on Aadum-Varde Bakkeø in West Jutland ($100 \times \frac{1}{10}$ sq. metre)

	Life-form.	Frequency number.	Leaf size in evergreen Phanerophytes and Chamaephytes with entire or lobed leaves.
Calluna vulgaris	Ch	98	l
Empetrum nigrum	Ch	90	l
Arctostaphylos uva ursi	Ch	11	n
Vaccinium uliginosum	N-Ch	1	
„ vitis idaea	Ch	3	n
Carex Goodenoughii	G	1	
„ panicea	G	1	

In the *Calluna vulgaris*-Formation (Table 1) for example, there are two species which are Nanophylls, viz. *Arctostaphylos* and *Vaccinium vitis idaea*, with altogether 14 points. There are 2 species, *Calluna vulgaris* and *Empetrum nigrum*, which are Leptophylls with altogether 188 points. Calculation reveals that the Nanophylls occurring in this formation amount to 71 per cent., and the Leptophylls to 93 per cent.

LEAF SIZE IN PLANT GEOGRAPHY

TABLE 2

Statistical-biological analysis of the *Thymus*-Formation on a range of hills about 2 km. east of Lerida (50 × $\frac{1}{10}$ sq. metre)

	Life-form.	Frequency number.	Leaf size in evergreen Phanerophytes and Chamaephytes with entire or lobed leaves.
Thymus hiemalis	Ch	100	1
Fumana glutinosa	Ch	4	1
Genista scorpius	Ch	2	
Helianthemum hirtum	Ch	6	1
Santolina chamaecyparissus	Ch	6	
Siderites hirsuta v. tomentosa	Ch	2	1
Teucrium polium	Ch	4	1

In the *Thymus hiemalis*-Formation all the species are Leptophylls (100 per cent.).

By treating the *Erica multiflora*-Formation (Table 3) in the same manner we obtain 21 per cent. of Nanophylls and 79 per cent. of Leptophylls.

In the *Arbutus unedo* + *Quercus coccifera*-Formation (Table 4) we have three classes of leaf size, viz. 59 per cent. of Microphylls, 27 per cent. of Nanophylls, and 14 per cent. of Leptophylls. It must, however, be observed that *Cistus salviaefolius* is here given as a Microphyll. As a matter of fact it is often intermediate between a Microphyll and a Nanophyll. If we reckon it both among the Microphylls and the Nanophylls our numbers become slightly different, viz. 49 per cent. Microphylls, 40 per cent. Nanophylls, and 11 per cent. Leptophylls. This, however, by no means masks the peculiarity of this formation compared with the other formation, namely the relatively large leaf size.

The results of this investigation of this leaf size in the four formations are placed in juxtaposition in Table 5, where the formations are arranged according to their height. From this it is seen that the resemblance is strongest between the *Calluna*-heath of west Jutland and the *Thymus*-heath of north Spain. This is an example of climates physically widely different determining the dominance of essentially the same life-form in a formation. In the *Calluna*-heath the leptophyllous condition, as indeed the xeromorphy taken as a whole, is determined by the physiological drought of winter, while the corresponding characters of the *Thymus*-heath are determined by the physical drought of summer. If we consider all the species in the formation, it is easily seen that the two formations belong to two different plant climates, the *Thymus hiemalis*-Formation developing many Therophytes during the favourable spring season. It should, however, be observed that for a single formation, especially when we investigate a circumscribed area, we cannot always

Table 3

Statistical-biological analysis of the *Erica multiflora*-Formation in the south of France ($50 \times \frac{1}{10}$ sq. metres)

	Life-form.	Frequency number.	Leaf size in evergreen Phanerophytes and Chamaephytes with entire or lobed leaves.
Erica multiflora	N	68	1
Amelanchier vulgaris	N	2	..
Daphne gnidium	N	2	n
Genista scorpius	N	28	..
Juniperus sp. (communis?)	N	4	1
Lavandula vera	N	18	n
Rosa sepium	N	2	..
Rosmarinus officinalis	N	46	n
Arenaria capitata	Ch	6	1
Artemisia campestris	Ch	2	..
Coris monspeliensis	Ch	26	1
Cytisus argenteus	Ch	12	..
Fumana procumbens	Ch	68	1
Helianthemum hirtum	Ch	2	1
„ polifolium	Ch	38	1
Helichrysum stoechas	Ch	20	1
Ononis minutissima	Ch	26	..
Satureia montana	Ch	10	..
Sedum anopetalum	Ch	2	1
„ nicaeense	Ch	12	n
Teucrium marum	Ch	6	1
Thymus vulgaris	Ch	50	1
Anthyllis vulneraria	H	10	..
Aphyllanthes monspeliensis	H	22	..
Asperula cynanchica	H	16	..
Avena bromoides	H	12	..
Carex Halleriana	H	14	..
Euphorbia serrata	H	10	..
Festuca ovina	H	56	..
Globularia vulgaris	H	2	..
Hieracium pictum	H	18	..
Hypochaeris radicata	H	2	..
Psoralea bituminosa	H	6	..
Schoenus nigricans	H	2	..
Stipa juncea?	H	20	..
Taraxacum gymnanthus?	H	2	..
Trinia dioeca	H	24	..
Brachypodium ramosum	G?	4	..
Cirsium (arvense?)	G	2	..
Aethionema saxatile	Th	2	..
Asterolinum stellatum	Th	2	..
Cuscuta sp.	Th	10	..

LEAF SIZE IN PLANT GEOGRAPHY

come to definite conclusion about the plant climate. For this purpose a local flora is needed, which may indeed well be the flora of a small area if it includes samples of the essential formations of that area.

TABLE 4

Statistical-biological analysis of the *Arbutus unedo*-Formation on the top of Tibidabo near Barcelona ($50 \times \frac{1}{10}$ sq. metre)

	Life-form.	Frequency number.	Leaf size in evergreen Phanerophytes and Chamaephytes with entire or lobed leaves.
Arbutus unedo	N	96	mic.
Calycotome spinosa	N	22	..
Cistus salviaefolius	N	62	n-mic.
Erica arborea	N	38	1
Lonicera caprifolium	N (-Ch)	14	..
Phillyrea media	N	2	mic.
Pistacia lentiscus	N	4	..
Quercus coccifera	N	76	n
,, ilex	M (-N)	2	mic.
,, pubescens	N	2	..
Smilax aspera	N	4	mic.
Dorycnium suffruticosum	Ch (-H)	2	..

The *Erica multiflora*-Formation is much richer in species than the *Calluna vulgaris*-Formation, but biologically the two formations do not differ widely. The *Arbutus unedo* + *Quercus coccifera*-Formation differs most markedly from the *Calluna*-heath, being much taller and having larger leaves.

TABLE 5

	N and Ch		Evergreen N and Ch with entire or lobed leaves.				
			Percentage of N and Ch with respect to		Percentage distribution in classes of leaf size on a basis of the valency of the species.		
	No. of species.	Points.	No. of species.	Points.	Micro- phyll.	Nano- phyll.	Lepto- phyll.
Thymus hiemalis-Formation	7	124	71	92			100
Calluna vulgaris-Formation	5	203	80	99·5		7	93
Erica multiflora-Formation	22	450	68	82		21	79
Arbutus unedo+Quercus coccifera-Formation	12	324	58	86	59	27	14

In Table 5 the formations, as already mentioned, are arranged according to their height, the highest being placed last. It can be seen that as the formation decreases in height the leaf size decreases, and this can, as shown here, be expressed in definite numbers, so that we can obtain a foundation making it possible to compare and criticize the relationships between a series of closely related formations. This is specially important in dealing with such regions as the Mediterranean, where one dominant formation rich in species, the maquis, is distributed over large areas, occurring floristically in many different patterns. By means of valency numbers these patterns can be determined and compared numerically in regard to each character, each adaptation, which in one way or another may be expressed numerically, as I have shown in dealing with leaf size. The same procedure can be used in dealing with such characters as stomatal covering, thick cuticle, hairiness, and other xeromorphic characters. Even if the difficulties encountered be great, this need not prevent us from striving after exact methods in plant geography. It is only by this means that it is possible to transcend the vague rough-and-ready determinations of 'tourist' plant geography.

XI
STATISTICAL RESEARCHES ON PLANT FORMATIONS
INTRODUCTION

In all vegetation which is dense enough to cover the soil, there arises between the different species which make it up a competition leading sooner or later to a state of things in which the species best fitted to live in the conditions of existence which the area in question presents become dominant over the species less well adapted, which suffer a diminution or are entirely suppressed. At the end of some time one finds that the vegetation has entered into a certain state of equilibrium, a state in which it will remain without undergoing appreciable alteration so long as the life conditions and the nature of the species remain the same. But if these change (and the vegetation itself may contribute to such a change) the equilibrium will come to an end; either the change in the conditions of existence does not exercise the same influence on all the species present, or the new conditions are sometimes more suitable for species arriving from the outside than for those already present which constituted the original population.

Wherever a vegetation has been abandoned to itself for a sufficiently long time, it will generally have attained a certain equilibrium, and by reason of the competition the species which compose it will be among those which are the best adapted to the environment. Such a vegetation is called a Natural Formation, by contrast with artificial formations arising under the continuous action of man. It may happen, though very rarely, that a single species succeeds over all its rivals to the point of exterminating them and of coming to constitute by itself the entire vegetation (for example, compact formations of *Picea abies* or of young Beeches); but, in general, besides the one or more dominant species we shall find a variable number of less fortunate rivals growing more or less sparsely. In traversing a formation composed in this way the relative proportions of the species will be found to change sooner or later, and we shall have to conclude that one or several new species have prospered to the point of taking the upper hand. If the vegetation considered is in a state of equilibrium it is clear that the altered composition means that the new area offers conditions of existence different from those of neighbouring areas: or, in other words, we have come upon a new formation. According to the laws which govern competition every modification in external conditions will necessarily bring with it a modification in the composition of the flora; the relative proportions of the species represented will change; perhaps even some of them will disappear—clearly those whose life-form does not suit the new conditions—and will be replaced by other species which represent a life-form more appropriate to them.

By 'life-form' or 'biological type' we understand, in speaking of a plant, the totality of its different adaptations to the conditions of the environment; in any vegetation the plants which are finally victorious in the struggle for life are precisely those which possess the life-form most appropriate to the conditions in which that struggle takes place. Wherever we perceive a marked difference in the specific composition of a flora of one and the same locality we perceive an expression of the fact that the physical conditions have changed—of course supposing that we are dealing with one and the same flora and with a vegetation left to itself for a sufficient time to enable it to attain equilibrium. Under these conditions the variation undergone by the vegetation from the point of view of the species which compose it always indicates a variation of the formation, if by 'formation' we understand that expression of life conditions which is produced by the adaptations of the plant covering—in other words, the life-form.

If we submit the plant covering to a more or less detailed analysis, the word 'formation' may be employed in a narrower or wider sense, just as in the classification of plants the designation 'species' is employed sometimes in its narrowest sense to express the smallest units which are hereditarily different (microspecies), or in a wider sense as 'Linnean' or collective species. With regard to the designation 'association', sometimes employed in a narrower sense than 'formation', to express simply a floristic difference (Third International Botanical Congress, Brussels, 14–22 May 1910, *Phytogeographic Nomenclature*, pp. 5–7), it is not only superfluous but false; for wherever in the interior of a limited area the vegetation has attained a certain equilibrium, one can verify the justice of the general observation that a difference of floristic order is always the expression of a difference of external conditions, and necessarily—as a result of competition—of a difference of adaptation to these conditions; in other words, we are concerned with a difference of life-form, and thus of formation, even according to the definition given in *Phytogeographic Nomenclature* cited above.

Every formation can be characterized under three different aspects:

(1) Floristic characterization, in which a list is given of the species of the formation, with an indication of their degree of frequency and of their distribution. This forms the basis of all later examination.

(2) Physiognomic characterization, which determines the relative proportions of the species present, a determination, which, as we shall see later, can be carried out at the same time as the floristic analysis.

(3) Biological characterization, which examines the manner in which the species are adapted to the conditions of life; here it is the life-form of the species that we have to determine. Meanwhile, it is not easy to

define the real life-form of a species, that is to say the sum of its adaptations to the conditions. In fact, these adaptations are multiple and of very different nature; they may be morphological, anatomical, or intracellular; some are sufficiently well known, while others seem more or less obscure, and others again are doubtless at present quite inaccessible to study. Every plant species presents a long series of adaptations, but we know no means of expressing them as simple quantities which can be calculated and which can express the degree of adaptation, that is to say the life-form: in order to characterize this last we are obliged, until we are better informed, to consider one essential aspect of the adaptations in order to be able to deal successively with other aspects. With regard to the basis on which we should construct this provisional system of life-forms, it is first of all indispensable that it should be essential in its nature; in other words, it ought to refer to that which is most fundamental in the adaptations of plants to the conditions of their environment. Further, for practical reasons, the basis sought ought to be easy to use, that is to say it ought to be easy to determine in the field to what life-form a given plant belongs. Finally, it is necessary that the basis employed should represent a single point of view and thus allow a statistical treatment of the vegetation of different regions. If the system of life-forms is constructed on a heterogeneous basis it cannot serve as a starting-point for exact comparative investigations. Beginning with these considerations, I have chosen as a basis of life-forms (which I have established and used in various publications) the adaptation which enables plants to survive the unfavourable season, especially in relation to the protection of the buds or extremities of the shoots. Within this framework, which should include the plant covering of the entire earth, it is possible to deal with any adaptation that one desires to examine (Raunkiaer: 1905; 1907; 1908; 1909, I; 1911. See bibliography, p. 423, and Chapters I, II, IV, V, and VII).

I. USE OF THE DEGREE OF FREQUENCY OF SPECIES FOR THE FLORISTIC CHARACTERIZATION OF PLANT FORMATIONS

The primitive line of demarcation of every formation is directly determined by observations relating to the floristic and physiognomic composition presented by the plant covering of the soil; the list of species of which the formation in question is composed forms the starting-point of every study which is more than superficial. Meanwhile, the specific nature of the formation is not solely determined by the enumeration of species, for the different species have not the same value in the composition of the plant covering; and, further, we may have two different localities presenting the same totality of species but nevertheless constituting two different formations determined by different conditions of the environment, the one favouring the development of one or many species which in the other only play quite a secondary role. Thus we have in the first place to determine the degree of frequency, the valency, of the species which make up a given formation.

As a rule the degree of frequency of the species of a formation is determined by an approximate estimation, following a procedure long employed by the floristic botanist when he has to determine the degree of frequency of the species which make up a given flora, a proceeding which consists in expressing the degree of frequency by adjectives or adverbs such as 'very rare', 'rare', 'scattered', 'common', 'very common', and similar terms. This method, nevertheless, has one great inconvenience: it does not give exact figures; one can, it is true, express the degrees of frequency by figures as one evaluates them, but these numerical evaluations, based as they are on subjective personal appreciation, form too uncertain a basis. It often happens that not only the judgements of different people differ between themselves, but even one and the same person may judge differently of one and the same thing at different times. For this reason I have tried to think out a method which in the determination of frequencies will enable one to obtain results more exact than those of subjective estimations and in which the investigation of formations can give to every species a mark expressing its degree of frequency by a number independently of subjective appreciation, so that the observation will be essentially the same whoever the person that makes it.

In this new method, the degrees of frequency of the species included in a formation—or, in other words, their valency—are determined by the floristic analysis of a certain number of units of area, 'samples' of definite extent, the degree of frequency of each species being finally expressed by the number representing the proportion of samples in which the species in question is found. Let us suppose that in 50 units of area we have encountered the species A, B, C in 47, 29, and 3 respectively: the degrees of frequency of these three species will then be expressed by the numbers 94, 58, and 6.

Before expounding in detail the mode of application of our method, it is necessary to examine more closely what should be

(1) the number of samples;
(2) the extent or size of these units of area;
(3) their distribution in the formation.

We shall have obtained the necessary number of samples when the result reached by the examination of a certain number has become constant, that is to say as soon as the figure expressing the degree of frequency does not undergo appreciable modification if one extends the examination to a greater number of samples. How many samples are requisite to enable this result to be obtained? That depends on the one hand upon the extent—more or less restricted—that has been given to the notion of formation, and, on the other hand, and especially, on the size of the samples. With regard to the relation existing between the size of the sample and the necessary number of samples, it goes without saying that the greater the size the smaller will be the minimum number enabling us to arrive at a constant result; but the more distant will be the relation found from the real relation between the degrees of frequency of the species; and, reciprocally, the smaller the samples are, the greater will be the number required to obtain a constant result, and the better this result will express in an adequate fashion the real state of things. We may add that the more extended the unit areas, the completer will be the floristic list resulting from their analysis. But, on the other hand, large unit areas are much more difficult to analyse than small ones.

It results from what we have said that in the choice of the size of the sample areas it is necessary to take many facts into account. The most suitable size will undoubtedly be that which will give results corresponding most closely to the work carried out. I have tried to determine this by experiment in a series of different sizes, that is to say 10 sq. metres, 1 sq. metre, $\frac{1}{10}$ sq. metre, $\frac{1}{100}$ sq. metre, with the result that I have fixed upon $\frac{1}{10}$th of a sq. metre as the most practical size (Chapter VI, pp. 207–9). I have therefore taken this size as a basis in my later work on the formations.

It is necessary to add that only results obtained by investigations carried out on a uniform size of sample area are directly comparable with one another: in fact the expression obtained by this method for the relative proportions of the species only corresponds approximately to reality; and the degree of concordance between the result obtained and the reality changes when the size of the sample is changed, the degree of concordance increasing as the size diminishes.

If we employed a very small unit of area such as 1/10,000 of a square metre (that is, 1 sq. cm.) on which grew only a single individual plant, one would be certain of obtaining a result coinciding exactly with reality; but an immense, practically unrealizable number of samples would be necessary to arrive at a constant result; and I have concluded that I need not work with so small a size, the more so that in comparative researches on formations the essential point is to obtain a constant expression of the proportions existing between the degrees of frequency of the species, even if this does not correspond with absolute truth.

In regard to the number of samples, I have been able to establish that if we adopt the size of $\frac{1}{10}$ sq. metre as I have proposed, it is only necessary to examine 25 to 50 samples to arrive at a constant result from the point of view of the proportions which the species characteristic of a formation bear to one another. When the word formation is taken in the narrow sense, in which it means a relatively homogeneous part of a flora, one would be quite content with 25 or even perhaps 20 samples.

Provided that one is dealing with a formation in the narrow (strict) sense the distribution of the samples has no importance: one can take them at random or at definite intervals along the length of a certain number of lines traced beforehand. In fact the word formation itself implies that we have to do with a vegetation whose composition is practically everywhere the same—if at the first glance one is not very sure that the vegetation in question has an almost homogeneous composition and that thus it does not constitute a single formation, it becomes evidently necessary to try to resolve this question; and this can be done, as we shall see in the sequel, by the determination of degrees of frequency.

The procedure which I first employed consisted in taking samples by means of a square frame enclosing an area of $\frac{1}{10}$ sq. metre, which I threw at random on the formation that was to be examined. Meanwhile, in certain conditions of vegetation, for instance when it is composed of herbaceous plants with tall and robust stems, or of Nanophanerophytes—for example, in maquis—it often happened that the frame remained hanging on the plants; for this reason in all my later work, in order to determine the samples, I always used a thin metal rod which was jointed at right angles to a stick and formed a radius of a circle enclosing $\frac{1}{10}$ sq. metre. In walking through the formation to be examined I then chose the samples by plunging the stick into the soil at successive intervals of two or more paces and rotating the stick on its axis so that the end of the metal rod described a circle whose circumference enclosed a sample with a surface of $\frac{1}{10}$ sq. metre.

On examining each sample, I marked each of the species found on the list of species noted. Then, having taken the number of paces selected for the distance between successive samples, I took the next sample in the same way, and so on up to the number desired.

For reasons which I shall explain immediately, it is necessary that the notes relating to the examination of a given formation should be arranged so that the results of the floristic analysis of each sample can be consulted in order. Consequently it is necessary to assign to each individual sample a definite place in the scheme employed, as is shown, for instance, in Table 1, where the degrees of frequency occupy the left-hand column and each of the five columns on the right contain 5 spaces, so that for each species there are 25 places in all and one can represent the numbers 1–25 by means of lines reading from left to right. Thus in the example represented in Table 1 the formation constituted by *Anemone nemorosa* + *Oxalis acetosella* occurring in the Dyrehave near Copenhagen, the presence of *Anemone nemorosa* has been established in all the units of area up to the number 25, and *Oxalis acetosella* only in 24 (this species being absent in No. 22); for the rest, *Asperula odorata* occurs in 4 samples, (Nos. 1, 12, 13, and 16), *Viola silvatica* in 2 (Nos. 8 and 10), *Rubus ideaus*, *Melica uniflora*, and *Ficaria verna* each in a single sample, Nos. 8, 17, and 19 respectively. Since we have only taken 25 samples, it is necessary to multiply each number by 4 to obtain the corresponding percentage. Thus we obtain an expression of valency or degree of frequency of each species entering into the formation examined, as set out in the first column on the left.

TABLE 1

Formation of *Anemone nemorosa* and *Oxalis acetosella* in the Dyrehave near Copenhagen

	Degree of frequency.	Samples or units of area.				
		1–5	6–10	11–15	16–20	21–25
Anemone nemorosa	100	|||||	|||||	|||||	|||||	|||||
Oxalis acetosella	96	|||||	|||||	|||||	|||||	|·|||
Asperula odorata	16	|····	·····	·||··	|····	·····
Viola silvatica	8	·····	··|·|	·····	·····	·····
Rubus idaeus	4	·····	··|··	·····	·····	·····
Melica uniflora	4	·····	·····	·····	·|···	·····
Ficaria verna	4	·····	·····	·····	···|·	·····

TABLE 2

Heath between oak coppice to the south of Varde (Jutland)

	Life-form.	Degree of frequency.	Samples or units of area.				
			1–5	6–10	11–15	16–20	21–25
Calluna vulgaris	Ch	100	|||||	|||||	|||||	|||||	|||||
Arctostaphylos uva-ursi	Ch	92	|||||	|||||	|·|||	|||||	|·|||
Empetrum nigrum	Ch	60	|||||	|||··	··|·|	|||·|	|····
Carex panicea	G	8	|····	·····	·····	|····	·····
Orchis maculata	G	8	·||··	·····	·····	·····	·····
Molinia coerulea	H	52	··|||	|·||·	·|||·	··|··	·|||·
Arnica montana	H	12	···||	|····	·····	·····	·¡···
Genista anglica	Ch	20	····|	||···	··|··	·····	··|··
Potentilla erecta	H	8	····|	·····	·····	·····	···|·
Majanthemum bifolium	G	8	····|	·····	·····	·····	···|·
Scirpus caespitosus	H	4	·····	|····	·····	·····	·····
Genista pilosa	Ch	8	·····	··|·|	·····	·····	·····
Aira flexuosa	H	4	·····	·····	···|·	·····	·····
Carex pilulifera	H	8	·····	·····	···|·	····|	·····
Lycopodium clavatum	Ch	4	·····	·····	·····	···|·	·····

Just as Table 1 gives the results obtained from the examination of a formation of *Anemone nemorosa–Oxalis acetosella*, Table 2 gives the results of an analogous examination of a formation of *Calluna vulgaris–Arctostaphylos uva-ursi*. Both these tables reproduce corresponding pages of the field notes; that is why the species appear to follow one another haphazard, because they occur in the order corresponding to that in which they appeared for the first time in the samples examined. The same applies to Tables 3–6.

It results from Tables 1 and 2 that the species which one has found only in a small number of samples are set out quite uniformly in the scheme, but this is not true of Table 3, which records the results obtained from the examination of a heath situated to the north of Varde (Jutland).

TABLE 3

Frøstrup Heath, to the north of Varde

	Degree of frequency.	Samples or units of area.				
		1–5	6–10	11–15	16–20	21–25
Calluna vulgaris . .	100	|||||	|||||	|||||	|||||	|||||
Empetrum nigrum .	32	.|...	||||.	.|.|.	....|	.....
Arctostaphylos uva-ursi	20	....|	|||.|	.....	.....	.....
Molinia coerulea . .	40	..|..	..|..	|||||	|...|	|....
Arnica montana . .	4	.....	..|..	.....	.....	.....
Erica tetralix . .	48	.....	.....	..|||	||||.	|||||
Carex panicea . .	8	.....	.....	...|.	.....	..|..
Vaccinium uliginosum .	4	.....	.....	....|	.....	.....
Scirpus caespitosus .	4	.....	.....	.....	|....	.....

From an examination of this table one sees that
(1) *Calluna vulgaris* occurs in all the samples.
(2) Among the rare species, Nos. 2, 3, 4, 5, 7, and 9 occur here and there in the formation.
(3) On the other hand, *Erica tetralix* does not occur in the first 12 samples but is found in all the remaining ones except No. 20.

It results from these facts that if the formation with which we started was formed essentially by *Calluna vulgaris*, this is not true of the later samples starting from No. 13, where one sees that the line on which the samples are situated is continued into another formation formed by *Calluna vulgaris* and *Erica tetralix*. If we examine the matter more closely we are able to establish that the modification in the vegetation thus recorded was the effect of a modification in the conditions of the environment and consisting of a very slight lowering of the soil surface, bringing with it an increase in the humidity of the soil. This example will serve to show how desirable it is to give each sample a definite place in the scheme. In this way we can immediately verify any variations in the composition of the vegetation of the different parts of the area adjoining the line along which the samples are situated.

In the great majority of cases a simple glance will suffice to determine if one is still in the same vegetation or if this is changed. It happens sometimes that a 'frequent' species which occurs for the first time escapes the attention owing to its small size, and thus the difference produced in the carpet of vegetation is not recognized immediately but only on an inspection of the results furnished by the analysis of the samples. Another reason which makes it desirable to assign each sample its proper place in the scheme is that it is necessary to avoid writing any species more than once under the heading of the same unit of area.

Whenever an analysis is found to contain elements belonging to two different formations—as we have seen, for example, in Table 3—it is evidently necessary to study the two formations separately in order to establish the difference or differences which distinguish them. This has been done in the case under discussion in Tables 4 and 5.

TABLE 4

Formation of *Calluna vulgaris*. Frøstrup Heath, to the north of Varde

	Degree of frequency.	Samples or units of area.				
		1–5	6–10	11–15	16–20	21–25
Calluna vulgaris . .	100	‖‖‖‖‖	‖‖‖‖‖	‖‖‖‖‖	‖‖‖‖‖	‖‖‖‖‖
Empetrum nigrum .	40	.|...	‖‖‖.	.|.|.	|....	|..|.
Arctostaphylos uva-ursi	36	....|	‖‖.|	....|	.....	..‖‖
Molinia coerulea . .	24	..|..	..|..	‖‖‖.	.....	.....
Arnica montana . .	4	.....	..|..	.....	.....	.....

TABLE 5

Formation of *Calluna vulgaris+Erica tetralix*. Frøstrup Heath, to the north of Varde

	Degree of frequency.	Samples or units of area.				
		1–5	6–10	11–15	16–20	21–25
Calluna vulgaris . .	92	‖‖‖‖‖	‖‖‖‖‖	‖‖.|	.‖‖‖	‖‖‖‖‖
Erica tetralix . .	88	‖‖‖‖‖	‖.‖‖	‖‖‖‖‖	‖‖‖‖‖	‖‖..|
Molinia coerulea . .	24	‖‖‖.	..‖.	.....	.....	.....
Empetrum nigrum .	24	.|...	..|..	...|.	‖|...	..|..
Carex panicea . .	16	.|...	.....	.....	|....	...‖	.....
Vaccinium uliginosum .	4	..|..	.....	.....	.....	.....
Scirpus caespitosus .	12	...|.	.....	.....	....|	|....
Salix repens . .	4	.....	.....	.....	.....	...|.

The first of these tables (Table 4) represents the formation of *Calluna vulgaris* and shows the 12 first samples of Table 3 together with 13 other samples belonging to the same area, in all 25 samples. In the same way Table 5 represents a formation of *Calluna vulgaris + Erica tetralix*, showing the 13 last samples of Table 3 and 12 others belonging to the same formation. Simple inspection of the two tables will enable one to see very clearly what constitutes the difference between the two formations.

Another example of a series of samples which crosses the limit between two different carpets of vegetation is given in Table 6. At first sight, it appears that one has simply a formation of *Calluna vulgaris* covering a flat, horizontal soil and quite homogeneous, this species occurring in all the samples. But in fact the scheme shows that *Empetrum nigrum* occurs fairly often in the first 14 samples, while it is totally absent from the last 11. Manifestly there is between the 14th and 15th samples some line of demarcation; what does this line mean? The soil is horizontal and entirely of the same nature on the two sides of the line. Meanwhile if one looks a little closer one can establish a slight difference in the height of *Calluna*, this being a little taller where *Empetrum* grows than it is in the areas where that plant is absent. Further, a closer examination reveals in this last area the remains of *Calluna* stems, probably survivals from a fire which had ravaged the area. In fact the conjecture that the difference established between the vegetation on the two sides of the line indicated was due to a fire was confirmed by information obtained from people acquainted with the locality, who said that the region where *Empetrum* did not grow had been devastated by fire about eight years previously. Thus this fire had entirely destroyed the *Empetrum*, which had not been able to re-establish itself on the burned area; the other species, on the contrary, showed themselves capable of rejuvenating by means of underground shoots; some perhaps had immigrated from the adjoining heath which had escaped the ravages of the fire. This heath was completely explored and the results are put together in Table 7, which comprises the first 14 samples of Table 6, with 11 other samples from the same area in addition.

Table 6

Heath near Grindsted

	Degree of frequency.	Samples or units of area.				
		1–5	6–10	11–15	16–20	21–25
Calluna vulgaris	100	|||||	|||||	|||||	|||||	|||||
Empetrum nigrum	56	|||||	|||||	||||.	.....	.....
Vaccinium vitis-idaea	20	||.||	.....	.....	.|...	.....
Carex Goodenoughii	24	|....	.....	|||..	.....	...||
„ panicea	28	.|...	..||.	|...|	..|.|	.....
Molinia coerulea	16	..||.	.....	.....	||...	.....
Scirpus caespitosus	4	.....	.|...	.....	.....	.....
Juncus squarrosus	4	.....	.....	.|...	.....	.....

Once we are assured that the examination has been carried out on one and the same formation, it becomes unnecessary to use schematic tables, which are always more or less complicated, in order to compare different formations: the valency numbers furnish all necessary information and brought together in a single table enable one further to compare a series of formations at the same time. Thus the four formations recorded in Tables 2, 4, 5, and 7 are brought together in Table 8 and can be directly

TABLE 7

Heath near Grindsted

	Degree of frequency.	Samples or units of area.				
		1–5	6–10	11–15	16–20	21–25
Calluna vulgaris . .	96	|||||	|||||	|||||	|||||	|.|||
Empetrum nigrum .	100	|||||	|||||	|||||	|||||	|||||
Vaccinium vitis-idaea .	36	||.||	.....	.....	....|	.||||
Carex Goodenoughii .	24	|....	.....	|||..	|..|.	.....
„ panicea . .	16	.|...	..||.	|....	.....	.....
Molinia coerulea . .	8	..||.	.....	.....	.....	.....
Scirpus caespitosus .	8	.....	.|...	....|	.....	.....
Juncus squarrosus .	4	.....	.....	.|...	.....	.....
Erica tetralix . .	4	.....	.....	.....	.|...	.....
Arctostaphylos uva-ursi	12	.....	.....	.....	.....	|||..
Aira flexuosa . .	4	.....	.....	.....	.....	.|...

TABLE 8

Summary of the formations described in Tables 2, 4, 5, and 7

	Life-form.	Degree of frequency of species.			
		Table 4.	Table 2.	Table 7.	Table 5.
Calluna vulgaris . .	Ch	100	100	96	92
Arctostaphylos uva-ursi .	Ch	36	92	12	..
Empetrum nigrum . .	Ch	40	60	100	24
Erica tetralix . . .	Ch	..	..	4	88
Aira flexuosa . . .	H (-Ch)	..	4	4	..
Arnica montana . . .	H	4	12	..	..
Carex Goodenoughii . .	G	..	..	24	..
„ panicea . . .	G	..	8	16	16
„ pilulifera . . .	H	..	8	..	..
Genista anglica . . .	Ch	..	20	..	..
„ pilosa . . .	Ch	..	8	..	..
Juncus squarrosus . .	H	..	..	4	..
Lycopodium clavatum . .	Ch	..	4	..	..
Majanthemum bifolium .	G	..	8	..	..
Molinia coerulea . .	H	24	52	8	24
Orchis maculata . .	G	..	8	..	..
Potentilla erecta . .	H	..	8	..	..
Salix repens . . .	N	..	..	..	4
Scirpus caespitosus . .	H	..	4	8	12
Vaccinium uliginosum . .	N	..	..	..	4
„ vitis-idaea . .	Ch	..	..	36	..

compared by means of the valency numbers. In these tables the species are recorded either in alphabetical order or according to their taxonomic position, excepting those of the highest degree of frequency, which are always placed at the top of the list so that the species characteristic of the formations occur together. The one or more species, which are dominant —either by the number or size of their individuals or by both—are, as is well known, commonly used to denominate formations. They can be brought into prominence by using heavy type for their valency numbers (Table 8) as in earlier publications (Chapters VI and VIII). In Chapter VIII I have thus brought into relief all the species whose degree of frequency equalled or surpassed 60, or, in other words, those which appeared in more than $\frac{3}{5}$ths of the total number of samples. All the species having a degree of frequency of more than 60 are here placed first in the table; but only those whose degree of frequency exceeds 80 have the corresponding numbers in heavy type. We shall deal later on with the motive which led to this usage.

II. THE LAW OF DISTRIBUTION OF FREQUENCIES

Since the frequency of a given species expresses itself by the ratio of the total number of samples examined to the number of those which contain the species in question, the percentage figure of the frequency will always be represented by a number between 1 and 100. We shall now inquire to what degree these percentage figures grouped into frequency classes will follow a general law.

Since formations are most often characterized by the predominance of one or several species, each formation will contain as a general rule at least one species of high frequency. But the examination of samples resulting from statistical analyses of one or many formations enables us to establish the presence of a sufficiently large number of species showing low percentage figures (compare Table 8); at first I was led to believe that a curve representing the distribution of the percentages in different classes would reach its maximum in the first group, that is to say, in the class containing the lowest frequencies, and that from that point it would fall more or less rapidly towards the last group containing the highest frequencies which would thus probably be the least numerous. But this turns out not to be the case.

In order to elucidate the question, I have put together all the frequency numbers which up to now have been established by my method, in which samples of $\frac{1}{10}$th of a square metre have been used for the statistical analysis of formations. Before considering the grouping of these figures it will be well to cite in chronological order the works hitherto published on this subject.

1. C. Raunkiaer. 1909, II. This memoir deals with a number of Danish plant formations, forests, fields, meadows, peat-bogs heaths, dunes. Most of them show quite a homogeneous composition; only a few have a collective character, especially among the meadows (Chapter VI).

The total number of samples in each formation is 50; the number of frequency figures about 1,350.

2. M. Vahl. 1911. Many formations of Phanerophytes and Chamaephytes, mostly in Sweden (Blekinge, Småland), some in Denmark (north-east of Zealand); in almost every formation the author took 50 samples. About 1,100 frequency numbers in all.

3. Hanna Resvoll-Holmsen. 1912. These researches are on coniferous forests of the high mountain regions of Norway, especially on the vegetation of the forest floor, on heaths ('Lyngmark'), plateaux ('Fjeldmark'), pastures, peat moors, of the regions in question. Number of samples in each formation: 50. Number of frequencies of vascular plants about 650. In many formations the author has also taken into consideration the mosses and lichens.

4. M. Vahl. 1912. Formations of Phanerophytes and Chamaephytes covering the little island of Notö situated in the lake Torsjö in Småland (Sweden). 50 samples in each formation; about 160 frequency numbers.

5. C. Raunkiaer. 1913. Formations of the northern point of Jutland (heath, marsh, meadow, cultivated land, dune). 25 samples in each formation; about 700 frequency numbers (Chapter VIII).

6. M. Vahl. 1913, I. Many formations of Chamaephytes and some of Phanerophytes, all in Lapland. 50 samples in each formation; about 550 frequency numbers.

7. M. Vahl. 1913, II. Various formations found in the Swedish marshes. 50 samples in each formation; about 550 frequency numbers.

8. Hanna Resvoll-Holmsen. 1914, I. Formations occupying the bottom and slopes of the valley Maalselvdalen in Norway about 69° N. lat. Principally forest floor formations, but also formations of meadow and marsh; and others inhabiting the river edge (Maalselven), where the vegetation is more or less open. Usually 50 samples in each formation; in some only 25. About 800 frequency figures relating to vascular plants. For some formations mosses and lichens are included.

9. Hanna Resvoll-Holmsen. 1914, II. Formations of the forest floor of coniferous and deciduous trees, and of heaths, 'Fjeldmark', 'snow patches'[1], pastures, peat-moors. In some cases 50 and in others 25 samples in each formation. About 1,550 frequency numbers relating to vascular plants, besides mosses and lichens.

10. C. Raunkiaer. 1914. Dune and sandy marsh of the French Mediterranean coast. 25 samples in each formation; about 60 frequency numbers (Chapter IX).

11. Carsten Olsen. 1914. Formations in *Sphagnum* bogs in the north-east of Zealand. Generally 25 samples in each formation. About 500 frequency numbers of vascular plants, besides mosses and lichens.

[1] Areas where snow lies late and which then support a characteristic vegetation.

From this short summary of the literature relating to the statistics of formations, it appears that some deal not only with vascular plants but also with mosses and lichens, while others are restricted to the former. For the moment we shall confine ourselves to these, and later on shall consider how far the cryptogams in question behave like the vascular plants from the point of view of the distribution of their frequency numbers.

In order to ascertain this distribution I have arranged the frequency numbers relating to the vascular plants in 5 classes in each of the eleven memoirs cited. Class I includes the frequency numbers (F) from 1 to 20; Class II, from 21 to 40; Class III, from 41 to 60; Class IV, from 61 to 80; and Class V, from 81 to 100. I should have preferred 10 classes to 5, but since for many formations the frequency numbers are only based on 25 samples in each formation it is clear that in such cases the 10 classes would not have the same value, because the 25 samples would have to be arranged in groups of 3 and 2 alternately. For this reason I have had to content myself with 5 classes, and this division has the advantage of enabling one to form very readily a picture of the state of affairs.

Table 9 gives a summary of the results obtained. The first and second columns contain references to the memoirs from which the data are taken. The number in the third column represents for the vascular plants the number of frequency figures established in the memoir concerned, and the numbers in the five last columns show in percentages the frequencies distributed in the 5 classes. At the bottom of the Table, under A, the total number of frequency figures (8,078) from the whole eleven memoirs is brought together, with the percentages represented in each of the 5 classes. Finally, B gives the mean of the percentages indicated under the numbers 1–11 of the Table.

Although the number of frequency figures is very different in the eleven memoirs, varying from 61 to 1,544, it is clear that the figures obtained are approximately the same whether one calculates the mean from the sum of all the frequency figures, as is done in the line A of the Table, or whether one takes the mean of the proportional values of these figures as they are distributed in the different classes, as in line B.

If the numerical relations of the frequencies of the species constituting the formation are represented by a curve, this is seen to have two peaks: it begins with a relatively high peak, corresponding to the figures of lowest frequency, while the second peak, a much lower one, occurs in the highest group of frequency figures. Starting from the first peak, the curve falls to the group preceding that represented by the second peak, which means, in a general way, that the least frequent species of a formation are the most numerous, and that as the frequency figures ascend the number of species diminishes, rising anew in the last group containing the species which are most frequent of all; while the smallest group is composed of those which are somewhat less frequent than these last.

By studying the frequency figures in each memoir separately, it can be shown that the result summarized in Table 9 is not an effect of pure chance. The lines 1–11 in Table 9 show that the march of the figures through the 5 classes is everywhere sensibly the same as in the mean figures given under A and B; the only exception is No. 3, where the second summit of the curve is lacking, because the size of the 5th group, containing the highest figures of frequency, is not greater than but equal to that of the 4th group. Further, in No. 10 the 4th group is larger than the 3rd and in Nos. 5 and 11 these two groups (3 and 4) are of the same size.

TABLE 9

Distribution, in percentages, of the frequency numbers in the 5 classes.
(Vascular plants only)

	References to Memoirs.	Frequency figure.	Frequency classes.				
			I 1–20 per cent.	II 21–40 per cent.	III 41–60 per cent.	IV 61–80 per cent.	V 81–100 per cent.
1	Raunkiær, 1909, II	1,355	65	11	7	6	11
2	Vahl, 1911	1,120	55	12	7	6	20
3	Resvoll-Holmsen, 1912	656	75	10	7	4	4
4	Vahl, 1912	162	46	16	11	8	19
5	Raunkiær, 1913	722	49	16	8	8	19
6	Vahl, 1913, I	555	43	15	13	9	20
7	Vahl, 1913, II	555	35	14	10	9	32
8	Resvoll-Holmsen, 1914, I	822	60	15	9	6	10
9	Resvoll-Holmsen, 1914, II	1,544	57	15	11	8	9
10	Raunkiær, 1914	61	57	16	7	8	12
11	Olsen, 1914	526	43	13	10	10	24
A	1–11	8,078	55	14	9	7	15
B	Mean of the percentages of 1–11	..	53	14	9	8	16

Let us now consider a little more closely the exceptional case of No. 3. If in this memoir the 5th class is not greater than the 4th, this is because a certain number of formations are included which do not contain any species whose frequency can enter the 5th class. The causes of such frequency distributions are various; sometimes it may depend especially on a phenomenon characteristic of tropical rain-forest, but a discussion of this we must leave on one side, since there are not in existence statistical researches relating to such forests, all the published work of this nature referring to the North Temperate Zone. In these countries the absence of very high frequency figures can arise especially from two causes: first, the formation in question may be too open to make it possible to obtain high frequency figures, however narrow the limits chosen to define the formation; this is the case, for example, in certain formations of dunes and deserts, whether these are due to dry climate or to cold; in the second place, the absence of high frequency figures may be due to giving too collective a meaning to the term formation. If, for example, one examines as a whole a heath made up of a formation of *Calluna vulgaris + Empetrum nigrum* and a formation of *Calluna vulgaris + Erica tetralix*, one will only have a single species of which the frequency figure comes into the 5th group, that is to say, *Calluna vulgaris*; but at the same time one would perhaps have two species of which the frequency figures belong to the 4th group, that is to say *Erica tetralix* and *Empetrum nigrum*. If, on the contrary, the two formations were examined separately—that is, true formations, not vegetations defined in rather a popular manner by their physiognomic aspect—each will show two species in the 5th group: in the one case, *Calluna vulgaris* and *Empetrum nigrum*, and in the other case, *Calluna vulgaris* and *Erica tetralix*; further, the formation *Calluna vulgaris + Erica tetralix* may present one species which will come into the 4th group, namely *Empetrum nigrum*.

Let us take another example. If we examine *en bloc* the floor of a beech forest bearing a formation of *Anemone nemorosa* and a formation of *Oxalis acetosella* with *Anemone nemorosa* intermixed, it may easily happen that the 5th class may not be represented at all; on the other hand, if the two formations are examined independently we shall probably find that they will both have one species at least which comes into the 5th class, that is to say *Anemone nemorosa* and *Oxalis acetosella* respectively.

In general, in any statistical examination of the formations of the North Temperate Zone the absence of frequency figures of the 5th class will ordinarily show that several formations have been examined together, in other words that the term 'formation' has been taken in too vague and collective a sense.

In order that the 5th class may contain frequency figures in greater number than the 4th, it is clearly necessary that all the individual investigations in question should establish frequency figures of the 5th class, that is to say figures exceeding 80; this condition being fulfilled, a distribution of the frequency figures like that given in line A of Table 9 will result, at least if the formations considered are not too few; in each formation taken separately the 5th class is not always more comprehensive than the 4th.

It is true that among the results summarized in Table 9 there are some which represent formations that have no frequency figures coming into the 5th class; but usually these isolated exceptions will be too insignificant to mask the law of distribution of frequencies which we have succeeded in establishing. Only in No. 3 in Table 9 are the two last classes equal. If one omits all investigations which do not show frequency figures in the

5th class, the remaining data will always show the number in the 5th class exceeding that in the 4th. This is clear from Table 10.

TABLE 10

Resvoll-Holmsen, 1912 (Table 9, No. 3).	No. of frequency figures.	Frequency classes.				
		I 1–20 per cent.	II 21–40 per cent.	III 41–60 per cent.	IV 61–80 per cent.	V 81–100 per cent.
A. All the formations examined	656	75	10	7	4	4
B. Formations which present frequency figures exceeding 80	488	74	9	6	4	6

Let us now examine the behaviour of the mosses and lichens, which we have omitted from the preceding account because most of the publications relate only to vascular plants and because one cannot tell *a priori* if the law of distribution which holds for these will also apply to the lower plants.

Only Hanna Resvoll-Holmsen (1912; 1914, I; 1914, II) and Carsten Olsen (1914) have determined the frequency figures for the mosses and lichens as well as those for the vascular plants in a series of formations. Since the formations are characterized and delimited by the vascular plants they cannot be considered as formations of mosses and lichens; in many formations the frequency figures of the species of these two classes never exceed 80. For this reason, in discussing the data in question it has seemed proper to calculate the frequency figures of the mosses and lichens respectively, both from the data relating to all the formations and also from those alone which show figures exceeding 80.

TABLE 11

Distribution, in percentages, of the frequency figures in the 5 classes among the mosses.

Resvoll-Holmsen, 1912; 1914, I; 1914, II.	No. of frequency figures.	Frequency classes.				
		I 1–20 per cent.	II 21–40 per cent.	III 41–60 per cent.	IV 61–80 per cent.	V 81–100 per cent.
A. All the formations . .	321	69	13	9	5	2
B. Formations which present frequency figures exceeding 80	73	49	23	6	4	18

The results relating to the mosses are brought together in Tables 11 and 12, the first giving the figures obtained with data from the three memoirs of Resvoll-Holmsen, while the second relates to those of Carsten

Olsen. It will be seen that the mosses obey the same law as the vascular plants.

TABLE 12

Distribution, in percentages, of the frequency figures in the 5 classes among the mosses.

Carsten Olsen, 1914.	No. of frequency figures.	Frequency classes.				
		I 1–20 per cent.	II 21–40 per cent.	III 41–60 per cent.	IV 61–80 per cent.	V 81–100 per cent.
A. All the formations	463	59	14	10	6	11
B. Formations which present frequency figures exceeding 80	227	50	16	8	5	21

As to the lichens, the memoir of C. Olsen has been left out of account because the frequency figures he gives for this class are too few. In Table 13 are given the results for the lichens from the memoirs of Resvoll-Holmsen, and these results tend to establish the same law which governs the higher plants.

TABLE 13

Distribution, in percentages, of the frequency figures in the 5 classes in the lichens.

Resvoll-Holmsen, 1912; 1914, I; 1914, II.	No. of frequency figures.	Frequency classes.				
		I 1–20 per cent.	II 21–40 per cent.	III 41–60 per cent.	IV 61–80 per cent.	V 81–100 per cent.
A. All the formations	505	57	13	9	7	14
B. Formations which present frequency figures exceeding 80	289	46	13	9	7	25

If we designate the 5 classes of frequency by the letters A, B, C, D, and E, the law of distribution of the frequency figures can be expressed thus:

$$A > B > C \gtreqless D < E$$

Usually one finds that $C > D$; but in Table 9 Nos. 5 and 11 show $C = D$, and No. 10, $C < D$.

In using the investigations in which 50 samples of each formation have been taken, I have employed a division into 10 classes to determine as far as possible the situation of the lowest point of the distribution curve. In this way I have established that in most cases this point coincides with the group 61–70; but it may occur in the group 71–80, or in the group 51–60.

The law of distribution of frequency numbers thus established can be explained, I believe, as follows. The dominant species in a formation which has attained a state of equilibrium are the best fitted to live in the conditions of existence presented by the formation of which they form part, and as the result of competition they hinder other species from equalling them in frequency. Meanwhile, although better equipped to sustain this competition with success, they cannot hinder less frequent and more scattered species from introducing themselves into the formation and occupying the lacunae left for any reason by the dominant species, and it is for this reason that we find the least frequent species are much the most numerous.

Besides the statistical researches quoted above, the results of which I have used to determine the distribution of frequency numbers, there is another investigation to which I have applied my method of valency but in which I have not been able to use the results at the same time as the others because the unit of area employed is not the same, being usually $\frac{1}{2}$ sq. metre instead of $\frac{1}{10}$ sq. metre. The investigation to which I allude is *Markflorans analys på objektiv grund* by Torsten Lagerberg (Lagerberg, 1914). Lagerberg is concerned with forest botany, the examination of wooded areas—sharply defined for the purpose of these investigations—in so exact a manner that by comparing the result with that of an analogous examination undertaken at the end of a certain number of years it will be possible to establish if in the interval there has been a notable modification in the vegetation as a consequence of a thinning undergone by the population of the forest. With this object Lagerberg proceeds as follows. He uses samples each having an area of $\frac{1}{2}$ sq. metre (in some cases only $\frac{1}{10}$ sq. metre) separated by intervals of 2, 4, or 8 metres; for each sample or unit of area he then estimates approximately the degree in which each species covers the soil, so as to determine both the frequency of the species and the relative areas they occupy. I have tested how far the frequency numbers obtained by this method correspond with those resulting from samples of $\frac{1}{10}$ sq. metre. Where Lagerberg has made two records of samples separated by different distances I have only used the result of the investigation of the greatest number of samples. The investigations limited in this way number 8 and comprise mosses and lichens besides the vascular plants, but the lichens are so few that I have omitted them.

Although Lagerberg was not concerned with studying sharply limited formations but rather wooded areas having a certain definite extent, one can establish that most of his investigations—7 out of 8—comprise species of which the frequency number exceeds 80, which would seem to indicate that the areas explored can be regarded as formations in the collective sense of the term; for the rest, it is evident that the larger size of the units he employs has a certain influence on the result. Each investigation comprises more than 100 (168–336) samples; consequently in many cases the frequency number of a given species is less than 1.

The results of my calculations for the vascular plants are assembled in Table 14, and for the mosses in Table 15. Under A all the investigations and all the species are included; in B all the species with frequency number less than 0·5 are excluded; in C only the investigations (7 out of 8) containing some frequency number exceeding 80 are included. It appears from the two tables that even in these researches the frequency figures conform to the law of distribution previously established.

TABLE 14

Distribution, in percentages, of the frequency figures in the 5 classes of vascular plants. (Unit of area: ½ sq. metre.)

Lagerberg 1914	No. of frequency figures.	Frequency classes:				
		I 1–20 per cent.	II 21–40 per cent.	III 41–60 per cent.	IV 61–80 per cent.	V 81–100 per cent.
A. All the researches and all the species	232	80	10	4	2	4
B. All the researches, but excluding the spp. with frequency figures less than 0·5	197	77	11	5	2	5
C. Only the researches (7 out of 8) containing some frequency figure exceeding 80 (for the others see B) .	164	79	10	4	2	5

TABLE 15

Distribution, in percentages, of the frequency figures of the mosses in the 5 classes. (Unit of area: ½ sq. metre.)

Lagerberg 1914.	No. of frequency figures.	Frequency classes.				
		I 1–20 per cent.	II 21–40 per cent.	III 41–60 per cent.	IV 61–80 per cent.	V 81–100 per cent.
A. All the researches and all the species	100	60	14	3	8	15
B. All the researches, but excluding the spp. with frequency figures less than than 0·5 . . .	90	55·5	15·5	3	9	17

We have already seen how the frequency numbers established from samples of small area behave; we shall now consider the behaviour of the distribution of frequency numbers when larger units are employed, either within a formation in the narrow sense or in populations presenting a more varied aspect.

If one confines oneself to any definite formation, it is clear *a priori* that the larger the unit of area adopted the more numerous the species with high frequency figures will be. In every region of some little extent which is explored by the method of units of area, the species having high frequency numbers are correspondingly less numerous as the nature of the ground and its flora are more varied; and in any given region the species with high frequency numbers will be correspondingly less numerous as the dimensions of the units of area are smaller. If, for example, we wish to determine the frequency numbers of the plants inhabiting Denmark and we divide the whole country into a certain number of regions—let us say 50—of equal or approximately equal extent, and then assign to each species a frequency number corresponding to the number of regions it inhabits, it is beyond doubt that the species with high frequency will be in the majority, but as the extent of the regions diminish the number of species of high frequency will also diminish, and finally there will be none at all.

Briquet has proposed to divide the surface of a country into a certain number of squares each of 100 sq. km. and to indicate for each species the number of squares in which it is found, in order to determine the degree of frequency shown by the plant species of a country.

But so far as I know this procedure has never been carried out for any country. It has been remarked that when we are dealing with units of area of relatively high order it is scarcely necessary for their size to be absolutely identical; and if we content ourselves with approximately equal dimensions we find that there now exist in floristic literature fairly good data which can be utilized for investigations enabling us to establish how the figures of frequency behave. Data of this nature have been published for Finland (Sælan, Kihlman, and Hjelt, 1889), for Great Britain (Watson, 1883), and for Ireland (More, 1898). I have determined the distribution of frequency figures for the first and the last of these three countries.

Finland. In the summary of the distribution of vascular plants in Finland, the country is divided into 29 departments or territories. In the exposition which follows, one of these regions, Lapponia enontekiensis, which is geographically quite isolated, has been omitted: I have thus been able to divide the frequencies into 28, and again into 14, 7, 4, and 2 groups, whenever it seemed convenient for the purpose I had in view.

Instead of determining at first the frequency figure of each species by the percentage of departments where it has been found, I have thought it better to establish how many species only occurred in a single department, how many occurred in two, how many in three, and so on. I have counted as species some forms which in the list quoted figure as subspecies; in this way the total number of species is raised to 1,206, which are distributed by percentages into the 28 groups, as indicated in Table 16. We can see that the most numerous of these groups is the first, composed of species which only occurred in one of the 28 departments; indeed, this

TABLE 16

Frequency figures of the vascular plants of Finland, determined by the number of departments in which the species were found; distribution, in percentages, of the frequency figures in the 28 classes. Number of species: 1,206.

Frequency class	1	2	3	4	5	6	7	8	9	10
No. of species, per cent.	13	8	6	5	5	4	3	3	3	4
Frequency class	11	12	13	14	15	16	17	18	19	20
No. of species, per cent.	2	2	3	3	3	1	3	2·5	2	2·5
Frequency class	21	22	23	24	25	26	27	28		
No. of species, per cent.	2	2	2	2	2	2	3	7		

group actually comprises 13 per cent. of the whole of the species considered. Next comes the 2nd group, comprising 8 per cent., then the 28th with 7 per cent.; the 3rd group contains 6 per cent.; the 4th and the 5th 5 per cent. The percentages of the other groups (to the number of 22) vary between 4 and 2, except group 16, which only includes 1 per cent. of the species. The curve which can be constructed from these proportional figures presents a rather irregular trace; but if one combines the groups in fours so as to reduce them to 7 in number, one obtains the result obtained in Table 17, and the curve becomes very regular, having its peak in the 1st group, falling to the 6th group, and rising again a little in the 7th and last group; so that in this case we have established practically the same trace as that presented by the frequency figures of the formations.

TABLE 17

The same data as in Table 16, but with the distribution of the frequency figures in 7 classes.

Frequency classes	1	2	3	4	5	6	7
No. of species, per cent.	31	16	12	10	10	8	13

Ireland. In the work entitled *Cybele hibernica* (More, 1898), Ireland is divided into 12 territories, and on pages lxxvii to xcvi there is a list of species indicating the territories in which each species occurs. If now we group the vascular plants of Ireland, to the number of 1,071, into 12 frequency classes, we shall obtain the result shown in Table 18 A, while if we divide them into 6 classes we shall arrive at the figures in Table 18 B. The curve which can be constructed with the help of these last values shows the same regular course with two peaks as that based on the figures of Table 17, but with this difference, that in the Irish figures the summit corresponding to the last frequency class is much higher than that on the first class, and that the lowest point of the curve does not correspond with the last class but one, as in Table 17, but with the second class. The territories into which Ireland has been divided are of much smaller area than those of Finland—in Table 18 B they are about one-quarter the size of those of Table 17—and the decrease in size of the regions gives rise, under conditions otherwise equal, to a displacement of the peaks from the right to the left within the frequency classes; if nevertheless the highest peak of the curve of distribution of the Irish frequency figures is still situated to the extreme right, differing in this respect from that of Finland, this is because the flora of Ireland is much less varied than that of Finland.

TABLE 18

Frequency figures of the vascular plants of Ireland, determined by the number of territories in which the species occur; distribution in percentages of the frequency figures; in A, 12 classes; in B, 6 classes.

(Number of species: 1,071)

Frequency classes		1	2	3	4	5	6	7	8	9	10	11	12
No. of species, per cent.	A.	8·1	7·5	4·2	3·9	4·5	4·5	4·7	4·6	5·7	5·1	7·5	39·2
	B.	16		8		9		9		11		47	

It is necessary to say that in determining the frequency figures of species I have only taken into consideration the works cited above, without using the more recent data concerning the wider distribution of certain species; it is evident that the establishment of such wider distributions of a certain number of species will give rise to a corresponding displacement from left to right in the series of frequency classes.

Following the discussion of the frequency figures based on our actual knowledge of the presence or absence of species in each territory forming part of a definite floristic domain, I shall give an example of the subjective estimation of frequency figures. As is well known, it has for a long time been customary to designate the extent of the distribution of any given species of a flora, the greater or less frequency with which it appears, by terms expressing an approximate personal estimate, such as 'very widespread', 'wide-spread', 'here and there', 'rare', &c. Some authors, notably S. Aubert in his work *La Flore de la Vallée de Joux* (Aubert, 1900), have also tried to express their estimates by numbers. Since the territory explored by this author—the valley of Joux in the Jura—has only an area of 260 sq. km., it was relatively easy to acquire a profound knowledge of the distribution of the species, and it is perhaps for this reason that the author believed himself justified in increasing the number of degrees of frequency and also the number of degrees of abundance to 10.

In the list of species found in this region (pp. 646 to 726 of the work cited), a list which comprises 827 species of vascular plants, there have been entered under the heading 'Degree of frequency', in respect of each species, two figures, of which the first represents the degree of frequency of the species, the second the degree of abundance; the smallest frequency is expressed by the number 1, the greatest by 10; and the same for the degrees of abundance. I have investigated how the 827 degrees

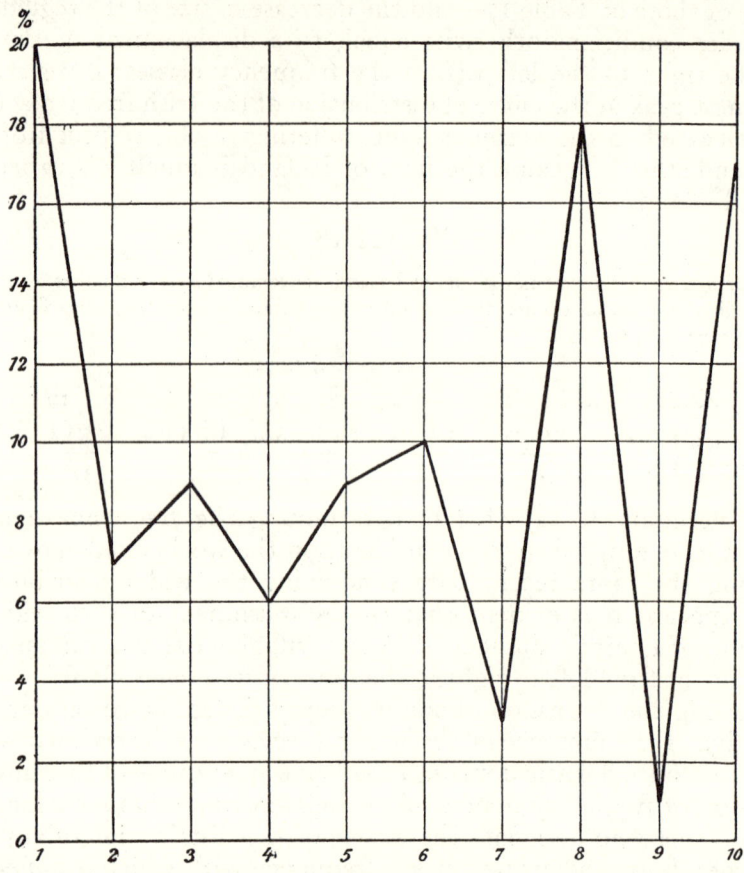

Fig. 140. Graphical representation corresponding with Table 19. Abscissae = frequency classes; Ordinates = percentages.

of frequency obtained by this method are distributed in the 10 frequency classes; and the results are shown in Table 19 (p. 403) and represented graphically in Fig. 140. The first thing which strikes us is the very low figure of the 9th class, in which only 1 per cent. of the whole number of species falls; secondly, in the 10th class (containing the most frequent species of all) the number of species is 17 times as great; and in the 8th class it is even 18 times as great. At the same time we remark that the number of species composing the 7th class is very low in comparison with the class which immediately follows it; in fact the species of the 8th class are $4\frac{1}{2}$ times more numerous than those of the 7th and the 9th taken together. It appears improbable that these proportions correspond to the reality of the facts and represent the true frequency of the species: the explanation must be found in the psychology of personal estimation. In fact as long as one tries to judge low degrees of frequency and can still

TABLE 19

Vascular Plants of the Valley of Joux (Aubert, 1900).
Percentage distribution of 10 classes of frequency numbers determined by estimation. Area: 260 sq. km. (Total number of species: 827.)

Frequency classes	1	2	3	4	5	6	7	8	9	10
Distribution of species, per cent.	20	7	9	6	9	10	3	18	1	17

to a certain extent include at a glance the proportional representation of different species, one is inclined to feel sufficiently sure of the fact, and this may have as a consequence that one will prefer uneven numbers, which express, so to say, a greater subjective certitude; at least it appears strange to me that in the 3rd class the number is greater than in the 2nd, and that in the 5th it is greater than in the 4th. On the other hand, one feels less certain in judging the higher degrees of frequency, so that, for example, when one has to decide whether a widely distributed species ought to be put in the 9th frequency class or in the 8th or 10th, one generally finishes by putting it in the 8th or 10th, the even numbers producing rather the effect of 'round numbers', and thus expressing better an estimate which is only approximate. In this way I should explain the deviations which the figures of Table 19 show compared with those of Tables 16 and 18, which do not depend on subjective estimation. Fig. 141 seems to confirm this point of view: the curve AB here represents simply the classes of Table 19 which bear uneven numbers; similarly, the curve MN is constructed from the classes with even numbers. It is easy to see that these two curves show almost diametrically opposite traces. The same thing applies to degrees of abundance, as is shown by Table 20, of

TABLE 20

Vascular plants of the Valley of Joux (Aubert, 1900).
Percentage distribution of the 10 classes of degrees of abundance determined by estimation.

Classes of abundance	1	2	3	4	5	6	7	8	9	10
Distribution of species, per cent.	10	8	15	5·5	10	6	3	10	0·5	32

which the figures make the same fact clear; but with this difference, that the number of species assigned to the 9th class is still less—only 0·5 per cent.—while the 10th class comprises 32 per cent. and the 8th class 10 per cent., there being 268 species in the 10th class, 84 in the 8th, and only 4 in the 9th.

This proves at the same time two things: the first is the well-known fact that we cannot trust personal estimates; and the second, that when we have to be content with a classification of this kind a scale with too great a number of degrees, such as 10, should never be employed; a restricted number of groups, such as 5, is greatly preferable.

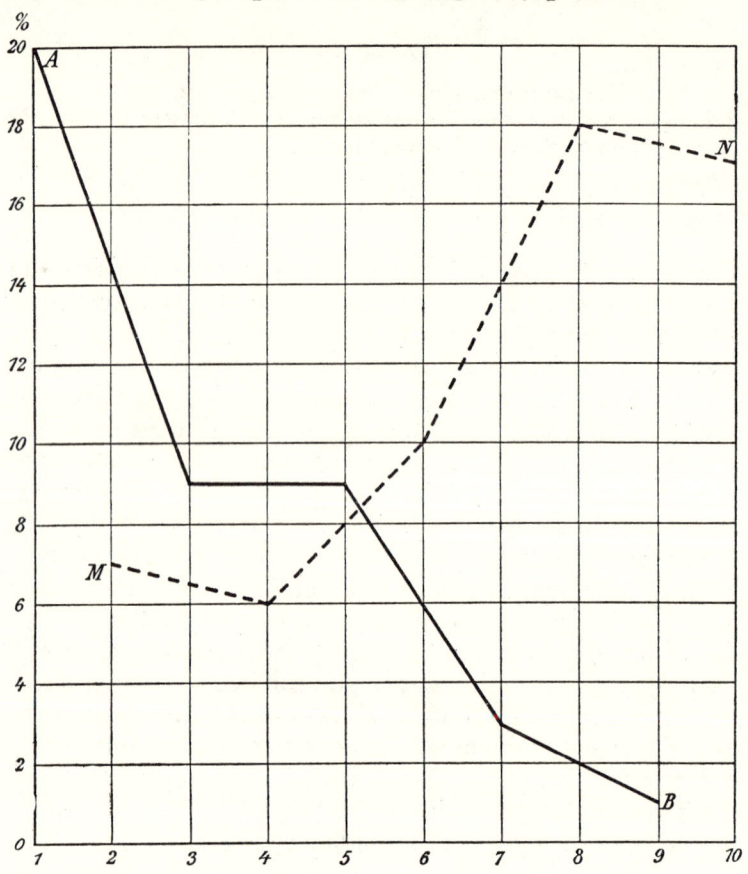

FIG. 141. Graphical representation corresponding with the elements of Table 19. The abscissae are frequency classes, the ordinates are percentages. *A—B*: classes with uneven numbers. *M—N*: classes with even numbers.

III. THE DEGREE OF FREQUENCY AS A BIOLOGICAL CHARACTERISTIC OF PLANT FORMATIONS

The climate of a country, in the phytological sense, is characterized by the adaptation of the plants to the climatic conditions, or in other words by the life-forms of the plants (see Chapter II, The Life-forms of Plants and their bearing on Geography).

In order to characterize the vegetational climate or plant climate of a country by means of the way in which plant species are distributed in the system of life-forms, it is necessary to determine first of all the life-form of each species represented: evidently a small number of species of a flora cannot suffice to give an exact expression of the adaptation of the plant carpet to the climatic conditions, even if for any reason certain of the species are to be regarded as particularly characteristic of the flora under consideration. If in order to make a choice of representative plants one takes up points of view which may be very definite and very essential, but are arbitrarily chosen, it will be difficult to avoid arriving at conclusions more or less false: for all the species which grow under the conditions of a given climate must evidently be adjusted to that climate—otherwise they would not exist in that position—and consequently they have essentially the same right to be included in the list. Thus if we wish to characterize the phytoclimate of a country with the help of a limited number of the species which constitute its flora, we must not set out from any preconception in making the choice, since a choice so determined will clearly be purely arbitrary. The surest method will naturally be to include all the species.

TABLE 21

Biological spectrum of the flora of Denmark.

Denmark.	No. of species.	S	E	M/M	M	N	Ch	H	G	H/H	Th
1. The whole flora	1,084	..	(0·1)	1	3	3	3	50	11	11	18
2. $\frac{1}{5}$th of the species, namely those whose last figure is 0 or 5	217	..	..	2	2	4	3	46	12	11	20
3. $\frac{1}{5}$th of the species, namely those whose last figure is 3 or 8	215	..	..	1	3	2	4	48	11	13	18
4. $\frac{2}{5}$ths of the species: those whose last figure is 0, 3, 5, 8 (= 2+3)	432	..	..	1	3	3	3	47	12	12	19

The considerations given above have led me to examine first of all the way in which the species making up the flora of a country are distributed in percentages among the classes of the system of life-forms, in order to characterize the phytoclimate. Thus I obtain a numerical expression which I call the 'phytobiological spectrum' of the region examined. As an example, the biological spectrum of Denmark is given in Table 21: first of all (1) that of the whole country, and then three examples (2–4) which show that a fraction chosen at random of the plant species of a country serves very well to establish the biological spectrum. In Nos. 2 and 3 the spectrum is based on $\frac{1}{5}$th of the species obtained as follows: in a complete and numbered list of the species of the Danish flora arranged systematically, I selected those whose numbers end in 0 or 5 (Table 21, No. 2), and those whose last figure is 3 or 8 (Table 21, No. 3). The spectra thus formed agree very well with that based on the entire list of species.

If the species of a flora are not known sufficiently well to determine to what class of life-form each species belongs, or if one has no need of a spectrum as detailed as that based on the life-form classes, one can be content with a spectrum indicating the percentage distribution of the species in the 5 series of life-forms: Phanerophytes (Ph), Chamaephytes (Ch), Hemicryptophytes (H), Cryptophytes (Cr), and Therophytes (Th). Table 22 contains 4 spectra of this kind, characterizing each of the 4 principal phytoclimates, first the tropical climate, constantly hot and constantly damp, that is to say the **phanerophytic climate**; secondly, the sub-tropical regions with winter rains, the **therophytic climate**; thirdly, the cold temperate climate, that of the **Hemicryptophytes**; and fourthly, the arctic climate, that of the **Chamaephytes**. By comparing these four spectra with the **normal spectrum**, that is to say the spectrum representing the entire flora of the world (see Chapter IV, pp. 115–23), it can be shown that each represents its own phytoclimate; a climate in the phytological sense is characterized by the life-forms of which the percentages exceed in marked degree those of the same life-forms in the normal spectrum.

TABLE 22
Biological Spectra

	No. of species.	Ph	Ch	H	Cr	Th
1. Seychelle Islands	258	**61**	6	12	5	16
2. Argentario	866	12	6	29	11	**42**
3. Denmark	1,084	7	3	**50**	22	18
4. Baffinland	129	1	**30**	51	16	2
5. Provisional normal spectrum (spectrum of the whole world)	..	47	9	27	4	12

By the aid of the biological spectra of a sufficient number of local floras, one can trace the limits of the phytoclimates by means of **isobiochores** analogous with the isotherms traced on the basis of observations made at meteorological stations. By **isobiochore** I understand a line traversing the regions of which the biological spectra are sensibly alike (see Chapter IV, pp. 117, 123 ff.).

In the determination of the phytoclimate of a country by means of its biological spectrum, the species which live under the conditions of this climate are of the same value, whatever their number or the stature of their individuals. If a given life-form predominates from the point of view of the number of species which belong to it, that is because this form is particularly suited to the climate under consideration; the number of individuals which any species possesses scarcely depends directly on the climate but primarily on its power of reproduction, and on the conditions of the ground on which it grows.

It is quite otherwise when one considers the flora from the point of view of the theory of formations: then it becomes the difference in the degree of predominance of the different species which is decisive. In order to arrive at a satisfactory biological spectrum for the flora of a country from this point of view, it would be necessary to take into consideration the differences of valency which the species present in the different parts of the country; but this is scarcely possible, since even for a country as small as Denmark one would not know how to determine, with however slight a degree of accuracy, the valency of each of the species which enter into the composition of its plant covering. Still less could this be done for the whole of the countries belonging to a given phytoclimatic region; for example the climate of the Hemicryptophytes of the North Temperate Zone, which extends from the Atlantic Ocean to the Pacific, both in the Old World and the New.

What can be done is this: among the formations which make up the domain of a phytoclimate we can determine the formation with the greatest extent, and in this way we can characterize the domain from the point of view of formations, provided that the area has not undergone changes brought about by man, for instance the felling of forests. Here a determination of the formation which predominates from the purely physiognomic point of view would have no fundamental importance. Physiognomic predominance depends especially on the dimensions of the plants. But the power of existence of a given species in a given climate is not a result of its larger size, rather the contrary is true: if the Phanerophytes, the trees, now form, or formed at some time in the past, the dominant formation in the forests of the North Temperate Zone, it is not that the phanerophytic life-form as such is particularly adapted to this climate: on the contrary, the predominance of trees is produced in spite of their phanerophytic characters, which are only moderately suitable, so that this life-form will succumb as soon as the surrounding conditions are changed for the worse—for example in places exposed to wind. The Hemicryptophytes will then show their superiority as the characteristic life-form of the climate under consideration. Consequently the dominant formation of a country is very often not at all identical with the life-form which characterizes the biological spectrum. Of the four principal phytoclimates indicated above there are only two, those of the Phanerophytes and of the Chamaephytes, where the life-form characteristic of the dominant formation is the same as that which characterizes the biological spectrum of the climate in question.

In the biological spectra of the phytoclimates, the life-forms dominate in number of species, while in the biological spectra of formations it is a question of the frequency of the species. The doctrine of the formations establishes the results brought about by the competition of the plant species, as conditioned by climate, in the different localities. A given locality will be occupied by the available species which are best fitted to the environment, and from this the formations result. Localities presenting sensibly identical conditions are often situated in countries very far removed from one another; also a formation generally comprises a number of small areas of more or less irregular form which it is impossible to trace on the map and whose extent cannot be measured. It is thus not possible to catalogue the plant species of a formation with the same certainty with which one can make a complete list of the species inhabiting a country which is well circumscribed. On the contrary, as we have seen in Section I, the valency of the species forming an isolated part of the formation can be determined with a high degree of precision. The figures of valency thus obtained make it possible to assign to each species an importance corresponding to the frequency with which it occurs in any given floristic sample; and the figures of valency offer an advantage still more important because when we are dealing with the biological characterization of formations, they serve as the means of translating the floristic units, species, into biological units, or life-forms. It is only thus that it becomes possible to make a comparison between formations which are close together from the biological point of view but very different floristically: for example, a Danish formation of *Empetrum nigrum*, an alpine formation of *Loiseleuria procumbens*, and an arctic formation of *Cassiope tetragona*.

To bring out more clearly the essence of the matter, we may refer to Table 2 (p. 385), which represents a formation composed of 15 species: 6 Chamaephytes, 6 Hemicryptophytes, 3 Geophytes (column 1). If we calculate the centesimal proportions existing between the life-forms of this formation, basing them exclusively on the floristic list, we shall obtain the spectrum of the formation given in Table 23, line 1, of which the figures give an entirely false idea of the real proportions. It will be quite otherwise if we convert the floristic units into biological units, that is to say into life-forms. Then we shall assign to the species the values of their valencies, that is their frequency numbers (see Table 2, column 2): the Chamaephytes show 284 units or points, while the Hemicryptophytes only show 88 and the Geophytes 24. Converted into percentages, these figures make the formation spectrum represented in Table 23, line 2, a result which approximates to reality, although the Chamaephytes have not yet attained the centesimal figure which rightly belongs to them.

Another example is furnished by the formation given in Table 5 and composed of evergreen Chamaephytes and deciduous Chamaephytes: 3 of the former (*Calluna vulgaris*, *Erica tetralix*, and *Empetrum nigrum*) and 2 of the latter (*Vaccinium uliginosum* and *Salix repens*), that is 60 per cent.

of evergreen Chamaephytes and 40 per cent. of deciduous Chamaephytes, figures far from reflecting the truth, if one wishes to take into account the physiognomic aspect as determined by the quantity of individuals. If on the contrary we calculate the value of the two groups by means of the valency figures, we shall have 96 per cent. of evergreens and 4 per cent. of deciduous plants, a result which is very close to the fact.

TABLE 23

The plants of Table 2.	Ch	H	G
	per cent.	per cent.	per cent.
1. According to the floristic list	40	40	20
2. According to the valency	72	22	6

As a last example of the use of valency figures, I take the relation between groups of species of which the leaves differ in size. The formation set out in Table 7 contains 5 evergreen Chamaephytes, of which 3 are Leptophylls (*Calluna vulgaris, Empetrum nigrum,* and *Erica tetralix*) and 2 Nanophylls (*Vaccinium vitis-idaea* and *Arctostaphylos uva-ursi*), that is respectively 60 per cent. and 40 per cent.; but if we calculate the relation by means of the frequency figures we arrive at 81 per cent. and 19 per cent., values which again approach closely to reality.

Nevertheless, the biological spectrum of the formation constructed on the basis of frequency figures does not always give a sufficiently correct expression of the formation in question. In determining the frequencies it is necessary also to attend to the degree of prosperity of each species, notably to that of the most frequent species. In fact, it may happen that a species which in two different localities shows the same very high frequency, even that expressed by 100, is in one of these localities on the point of succumbing, while in the other it is in full vigour: it is not only in a luxuriant and freely flowering formation of *Anemone nemorosa* that this species may show a frequency of 100 or nearly: the same frequency may be attained on a peaty soil exposed to wind where the individuals grow badly, so that only some succeed in flowering. This difference could not be expressed in the biological spectrum; it would be necessary to emphasize it in the description of the species examined, noting the prosperity, the stature, the flowering, the growth in more or less close masses —all in comparison with that which is normal for each species.

IV. USE OF THE DEGREE OF SOIL COVER OF SPECIES FOR THE PHYSIOGNOMIC CHARACTERIZATION OF PLANT FORMATIONS

Under the name of quality of a formation I have designated elsewhere (Chapter VIII, pp. 306–7) the whole of its floristic and biological characters, in other words the specific composition of the formation and the imprint imposed on the species by the conditions of the environment (the degree of frequency is nothing but one of the manifestations of this imprint). **Physiognomic** character is chiefly established by the aid of **quantity**. Meanwhile, even in units of area of very slight extent, the proportion in which each species occurs cannot be established in any absolute fashion: one cannot determine the annual production of each species either in weight or in volume. If the evaluation is made in the middle of a period of vegetation, the plant substance produced during the remainder of the period will necessarily be omitted; and on the other hand, if the determination is made at the end of the season, part of the annual production will have already disappeared. Furthermore, even if one is content with determining the relative proportion of the species as it presents itself at one or several given moments during the period in question, the work is so complicated that the procedure can scarcely be applied to ordinary investigations of the formations.

The preceding considerations led me already in 1913 (p. 307) to propose the application of a combined method of valency and judgement in order to obtain an adequate expression of quantity. This method consists in determining the quantity of each species in each of the samples employed to determine their valency or degree of frequency by an approximate evaluation made on a fixed scale from 1 to 5. In proposing this new procedure I started from the consideration that if it is difficult to evaluate even approximately the relative proportions of the plant species making up the whole of a formation and consequently occupying a more or less extensive area, it should be, on the contrary, very easy to arrive at this approximation if we only consider subdivisions of the area, each of $\frac{1}{10}$ sq. metre, one after the other. This method of division was afterwards used by Lagerberg in 1914, whose procedure consisted in estimating for each species its relative proportion expressed as the area covered by the species in question. This area was determined as the sum of the vertical projections of the aerial organs on the soil, the percentage of the whole area thus obtained for each species being afterwards determined. This proportion may be called 'degree of soil cover', or 'percentage area' (A per cent.)

In order to be included in frequency determinations, the species in question must have either rooted shoots or buds hibernating above the soil within the limits of the unit of area considered. In regard to perennial species, the determination only includes organs on which depends the persistence of the individual, its rejuvenation; frequency is independent of the degree of development which the organs of assimilation may have attained. It is not, however, the same for the determination of the degree of soil cover in relation to the extent of the whole area: here it is evidently necessary to include all the species which have organs situated above that part of the area limited in the manner indicated above, and that even if these organs are neither rooted in the soil nor furnished with perennating buds within the unit area; we are here dealing only with soil cover. Thus it is not necessary that the percentage of frequency calculated by means of the number of samples in which a given species covers the soil or part of the soil should coincide with that which represents the actual frequency of the species. This must be represented in another way, either by frequency calculated according to area, AF per cent., while the percentage of frequency in the strict sense is designated by F per cent. These two values may show considerable deviations. Thus a formation of *Petasites ovatus* showed in July AF per cent. = 100, while F per cent. = 61, the same frequency figure recorded in April, when the value AF per cent. did not exceed 61. One may, in fact, state generally that F per cent. remains sensibly the same for the whole period of vegetation in the same formation, while AF per cent. sometimes shows considerable variation. Nevertheless, the two quantities are in some cases approximately the same.

As regards the distribution and number of the unit areas, it will be necessary to follow the rules already adopted for the determination of frequency. The form and extent of the units, on the contrary, employed for the determination of the frequency numbers according to area, have a practical rather than a theoretical importance. In other words, the most convenient forms and sizes of the unit areas are the best; for, as opposed to degrees of frequency determined by means of the sizes of different samples, the absolute values which represent the degrees of soil cover can be directly compared. Meanwhile, since it is convenient to establish frequency at the same time as degree of cover, and since the size, $\frac{1}{10}$ sq. metre, adopted for my frequency determinations seems to be equally suitable for determining degrees of cover, I have employed a unit of area of the same form and the same extent as that previously used for the estimation of frequency, that is to say a circle with an area of $\frac{1}{10}$ sq. metre described by means of a rotating rod attached to a stick. The use of a rigid frame for the delimitation of the sample is inconvenient, especially when we have to determine the degree of cover; for the plants will be bent by the frame which, besides, easily becomes shifted in the course of the work, disturbances which may falsify the result. On the other hand, the delimitation of the circular sample with the aid of the rotating rod is constant owing to the stick being situated at the centre of the circle; and the rod can be fixed at any convenient height so as not to cause any disturbance of the vegetation. It is true that in estimating the degree of cover of each species the use of a square frame has the advantage that the area enclosed can easily be divided at will into two, four, or eight parts; perhaps it is for this reason that Lagerberg, who used a square frame, distinguished four degrees of cover, while I originally proposed a scale of 5. It is clear that for purposes of calculation a scale of 5 is preferable; and the difficulties experienced in judging the fifths and the tenths of a circle disappear if the apparatus is furnished with two radii marking off a sector corresponding to the area of the smallest degree of cover which one proposes to use. A convenient area for this purpose would be 0·01 sq. metre, that is to say $\frac{1}{10}$th of the unit represented by the whole area. Thus the apparatus for measuring these units should be constructed as shown in Fig. 142. The ring clamped to the stick has two

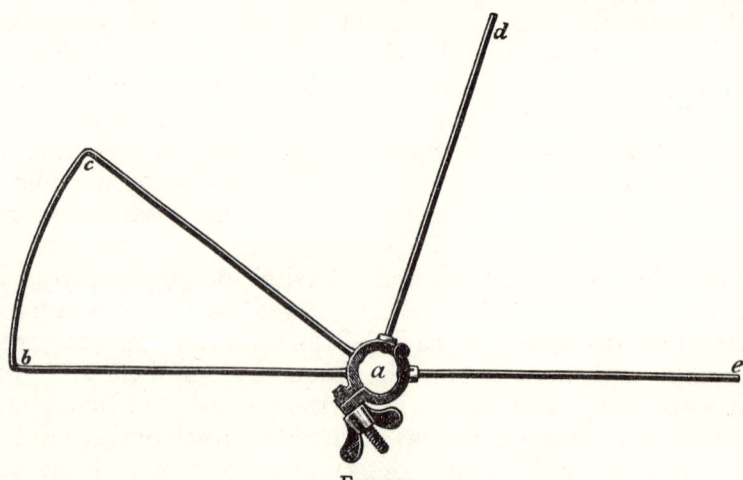

FIG. 142.

radial arms, ab and ac, rigidly soldered to it, and these, together with the arc bc, include $\frac{1}{10}$th of the area of the circle; by further jointing to the ring the two radial arms d and e, separated by $\frac{1}{5}$th of the circle, a division is obtained which is capable of furnishing measures corresponding to each of the ten degrees of cover. This arrangement facilitates the work of evaluation very markedly.

Thus we have 10 classes of degree of cover, constituting fractions of the circular unit of area, with the limits of classes and the values of classes as follows:

$$\text{Limits of classes} \quad \frac{0-1-2-3-4-5-6-7-8-9-10}{10}$$

$$\text{Values of classes} \quad \frac{|\ \ |\ \ |\ \ |\ \ |\ \ |\ \ |\ \ |\ \ |\ \ |}{1\ \ 3\ \ 5\ \ 7\ \ 9\ \ 11\ \ 13\ \ 15\ \ 17\ \ 19}$$
$$\phantom{\text{Values of classes} \quad}\overline{20}$$

Nevertheless it would not be at all convenient to use the numbers 1, 3, 5, ... 17, 19 to designate the degrees of cover. For this reason, in that which follows I have substituted for these numbers those which designate the limits of classes, so that a given degree of cover will be constantly represented by the figure indicating the limit of the class above the

STATISTICAL RESEARCHES ON PLANT FORMATIONS 413

degree in question. Thus a species of which the degree of cover is situated between 0 and 1—that is to say it covers less than $\frac{1}{10}$th of the area—and the mean value of whose class is consequently $\frac{1}{20}$th, will be represented by the figure 1 (that is to say $\frac{1}{10}$th); a species which in a given case covers more than $\frac{6}{10}$ths and less than $\frac{7}{10}$ths of the area and of which, consequently, the mean value is $\frac{13}{20}$ths, will be designated by 7 (that is to say $\frac{7}{10}$ths of the sample), and so on. Each determination will thus be raised $\frac{1}{20}$th, and it will eventually be necessary in calculating the frequency figures expressed in area of cover to use a consistent correction by deducting from the sum of the degrees of cover of each species as many twentieths as there are observations relating to the cover exercised by the species in question.

In order to avoid the use of numbers of two figures, the degree of cover 10 can be represented by ø or simply by 0. And to denote that a species which cannot be included in the determination of degree of frequency nevertheless has a certain degree of cover, one can add a sign, such as o, above the number representing its degree of cover.

As an example of the procedure that has been followed in determining the frequency numbers related to the area covered, let us choose the examination, summarized in Table 24, of the formation of herbaceous plants covering the humus soil of a dense beechwood situated in the northern part of the 'Dyrehave' near Copenhagen. We are dealing here essentially with a formation of *Anemone*, *Asperula*, and *Oxalis*. Each of the species which compose it, *Anemone nemorosa*, *Asperula odorata*, *Oxalis acetosella*, has a high frequency figure, that is to say 100; the other species have quite secondary importance. The formation was examined on the 4th June 1916, and 50 samples were taken. Each of these has its number marked in the Table; they are counted from left to right. The degree of cover for each species in any given sample is expressed by a number which represents tenths of this unit of area: the proportional extent which its aerial organs cover in vertical projection is determined by an approximate estimation for each species; thus, for example, in the 6th sample, *Asperula odorata* covers $\frac{2}{10}$ths, *Anemone nemorosa* $\frac{3}{10}$ths, and *Oxalis acetosella* $\frac{4}{10}$ths of the unit of area. As regards *Anemone nemorosa*, the total number of tenths covered in the 50 samples amounts to 109. Meanwhile, in calculating the surface that a given species occupies, that is its percentage area, in relative figures, we must remember that in using the limit of the next higher class as the designation of the degree of cover we obtain in each particular case an average value $\frac{1}{20}$th too high; consequently, in the present case, when the species occurs in each of the 50 samples, the sum—$\frac{109}{10}$ths— is too high by $\frac{50}{20}$ths. The frequency figure (A per cent.) of *Anemone nemorosa* will then be: A per cent. $= 2 \left(\frac{109}{10} - \frac{50}{20} \right) = 16 \cdot 8$. In the same way one has for *Asperula odorata* A per cent. $= 2 \left(\frac{156}{10} - \frac{50}{20} \right) = 26 \cdot 2$, and for *Oxalis acetosella*, A per cent. $= 2 \left(\frac{119}{10} - \frac{50}{20} \right) = 18 \cdot 8$. The other species play quite a secondary part: for *Melica uniflora*, A per cent. is $2 \left(\frac{7}{10} - \frac{7}{20} \right) = 0 \cdot 7$;

TABLE 24

Scheme for determination of the percentage area of a formation of *Anemone–Asperula–Oxalis*.

The figures in parentheses refer only to the determination of percentage frequency.

												Total tenths.	A %	F %	AF %
Anemone nemorosa	65322	33223	24131	21231	21222	32122	13132	12422	23241	11112		109	16·8	100	100
Asperula odorata	34524	23321	12241	44339	23315	23433	31341	35332	32522	24935		156	26·2	100	100
Oxalis acetosella	41132	44455	51512	11231	11451	23222	13122	42213	31214	11142		119	18·8	100	100
Melica uniflora	.....	.....	1...1	.1...	.....	.111	.....	...1	.....	.....		7 (3)	0·7	6	14
Urtica dioeca	.....	.....	...1	.....	.....	..1	.....	.....	.....	.....		2 (1)	0·2	2	4
Stellaria holostea	.....	.....	.....	1....	1..1	1....	...1	.....	1....	1....		7 (6)	0·7	12	14
Viola silvatica	.....	.....	.....	1....	.....	.....	..1	.....	.....	.....		2 (0)	0·2	0	4
Aira caespitosa	.....	.....	.....	.1...	.....	.....	.....	.....	.....	.....		1 (0)	0·1	0	2
Poa nemoralis	.....	.....	.....	.....	.....	.....	1....	.....	..1..	.....		2 (1)	0·2	2	4

for *Urtica dioeca*, A per cent. = 0·2, *Stellaria holostea*, 0·7, *Viola silvatica*, 0·2, *Aira caespitosa*, 0·1, *Poa nemoralis*, 0·2.

In the last column and that which precedes it, there are indicated respectively the AF per cent. and F per cent. of the different species. For the three dominant species the two quantities are identical; while those which are sporadic show divergences.

In the example cited above (Table 24), the sum of the percentage areas of all the species is 63·9, that is to say only about 64 per cent., or scarcely more than ⅝ths of the soil is covered by plants. If we had added the phanerophytic population, we should have obtained a degree of cover for the formation (AF per cent.) much higher, exceeding 150; that is to say the soil would be shown as covered something between once and twice. Many-layered formations may show an AF per cent. notably higher than this.

The determination of the degree of cover acquires quite a special importance when we are concerned with a deeper investigation of the question whether, on a given soil, the proportions of the plant species are changing in the course of time (succession of formations). In this kind of research, the application of degree of cover is easier in the formations of Chamaephytes, and more particularly of evergreens, where the annual variations undergone are least considerable. When one is dealing with formations of herbaceous plants, matters are somewhat complicated because, as a result of the difference of seasons marking the culminating points of development, there are many fluctuations in the relative proportions during the course of the vegetative period. Consequently, to arrive at a result which can be used in comparative researches, it is necessary to make the determinations of degrees of cover on the same aspect of development in different years.

In the present state of this line of work, the establishment of degrees of cover will be as a rule of little importance in ordinary researches on formations, because the determination of A per cent. requires too much time in relation to its value for the physiognomic characterization of formations. In fact, the degree of cover is not necessarily proportional to the mass: the same A per cent. may, for one and the same species, correspond to a very different quantity; a meagre covering formed, for example, by *Calluna vulgaris*, only 10 to 15 cm. high, may show the same A per cent. as a luxuriant vegetation of two or three times the stature. I hold that in such a case it is more practical to content oneself with adding to the frequency determination simply a note descriptive of the physiognomic aspect.

In the use of diverse physiognomic formations for the characterization of a landscape, I find it natural to employ the system of life-forms used for the biological characterization of formations, the system which I have also employed in the characterization and delimitation of phytoclimates.

This system of life-forms, while it is based on purely biological considerations, the adaptation of plants for passing the unfavourable season, is in fact clearly a physiognomic system—although that was no part of the original design—including the principal types which characterize the physiognomic aspect of the landscape.

SUMMARY

1. I use the word 'formation' to mean the fundamental unit of the science of plant formations, corresponding to some extent to the unit 'species' in the morphological classification of plants. Just as the species of the natural system are grouped into genera, and these into families, &c., so one can assemble more or less similar formations into a hierarchy of categories: groups of formations, classes of formations, branches of formations, &c.

2. A formation is a vegetation which is sensibly homogeneous in regard to its floristic composition and to the general character impressed upon it by the conditions of the environment. As a result of the struggle for existence, the formation is the biological expression of these conditions.

3. Wherever nature has been abandoned to herself for a sufficiently long space of time and where no marked modification has taken place in the external conditions, the vegetation will present a certain state of equilibrium, and as a result of the struggle for existence the different parts of the area will be occupied by those species inhabiting the region which are best qualified to live in the existing conditions and which thus come to constitute a formation.

4. No case has ever been established in which two or more species are completely alike from the point of view of their adaptation to conditions, and there is no reason to suppose that such a thing ever exists. Consequently, wherever vegetation is in a state of equilibrium, any difference that we can establish in the floristic composition will indicate a difference of conditions, and as a result, a difference of formation. Meanwhile, modifications of the conditions generally work in a continuous and progressive fashion, and correspondingly the change in the vegetation will, as a general rule, occur by gradual and imperceptible transitions, so that the number of formations is, practically speaking, infinite. But in order to recognize and take account of anything, so that we can compare it with something else, we are obliged to define it, in other words to delimit it; and that is why in my definition of the notion of 'formation' it has been necessary to add to the adjective 'homogeneous' the adverb 'sensibly', and by 'sensibly homogeneous' from the floristic point of view I understand that the vegetation with which we are dealing should be homogeneous in regard to its dominant species, that is those with the highest frequency.

STATISTICAL RESEARCHES ON PLANT FORMATIONS

The determination of frequency is carried out by means of the method of valency.

5. The **method of valency** is employed in researches on formations in order to assign objectively to each plant species a number expressing the valency which belongs to it in the formation. This number will serve, on the one hand, as a basis of direct comparison of formations which are similar from the floristic point of view, and on the other hand as a means to the conversion of systematic units, or species, into units of a different order (that is to say biological or physiognomic units). In this way we can obtain an exact comparison of formations which are essentially different.

6. In the **biological characterization** of formations we are principally concerned with the frequency and degree of prosperity of each of the species composing it, while the **physiognomic characterization** depends especially on their quantities and their relative proportions (taken together with their frequency).

A. Frequency

7. **In order to determine frequency**, a certain number of samples are selected from a definite area; the frequency is then expressed by the percentage number of samples in which the species in question is found, i.e. **percentage frequency (F per cent.)**.

8. The **necessary number of samples** will have been reached as soon as the result obtained becomes constant, that is to say when the F per cent. does not undergo any further sensible modification if one takes a greater number of samples. This result will depend on the area of each sample: the smaller they are, the greater will be the number that has to be taken; but, on the other hand, the more precise will be the degree of frequency established.

9. The most convenient **size or area of the sample**, i.e. unit of area, is that which will give the best result in the least time. As a result of preliminary trials, I have fixed on a unit of $\frac{1}{10}$th sq. metre, an area which seems to me sufficiently small to ensure results which are sensibly exact and at the same time not necessitating too great a number of samples to obtain constant results: 25 to 50 samples are sufficient.

10. When investigating a formation, that is to say a vegetation of sensibly homogeneous composition, the limits of the area as well as the method of distribution of the samples are indifferent: the samples can be taken at random or at definite intervals along a certain number of lines traced beforehand. If one is not sure that the composition of the vegetation is really sensibly homogeneous, in other words that one is actually dealing with a single formation, it becomes evidently necessary to determine exactly what one is dealing with, and this is most easily done

at the same time as the determination of frequency itself. For this purpose the samples are taken at regular intervals along lines traced beforehand, and a separate place in the scheme of notation is assigned to each sample. In this way it is possible to see immediately if in the course of sampling one has passed from one formation to another: such a passage will result in change of the frequency of species or in the occurrence of new species of high frequency.

When a vegetation composed of many formations is examined as a whole, the samples should be distributed uniformly over the whole surface of an area circumscribed beforehand.

11. The location and delimitation of the samples are carried out as follows: A ring is attached to a stick, and a thin metallic arm is attached to the ring perpendicularly to the length of the stick. This arm is of such a length that when the stick is fixed vertically in the soil and the arm is rotated about the axis of the stick, the free extremity of the arm will describe a circle with a surface of $\frac{1}{10}$th sq. metre.

12. The analysis of each sample is applied only to those species of which the living basal shoots are included within the limits of the sample and to those which possess perennial shoots or parts of shoots above the surface of the soil of the sample, even if they are not rooted within the area.

B. Degree of Prosperity

13. The imprint left by the conditions of existence is determined descriptively as well as statistically by the use of frequency figures as a link between the systematic and biological units to establish the valency of each adaptation of the formation. In this way one is also establishing the spectra of the formation, not only from the point of view of the adaptations of the species which enable them to survive the unfavourable season, but also of the other biological factors, such as xeromorphy (dimensions, structure, &c., of the leaves), pollination, dispersal of seeds, &c.

C. Height and Mass

14. The height or stature and the mass or quantity of the species of a formation are two very important elements from the point of view of its physiognomic aspect. As regards Chamaephytes and Phanerophytes, the height is expressed by the life-form (Nanophanerophytes, Microphanerophytes, &c.), and when we are dealing with formations of herbaceous plants, the height can easily be determined directly. On the other hand, it is difficult to obtain an exact expression of quantity, since the determination of weight and volume is impracticable, and, moreover, not in accordance with the physiognomic point of view. Consequently, in default of a better method, we have to be content with determining the mass by the help of the methods of valency and of

estimation combined, that is to say by evaluating the quantity presented by each sample taken for the determination of the frequency of the species. An important element from the physiognomic point of view of the quantity of a species is its degree of soil cover.

15. The degree of soil cover is determined by vertical projection and expressed by the percentage proportion of the area under consideration, i.e. percentage area (Lagerberg).

16. For the determination of percentage area (A per cent.) the extent of the units of area has no theoretical importance: the most convenient size of area is the best. While the percentage frequencies established on areas of different sizes are not directly comparable, percentage areas are comparable because they are absolute quantities. Nevertheless, even for the determination of percentage area, $\frac{1}{10}$th sq. metre is sensibly the best size, the more so as it is most convenient to determine frequency at the same time as degree of cover.

17. In regard to the distribution and number of samples, we need only repeat what has been said on the subject of the determination of frequency.

18. In the determination of percentage area, it is necessary to include all the species which have living aerial organs within the unit of area under consideration, even if they are neither rooted nor furnished with shoots or perennating parts of shoots. Contrasting this rule with those proposed on the subject of frequency determination, it will be seen that the percentage frequency which we can deduce from the number of units of area in which a particular species shows a certain cover does not necessarily coincide with the percentage frequency proper (F per cent.) of the same species and ought thus to have a different designation, such as areal frequency (AF per cent.).

19. To determine the percentage area (A per cent.), a certain number (such as 10) of classes of degrees of cover are adopted, and a given unit of area is represented as if it were divided into 10 equal parts. The number of these parts which are covered by the projection of a given species is then estimated approximately. The sum of the individual estimations will easily give the A per cent.

20. For the location and delimitation of the units of area, the same apparatus is employed as for the determination of frequency but with this addition, that the ring attached to the stick is furnished with two radial arms cutting off a sector whose area indicates the smallest degree of cover which it is desired to evaluate, that is to say the interval between successive classes of degrees of cover, or $\frac{1}{10}$th of the unit of area (= 0·01 sq. metre). If then two other radial arms enclosing $\frac{1}{5}$ of the circle are attached to the ring, one can obtain a measure of each of the 10 degrees of cover by the division of the circle, thus greatly facilitating the estimation (see Fig. 142, p. 412).

21. Thus we arrive at the following limits and values of the classes, determined as fractions of the circular unit of area:

Limits of classes $\dfrac{0 \quad 1 \quad 2 \quad 3 \quad 4 \quad 5 \quad 6 \quad 7 \quad 8 \quad 9 \quad 10}{10}$

Values of classes $\dfrac{1—3—5—7—9—11—13—15—17—19}{20}$

Nevertheless, both for notation and calculation, it is more convenient, in expressing the degree of cover, to use the numerical expressions of the limits of classes rather than those of the values of classes. For this reason, in the scheme of sampling a given degree of cover is expressed by the number indicating the limit of the class a b o v e the degree in question. Thus a species whose degree of cover is situated between 0 and 1—that is it covers up to $\frac{1}{10}$th of the unit of area—and of which consequently the mean value is $\frac{1}{20}$th, is expressed by 1 (that is to say $\frac{1}{10}$th); another species which in a given case covers between $\frac{6}{10}$ths and $\frac{7}{10}$ths of the unit and whose mean value is consequently $\frac{13}{20}$ths, is expressed by 7 (that is to say $\frac{7}{10}$ths of the unit), and so on. The numerical result of each determination will thus be $\frac{1}{20}$th too high, so that in calculating the percentage areas, it will be necessary by way of correction to make a uniform deduction from the sum of the degrees of cover of each species of as many 20ths as one has made observations of the cover exercised by the species in question. Let us suppose that one has examined 25 samples, and that the presence of a given species has been established in 22 of them; if the sum of its degrees of cover is, for example, 154, the percentage area of the species will be $4\left(\frac{154}{10} - \frac{22}{20}\right) = 57 \cdot 2$.

22. A very high percentage area means that the species in question not only presents a high degree of frequency in the formation under consideration, but that its growth is very thick. On the contrary, a very low percentage area does not enable us to draw conclusions as to the closeness or otherwise of the growth. Thus, for example, a percentage of 12 may be a result not only of high frequency with low soil cover in each particular sample, but also may be a consequence of low frequency associated with high soil cover. For this reason the percentage area (A per cent.) ought to be accompanied by the frequency area (AF per cent.), thus giving the mean degree of cover in each sample: p e r c e n t a g e p r e s e n c e (P per cent.) $= \dfrac{\text{A per cent.} \times 100}{\text{AF per cent.}}$. If A per cent. = 12 and AF per cent. 15, one will have P per cent. $= \dfrac{12 \times 100}{15} = 80$, which means that the species in question, while only appearing here and there, nevertheless presents a thick growth wherever it occurs.

23. For ordinary investigations on formations the percentage areas

will have comparatively little importance, since their value from the physiognomic point of view is too slight in relation to the time required to establish it. For the present we shall have to content ourselves with adding to the determination of frequency an indication descriptive of the value of each species from the point of view of the physiognomic aspect of the formation. On the other hand, it is of great importance to determine the percentage area when we are seeking the means of verifying if in the course of years slight modifications are occurring in the relative proportions of the species in the population of a limited and sharply circumscribed area.

D. The Importance of Formations for the Establishment of the Features of the Landscape

24. Climates in the phytological sense (phytoclimates) are characterized and delimited by means of the biological spectrum of the flora, which is based on the system of life-forms established on the element which is the most essential for the adaptation of the plants to the conditions of the environment, namely their adaptation for withstanding the unfavourable seasons.

25. The biological spectrum is an important side of the biological description of formations. With the frequency figures as mean terms, it is established on the same system of life-forms which serves to characterize the phytoclimates.

26. Besides having a fundamental biological importance, this system of life-forms also includes, though it was not originally intended to do so, the physiognomic types which characterize the landscape.

27. As a consequence, the spectra of formations have an importance from both points of view, the biological and the physiognomic: **biologically**, the formations are characterized by the life-form which numerically predominates, that is to say by its frequency; **physiognomically**, by the form which dominates by its mass.

28. In taking the system of life-forms as a basis, we shall obtain the classes of formations which are enumerated below in a serial order in which the adaptation for passing the unfavourable season increases, while, so far as terrestrial plants are concerned, the physiognomic predominance decreases. The Hemicryptophytes, Geophytes, and Therophytes, very different from the biological point of view, form a single class in respect of physiognomy.

1st Class. Formations of Megaphanerophytes.

A. Evergreen.   Examples.

   a. With large leaves (mega-microphylls)   The loftiest forests of tropical and subtropical rainy regions.

   b. With small leaves (micro-leptophylls)   The loftiest coniferous forests.

B. With deciduous leaves . . .   The loftiest Monsoon forests; forests of leafy trees in the temperate zone.

2nd Class. Formations of Mesophanerophytes.

A. Evergreen.
   a. With large leaves (mega-microphylls) — Tropical and subtropical rain forests of medium height (8–25 m.).
   b. With small leaves (micro-leptophylls) — Coniferous forests of medium height.
B. With deciduous leaves . . . Monsoon forests of medium height. Leafy woods of the temperate zone, of medium height.

3rd Class. Formations of Microphanerophytes.

A. Evergreen.
   a. With large leaves (mega-microphylls) — Low evergreen forests in the tropical, subtropical and temperate zone.
   b. With small leaves (micro-leptophylls) — Low forests of conifers, &c.
B. With deciduous leaves . . . Low forests in the tropical, subtropical and temperate zones; high scrub or brushwood.

4th Class. Formations of Nanophanerophytes.

A. Evergreen.
   a. With large leaves (mega-microphylls) — Palm scrub; certain forms of maquis.
   b. With small leaves (micro-leptophylls) — Maquis.
B. With deciduous leaves . . . Low scrub.

5th Class. Formations of Chamaephytes.

A. Evergreen.
   (Subdivided according to xeromorphy, including size of leaves) . . . Heaths.
B. With deciduous leaves . . . Formations of chamaephytic *Salix*, *Vaccinium*, &c.

6th Class. Herbaceous Formations (Hemicryptophytes, Geophytes, and Therophytes).

Subdivided principally according to the height of the perennial plants and according to xeromorphy . . . . . Hygrophilous and mesophilous meadows, vegetation of the forest floor, grassy steppes, &c.

7th Class. Formations of Helophytes.

A. High . . . . . . Formations of *Typha*, *Scirpus lacustris*, *Phragmites*, &c., &c.
B. Low . . . . . . Formations of *Carex*, *Equisetum*, &c., &c.

8th Class. Formations of Hydrophytes.

A. With floating leaves.
   a. Rooted . . . . . Formations of Nymphaeaceae, *Potamogeton*, &c.
   b. Free floating . . . . Formations of *Hydrocharis*, *Lemna*, &c.
B. Submerged.
   a. Rooted . . . . . Formations of *Potamogeton*, *Myriophyllum*, Characeae, &c.
   b. Not rooted . . . . Formations of *Ceratophyllum*, *Utricularia*, *Lemna*, &c.

The foregoing Table only gives the main divisions of the system of formations, and only relates to cases in which the biological and physiognomic aspects of the formation coincide, and are both well marked. We can also include in this framework formations characterized especially by distinctive features of a biological order and which demand rather a specially biological or ecological description of the fundamental units of the system in question, or, in other words, formations in the narrowest sense of the term.

29. Plant ecology has for its goal the explanation of the way in which each plant species behaves in the formation. It seems to attain this end on the one hand by deeper and deeper studies of the climatic conditions, as well as of edaphic conditions in their chemical, physical, and biological aspects; and, on the other hand, by not less detailed investigations of the nature and properties of plants in their morphological, anatomical, and physiological aspects.

BIBLIOGRAPHY

AUBERT, S. (**1900**), La flore de la Vallée de Joux. Bull. de la Soc. Vaudoise des Sciences naturelles, vol. **36**, 1900.

BRIQUET, J. (**1893**), Les méthodes statistiques applicables aux recherches de floristique. Bull. Herb. Boiss. **1**, 1893, p. 133 &c.

LAGERBERG, TORSTEN (**1914**), Markflorans analys på objektiv grund. Meddelanden från Statens Skogsförsöksanstalt, H. **11**, Stockholm, 1914.

MORE, A. G. (**1898**), Cybele Hibernica. 2nd edition, by N. Colgan and R. W. Scully. Dublin, 1898.

OLSEN, CARSTEN (**1914**), Vegetationen i nordsjællandske Sphagnummoser. Botanisk Tidsskrift, **34**. Copenhagen, 1914.

RAUNKIÆR, C. (**1905**), Types biologiques pour la géographie botanique. Académie Royale des Sciences et des Lettres de Danemark. Bulletin de l'année 1905, No 5. Copenhagen, 1905.

—— (**1907**), Planterigets Livsformer og deres Betydning for Geografien. Copenhagen, 1907. (Chapter II.)[1]

—— (**1908**), Livsformernes Statistik som Grundlag for biologisk Plantegeografi. Botanisk Tidsskrift, **29**. Copenhagen, 1908. (Chapter IV.)

—— Statistik der Lebensformen als Grundlage für die biologische Pflanzengeographie. Beihefte zum Bot. Centralbl. **27**, 2. Abt. 1910.

—— (**1909**, I), Livsformen hos Planter paa ny Jord. D. Kgl. Danske Vidensk. Selsk. Skrifter, 7. Række. Naturvidensk. og mathem. Afd. VIII. Copenhagen, 1909. (Chapter V.)

—— (**1909**, II), Formationsundersøgelse og Formationsstatistik. Botanisk Tidsskrift, **30**. Copenhagen, 1909. (Chapter VI.)

—— (**1911**), Det arktiske og det antarktiske Chamaefytklima. Biologiske Arbejder tilegnede Eug. Warming. Copenhagen, 1911. (Chapter VII.)

—— (**1912**), Measuring-apparatus for statistical Investigations of Plant-formations. Botanisk Tidsskrift, **33**. Copenhagen, 1912.

—— (**1913**), Formationsstatistiske Undersøgelser paa Skagens Odde. Botanisk Tidsskrift, **33**. Copenhagen, 1913. (Chapter VIII.)

—— (**1914**), Sur la végétation des alluvions méditerranéennes françaises. Mindeskrift for Japetus Steenstrup. Copenhagen, 1914. (Chapter IX.)

—— (**1916**), Om Bladstørrelsens Anvendelse i den biologiske Plantegeografi. Botanisk Tidsskrift, **34**. Copenhagen, 1916. (Chapter X.)

RESVOLL-HOLMSEN, HANNA (**1912**), Om Vegetationen ved Tessevand i Lom. Videnskabsselskabets Skrifter. I. Mat.-naturvid. Klasse, 1912, No. 16. Christiania, 1912.

[1] The Chapter numbers in parentheses refer to the Chapters of the present work.

RESVOLL-HOLMSEN, HANNA **(1914,** I), Statistiske Vegetationsundersøgelser fra Maalselvdaleni Tromsø Amt. Ibidem, 1913, No. 13. Christiania, 1914.

—— **(1914,** II), Statistiske Vegetationsundersøgelser fra Foldalsfjeldene. Ibidem, 1914, No. 7. Christiania, 1914.

SÆLAN (TH.), KIHLMAN (A. Osw.), HJELT (HJ.) **(1889),** Herbarium Musei Fennici. Sec. Edit. I. Plantae vasculares. Helsingfors, 1889.

VAHL, MARTIN **(1911),** Les types biologiques dans quelques formations végétales de la Scandinavie. Académie Royale des Sciences et des Lettres de Danemark, Extrait du Bulletin de l'année 1911, No. 5. Copenhagen, 1911.

—— **(1912),** The Vegetation of the Notø. Botanisk Tidsskrift, **32.** Copenhagen, 1912.

—— **(1913,** I), The Growth-Forms of some Plant Formations of Swedish Lapland. Dansk Botanisk Arkiv. Bd. **1,** Nr. 2. Copenhagen, 1913.

—— **(1913,** II), Livsformerne i nogle svenske Moser. Mindeskrift for Japetus Steenstrup. Copenhagen, 1913.

WATSON, H. C. **(1883),** Topographical Botany: The Distribution of British Plants. 2nd edition. London, 1883.

XII

ON THE BIOLOGICAL NORMAL SPECTRUM

THE existence of every plant is determined by the fortunate correspondence of its needs with the given conditions, that is the plant must be adapted to the conditions: the adaptation of a plant is a biological expression of the existing conditions of life. This is the starting-point of biological plant geography, which considers the adaptational phenomena of plants, that is the life-forms, as reactions to the conditions of life, and whose aim is, by this means, to delimit and to characterize regions within which the conditions of life (from the standpoint of the plant) remain in essentials the same.

In consequence of the complexity of biological relations great difficulties have to be overcome: conditions of life are variable both in space and time: composition and moisture of the soil differ from place to place, even within very small areas; and throughout the year the conditions vary in one and the same place, as do the essential elements of climate, viz. precipitation and temperature. In order to maintain itself the plant must be adapted to every change of conditions to which it may be exposed. Therein lies the cause of the separation of the two disciplines: the study of plant formations and the climatology of plants (biological plant geography in the narrower sense).

In the study of plant formations we investigate how plants are associated into communities (formations) in correspondence with the conditions of the habitat.

In the climatology of plants, on the other hand, we investigate how plants are adapted to the climate, changing as it does in the course of the year; and how they set about the task of maintaining themselves through the season unfavourable to growth—because this primarily determines their existence or non-existence in a given climate. Since the unfavourable seasons of various regions differ from one another far more than the favourable, the plant world also bears the stamp of adaptation to the former far more strongly impressed than to the latter. I have therefore chosen adaptation to the unfavourable season as the best method of limiting and characterizing plant climates.

The adaptations we have to consider in this connexion are of very diverse kinds, and even in one and the same plant there appear a whole series of adaptations which enable it to survive the unfavourable season: for example, the stem may be adapted in one way, the root in another, &c. Since we are dealing with a comparative investigation of the plant growth of various regions it is necessary to obtain a common basis of comparison: it is useless to employ in one case one condition and in another, another: the system of life-forms serving as a basis must therefore be constructed on a single but essential aspect of adaptation, and the uniform basis so chosen must be consistently carried through and must possess both essential significance and practical applicability. Not all parts of the plant are equally important for the maintenance of the existence of the individual, and not all are equally sensitive. Most important, and at the same time most sensitive, are embryonal tissues in the apices or buds of the shoot, for on their existence the maintenance of the individual rests, and we can perceive that there exist in the plant world a number of different methods by which these buds are protected and are thus able to survive the unfavourable season. In the first place a series of main types will be distinguished according to the position of the buds in relation to the soil and the amount of protection thus obtained. Thus they may be borne aloft in the air (Phanerophytes), near the soil (Chamaephytes), lodged in the earth (Hemicryptophytes), concealed below the surface of the earth (Geophytes), or at the bottom of water (Helo- and Hydro-phytes). In correspondence with these facts I have constructed a system of life-forms on the basis of the adaptation of the plant to survive the unfavourable season, that is to say with reference to the protection of the buds, on whose life the further existence of the individual depends (Chapters I and II).

If we now investigate the plant world in various climates with reference to the adaptation of the species to the unfavourable season, it appears that one life-form is predominant in one climate, and another in another, but it never happens that all the species belong to the same life-form. It is therefore necessary for comparative purposes to use statistical methods; and, since all the plant species of a definite region must be adapted to the climate of this region, all have the same right to be considered if we are seeking to characterize the region by means of the adaptation of its plant world to the climate. Thus one must first determine the life-forms of the single species and then show how the totality of species are distributed among the different life-forms. This percentage distribution of species among various life-forms I have named the biological spectrum, and I have shown that this spectrum actually gives a picture of the relation of the plant world to the climate, various regions with the same climate showing the same biological spectrum even if the floristic composition of their vegetation is wholly different. Correspondingly floras of different climates yield different biological spectra. I have therefore established the biological spectrum as a basis for the characterization and delimitation of plant climates (Chapters IV and VII).

The fact that the main principle of division of the system of life-forms is, and must be, established from a single point of view naturally does not exclude the possibility of taking into consideration, when so desired, any other adaptation within the limits of the framework constructed; but we must deal with the various adaptations single and methodically.

In other places (Chapters II and VI) I have given a synopsis of the chief types constituting the system of life-forms, and have argued that it is necessary on practical grounds and also on account of our defective knowledge of the adaptations of individual species in various climates, to classify the types of life-forms in larger groups and thus to simplify the system so that it can be used in the present state of our knowledge. In the delimitation of the ten classes of life-forms which can be used for the construction of the biological spectra of single regions this simplification has found various expressions. I have not for example been able to consider the distinction between deciduous and evergreen Phanerophytes. This distinction can only be employed in special researches relating to particular plant climates.

In Table 1 the classes of life-forms are set out as follows: with the exception of Stem Succulents (S), Epiphytes and phanerogamic parasites (E) and Therophytes (Th) the name of each class is given by the position in relation to the surface of the soil in which the life-form in question has its buds during the unfavourable season. On the right of the Table are the abbreviations which serve to distinguish the life-forms in the biological spectrum.

Table 1

	(Stem Succulents)	S
	(Epiphytes)	E
Above 30 metres.	Megaphanerophytes }	MM
,, 8 metres.	Mesophanerophytes }	
,, 2 metres.	Microphanerophytes	M
,, 0·25 metre.	Nanophanerophytes	N
Below 0·25 metre.	Chamaephytes	Ch
At surface of soil.	Hemicryptophytes	H
	Geophytes	G
	Helophytes }	HH
	Hydrophytes }	
	(Therophytes)	Th

If now, using these classes of life-forms as a starting-point, we determine the biological spectrum for a series of regions and then assemble the spectra so obtained as shown in Table 2, we immediately perceive the differences and similarities.

The eight spectra clearly fall into four groups so that they are associated in pairs, and each of these pairs, A, B, C, D, is entirely different from the others: A is conspicuous for a high percentage of Phanerophytes, that is of Micro- and Nanophanerophytes; B is characterized by a high percentage of Hemicryptophytes; C by a preponderance of Therophytes; and D by a high percentage of Chamaephytes together with a considerable proportion of Hemicryptophytes.

TABLE 2

		No. of species.	S	E	MM	M	N	Ch	H	G	HH	Th
A	St. Thomas and St. Jan	904	2	1	5	23	30	12	9	3	1	14
	Seychelles	258	1	3	10	23	24	6	12	3	2	16
B	Altamaha, Georgia, N. Am.	717	0·1	0·4	5	7	11	4·	55	4	6	8
	Denmark	1,084	..	0·1	1	3	3	3	50	11	11	18
C	Death Valley (N. Am.)	294	3	..	..	2	21	7	18	2	5	42
	Argentario (Italy)	866	..	..	2	4	6	6	29	9	2	42
D	Spitsbergen	110	..	..	..	..	1	22	60	13	2	2
	St. Lawrence I. (Alaska)	126	..	..	..	..	..	23	61	11	4	1

In the representation of these four typical plant climates which are illustrated in the biological spectra shown in Table 2, it is not necessary to use all ten classes of life-forms: the following four will suffice: Ph = Phanerophytes (S, E, MM, M, N), Ch = Chamaephytes, H = Hemicryptophytes, Cr = Cryptophytes (G and HH), and Th = Therophytes. The relations thus become considerably easier to perceive, as Table 3 shows.

TABLE 3

	No. of species.	Ph	Ch	H	Cr	Th
St. Thomas and St. Jan.	904	**61**	12	9	4	14
Seychelles	258	**61**	6	12	5	16
Altamaha, Georgia (N. Am.)	717	23	4	55	10	8
Denmark	1,084	7	3	50	22	18
Death Valley (N. Am.)	294	26	7	18	7	**42**
Argentario (Italy)	866	12	6	29	11	**42**
Spitsbergen	110	1	22	60	15	2
St. Lawrence Island, Alaska	126	..	23	61	15	1

Here it is easy to see what belongs together, but if we take into consideration a larger number of floras we shall certainly come across biological spectra which form transitions between the four groups of Table 3, and we have no criterion which would enable us to determine which of these transitional spectra are to be regarded as limiting spectra—e.g. between the Phanerophyte and Hemicryptophyte climates. We might take arbitrary limits—e.g. 40 per cent. of Phanerophytes, but we could not assert that the number 40 is more natural than, for example, 38 or 42, or some other number. We are in the same position as if we were dealing with isotherms. All isotherms are, as such, equally significant. If we choose a definite annual isotherm as a limit between two zones—e.g. between the tropical and subtropical zones—the grounds of our choice do not lie in the consideration of the isotherms as such, but in quite other circumstances, and our decision is arbitrary and lacks precision. In the delimitation of plant climates with the help of a biological spectrum it is very desirable to avoid arbitrarily estimated determinations. In order to do this we must seek a measure, a norm, with which the biological spectrum of each individual local flora can be compared and by means of which we can objectively decide where the limits between various plant climates should be drawn. It is clear that this common measure with which the biological spectra of individual parts of the earth's surface, that is of the individual local floras, may be compared can be nothing else than the biological spectrum of the entire flora of the surface of the earth, and this spectrum, that is the percentage relations between the life-forms of all the Phanerogams of the world, I call the normal spectrum.

We must now consider how such a normal spectrum can be established. More than ten years ago I took this work in hand and obtained a provisional normal spectrum which has already rendered good service (Raunkiær, 1908). Naturally it was not to be expected that all the Phanerogams of the earth's surface could be investigated and assigned to their proper life-forms within a measurable time; it was therefore necessary to adopt a method such as is used in similar cases where the whole of the material involved cannot be investigated but one has to be content with samples— e.g. in the determination of the percentages of various kinds of seeds in a large sample of a seed mixture of unknown composition. Here, however, I came upon the difficulty that there was no comprehensive descriptive list of Phanerogams of the whole earth which represented a conglomeration approximately as uniform and homogeneous as a mass of seeds which one can thoroughly mix by shaking. If there were available a catalogue of all the Phanerogams in which the different species were arranged alphabetically according to their specific names without reference to the genus it might be assumed that we should have such a mixture; but a catalogue of this sort does not exist. Nevertheless, we have in the *Index Kewensis* an alphabetical list of genera and the species arranged alphabetically under the genera; but, since the species of the same genus very often belong, all or in great part, to the same life-form, such a catalogue as the *Index Kewensis* is more like a mixture of seeds in which the species are on the whole scattered among one another, but nevertheless the seeds of one species often form larger or smaller clumps. If now the seeds of such a clump were stuck together so that we could not bring about a more homogeneous mixture of the material, we should have to take samples from various parts of the whole mass and investigate these separately. We have to proceed in a somewhat similar manner in taking samples from a catalogue of species like the *Index Kewensis*, when we are trying to use the life-forms of these species as the foundation for the construction of a normal spectrum which shall represent the percentage composition of the entire mass of species.

I decided, as a preliminary, to take one thousand species—that is, approximately one one-hundred-and-fortieth part of the entire number. If the species had been alphabetically arranged without reference to their genera, and if one had collected the thousand species at equal intervals through the entire catalogue, one would certainly have obtained a very good picture of the whole, but in the application of such a method to the *Index Kewensis* the great genera, all, or a large part, of whose species belong to one and the same life-form, would in this way scarcely obtain their due representation.

I therefore took groups of species with certain intervals between them and between the species of each group. I chose ten groups, each of one hundred species, the ten groups being so distributed in the *Index Kewensis* that the first began on p. 150, the second on p. 400, the third on p. 650, &c., that is at intervals of 250 pages, and I chose in each of the hundred columns beginning at the above designated starting-points the last cited species. For the sake of accuracy it may be remarked that by 'last cited species' is to be understood the last valid species according to the rules of the *Index Kewensis*: thus where the last species was cited as doubtful or of unknown origin then the last species but one was taken. It is very possible that another method of selection would have been better—e.g. one hundred groups each of ten species—but I must content myself provisionally with my first method. In most cases it was very easy to determine the life-form of the species selected, but for a certain number of species this could not be done conclusively since the data concerned were defective: thus, for example, it is possible that a Geophyte might be cited as a Hemicryptophyte, or vice versa. But these possible errors can scarcely have exercised a marked influence on the final result. We have no ground for supposing that they would have more effect in one direction than in the opposite; on the whole they would cancel out.

In the year 1908 I determined the life-forms of the first 400 species, and in the same year published the normal spectrum calculated from them (Chapter IV). This was later used as the provisional normal spectrum (Chapters V, VII, and IX), and is given in Table 4.

In the autumn of 1916 I determined the life-forms of the remaining 600 of the thousand species which I had originally chosen as the foundation for the construction of the normal spectrum. In this way I obtained the life-forms of the whole thousand species on the original plan and the result is given in Table 4 below that relating to the first four hundred species. Thus in Table 4 the differences and correspondences between the provisional normal spectrum of 1908 and the normal spectrum based on the whole thousand species can easily be seen.

In most points the correspondence is good. For E, Ch, and Th the numbers are the same in both spectra. In S, M, H, G, HH, the difference is only 1, in MM 2. On the other hand, the Nano-phanerophytes show

Table 4

		Epigeal 56						Hypogeal 31			Th 13
		Ph 47					Ch 9	H 27	Cr 4		Th 13
		S	E	MM	M	N	Ch	H	G	HH	Th
1908 400 species		1	· 3	6	17	20	9	27	3	1	13
1916 1,000 species		2	3	8	18	15	9	26	4	2	13
		Ph 46					Ch 9	H 26	Cr 6		Th 13
		Epigeal 55						Hypogeal 32			13

a greater deviation, namely 5; the spectrum of 1908 gives 20 per cent. N: the final spectrum, on the other hand, only 15 per cent. The higher figure depends among other things on the fact that the first century contained the genus *Croton*, which consisted mainly of Nano-phanerophytes.

If we confined ourselves to the series of life-forms Ph, Ch, H, Cr, Th, which is sufficient for the determination and delimitation of the main plant climates, the numbers of the two spectra are approximately equal; and the correspondence is the more complete if we unite H and Cr into one group (the group of plants passing the winter below ground). This is not unpermissible since, on account of the relatively small number of Cryptophytes, it appears that no main plant climate exists characterized by Cryptophytes to such a high degree that it deserves the name of Cryptophyte climate. In this way the determination of the life-forms of species, absolutely necessary for the determination and delimitation of the main plant climates, is greatly facilitated, since it is relatively easy to decide if a species is Ph, Ch, hypogeal (H and Cr), or Th.

Setting aside the limiting floras through whose biological spectra the limits between the plant climates can be established, it is not necessary in the determination of the plant climate to settle what life-form every single species belongs to; it suffices here to determine for each single species if it can be assigned to a definite life-form or not. For example, supposing we ask if and how far the Bahama Islands belong to the Phanerophyte climate, it is only necessary to know for each species if it is a Phanerophyte or not—in a word, it is sufficient in a preliminary way, to be able to determine the percentage of Phanerophytes; if this appears as 50 per cent., for example, the region belongs to the Phanerophyte climate. If the Ph percentage of a local flora exceeds that of the normal spectrum (that is the Ph percentage of the entire earth) then we have a Phanerophyte climate; if it sinks below the Ph percentage of the normal spectrum then naturally there must be some other life-form, or several, whose percentage will exceed the corresponding percentage of the normal spectrum and in this way enable us to characterize the plant climate. In order to determine what life-form that is, we must determine the life-form of the species in the limiting floras to such a point that the spectrum representing the complete series of life-forms can be constructed.

If in a given plant climate we find that the percentages of several life-forms are greater than in the normal spectrum, it is practically always the case that one given life-form will be predominant and in a specially high degree characteristic of the plant climate in question. The fact appears to be that on the one side of a boundary line the vegetation is characterized by a definite life-form, while on the other side another life-form takes over the characteristic role. If, therefore, in the establishment of the limit between two plant climates we hold strictly to the normal spectrum, it is very easy to calculate at what number the limiting line is to be drawn: this will be where the ratio of the two competing life-forms is equal to that of the normal spectrum. If we designate the life-forms in question by A and B and their numbers in the local spectra and normal spectrum respectively as l_A, l_B and n_A, n_B, then the limit between the two plant climates lies at the point where $\dfrac{l_A}{n_A} = \dfrac{l_B}{n_B}$; if, on the other hand, $\dfrac{l_A}{n_A} > \dfrac{l_B}{n_B}$, we have the A climate; while if $\dfrac{l_A}{n_A} < \dfrac{l_B}{n_B}$ we have the B climate.

Even if we neglect this formula on grounds which I cannot here discuss, nevertheless a boundary line, an isobiochore, must naturally be so drawn that all local floras through which it passes show the same steady relation to the normal spectrum in regard to the characteristic life-forms. Thus, as the limit between the hemicryptophytic climate and the arctic chamaephytic climate the line of 20 per cent. Chamaephytes is taken. I will not discuss here how such a limiting line is constructed but will refer to my earlier work (Chapter IV). Further, in another place (Chapter VII) I have shown the relation which a limiting line drawn according to this principle bears to the formula given above.

Finally, I will here discuss an attempt which I have made to control the accuracy of the normal spectrum based on the thousand species.

For this purpose, with the help of Engler and Prantl's *Natürlichen Pflanzenfamilien* and other sources, I have first of all counted how many species of Phanerogams were known about the year 1900, and how many of these belong respectively to Gymnosperms, Monocotyledons, Choripetalae, and Gamopetalae. The numbers for the plant groups in question, with the corresponding percentages, are given in Table 5 on the right. The number of Phanerogams comprised 139,953 species—that is to say approximately 140,000, so that I can use this last number in the following calculations. In this way I calculated how many species each of the four groups contained if the thousand species used in the establishment of the normal spectrum presented an accurate picture of the proportions. The numbers so found (Table 5, col. 2) were also calculated in percentages which appear in the fourth column of Table 5.

TABLE 5

	In 1,000 species.	In 140,000 species.	Per cent.	Per cent.	By summation.
Gymnosperms	3	420	0·3	0·3	471
Monocotyledons	198	27,720	19·8	17·2	24,083
Choripetalae	464	64,960	46·4	49·8	69,681
Gamopetalae	335	46,900	33·5	32·6	45,718
				Total	139,953

The correspondence between the actual and the calculated numbers in the two series of percentages must be considered very good. The greatest difference is shown by the Choripetalae, namely 3·4. If we reckon the mean error m according to the formula

$$m = \sqrt{\left(\frac{P_1 \times P_2}{n}\right)},$$

we have

$$m = \sqrt{\left(\frac{46·4 \times 53·6}{1,000}\right)} = \pm 1·6.$$

The mean error doubled is thus ±3·2, trebled ±4·8; the difference between the calculated and the actual percentage (49·8−46·4) amounts to 3·4, and is thus greater than twice but smaller than three times the mean error, quite a satisfactory approximation. In any case I shall keep the normal spectrum based on these thousand species; it would be no small task to obtain a better one based on a larger mass of material.

Furthermore, if in this spectrum certain figures deviate from accuracy by a small percentage, the usefulness of the spectrum as a norm is not markedly diminished; what matters is not so much the absolutely accurate number but a common measure with which the biological spectrum of each local flora may be compared and measured from the same standpoint; all that is necessary is that the figures of this measure should not deviate too much from the actual figures. The result of the test just described appears to show that a very far-reaching correspondence has been obtained.

In conclusion, I have used the foregoing material, that is the thousand species which have formed the basis of the normal spectrum, in order to obtain provisional spectra of each of the four plant groups, Gymnosperms,

Monocotyledons, Choripetalae, and Gamopetalae. Although the life-forms of these groups have not yet been statistically investigated, it is nevertheless well known that the numerical relations of the single life-forms in various plant groups are not the same; for example, Warming (*Frøplanterne*, p. 347) writes that the Gamopetalae contain relatively fewer woody plants than the Choripetalae, and the spectra obtained, which are given in Table 6, show that for Gymnosperms and Choripetalae the phanerophytic life-form, for the Gamopetalae the hemicryptophytic, and for the Monocotyledons the hemicryptophytic and cryptophytic life-forms are characteristic.

TABLE 6

The biological spectra of the Gymnosperms, Monocotyledons, Choripetalae and Gamopetalae of the thousand investigated species compared with the normal spectrum.

	Ph	Ch	H	Cr	Th
Gymnosperms	**100**	..	..	..	..
Monocotyledons	27	5	36·9	21·3	9·6
Choripetalae	**59·8**	9·7	15·1	1·5	13·8
Gamopetalae	34·9	11·6	**36·4**	3	14
Normal spectrum	46	9	26	6	13

XIII

ON THE SIGNIFICANCE OF CRYPTOGAMS FOR CHARACTERIZING PLANT CLIMATES

IN my earlier investigations I have based the biological spectrum upon flowering plants alone. Though it is true that actually there is nothing to prevent us from including Cryptogams and Phanerogams together in a system of life-forms, yet it would be entirely misleading to allow the biological spectrum to include Thallophytes as well as Phanerogams, quite apart from the fact that the number of Thallophytes in the individual local floras is usually not even approximately as well known as the species of Phanerogams. The plant covering of the world is made up in the main of Phanerogams; in comparison with them the bulk of Algae and Fungi is infinitesimal, while the number of their species is very great. It is true that in the plant covering, Lichens and Mosses, at any rate in some places, bulk very large; but in comparison with the majority of Phanerogams the individuals are small, while the number of species is great. From this it follows that if we included them along with Phanerogams as a basis for the biological spectrum, the spectrum would entirely lose its character of being an intelligible expression of the plant covering In practice the inclusion of Lichens and Mosses with Phanerogams is ruled out by the fact that the lists of plants available for the individual districts do not usually include Lichens and Mosses as well as Phanerogams. It goes without saying that in a comparative investigation the spectra to be compared must always contain the same groups of plants. We cannot place side by side biological spectra including only Phanerogams and those which include Phanerogams and also other groups such as Lichens and Mosses.

For the Pteridophytes the case is rather different. They resemble Phanerogams both morphologically and physiognomically so closely that it would not be unnatural to treat the two groups as one. At first I attempted, therefore, to include the Pteridophytes in the biological spectrum, but it soon became apparent that this could not be done, since they are absent from too many lists of plants. Later, therefore, I always formed the biological spectrum of Phanerogams alone. This is not merely a necessity arising from the fact that lists often include only Phanerogams; for I consider that the important part played by Pteridophytes, as also by Lichens and Mosses, in the plant covering is expressed more justly when each of these groups is treated separately. By doing this a means is obtained of further characterizing regions themselves based upon a spectrum of Phanerogams, and in several cases we succeed in subdividing such regions by means of the behaviour of Cryptogams. By treating separately the individual groups of Cryptogams, especially Lichens, Liverworts, Mosses, and Pteridophytes, we have the further advantage of not being confined to the system of life-forms used for the Phanerogams, but are able to use for each group the system of life-forms that is most appropriate. Further, by treating each group separately the mutual relationship of the groups and their distributional relationship to the Phanerogams in the world can be submitted to a comparative investigation, which is demanded because the conditions for distribution of Cryptogams are not the same as those of the Phanerogams.

The first question to be investigated in dealing with the geographical relationships of any group of Cryptogams is the number of species of that group in the individual floras. By this means the various floras are made comparable by using the flora of the world as a common measure. Thus if we take the Pteridophytes as an example we must determine the relationship between the number of species of Phanerogams and Pteridophytes in the flora investigated, and this relationship is then measured by the already determined relationship between the number of species of Phanerogams and Pteridophytes in the whole world. The number thus obtained I call the Pteridophyte-Quotient (Ptph.-Q.), which expresses how many times more (or fewer) Pteridophytes the flora includes than it would comprise if the proportion were the same as that in the flora of the whole world. In determining the number of Phanerogams and Pteridophytes in the world in order to decide the common measure for all floras, it is of course very important to approach the correct number as closely as possible; but on the other hand the significance and feasibility of the measure does not stand or fall with its absolute correctness; for if the same measure be always used, the mutual relationship of the numbers dealt with will remain the same whether the measure used is absolutely exact or whether it deviates more or less from exactness.

For many reasons, amongst others in order to synchronize approximately the floral lists and the number of species derived from them in constructing the flora of the world, I have taken as my starting-point the number of species known about the year 1900. The number, which is of course only approximate, is so adapted that the number of Phanerogams in the world is divided by the number of species in the individual groups of Cryptogams so as to give a whole number. By this means the computation of the quotient is made much easier. These are the numbers which I have taken as my starting-point:

Phanerogams	140,000 species	
Pteridophytes	5,600 „	(25)
Mosses	about 12,700 „	(11)
Liverworts	4,000 „	(35)
Lichens	6,350 „	(22)

The numbers given in brackets show how many times the Phanerogams of the world exceed the number of species of the individual groups of Cryptogams.

Let us take the flora of Ireland as an example. It contains 1,026 Phanerogams and 50 Pteridophytes.

From the equation

$$\frac{\text{The Phanerogams of the world, 140,000}}{\text{The Pteridophytes of the world, 5,600}} = \frac{25}{1} = \frac{\text{The Phanerogams of Ireland, 1,026}}{x}$$

we get $x = \frac{1 \times 1,026}{25} = 41$, so that Ireland should have 41 species of Pteridophytes if the relation between the Pteridophytes and Phanerogams in Ireland were the same as that of the whole world. But since the flora of Ireland includes 50 Pteridophytes the Ptph.-Q. = 50/41 = about 1·2, expressing the fact that the number of Pteridophytes in Ireland is 1·2 times as large as it would be if the proportion in the flora between Phanerogams and Pteridophytes were the same as that in the whole world.

Starting with the fact that in the flora of the whole world there are 25 times as many Phanerogams as Pteridophytes, the Ptph.-Q. of any given flora can be obtained by multiplying the number of Pteridophytes in that flora by 25 and dividing the result by the number of Phanerogams present.

For comparison with Ireland we can take Formosa with 1,297 species of Phanerogams and 149 Pteridophytes. Starting with these numbers the Ptph.-Q. of Formosa is $\frac{25 \times 149}{1,297}$: about 2·9, showing that the flora of Formosa has nearly three times as many Pteridophytes as it should have if the relationship between Phanerogams and Pteridophytes were the same as that for the whole world.

Thus in order to determine the Pteridophyte, Moss, Liverwort, or Lichen Quotient in a local flora the number of Pteridophytes, Mosses, Liverworts, or Lichens is multiplied by 25, 11, 35, and 22 respectively, and the result is divided by the number of Phanerogams in the flora. The numbers thus obtained express directly the relationship of the flora in question to that of the whole world. It is the flora of the world which is used as a common measure for all local floras, and to it they must be made directly comparable.

I have determined the Ptph.-Q. of about 100 local floras from different regions of the world, and the result of my investigations is that, if these floras can be looked upon as a valid expression for the whole, the Ptph.-Q. of most local floras will be larger than 1. This can be assigned to several causes. It may be because I have made the number of Pteridophytes in the world too low in comparison with the Phanerogams; or it may arise from the fact that the species of Pteridophytes have on an average wider areas of distribution than the Phanerogams. It is possible that both these causes may work together.

The question whether the species of Pteridophytes have on an average a wider distribution than the Phanerogams can be investigated by means of this method independently of whether or not the numerical relationship between the Phanerogams and Pteridophytes of the world is right, i.e. independently of whether or not the calculated Ptph.-Q. is absolutely correct.

If the species of Pteridophytes have on an average a wider distribution than the species of Phanerogams, then among the Pteridophytes more often than among the Phanerogams will the same species be common to different floras when the floras have any species at all in common. This amounts to saying that the Ptph.-Q. of the common species will on an average be greater than the Ptph.-Q. of the floras under comparison taken as one unit. This is just what is shown by the investigations, as far as they go.

The relationship can be illustrated by comparing Denmark with such diverse and mutually distant regions as Argentario (on the west coast of Italy), Racine (Wisconsin), the Alpine region of Shirouma (west of Tokio), and the Island of Formosa.

TABLE I

	No. of species.		
	Phanero-gams.	Pterido-phytes.	Ptph.-Q.
1. Denmark and Argentario as one unit	1,703	57	0·84
Common species	247	13	1·32
2. ,, ,, Racine as one unit.	1,734	52	0·75
Common species	147	17	2·89
3. ,, ,, the Alpine Region of Shirouma as one unit	1,279	45	0·38
Common Species	26	2	1·92
4. ,, ,, Formosa as one unit	2,697	299	2·77
Common species	19	5	6·58

Table I shows that in all these examples the species common to Denmark and the individual regions have a much higher Ptph.-Q. than Denmark or than any of the other floras taken alone.

Within the individual flora the same relationship is shown by the average of the Ptph.-Quotients of the individual districts being greater than the Ptph.-Q. of the flora as a whole. In More's *Cybele Hibernica* Ireland is divided into 12 districts. If we determine the Ptph.-Q. for each of these 12 local floras, and calculate the average of the 12 numbers, we obtain a higher local Ptph.-Q. than for the flora of Ireland taken as a whole, viz. 1·42 instead of 1·22.

Another question that we can try to answer by means of this method is in what degree the environment of the various climatic regions is favourable to Pteridophytes. In Table 2 I have given the Pteridophyte quotient for a series of floras in different climates. It is clear from this Table that it is the regions poor in rain that have the lowest Ptph.-Q. Within the Therophyte climate the Ptph.-Q. is thus usually lower than 1, often indeed a great deal lower. The Hemicryptophyte climate too is not particularly favourable for Pteridophytes. Damp and especially damp hot regions have the highest Ptph.-Q.

It cannot always be taken for granted that a high Ptph.-Q. is a proof that the environment of any given region is correspondingly highly favourable to Pteridophytes. It is reasonable to suppose that the average large areas of distribution of Pteridophytes is connected with the fact that the Pteridophytes have on an average better means of distribution than the Phanerogams. It may thus happen that where we are dealing with regions, such as oceanic islands, which plants reach with difficulty, it may well be that the species forming the plant covering are not necessarily those best adapted to the district, but those able to get there, provided of course they are able to grow there. In circumstances such as these a high Pteridophyte quotient cannot immediately be taken as a proof that the environment is correspondingly well adapted to Pteridophytes; it is just as much a question of means of distribution. Let us now investigate this matter further by a study of the oceanic islands.

Table 3 gives a conspectus of the Ptph.-Q. of a series of oceanic islands. The Ptph.-Q. may seem to be high for most of the entries, but it varies widely on the different islands and groups of islands. There are doubtless several reasons for these differences, the details of which cannot be explained. Such an explanation would demand a knowledge of the oceanic currents, the direction of the wind, of the hydrotherm figures, and of the soil conditions not only prevailing at present, but also in the past. The actual composition of the soil scarcely plays a very important part for the Ptph.-Q., provided the different kinds of localities are not too few. Where, however, an island offers few or even only one kind of habitat, which for most Pteridophytes may be very unfavourable, as for example coral islands, then the soil conditions may determine a low Ptph.-Q., as seen in Table 3, Nos. 1–5.

The precipitation exercises a marked influence on the Ptph.-Q., as is clearly seen from Table 2; but it will not do to treat the individual islands from this point of view. I shall content myself here merely with comparing those islands whose hydrotherm figures and other relationships determine the presence of formations of tall Phanerophytes. It is clear from Table 3, Nos. 6–21, that these islands have a high, often indeed very

Table 2

Pteridophyte-Quotients of different regions.

Tunis and Algeria	0.3
Tripoli	0.2
Spain	0.3
Italy	0.4
Malta	0.3
Cyprus	0.4
Samos	0.5
New Mexico	0.3
Alabama	0.7
Calcutta	0.8
Ganges Delta	1.3
Lagoa Santa	1
Racine	0.8
Connecticut	1.3
North Germany	0.8
Stuttgart	0.8
Denmark	1
West Lancashire	1.2
Ireland	1.2
Misaki (south of Tokio)	1.5
Ceylon	2.4
Ceylon: summits of Naminakulikunda	3.8
Ceylon: summits of Ritigala	5
Formosa	4

Table 3

Pteridophyte-Quotients of oceanic islands.

Jaluit	0.9
Marshall Islands	1.6
Laccadive Islands	1.1
Minikoi	1
Maldive Islands	0.9
Hongkong	2.4
Formosa	4
Polillo (near Luzon)	3.2
Guam	5.1
Yap	3.3
Fiji	5.5
Norfolk Island	7.4
Lord Howe Island	7.6
Tonga	3.2
Roratonga	9.3
Hawaii	5.6
Tahiti	12.2
Reunion	4.8
Mauritius	5.8
Rodriguez	3.8
Seychelles	7.8

high Ptph.-Q. It is, however, very far from true that the islands mentioned are particularly highly favoured in their precipitation; and the environment taken as a whole cannot by itself give a satisfactory explanation of the high Ptph.-Quotients. If it were particularly the environment favourable to Pteridophytes that had determined the high Ptph.-Quotients, then we might expect the same circumstances to have determined the occurrence of many species peculiar to the particular islands, and that the endemic flora of the island would therefore show a higher Ptph.-Q. than the flora as a whole. But that this is not so will be seen from Table 4, which shows what is found on five islands and groups of islands I have investigated. We see here that the Ptph.-Q. of the endemic species is much lower than that of all the other species. If, therefore, the copious occurrence of new species can in general be regarded as an indication that the group of plants to which they belong particularly prefers the climate, we cannot always conclude that the climate of these oceanic islands is more favourable for Pteridophytes than for Phanerogams.

I have shown that the species common to the different floras have a higher Ptph.-Q. than the floras taken as a whole. It follows from this that the species of Pteridophytes must cover on an average a larger area

Table 4

	Ptph.-Q. for	
	all the indigenous species.	the endemic species alone.
Guam	5·1	0·4
Lord Howe Isl. . .	7·6	5·6
Roratonga . . .	9·3	1·5
Hawaii. . . .	5·6	3·2
Tahiti	12·2	2·9

than the Phanerogams; and this corresponds with the fact that the units of distribution of Pteridophytes, the spores, can be distributed by means of the wind over wide areas more easily than the much heavier fruits and seeds of Phanerogams. This it seems to me is the most important reason why isolated floras, especially those of oceanic islands, have such a high Ptph.-Q. In harmony with this explanation is the fact that the highest Ptph.-Quotients are found on the islands farthest from the mainland. A regular increase in the Ptph.-Q. as the distance of large islands from the mainland increases cannot indeed be observed, and could not be expected, since the conditions of distribution differ in different regions. Beside this we know very little about the conditions for distribution at the present time, to say nothing of the past, which is of great importance when we are dealing with species which have migrated a long time ago. It is at any rate certain that the highest Ptph.-Quotients are found on those islands situated farthest away from the land, such as the Seychelles (7·8), Mauritius (5·8), Norfolk Island (7·4), Fiji (5·5), Hawaii (5·6), and Roratonga (9·3); not to speak of the great group of islands lying farthest away of all, the Society Islands, which have the highest Ptph.-Q. The island of Tahiti belonging to this group has a Ptph.-Q. of 12·2. This island thus has 12·2 times as many Pteridophytes as it should have if the proportion there were the same as in the world as a whole.

We might then expect to find, even in large areas with a dry climate unfavourable to the Pteridophytes, isolated floras with a comparatively high Ptph.-Q. This is actually so, examples being the Galapagos Islands with a Ptph.-Q. of 4·4 and the Azores of 2·2. I do not for one minute doubt that the Azores, which have the same latitude as the south of Portugal, would have a much lower Ptph.-Q. if they were situated in a similar environment near the coast of Algarve instead of at a distance of 1,300 km. from it.

I shall not enter here into further details about the quotients of the other groups of Cryptogams, but shall only remark that the Mosses, Liverworts, and Lichens also show widely different quotients in different climates. In these groups too the quotient seems on an average to be higher than 1; and here too there is no reason to doubt that in these groups the areas covered by species are on an average larger than those of the Phanerogams. The floras investigated show particularly high Lichen quotients. Whether this arises from the fact that the Lichens of the world have been computed at too low a figure, or whether the area of the species of Lichens are on an average very large, are questions that can be investigated by the method used in dealing with the Pteridophytes.

In Table 5 will be found some floras in which the quotients of the various groups of Cryptograms have been determined as far as the material will allow.

TABLE 5

	Ptph.-Q.	Musc.-Q.	Hept.-Q.	Lich.-Q.
Faroe Islands	2·4	10·6	13·1	16·9
West Lancashire	1·2	4·4	3·9	··
Alabama	0·7	0·7	0·6	2·3
Azores	2·2	2·4	2·4	3·7
Lagoa Santa	1	0·3	··	0·9
Connecticut	1·3	2·6	3·3	··
Reunion	4·8	2·1	4·3	2·4
Cape Horn	3	4·4	16	10·1

In conclusion let us present an example of the comparatively large quotients of the cryptogamic species in two widely separated floras. According to J. Stirling (*Trans. and Proc. of the Bot. Soc. Edinb.*, xx, pp. 348–9) the Australian Alps and the Island of Arran off the coast of Scotland have the following species in common: 27 Phanerogams, 10 Pteridophytes, 11 Mosses, and 16 Lichens. Table 6 shows that the quotients based on these species are very high. For comparison the quotients for the flora of the Australian Alps alone are added.

TABLE 6

	Ptph.-Q.	Musc.-Q.	Lich.-Q.
Australian Alps	1·9	2·8	3·1
Common to Australian Alps and Arran	9·3	4·5	13·0

XIV

THE DIFFERENT INFLUENCE EXERCISED BY VARIOUS TYPES OF VEGETATION ON THE DEGREE OF ACIDITY (HYDROGEN-ION CONCENTRATION) OF THE SOIL

INTRODUCTION

WHEREVER new soil arises in the neighbourhood of land that is covered by plants, as it may do on dunes or salt marshes, through the agency of water deposits or by other means, or wherever ground formerly covered with plants is laid bare, seeds will soon be carried on to the new soil from the surrounding country by wind, animals, &c., and those able to germinate and grow in the new environment will produce plants. As long as the plant covering remains open and competition is not keen, the vegetation will appear motley and haphazard because of the irregularities of the factors determining migration; but as the covering becomes gradually denser and the competition for the essentials of life becomes correspondingly keener, one species after another succumbs, so that after a varying number of years the result will be a denser, relatively stable vegetation consisting of those migrants from the surrounding territory that are best adapted to grow in the environment they have found. Where this environment is uniform, not merely the species composition, but also the frequency and mutual mass relationships of the species will be essentially uniform, while even apparently small deviations in the factors essential for plants will turn the scales, producing corresponding deviations in the composition and character of the vegetation.

Though seeds of different species are constantly introduced, yet a stabilized vegetation will not be markedly altered as long as the environment remains the same. But if the environment be altered essentially, either by influence of the vegetation itself on the soil or by any other means, then species other than the first conquerors can gain access and enter the competition, causing the composition of the vegetation to undergo changes, which in their turn, after the lapse of time, will again lead to a certain state of equilibrium. Where, then, the vegetation has had time to arrive at a state of equilibrium, every habitat will be occupied by those species of the region that are best adapted to live in the given environment. In accordance with this principle experience shows us that wherever we find a series of habitats determined by the form and nature of the ground, and the units of this series are determined by different influences and by the environmental factors of importance for the individual plants, there we find a corresponding series of plant formations, whose individual members are the same, and follow each other in the same invariable series, wherever the same series of localities is encountered in the district. As an example may be cited the varying degree of soil humidity dependent on the height of the ground above the water table. In numerous places in Denmark, especially in the dune region of West Jutland, and on our heaths and salt marshes in thousands of localities, this law-determined connexion between series of habitats and series of plant formations may be demonstrated.

Thus it is easy to observe that the zonation of the vegetation surrounding every hollow on heaths, dunes, &c., is determined by varying humidity; and it is easy to recognize the zones in these different places. But when we undertake a more thorough statistical analysis of the composition of the vegetation, differences are soon discovered which cannot always be regarded as determined by difference of humidity. Sometimes, perhaps, the difference may be due to a difference in the history of the migration of the plants, merely signifying that the vegetation is not yet stabilized. But where we are dealing with regions undisturbed for a long time by cultivation, other causes must be supposed to be at work.

Besides humidity there are many other soil factors which may co-operate in determining what combination of species covers any given habitat. There is no doubt that one very important factor is the reaction of the soil, i.e. its hydrogen-ion concentration. Now that S. P. L. Sørensen, by his colorimetric method, has given us a means which has made feasible the determination of the hydrogen-ion concentration, this factor will certainly in future be employed with great gain in ecological investigations. The method gives us a practicable means of grading the reaction of the soil far more accurately than was hitherto possible.

It is especially American investigators who have worked on these lines; but the subject has recently aroused interest in Denmark, where it has been treated on a broader basis than hitherto, especially by C. Olsen in his work entitled: *Studier over Jordbundens Brintionkoncentration og dens Betydning for Vegetationen særlig for Plantefordelingen i Naturen*.[1] By comparing statistical investigations of formations in a series of meadow and woodland localities with the determination of the hydrogen-ion concentration of the soil of these localities, C. Olsen has shown that in natural formations the individual species occur only on the soil 'whose hydrogen-ion concentration lies within limits characteristic for each species' and that within these limits 'there occur narrower limits within which the species has its greatest average frequency'. By means of a series of water culture experiments C. Olsen has shown 'that species which in nature occur on very acid soil approach their most vigorous growth in culture solution whose P_H value is somewhere about 4, while species that in nature occur only on weakly acid, neutral, or basic soil, approach their most vigorous growth with a P_H value between 6 and 7. In the weakly acid culture

[1] Meddelelser fra Carlsberg Laboratoriet 15, No. 1. Copenhagen, 1921.

solutions in which plants of basic soils approach their most vigorous growth the plants of acid soils did very badly and became chlorotic (C. Olsen, l.c., p. 144).

There is scarcely then any doubt that the hydrogen-ion concentration of the soil is very important if we wish to understand the occurrence in nature of individual species. This subject presents a very wide field for future investigations. Another question is whether the plants affect the hydrogen-ion concentration of the soil, and, if so, in which direction and in what degree. If it can be proved that a species or a formation or a type of formation affects the P_H value of the soil in a definite direction we shall then have demonstrated one of the factors which may be co-determinant when a given formation or type of formation in course of time alters its environment, and by that means brings about its own downfall; it alters in fact the environment in favour of another combination of species.

Before giving any more details about my investigations of the influence which various types of formation exercise on the hydrogen-ion concentration of the soil I must say a few words about certain aspects of the method itself.

My thanks are due to Professor S. P. L. Sørensen, the director of the chemical department of the Carlsberg Laboratory, for so readily placing at my disposal the standard solutions and indicators for my investigations; and I am grateful to Dr. C. Olsen of that laboratory, who has made me familiar with his procedure for determining the hydrogen-ion concentration in soil samples. My thanks are also due to Kgl. Skovrider N. Vestergaard for explaining several points about the history of the Dyrehave which were important for my work, and to Dr. Boysen Jensen for good advice about the construction of the atmometer described on pp. 481–3.

SOME POINTS ABOUT THE METHOD

Concerning the procedure necessary for the colorimetric determination of the hydrogen-ion concentration of the soil I must here be content to refer to the descriptions of C. Olsen. Individual details must however be discussed more fully, especially the question of the extent of variation in the hydrogen-ion concentration of the soil at the same locality, partly at the same depths at different places, and partly at different depths at the same place.

Besides this there is the question of the length of time spent on making the extract, and of the kind of water used for it, and the desirability of doing without standard solutions coloured by indicators.

The difference in the hydrogen-ion concentration of the soil at the same depth in different localities. The investigation is based on soil samples taken from a depth of from 7 to 10 cm. There is no a priori reason to believe that the P_H value of soil extract will be uniform everywhere within the same region, however uniform that region may appear. The investigations showed that the variations can be very great. Neither is there any reason to suppose that the factors, which are presumably very complicated but are little understood, determining hydrogen-ion concentration, should be uniform everywhere in the same locality, however uniform that locality may appear. Vegetation forms a mosaic, and animal life is heterogeneous and mobile. The relationship of the vegetation to the hydrogen-ion concentration of the soil is not then such that the species present are in equilibrium with a definite hydrogen-ion concentration, and is not such that the plant growth itself in the most stabilized formation is an expression of definite environmental factors at any moment. The environment alters from year to year, from day to day, from hour to hour; but the species composition for all that may remain essentially the same. One year the precipitation may be so copious that a given locality is completely saturated with water throughout almost the whole of the summer, whilst the same locality another year may be so dry during most of the summer that one can lie on the ground without noticing any moisture. If these two conditions were each permanent throughout a long number of years on two parts of the locality in question, these two parts eventually would inevitably bear a different vegetation, each with its definite stabilized formation. In all the places where a piece of ground becomes permanently drier because of draining, or permanently more humid because of damming, we have sufficiently convincing examples of this. But where the differences in the environment are due to the climate varying from year to year, while the average climate remains the same, then the vegetation remains essentially the same in its species composition, and usually in its species frequency. But it goes without saying that the luxuriance of the species will vary in different years. In other words: the species composition and the species frequency of the formation are not in general a stabilized expression of environment at any one moment, but a constantly oscillating, only relatively stabilized, expression of the average environment prevailing throughout a number of years.

Something similar to that which is true of the relationship of time is true also of the relationship of space, as for example the significance of the slight differences in the soil of the same locality, these including the differences in the hydrogen-ion concentration. A stabilized vegetation is not stabilized in its relationship to one definite P_H value throughout its extent, but it is stabilized in relationship to a mosaic of areas varying in extent and varying also in their P_H value, which, though it may vary, yet, when its average value is determined statistically, is found to remain essentially uniform from year to year. It is therefore necessary in determining the hydrogen-ion concentration of the soil to use the statistical method.

It is very probable that the variation of the P_H values differs somewhat in different localities. C. Olsen found in the meadows he investigated that the variation in the same small plot did not amount to more than 0·3 in P_H value. It is indeed possible that the meadows in question were very uniform in the P_H value of their various portions; but I think it most probable that an investigation of more samples taken in different places would have shown a considerable difference between the highest and lowest P_H value. At all events my investigations have shown a much wider variation within single restricted plots measuring only a few square metres. For the purpose of this study I have investigated a series of localities on pastured common especially on the Plain of the Eremitage in the forest of 'Dyrehave' near Copenhagen, and also a series of woodland localities in spruce, beech, and oak woods. A hundred localities in all were investigated, and in each of these localities 5 soil samples were taken from a depth of from 7 to 10 cm. The P_H value of each soil sample was determined separately; that of the 500 samples fluctuated between 3·7 and 7·6, figures which include the P_H values of the majority of Danish soils. In the group of 5 samples which showed the widest difference between the highest and the lowest P_H value this difference was 2·6. In 30 per cent. of the groups of 5 the greatest difference in an individual group was more than one degree. In none of the 100 groups of 5 was the greatest difference less than 0·3; on an average it was 0·89.

As far as can be judged from the material before us, there does not appear to be any noteworthy difference in the extent of the variation of hydrogen-ion concentration between localities in the woods and localities on the common. In the 56 groups of 5 taken from woodland localities the greatest average difference was 0·92; in the 44 groups of 5 taken from the common the corresponding number was 0·86.

Whether there are perceptible differences in the variation of the P_H value of soils with different degrees of acidity is a cognate problem of considerable interest. Since the numbers by which the P_H value is expressed do not form an equidistant series when they are converted into figures giving the absolute weight of hydrogen-ions expressed in grammes per litre, one might suppose it possible that there is a difference in the variation of the P_H value in soils with different hydrogen-ion concentrations. But this does not appear to be so; at any rate my material shows nothing of the kind. If the 100 localities investigated by me be divided into two groups, the one including the most acid and the other the less acid and basic localities, and if the average amount of variation of P_H value be determined for each group, then, as seen by these figures, no perceptible difference is apparent. The amount of the variation of P_H value in the different localities appears to be independent of the absolute P_H value. Thus in 51 localities with P_H 3·6–5·4, the average variation was 0·87, and in 49 localities with P_H 5·5–7·6 the average variation was 0·91.

The rather large variation of P_H value shown in the same apparently uniform localities makes it necessary to determine the P_H value of several soil samples if we wish to state the actual degree of acidity of a locality. Where the numbers, as those following, have to serve as a basis for a comparison of different localities I have therefore made it a rule to investigate 5 samples from each locality. Sometimes more will have to be investigated; the more samples investigated the closer we shall get to a constant average number. Of course it would be much easier to mix the samples taken from one locality, and to ascertain the P_H value, as C. Olsen has done, from the mixture. I cannot however recommend this procedure; especially as by using it we obtain no information about the extent of variation, which may differ very widely in different localities, and which must not be overlooked if we wish to obtain a proper understanding of the localities.

The difference in amount of hydrogen-ion concentration at different depths in the same locality. On this question C. Olsen (l.c., p. 25) writes: 'there was no essential difference whether the soil samples were taken at a depth of 5, 10, or 20 cm. The deviations never exceeded 0·3 in P_H value.' Olsen further points out that according to Arrhenius the hydrogen-ion concentration decreases with increasing depth, while Plummer, dealing with agricultural soils, observed that the hydrogen-ion concentration rises with increasing depth. There seems good reason for pursuing this subject farther, basing the investigation of localities on several samples. I have undertaken no exhaustive investigations of this kind; they have not been necessary in dealing with the problem confronting me. In order however to obtain a preliminary idea I have determined the P_H value of some soil samples taken from a few places at different depths.

In a 75-year-old spruce wood in Fortun-Indelukke I took three soil samples from a plot 0·25 sq. metre in extent from depths of 7–10, 17–20, and 27–30 cm. respectively. Five samples were taken from each depth. Table 1 shows the result of determining the P_H value of these samples.

Table 1

The P_H value of the soil at three different depths in a 75-year-old spruce wood.

Depth in cm.	P_H value of the single samples.					Average P_H value.	Greatest deviation.
7–10	4·7	4·9	4·9	5·1	5·3	4·98	0·6
17–20	4·9	4·9	5·1	5·8	5·9	5·32	1·0
27–30	4·8	4·8	5·4	5·4	5·9	5·26	1·1

The soil is here seen to become slightly less acid downwards; but the difference is not great, much less in fact than the greatest difference between the individual samples taken at the same depth.

TABLE 2

The P_H value of the soil at three depths of the common at 'Nørrefælled' near Copenhagen.

Depth in cm.	P_H value of the single samples.					Average P_H value.	Greatest deviation.
7–10	6·6	6·6	6·7	6·7	7·1	6·74	0·5
17–20	5·7	5·8	5·9	6·0	6·2	5·92	0·5
27–30	5·5	5·6	5·7	6·1	6·2	5·82	0·7

Table 2 shows the result of an investigation of the P_H value of the soil at different depths in an area of 0·25 sq. metre at Nørrefælled near Copenhagen. Just as in the former case we see here only a slight difference in the degree of acidity between the depths of 17–20 and 27–30. If these two depths are taken together we obtain a P_H value of 5·87. The deeper layers then are considerably more acid than the soil at a depth of 7–10 cm. This is the reverse of what we saw in the considerably more acid soil of the spruce wood.

These facts, together with the already mentioned apparently contradictory statements in the literature, may perhaps be explained by supposing that where we are dealing with localities covered with a vegetation like that of a dense spruce wood, which determines a greater acidity of the soil, the soil will always, at any rate in a comparatively young wood, be less acid downwards, since the lower layers will not be so much influenced as the upper ones. The reverse will be seen in localities where, by cultivation or other means, the soil has been so affected that its upper layers have had their P_H value increased, showing the phenomenon demonstrated by Plummer, that the lower soil is more acid than the upper more strongly affected layers.

The facts thus brought to light show that it is desirable to investigate this problem much more thoroughly than has hitherto been done. The investigation should include a series of soils of different kinds and bearing different types of vegetation, and should be based on 5 or even more determinations of the P_H value at each depth from each single circumscribed plot.

The treatment of the soil samples. Using the colorimetric method it is necessary to filter the soil extracts; and since these are almost free of the buffer, and therefore easily alter their hydrogen-ion concentration, it is important that the procedure should be as uniform as possible, especially where we are dealing with a comparative investigation in which even small deviations in hydrogen-ion concentration may have significance.

For collecting the soil samples I have always used cylindrical glass tubes open and fitted with corks at each end. It is easy to press the entire soil sample all at once from one end of these tubes into a glass in which it is to be extracted with water. The procedure is particularly simple if the glass used has a mouth wide enough to fit the tube in which the soil sample has been kept. The tubes used contained about 70 cubic cm. of soil; and for extracting this amount of soil about 80 cubic cm. of water or slightly more were necessary. If one restricts oneself to a volume of water equal to the volume of the soil sample, it may sometimes occur, especially where several indicators have to be used, that too little of the filtered solution is obtained. As C. Olsen emphasizes, and as I myself can confirm, there is no demonstrable difference due to the amount of water used for the extract. It is important of course that the procedure should be as uniform as possible.

The duration of the process of extraction. C. Olsen unfortunately gives us no information about the influence which the length of time spent on making the extract has upon the strength of the hydrogen-ion concentration of the filtered solution. He merely says (l.c., p. 14) that the soil samples remained in water for 24 hours, during which time they were repeatedly stirred with a glass rod, before the determination of the hydrogen-ion concentration was made.

If the place where the investigation is made is not too distant a set of soil samples can be collected during the afternoon and put into water on returning home in the evening. They can be left overnight, and their P_H values determined the next morning, after shaking the samples briskly several times both in the evening and in the morning. In this way one can manage to fetch and determine a set of from 20 to 30 soil samples each day. Thus the soil samples can be extracted in as short a time as 15 hours; this period was used for all the samples whose P_H values were determined for this work. It makes scarcely any difference whether the extraction takes 15 or 24 hours. Since it would be of interest, however, to know whether a period still shorter than 15 hours could be used, so that the P_H values could be determined the same day as the samples were collected, I performed an experiment, using the same well-mixed soil for making 15 extracts, of which the first 5 were extracted in 3 hours, 5 in 6 hours, and 5 in 15 hours. The result of this experiment is shown in Table 3, which shows no marked difference whether the extracts are made

Table 3

The significance of the duration of the process of extraction.

Period of extraction.	P_H value of the individual soil samples.					Average P_H value.
3 hours	7·4	7·4	7·4	7·5	7·5	7·44
6 "	7·6	7·6	7·7	7·7	7·7	7·66
15 "	7·6	7·7	7·7	7·8	7·9	7·74

in 6 or 15 hours; but there is a noticeable difference if they are extracted in 6 or in 3 hours. The question deserves a thorough investigation, especially as it affects the planning of future work.

The kind of water used for making the extracts. Since it is of course important that our technique should be as uniform as possible it might be supposed that there would be no question of using anything but distilled water; C. Olsen mentions no alternative. For practical reasons, however, it is desirable to investigate this question in order to find out whether it is not possible to use other kinds of water, e.g. rain-water or filtered and boiled tap- or well-water. Where distilled water is accessible it must of course be used, but it is sometimes very difficult to obtain. One may for example be investigating soils in a sequestered part of the country. I consider therefore that it is worth making a thorough investigation of the differences resulting from using distilled water or some other kind of water. In the problem confronting us this question has indeed no significance, as I have always used distilled water. It cannot, however, be denied that the question may have significance; to make myself familiar with it I have carried out a small experiment, the result of which I will now give. In this experiment three kinds of water were tried: distilled water, unboiled Copenhagen tap-water, and boiled and filtered Copenhagen tap-water. The P_H value of the water was determined by using 5 samples of each kind, with the results shown in Table 4.

TABLE 4
The P_H value of the water used in the experiment given in Table 5.

	P_H value of the individual water samples.					Average P_H value.
Distilled water	4·8	4·8	5·0	5·2	5·3	5·02
Unboiled tap-water	7·6	7·6	7·6	7·6	7·6	7·60
Boiled and filtered tap-water	8·2	8·2	8·2	8·2	8·3	8·22

In order to try the effects of these different kinds of water a sifted and well-mixed peat soil from a *Calluna*-heath was used. Thirty samples of the soil were taken. Ten of them were treated with distilled water, 10 with unboiled tap-water, and 10 with boiled and filtered tap-water (cf. Table 4). The determination of the P_H values of the extracts gave the result shown in Table 5.

TABLE 5
The different kinds of water used for extracting.

Peaty soil from Calluna-heath extracted with:	P_H value of the individual samples.										Average P_H value.
Distilled water	4·3	4·3	4·3	4·4	4·5	4·6	4·8	4·8	4·8	4·8	4·56
Unboiled tap-water	4·4	4·4	4·6	4·7	4·7	4·7	4·8	4·8	4·8	4·8	4·67
Boiled and filtered tap-water	4·2	4·2	4·3	4·3	4·4	4·4	4·4	4·8	4·8	4·3	4·46

It is seen from this table that although there was actually a considerable difference in the hydrogen-ion concentrations of the water used the extracts investigated show almost identically the same P_H values.

It seems then that boiled and filtered tap-water is just as good as distilled water, the samples extracted with the latter being 0·1 less acid than those extracted with the former; this is about the same difference as C. Olsen found between the P_H value of the samples extracted with distilled water and that of the water pressed out of the soil, which was on an average slightly more acid than the extract (l.c., p. 23).

This experiment is designed merely to direct attention to this question, and to show that there appears no reason to doubt that distilled water can be dispensed with if necessary in determining the P_H values of soil samples; but it goes without saying that it is necessary first to compare the behaviour of the water used with that of distilled water by using both kinds for treating a series of samples of the same uniform mixture.

On the possibility of using a comparimeter instead of standard colour solutions. A question of even greater practical importance than whether one can dispense with distilled water when making soil extracts is that of the possibility of doing without the standard solutions used in the colorimetric determination of hydrogen-ion concentration. If a soil investigation has to be done at a great distance from a laboratory, the transport of an extensive series of bottles containing the different standard solutions presents great difficulty; if any of the bottles break, as they easily may, the investigator is placed in an awkward situation. I have therefore from the beginning of my investigations carefully considered whether it would not be possible to substitute the indicator solutions by objects with the same tint, e.g. coloured glass, permanently coloured solutions, or coloured paper. By the mixing of various coloured substances, one can obtain, it is true, solutions with the same tint as the standard solutions coloured by indicators. But it will be certainly difficult, perhaps impossible, to obtain sufficiently permanent colours, and since the use of such solutions is much less practicable than the use of coloured glass or coloured paper I have pursued this question no farther. It is possible that coloured glass will be found to give the best result;[1] but I have had no opportunity of carrying out experiments with it. I shall therefore confine myself exclusively to my experiments with coloured paper. These experiments have taught me that it is possible to make coloured paper which, seen through a test tube of water placed in a comparimeter, gives a colour indistinguishable from that of a standard solution with an indicator in an adjacent test-tube. The difficulty consists

---

[1] Cf. Klas Sondén, *Zur Anwendung gefärbter Gläser statt Flüssigkeiten bei kolorimetrischen Untersuchungen* (Arkiv för Kemi, Mineralogi och Geologi. Utgivet av K. Svenska Vetenskapsakademien, Band 8, No. 7, 1921).

merely in obtaining a paper whose colour is sufficiently permanent. It is true that the colour need not be absolutely permanent, since the paper can be renewed; but in order to be practicable the colours must be sufficiently permanent to stand use for some time without altering more than a shade corresponding to 0·2 in P_H value.

The method of procedure is as follows. The coloured paper is placed horizontally inside the windows of a comparator which slope at an angle of 45 degrees; it must be close to these windows and so arranged that the light from the surfaces of the pieces of paper is projected through a test-tube of water placed in the comparator. By the side of this test-tube another test-tube is placed containing a standard solution coloured by an indicator. The coloured paper is changed until the colour is found which, when seen through the test-tube of water, gives to this test-tube exactly the same tint as the standard solution concerned, which has been coloured by an indicator. By this method the necessary number of different coloured papers is obtained. It is of course not necessary to have coloured paper for each P_H value; a series showing intervals of 0·4 in P_H value suffices; and if difficulty be found in obtaining a colour, say for example the colour corresponding to 4·4 in the series used by C. Olsen, while it may, however, be easy to obtain coloured paper corresponding to 4·3 or 4·5, one of the latter numbers may well be used, since it is not absolutely necessary that the distance between the units in the scale should always be the same.

Having prepared the necessary number of coloured papers those which correspond with a definite indicator are placed in pieces 2·5–3 cm. long and about 2 cm. broad on a ruler about 4 cm. broad. The sample papers are arranged in a series corresponding with the progressive P_H values, and the distance between the coloured fields is such that one of them is seen through every alternate window of the comparator when the ruler is brought up to them. In this way the soil extract whose P_H value is to be determined with the indicator added can always be placed in such a position that by moving the ruler backwards and forwards at the windows it is brought to a stand between the two fields of colour on the ruler which show the numbers between which the P_H value of the soil extract lies; or it may turn out that it is identical with one of the colours. If there is one comparator ruler for each indicator the determination of the P_H value of the solution takes only a few seconds. It is then possible to work much more rapidly than when using the coloured standard solutions. These solutions moreover are not permanent, and have to be renewed often, some of them every day. Their instability and the great difficulty of carrying the numerous bottles of standard solution makes it very desirable to obtain permanently coloured paper to use for the comparator ruler. Then, even on a journey, exact determinations of the P_H value of soil samples can be undertaken.

THE PLAN AND SCOPE OF THE WORK

Even if, as C. Olsen has shown, the P_H value is one of the factors determining which species occupy a given locality, and even if the P_H value also codetermines the frequency of these species in that locality, these facts by no means exclude the possibility that the vegetation of a habitat can itself directly or indirectly determine changes in the original P_H value of the soil, and thus seal its own doom. It is this problem that I shall now discuss.

The most scientific procedure would be to start with a uniform locality everywhere having essentially the same P_H value, and then to allow part of the area to bear one kind of vegetation and another part an essentially different kind, e.g. herbaceous vegetation and woodland respectively, and then at longer or shorter intervals of years to determine the P_H values of the soil dominated by the different kinds of vegetation. We could then find out whether any demonstrable change in P_H value had taken place, and whether the two different kinds of vegetation showed any difference in this respect. To obtain decisive results in this manner the experiment would probably have to extend over years. As this was not feasible I looked for places which during the process of time have been subject to changes corresponding to those mentioned above. Such changes occur not only as a result of cultivation, but they also take place in nature apart from man's interference.

From the very outset it was clear to me that in some ways the Dyrehave was particularly suitable for such an experiment. Partly because, compared with other forests of similar size, it has been allowed to remain undisturbed for a long period, and partly because fenced plantations have from time to time been made on the grass areas which have arisen from the death of the old trees, and from the herds of animals preventing rejuvenation. We thus find here side by side on what was originally the same soil young woodland, old woodland, and herbaceous vegetation. We find moreover in other places herbaceous vegetation (pasture), where old woodland was present before, and where thus the results obtained by following the development from pasture to wood can be further put to the test by following the opposite process, i.e. from wood to pasture. All the investigations here described of the different influence on the different kind of formations of the hydrogen-ion concentration of the soil have been undertaken exclusively in the Dyrehave and adjacent woods, i.e. Stampeskov, Chr. IX's Hegn, and Jægersborg Hegn. These woods are conveniently near Copenhagen and thus very easy to study.

The north side of the Eremitageslette.[1] The western portion of Eremitageslette is bounded on the north by Stampeskov and its eastern portion by Chr. IX's Hegn. Both these woods date from the middle of last century. The boundary between these and the now existing plain (the

[1] *Slette*, plain.

Eremitageslette) was drawn in 1853 as an arbitrary line running across the pasture which had arisen in course of time from the disappearance, through felling or old age, of the trees in the forest. Scattered here and there on this plain trees stood isolated and in small groups; and some of these are still standing, some on Eremitageslette, some in Stampeskov and in Chr. IX's Hegn. An opportunity is here presented of comparing the P_H values in beech- and oak-wood 60 to 70 years old with that of the original pasture at the same level. Further we can compare the oakwood 60 to 70 years old with the beechwood of the same age and this beechwood with older beechwood.

Fortunens Indelukke[1] includes woods of different kind and of different age. The southern portion was enclosed in 1831, and planting was completed between 1840 and 1850, with the exception of some parts which are younger. Towards the south and east the wood is bounded for some of its extent by old pasture, affording an opportunity of comparing this pasture with beechwood and sprucewood 70 to 80 years old. We can here also compare sprucewood and beechwood. The middle part of Fortun-Indelukke dates from the years 1865 to 1870. Its northern portion includes areas which have been cultivated at very different periods. Some parts date from near the middle of the last century, so that they are about 70 years old; some parts are much younger. This portion of the wood, too, abuts upon what was originally pasture; but it should be noted that the pasture has not everywhere been left undisturbed: towards the north it is at present under cultivation, and towards the east certain portions have been cultivated for a number of years. But this cultivation does not seem to have had a lasting effect on the hydrogen-ion concentration of the soil. Further, there are places where the old pasture has not been disturbed, and these places give us an opportunity of comparing sprucewood and beechwood with original pasture.

The middle and northerly portions of Fortun-Indelukke are of particular interest because they give the opportunity of comparing the P_H value in the soil of the wood with that of a recent pasture which owes its origin to the felling of the wood and to the ground being allowed to go to grass. I refer to the broad portion of wood running through Fortun-Indelukke which for military reasons was felled in 1914 and during the following years rapidly became pasture. This pasture is bounded by arbitrary lines which in different places traverse different types of wood, whose P_H value can be compared with that of the recent pasture.

Dating from about the same period (1914–15) are several narrow strips of pasture between rows of spruce trees in the most northerly part of Fortun-Indelukke.

Then we have the small enclosures on Eremitageslette made at different times for the purpose of growing groups of trees on the plain. First of all

[1] *Indelukke*, enclosure.

we have four groups planted about 1840; one of these is west-north-west and three are south and south-west of the hunting-lodge ('Eremitagen'). They consist principally of beech. As long ago as 1913 the fence surrounding them was taken away. Extensive portions of the plain surrounding these islands of woodland have from time to time been cultivated for a few years; but we have sufficient places where the islands of wood abut on portions of the plain that have not been cultivated, at any rate after the trees were planted.

Again towards the north-west and the north of Eremitagen there are four small enclosures of oak having an undergrowth of hawthorn. They date from 1885 to 1895.

Finally between the years 1885 and 1910 some enclosures were planted, especially with beech and oak scattered in the old forest, where, owing to the death of the old trees, large open places had appeared in the course of time. Here we have an opportunity of studying the influence of young wood on the P_H value of the soil, comparing it with the P_H value in the part of the glade that has not been planted.

In the following account I shall consider those places where for varying periods parts of the old pasture have been planted with wood—spruce, beech, and oak—while the corresponding parts of the pasture immediately abutting on the wood have lain undisturbed, giving us an opportunity of studying the influence the different kinds of wood have had on the hydrogen-ion concentration of the soil. Afterwards I shall give some observations of the hydrogen-ion concentration of the soil in different types of wood of the same age and originally on the same soil. Finally will come investigations of the hydrogen-ion concentrations in old and new parts of the wood compared with those of the pasture which has taken the place of the portions of the wood that have been cleared by felling or old age, thus carrying us back to our starting-point, pasture.

In each locality the hydrogen-ion concentration has been determined by 5 soil samples taken from a depth of from 7 to 10 cm. Since we are here only dealing with a comparison of different types of vegetation and not with narrowly defined formations, I have undertaken no statistical analyses of the formations, but have contented myself with giving a short description of the vegetation, mentioning those species which were observed within the narrower confines from which each soil sample was taken.

PASTURE—WOOD

A. Pasture—Sprucewood (Table 6)

The relationship between the hydrogen-ion concentration in the pasture and in the sprucewood planted on portions of that pasture was investigated in six localities, which in the first column of Table 6 are given numbers corresponding to the descriptions of the localities cited in the text. In the second and third columns are given the P_H values of the pasture (*a*) and the wood (*b*). Column 4 gives the number by which the P_H value of the pasture exceeds that of the sprucewood. The tables following are arranged in the same way and the numbers of localities are consecutive.

1. The south end of a narrow pasture to the east of the southern portion of Fortun-Indelukke compared with the part of the sprucewood lying to the south of the pasture.
    a. Pasture: Luxuriant *Agrostis tenuis–Cynosurus cristatus*-Formation with much moss, containing the following species, which, in all the following descriptions, are given in alphabetical order: *Cerastium caespitosum, Deschampsia caespitosa, Phleum pratense, Rumex acetosa, Trifolium repens, Veronica chamaedrys*; besides members of the *Hypnaceae*, especially *Hylocomium squarrosum*. $P_H = 6.76$.
    b. Sprucewood 70–80 years old: The ground in parts is covered by spruce needles and by wind-carried beech leaves; in parts it is bare. Recent felling has admitted light into the wood, so that we now see some scattered *Oxalis acetosella* and very occasional weak specimens of *Urtica dioeca* and *Lactuca muralis*. $P_H = 4.10$.
2. The same locality as No. 1 but a little farther north.
    a. Pasture: Luxuriant *Agrostis tenuis*-Formation, with *Anthoxanthum odoratum, Carex hirta, Cynosurus cristatus, Poa pratensis, Rumex acetosa, Stellaria graminea, Trifolium repens*. $P_H = 6.32$.
    b. Sprucewood 70–80 years old with a layer of needles and a few beech leaves. No ground flora. $P_H = 4.16$.
3. The east side of Fortun-Indelukke, north of the road between Eremitagen and Fortunen.
    a. Pasture: *Agrostis tenuis–Achillea millefolium*-Formation with *Anthoxanthum odoratum, Phleum pratense, Plantago lanceolata, Poa pratensis, Rumex acetosa, Taraxacum* sp., *Trifolium repens*. $P_H = 5.76$.
    b. Sprucewood 50–5 years old with a ground covered with needles and no ground flora. $P_H = 3.92$.
4. The northern portion of Fortun-Indelukke; towards the west side of the uneven portion planted a few years ago (1914) with spruce.
    a. Strip of pasture between the young spruce plantation which is 0·25–1·5 metres high and the older sprucewood to the west of it: *Agrostis tenuis–Festuca ovina*-Formation with *Agrostis canina, Calluna vulgaris* (very small plants), *Campanula rotundifolia, Festuca rubra, Hieracium pilosella, Luzula campestris, Plantago lanceolata, Poa pratensis, Sieglingia decumbens, Trifolium repens*. $P_H = 6.48$.

458   INFLUENCE OF TYPES OF VEGETATION ON SOIL ACIDITY

    b. Sprucewood, very dense and dark, with a thick layer of needles; incipient acid humus formation; no ground flora. $P_H = 3·98$.

5. The same locality as No. 4, but to the east side of the young spruce plantation.

    a. Strip of pasture between the young spruce plantation and the older spruce to the east of it: poor *Agrostis tenuis–Hieracium pilosella*-Formation with *Anthoxanthum odoratum*, *Calluna vulgaris* (very small plants), *Festuca ovina*, *Galium verum*, *Leontodon autumnalis*, *Luzula campestris*, *Plantago lanceolata*, *Poa pratensis*, *Polygala vulgaris*, *Sieglingia decumbens*, *Trifolium repens*. $P_H = 5·16$.

    b. Sprucewood with a thick layer of needles. No ground flora. $P_H = 4·04$.

6. Jægersborg Hegn; north-west of Skodsborg Station.

    a. Pasture: *Agrostis tenuis–Festuca rubra*-Formation with *Achillea millefolium*, *Brunella vulgaris*, *Campanula rotundifolia*, *Cynosurus cristatus*, *Holcus lanatus*, *Hypericum perforatum*, *Lathyrus montanus*, *Leontodon autumnalis*, *L. hispidus*, *Potentilla erecta*, *Ranunculus acer*, *Rumex acetosa*, *R. acetosella*, *Stellaria graminea*, *Taraxacum* sp., *Trifolium pratense*, *Veronica chamaedrys*. $P_H = 5·80$.

    b. Sprucewood with layer of humus 4–6 cm. thick. Recent felling has admitted so much light that *Oxalis acetosella* has begun to come in; otherwise there is no ground flora. $P_H = 3·96$.

Table 6

Pasture—Sprucewood

Locality.	P_H value of the soil in:		P_H value of the pasture exceeds that of the sprucewood by:
	a Pasture.	*b* Sprucewood.	
1. The east side of the southerly portion of Fortun-Indelukke: near the south end of the narrow pasture	6·76	4·10	2·66
2. The same locality. By the west side of the pasture.	6·32	4·16	2·16
3. The east side of Fortun-Indelukke near Eremitageslette: north of the road between Eremitagen and Fortunen	5·76	3·92	1·84
4. The northern part of Fortun-Indelukke near the west side of the young spruce plantation	6·48	3·98	2·50
5. The same locality. Near the east side of the young spruce plantation.	5·16	4·04	1·12
6. Jægersborg Hegn: north-west of Skodsborg Station.	5·80	3·96	1·84
Average	6·05	4·03	2·02

INFLUENCE OF TYPES OF VEGETATION ON SOIL ACIDITY 459

Table 6 shows that in all six localities investigated the sprucewood has greatly increased the acidity of the soil. This increase is from 1·16 to 2·66 expressed in P_H value and the average of all six localities is about 2.

B. Pasture—Beechwood (Table 7)

A comparison between the P_H value of the soil of the pasture with that of the beechwood was undertaken in the following localities:

7. The middle portion of the southern edge of Fortun-Indelukke.
   a. Pasture: *Agrostis tenuis*-Formation with *Anthoxanthum odoratum, Dactylis glomerata, Deschampsia caespitosa, Festuca rubra, Poa pratensis, Ranunculus acer, Rumex acetosa, Stellaria holostea, Veronica chamaedrys, Viola silvestris.* $P_H = 5·48$.
   b. Beechwood 70–80 years old: wind-swept almost naked soil with bud-scales and husks of beech; some *Oxalis acetosella.* $P_H = 4·80$.
8. The same locality as No. 7 but farther east near 'Kjøbenhavns Allé'.
   a. Pasture: *Agrostis tenuis*-Formation with *Achillea millefolium, Anthoxanthum odoratum, Campanula rotundifolia, Carex pallescens, Cynosurus cristatus, Dactylis glomerata, Festuca rubra, Hieracium auricula, Luzula campestris, Plantago lanceolata, Potentilla erecta, Ranunculus acer, Rumex acetosa, Trifolium* sp., *Veronica chamaedrys, Viola* sp. $P_H = 6·32$.
   b. Beechwood 70–80 years old with some *Larix*; the ground partly bare or with bud-scales and husks of beech, partly covered with leaves; apart from some *Oxalis acetosella* there is no ground flora. $P_H = 5·42$.
9. The east side of Fortun-Indelukke, north of the road between Eremitagen and Fortunen.
   a. Pasture (same locality as 3 a): *Agrostis tenuis–Achillea millefolium*-Formation. $P_H = 5·76$.
   b. Beechwood 50–5 years old without ground flora; ground partly bare and partly covered with bud-scales, husks, and leaves of beech. $P_H = 4·56$.
10. The north-east corner of Fortun-Indelukke.
    a. Pasture. This part of the plain has however not lain entirely undisturbed since the bordering portion was planted with beech in about 1880: it was cultivated in 1895–7, since when it has not been touched. $P_H = 6·04$.
    b. Dense beechwood about 40 years old with loose soil rich in humus and a thick layer of leaves. No ground flora. $P_H = 5·44$.
11. Near the western border of a group of oaks on a range of hills east of Hjortekær.
    a. Pasture ('Eremitageslette'): Luxuriant *Agrostis tenuis*-Formation. $P_H = 6·34$.

460 INFLUENCE OF TYPES OF VEGETATION ON SOIL ACIDITY

    b. Dark beechwood about 65 years old immediately to the north of the fence between Stampeskov and Eremitageslette, near the locality *a.* The ground is fairly loose and there is a considerable layer of leaves. No ground flora. $P_H = 5.78$.

12. Near the stile between Eremitageslette and Stampeskov.

    a. Pasture. Luxuriant *Agrostis tenuis–Anthoxanthum odoratum-*Formation with *Achillea millefolium, Campanula rotundifolia, Cirsium arvense, Crataegus* sp. (scattered individuals a few cm. high), *Dianthus deltoides, Festuca ovina, F. rubra, Holcus lanatus, Leontodon autumnalis, Plantago lanceolata, Poa pratensis, Polygala vulgaris, Potentilla reptans, Veronica chamaedrys, Viola canina.* $P_H = 5.82$.

    b. Beechwood about 65 years old with some *Larix*; fair layer of leaves; no ground flora. $P_H = 4.22$.

13. The south-eastern corner of Chr. IX's Hegn.

    a. Pasture (Eremitageslette): *Agrostis tenuis-*Formation with *Anthoxanthum odoratum, Campanula rotundifolia, Cynosurus cristatus, Galium verum, Hieracium pilosella, Holcus lanatus, Luzula campestris, Ranunculus bulbosus, Plantago lanceolata, Poa pratensis, Rumex acetosa, Taraxacum* sp., *Trifolium repens.* $P_H = 6.54$.

    b. Beechwood about 65 years old on hard, almost bare ground. Beech husks and bud-scales but few leaves. Ground flora almost absent; only a few poor individuals of *Arenaria trinervia, Dactylis glomerata, Poa nemoralis.* $P_H = 5.80$.

14. North-westerly group of beeches on Eremitageslette.

    a. Pasture to the south of the east end of the group: *Agrostis tenuis-*Formation with *Achillea millefolium, Carex hirta, Cerastium caespitosum, Hieracium pilosella, Lolium perenne, Plantago lanceolata, Poa pratensis, Ranunculus bulbosus, Stellaria graminea, Trifolium repens.* $P_H = 6.1$.

    b. Beechwood 80 years old with some oak. Nearly bare, very windswept soil with beech husks, bud-scales, and small twigs. Scattered *Poa nemoralis*; no other ground flora. $P_H = 5.38$.

15. Groups of beeches immediately to the south of the Eremitage.

    a. Pasture: *Cynosurus cristatus–Lolium perenne–Agrostis tenuis-*Formation with *Achillea millefolium, Phleum pratense, Plantago lanceolata, Polygonum aviculare, Potentilla reptans, Ranunculus bulbosus, Taraxacum* sp., *Trifolium repens.* $P_H = 6.42$.

    b. Beechwood 80 years old: windswept ground with beech husks, bud-scales, and twigs. Scattered *Dactylis glomerata, Poa annua, Poa nemoralis.* $P_H = 5.86$.

16. Jægersborg Hegn, north-west of Skodsborg Station.
   a. Pasture (in the same locality as 6 a): *Agrostis tenuis–Festuca rubra*-Formation. $P_H = 5.80$.
   b. Beechwood with poor *Oxalis acetosella*-Formation. The soil partly bare and partly covered with beech husks, bud-scales, leaves, and twigs. $P_H = 5.36$.

From this and from Table 7 it will be seen that in ten localities investigated the soil of the beechwood is more acid than that of the corresponding pasture, but the difference is much less than that between the pasture and the sprucewood. It is in fact on an average 0.8 in P_H value, while the sprucewood showed a difference of about 2.

TABLE 7

Pasture—Beechwood

Locality.	P_H value of the soil in:		P_H value of the pasture exceeds that of the beechwood by:
	a Pasture.	b Beechwood.	
7. South edge of Fortun-Indelukke	5.48	4.80	0.68
8. The east side of the southern portion of Fortun-Indelukke.	6.32	5.42	0.90
9. The east side of Fortun-Indelukke north of the road between the Eremitage and Fortunen	5.76	4.56	1.20
10. The north-east corner of Fortun-Indelukke (beechwood 10–15 metres high)	6.04	5.44	0.60
11. The southern edge of Stampeskov near the western boundary of the group of oaks east of Hjortekær	6.34	5.78	0.56
12. Near the stile between Eremitageslette and the south-east corner of Stampeskov	5.82	4.22	1.60
13. South-east corner of Chr. IX's Hegn	6.54	5.80	0.74
14. The north-westerly group of beech trees on Eremitageslette	6.10	5.38	0.72
15. The group of beech trees immediately to the south of the Eremitage	6.42	5.86	0.56
16. Jægersborg Hegn: north-west of Skodsborg Station	5.80	5.36	0.44
Average	6.06	5.26	0.80

C. Pasture—Oakwood

I have had the opportunity of comparing in three places the hydrogen-ion concentration of the soil of the pasture with that of the oakwood planted on the pasture. These oakwoods however differ widely in respect of light and shelter; and since they also show noticeable differences in hydrogen-ion concentration when compared with that of the original pasture we must here treat each wood separately.

The Oak enclosures on Eremitageslette (Table 8). On the plain to the north of the Eremitage there are four small enclosures which during the decade 1885–95 were planted with oak. There is now an undergrowth of hawthorn. The oaks are very close together, and since the hawthorns fill up all the intervening space the shade cast is very deep. And since these groups of trees are small and lie isolated on the plain where they are exposed to all the winds, most of the fallen leaves are swept away and the surface of the soil becomes dried out. The conditions for a ground flora are therefore very bad.

17. The most westerly oak enclosure north-west of the Eremitage.
    a. Pasture immediately to the west of the enclosure: *Agrostis tenuis*-Formation rich in species. $P_H = 5.38$.
    b. Westerly portion of the oak enclosure: the windswept ground is almost bare; here and there are poor specimens of *Holcus mollis* and *Agrostis tenuis*. $P_H = 4.74$.

18. The same enclosure as No. 17.
    a. Pasture immediately to the east of the enclosure: *Agrostis tenuis*-Formation with *Achillea millefolium, Anthoxanthum odoratum, Cynosurus cristatus, Dactylis glomerata, Leontodon autumnalis, Lolium perenne, Phleum pratense, Plantago lanceolata, Ranunculus bulbosus, Rumex acetosa, Trifolium repens, Veronica chamaedrys.* $P_H = 5.26$.
    b. The easterly portion of the oak enclosure; in places the ground is bare and in places there are weak plants of *Holcus mollis, Agrostis tenuis, Urtica dioeca.* $P_H = 4.26$.

19. The oak enclosure farthest to the north-west, north of the Eremitage.
    a. Pasture immediately to the west of the enclosure: *Achillea millefolium, Agrostis tenuis, Anthoxanthum odoratum, Avena pratensis, Cerastium caespitosum, Cirsium arvense, Deschampsia caespitosa, Phleum pratense, Plantago lanceolata, Poa pratensis, Rumex acetosa, Trifolium repens.* $P_H = 6.80$.
    b. The westerly part of the enclosure; ground windswept, partly bare and partly with a few leaves; a small quantity of *Agrostis tenuis* and scattered low *Urtica dioeca.* $P_H = 5.44$.

20. The same enclosure as No. 19.
    a. Grass plain immediately to the east of the enclosure: the vegetation essentially as in 19 a; yet I saw no *Avena pratensis.* $P_H = 6.12$.
    b. The east portion of the enclosure essentially the same as 19 b. $P_H = 4.88$.

INFLUENCE OF TYPES OF VEGETATION ON SOIL ACIDITY

21. The south-easterly oak enclosure north-east of the Eremitage.
    a. The pasture immediately to the north of the enclosure: *Cynosurus cristatus–Agrostis tenuis*-Formation. $P_H = 5\cdot88$.
    b. The northerly part of the enclosure: ground partly bare and partly covered with leaves; here and there the layer of leaves is rather thick. $P_H = 4\cdot92$.
22. Same enclosure as No. 21.
    a. Pasture immediately to the south of the enclosure. $P_H = 6\cdot02$.
    b. Southern portion of the enclosure, partly bare, partly covered with leaves; here and there poor *Agrostis tenuis*. $P_H = 4\cdot96$.

TABLE 8

Pasture—Oakwood. Oak enclosures on Eremitageslette

Locality.	P_H value of the soil in:		P_H value of the pasture exceeds that of the oakwood by:
	a Pasture.	b Oakwood.	
17. The west side of the most westerly enclosure	5·38	4·74	0·64
18. The east side of the most westerly enclosure	5·26	4·26	1·00
19. The west side of the north-western enclosure	6·80	5·44	1·36
20. The east side of the north-western enclosure	6·12	4·88	1·24
21. The north side of the south-eastern enclosure	5·88	4·92	0·96
22. The south side of the south-eastern enclosure	6·02	4·96	1·06
Average	5·91	4·87	1·04

It follows from this and from Table 8 that the soil of the oakwood is here considerably more acid than that of the pasture, the difference amounting on an average to about 1 in P_H value. It also follows that it is on an average more acid both absolutely and relatively than that of the beechwood localities given in Table 7. The decisive factor for the soil of these small oak enclosures is the shade caused by the density of the wood in connexion with the complete absence of shelter, which allows the wind to blow freely through and to stop the development of the low woodland plants which can endure shade but which demand shelter. This results in the drying out of the upper layers of the soil.

Egesaaten (Table 9). The enclosure immediately east of the southern portion of Ulvedalene was planted with oak 20 or 30 years ago; at all events in the localities we are treating beech was planted (later?) among the oak, so that the ground in these places is deeply shaded. Scattered ancient beeches and oaks are found in the enclosure, and immediately outside it there are also some old trees which cast some shade on to the soil within.

23. The north-western portion of the enclosure and the pasture to the north of it.
   a. Pasture: *Agrostis tenuis*-Formation with *Achillea millefolium, Anthoxanthum odoratum, Campanula rotundifolia, Cynosurus cristatus, Deschampsia caespitosa, Festuca ovina, Leontodon autumnalis, Luzula campestris, Plantago lanceolata, Potentilla erecta, Rumex acetosa, Sieglingia decumbens, Trifolium repens, Veronica chamaedrys, Viola canina, Hylocomium squarrosum*, and other *Hypnaceae*. $P_H = 5·86$.
   b. Oakwood with undergrowth of low beech: dark, without ground flora; good layer of leaves; the ground is friable and harbours earth-worms. *Oxalis acetosella* and *Stellaria media* were entering the wood in this neighbourhood. $P_H = 5·10$.

24. The north-easterly portion of the enclosure and the pasture and the neighbouring pasture.
   a. Pasture: *Agrostis tenuis–Deschampsia caespitosa*-Formation with *Anthoxanthum odoratum, Carex hirta, Cynosurus cristatus, Festuca rubra, Juncus conglomeratus, Lotus corniculatus, Plantago lanceolata, Ranunculus acer, Rumex acetosa, Veronica chamaedrys*. $P_H = 6·40$.
   b. Oakwood with low beech as undergrowth; layer of leaves; no ground flora. $P_H = 5·26$.

25. The east side of the enclosure and the pasture to the east of it.
   a. Pasture in a glade among old beech trees, partly (1) shaded *Agrostis tenuis*-Formation with much moss and with *Anthoxanthum odoratum, Dactylis glomerata, Deschampsia caespitosa, Ranunculus acer*; the chief moss here is *Hylocomium squarrosum*; and partly (2) *Urtica dioeca–Carex remota*-Formation with *Dactylis glomerata, Deschampsia caespitosa, Juncus effusus, Melica uniflora, Oxalis acetosella, Rumex nemorosus*. (*Agrostis tenuis*-Formation: $P_H = 6·5, 6·2$; *Urtica–Carex*-Formation: $P_H = 6·6, 6·3$; the boundary region between both: $P_H = 6·5$.) $P_H = 6·42$.
   b. Oakwood with undergrowth of low beech surrounded by old beech trees: very dark; good layer of leaves; earth-worms present; no ground flora except *Anemone nemorosa* and *Oxalis acetosella*. Near this locality species begin to come in from the better lighted pure oakwood lying to the west, viz. *Carex remota, Deschampsia caespitosa, Melica uniflora, Urtica dioeca*. $P_H = 5·58$.

26. The south-easterly portion of the enclosure and the pasture to the south of it.
   a. The pasture in a glade among old beech trees: partly (1) shaded *Agrostis tenuis*-Formation with *Dactylis glomerata, Deschampsia caespitosa, Melica uniflora, Oxalis acetosella, Ranunculus acer*; and

partly (2) *Urtica dioeca–Melica uniflora*-Formation with *Deschampsia caespitosa, Festuca gigantea, Geranium robertianum, Juncus effusus, Rumex nemorosus*. $P_H = 5 \cdot 66$.

b. Oakwood with undergrowth of low beech: thick layer of leaves; very dark; no ground flora. $P_H = 5 \cdot 10$.

TABLE 9

Pasture—Oakwood. Egesaaten (east of Ulvedalene)

Locality.	P_H value of the soil in:		P_H value of the pasture exceeds that of the oakwood by:
	a Pasture.	b Oakwood.	
23. Egesaaten: north-western part	5·86	5·10	0·76
24. " north-eastern part	6·40	5·26	1·14
25. " east side	6·42	5·58	0·84
26. " south-east corner	5·66	5·10	0·56
Average	6·09	5·26	0·83

Just as in the oak enclosures on Eremitageslette the parts of Egesaaten which have been investigated are almost devoid of ground flora. But at Egesaaten there is much better shelter and the ground is for the most part covered by a layer of leaves. Corresponding with this the difference between the acidity of the soil in the oakwood and pasture is not so great at Egesaaten (Table 9) as it is in the oak enclosures on Eremitageslette. Because of the beeches which have been planted beneath the oaks and because of the large shade-casting beech trees at the side of the wood, the part of Egesaaten investigated behaves, as far as shade is concerned, exactly like a young beechwood; the relationship between the P_H value of the soil of the wood and that of the corresponding pasture is here also seen to be essentially the same as that of the beechwood and that of the corresponding pasture. The acidity of the soil of the beechwood was 0·80 higher in P_H value than that of the pasture, while the corresponding number for Egesaaten was 0·83.

Stampeskov (Table 10). The southern portion of Stampeskov, which abuts upon Eremitageslette, consists for the most part of an oakwood 65 years old. In several places beech was planted some years ago among the oak trees on the outskirts of the wood; but these beeches are only about the height of a man and have not yet been able to shade out the ground flora that was there before them. The wood is well illuminated, the canopy being sufficiently open to permit enough sunlight at various times to reach the ground and allow the growth of a luxuriant tall ground flora. This ground flora is usually not dense enough to exclude light from the ground, over which it casts a flimsy veil of shade. This ground flora consists in most places of *Rubus idaeus*, among which are found scattered other species, especially: *Avena elatior, Dactylis glomerata, Deschampsia caespitosa, Fragaria vesca, Hypericum perforatum, Lampsana communis, Mercurialis perennis, Oxalis acetosella, Stachys silvatica, Stellaria holostea*.

The favourable illumination and shelter determine a rich and manifold fauna both above and beneath the ground; and the ground, because of the sunlight that reaches it, becomes warmer than that of the corresponding deeply shaded beechwood. This presumably fosters the microflora which assists in the speedy conversion and use of dead organic substances. At any rate in the summer time, in spite of the fact that a large quantity of leaves are retained by the ground flora, we find only a thin layer of leaves and dead portions of herbs compared with what obtains in the dark neighbouring beechwood. Here and there bare ground is seen. This is partly due to the activity of digging animals. *Catharinea undulata* often occurs on these bare patches. Because of the rich fauna the soil is loose and friable, much more so than under the beech canopy, which is so dense that no summer ground flora can thrive there.

In six places along the boundary between this oakwood and Eremitageslette I have determined the relationship between the acidity of the soil in the oakwood and of the corresponding portion of the pasture.

27. Low ground near Hjortekær.
    a. Pasture: *Achillea millefolium, Agrostis tenuis, Anthoxanthum odoratum, Calluna vulgaris, Campanula rotundifolia, Dianthus deltoides, Festuca ovina, Galium verum, Hieracium pilosella, Leontodon autumnalis, Lotus corniculatus, Plantago lanceolata, Potentilla erecta, Polygala vulgaris, Sieglingia decumbens, Thymus serpyllum, Trifolium repens, Viola canina*. The frequency percentage of *Calluna vulgaris* is here 100, but the plants are very small. $P_H = 6.50$.
    b. Oakwood: Originally there was certainly beech among the oak here, but the wood in the part investigated contained only oak. There was practically no raspberry. The ground flora consisted especially of *Anthriscus silvester, Avena elatior, Carex hirta, Dactylis glomerata, Deschampsia caespitosa, Oxalis acetosella, Stellaria holostea*. $P_H = 5.68$.

28. The westerly portion of the range of hills east of No. 27.
    a. Pasture: *Agrostis tenuis*-Formation with *Achillea millefolium, Anthoxanthum odoratum, Calluna vulgaris* (individuals small and low), *Campanula rotundifolia, Festuca ovina, Galium verum, Hieracium pilosella, Holcus lanatus, Lathyrus montanus, Luzula campestris, Poa pratensis, Potentilla erecta, Rumex acetosa, R. acetosella, Trifolium repens, Veronica chamaedrys, Viola canina*. $P_H = 5.42$.
    b. Oakwood with young beeches as high as a man, not covering the ground; layer of leaves: *Agrostis tenuis, Avena elatior, Dactylis*

glomerata, Deschampsia caespitosa, Hypericum perforatum, Oxalis acetosella, Rubus idaeus (weak), *Stellaria holostea*. Of mosses there were *Catharinea undulata* and some *Hypnum* (*purum*?). $P_H = 4·74$.

29. Slightly farther to the east than No. 28, near the western edge of the group of old oaks on Eremitageslette east of Hjortekær.
    a. Pasture: *Agrostis tenuis*-Formation luxuriant and rich in species. $P_H = 6·34$.
    b. Oakwood with beeches up to the height of a man, which, however, do not cover the ground; scrub of raspberry bushes with the accompanying species mentioned above (p. 466). $P_H = 5·86$.

30. Farther east than No. 29, outside the space between the westerly and easterly oaks in the group to the east of Hjortekær.
    a. Pasture: Luxuriant *Agrostis tenuis*-Formation with *Agrostis canina, Anthoxanthum odoratum, Campanula rotundifolia, Carex muricata, Deschampsia caespitosa, Galium verum, Hieracium pilosella, Holcus lanatus, Hypericum perforatum, Leontodon autumnalis, Lotus corniculatus, Phleum pratense; Plantago lanceolata, Poa pratensis, Potentilla erecta, Ranunculus acer, Rumex acetosa, R. acetosella, Veronica chamaedrys*. $P_H = 6·42$.
    b. Oakwood: scrub of raspberries with the species mentioned above. $P_H = 6·30$.

31. East of the group of oak trees mentioned in Nos. 29–30.
    a. Pasture: Luxuriant *Agrostis tenuis–Anthoxanthum odoratum*-Formation (= 12 *a*). $P_H = 5·82$.
    b. Oakwood between the fence and the road running between Raavad and Hjortekær. $P_H = 5·92$.

32. The same place as No. 31.
    a. Pasture (= 31 *a*). $P_H = 5·82$.
    b. Oakwood immediately to the north of the road between Raavad and Hjortekær. Ground flora with raspberries and the accompanying species. $P_H = 5·46$.

Table 10 gives the results of investigations in the well-illuminated parts of the oakwood compared with the corresponding parts of the pasture. By comparing this table with the earlier ones we see that of the types of wood investigated the oakwood is the type that increases the acidity of the original pasture less than any other type. An average of six localities gives an increase of acidity amounting to only 0·39. This difference is so slight that we need not be surprised that in one locality (No. 31) the soil of the oakwood was slightly less acid than that of the pasture.

By comparing Tables 6–10 it is apparent that within the district investigated the wood increases the acidity of the soil of the original

Table 10

Pasture—Oakwood. Stampeskov

	P_H value of the soil in:		P_H value of the pasture exceeds that of the oakwood by:
	a Pasture.	*b* Oakwood.	
27. Low ground near Hjortekær	6·50	5·68	0·82
28. The western portion of the range of hills east of No. 27	5·42	4·74	0·68
29. Slightly to the east of No. 28, near the western margin of the group of old oak trees east of Hjortekær	6·34	5·86	0·48
30. Slightly to the east of No. 29, outside the space between the westerly and the easterly oaks in the group east of Hjortekær	6·42	6·30	0·12
31. East of the above-mentioned group of oaks	5·82	5·92	−0·10
32. The same place as No. 31 (see text)	5·82	5·46	0·36
Average	6·05	5·66	0·39

pasture, and that it does so in varying degrees, the soil becoming more acid as the wood gives more shade. The well-illuminated oakwood makes the soil least acid; the dark beechwood makes it rather acid, as does the oakwood when it contains undergrowth of beech, hawthorn, &c. The dense dark sprucewood makes the soil more acid than any, the difference between the soil of this wood and that of the original pasture on which the wood was planted amounting to even more than 2 degrees in P_H value.

In perfect correspondence with this we have the result of the following investigations of the P_H value of the soil in different types of wood of the same age growing on what was originally the same soil.

Different types of wood on the same soil and of the same age.

A. Sprucewood and beechwood on what was originally the same soil and of the same age (Table 11).

33. The east side of the southern portion of Fortun-Indelukke; wood 70–80 years old.
    a. Sprucewood (= 2*b*). $P_H = 4·16$.
    b. Beechwood (= 8*b*). $P_H = 5·42$.
34. The east side of Fortun-Indelukke north of the road between Eremitagen and Fortunen: wood 50–5 years old.
    a. Sprucewood (= 3*b*). $P_H = 3·92$.
    b. Beechwood (= 9*b*). $P_H = 4·56$.
35. Jægersborg Hegn: north-west of Skodsborg Station.
    a. Sprucewood (= 6*b*). $P_H = 3·96$.
    b. Beechwood (= 16*b*). $P_H = 5·36$.

TABLE 11

Sprucewood and beechwood on what was originally the same ground and of the same age.

Locality.	P_H value of the soil in:		P_H value of the beechwood exceeds that of the sprucewood by:
	a Sprucewood.	b Beechwood.	
33. The east side of the southern portion of Fortun-Indelukke.	4·16	5·42	1·26
34. The east side of Fortun-Indelukke north of the road between Eremitagen and Fortunen	3·92	4·56	0·64
35. Jægersborg Hegn: north-west of Skodsborg Station	3·96	5·36	1·40
Average	4·01	5·11	1·10

All three localities show that the soil of the sprucewood is considerably more acid than that of the beechwood, the average difference being 1·1.

B. Beechwood and oakwood on what was originally the same soil (Table 12). All the localities belonging here are situated to the south side of Stampeskov. The oakwood localities are partly the same as those in Table 10, and have the ground flora already mentioned.

36. Near the western margin of a group of old oak on the hill east of Hjortekær.
    a. Beechwood (= 11 b). $P_H = 5·78$.
    b. Oakwood (= 29 b); young beeches planted amongst the oak. $P_H = 5·86$.

37. South-east of the woodland marsh on the south side of Stampeskov.
    a. Beechwood planted around a group of old oaks and presumably somewhat younger than the adjacent oakwood: good layer of leaves; rather loose soil; deep shade; no ground flora. $P_H = 5·82$.
    b. Oakwood. $P_H = 6·22$.

38. Near the middle part of the north side of the oakwood.
    a. Beechwood: good layer of leaves; soil rather loose; deep shade; no ground flora. $P_H = 5·56$.
    b. Oakwood. $P_H = 6·32$.

39. Slightly to the east of No. 38.
    a. Beechwood: good layer of leaves; rather loose soil; no ground flora. $P_H = 5·06$.
    b. Oakwood. $P_H = 5·90$.

40. In the easterly part of the oakwood.
    a. Beechwood: little group (10–12 metres in diameter) of beech trees planted around an old oak. Good layer of leaves; here and there

isolated plants of *Oxalis acetosella* and *Anemone nemorosa*; no other ground flora. $P_H = 5 \cdot 28$.
    b. Oakwood. $P_H = 6 \cdot 06$.
41. Near the eastern margin of the oakwood.
    a. Beechwood: a good layer of leaves; no ground flora. $P_H = 4 \cdot 22$.
    b. Oakwood. $P_H = 5 \cdot 46$.

TABLE 12

Beechwood and oakwood on what was originally the same soil, and of the same age. (Stampeskov)

	P_H value of the soil in:		P_H value of the oakwood exceeds that of the beechwood by:
	a Beechwood.	*b* Oakwood.	
36. Near the western margin of the group of old oaks on the hill east of Hjortekær . .	5·78	5·86	0·08
37. South-east of the woodland marsh on the south side of Stampeskov . . .	5·82	6·22	0·40
38. Near the middle part of the north side of the oakwood	5·56	6·32	0·76
39. Slightly to the east of 38	5·06	5·90	0·84
40. Eastern part of the oakwood . . .	5·28	6·06	0·78
41. Near the east edge of the oakwood . .	4·22	5·46	1·24
Average	5·29	5·97	0·68

Again we see from this and from the summary in Table 12 that the soil of the beechwood is considerably more acid than that of the oakwood, the average difference being 0·68.

WOODLAND—PASTURE

The foregoing investigations of soil acidity in a series of different woods show that the wood makes the soil more acid, and that it is the dense, deeply shading wood that determines the greatest acidity, whether the wood be of spruce, of beech, or of oak whose shade is rendered more dense either by undergrowth or by adjacent beechwood. We have reached this result in our investigations of a series of different places by examining level areas situated on what was apparently originally the same kind of soil, and was originally covered by pasture. Part of this pasture has for a varying number of years been planted with wood, while other parts have remained as pasture to the present day.

The next thing to be done was to find out to what extent the increasing acidity which results when wood is planted on the pasture can be reversed when the wood is removed and the ground again allowed to go to grass. For the investigation of this problem too the Dyrehave, as mentioned

INFLUENCE OF TYPES OF VEGETATION ON SOIL ACIDITY

above, is well suited, since there are places in the northern part of Fortun-Indelukke, in a line with Fortun Fort, along the margins of most of the young mixed wood felled in 1914 for military reasons, where individual trees of the old wood have disappeared and the ground has gone back to grass. I shall now give an account of my investigations of the changes in the acidity of the soil occurring when the wood gives place to pasture.

A. Sprucewood—Pasture (Table 13)

42. Fortun-Indelukke: the south side of the eastern end of the portion cleared in 1914.
    a. Sprucewood 50–5 years old, dense, dark, and without ground flora. $P_H = 3.92$.
    b. Pasture 7 years old occupying ground originally like that of a. $P_H = 4.96$.
43. The same place as No. 42, on the north side of the cleared portion.
    a. Sprucewood 50–5 years old with thick layer of needles and without ground flora. $P_H = 4.18$.
    b. Pasture 7 years old: *Agrostis tenuis*-Formation. $P_H = 5.30$.
44. North of the east end of the cleared portion.
    a. Sprucewood 50–5 years old with thick layer of needles and no ground flora. $P_H = 4.08$.
    b. Pasture 7 years old: in 1914–15 some of the rows of spruces were cut down, their places being taken by small areas of pasture. These strips of pasture which run from west to east are only about 7 metres broad and are covered with an *Agrostis tenuis*-Formation rich in mosses. This formation is noticeably affected by the shade of the tall spruce trees between which it lies. List of species: *Achillea millefolium, Agrostis canina, A. tenuis, Anthoxanthum odoratum, Calluna vulgaris, Campanula rotundifolia, Cerastium caespitosum, Cirsium arvense, Dactylis glomerata, Deschampsia caespitosa, Festuca rubra, Fragaria vesca, Galeopsis tetrahit, Galium harcynicum, Gnaphalium silvaticum, Holcus lanatus, Hypericum perforatum, Knautia arvensis, Leontodon autumnalis, Plantago lanceolata, Poa pratensis, Ranunculus repens, Rumex acetosella, Stellaria graminea, Taraxacum sp., Urtica dioeca, Veronica chamaedrys, V. officinalis, Vicia cracca, Viola canina, V. silvestris.* $P_H = 5.42$.

The summary in Table 13 shows us that the acidity of the pasture is considerably lower than that of the sprucewood; in the seven years that have elapsed since the felling of the sprucewood the acidity of the soil has diminished on an average 1·17 in P_H value.

TABLE 13

Sprucewood—Pasture. The northern portion of Fortun-Indelukke: the part cleared in 1914.

Locality.	P_H value of the soil in:		P_H value of the pasture exceeds that of the sprucewood by:
	a Sprucewood.	b Pasture.	
42. Near the east end of the southern boundary of the felled portion	3·92	4·96	1·04
43. On the north side of the felled portion .	4·18	5·30	1·12
44. North of the east end of the felled portion (see text)	4·08	5·42	1·34
Average	4·06	5·23	1·17

B. Young and Old Beechwood—Pasture (Table 14)

45. Fortun-Indelukke: near the west end of the north side of the part cleared in 1914.
   a. Beechwood 50–60 years old with oak and larch; many twigs and bud-scales but only a few leaves on the ground; scattered weak *Agrostis tenuis*. $P_H = 4·44$.
   b. Pasture 7 years old: *Agrostis tenuis*-Formation with *Anthoxanthum odoratum, Campanula rotundifolia, Holcus mollis, Hypericum perforatum, Potentilla erecta, Rumex acetosa, R. acetosella*; the commonest moss is *Hylocomium squarrosum*. $P_H = 5·50$.

46. The same place as No. 45 but farther to the east.
   a. Beechwood 50–60 years old; a slightly better illumination allowing *Agrostis tenuis* to enter. Numerous bud-scales, beech husks, and twigs; only few leaves; most of the ones present have been held by *Agrostis tenuis*. $P_H = 4·92$.
   b. Pasture 7 years old with *Agrostis tenuis*-Formation containing *Achillea millefolium, Anthoxanthum odoratum, Campanula rotundifolia, Carex muricata, Cirsium arvense, Hypericum perforatum, Juncus effusus, Linaria vulgaris, Rumex acetosella, Stellaria graminea, Taraxacum* sp., *Veronica chamaedrys, Viola canina*. $P_H = 5·44$.

47. The same place as Nos. 45 and 46, but towards the east end of the south side of the cleared portion.
   a. Beechwood 50–60 years old without ground vegetation. Bud-scales, beech husks, and twigs. Very few leaves. $P_H = 4·56$.
   b. Pasture 7 years old: *Agrostis tenuis*-Formation with *Agrostis canina, Campanula rotundifolia, Carex muricata, Cirsium arvense, C. lanceolatum, Deschampsia caespitosa, Galium harcynicum, Holcus*

lanatus, Juncus conglomeratus, J. effusus, Linaria vulgaris, Luzula campestris, Rumex acetosella, Stellaria graminea, Taraxacum sp., *Veronica chamaedrys*. $P_H = 5.38$.

48. Near the stile at the south-east corner of Stampeskov.
    a. Beechwood 60–70 years old: good covering of leaves; no ground flora. $P_H = 4.22$.
    b. Pasture 8 years old: small plot cleared of wood in 1913 and now covered with *Agrostis tenuis*-Formation containing *Campanula rotundifolia, Carex muricata, Dactylis glomerata, Dianthus deltoides, Gnaphalium silvaticum* (weak), *Hieracium auricula, Juncus effusus* (weak), *Rubus idaeus* (very small and weak), *Trifolium (medium?), Urtica dioeca* (weak), *Veronica chamaedrys*. $P_H = 5.36$.

49. Jægersborg Hegn south of Bøllemose.
    a. Beechwood with a poor open ground flora of *Asperula odorata, Carex pilulifera, Luzula pilosa, Melica uniflora, Oxalis acetosella, Veronica chamaedrys, Viola silvestris*. $P_H = 5.24$.
    b. Pasture in all probability about 10 years old with luxurious *Juncus effusus–Agrostis tenuis*-Formation containing among other plants *Chamaenerium angustifolium, Cirsium arvense, Deschampsia caespitosa, Digraphis arundinacea, Holcus lanatus, Plantago lanceolata, Rubus idaeus*. $P_H = 6.12$.

TABLE 14

Young and old beechwood—Pasture.

Locality.	P_H value of the soil in:		P_H value of the pasture exceeds that of the beechwood by:
	a Beechwood.	b Pasture.	
45. Near the west end of the north side of the portion felled in 1914	4·44	5·50	1·06
46. The same place as 45 but farther to the east	4·92	5·44	0·52
47. Near the east end of the south side of the felled portion	4·56	5·38	0·82
48. Near the stile at the south-eastern corner of Stampeskov	4·22	5·36	1·14
49. Jægersborg Hegn: south of Bøllemose	5·24	6·12	0·88
Average	4·68	5·56	0·88

Here too the acidity of the soil has decreased after the felling of the wood has allowed pasture to occupy the ground (N.B. The area has now been planted with spruce.) The average difference in P_H value is, as seen in Table 14, 0·88. This is about the same difference as was found between the soil of the beechwood and that of the original pasture (cf. Table 7).

C. OLD BEECHWOOD—NETTLE GLADE (Table 15)

Every year some of the venerable Dyrehave beeches fall, allowing glades to appear, in which the ground, even where vegetation was absent before, often gradually becomes covered with nettles (*Urtica dioeca*). A great many leaves are caught in the nettle stems, favouring a more abundant fauna and a microflora richer and different from that which thrived in the acid soil under the old beech trees, which was deeply shaded, and often wind-swept. With suitable illumination and humidity the nettles may for a long time dominate the glade, completely covering the ground in summer, even though they may attain to no great height. If the light becomes stronger, as it may for example where several trees fall, then a series of other herbs get a foothold, e.g. *Mercurialis perennis*, broadleaved grasses as *Festuca gigantea* and *Dactylis glomerata*; further, *Deschampsia caespitosa* and others. The nettles are then repressed, and, unless the ground be too wet, it will be gradually occupied by the ordinary *Agrostis tenuis*-Formation of the pasture with various species determined by the varying conditions. This new kind of pasture will be discussed in the next section. Here we shall only deal with the nettle glade.

50. North-west of 'Første Tøjreslag'.
    a. Beneath the old beech trees: ground almost without leaves, but with bud-scales, beech husks, and twigs; no moss but the uppermost 1–2 cm. of the soil rich in humus. Scattered *Oxalis acetosella*, here and there *Melica uniflora* and some weak individual plants of *Deschampsia caespitosa* and *Dactylis glomerata*. $P_H = 4.96$.
    b. Nettle glade with fairly good covering of leaves; abundant earthworms; besides *Urtica dioeca* there were found *Aspidium filix foemina, Dactylis glomerata, Deschampsia caespitosa, Nepeta glechoma, Ranunculus repens*. $P_H = 6.12$.

51. South of 'Første Tøjreslag'.
    a. Beneath the old beech trees: here and there a few leaves on the ground; numerous bud-scales, twigs, and beech husks; the upper layer of soil rich in humus; no ground flora. $P_H = 4.84$.
    b. Nettle glade, with *Festuca gigantea* and other species scattered about. Fairly good layer of leaves. $P_H = 5.90$.

52. Ørnekuldsbakke to the west of Kildesø.
    a. Under an old beech tree: no leaves on the ground, but bud-scales, twigs, and beech husks; the top 1–2 cm. rich in humus. Scattered weak plants of *Dactylis glomerata* and *Deschampsia caespitosa*; no other ground flora. $P_H = 4.42$.
    b. Nettle glade: very dense, almost a pure community of nettles, which reach about to the knees; good layer of leaves. $P_H = 6.56$.

53. Near the stile by the south-east corner of Stampeskov.
   a. Beneath old beech trees: few leaves, but bud-scales and beech husks. Slight lateral illumination; scattered *Oxalis acetosella*, *Stellaria holostea*, and *Viola silvestris*. $P_H = 4.26$.
   b. A fairly new glade in which *Rubus idaeus* and *Urtica dioeca* bulk large. There also occur *Anemone nemorosa*, *Cirsium arvense*, *Festuca gigantea*, and *Oxalis acetosella*. $P_H = 4.70$.

TABLE 15
Old beechwood—Nettle glade.

Locality.	P_H value of the soil in:		P_H value of the nettle glade exceeds that of the beechwood by:
	a Beechwood.	b Nettle glade.	
50. North-west of 'Første Tøjreslag'	4.96	6.12	1.16
51. South of 'Første Tøjreslag'	4.84	5.90	1.06
52. Ørnekuldsbakke	4.42	6.56	2.14
53. Near the stile by the south-eastern corner of Stampeskov	4.26	4.70	0.44
Average	4.62	5.82	1.20

We see then (cf. Table 15) that the soil of the nettle glade is less acid than the soil beneath the beech trees, the average difference being P_H 1.2. But while the acidities of the four beechwood localities show only slight differences varying from 4.26 to 4.96, the differences in the corresponding nettle glades is much greater, the P_H value varying from 4.70 to 6.56, which gives a difference of 1.86 against 0.70 of the beechwood localities. In all probability the cause of this discrepancy is the different ages of the nettle glades. The nettle glade in locality No. 53 is certainly the most recent, showing the least difference from the P_H value in the beechwood. The glade in locality No. 52 is, as far as I know, the oldest, and, corresponding with its greater age, shows the widest deviation from the soil of the corresponding beechwood.

D. OLD BEECHWOOD—PASTURE (Table 16)

54. Ørnekuldsbakke west of Kildesø.
   a. Under an old beech tree (same locality as No. 52a). $P_H = 4.42$.
   b. Pasture beside the nettle vegetation of locality No. 52b. *Agrostis tenuis*-Formation with *Cynosurus cristatus*, *Deschampsia caespitosa* (scattered), *Poa pratensis*, *Ranunculus acer*, *Rumex acetosa*, *Trifolium repens* (abundant), *Veronica chamaedrys*. $P_H = 5.98$.
55. Immediately to the north of Egesaaten.
   a. Under the edge of the canopy of old beech trees: many leaves; because of the lateral illumination scattered plants make their

appearance, such as a few individuals of *Carex hirta*, *Dactylis glomerata*, *Oxalis acetosella*. $P_H = 4.88$.

    b. Pasture with *Agrostis tenuis–Deschampsia caespitosa*-Formation. $P_H = 6.40$.

56. To the west side of the plain south of Schimmelmanns Vildthus.

    a. Under the old beech trees: bud-scales, twigs, and beech husks, but few leaves. No ground flora. $P_H = 5.22$.

    b. Pasture immediately outside the canopy of the old beech trees. The beeches which stood here disappeared long ago. *Festuca rubra–Cynosurus cristatus–Agrostis tenuis*-Formation with *Achillea millefolium*, *Dactylis glomerata*, *Galium verum*, *Leontodon autumnalis*, *Plantago lanceolata*, *Rumex acetosa*, *Trifolium* sp., various mosses. $P_H = 6.32$.

57. South-west of Schimmelmanns Vildthus.

    a. Under old beech trees: Bud-scales, twigs, and beech husks; few leaves. A few plants are entering the outermost part. $P_H = 5.31$.

    b. Pasture outside the canopy: *Agrostis tenuis*-Formation with *Achillea millefolium*, *Brunella vulgaris*, *Cerastium caespitosum*, *Cynosurus cristatus*, *Dactylis glomerata*, *Festuca rubra*, *Juncus conglomeratus* (scattered weak individuals), *Plantago lanceolata*, *Poa pratensis*, *Rumex acetosa*, *Stellaria graminea*, *Trifolium repens*, *Veronica chamaedrys*. $P_H = 5.82$.

58. Near the north-west corner of Rødelyngen.

    a. Under old beech trees: scales and beech husks; very few leaves; scattered, weak, shaded specimens of *Dactylis glomerata*. $P_H = 4.82$.

    b. Pasture immediately outside the canopy: luxurious *Agrostis tenuis*-Formation, rich in species. $P_H = 5.82$.

TABLE 16

Old beechwood—Pasture.

Locality.	P_H value of the soil in:		P_H value of the pasture exceeds that of the beechwood by:
	a Beechwood.	*b* Pasture.	
54. Ørnekuldsbakke	4·42	5·98	1·56
55. Immediately to the north of Egesaaten .	4·88	6·40	1·52
56. On the west side of the plain south of Schimmelmanns Vildthus . . .	5·22	6·32	1·10
57. South-west of Schimmelmanns Vildthus .	5·31	5·82	0·51
58. Near the north-west corner of Rødelyngen .	4·82	5·82	1·00
Average	4·93	6·07	1·14

INFLUENCE OF TYPES OF VEGETATION ON SOIL ACIDITY

It must be a long time ago that the portions of pasture given in Table 16 were covered with old beechwood, and the vegetation now covering them gives the impression of being in an advanced stage of stability. Naturally the acidity of the soil is not quite the same in the investigated pastures; but by comparing Table 16 with Table 7 we see that the average degree of acidity is the same in the relatively primitive pasture (Table 7) as it is in the pasture (Table 16) that in course of time has arisen on the ground formerly occupied by beechwood. On the contrary the soil of the old beechwood (Table 16) is considerably more acid than that of the young beechwood and that of the wood of intermediate age (Table 7).

Summarizing the result of all these investigations, we may say that, **compared with pasture, woodland makes the soil more acid, and the deeper the shade of the wood the greater is the degree of acidity. Further, the soil returns to its original degree of acidity if the wood again gives place to pasture which is allowed to occupy the ground for a long period.**

Supposing nature be left to herself, woodland in one form or another will probably in most parts of the world be the ultimate vegetation; only here and there and before the final stages are reached will the woodland give place to grassland. The relationship between acidity of woodland and grassland soil in the warmer parts of the earth have to be investigated more closely. In the temperate zones woodland belongs to the chief type of vegetation which we have shown to increase the acidity of the soil. And since woodland in these regions, supposing nature be left to herself, will be the ultimate vegetation, we are justified in concluding that these parts of the world become progressively more acid, apart, that is, from the interference of cultivators.

THE REASONS WHY WOODLAND SOIL BECOMES MORE ACID THAN THE SOIL OF PASTURE

There can be scarcely any doubt that we are here confronted with a very intricate problem. In order to get as near as possible to its solution we must first resolve it into its component parts, and investigate each of these separately. The results then obtained can be used as a foundation on which a satisfactory explanation of the phenomenon may one day be built.

The first question is this: Do flowering plants directly alter the acidity of the soil, or do they do so indirectly by changing the conditions which determines the alteration in acidity, or do they do both these things?

We know that plants can alter the acidity of a culture solution, so that we may suppose that they are able to some extent to affect the acidity of the soil in which they grow. And since it is quite possible that different species behave differently in this respect, and that even the same species may behave differently under different conditions of soil, it is very desirable to make experimental trials. But the hydrogen-ion concentration may not only differ widely in different places within the same small region, but it can even differ in different samples of the same well-mixed soil mass. It is clear then that in making an experiment of this kind we cannot rely upon the determination of the P_H values of isolated samples, but we must undertake the determination of a series of samples of the soil used in each experiment. Of work which satisfies this demand I will avail myself only of the results of one small series of experiments. I shall mention these experiments although they extended only over two months, and therefore cannot be considered decisive.

At the beginning of October 1921 some oats, barley, rye, and wheat were sown in different kinds of soil, namely a grassland soil from Nørrefælled near Copenhagen and peat soil from *Calluna*-heath. Each kind of soil was carefully sifted and mixed, of each there were taken two sets of five samples, and their hydrogen-ion concentrations were determined, with the results given in Table 17.

TABLE 17

	Average P_H value of		
	5 samples.	5 samples.	all 10 samples.
Peat soil from *Calluna*-heath	4·48	4·56	4·52
Grassland soil	4·76	4·8	4·78

On the 6th of October flower-pots of equal size were filled with soil, ten with heath soil and ten with grassland soil. Two pots of each kind were sown with oats; and in the same way four pots were sown with barley, four with rye, and four with wheat. In each of the pots 100 selected grains were sown. Nothing was sown in the four remaining pots, two of which contained heath soil and two grassland soil; they were treated during the period of the experiment exactly like the rest, and were kept suitably moist by watering with distilled water.

All four species germinated between the 9th and the 12th of October, and they all germinated well; first the rye germinated, then the wheat, then the barley, last of all the oats. The germination always began first in the peat soil. During the whole period of the experiment they were nearly all equally vigorous, and there was no noticeable difference in vigour between plants growing in the two kinds of soil.

Between the 6th and the 11th of December the P_H value of the soil was determined in ten pots belonging to each set. Each of eight pots with plants growing in them had their earth freed as far as possible from roots and then well mixed. Similarly the earth from each of the pots containing no plants was well mixed. The P_H value was determined in ten samples from the soil from each of the ten pots. These samples included about

INFLUENCE OF TYPES OF VEGETATION ON SOIL ACIDITY

two-thirds of the whole mass of soil. The result is seen in Table 18, in which are also given the P_H values of the soil shown at the beginning of the experiment on the 6th of October.

TABLE 18

	P_H value of soil without plants.		P_H value of soil with plants.			
	Oct. 1921.	Dec. 1921.	Oats.	Barley.	Rye.	Wheat.
Heath soil	4·52	4·61	4·72	4·84	4·97	4·7
Grassland soil	4·78	4·88	4·93	5·03	4·95	4·92
Average	4·65	4·75	4·83	4·94	4·96	4·81

It is seen from this table that in comparison with the soil bearing no plants the soil bearing the plants has its P_H value raised by the numbers given in Table 19.

TABLE 19

	Oats.	Barley.	Rye.	Wheat.
Heath soil	0·11	0·23	0·36	0·09
Grassland soil	0·05	0·15	0·07	0·04
Average	0·08	0·19	0·22	0·07

The question then turns upon minute differences, so minute in fact that it is open to doubt how far they represent evidence of a change in the P_H value of the soil. The fact that the change was in the same direction in all the flower-pots shows, however, that a real change had taken place. If we take for granted that the numbers found express correctly what really happens, these numbers show us firstly that both kinds of soil with no plants growing in them decrease their acidity slightly during the experiment, the decrease being about 0·1 in P_H value. Secondly we see that the acidity of the soil containing plants has also decreased, and that the decrease in this soil is greater than in the soil without plants, and further that it is greater in the heath soil than in the grass soil, especially in the soil containing barley and rye.

It is evident that since the changes found in the P_H values of the soil are minute, the experiments should be repeated and extended over a longer period and made to include more species and different kinds of soil, before we can be certain to what extent the green plants are able directly to alter the acidity of the soil.

But even if individual species are able to some extent directly to alter the P_H value of the soil, it still seems difficult to explain sufficiently by this fact the statements on p. 477. At all events it will be necessary to investigate the significance of the influence of the various formations on the component factors of the environment, such as light, temperature, and humidity, and to ascertain whether these factors determine the ability of flowering plants to alter the P_H value, or whether their essential significance lies in the fact that they determine the influence of micro-organisms on the P_H value, or in the fact that they influence the course of these processes by purely chemical means.

There are perhaps no a priori reasons which would prevent us from supposing that the spruce is able directly to make the soil more acid than the beech or the beech than the oak. But when we observe that the different degrees of acidity gradually increase in correspondence with the shade cast by the species in question, and that the degree of acidity in the oakwood is perceptibly higher, not merely where the oakwood casts a deep shade because of its density, but also where the deep shade is not only due to the oak trees, but to adjacent tall beech trees, it is difficult to avoid the supposition that the degree of acidity is not so much due to the influence of the kind of tree, but rather that it is connected with a difference in environment determined by the different trees.

In the dense sprucewood a ground flora of green plants is absent; only when trees are cut down and light reaches the soil does a ground flora enter the wood. The soil has by that time no doubt been for a long time as acid as it ever becomes.

In the dense young beechwood, especially as long as the dead leaves persist until the following spring, lack of light prevents the development of a ground flora of green plants. Later, when the leaves no longer persist, the floor of the wood is so light in the spring that a ground flora can flourish at that time of the year. The epigeal organs of those plants, however, perish when the wood comes into leaf; the floor of the wood then resembles a plantless desert. Only later, when the trees have grown high, and a good many of them have been cut down, does the ground become light enough to allow a summer flora. But from this time onwards the soil hardly increases in acidity, supposing that the environment, especially shelter, has not altered for some other reason.

In contrast with spruce and beech, the oakwood is from the very start so light that in the spring and summer a ground flora can luxuriate. As long as the wood remains closed, this ground flora consists for the most part of comparatively tall Proto-hemicryptophytes, which do not wholly exclude the light from the ground, but draw a translucent veil over it.

There are, it is true, a number of Rosette Hemicryptophytes; but these are scattered, and, because of the form and position of their leaves, they give comparatively little shade. It would appear a priori that because of the favourable light in the oakwood, and because the ground flora with its different species can give both shelter and food, the conditions here would be much more favourable for animal life than in the dark spruce and beechwood. There is indeed scarcely any doubt that the fauna in the oakwood both above and below the ground is richer; the soil is consequently looser and more perforated by animals. The transformation of organic substances is hastened and made more complete than in the sprucewood and beechwood, in which no summer ground flora can live. This absence of a summer ground flora is probably at any rate one of the causes of the soil being more acid than that of the oakwood. Unfortunately we still know little about the microflora and its connexion with soil acidity; but we can at any rate see that, for the microflora of the upper layers of the soil, conditions in the oakwood may differ widely from those of the spruce and beechwood, and it is very probable therefore that these two kinds of wood have a different microflora, which, in its turn, determines or helps to determine different degrees of acidity. I am here thinking especially of the different conditions affecting evaporation from the upper layer of soil in the two types of wood. The ground flora of the oakwood, and of course of other woods light enough to allow a ground flora to succeed, robs the soil of some of its moisture. The evaporation from the leaves of the plants composing the ground flora may be considerably greater than the evaporation from the bare ground. But we must bear in mind the important fact that the moisture which evaporates from the leaves of these plants is principally drawn up from a depth in the soil where there is usually a good deal of water, while the ground flora, by giving shelter, protects the upper layers of soil from desiccation. In the dark spruce and beechwood there is no ground flora to give shelter. The conditions therefore are more favourable to evaporation from the uppermost layers of soil, so that these layers, together with the dead plant remains lying on them, are dried up by the draught which blows unhindered along the ground.

I shall give a few examples to show how great the difference in evaporation can be in two places close together in the same wood, the one without and the other with a ground flora. I must first give a short description of the atmometer used in these investigations.

It is of course very interesting to follow and compare throughout the whole growing period, or even for a longer time, the conditions affecting evaporation in different localities. But it seems to me most important to know what the factors are in the critical period when evaporation almost holds the plant's life in the balance. It is the factors of such a period as this that decide what community of plants will find its home in a given locality. It was, then, my object to make an atmometer so finely graduated that even in the course of a short experiment it would give a sufficiently exact expression of the evaporation in the localities compared. Following Mitscherlich and Livingston I have used for my measurements the evaporation from the surface of a porous receptacle filled with water. And since the precision of the measurement needs only a small evaporating surface, I was able to use Chamberland's porcelain filter with a porous portion only about 4·5 cm. long. After being filled with distilled water the filter-candle (Fig. 143*f*) was connected with the measuring apparatus by means of a rubber tube (*g*) 4 cm. long, one end of which embraced the glazed opening of the filter and the other end fitted over a glass tube (*a*) attached to the measuring apparatus. As Fig. 143 shows, the apparatus consists of a three-way cock. To the ascending branch (*a*) is attached the evaporating Chamberland filter (*f*) as described above.

One of the two other branches (*b*) is connected with a funnel (*t*) for pouring in the distilled water. The third branch is connected with a

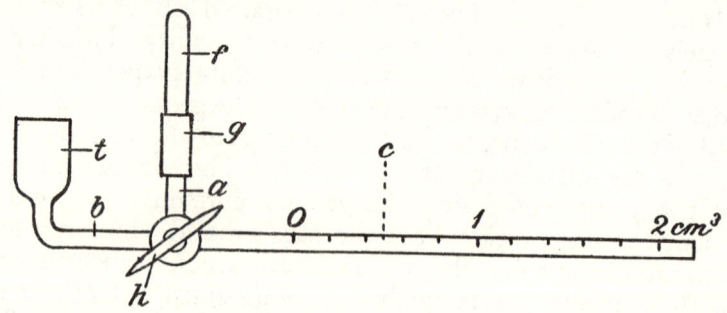

FIG. 143.

graduated pipette (*c*). The whole apparatus can easily be attached with a piece of elastic to a notch in the top of a slender stand.

Care must of course be taken that no air gets into the filter-candle. Before the candle is attached to the apparatus the tap (*h*) is turned so as to connect the funnel with branch *a*, while it is closed to the pipette. Then water is poured into the funnel, and the apparatus is held so that the branch *a* becomes completely filled with water. Then the tap is turned sufficiently to shut the water out of the branch *a*. Next the filter-candle, which, with its rubber tube, has been previously soaked in water, is connected with the branch *a*, care being taken that no air is admitted. By a little practice this is easily avoided. Then the tap is turned so that *b* and *c* are connected and the pipette can be filled with water from the funnel. Any bubbles in the pipette must be got rid of by turning the apparatus to allow the water to rise and sink in the tube until they disappear. When the pipette is full of water it is shut by turning the tap to exclude it from *b*. The pipette is now connected with the filter-candle. Loss of water from the filter-candle by evaporation from its surface is now replaced by suction from the pipette, in which the water therefore is gradually withdrawn from the apex. The extent of the evaporation in a given time can then be read directly from the graduations. Of course the reading must not begin before the adhering water has evaporated off, and the apparatus has come to equilibrium. When this has taken

INFLUENCE OF TYPES OF VEGETATION ON SOIL ACIDITY

place, and the apparatus is fixed in its place, a note is made of the time, and of the position of the fluid in the pipette. At the end of the experiment, and, if desired, at definite intervals during the experiment, a note is made of the number of cubic millimetres of water which have evaporated. If the evaporation is very extensive the pipette may be empty after about two hours; but the experiment can easily be extended over a longer period by instantly filling the pipette again, wholly or partially, with water from the funnel.

If it is desired to compare the evaporation at several places at the same time, and this is what one usually wishes to do, a corresponding number of instruments must be used. Since however it cannot be taken for granted that the filter-candles are uniform, so that the results can be directly compared, the instruments must be standardized before the experiments take place. This can be done by setting them going for the same period under uniform conditions, so that their behaviour may be easily compared. It is then easy to compute the correct results of the instruments in different localities.

If we wish to know not merely the relationship between the evaporation in different localities, but also the absolute amount measured by evaporation from a free surface of water during the same period, this can be done by attaching to each instrument, while it is being standardized, bowls of water the extent of whose surface is known. The bowl and its water are weighed at the beginning and end of the experiment, and from the loss of weight, the extent of the surface of water, and the evaporation measured by the atmometers during the time of the experiment, it is easy to calculate the relationship between the evaporation from each filter-candle and to compare it with that of the free water surface under the same conditions.

In the course of time some change may take place in the evaporating power of the filter-candles, and since this change may not be the same for all the candles, they must be compared and standardized at intervals.

By means of the atmometer[1] here described I have made some experiments to find out the evaporation in different localities. I shall now describe some of these experiments which relate to the influence of the ground flora on evaporation from the soil. These experiments were made in Allindelille Fredskov at the beginning of August 1920. In the first experiment the evaporation was measured at the same time in the four localities mentioned, all of which are situated close together near a little bog, Thomaspark, nearly in the middle of the wood.

[1] In Eduard Rübel's book published this year (1922) entitled *Geobotanische Untersuchungsmethoden*, p. 81, I read that E. S. Johnston and B. E. Livingston as long ago as 1916 had modified Livingston's Atmometer, making it capable of determining the extent of evaporation during a short time. The difference between the apparatus described above, which I used, and that of Johnston and Livingston will be easily understood by comparing Fig. 143 in this work with the figure on p. 81 of Rübel's book. Unfortunately I have not had access to Johnston and Livingston's work (*Plant World*, xix. 1916).

A. Firm ground exposed to the sun, near the north edge of the bog. Dense herbaceous vegetation rich in species, among other plants *Selinum carvifolium* was very prominent.

B. Tall beechwood south of the bog. The ground, which slopes towards the north, was partly bare and partly covered with a thin layer of leaves and bud-scales. Vegetation was absent or consisted of scattered weak *Asperula odorata* with a few individuals of *Viola silvestris*, *Oxalis acetosella*, *Anemone hepatica*, *Poa nemoralis*, and *Vicia sepium*. The soil was acid. From about 5 cm. depth and downwards it consisted of a hard clay mixed with sand.

C. A few metres to the west of B, with beech of the same age as B and the ground sloping towards the north. The soil however is strongly calcareous. Luxuriant ground flora of *Actaea spicata*, *Brachypodium silvaticum*, *Hordeum europaeum*, *Asperula odorata*, *Convallaria majalis*, *Mercurialis perennis*, *Bromus ramosus*, *Hedera helix*, *Rubus saxatilis*, *Fragaria vesca*, *Viola silvestris*, *Pulmonaria officinalis*. A little undergrowth was present consisting of *Viburnum opulus* and other plants.

D. Dense willow thicket on the south side of the bog.

The experiment lasted from 1.40 p.m. to 4.5 p.m. on the 5th of August. The sun was shining at the time, but now and then it was covered with white clouds. At 2 o'clock the light intensity was determined in all four localities by Dr. Boysen Jensen's actinometer[1] in two different ways. Firstly the actinometer was protected from direct sunlight by means of a shade placed at a suitable distance. Secondly the actinometer, kept in a horizontal position, was carried backwards and forwards over the area without being directly shaded, so that in locality A it was exposed to the sun all the time, while in the other localities it only encountered the direct rays which happened to penetrate the canopy. Table 20 gives the per-

TABLE 20

Locality.	Percentage of evaporation.	Percentage of light.	
		Actinometer not protected from direct sunlight.	Actinometer protected from direct sunlight.
A	100	100	100
B	82.7	2.3	6.9
C	39.2	3.8	11.4
D	16.9	2.1	6.3

centage of light and the amount of evaporation. The evaporation was actually greatest in the meadow exposed to the sun, where the exposed

[1] P. Boysen Jensen: 'Studies on the production of matter in Light- and Shadow-Plants', *Botanisk Tidsskrift*, vol. xxxvi, 1918 (p. 233).

filter-candle gave off 900 cubic millimetres of water during the 2¼ hours (*sic*) which the experiment lasted. This number is reduced in the table to 100, and the evaporation of the other localities is given proportionally.

The figures for localities B and C are the most important part of these experiments, and these figures must be compared. These localities are separated by only a few metres; both are situated on the same inclination in the same beechwood; the only difference in them being that determined by the soil, viz. that B has no shelter worth mentioning while C possesses a ground flora giving good shelter. The conditions however in other ways were not particularly favourable for the locality C on the day the investigation was made. The wind, though quite slight, blew from the south-east obliquely inwards towards the boundary of B and C. The atmometer at C stood only a few metres distant from the open floor of the wood surrounding B, which was almost devoid of vegetation and over which the wind was blowing towards C. All the same, as seen from the numbers in the table, the evaporation at C is only half what it is at B. We cannot doubt that it is the rich sheltering ground flora at C that has so considerably reduced the evaporation on the surface of the ground. As seen in the table the difference cannot depend upon the illumination of the two places. There is, it is true, a slight difference in illumination; but the tendency of this is to determine more evaporation at C than B, since the percentage of light at C is slightly higher than it is at B. In order to elucidate further the significance of the ground flora in the reduction of evaporation from the surface of the soil and the resulting counteraction of the drying out of the soil I made on the 6th of August the following experiment on these two localities (B and C). This experiment lasted only one hour (from 2 to 3 o'clock). In order to eliminate the differences which might be thought to be due to a difference in the sun's rays impinging on the filter-candle, a screen was placed at a suitable distance from the candle of each apparatus. Further, at B, the locality with no sheltering ground flora, two instruments were placed immediately on the ground, one of them with, the other without, a shelter. The shelter used was, in the absence of anything better, a flat travelling trunk 65 cm. long and 40 cm. high. At C too we used two instruments; one was placed on the ground in the shelter of the ground flora and the other at a height of about 40 cm., immediately above the main mass of the ground flora, from which only scattered stems stood out. The result of the experiment, which is given in Table 21, shows in a striking manner the significance of the ground flora as a factor reducing evaporation from the soil. At C the evaporation from the instrument immediately above the ground flora is more than twice as great as that from the instrument on the ground in shelter of the vegetation. At B the evaporation from the instrument on the ground without a shelter is twice as great as that from the instrument on the ground but protected by a shelter.

TABLE 21

Locality.	Atmometer.	Percentage of evaporation.	
B	a. Near the ground, without shelter	100	
B	b. Near the ground, artificially sheltered	49·3	
C	c. Near the ground, sheltered by the ground flora	37·7	(46
C	d. At the height of 40 cm., just above the ground flora	81·9	100)

In Table 22 I have placed together the results of the experiments made in localities B and C with the atmometer placed near the ground without artificial shelter. It will be remembered that the two series of experiments differed in this respect. On the 5th of August, the day of the first experiment, the filter-candle was not sheltered from the rays of the sun, which were probably of different intensity in the two localities, and thus might be supposed to affect the extent of the evaporation. On the 6th of August all the filter-candles were screened from the sun. It will be seen from Table 22 that the difference due to the light being stronger at C than at B is partially reduced by excluding the direct rays of the sun in both localities so that the evaporation at C falls still more in proportion to that at B.

TABLE 22

Locality.	Percentage of evaporation.	
	5 August.	6 August.
B	100	100
C	47·4	37·4

It is true that the ground flora of the oakwood at Stampeskov described on pp. 465–6 floristically differs widely from that of the locality C described above; but in the capacity of protecting the soil from desiccation there is no great difference between these two ground floras. Looking through the lists of species in the two localities we find in both that the leaves, which are the organs that shelter and protect the ground, are situated at varying heights above the ground in the different species. It is sufficient for us to differentiate between two layers: (1) a ground layer consisting chiefly of rosette Hemicryptophytes, in which the majority of leaves are found on the surface of the ground, and (2) an upper layer with leaves borne at a greater elevation above the ground. The leaves of these plants are long-petioled or else the plants have elongated internodes.

If we confine ourselves to woodland sufficiently well illuminated for the success of a protecting summer ground flora, and if we begin with the type of wood that has the most deeply shaded floor, we shall then see that apart from wind-swept localities the species forming the upper layer are dominant, and that they, together with the few species of the lower layer, are able to protect the surface of the ground against excessive desiccation. But the species of the upper layer are more or less mesophilous plants with rather broad leaves. These species recede if the conditions become more favourable to evaporation. Where, therefore, the floor of the wood becomes better illuminated the species of the upper layer gradually disappear; but at the same time increasing numbers of new rosette plants enter the lower layer, which eventually becomes a continuous carpet investing the soil and protecting its upper layers from desiccation by wind and sun. The vegetation has become pasture-like.

As already described the hydrogen-ion concentration of pasture soil is lower than that of the shaded soil of woodlands. If a low hydrogen-ion concentration, a low degree of acidity, is an advantage, it must therefore be looked upon as a good thing, as far as the reaction of the soil is concerned, that the illumination of the wood is sufficient to allow the ground flora to become a dense coherent covering. There are, however, other important factors beside the degree of acidity. There is no doubt the species composing the dense flora of the well-illuminated wood make considerable demands on the food materials in the soil, and that they take water from all layers of soil but the uppermost, and thus deprive the trees of large quantities of water. It is therefore probable that the most favoured type of wood in our climate is the type that is sufficiently illuminated, or sufficiently shaded, to allow the success of a summer ground flora consisting essentially of an upper layer of broad-leaved species which do not wholly exclude the light from the ground, but yet give so much shelter that the upper layer of the soil does not become dried out, retaining sufficient moisture for a rapid and perfect breaking down and elaboration of materials, and at the same time creates favourable conditions for an abundant microflora and an abundant fauna.

XV
THE NITRATE CONTENT OF *ANEMONE NEMOROSA* GROWING IN VARIOUS LOCALITIES

Wherever nature remains unmolested for a long time, the ground that is capable of supporting plants becomes covered by a series of communities (formations), which appear according to natural laws, in correspondence with differences of environment. Both the formations and the places they occupy show endless variations, and it is the task of the plant ecologist to investigate the causes which bring about differences in the composition and distribution of these communities. We are confronted with two problems whose solutions are found by analysing respectively the formations and their habitats. We have means of carrying out both qualitative and quantitative analyses of the formations; but the analysis of the habitats presents difficulties. The links in the chain of factors are numerous and often difficult to investigate; we know neither the value of these factors nor how they react on each other; indeed there are probably factors of which we are entirely ignorant.

Since there are very numerous habitats which are not equiconditional, and numerous plant communities corresponding to them, it is of great importance to botanists to obtain some easily used methods of investigating the single factors. We have some such methods at our disposal. We can measure, for example, light, transpiration, the water content of the soil, and its degree of acidity (the hydrogen-ion concentration), &c.; but the investigation of many of the chemical, physical, and microbiological properties of the soil are so detailed and occupy so much time that they can scarcely be employed by botanists in comparing a large number of habitats.

An important environmental factor is the nitrifying properties and the nitrate content of the soil. But a satisfactory investigation of this factor is so laborious that it is easy to understand why attempts have been made to obtain by a ready means sufficient information of this kind to serve for a preliminary survey. Such a method is the one introduced by Molisch for demonstrating the presence of nitrates in plants by means of diphenylamine sulphate. Molisch showed in 1883 that a section of plant tissue placed in diphenylamine sulphate shows a blue coloration if the tissue contains nitrates; and since plants, as far as we know, are not able to elaborate nitrates, their presence in plants must be dependent upon their presence in the soil. We thus have a means of showing by means of nitrates in plants that nitrates were present in the soil in which the plants grew.

In plant physiology the method was used later by various investigators, especially by Frank (1887, 1888), Schimper (1890), Stahl (1900), and

Klein (1913). In plant sociology the method was used specially by Hesselman (1917), who, in his sylvicultural work, investigated the nitrate content of an extensive series of plant habitats. Judging by the deepness of the tint Hesselman separates three degrees of nitrate content, calling them weak, distinct, and strong; and in describing habitats the species which occur are placed in groups according to the strength of the reaction.

Investigations of the nitrate content of plants by diphenylamine sulphate have been carried out in Denmark by Carstén Olsen and by Bornebusch (1923). Weis (1924) has used the reaction in investigating the nitrate content of the soil in a series of Danish beechwoods.

The facts most important for plant sociology can be summed up as follows:

1. The presence of nitrates in plants can be shown by diphenylamine sulphate (Molisch, 1883). If a small quantity of the reagent be applied to the tissue containing nitrates, a blue coloration results. Though the colour is deeper according to the amount of nitrate present, yet the deepness of the tint has little quantitative significance, since a small quantity of nitrate strikes a deep colour.

2. Since plants, as far as it is known, are not able to elaborate nitrates, but obtain them from the soil, it follows from the presence of nitrates in plants that nitrates are present in the soil on which the plants grow.

3. The absence of a blue colour from a tissue to which the reagent has been applied does not necessarily indicate absence of nitrates either in the plant or the soil. The reaction may be suspended for any of the following reasons:

    a. The nitrate may be gradually assimilated as soon as it is being taken up.

    b. For some time immediately before the application of the reagent particularly good conditions for the assimilation of nitrates may have prevailed, so that the nitrate may be consumed. Conditions that favour carbon assimilation cause a rapid consumption of nitrates. According to Schimper nitrates are often more easily demonstrated in plants in dark and cold than in light and hot weather.

    c. The plant may be at a stage in its development at which no storage of nitrates occurs, or perhaps even no nitrates are taken up. Young plants usually contain larger amounts of nitrates than older ones. In the locality where young individuals of a species show a strong reaction, the same species later in the year may be entirely devoid of nitrates.

    d. Substances may be present which hinder the reaction, although the plant contains nitrates. Thus Molisch (1883) and later Schimper (1890) have shown that the presence of lignin may prevent the reaction. It is for this reason amongst others that in my investigation

of the occurrence of nitrates in plants I have not used the ordinary method, which consists in placing a section of the tissue in the reagent, but have always used the juice expressed from the tissues.

e. Finally it might be thought that some plants do not take up nitrates at all. The presence of nitrates in various species has not yet been demonstrated, although these species grow on soil containing nitrates. These plants are for the most part trees, especially those having mycorrhiza (e.g. many *Cupuliferae*). Frank (1888) suggested that such plants do not absorb nitrates, but that they obtain their nitrogen partly or entirely from organic nitrogen compounds which are presumably taken up by the mycorrhiza.

4. Our knowledge of the presence of nitrates in Danish plants is still very limited. Investigations hitherto carried out seem to justify a classification of our plants as follows:

    a. Some species have nitrates in all the individuals, at any rate at a certain stage of their development. The presence of nitrates in these plants in their various habitats may be regarded as a proof of the occurrence of nitrate in the soil.

    b. Some species are rich in nitrate in some localities, poor in nitrate in other localities, and in other localities again some or all individuals are devoid of nitrate, even if nitrate be present in the soil. In this respect however there are wide differences between the different species, some being rich in nitrate in the same locality in which others are poor in nitrate or free from nitrate. A corresponding difference is seen among the individuals of the same species.

    c. Certain species only exceptionally contain nitrates.

    d. In some species the presence of nitrate has not been demonstrated (3*e*), although they grow with species which contain nitrate.

A consideration and comparison of the individual clauses in the above summary makes it obvious that we are as yet unable to determine the amount of nitrate in the soil by investigating the nitrate contents of the plants; but much can be gained by obtaining a relative determination by a comparative investigation of a species in a series of localities. Various factors however are in operation. For example the habitats in question may have different relationships to light (3*b*), and even if we overcome this objection by investigating and comparing habitats with the same relationship to light, we encounter the difficulty that our knowledge of the capacity of the individual species to accumulate nitrate is very imperfect. It is easiest to carry out a comparison by investigating one or more species growing in a series of habitats. It will always be of interest to investigate the nitrate content of an individual species in all its habitats. In so doing we can use the same test, i.e. the plant's capacity of accumulating nitrates. In this manner I have investigated *Anemone nemorosa* in all its habitats within a circumscribed region, i.e. the Dyrehave and Jægersborg Hegn.[1] I have chosen this species partly because it can grow in very different habitats and partly because a preliminary investigation has shown that its nitrate content can vary from entire absence to the greatest possible abundance of nitrate, as far as can be determined by diphenylamine sulphate. In using this reagent for demonstrating nitrates in plants I have, as already mentioned, always used the expressed juice of the plant instead of a section, thus avoiding the reaction brought about by lignin (3d). Besides having this advantage the method is more rapid, and this is very important when we wish to make many determinations, for the sake of comparison, in as short a time as possible, in order that the environmental factors may be as uniform as possible. In using the expressed juice the technique is more uniform, as in each test approximately the same quantity of fluid is used.

For expressing the juice it is best to use a pair of forceps with slightly bent points, concave on the inner side so that the portion of the plant can be easily gripped. The tissue is held between these points and the projecting part is cut away. Through the cut surface the contents of the cells are pressed out by manipulating the forceps. Since the ends of the forceps are bent it is easy to apply them to the bottom of a shallow white porcelain dish, putting on to it a suitable quantity of the juice; a fraction of a drop suffices. To this is applied a drop of diphenylamine sulphate. If nitrate be present there at once appears a blue colour the intensity of which is easily determined, as it is not masked by the presence of tissue. Unfortunately not many degrees of intensity can be recognized because the fluid soon becomes as blue as blue ink. In the following investigation I have discriminated four degrees of intensity:

1. Very weak reaction, a pale transparent blue cloud.
2. Rather weak reaction; the blue cloud denser and more or less opaque.
3. Rather strong reaction and
4. Very strong reaction; the fluid at once becomes as blue as blue ink throughout.

For these four degrees of concentration I have used the above numbers (1–4); absence of reaction can be expressed by 0. The boundaries separating the different degrees vary according to the opinion of the observer. The boundary between degrees 3 and 4 is the most difficult one to draw; but if the reagent be applied to the expressed juice it is so easy to differentiate slight shades of colour that it seems insufficient to distinguish less than four degrees. If the individual numbers found by investigating a series of plants be specified, 3 and 4 can always be united if thought desirable, and we can follow Hesselman in recognizing only three degrees. Since the same species contains different amounts of nitrates at different

[1] Woods in the vicinity of Copenhagen.

periods of development I have undertaken the investigation of the nitrate content of *Anemone nemorosa* within a very short period, i.e. the last half of May. In 1924, when the investigation was carried out, this was the period within which *Anemone nemorosa* was in full flower. During this time two or three determinations were often made of plants growing in the same place, partly in the same kind of weather, and before and after heavy rain. The rain made no difference to the nitrate content. Later on I shall mention the significance of the influence of light in the various habitats.

Since the different parts of the plant have different nitrate contents, I have in this comparative investigation always used the same part, i.e. the middle of the flowering stem. The flowering shoots of plants growing in one habitat are more uniform in development and in their position relative to light than the radical foliage leaves, which, however, would possess the advantage that by their means the behaviour of the species could be followed over a wider area, since *Anemone nemorosa* can continue to live vegetatively in localities where the environmental factors seldom or never allow it to produce flowering shoots.

Anemone nemorosa occurs as a component of various types of vegetation. In classifying these we must first consider the plant's relationship to two important factors, water and light. Firstly then, making use of the relationship to water and of the words used in the language of the people for the main types of vegetation which characterize the landscape, we get on the one hand water and marsh, including a series of formations of Hydrophytes and Helophytes, and on the other hand, wood, scrub, heath, bog, meadow, pasture,[1] dune, and beach. In the district investigated wood, meadow, and pasture are the only localities in which *Anemone nemorosa* grows.

It is true that the chief flowering period of *Anemone nemorosa*, when the nitrate investigation was made, is the season when the wood has not yet fully unfolded its leaves, and still gives but little shade; but on the other hand a wood from its very nature casts such deep shade, which varies with the kind of woodland, that we cannot simply compare the behaviour of the plant in different types of wood, but must consider these varying degrees of shade. It is this intensity of shade, varying somewhat during spring in the different kinds of woods, and more particularly the variations of shade in the summer, in combination with the composition, humidity, degree of acidity, nitrifying properties and other varying factors of the soil, that bring about a series of more or less well-marked types of vegetation in which *Anemone nemorosa* is often an important component. Somewhat the same is true of meadow and pasture, but in these habitats, except where they abut on wood (in which, indeed, *Anemone nemorosa* is very likely to grow), the illumination is essentially uniform. These remarks apply only

[1] Danish *Ore*.

to those parts of the meadow and pasture on which *Anemone nemorosa* is most likely to grow.

In order to be able to compare possible variations in the nitrate content of *Anemone nemorosa* with differences in the composition of the flora of which it is a component, I have undertaken a statistical analysis of the vegetation concerned. In such a comparative treatment we encounter this difficulty, especially in some of the woodland localities, that it is not known whether or not the vegetation is stable; the illumination may have altered recently, either because of the interference of foresters or of damage due to storms. After such disturbing factors the vegetation may not be in a state of equilibrium with the altered illumination.

The Formations. Particularly where the ground is flat the conditions may be approximately uniform over expanses of varying extent, so that a stable vegetation will have a uniform appearance. But in investigating an area not too restricted the conditions will generally be found to vary in many ways, in harmony with the individual factors which determine the vegetation. The transitions are seldom abrupt. For example, in passing from lighter to darker regions or from damp to dry soil the composition and character of the vegetation gradually change. We must of course first investigate those large expanses on which the vegetation is uniform. But in neglecting transitional types we must not be led into the error of regarding the social units as being anything like the genotypic species of Systematic Botany. These units are in two senses phenotypes; firstly they are groups of systematic species, constellations determined by the environment, and secondly they depend upon the phenotypes of the component species.

When we are faced with a problem of investigating such an endless series of graded phenomena as plant formations we must have recourse to classification. The manner of carrying out this classification is a practical problem. But whether we approach this classification by restricting the concept of the formation, or whether we make use of a wider concept, employing the word facies as a designation for a subordinate group, we must first of all investigate the individual habitats and their vegetation by exact methods, which must be as objective as possible. For example, by determining the species composition, the dominant species, the reaction to environment, the density of the species, &c., we obtain material which will serve for a properly based limitation of the social units and for the determination of their constants.

The Frequency-Dominants are dominant species with the highest degree of frequency. In my first investigation I included among them those species which on random plots of 0·1 sq. metre occurred in over three-fifths of all the samples, and I placed these species at the beginning of floral lists in the formation-tables. Later I adopted the plan used in this paper of including among the frequency-dominants only those species which have the highest degree of frequency, i.e. a frequency percentage of more than 80. I was formerly inclined to believe that one should demarcate as many formations as there are combinations of frequency-dominants. Theoretically this is justifiable; but it is often not practicable, especially when differentiating units within vegetation where the species are dense, and where there are many frequency-dominants. But also in dealing with vegetation having more scattered species with few frequency-dominants we may encounter difficulties, especially because the physiognomy of the vegetation again and again forces its claim to be considered. For example, there may be no physiognomical differences between units of vegetation where *Anemone nemorosa* is the only frequency-dominant and units where both *Anemone nemorosa* and *Oxalis acetosella* have a frequency percentage of over 80, but where *Anemone* is physiognomically as dominant as in the former unit, while *Oxalis* occupies a subordinate position which first becomes apparent when the formation is analysed statistically. Near such communities we may find others in which the two species have physiognomically approximately the same value or *Oxalis* may dominate over *Anemone*, even when the latter is the frequency-dominant. Thus we are obliged to make use of the concept 'physiognomical dominant'.

Physiognomical Dominants are species which dominate physiognomically in a given piece of vegetation. They are usually also, but not necessarily, frequency-dominants. They need not necessarily occupy the highest percentage of area, though as a matter of fact they often do.

While frequency-dominants must be determined objectively, physiognomical dominants are computed by ocular estimate. This drawback can scarcely be avoided if we wish to give to the physiognomy of the vegetation the attention which is due to it. Our tendency to name formations after their physiognomical dominants forces the physiognomical principle upon us.

In statistical tables it should always be made clear which of the species, if any, are physiognomical dominants. I have shown the physiognomical dominants by underlining their frequency percentage.

Species constants are the characteristic species of a social unit, i.e. those species which must invariably be present in communities which belong to the same unit.

The constant of species density (Chapter VIII, p. 307) is the designation for the average number of species on each sample (0·1 sq. metre) in a given vegetation. During statistical analysis the lists of species increase in length, first rapidly, then more slowly, until all the species are included. With species density, on the contrary, it is usually only necessary to investigate a few samples in order to arrive at a figure so constant that it varies by less than 1 per cent., however many samples are subsequently investigated.

High density of species shows that the environment is suitable for many of the species in the district; it also shows that no single species is able to get the better of the others in competition. A diminution in activity of one or more of the environmental factors causes a fall of the species density, which decreases in proportion to the deviation of the environment from the demands of the majority of the species.

The plants themselves, however, also affect their environmental factors, e.g. the properties of the soil. But such changes usually take place slowly, and are succeeded by correspondingly slow changes in the species composition and density. There is on the other hand one factor upon which plants are dependent which can change and become decisive very rapidly. This factor is light. Certain species, in virtue of their height and of the density of their foliage, have a decisive effect on the existence of other species. In this respect plants dominate one another in proportion to their height: the tall Phanerophytes dominate Micro- and Nanophanerophytes, which in turn dominate Chamaephytes, and these again usually dominate Hypogeophytes.[1]

It is manifest that even if the soil is essentially uniform, but a portion of the space is occupied by a vegetation of Phanerophytes (woodland), while another part is pasture, with a vegetation of Hypogeophytes, we shall find in these two localities an entirely different species density, determined by the fact that in the woodland one or two species have shaded their competitors out of existence. Thus in Fortun-Indelukke in the Dyrehave the species density in thick beechwood was only about two, while in the adjacent common, which originally had the same kind of soil, the species density was about 19.

The difference found on calcareous soil in Allindelille Fredskov was still greater. Here on pasture only 4 or 5 metres distant from the almost bare floor of the forest under the dense canopy of the beech trees there was found a species density slightly over 24, the greatest species density I have ever observed. The number of species too was very great; on 2 sq. metres no less than 52 species of vascular plants were found.

The comparison of the species density under the varying shade of the canopy of Phanerophytes is very interesting. Subject to exceptions to which I shall soon revert, the rule is that, other factors being equal, the species density increases with the intensity of the light. This is seen by comparing the ground flora of a beechwood with that of an oakwood. In comparing the average species density in these two kinds of wood I have made use of all the statistics available which are made on the basis of random samples with an area of 0·1 sq. metre. Besides my own observations, some only of which have been published, I have availed myself of material used by Vahl (1911), Olsen (1921), and Bornebusch (1920, 1921,

[1] This word is used as a collective term for Hemicryptophytes and Geophytes taken together. It is often convenient to use a term which includes these two life-forms.

1923). The number of analyses of the ground flora of the beechwood is 260, those of the oakwood only 21.

TABLE I

Species density in the ground flora of oak- and beech-woods

Ground flora in	No. of observations.	Average species density.	Distribution in percentage in classes of species density.									
			1	2	3	4	5	6	7	8	9	10
Beechwood	260	2·57	12	27	28	20	9	2	1	1	..	..
Oakwood	21	4·46	..	..	19	24	24	14	9	5	..	5

From Table I it will be seen that the material used yields an average species density for the ground flora of the beechwood of 2·57 and for the oakwood 4·46. Columns 4–13 (numbered 1–10) illustrate how the percentages of species density fall into classes with an interval of one between each. The number of the boundary of a class belongs to that of the class before it. For example a species density of 3 is reckoned as belonging to the class between 2 and 3.

It is not however only the Phanerophytes that markedly affect the species density by their shade; the Hypogeophytes also compete with one another for the occupation of space, and in this competition shade is an important weapon. Apart from the phanerophytic formations we encounter certain gregarious large-leaved Hypogeophytes that are able to shade out competitors, e.g. *Tussilago farfarus* and notably *Petasites ovatus*. These plants may reduce a vegetation originally very rich in species to a species density of one.

In the ground flora of the phanerophytic formations shading plays a very large part in competition, since the light already approaches the minimum for the species present. In the ground flora too we often find species with comparatively broad leaves. A thick carpet of such low plants as *Anemone nemorosa* or *Asperula odorata* is capable of shading out species of lower stature, e.g. *Ficaria, Oxalis*, &c. But this takes place more markedly amongst taller plants with broader leaves such as *Mercurialis, Aegopodium, Urtica*, &c. When these plants grow thickly they are able eventually to shade out all the other plants of the ground flora. In Table 2 I have given a conspectus of 31 determinations of species density in different parts of a *Mercurialis perennis*-Formation, that is a formation in which *Mercurialis perennis* is the only physiognomical dominant.

The large numbers of the lowest species densities is not due to any dense shade cast by the trees, but is the result of the successful competition of the *Mercurialis* community. For the sake of comparison I have included in Table 2 a conspectus of 36 observations in the *Melica uniflora*-Formation. Another example of the effect of the shade on species density

Table 2

Species density in formation of *Mercurialis perennis* and *Melica uniflora*

	No. of observations.	Average species density.	Distribution in percentage in classes of species density.						
			1	2	3	4	5	6	7
Mercurialis perennis-Formation	31	2·23	13	39	29	10	3	3	3
Melica uniflora-Formation	36	3·23	..	12	30	47	5	3	3

is seen in Table 12 (p. 507), Nos. 3–4. These observations are from vegetation occupying the same soil in a light alderwood; over large areas (3) none of the species were dominating entirely by shading; but in some places (4) *Impatiens* and *Urtica* grew so thickly that they shaded the ground completely, so that the other species lingered on as starved, moribund individuals. It is seen in Table 12 that the species density in the latter example is only half as great in the former.

Finally the dense *Allium ursinum*-Formation is a particularly instructive example. In 14 observations the species density varied only between 1 and 1·4, and in 9 of the observations *Allium ursinum* was the only plant found. In the 5 remaining observations the following plants were found: *Corydalis cava, C. intermedia, Dryopteris filix mas, Ficaria verna, Melica uniflora, Mercurialis perennis*, and *Urtica dioeca*.

The presence of these species and of others that were found outside the samples shows that the conditions of soil and light permit the occurence of various woodland species in no inconsiderable species density, if the broad-leaved, densely growing *Allium ursinum* does not overwhelm all competitors.

Apart from the special factors which prevail in formations dominated by the dense shading of a single species, the species density in stabilized woodland formations will increase with the light that has significance for the plants, other factors being equal, so that the species density becomes a biological measure of the light necessary for plant growth. This light is not exactly the same as that measured by our photometers.

The relationship between species density and the hydrogen-ion concentration of the soil. Other factors being equal the species density continues to increase with decreasing hydrogen-ion concentration until a point is reached which lies near neutrality (Olsen, 1921).

The relationship between species density and soil humidity. It is almost certain as a general rule that other factors being equal in a given area the species density increases with increasing soil humidity.

If we walk over an evenly sloping heath which has suffered little interference, passing from the highest and driest parts to the lower and damper portions, we meet an increase in species density, as is seen in Table 3, which shows the species density of (1) *Calluna*-heath, (2) the uppermost portion of *Erica*-heath, and (3) the lowermost portion of *Erica*-heath.

TABLE 3

Species density in *Calluna*-heath (1) and in *Erica*-heaths, uppermost (2) and lowermost portion (3). (Most of the observations were made by Magister Mølholm Hansen.)

	No. of observations.	Average species density.	Classes of species density.						
			1	2	3	4	5	6	7
1. Calluna-heath . . .	36	2·79	..	11	64	8	17	..	..
2. Upper part of Erica-heath .	14	3·15	..	..	43	50	7	..	..
3. Lower part of Erica-heath .	19	5·08	..	..	5·3	5·3	21	63	5·3

But the plant's relationship to humidity in different localities is probably very complicated, since the humidity is liable to change in very different ways in each locality in the course of the year. I believe that it would be well worth while to study the species density in relationship to different types of humidity.

THE NITRATE CONTENT OF *ANEMONE NEMOROSA*

I. Shaded Places

A. Beechwood. In the Jaegersborg Hegn and Stampeskov, where the great herds of deer of the Dyrehave do not obtain access, the ground flora of the beechwood is generally very luxuriant, differing widely from that seen in the Dyrehave, which is greatly altered by the grazing of the deer. In the old forest the original ground flora of the shaded beechwood has been almost obliterated, and over wide areas, where the wind has swept away the dead leaves, the ground has become so impoverished that a luxuriant ground flora can scarcely develop, even if animals are kept away. The nitrate content of the *Anemone* in localities like this will be discussed separately. In other situations in the Dyrehave yet other conditions prevail, as for example in the district between Fortunen and Fuglesangssøen, the most southerly, oldest part of Fortun-Indelukke, and in Chr. IX's Hegn. The ground flora in these localities has certainly been much damaged, and it is devoid of the physiognomical features of a flora unaltered by grazing. But on the other hand a statistical analysis shows one or more frequency-dominants, and the species density is essentially the same as that in the corresponding Jægersborg Hegn, which, however, is more luxuriant because it is fenced in.

1. Enclosed Beechwood whose floor is protected from the wind and more or less covered with leaves. This is for the most part occupied by diverse kinds of *Anemone nemorosa*-Formation (Table 4,

TABLE 4

Ground flora of the beechwood

Oxalis acetosella-Formation: 1–2 (Chr. IX's Hegn).
Anemone nemorosa-Formation: 3–4 (Fortun-Indelukke), 5–6 (Stampeskov), 7–13 (Jægersborg Hegn).
Asperula odorata-Formation: 14 (Jægersborg Hegn).

	1	2	3	4	5	6	7	8	9	10	11	12	13	14
Oxalis acetosella	100	100	.	100	15	90	100	100	100	100	100	100	100	100
Anemone nemorosa	.	90	100	90	100	90	100	100	100	100	100	100	100	100
Asperula odorata	.	.	.	.	.	.	.	20	.	.	.	30	.	100
Carex remota	.	.	.	.	.	.	5	.	.	.	.	.	.	.
„ silvatica	.	.	.	.	.	.	.	.	.	.	.	.	.	5
Dactylis glomerata	.	.	.	.	.	.	.	.	.	.	.	5	.	.
Dryopteris pulchella	.	.	.	.	.	.	.	.	.	.	50	50	.	.
Ficaria verna	.	.	.	.	.	.	.	5	20	20	.	.	.	.
Melica uniflora	.	.	.	.	.	.	.	5	20	5	20	15	.	.
Milium effusum	.	.	.	.	.	.	.	.	.	.	.	.	.	.
Poa nemoralis	.	.	.	.	.	.	.	.	.	.	.	.	.	5
Rubus idaeus	.	.	.	.	.	.	.	10	.	5	.	.	.	.
Stellaria holostea	.	40	.	.	.	.	.	.	.	5	.	.	.	.
Urtica dioeca	.	.	5	.	.	.	.	.	.	5	.	10	.	.
Viola silvatica	.	.	.	.	.	.	.	10	.	.	.	.	.	.
Species density	1	2·3	1·1	1·9	1·2	1·8	2·1	2·5	2·4	2·4	2·7	3·1	2	3·1

Nos. 3–13); besides *Anemone nemorosa*, *Oxalis acetosella* is also a common frequency-dominant (Table 4, Nos. 4 and 6–13). Here and there the *Oxalis acetosella*-Formation occurs (Table 4, Nos. 1–2). Especially in the northern portion of Jægersborg Hegn there are found expanses covered with *Asperula odorata*-Formation (Table 4, No. 14), in which *Anemone* and *Oxalis*, besides *Asperula*, occur as frequency-dominants, but are physiognomically repressed. Here and there a more or less well-marked *Melica uniflora*-Formation occurs (Table 5), in which both *Anemone nemorosa* and *Oxalis* are usually found as frequency-dominants; *Anemone ranunculoides* may also be abundantly present, and may even be a frequency-dominant (Table 5, No. 4).

Table 5

Ground flora of beechwood

Melica uniflora-Formation: 1–4 (Dyrehave, between Fortunen and Fuglesangssøen), 5–8 (Jægersborg Hegn)

	1	2	3	4	5	6	7	8
Melica uniflora	**100**	**100**	**100**	**100**	**100**	**100**	**100**	95
Oxalis acetosella	100	100	100	70	20	5	70	100
Anemone nemorosa	90	100	5	..	100	100	100	95
„ ranunculoides	..	..	65	90	..	..	..	..
Asperula odorata	..	..	..	..	20	70	70	20
Dactylis glomerata	..	..	..	..	..	..	..	5
Festuca gigantea	..	..	..	..	..	..	..	5
Ficaria verna	10	5	55	20	..	5	30	..
Mercurialis perennis	..	5	..	..	..	..	..	5
Milium effusum	..	..	..	..	..	..	..	15
Stellaria holostea	..	..	..	..	5	..	..	60
„ neglecta	..	..	25	5	..	..	..	..
Urtica dioeca	..	..	20	..	..	..	..	5
Veronica hederifolia	..	..	..	5	..	..	..	..
Viola silvatica	..	..	..	25	..	..	..	10
Species density	3	3·1	3·7	3·2	2·5	2·8	3·7	4·2

In certain places where the wood is more open, and especially where there is lateral illumination under the beech trees, there is found a Hypogeophyte Formation richer in species and with a greater species density; this community as a rule has no outstanding physiognomical dominants (Table 6).

I have made in all 300 determinations of the nitrate content of *Anemone nemorosa* in the part of the beechwood characterized by the above-named formations. Sixty different places were investigated, and 5 determinations made in each place. Table 7, *a* shows the percentage distribution of the different grades of reaction of the 300 determinations. In all the localities one or several of the plants, usually all of them, contained

Table 6

Ground flora of beechwood

1. Light beechwood in Stampeskov. 2. Under the edge of beeches surrounded by oaks
(Langens Plantage)

	1	2		1	2
Anemone nemorosa	**100**	100	Lampsana communis	10	..
Oxalis acetosella	**100**	100	Mercurialis perennis	10	..
Dactylis glomerata	40	85	Milium effusum	45	15
Deschampsia caespitosa	..	85	Polygonatum multiflorum	5	..
Agropyrum caninum	..	10	Poa nemoralis	35	10
Agrostis tenuis	..	5	,, trivialis	5	10
Arenaria trinervia	5	..	Ranunculus acer	..	15
Anthoxanthum odoratum	..	5	,, auricomus	5	..
Brachypodium silvaticum	..	5	Rubus idaeus	5	..
Carex silvatica	..	35	Stellaria holostea	15	..
Epilobium montanum	10	..	,, media	..	5
Festuca gigantea	..	25	Taraxacum vulgare	5	..
Ficaria verna	10	30	Urtica dioeca	10	..
Fraxinus excelsior, seedlings	..	20	Veronica chamaedrys	20	15
Hieracium silvaticum	5	..	,, officinalis	5	..
Holcus mollis	..	10	Viburnum opulus	..	5
Lactuca muralis	25	..	Viola silvatica	10	45
			Species density	4·8	6·4

nitrates, and only 6 per cent. of the plants investigated were free from nitrates. In more than half of the localities one or more individuals showed the highest degree of reaction, but it was possible only on small expanses to find any demonstrable connexion between the degree of the reaction and the different vegetation. A few points must be discussed in greater detail.

Table 7

Nitrate content of *Anemone nemorosa* in beechwood

	No. of plants investigated.	Distribution in percentage of classes of reaction.				
		0	1	2	3	4
a. Dense beechwood with floor protected from the wind and more or less covered with dead leaves	5×60	6	20	22	20	32
b. Melica uniflora-Formation	5× 4	20	40	30	10	..
c. The north-eastern part of Jægersborg Hegn	5× 4	65	25	10	..	..
d. Wind-swept ground	5×18	52	41	6	1	..
e. Deschampsia flexuosa-Formation	5× 9	67	33	..	..	..
f. Nettle glades	5× 5	..	8	32	32	28

Melica uniflora-Formation. In all 4 localities investigated the nitrate content of the *Anemone* was very low (Table 7, b). None of the plants showed the highest degree of the reaction, while, as Table 7, *a* shows, as many as 32 per cent. of the individuals investigated which were growing in a more or less luxuriant beechwood flora showed the highest reaction. The investigations are few and it is scarcely fair to draw general conclusions from them.

Another point that should be discussed is the peculiarly low nitrate content of *Anemone* in a large area in the north-east part of Jægersborg Hegn. Investigation of a luxuriant *Asperula*-Formation south-west of Kørom showed the nitrate content of *Anemone* to be strikingly low. Since there was no a p r i o r i reason to suppose that the ground of the *Asperula*-Formation was in general strikingly deficient in its power of nitrification I investigated the nitrate content of the anemones found in luxuriant *Anemone nemorosa*-Formations adjacent to the *Asperula*-Formation. Of 20 determinations 13 showed no trace of nitrate; of the 7 others only 5 showed the weakest and 2 showed the weakest reaction but one (Table 7, *c*). The night before the investigation however it had rained heavily and it is possible that the rain had leached the soil, leaving but little nitrate available for the plants. I therefore at once investigated the nitrate content of the anemones in some localities where I had determined their nitrate contents on days before rain. The result is seen in Table 8, which shows that no perceptible difference is demonstrable before and after rain.

TABLE 8

Nitrate content of the *Anemone* in the same locality the day before and the day after heavy rain

	No. of plants investigated.	*Percentage distribution among classes of reaction.*				
		0	1	2	3	4
Days before heavy rain	5×5	40	12	12	16	20
Days after heavy rain.	5×5	32	24	12	12	20

2. Wind-swept soil in the beechwood. Because of the lack of shelter-giving undergrowth the soil beneath the old beeches in the ancient forest of Dyrehaven is in many places exposed to the wind, which carries away the fallen leaves, so that the ground becomes quite bare or covered only with dead twigs, beech husks, bud-scales, and similar material that is able to form a thin layer of humus. In the younger beechwood too wind-swept soils of this kind are found in exposed situations. The soil gradually becomes more acid than that covered with leaves, and the vegetation is exceedingly poor, being formed of scattered individuals of the common woodland species. They are usually flowerless and depauperate from lack of light and nourishment, so that from a distance the ground appears to be entirely devoid of vegetation. Among the few plants able to persist for a long time is *Anemone nemorosa*, which, however, seldom flowers. In 18 localities of this kind I have investigated the nitrate con-

tent of *Anemone*; in two of these places, of which one was a wind-swept slope near the mill-pond at Stampen, and in the neighbourhood of 'Øhlenschaeger's Beech', I found no trace of nitrate in any of the five plants I investigated in each place. In the other 16 localities nitrate was demonstrated in one or more individuals; only one plant out of ninety showed the degree of reaction No. 3; quite half of them were free of nitrate, and in the rest only a trace of nitrate was found. Table 7, *d* shows the percentage distribution of the degrees of reaction of the plants investigated.

3. *Deschampsia flexuosa*-Formation. Where sufficient light reaches the wind-swept, naked, strongly acid soil to allow *Deschampsia flexuosa* to thrive this species will, if it gets the opportunity, invade and gradually cover the soil, forming a *Deschampsia flexuosa*-Formation. This formation is sometimes pure, and sometimes, where there is more light, several other species enter (Table 9, Nos. 1–4). On wind-swept slopes along the mill-stream and in some other places, as in the region about Bøllemosen, there are a number of usually well-defined localities bearing this formation.

TABLE 9

Deschampsia flexuosa-Formation

	1	2	3	4
Anemone nemorosa	10	5	60	100
Deschampsia flexuosa	**100**	**100**	**100**	**100**
Luzula pilosa	30	..	80	100
Majanthemum bifolium	95	15	15	100
Oxalis acetosella	..	100	..	40
Stellaria holostea	..	100	..	40
Agrostis tenuis	..	..	35	25
Anthoxanthum odoratum	..	..	..	25
Calluna vulgaris	..	..	20	..
Carex pilulifera	5	..	30	..
Convallaria majalis	20	..	..	..
Dactylis glomerata	..	..	..	25
Dryopteris pulchella	..	..	..	30
Holcus mollis	15	..	..	..
Luzula multiflora	..	..	..	10
Milium effusum	..	..	..	25
Poa nemoralis	..	..	..	10
Rubus idaeus	..	10	5	..
Trientalis europaea	..	..	5	..
Viola canina	..	..	..	5
Species density	2·8	3·3	3·5	6·4

Of all the formations in which *Anemone nemorosa* occurs the *Deschampsia flexuosa*-Formation was the one in which the *Anemone* showed the lowest grade of nitrate content. In three of the nine localities all of the individuals investigated were free from nitrate; and the positive reaction all belonged to the lowest grade (Table 7, *e*).

4. **Nettle glades.** Where either felling or storms have made openings in the canopy sufficiently large to form glades in the forest, an *Urtica dioeca*-Formation will often be found. Since the stems of the nettles gradually collect a number of wind-carried leaves, improvement takes place in the soil, and the degree of acidity decreases (Chapter XIV). Where *Anemone* occurs in this formation it shows a rich nitrate content (Table 7,*f*), richer, it seems, than in all other localities with the exception of alder wood on humus (Table 11, *a*).

B. **Oakwood.** We must here consider the small portions of the younger oakwood, especially Langen's Plantation and the oakwood on the south side of Stampeskov. We must also deal with the ground surrounding the huge old oak trees scattered throughout the beech forest, as, for example, in Stampeskov and in Chr. IX's Hegn. These oaks may occur in groups, as in Stampeskov and especially in the middle portion of Fortun-Indelukke, where there are considerable remains of oak forest representing the wood districts, which, in a map dating from the time of Langen, are marked on the southerly part of the grass plain then covering that part of Dyrehaven where Fortun-Indelukke is now situated (Lütken, 1899). Among the hundreds of more or less isolated old oaks in the ancient forest no anemones are usually seen, because the light is sufficiently strong to allow the ground to be covered by a herbaceous vegetation of considerable density in which grasses predominate.

Under the old trees which are surrounded by beech forest, where the weaker light checks the invasion of the grass, we see, especially in Stampeskov, a *Rubus idaeus*-Formation with a number of broad-leaved herbs, e.g. *Mercurialis perennis*, *Anemone nemorosa*, *Oxalis acetosella* (Table 10, Nos. 6–7). This is seen also over large extents of the oakwood on the south side of Stampeskov (Table 10, No. 5). In other parts only isolated plants of *Rubus idaeus* are seen, or this species may be altogether absent, but we find the *Mercurialis perennis*-Formation (Table 10, Nos. 2–4), *Melica–Anemone*-Formation (Table 10, No. 1), *Urtica dioeca*-Formation (Table 10, No. 8), or where the soil is better illuminated a Hypogeophyte vegetation rich in species and with a high species density without well-marked physiognomical dominants (Table 10, No. 9).

I have made altogether 215 single determinations of the nitrate content of *Anemone* in the oakwoods with the results shown in Table 11, *c*, which show no marked difference from the results obtained in the beechwood (Table 11, *b*). But there appears to be a slight difference between the anemones under the old oaks and those beneath the younger oaks, the former being slightly richer in nitrates than the latter. Out of 100 plants investigated under the old oaks none were entirely devoid of nitrate and 31 (31 per cent.) showed the highest degree of reaction, while out of 115

Table 10

Ground flora of the oakwood

1. *Melica–Anemone*-Formation between Fortunen and Rødeport.
2-4. *Mercurialis perennis*-Formation. 2, between Fortunen and Rødeport; 3, young oakwood in Stampeskov; 4, under old oak trees in Stampeskov.
5-7. *Rubus idaeus*-Formation. 5, young oakwood in Stampeskov; 6-7, beneath old oak trees in Stampeskov.
8. *Urtica dioeca*-Formation, under oak trees in the Fortun-Indelukke.
9. Hypogeophyte-Formation rich in species beneath an oak in Langen's Plantation.

	1	2	3	4	5	6	7	8	9
Anemone nemorosa	**100**	90	85	70	85	95	100	40	100
Ficaria verna	55	95	..	5	..	25	..	5	60
Melica uniflora	**100**	15	..	..	..	..	..	..	35
Oxalis acetosella	75	10	..	90	25	85	100	80	95
Mercurialis perennis	5	**100**	**100**	**100**	35	25	10	70	40
Rubus idaeus	..	..	25	5	**90**	**100**	**90**	10	..
Urtica dioeca	45	55	..	20	5	40	5	**100**	75
Poa trivialis	..	..	..	..	..	..	..	5	**95**
Aegopodium podagraria	..	..	..	..	..	10	..	..	..
Agrostis tenuis	..	..	..	..	5	..	..	..	15
Arenaria trinervia	..	..	..	5	..	..	5	..	..
Avena elatior	..	..	..	..	25	..	..	..	..
Calamagrostis lanceolata	..	..	..	..	25	..	5	..	..
Carex remota	..	..	..	..	..	..	..	..	5
„ silvatica	..	..	..	..	..	..	..	..	5
Circaea lutetiana	..	5	..	..	10	..	..	..	10
Dactylis glomerata	..	..	..	..	50	..	..	..	65
Deschampsia caespitosa	5	..	5	10	60	..	35	..	65
Equisetum silvaticum	..	..	5	..	..	..	..	..	..
Festuca gigantea	..	..	..	..	..	..	..	..	35
Gagea lutea	..	5	..	..	..	..	..	..	..
Galium aparine	..	..	..	5	30	25	5	..	5
Geum urbanum	..	..	..	..	5	..	..	5	..
Holcus mollis	..	..	..	..	15	..	..	..	15
Lampsana communis	..	..	..	..	5	..	..	..	..
Milium effusum	..	..	..	15	..	15	..	..	15
Nepeta hederacea	..	..	..	..	..	15	..	..	..
Poa annua	5	..	..	..	..	..	..	..	..
Polygonum dumetorum	..	..	..	..	..	5	..	..	..
Ranunculus acer	..	..	..	..	..	..	..	..	30
Stachys silvatica	..	..	..	..	20	..	5	15	..
Stellaria holostea	5	..	..	20	60	15	..	..	35
„ media	..	..	..	..	..	..	10	..	..
Veronica chamaedrys	..	..	..	..	10	..	..	..	10
„ hederifolia	15	..	..	..	..	35	..	..	..
Vicia sepium	..	..	..	..	..	..	..	..	15
Viola silvatica	..	..	..	10	..	..	..	..	40
Species density	3·1	3·8	2·2	3·6	5·6	4·9	3·7	3·3	8·6

plants investigated in the younger oakwood 7 per cent. were free, and only 9 per cent. showed the highest degree of reaction (Table 11, *d* and *e*).

TABLE 11

The nitrate content of *Anemone nemorosa* in woods of alder, beech, and oak

	No. of plants investigated.	Percentage distribution among classes of reaction.				
		0	1	2	3	4
a. Alderwood	5×12	..	13	20	30	37
d. Total of beechwood localities . (Tab. 7, *a, d, e, f*)	5×92	21	25	17	15	22
c. Total of oakwood localities .	5×43	4	27	28	22	19
d. Under old oaks	5×20	..	13	31	25	31
e. Young oakwood	5×23	7	39	25	20	9

C. **Alderwood.** Of all the investigated woodland ground floras in which anemones were found that of the alderwood on humus was the richest and at the same time the one that showed the greatest species density (Table 12, No. 3), apart from localities where broad-leaved species such as *Urtica, Impatiens*, and *Mercurialis* had invaded the ground in sufficient numbers to shade most other species out of existence (Table 12, No. 4). *Anemone nemorosa* showed a higher nitrate content here (Table 11, *a*) than in any other locality, although it might be supposed that the stronger light would favour the consumption of the stored nitrate.

II. SITUATIONS OPEN TO THE LIGHT. MEADOW AND PASTURE

By places open to light I mean situations where sufficient light penetrates to allow the ground to become covered with a dense closed vegetation. In our climate grass-like plants dominate such localities. We need not here dwell on the fact that this vegetation is not a natural climax, but is brought about by a cultivation which keeps away the Phanerophytes. Illuminated ground of this kind includes all grades from the dampest to the driest soils, so that a very varied series of formations occurs. It will here suffice to discriminate two main types which are dependent essentially on soil humidity, that is me a d o w and p a s t u r e. By meadow I understand a dense herbaceous vegetation growing where the conditions, especially moisture, are sufficiently favourable to allow the plants to be profitably cut for hay. By pasture I mean land where the conditions are such that the vegetation remains so poor that it cannot profitably be used for hay, but is useful for grazing. The division between the two kinds of vegetation is dictated by practical utility, and is not a sharp one, so that actually the two types are not always distinct. By cultivation, especially by irrigation, pasture can be turned into meadow;

NITRATE CONTENT OF *ANEMONE NEMOROSA*

Table 12

Ground flora on humus

1. Under ash (Jægersborg Hegn); 2, under birch (Jægersborg Hegn); 3–4, under alder (Duschbad Mose in Dyrehaven)

	1	2	3	4
Anemone nemorosa	**100**	**100**	85	10
Ficaria verna	**100**	75	10	10
Impatiens nolitangere	..	..	**90**	100
Mercurialis perennis	..	**85**	30	..
Oxalis acetosella	..	95	95	15
Poa trivialis	..	..	100	95
Stellaria neglecta	..	..	95	65
Urtica dioeca	20	40	15	**100**
Agropyrum caninum	..	..	5	..
Agrostis alba	..	..	25	..
Asperula odorata	..	15	..	..
Arenaria trinervia	..	5	..	..
Baldingera arundinacea	..	..	55	..
Brachypodium silvaticum	..	..	15	..
Calamagrostis lanceolata	..	..	20	..
Carex acutiformis	..	..	15	..
Circaea lutetiana	..	..	5	5
Dactylis glomerata	5	..	45	5
Deschampsia caespitosa	5	..	65	..
Equisetum silvaticum	..	5	..	..
Euonymus europaeus	..	..	5	..
Festuca gigantea	..	..	40	..
Filipendula ulmaria	..	..	10	..
Fraxinus excelsior	..	..	10	..
Galium aparine	..	70	..	50
Geranium robertianum	5	..	..	..
Geum rivale	..	..	10	..
„ urbanum	..	5	..	..
Holcus lanatus	..	15	..	..
Humulus lupulus	..	..	40	5
Lysimachia vulgaris	..	..	25	..
Milium effusum	..	35	..	..
Nepeta hederacea	..	80	5	15
Polygonum hydropiper	..	..	25	..
Ranunculus repens	5	..	30	..
Rubus idaeus	..	35	..	..
Rumex acetosa	..	..	10	..
Stellaria holostea	15	40	..	..
„ media	..	..	70	10
Veronica hederifolia	..	..	..	55
Viola palustris	..	..	5	..
„ silvatica	5	..	..	..
Species density	2·6	7	10·6	5·4

similarly too much drainage of the land can transform meadow into pasture.

Meadow. Within the district investigated *Anemone nemorosa* grows but sparingly in meadows. It is found in one locality at the west end of Fuglesangssøen, in a *Molinia coerulea*-Formation in a hollow in Chr. IX's Hegn, and in a meadow north of Mølleaaen between Raavad and Stampen. Only few of the anemones were found in flower, and their nitrate content was very low (Table 13, *a* and *b*), especially on the meadow near Raavad, which I shall mention again later.

TABLE 13

Nitrate content of *Anemone nemorosa* in Meadow (*a–c*) and on common (*d–e*)

	No. of plants investigated.	Percentage distribution among classes of reaction.				
		0	1	2	3	4
a. Meadow near Raavad	5× 5	76	24	..	..	..
b. *Molinia coerulea*-Formation	5× 5	12	68	20	..	..
c. Meadow near Fuglesangssøen	5× 2	..	40	40	20	..
d. *Nardus* pasture	5× 5	72	24	4	..	..
e. Pasture rich in species and species density	5×20	50	39	9	2	..

Pasture. The nitrate content of *Anemone nemorosa* was investigated partly in a *Nardus strictus*-Formation at 'Andet Tøjreslag' in Dyrehaven (Table 14), and partly on a pasture richer in species outside the edge of the wood.

As Table 15 shows, the vegetation in the latter place is both rich in species and has considerable species density: in the 20 random samples analysed, each of 0·1 sq. metre, amounting altogether to 2 sq. metres, 40 species were found. Ten of these occurred as frequency-dominants. The large number of species shows that the conditions suit a large number of our species, enabling them to maintain their place in spite of keen competition. This perhaps explains the fact that the anemones in such localities were almost as poor in nitrate (Table 13, *e*) as in the poor *Nardus*-Formation (Table 13, *d*). In order to draw conclusions about the relationship between the amount of nitrate stored by any species (e.g. *Anemone nemorosa*) and the nitrifying property of the soil on which it grows, it is always necessary to take into consideration the demand for nitrate according to the species density and the number of individuals.

In Dyrehaven and in Jægersborg Hegn I have investigated the nitrate content of *Anemone* in all the different types of habitat occupied by this species. In the habitats that make up the largest area a great many more individuals have been investigated than in the other localities, so that we may look upon the total result as an expression of the entire district, even

Table 14

Nardus-Formation

(1. Nardus–Molinia–Agrostis tenuis-Facies; 2, Nardus–Agrostis tenuis-Facies)

	F per cent.		
	1	2	3
Molinia coerulea	95	..	..
Nardus strictus	100	100	100
Agrostis tenuis	100	100	80
Potentilla erecta	50	95	5
Achillea millefolium	..	50	..
Agrostis canina	35	10	..
Anthoxanthum odoratum	30	80	..
Avena pratensis	..	5	..
„ pubescens	..	5	..
Campanula rotundifolia	..	80	25
Cardamine pratensis (very weak)	5	..	..
Carex caryophyllea	10	..	..
„ hirta	20	10	..
„ leporina	10	..	20
„ pilulifera	..	65	..
Deschampsia caespitosa	30	..	..
„ flexuosa	5	25	10
Festuca ovina	..	..	10
Galium harcynicum	45	20	5
Hieracium auricula	..	..	5
„ pilosella	..	30	..
Holcus lanatus	25	40	..
Hypericum perforatum	..	5	..
Juncus effusus	15	..	..
Luzula campestris (coll.)	40	80	5
Lysimachia vulgaris (very weak)	10	..	..
Plantago lanceolata	..	5	5
Poa pratensis	35	20	..
„ trivialis (very weak)	25	..	..
Ranunculus acer	..	5	..
„ repens (very weak)	5	..	..
Rumex acetosa	55	50	10
„ acetosella	..	..	10
Sieglingia decumbens	40	65	..
Stellaria graminea	25	20	5
Trifolium medium	..	10	..
Veronica chamaedrys	5	35	15
„ officinalis	..	5	5
Viola canina	..	60	5
„ palustris	65	5	..
Species density	8·8	10·8	3·2

Table 15
Pasture rich in species and with a high species density

	F%		F%
Agrostis tenuis	95	Deschampsia flexuosa	10
Anemone nemorosa	100	Festuca ovina	40
Festuca rubra	100	Galium harcynicum	5
Holcus mollis	100	Hieracium auricula	15
Luzula campestris	85	Hypericum perforatum	60
Oxalis acetosella	90	Lathyrus montanus	15
Poa pratensis	95	Listera ovata	10
Potentilla erecta	95	Plantago lanceolata	15
Rumex acetosa	95	Poa trivialis	35
Veronica chamaedrys	100	Populus tremula	10
Achillea millefolium	50	Ranunculus acer	75
Alchimilla minor	5	Stellaria graminea	55
Anthoxanthum odoratum	65	,, holostea	20
Avena pubescens	20	Trifolium medium	65
Betula pubescens	5	,, repens	45
Campanula rotundifolia	75	Veronica serpyllifolia	10
Carex leporina	5	Vicia cracca	20
,, pilulifera	55	,, sepium	15
,, caryophyllea	35	Viola silvatica	45
Cerastium caespitosum	10	Species density	18·8
Deschampsia caespitosa	10		

if the investigated plants do not represent a uniform system embracing the whole area.

The 1,040 plants investigated are distributed as follows among the different degrees of reaction.

Table 15 A

	No. of plants of Anemone investigated.	Percentage distribution in classes of reaction.				
		0	1	2	3	4
Dyrehaven and Jægersborg Hegn.	1,040	21	27	21	15	16

This table shows that about four-fifths of the individuals contain nitrate, a result not in good accord with other investigations of the nitrate content of *Anemone nemorosa*. For example Hesselman (1917, p. 387) states that this species is practically always devoid of nitrate, and that the soil on which it grows usually elaborates no nitrate. Bornebusch (1923, p. 103) also finds that in Denmark *Anemone nemorosa* is almost always devoid of nitrate. I suppose that the great difference between my result and those of other investigators is due to the fact that Hesselman and Bornebusch used sections of the plant's tissue, whereas I have always made use of the expressed juice, a technique better adapted for demonstrating nitrates. I have investigated groups of five plants 208 times, and only in nine of these observations were all five plants devoid of nitrate.

Next we must try to compare the results of my investigations of the nitrate content of *Anemone* with the results obtained by Weis (1924) in his investigations of the nitrate content of the soil in a series of Danish beechwoods. The soils investigated were very probably almost all of them taken from *Anemone* localities. Weis used as his reagent diphenylamine sulphate, the same reagent that I have employed. In each of the 779 soil samples 25–30 grammes of soil were shaken together with 25 c.c. of water and shaken continuously for 2 to 3 hours. Of the filtered extract a drop was taken in a pipette and placed in the hollow of a porcelain dish which already contained 10 drops of diphenylamine sulphate. If any nitrate was present in the extract a blue coloration of varying intensity was produced, and Weis differentiates 6 degrees of reaction which he designates as follows:

	Nitrate reaction.
Trace of blue	0·5 a trace
Very pale blue	1·0 very weak
Pale blue ring	1·5 weak
Distinct blue ring	2·0 distinct
Strong blue ring which broadens	2·5 strong
The whole drop blue black	3·0 very strong

It must be borne in mind that we are dealing with a very sensitive reagent whose action is rapid, and that we have no right to assume that the degrees of reaction are uniform classes. The last of these degrees embraces a series of concentrations much greater than that embraced by all the other degrees put together. It seems to me that Weis differentiates too many classes; the two first classes in particular cannot be relied upon; they should be united into a single class, which would coincide more or less with a class I have called No. 1. Similarly Weis's reactions Nos. 3 and 4 can be united into one which would correspond roughly with my class No. 2. Weis's two last degrees of reaction correspond most probably with the two degrees that I have called respectively 3 and 4. It is on this basis that I will attempt my comparison.

The distribution of Weis's determinations among the different classes of reaction seems to some extent to justify my proposed reduction in the number of classes. If we investigate the distribution in the various classes of reaction of the 547 determinations of nitrate content carried out by Weis on superficial samples of soil, we find the result given in Table 16. We see in this table the striking fact that the class of reaction designated 1·5 (pale blue ring) is only found in 4 samples, while the previous class has 63 and the following class 109 samples. Of course this does not exclude the possibility that the very low number in class 1·5 is an expression of the fact that soil with a correspondingly low nitrate content is seldom found in our woods, while soil a little richer or a little poorer in nitrate occurs more commonly. But it is very much more probable that the low number in class 1·5 expresses the fact that the class is not a good one. If my reduction of Weis's classes be accepted we then obtain the percentage distribution seen in Table 16 and in Table 17, b.

TABLE 16

Distribution among classes of reaction of Weis's determinations of nitrate content in the superficial soil of Danish beechwoods

No. of determinations.	Classes of reaction.						
	0	0·5	1	1·5	2	2·5	3
547	228	39	63	4	109	88	16
%	42	7	11	1	20	16	3
	42	18		21		16	3

In Table 17, b, for the sake of comparison the numbers are placed together with my own numbers (Table 17, a) according to the method I have already described. The difference is apparent especially in two points: firstly the percentage of nitrate-free localities in Weis's material is twice as great as the percentage of nitrate-free *Anemones* in my investigations; secondly in the highest class of reaction (No. 4) my number exceeds Weis's number greatly, being 16 per cent. against his 3 per cent.

TABLE 17

	No. of determinations.	Percentage distribution among classes of reaction.				
		0	1	2	3	4
a. My investigations of the nitrate content of *Anemone nemorosa* in Dyrehaven and Jægersborg Hegn	1,040	21	27	21	15	16
b. Weis's determinations of the nitrate content of the superficial layer of soil in a series of Danish beechwoods	547	42	18	21	16	3
c. Weis's determinations of the nitrate content in Dyrehaven alone	55	49	31	15	5	..

Since Weis also made a series of investigations in Dyrehaven it might be supposed that if one wishes to make a comparison, my observations should be compared with these. I have therefore in Table 17, c compiled the results of Weis's observations in Dyrehaven. In the two points mentioned above these numbers deviate even more than do the numbers in Weis's material taken as a whole. The nitrate content of the soil in the part of Dyrehaven investigated by Weis is considerably lower than the average nitrate content of all the woods investigated by him. The part of Dyrehaven investigated by Weis may be considered the poorest in nitrifying properties of all the parts I investigated. At the same time it includes but a small area of the localities I investigated in which *Anemone* grow. I think therefore that it is more in harmony with the actual facts to compare my investigations with the results of Weis's observations of all the woods that he investigated. The difference demonstrated between the two series of investigations presumably depends upon the fact that the tissue of the plant is more concentrated than the greatly diluted soil extract employed by Weis. It seems to me that this dilution is a slight defect in Weis's method. It is true that diphenylamine sulphate is a very delicate reagent, but when there are only minute traces of nitrate in the soil, and when such large quantities of water are used as are necessary to obtain the filtrate, it may well happen that the nitrate content of the solution becomes so dilute that no reaction occurs. That this may occur I have proved on several occasions actually in one of the localities in Dyrehaven which Weis investigated. From this locality I took soil samples and treated them by Weis's method, and the filtrate showed no trace of nitrate-reaction; but after evaporating the solution, thus making it more concentrated, the reaction was clearly discernible.

Instead of investigating the nitrate content of soil extract by means of diphenylamine sulphate in order to find out the nitrifying properties of the soil, it seems both easy and more certain to find out the nitrate content of the plants which grow on the soil in question. By this means we avoid also the possible error caused by the presence of substances in the soil other than nitrate which give a blue coloration with diphenylamine sulphate.

I mentioned earlier that in the attempt to use the nitrate content of plants as a preliminary means of determining the nitrifying properties of the soil, we must consider the illumination and the competition caused by the density of the vegetation, including density of individuals and species density. In the meadow near Raavad (Table 13, *a*) the anemones were partly free from nitrate and partly very poor in nitrate; but individuals of *Cirsium palustre* and *Cirsium oleraceum* were found to have the degree of reaction of 3 and 4. The same state of affairs was found in several places. Doubtless nitrification proceeds actively in these places, but competition is so great that only those species which take up and store nitrate very efficiently show a high degree of reaction, while species such as *Anemone nemorosa*, in which these properties are less active, show only a small nitrate content. But for this very reason such species are particularly useful in the study of the nitrifying properties of the soil.

In order to illustrate the different capacity for storing nitrate in different species I have compared in some localities *Anemone nemorosa* with a species often growing in company with it, *Mercurialis perennis*. In each of the four localities 25 individuals of each species were investigated. Table 18 shows that *Mercurialis perennis* has a distinctly higher nitrate content than *Anemone* both in the average of all the plants investigated and in the average of plants in each of the localities.

TABLE 18

Nitrate content of *Anemone* and *Mercurialis perennis* growing in the same locality

Locality.	Species.	No. of plants investigated.	Percentage distribution among classes of reaction.				
			0	1	2	3	4
A. Oakwood	Anemone	25	52	44	4	..	..
	Mercurialis	25	52	20	8	16	4
B. Alderwood	Anemone	25	..	36	40	20	4
	Mercurialis	25	8	4	20	40	28
C. *Rubus idaeus*-Formation beneath old oak trees	Anemone	25	..	24	64	12	..
	Mercurialis	25	..	4	28	24	44
D. Beechwood	Anemone	25	..	..	56	44	..
	Mercurialis	25	..	..	4	52	44
A–D	Anemone	100	13	26	41	19	1
	Mercurialis	100	15	7	15	33	30

Most of the more or less shaded formations on the floor of the forest are as a rule neither rich in individuals nor dense in species. Competition in these formations cannot be nearly so keen as it is in fully illuminated, dense formations on meadow and common. But the other determining factor for the storage of nitrates, i.e. light, is here active. In order to investigate the influence of light upon the nitrate content of the plant I have determined the nitrate content of shaded and illuminated individuals of the same species. For this purpose *Mercurialis perennis* is well suited, because, when it grows in dense communities the taller plants form a well-illuminated upper layer which casts a deep shade on to the shorter individuals. I investigated 50 individuals in each of these layers, and the results are shown in Table 19, A. After that I compared the nitrate content of individuals shaded by dense young beech trees with that of individuals growing near them in full illumination (19, C). Again I compared the individuals of the dense community of *Mercurialis perennis* growing on the north side of a long pile of firewood running from east to west with fully lighted individuals from its south side (19, B).

It is seen from the table that all three cases show a distinct difference in nitrate content between the illuminated and the shaded individuals.

Finally I investigated lighted and shaded leaves in the same individual, using the upper part of the leaf stalk in *Aesculus hippocastanum*. Five individuals were investigated, and in each an equal number of

TABLE 19

The nitrate content of *Mercurialis perennis* in shade (*a*) and in light (*b*)

		No. of plants investigated.	Percentage distribution among classes of reaction.				
			0	1	2	3	4
A	*a.* Shaded layer	50	..	16	26	22	36
A	*b.* Illuminated layer	50	16	32	28	14	10
B	*a.* North side of wood stack	25	..	20	24	20	36
B	*b.* South side of wood stack	25	36	40	20	4	..
C	*a.* In shade of young beeches	50	..	2	6	16	76
C	*b.* In light about 1 metre from *a*	50	24	36	18	14	8
A–C	*a.* In shade	125	..	11	18	19	52
A–C	*b.* In light	125	23	35	23	12	7

illuminated and shaded leaves were taken. In one of the individuals the shaded leaves were distinctly poorer in nitrogen than the lighted leaves; but, taken as a whole, as is seen in Table 20, the shaded leaves were considerably richer in nitrate than the lighted ones. As a result, then, of these investigations, scarcely any doubt can remain that, other factors being equal, shade encourages the storage of nitrate in plants while light encourages its consumption.

TABLE 20

The difference in nitrate content in the upper portion of the leaf stalk of illuminated and shaded leaves of the same individuals of *Aesculus hippocastanum*

	No. of leaf stalks investigated.	Percentage distribution among classes of reaction.				
		0	1	2	3	4
Shaded leaves	30	10	37	30	23	..
Illuminated leaves	30	47	37	13	3	..

Different species of plants have different capacities for storing nitrates and thus, other circumstances being equal, contain different amounts of nitrate. It is therefore extremely important to base a comparative investigation on the behaviour of a single species or of a few species whose behaviour can be followed accurately in different localities.

The nitrate content of the individual species depends, apart from its capacity for storing nitrates, upon the amount of nitrate available in the soil, which in turn is not only dependent upon the nitrifying properties of the soil but, as investigation shows, upon the density of the vegetation both in individuals and species.

Only if in the places compared we have essentially the same illumination and essentially the same density of individuals and species, and only if the investigations take place at the same time, can we expect to find that the difference in the nitrate content of a plant in any way reflects a corresponding difference in the nitrifying properties of the soil.

In order to use the nitrate content of a plant, e.g. that of *Anemone*, as a measure of the state of nitrification in the soil in any locality it is necessary, at any rate in a number of typical cases, to carry out an investigation of the nitrifying properties of the soil to compare with the nitrate content of the plant. This I have unfortunately never had the opportunity of doing.

LITERATURE
to which reference is made.

BORNEBUSCH, C. H. (1920), Om Bedömmelse af Skovjordens Godhed ved Hjælp af Bundfloraen. Dansk Skovforenings Tidsskrift. 1920.

—— (1921), Objektiv Beskrivelse af et Skovdistrikts Urteflora. Ibid. 1921.

—— (1923), Skovbundsstudier. Det forstlige Forsøgsvæsen i Danmark, viii, 1923.

FRANK, A. B. (1887), Ueber Ursprung und Schicksal der Salpetersäure in der Pflanze. Ber. Deutsch. bot. Ges. Bd. **5**, 1887, pp. 472–87.

—— (1888), Ueber die physiologische Bedeutung der Mycorhiza. Ibid. Bd. **6**, 1888, pp. 248–69.

HESSELMAN, H. (1917), Studier över salpeterbildningen i naturliga jordmåner och dess betydelse i växtekologiskt avseende. Meddelanden från Statens Skogsförsöksanstalt. Bd. **1**, pp. 297–528. Stockholm 1917.

KLEIN, R. (1913), Ueber Nachweis und Vorkommen von Nitraten und Nitriten in Pflanzen. Beihefte z. Bot. Centralblatt. Bd. **30**, 1. Abt., 1913, pp. 141–66.

LÜTKEN, CH. (1899), Den Langenske Forstordning. Et Bidrag til det danske Skovbrugs Historie. København 1899. 4. Med Atlas i Fol.

MOLISCH, H. (1883), Ueber den mikrochemischen Nachweis von Nitraten und Nitriten in der Planze mittels Diphenylamin oder Brucin. Ber. Deutsch. bot. Ges. Bd. **1**.

OLSEN, CARSTEN (1921), Studier over Jordbundens Brintionkoncentration og dens Betydning for Vegetationen, særlig for Plantefordelingen i Naturen. Copenhagen 1921.

RAUNKIÆR, C. (1909), Formationsundersøgelser og Formationsstatistik. Bot. Tidsskrift. **30**, 1909. (Chapter VI of the present book.)

—— (1913), Formationsstatistiske Undersøgelser paa Skagens Odde. Bot. Tidsskrift. **33**, 1913. (Chapter VIII.)

—— (1918), Recherches statistiques sur les formations végétales. Det Kgl. Danske Videnskabernes Selskab. Biologiske Meddelelser. **1**, 3, 1918. (Chapter XI.)

—— (1922), Forskellige Vegetationstypers forskellige Indflydelse paa Jordbundens Surhedsgrad (Brintionkoncentration). Ibid. iii. 10. (Chapter XIV.)

SCHIMPER, A. F. W. (1890), Zur Frage der Assimilation der Mineralsalze durch die grüne Pflanze. Flora, Bd. **48**, 1890, pp. 207–61.

STAHL, E. (1900), Der Sinn der Mycorhizenbildung. Pringsh. Jahrb. Bd. **34**, 1899–1900, pp. 539–668.

VAHL, M. (1911), Les types biologiques dans quelques formations végétales de la Scandinavie. Académie Royale des Sciences et des Lettres de Danemark. Bull. de l'année 1911.

WEIS, FR. (1924), Undersøgelser over Jordbundens Reaktion og Nitrifikationsevne. I. Typiske danske Bøgeskove. Meddelelser fra Dansk Skovforenings Gødningsforsøg, iv, 1924.

XVI

THE AREA OF DOMINANCE, SPECIES DENSITY, AND FORMATION DOMINANTS

Wherever we travel we find in nature that the environment alters from place to place, indeed often from step to step. Correspondingly the vegetation alters qualitatively, i.e. in its species composition, and quantitatively, i.e. in the frequency of the component species. The qualitative (floristic) changes can be fairly easily observed and followed; but this is not true of the quantitative relationships, which change continually, giving no fixed points for our observations. Yet it is these changes that are of fundamental importance in the science of plant formations, in the study of the changes undergone by vegetation behaving as a biological expression of the changing environment.

Since we cannot immediately observe and determine a continuous quantitative change, we must seek some device, some definite method, of reducing these changes to a series of discontinuous units, which can be treated like the quantitative units, the species of the flora. This may be done by the valency method, which aims at allotting to each of the species occurring in a definite vegetation a number expressing its frequency. This number is determined by measuring the extent of the area covered by the individuals of the species. The figure can be computed to any degree of fineness, the degree used depending upon practical needs, and upon the relationship between the extent of the observations made and the value of the results for the object in view.

I shall at present enter into no further details about this question. It is enough to mention that in my investigation of Danish plant communities I have been satisfied with a method of determining valency based on the computation of the number of spaces between the individuals of a species, each space being large enough to contain a circular area of 0·1 sq. metre. Let us imagine a given vegetation to be investigated by placing upon it an infinite number of such circular areas in such a way that the centre of the circle had fallen once upon every point within the area, then the centre of the circle would have encountered all the points so situated that the circumference of the circle neither included nor touched a single individual of the species dealt with; provided of course that large enough intervals were present. We should then have for our species in the area investigated an absolute measure of the relationship between the area occupied and that unoccupied by it, the unit of measurement being the size of the sample circle. We must, however, content ourselves with a very restricted number of sample areas. The actual number will be discussed later.

Suppose, for example, that we are investigating 100 sample areas, and that the vegetation is uniform, so that our samples may be taken at random, i.e. according to the principle of chance. If we then find that the species to be investigated is absent from 20 of these 100 sample areas we see, allowing the 100 areas to be unity, that the extent of the area within which the sample area of 0·1 sq. metre can be put down without encountering the species, is 20 per cent. of the whole area. But since it is a positive and not a negative determination that we require, a determination of the area occupied by the species, i.e. the number of sample areas which include or touch one or more individuals of the species, we express the relationship as 80 per cent. of the area. I have called this number the **frequency percentage** (F per cent.) of the species in the vegetation here dealt with.

Let us take as our starting-point 100 samples in which 10 species are found. Each of these species will be encountered a number of times, the number usually differing in all 10 species. Each of the species then has obtained an F per cent. corresponding to its distribution in the vegetation investigated. If the experiment be repeated the result will certainly be, that of all those species whose F per cent. were lower than 100 not one will be found to have the same F per cent. as in the first experiment. Only those species whose F per cent. in the first experiment was 100 will as a rule obtain the same F per cent. again. This is because they are almost always species whose individuals are so numerous that their F per cent. would be 100 even if the sample area were much less than 0·1 sq. metre.

If we restrict our concept of a plant formation by stating that two vegetation units cannot belong to the same formation unless they agree in the species present and in their valency (F per cent.), we shall scarcely ever find two areas which belong to the same formation. We must therefore abandon this restricted concept of the uniform composition of plant formations. In dealing with this question we must first, however, examine a very important constant in characterizing a formation, the **species density**, i.e. the average number of species per sample plot. This number becomes constant even after the investigation of quite a small number of samples, and does not alter to any extent however many more samples are examined. If we divide the species found into five classes of frequency, viz. F per cent. 1–20, 21–40, 41–60, 61–80, and 81–100, we then find that the number of species in the classes decreases with ascending frequency. We must, however, exclude from the fifth class species whose F per cent. is 100, since these species, as before mentioned, will nearly always have an F per cent. of 100 when a smaller sample plot is used, and so may be said to have an F per cent. of more than 100.

As we shall see later the sum of the frequencies of the fifth class is almost as great as that of all the other classes together; from this it follows that the fifth class has a share in the species density almost as great as the four first frequency classes together. If we then have to restrict the concept of unity which must characterize vegetation in order that it may belong to the same least unit, i.e. to the same formation in the strictest sense, it is reasonable, first of all, to renounce the necessity for agreement in the four lower classes of frequency. Thus these classes are important only because they help us in the determination of constancy of species density. The demand for unity is insisted upon only in connexion with the fifth class of frequency and is insisted upon only among types of vegetation which agree in respect to the occurrence of species in the fifth class. It is only kinds of vegetation which thus agree that belong to the same smallest unit, the same formation in the strictest sense. I call these species frequency dominants; from the point of view of frequency these are the only species which are constant in the formation.

In the investigation of formations that are poor in species it is comparatively easy to maintain the demand of unity with respect to frequency dominants and essential unity with respect to species density. But this is not so in kinds of vegetation rich in species and with a high species density and many frequency dominants. It is here questionable to what extent it is possible or desirable to use the demarcation of formations defined above. At all events it will be very difficult to use, and we shall see later on that no use of it has been made in the investigations already undertaken.

Finally it might be doubted whether it were best to place the lowest boundary of the dominants at F per cent. 80, or whether one should not restrict the dominants to species with F per cent. 100. I think, however, that in order to have an interval to allow for chance variations it is best to consider all the species in the fifth class as dominants.

The dominance-area of the species, and co-dominants. By dominance-area of a species I understand the region or regions throughout which it occurs as a frequency dominant (i.e. with F per cent. over 80 determined by 0·1 sq. metre plots). By the co-dominants of a given species I understand those species that may occur as frequency dominants within the dominance area of that species. The degree of co-dominance can be expressed as a percentage of the observations, which must, of course, be as uniformly distributed as possible in the region in which the degree of dominance of the co-dominants is investigated. Co-dominance is an expression of ecological relationship, but it does not assert on which side the relationship is; this is only perceived by determining what is the co-dominance of the one or more co-dominant species in question. The ecological relationship between species is much better expressed by co-dominance than merely by co-occurrence. It is, therefore, of interest to determine within the region under investigation the grade of the co-dominance of the species accompanying the dominants.

There scarcely exist two species with exactly the same range of habitat, with exactly the same distribution; and species vary widely in the abundance with which they occur. Even if it be true that every species may in one place or another, on an area of varying size, occur as a frequency dominant, yet only the minority of our species occur as frequency dominants over extensive areas.

Where we are dealing with a stabilized plant community we may suppose that the species occurring as frequency dominants are to be found on a soil particularly suited to them, even though their success in the company of competitors bears no comparison to their success in places where competition is for any reason excluded. It is therefore of particular interest to know the nature and the boundaries of those regions within which the individual species occur as frequency dominants. It is also of interest to know which other species can occur as frequency dominants in their company. It has been customary to state what species grow in company with a given species; but we obtain a much truer picture by learning the frequency grade of the species in question in the given locality and the dominant species with which it grows. We ought therefore to strive to obtain such information about the individual species as will enable us to render an account of the species in a flora, stating, for each species, in which community or communities it is chiefly distributed; and, if the species occurs as a frequency dominant, we should be able to state its commonest co-dominants.

The regions within which the species occur as a frequency dominant are scarcely the same for any two species, still less for more. The overlapping of the dominance areas of species gives us a means of determining ecological affinity. The simplest and easiest behaviour to study is where there is a gradual change of one of the most important factors determining plant distribution, while the other factors remain essentially uniform. In certain parts of the dune regions, on certain heaths, and in many seaside meadows, where the composition of the soil is fairly uniform, if we proceed along the evenly sloping surface from a lower to a higher level, in the main only one of the chief factors regulating plant distribution alters, and that factor is moisture. By determining in such a situation the boundaries of the dominance of the species, and how the areas of dominance overlap one another, we are able to find out the succession of the species with respect to the given factor. And what we here discover may be important in investigating distribution in regions where the environment is less regular. If we have no opportunity of using a comparative investigation of series of this kind we must begin by determining the status of the individual species in relationship to the other species by determining the co-dominants of this species within its dominance area. In order to illustrate this for an individual species I will give the result of some analyses of the statistics of formations which I undertook in 1915 in some *Gnaphalium arenarium*[1] communities in the neighbourhood of Frederiksværk.

[1] *Helichrysum arenarium*, DC.

TABLE I

1–5. Five analyses, each of 25 plots of 0·1 sq. metre within the dominance area of *Gnaphalium arenarium*.
6. An analysis (25×0·1 sq. metre) in the *Trifolium arvense–Jasione montana*-Formation, in which *Gnaphalium arenarium* is recessive.
B. Analyses 1–5 taken together as one analysis on 125 sample plots.

No. of analysis.	B	1	2	3	4	5	6
Species density	11·50	12·52	14·68	15·52	9·20	6·52	7·64
Number of dominants	1	5	4	8	6	3	2
Number of species	68	30	43	31	23	18	20
Gnaphalium arenarium F%	100	100	100	100	100	100	76
Festuca ovina	17	84	..	..	..	..	..
Galium verum	18	88	..	..	..	..	..
Phleum Boehmeri	18	92	..	..	..	..	..
Thymus serpyllum	19	96	..	..	..	..	..
[Lolium perenne]	19	..	96	..	..	..	..
Scleranthus annuus	25	..	88	8	20	8	..
Rumex acetosella	55	..	100	88	44	44	64
Festuca rubra	38	72	20	92	8	..	60
Chrysanthemum leucanthemum	22	..	8	88	12	..	..
Aira praecox	47	..	24	100	84	28	44
Trifolium arvense	50	16	36	100	84	16	100
Corynephorus canescens	69	4	52	88	100	100	80
Jasione montana	51	..	68	92	96	100	88
Teesdalea nudicaulis	51	..	60	56	92	44	4
Achillea millefolium	22	..	20	68	8	16	4
Agrostis alba	6	16	..	..	..	12	4
„ spica venti	1	..	4	..	..	..	..
Aira caryophyllea	31	..	48	44	60	..	..
Alchimilla aphanes	2	..	12	..	..	..	..
Anchusa officinalis	1	..	4	..	..	..	..
Anthemis arvensis	4	..	20	..	..	..	..
Anthoxanthum odoratum	13	..	..	64	..	..	56
Anthyllis vulneraria	9	4	4	24	12	..	..
Arenaria serpyllifolia	17	52	32	..	..	..	..
Artemisia campestris	53	64	32	72	60	36	76
Avena pratensis	14	72	..	..	..	..	..
Bromus mollis	1	4	..	..	..	..	..
Calamintha acinos	8	40	..	..	..	..	..
Campanula rotundifolia	7	36	..	..	..	..	4
Carex arenaria	13	64	..	..	..	..	..
„ caryophyllea	10	52	..	..	..	..	..
Cerastium semidecandrum	35	..	80	28	44	24	..
Convolvulus arvensis	1	..	4	..	..	..	..
Deschampsia flexuosa	1	4	..	..	..	..	..
Echium vulgare	2	..	12	..	..	..	..
Erigeron acer	5	..	20	..	..	4	8
Erodium cicutarium	14	..	72	..	..	..	..
Euphrasia officinalis	1	4	..	..	..	..	..
Filago minima	30	..	72	4	28	44	8

TABLE I—continued.

No. of analysis.	B	1	2	3	4	5	6
Filipendula hexapetala	1	4	..	..	..	..	..
Gnaphalium silvaticum	1	..	..	4	..	..	..
Hieracium pilosella	38	20	56	68	24	24	48
Holcus lanatus	19	..	44	52	..	..	4
Hypericum perforatum	2	..	..	8	..	..	12
Hypochaeris radicata	10	..	28	20	..	4	..
Knautia arvensis	12	..	32	16	12	..	..
Koeleria cristata	6	28	..	..	..	..	..
Linaria vulgaris	4	..	8	8	4	..	..
Luzula campestris	2	8	..	..	..	..	..
Medicago lupulina	5	24	..	..	..	..	..
Ononis repens	4	16	..	..	..	4	..
Plantago lanceolata	11	..	8	48	..	..	..
Poa pratensis	11	..	..	52	4	..	..
Polygonum convolvulus	2	..	8	..	..	..	..
Potentilla argentea	7	..	4	16	16	..	..
Pulsatilla pratensis	14	68	..	..	..	..	..
Sedum acre	15	76	..	..	..	..	..
„ telephium	..	..	..	..	..	..	20
Solidago virga aurea	36	..	64	68	4	44	..
Spergula arvensis	7	..	36	..	..	..	4
Thalictrum minus	5	24	..	..	..	..	..
Trifolium minus	6	20	12	..	..	..	..
„ pratense	1	..	4	..	..	..	..
„ procumbens	9	..	4	36	4	..	..
Veronica arvensis	6	..	32	..	..	..	..
Vicia (Ervum ?), seedlings	7	..	16	20	..	..	..
Viola tricolor	3	..	16	..	..	..	..
Vulpia dertonensis	6	..	8	20	..	..	..

In Table 1 I have given the results of the analyses of five different places (1–5) in which *Gnaphalium arenarium* was the frequency dominant as well as the physiognomical dominant, and in one place (6) where the *Gnaphalium* was recessive both in frequency and physiognomically. In each place 25 sample plots were analysed, each plot being 0·1 sq. metre. No. 1 is from 'Hvide Klint', a hill sloping abruptly towards the fiord south of Frederiksværk. The hill was planted with Conifers; but on the southern part were unplanted areas with *Gnaphalium arenarium*-Formation, containing, besides *Gnaphalium*, *Festuca ovina*, *Galium verum*, *Phleum Boehmeri*, and *Thymys serpyllum* as frequency dominants. The number of species and the species density were fairly large, 30 and 12·52 respectively. In 1920 the reaction of the soil was determined in two samples, which were both alkaline, both having the P_H value of 7·5.

Observations Nos. 2–6 are all from a sandy ridge immediately to the east of Frederiksværk. Nearest to the town the hills composing this ridge are covered with wood, principally beech; but to the east there were in

1915 portions without wood, and these, at any rate in part, had not been cultivated for many years, as evinced by the scattered, presumably self-sown, pines and birches, some of which attained 4·5 metres in height. In certain places *Gnaphalium arenarium* dominated both physiognomically and in frequency. Wherever this plant grows it easily becomes a frequency dominant because of its capacity for spreading by its rootstocks.

Even if the vegetation investigated had been left to itself for a long period it had certainly not yet come to equilibrium. Its final state will be scarcely as favourable to *Gnaphalium arenarium* as when the investigation took place. When I again investigated the locality in 1920 this species was recessive.

The degree of soil acidity in the *Gnaphalium*-Formation differed somewhat from that found in the abutting beechwood. In 1920 the reactions of 4 samples in the *Gnaphalium*-Formation were determined. The P_H value was 6 (varying between 5·9 and 6·1). At the same time thirteen samples from the beechwood showed on an average a P_H value of 4·3 (varying between 4 and 4·7).

Observation No. 2 is from a lower part of the locality, which apparently had been cultivated not so very long ago, as was evinced by *Lolium perenne*, whose occurrence as a frequency dominant must have been due to sown seed. Physiognomically the community was a *Gnaphalium arenarium–Rumex acetosella*-Formation; but *Gnaphalium* was the more conspicuous of the two components. The soil could be seen between the plants, which grew luxuriantly. Both the number of species and the species density were high, 43 and 14·68.

Observations Nos. 3–5 were made on a portion of ground which had been uncultivated for many years. In Table 1 the observations are arranged according to the decrease in richness of the vegetation; this is most easily seen by looking at the number of species and the species density. The behaviour of the dominants shows the same thing; in No. 3 there are, besides *Gnaphalium*, 7 frequency dominants, in No. 4 there are 5, and in No. 5 there are only 2.

In a little spruce plantation some open places and hollows were found with a poor vegetation of scattered Phanerogams between which the soil was visible, where it was not covered with *Polytrichum piliferum* and *P. juniperinum*. *Gnaphalium* had doubtless formerly been a frequency dominant, but it is now recessive. No. 6 shows the composition of the vegetation.

We see from Table 1 that of the 68 species which occurred in the five localities (1–5) in which *Gnaphalium arenarium* was the frequency dominant, only 4 species besides *Gnaphalium* are common to all; and we see that more than half of the species are found in only one analysis. Fourteen species occur as co-dominants, but not one of them is common to the five analyses; 9 of them are found in only one analysis,

3 in two, and 2 (*Corynephorus canescens* and *Jasione montana*) in three analyses.

In Table 2, A, we see how the frequency numbers in the analyses arrange themselves in five classes of percentage frequency, and we notice the rise in the number in the fifth frequency class usually seen in narrowly demarcated formations. If, however, we take the five analyses together, making one complete analysis on 125 plots within the dominance area of *Gnaphalium arenarium*, we do not get a single dominant beside *Gnaphalium*, the dominance of which was recognized to begin with; and the distribution of the frequency numbers among the frequency classes becomes quite different (Table 2, B). Although the analyses 1–5 each represents a narrowly demarcated formation, and though they all lie within the dominance area of the same species, yet they differ mutually so widely that when they are taken together as a single analysis they give a frequency distribution curve showing no rise in the fifth frequency class.

TABLE 2

A. Percentage distribution (in 5 frequency classes) of the frequency numbers in analyses 1–5 in Table 1.
B. The same for B in Table 1.

	No. of frequency numbers.	Frequency classes.				
		1	2	3	4	5
A	145	% 42	% 15	% 14	% 11	% 18
B	68	75	13	9	1·5	1·5

It is clear that only when an investigation is extended in this manner to include a large number of localities within the dominance area of a species can we find out which species are the chief co-dominants of the species in question, and with which species the dominant has the closest ecological relationship. As long as investigations are not undertaken with this definite object in view we must content ourselves with materials procured for another purpose. This may well be done where, considering the object in view, the material does not appear to be one-sided. As an example I have chosen *Anemone nemorosa*. I have had access to 140 observations within the dominance area of this species; but they were confined principally to the island of Zealand. In these 140 observations 36 species in all occurred as co-dominants; but of these 22 were found only once as co-dominants, 5 species twice, 1 species four times, and 1 species five times. Of the remaining species *Oxalis acetosella*, *Mercurialis perennis*, *Melica uniflora*, *Asperula odorata*, *Ficaria verna*, *Lamium galeobdolon*, and *Hordeum europaeum* were co-dominants in 7–10 per cent. of the observations.

As another example I will describe what happens in a narrowly demarcated locality, i.e. the dominance area of *Hypochaeris radicata* in the dune region of the island of Fanø. In this cultivated area, which corresponds to the *Nardus* level and the higher portion of the *Erica* level, *Hypochaeris radicata* and *Leontodon autumnalis* grow together. The fields, which are in the grass stage, are covered in summer with the shining yellow blossoms of both these species. In the parts of the *Nardus*-Formation which still persist both species occur, but they are scattered, being unable to compete with the thick carpet of *Nardus*. In contradistinction to what happens in the cultivated area where *Leontodon autumnalis* and *Hypochaeris radicata* occur together over wide extents, on the uncultivated ground their areas of dominance are quite distinct; *Leontodon autumnalis* being dominant on the seaside meadows below the *Nardus*-Formation, and *Hypochaeris radicata* being dominant above the *Nardus*-Formation on the dry soil of the so-called 'Grey Dunes', which are covered with herbs. Here *Hypochaeris radicata* can occur as a frequency dominant in a number of formations which are each characterized by one or more of the following species: *Corynephorus canescens, Festuca ovina, F. rubra, Anthoxanthum odoratum, Agrostis tenuis,* and *Carex arenaria*. At the height of summer *Hypochaeris* is the physiognomical dominant.

In 33 different places within the dominance area of *Hypochaeris radicata* the vegetation has been analysed by taking in each place 20 samples, each 0·1 sq. metre in area. Besides *Hypochaeris* there were 13 other dominants with 72 frequency numbers in all, thus on an average 2–3 co-dominants in each observation, and since both the number of species and the species density are rather low (on an average 8·8 and 3·65 respectively) the large number of co-dominants shows that we are here dealing with narrowly demarcated formations, and this also follows from the frequency classes given in Table 3, A.

TABLE 3

Dominance area of *Hypochaeris radicata* in the 'Grey Dune' of Fanø.

	Frequency classes.				
	1	2	3	4	5
	%	%	%	%	%
A. Thirty-three observations in 6 narrowly demarcated formations: 289 frequency numbers	38	11	8	7	36
B. The single observations in each formation taken together. The whole material, therefore, is as from 6 observations with 97 frequency numbers	60	12	6	4	18
C. All the 33 observations treated as one within the dominance area of *Hypochaeris radicata*: 35 frequency numbers	83	8	0	6	3

Demarcating formations very narrowly, though not as narrowly as it is possible to demarcate them, the 33 observations represent six distinct formations. If these formations are so treated that the observations made within the same formation, but on different dunes, are regarded as a single observation, in which the F per cent. for each species is determined by the usual means, we then obtain the distribution of frequency seen in Table 3, B. It is true that the number in the fifth frequency class is still very high, but taken as a whole the series shows, when compared with Table 3, A, a marked displacement to the left. If we finally take all 33 observations as one, treating them as an observation of the dominance area of *Hypochaeris radicata* on the dunes of Fanø, then all trace of uniformity disappears except, as we already know, that *Hypochaeris radicata* is the frequency dominant. The 33 observations include in all 35 species, so that we have 35 frequency numbers with the percentage distribution in 5 frequency classes expressed in Table 3, C. The numbers show the marked displacement to the left characteristic of the statistical analyses of a heterogeneous vegetation. Since there are only 35 frequency numbers, one frequency number here corresponds to about 3 per cent. The number 3 in the fifth class, therefore, represent *Hypochaeris* alone. Were it not for this species there would be no species at all in the fifth class; indeed although the vegetation represented by the 33 observations would be considered by many to be a single formation, yet its lack of uniformity is so considerable that no single common co-dominant is found: moreover the 33 observations have not one single species common to them all apart from *Hypochaeris radicata*.

Since then even within so confined a region as the dominance area of *Hypochaeris radicata* on the dunes of Fanø there are not merely no co-dominant species, but there is actually no species common to all the observations, the question arises, which co-dominants are found in the single observations and what degree of co-dominance each of these species shows? As already mentioned 13 of the 35 species occur as dominants on one or more of the small localities investigated. The state of affairs was as follows:

Co-dominants in 1 observation: *Aira praecox, Jasione montana, Thymus serpyllum.*
„ 2 observations: *Festuca rubra, Galium verum, Sedum acre, Trifolium arvense.*
„ 3 „ *Rumex acetosella.*
„ 4 „ *Anthoxanthum odoratum.*
„ 9 „ *Corynephorus canescens, Festuca ovina.*
„ 17 „ *Carex arenaria.*
„ 19 „ *Agrostis tenuis.*

The co-dominants of *Hypochaeris radicata* are thus first and foremost *Agrostis tenuis* and *Carex arenaria*, whose co-dominance percentages are 58 and 52 respectively. After them come *Corynephorus canescens* and

Festuca ovina with a much lower percentage of about 27. The region of dominance of *Hypochaeris radicata* in Fanø extends over the lowest portion of that of *Corynephorus canescens* and especially over the uppermost portion of the dominance area of *Agrostis tenuis*. At essentially the same level but in scattered patches only, we find the dominance area of *Festuca ovina*, in which *Hypochaeris radicata*, as far as my observations extend, always occurs, but not always as a frequency dominant, since *Festuca ovina* by its capacity for forming a continuous mat often competes successfully with other plants, especially rosette plants, reducing them to a low frequency grade.

Formation dominants. By the dominants of a formation I understand those species which occur everywhere within the formation with an F per cent. of more than 80. In narrowly demarcated communities of comparatively uniform composition we can determine even by the investigation of a small number of plots which species are dominant. The number of dominants may differ in different formations, and may vary within certain limits with the number of species and species density.

The number of dominants, the percentage of dominants, and the number of species. The number of dominants expressed as a percentage of the number of species I shall call the numerical percentage of dominants or simply the dominant percentage. In the above determination the percentage of dominants is given by the number in the fifth frequency class. But the number of species is not constant in the same way as the species density and the number of dominants. In dealing with the two latter the number becomes constant after analysing a comparatively small number of plots; but the number of species goes on increasing with the number of samples examined until all the species of the formation have been encountered. Since the number of dominants remains the same while the number of species is rising, it necessarily follows that the percentage of dominants falls. As an example let us take an observation made in a *Poa nemoralis*-Formation in Allindelille Fredskov. Twenty-five sample plots were taken, and in the result given in Table 4 they are divided into areas numbered 1–5, 1–10, 1–15, 1–20, and finally

TABLE 4

A *Poa nemoralis*-Formation in Allindelille Fredskov. Species density, number of dominants, number of species, and dominant percentage according to the analyses of 5, 10, 15, 20 and 25 sample plots

	No. of samples.				
	5	10	15	20	25
Species density.	4·40	4·50	4·27	4·30	4·58
Number of dominants	2	2	2	2	2
Number of species	9	12	12	14	16
Dominant percentage	22	18	18	14	13

for the whole 25. It is seen from this Table that while the number of dominants and the species density remains constant, yet after taking only 5 sample plots the number of species rises from 9 to 16, an increase of almost twice the number in the first group of 5. Although the number of dominants remains unaltered, the dominant percentage falls from 22 to 13.

In order to show what happens in a more extensive investigation of formation statistics I shall make use of a series of analyses undertaken on the island of Fanø in the year 1926. In each of the 100 communities, which represented a series of the formations occurring on the island, there were taken 20 sample plots of 0·1 sq. metre, in such a way that for each observation the first 10 samples were kept to themselves (A) and the last 10 to themselves (B). Thus were obtained three series, i.e. besides the chief series (C) consisting of 100 analyses, each of 20 × 0·1 sq. m. plots, there were two parallel series, A and B, each consisting of 100 analyses of 10 0·1 sq. metre plots. In Table 5 I have given a conspectus of the range of frequency numbers of this material set out so that the numbers give the percentage distribution in frequency classes of the frequency numbers (the species), first according to a 10-fold and then according to a 5-fold scale.

The correspondence between the two parallel series A and B is as good as any one could wish. The number of species in the single observations varied in A from 1 to 28, and in B from 2 to 28.

Table 5

The percentage distribution of the frequency numbers in 10 and 5 frequency classes respectively for:

A. 100 analyses of 10 × 0·1 sq. metre plots, and
B. 100 analyses parallel with A, also each on 10 × 0·1 sq. metre plots.
A and B. The parallel series A and B taken together as 200 analyses of 10 × 0·1 sq. metre plots.
C. The analyses corresponding to A and B embodied in one analysis, made up of 100 analyses of 20 × 0·1 sq. metre plots.

	No. of frequency numbers.	Average F%	Frequency classes.									
			1	2	3	4	5	6	7	8	9	10
A	797	56·3	21·6%	10·8%	6·4%	5·8%	5·8%	4·2%	4·0%	6·6%	5·7%	29·1%
B	830	56·0	32·4		12·2		10·0		10·6		34·8	
			20·2	11·4	7·6	7·4	3·9	4·9	5·2	4·2	5·7	29·5
A+B	1,627	56·2	31·6		15·0		8·8		9·4		35·2	
			20·9	11·1	7·0	6·6	4·8	4·6	4·6	5·4	5·7	29·3
C	962	47·7	32·0		13·6		9·4		10·0		35·0	
			28·0	12·5	7·4	4·9	4·3	4·7	4·2	4·3	4·6	25·1
			40·5		12·3		9·0		8·5		29·7	

The average number of species per observation is in both cases about 8; B shows a slight excess over A, on the average 0·33 per observation. The average frequency of the species in A is 56·3 against 56·0 in B. The correspondence in the distribution into frequency classes is at once evident from the Table. A+B consists of two parallel series of analyses taken together, thus forming a single series of 200 analyses, each of 10 sample plots.

If we compare this with series C, which includes the same material as A and B, but which is taken as 100 analyses each of 20 sample plots, the consequence of the increase of sample plots in the individual analysis on the distribution of the frequency numbers is seen at once. The number in the first frequency class shows a rise and at the same time there is a fall in the percentage of dominants, i.e. the number in the last frequency class. By increasing the number of sample plots in the analyses several species of low frequency have been added, on an average 1·5 per analysis. This causes especially a rise in the number of the lowest frequency class, contributing to a fall in the percentage of dominants, because the number of dominants is unaltered. When the number in the first frequency class rises the average frequency of the species necessarily falls. In this example the fall is from 56·2 to 47·7.

It is therefore necessary, if we wish to use the percentage of dominants in comparative investigations, to take as our basis the same number of sample plots; and this leads to the question whether the percentage of dominants taken as a whole is essentially the same in different formations, or whether it is different in different types of formation. It is especially necessary to investigate the behaviour of the percentage of dominants in formations with different numbers of species. Since the number of species in the same formation rises and the percentage of dominants falls with an increase in the number of samples, it is obvious to the meanest intellect that a comparative investigation of what takes place in different formations can be carried out only by using the same number of samples.

In the investigation of the relationship between the percentage of dominants and the number of species in different formations I shall begin by making use of the material published by Carsten Olsen.[1] This material has the advantage of extending over formations widely different in the number of species and including a large number, in all 266, of analyses of vegetation, all carried out on the same principle: 10 samples each of 0·1 sq. metre were taken in each of the narrowly demarcated areas in which the determination of hydrogen-ion concentration of the soil was made. In Table 6 I have arranged this material according to the rising number of species in five groups (I–V), of which, however, only the first four are of uniform extent. The fifth group, because the number of plots was smaller and taken from formations particularly rich in species, extends over an

---

[1] Carsten Olsen, *Studier over Jordbundens Brintionkoncentration og dens Betydning for Vegetationen, særlig for Plantefordelingen i Naturen.* Copenhagen, 1921.

TABLE 6

The number of species and the percentage distribution of the frequency numbers into frequency classes (see text)

	No. of species.	No. of frequency numbers.	Frequency classes.				
			1	2	3	4	5
			%	%	%	%	%
I	1-5	410	30	12	8	4	46
II	6-10	518	34	18	12	7	29
III	11-15	362	33	19	13	11	24
IV	16-20	429	34	19	13	13	21
V	21-30	489	35	23	13	10	19
VI	31-55	917	44	19	14	10	13
VII	Average	..	35	18·3	12·2	9·2	25·3

interval of double each of the first four. In a sixth group (Table 6, VI) are added the numbers for a series of formations very rich in species. For these formations the number of species found in the analysis was between 30 and 55. This group includes in all 20 observations, of which 4 are taken from C. Ferdinandsen[1] (l.c., Tables 1, 2, 57, and 63), each on 25 sample plots. Three are from C. Olsen (l.c., Nos. 296-8), each on 10 sample plots. The remaining 13 observations are from my own observations in Allindelille Fredskov; 1 of them on 10 sample plots, 3 on 20, and 9 on 25. From the conspectus given in the Table of the percentage distribution of frequency numbers into frequency classes it is clear that while the number of species rises from 1 to 55, the percentage of dominants (see last column) falls from 46 to 13. Since the sixth group, however, differs somewhat from the others in the number of sample plots, it is perhaps sufficient to say that while the number of species rises from 1 to 30, the percentage of dominants falls from 46 to 19.

As another example I shall make use of the material above described from Fanø, viz. the two parallel series A and B, which agree with C. Olsen's material in that each analysis consists of ten sample plots. The proportion between the percentage of dominants and the number of species in these 200 analyses is made plain in Table 7, where the material is arranged exactly in the same way as in Table 6. From Table 7 it is clear that here too the percentage of dominants falls with increasing numbers of species in the formations. But comparing the numbers in the two Tables we shall see that in my material from Fanø the percentage of dominants does not fall as much as it does in C. Olsen's material. I shall come back to this later on; but shall first try to obtain a dominance value expressed by a proportion that does not vary with the number of sample plots.

[1] C. Ferdinandsen, *Undersøgelser over danske Ukrudsformationer paa Mineraljorder.* Copenhagen, 1918.

TABLE 7

The number of species and the percentage distribution of the frequency numbers into frequency classes (see text)

No. of species.	No. of analyses.	No. of frequency numbers.	Average. F%	Average. Species density.	Frequency classes. 1	2	3	4	5
					%	%	%	%	%
1– 5	53	220	64	2·30	27	10	8	7	48
6–10	105	774	56	4·10	34	12·5	8·5	9	36
11–15	27	331	56	6·82	32	14	10	10	34
16–20	10	181	52	9·43	29	22	11	15	23
21–30	5	121	51	12·40	32	17	15	12	24
	200	1,627	56	4·57	32	14	9	10	35

The number of dominants and the species density. As already mentioned the dominant percentage alters, and as a whole the percentage number in the frequency classes also alters, with the number of species. And since the number of species varies with the number of sample plots, a comparison can only take place between analyses with the same number of sample plots. This difficulty is avoided by using instead of the number of species, either the species density, or what comes to the same thing, the sum of frequency (= the species density $\times 100$), which, as shown above, comparatively quickly becomes a constant figure. In determining the frequency of dominants as a percentage of the total sum of frequency we not only manage to obtain the relationship of the dominants to a constant already determined, but at the same time we obtain only by this means a numerically true expression of the importance of the dominants. This the dominant percentage in the curve of frequency distribution does not tell us, or at any rate only gives very imperfect information about it, because the number of low frequency classes predominates, and this preponderance in no way corresponds to the participation in the vegetation of the species in question.

Just as I showed in Table 6 the relationship of the dominant percentage to the number of species, I have in Table 8 made use of the material of Table 6 to show the proportion between the sum of the frequency of dominants and the total sum of frequency (= species density $\times 100$). I have shown at the same time how the frequency sums of the other frequency classes behave when expressed in the same manner. About the four first frequency classes I shall here only mention that all the numbers are low in proportion to the number in the fifth class (the sum of the frequency of dominants) whose average figure for all the investigated formations is about equal to the sum of the number of the four first classes (Table 8, VII), while this last sum in Table 6, VII, which shows the number of the frequency classes expressed as the percentage of the number of species taken together, is three times as large as the number in the dominants' class (fifth class).

TABLE 8

The number of species and the sum of the frequency distribution for the same material as Table 6

	No. of species.	Frequency classes.				
		1	2	3	4	5
		%	%	%	%	%
I	1– 5	7	7	8	5	73
II	6–10	10	12	13	10	55
III	11–15	10	13	14	17	46
IV	16–20	10	14	15	20	41
V	21–30	11	18	15	16	40
VI	31–55	13	17	19	20	31
VII	Average	10	13	14	15	48

It is, however, the behaviour of the dominants, expressed in the figures in the fifth class, that is here of most interest. It must first be borne in mind that in the materials used, with the exception of group VI, the dependence of the number of species on the number of sample plots plays no part, the number of plots being here everywhere the same. If we now compare the numbers in the dominant class with the groups of species, we see that while the number of species in the formations rises from 1–5 in the first group to 21–30 in the fifth group, the frequency percentage of the dominants falls from 73 to 40. If we include group VI the number falls to 31, thus from about three-quarters to less than a third of the whole sum of frequency.

Essentially the same state of affairs occurs if we use the material I procured from a series of formations on Fanø. Just as Table 8 corresponds with Table 6, so here Table 9 corresponds with Table 7. On the basis of 200 analyses there was given in Table 7 a conspectus of the proportion of the percentage of dominants to the number of species; in Table 9 the same material is used to show the relationship between the percentage of the sum of the frequency of the dominants and the number of species. From the numbers in the fifth frequency class in Table 9 it is clear that here also the percentage of the sum of the frequency of the dominants falls with a rising number of species (first column). In the first group of species (1–5) the percentage of the sum of the frequency is essentially the same as in the corresponding group in Table 8; but in the following groups the percentage of the sum of the frequency of the dominants does not fall as much, not descending to the level seen in Table 8, the number in the fifth group of species in Table 9 descending only to 46 compared with 40 in Table 8.

TABLE 9

The number of species and the distribution of the sum of frequency for the material in Table 7

No. of species.	No. of analyses.	No. of frequency numbers.	Sum of frequency.	Average. F%	Average. Species density.	Frequency classes. 1	2	3	4	5
						%	%	%	%	%
1– 5 . .	53	220	14,200	64	2·30	6	5	7	8	74
6–10 . .	105	774	43,090	56	4·10	8	8	8	12	64
11–15 . .	27	331	18,420	56	6·82	8	9	10	13	60
16–20 . .	10	181	9,430	52	9·43	7	15	11	23	44
21–30 . .	5	121	6,200	51	12·40	9	12	16	17	46
Whole material	200	1,627	91,340	56	4·57	8	9	9	13	61

In connexion with the use just made of the percentage of the sum of the frequency I will here take the opportunity to illustrate directly that the percentage of the sum of the frequency has a far wider significance in the science of formations than the percentage of the frequency numbers. The relationship between the percentage of the number of frequency of the frequency classes and the percentage of the sum of the frequency is most easily seen by comparing Table 5 and Table 10. In both Tables exactly the same materials are used, and in both Tables the materials are first divided into ten frequency classes, which are then united in pairs to form 5 classes. The numbers in the frequency classes in Table 5 state what percentage of the species found in the analyses the frequency class in question includes. The distribution is seen to be the general one. That the number in the last class, whether the classes are made 10 or 5, is high, is a direct result of dealing with the dominants whose presence has already been used for demarcating the formations. The high number in the first class expresses the fact that the species of low frequency exceed by far those of high and medium frequency. It is thus true that the high number in the first frequency class is an expression of actual fact, but it should be emphasized that this is only true of the number of species. On the contrary the percentage distribution of the frequency numbers into classes is not an expression of the importance that the species in the individual frequency class possess in the composition of the vegetation. This importance can only be shown by the percentage sum of the frequency, in so far as the sum of the species in a formation can be expressed in terms of frequency.

In Table 10 is shown, for the same material as Table 5, how the relative strength of the frequency classes works out when expressed by the sum of frequency for the species of each individual frequency class. The most conspicuous difference between Table 5 and Table 10 is that the high

TABLE 10

The distribution of the sum of frequency in 10 and 5 frequency classes respectively for the same material as that of Table 5 (p. 528) where further details about the material will be found.

	No. of frequency numbers.	Average species density.	Frequency classes.									
			1	2	3	4	5	6	7	8	9	10
			%	%	%	%	%	%	%	%	%	%
A	797	4·49	3·8	3·8	3·4	4·1	5·1	4·6	5·0	9·5	9·0	51·7
			7·6		7·5		9·7		14·5		60·7	
B	830	4·65	3·6	4·1	4·1	5·2	3·4	5·3	6·5	6·0	9·1	52·7
			7·7		9·3		8·7		12·5		61·8	
A+B	1,627	4·57	3·7	4·0	3·7	4·7	4·3	4·9	5·7	7·7	9·1	52·2
			7·7		8·4		9·2		13·4		61·3	
C	962	4·59	4·0	4·5	4·2	3·7	4·3	5·6	5·9	7·1	8·4	52·3
			8·5		7·9		9·9		13·0		60·7	

number in the first frequency class in Table 5 is reduced very considerably in Table 10. This low figure corresponds with the slight participation of the group of species in the plant covering. Next in importance is that the number in the last class has risen considerably so that the tenth class includes about a half of the entire sum of frequency. When divided into 5 the last class includes as much as three-fifths of the whole sum of the frequency. This great excess in the last class gives a much better picture of the overwhelming importance of the dominants in the vegetation than the percentage frequency number. For the rest the numbers divided into 10 in Table 10 show that with the exception of the tenth class the numbers in the individual classes are very low, and that they are almost uniform in all the classes with the exception of a slight rise in the classes next to the last class. The state of affairs is most easily seen by glancing at Fig. 144 and Fig. 145. In Fig. 144 the columns give the figures in the scale divided into 10 of Table 5, C, while the curve gives the numbers in the scale divided into 5. In exactly the same way the columns in Fig. 145 give the scale divided into 10 in Table 10, C, and the curve gives the numbers on the scale divided into 5.

From what has been described it is clear that whether the proportion of the dominants was determined by means of the percentage number of dominants, i.e. the percentage participation of the dominants in the number of species, or whether we used the percentage of the sum of the frequency of the dominants, i.e. the participation of the dominants in the total sum of frequency, in the material used the importance of the dominants fell as the number of species in the formations rose. In seeking an explanation for this phenomenon we find three circumstances which may be considered causal, (1) the different number of samples in the analysis used, (2) the different demarcation of the formations, and (3) the difference in the distribution of the species in the communities, which varied in the number of species they contained.

As already shown the first of these three circumstances has no significance if the mass of the dominants be measured by the percentage of the sum of frequency, because this number becomes constant in the observations as soon as the species density becomes constant, and this takes place very soon. On the other hand the number of sample plots is of very great

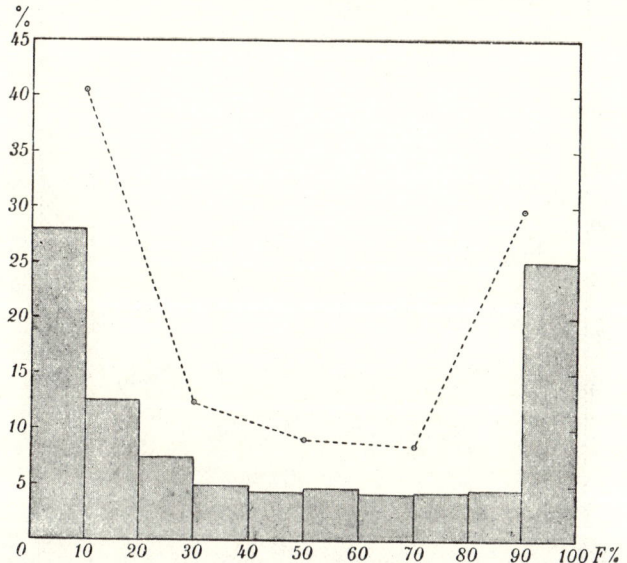

FIG. 144. Graphic representation of the material in Table 5, C (100 analyses of 20 × 0·1 sq. metre plots). The columns give the value of the single frequency classes expressed as a percentage of the total frequency numbers divided into 10 classes. The curve gives the same, but the frequency numbers are divided into 5 classes.

importance when making use of the percentage number of dominants, because this number in the same observation decreases with an increase of the number of the samples, and goes on decreasing until all the species have been encountered. A comparative investigation cannot, therefore, be made except on a basis of the same number of samples. This demand is, however, satisfied both in Table 6, I–V, and in Table 8; nevertheless, there is a marked fall in the percentage number of dominants with a rising number of species in the formations investigated. Thus in the material before us the first of the three explanations will not hold. As said above, it never holds for the percentage sum of the frequency of the dominants, supposing sufficient samples be taken to allow the frequency density to become constant. I shall, therefore, in what follows, leave out of consideration the percentage number of dominants, pursuing the problem only by means of the percentage of the sum of the frequency, which is the natural expression of the relative strength of the frequency classes, as far as this can be expressed by statistical analyses of the formations.

So far as concerns the second explanation, the varying demarcation of the formations, what occurs can be expressed as follows. Supposing that in procuring the materials used the same principle is always employed in

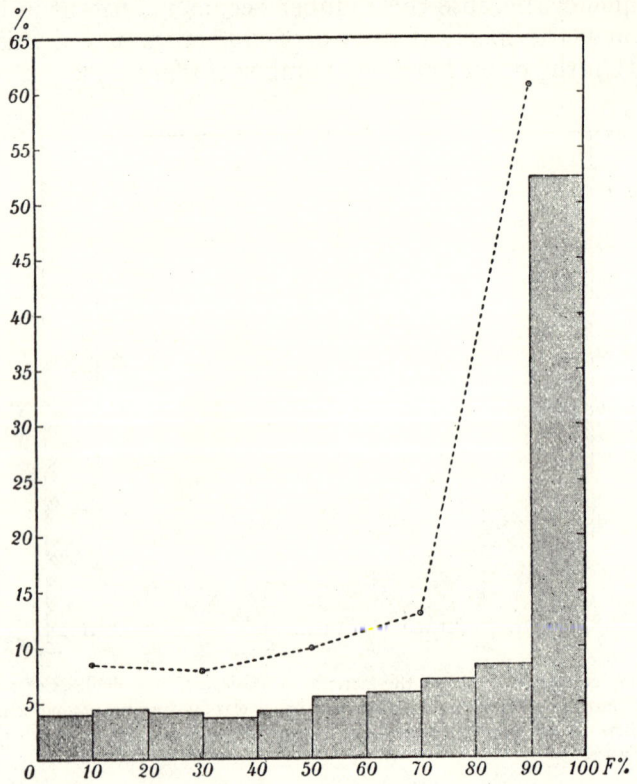

FIG 145. Graphic representation of the material in Table 10, C, which is the same as in Table 5, C and in Fig. 144. But while the material in Fig. 144 is arranged on a basis of the frequency number (F%) of the species, in Fig. 145 it is arranged on a basis of the sum of the frequency of the species. The columns give the value of the individual frequency classes expressed as a percentage of the total sum of frequency divided into 10 classes. The curve gives the same but in such a way that the species are divided into 5 frequency classes.

demarcating the formations, then the numbers in the dominant class in Tables 8 and 9 (and also in Tables 6 and 7) signify that the richer the formation is in species the greater is the part played by the less frequent species when compared with the dominants. The question is whether the proposition is exact or how far it is exact. In dealing with formations poor in species it is comparatively easy to demarcate the formations narrowly, trusting to the determination of the dominance area of the species. To a certain extent the principle stated above holds here, viz. that with a rising number of species the percentage sum of the frequency of the dominants falls. Observations made in pure communities of a single species must necessarily show 100 per cent. in the dominant class; and even if other species be thinly dispersed in the community yet the percentage of the sum of the frequency of the dominant class will as a rule be comparatively high. We need only consider such formations as the *Anemone nemorosa*-Formation, *Anemone nemorosa+Oxalis acetosella*-Formation, *Allium ursinum*-Formation, *Calluna vulgaris*-Formation, *C. vulgaris+Empetrum nigrum*-Formation, &c., &c. It does not, however, follow from this that the continued decrease of the percentage sum of the frequency of the

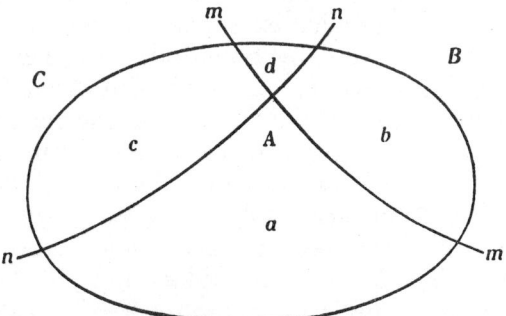

FIG. 146. (For explanation see text)

dominants with a rising number of species, which the above material shows, expresses what actually occurs in nature. The decrease of percentage of the sum of the frequency of the dominant class with a rising number of species is at any rate essentially the outcome of the fact that **the richer a piece of vegetation is in species the more difficult it is to be satisfied that we are working with as narrow a formation concept as is used in vegetation poor in species.** This is because in vegetation rich in species it is much more difficult to observe the necessary precaution of taking the sample plots in the same system of dominance areas; and if this is not done, it may easily be that one or several of the species, which otherwise would have been included in the class of dominants, will now sink into one of the lower classes, and as a result of this the sum of the frequency of the dominants' class will fall. In Fig. 146 the oval represents the dominance area of a given species A. This area of dominance is cut by the boundaries of the dominance areas of B and C in such a way that to the right of $m-m$ B is dominant, while to the left of the line it occurs with an F per cent. of under 80. Furthermore the dominance area of C extends from the left to the line $n-n$. To the right of this line C occurs with a lower frequency. If we investigate A's dominance area as a whole we obtain only one dominant (A) and 2 species of intermediate frequency (B, C). If we investigate region *a* by itself, essentially the same result is obtained, except that B and C do not get such a high frequency percentage as in the first case. If *b* is investigated by itself and *c* by itself then in both cases we get two dominants A+B and A+C respectively. Finally in *d* by itself we get three dominants A, B, and C, so that the sum of the frequency of the dominant class becomes comparatively very high.

As long as one has to deal with such simple cases as this, it is quite easy to draw proper boundary lines for the dominance areas of the individual species. But this is not so in formations that are rich in species, where we may have up to 10 or even more dominants. In the majority of observations made hitherto in such formations rich in species attention has scarcely been paid to determining to what extent the boundaries of the dominance area in question is intersected by boundaries of other dominance areas such as those determined above. As mentioned above I believe that we have here the essential cause of the fact that in the material described the percentage of the sum of the frequency of the dominant class falls with an increasing number of species—apart of course from the case mentioned of formations very poor in species.

Even if we disregard this latter special case of formations poor in species, e.g. the first group of species (1–5) in Tables 8 and 9, and disregard also the formations very rich in species, e.g. groups 4–6 in Table 8 and 4–5 in Table 9, where the demarcation of the boundary of the formations was not perhaps so narrow as in the communities with fewer species, even then we have remaining group 2 (6–10 species) and 3 (11–15 species). In these groups, in the material in Table 9, I am conscious of having striven to make use of the same narrow demarcation of formations as in the formations in the first group. Yet here too a fall occurs in the percentage of the sum of the frequency of the dominant group. The difference indeed is not great; in Table 9 the number falls only from 64 to 60. In C. Olsen's material (Table 8) the difference is greater, the number falling from 55 to 46. I think, therefore, in all these cases we should direct our attention to the question of whether the distribution of the plants in nature may not play a part in the phenomenon of the fall of the percentage of the sum of the frequency in the dominant class with the increasing richness in species of the formations. But whether this is so or not the fall is doubtless connected, as any rate in part, with the less narrow delimitation of the formation which we unconsciously use when dealing with a community rich in species.

The question then arises whether we must always use a narrow demarcation of formations, or, if not, to what extent we should delimit the formations.

So long as the analyses serve an ecological aim we must of course always, even in vegetation rich in species, delimit our formations with a narrowness based upon unity of species density and of dominants. But in a general preliminary view of the vegetation of a district it will be scarcely practicable in formations rich in species to work with such a narrow delimitation. Here we must first make our divisions by means of individual physiognomical dominants; so that the demand of unity with respect to all the dominants is for the time put aside. We have remaining a constancy of species density and of individual dominants, which should if possible be physiognomical dominants.

A statistical analysis of a vaguely demarcated vegetation which is not uniform has, however, but little interest. The investigation should, therefore, be made in such a way that it does not yield merely an average of the vegetation as a whole, but above all reproduces the actual state of affairs in a series of more accurately investigated places. Instead of basing the investigation on a representative system of sample plots, the investigation should be made by means of a system of small groups of sample plots, paying attention to a narrow delimitation of formations within each of these groups. This narrow delimitation is comparatively easy within the small area in which each group of samples is taken. Four such groups of 5 (or two groups of 10), each by itself giving a picture of the actual state of affairs at the points in question, yield more satisfactory information about the vegetation than 20 samples scattered over the entire district and only giving an average which is realized at no point within the district.

The narrower the formation concept used, the greater will be the significance of the statistical analysis of the formation. As an example I shall take the investigation of the ground flora of a little ashwood growing on much altered bog (Table 11). Ten analyses were made, of which 9 (Table 11, Nos. 2–10) are within narrowly demarcated formations; but this was not so with No. 1, which I shall discuss later. Each analysis included 20 samples, so that 200 samples were investigated altogether, representing different parts of the area. If we now contemplate these 200 samples as a single analysis of the ground flora of the ash swamp we get the result shown in Table 11, No. 11, which shows 25 species of which none is dominant. Table 12 shows the distribution into frequency classes, both with regard to the number of species and to the sum of the frequency. These numbers may of course be important in comparing the floristic condition of other ash swamps, but they tell us nothing about the individual species and of how far they belong together ecologically.

If in each of the 10 analyses (Table 11, Nos. 1–10) we take the 20 samples together, thus making a single sample plot of 2 sq. metres, we then obtain an analysis of 10 sample plots each of 2 sq. metres. The species density, which is connected with, and increases in proportion to, the size of the sample plot now of course alters, becoming 9·2 instead of 3·44. At the same time it is obvious that the numbers in the frequency class must become displaced to the right (Table 13). Two species, *Circaea lutetiana*

TABLE II
The ground flora of a Fraxinus swamp (see text)

No.		1	2	3	4	5	6	7	8	9	10	11
Number of species		18	10	8	12	9	7	7	9	4	8	25
Species density		6·50	2·75	2·35	4·60	2·50	1·95	2·90	3·20	2·75	4·85	3·44
The Dominants' { Number		2	1	1	1	1	1	2	2	2	4	0
Percentage of the sum of the frequency		28	36	43	22	40	51	70	63	73	80	0
Geum rivale	F%	100	. .	. .	. .	. .	. .	. .	. .	. .	. .	19
Galium aparine		85	5	. .	30	10	15	15	15	. .	. .	28
Mercurialis perennis		25	100	100	55	20	5	25	10	70	. .	55
Impatiens nolitangere		30	45	. .	100	100	100	. .	10	. .	15	42
Ficaria verna		80	50	40	40	10	30	100	60	5	100	65
Aegopodium podagraria		. .	. .	. .	80	. .	. .	100	100	100	100	30
Anemone nemorosa		5	20	25	15	35	. .	. .	. .	. .	90	19
Adoxa moschatellina		20	. .	. .	. .	. .	. .	. .	. .	. .	. .	2
Alliaria officinalis		50	10	35	15	. .	. .	. .	. .	. .	. .	11
Anemone ranunculoides		. .	. .	10	. .	. .	. .	. .	. .	. .	15	3
Circaea lutetiana		45	25	. .	20	35	. .	40	15	. .	60	25
Cirsium oleraceum		10	. .	. .	. .	. .	. .	. .	5	. .	. .	1
Crepis paludosa (Aracium)		45	. .	. .	. .	. .	5	. .	. .	. .	. .	5
Filipendula ulmaria		. .	5	. .	5	. .	. .	. .	5	. .	. .	2
Galeopsis tetrahit		20	. .	. .	. .	. .	. .	. .	. .	. .	. .	2
Geranium robertianum		70	. .	. .	. .	. .	. .	. .	. .	. .	. .	7
Lysimachia thyrsiflora		5	. .	. .	. .	. .	. .	. .	. .	. .	. .	1
Majanthemum bifolium		. .	. .	10	. .	. .	. .	. .	. .	. .	. .	1
Paris quadrifolia		. .	5	5	. .	. .	52	. .	. .	. .	5	2
Polygonatum multiflorum		. .	. .	. .	30	25	15	5	. .	. .	. .	3
Ranunculus auricomus		10	. .	. .	5	5	. .	. .	. .	. .	. .	1
” repens		15	. .	10	. .	. .	. .	. .	. .	. .	. .	9
Rubus idaeus		5	. .	. .	. .	. .	. .	. .	. .	. .	. .	2
Stachys silvatica		. .	10	. .	. .	. .	. .	. .	. .	. .	. .	1
Urtica dioeca		30	. .	. .	65	10	. .	5	. .	. .	. .	12

and *Impatiens nolitangere*, will now be found in the dominant class, but the result expresses no better the difference in the vegetation.

TABLE 12

No. 11 in Table 11. The distribution in frequency classes of the species and of the sum of the frequency

	Frequency classes.				
	1	2	3	4	5
Number of species	% 76	% 12	% 8	% 4	% 0
Sum of frequency	29	24	28	19	0

TABLE 13

(See text)

	Frequency classes.				
	1	2	3	4	5
Number of species	% 44	% 20	% 16	% 12	% 8
Sum of frequency	13	17	24	26	20

Let us now proceed to the special investigations taken on the basis of a narrow demarcation of formations. Four different formations in all are represented: *Mercurialis perennis*-Formation (Nos. 2–6), *Impatiens* + *Ficaria*-Formation (No. 7), *Aegopodium* + *Ficaria*-Formation (Nos. 8–9), and *Aegopodium* + *Anemone* + *Impatiens* + *Ficaria*-Formation (No. 10). The analyses of the *Mercurialis*-Formation show two sections through the south side of the ash swamp, where the ground was more deeply shaded than in its other parts. Nos. 2 and 3 are from a part where the deeper shade was due to the presence in the wood of elm and maple. No. 2 was nearest to the better illuminated portion, and No. 3 was more in the shade of the elm and maple. Corresponding to this shade relationship the species density in No. 2 is slightly higher than in No. 3. In Nos. 4–6 the shade from the beechwood south of the swamp was rather dense. No. 4 was nearest to the better illuminated part of the middle portion of the ash swamp, and No. 5 was the next nearest. No. 6 was nearest to the beechwood and therefore deeply shaded. In this series the number of species decreases 12–9–7 and similarly the decrease in species density is 4·60, 2·50, and 1·95.

The analyses Nos. 7–10 are from the northern, well-illuminated side of the ash swamp, where there is some mineral matter in the soil. No. 9 is from the highest portion where the soil contains most mineral matter. Here *Aegopodium* dominates and has shaded out all competitors except *Ficaria* and *Galium aparine*. The former escapes destruction because it is a plant of early spring, and can complete its aerial life before *Aegopodium* completely shades the ground. The latter escapes by climbing up to the light. No. 8 is close to No. 9 but on slightly lower ground, where *Aegopodium* does not make such a complete cover, allowing the species density to rise slightly. No. 7 is from still lower ground but quite close to No. 8. In May the ground of No. 7 is green and golden with a dense carpet of *Ficaria*: seedlings of *Impatiens* are present, a plant that covers the ground in the summer. Finally we come to No. 10, in which the soil is rather damp, containing but little mineral matter. It is true that *Aegopodium* is dominant here, but it does not thrive sufficiently to cover the ground completely. Besides *Ficaria* both *Impatiens* and *Anemone nemorosa* may be dominants; this gives the comparatively high species density of 5.

While Nos. 2–10 are taken in narrowly demarcated formations, No. 1 is different, representing the large light middle part of the ash swamp. In this analysis no sharp demarcation of the formation is undertaken; but because of a certain uniformity in the environment, two dominants occur in this observation. Since, however, the species density is 6·5, we might expect, if we were dealing with a narrowly demarcated formation, to find 3–4 dominants. Many of the species, in particular *Alliaria, Mercurialis, Impatiens, Aracium, Cirsium, Ficaria, Circaea,* and *Geranium robertianum* occur as dominants over varying extents, but do not assert themselves as dominants when the whole area is taken together as one. There is little doubt that the difference is due principally to illumination; but that is not to say that the present illumination alone will explain the differences. Where felling of trees and development of undergrowth take place, the ground flora cannot be expected to be stabilized in respect to the present illumination. That we are not here dealing with a narrowly demarcated formation is decided by the fact that there is no rise in the number of species in the fifth frequency class, and that the sum of the frequency of this class amounts to scarcely a third of the total sum of frequency (Table 14), while in narrowly demarcated formations it is about a half of the total

TABLE 14

Observation No. 1 in Table 11

	Frequency classes.				
	1	2	3	4	5
Number of species .	% 44	% 17	% 17	% 11	% 11
Sum of frequency .	14	13	22	23	28

sum of frequency. This is seen by a comparison of observations Nos. 2–10, which come from narrowly demarcated formations. Here, on an average, and indeed also in the single observations, the number of species in the fifth class is higher than that in both the foregoing classes, and on an average the sum of the frequency of the dominants is just as high as the sum of the frequency in the 4 first classes taken together (Table 15).

TABLE 15
Observations Nos. 2–10 in Table 11

	Frequency classes.				
	1	2	3	4	5
	%	%	%	%	%
Number of species . .	50	19	7	4	20
Sum of frequency . .	13	16	10	8	53

The number of species and the species density. A high number of species in a formation analysis does not necessarily entail a high species density, and conversely a high species density is not necessarily connected with a high number of species. We must, moreover, bear in mind that the number of species depends upon the number of samples taken. But if we take as our basis the same number of samples, we obtain, as we should expect, in the main features, a parallelism between the number of species and the species density. This is seen clearly in Table 16, where I have put together 260

TABLE 16
The relationship between species density and species number in 260 analyses of 10 sample plots (0·1 sq. metre)

Classes of no. of species.	Classes of species density.							
	2	4	6	8	10	12	14	
4	70	35	..	..	..	..	..	105
8	9	56	8	..	..	..	..	73
12	..	3	24	5	..	..	..	32
16	..	..	4	8	4	..	..	16
20	..	..	..	2	12	3	..	17
24	..	..	..	2	3	3	2	10
28	..	..	..	..	1	1	5	7
	79	94	36	17	20	7	7	260

of the analyses (each of 10 samples) undertaken by C. Olsen in woodland and meadow formations. It is apparent that what happens is approximately this: while the number of species rises by 4 the species density rises by 2, which means that the average percentage frequency lies somewhere near 50, working with analyses of 10 sample plots. The result of the 200 analyses of 10 samples, which I took on the Island of Fanø, corresponds with this, as is apparent in Table 7, if we compare the numbers for the species density in column 5 with the number of species in the first column.

If more samples are taken the number of species increases with no important rise in the species density, so that the numbers in the table of correlation are displaced to the left of the diagonal, as seen in Table 17, which gives a conspectus of what occurs in 130 analyses, each made on 25 sample plots. My thanks are due to Magister Joh. Gröntved for a considerable portion of the material used in this Table.

TABLE 17

The relationship between the species density and the number of species in 130 analyses each of 25 sample plots (0·1 sq. metre)

Classes of no. of species.	Classes of species density.													
	2	4	6	8	10	12	14	16	18	20	22	24	26	28
4	..	..	..	..	..	..	..	..	..	..	..	..	..	..
8	1	9	..	..	..	..	..	..	..	..	..	..	..	10
12	..	17	12	2	..	..	..	..	..	..	..	..	..	31
16	..	2	7	2	2	..	..	..	..	..	..	..	..	13
20	..	..	4	9	1	..	1	..	..	..	..	..	..	15
24	..	..	..	5	5	..	..	..	..	..	..	..	..	10
28	..	..	1	1	4	1	..	..	..	..	..	..	..	7
32	..	..	..	..	2	1	3	..	..	..	..	..	..	6
36	..	..	..	..	2	3	2	4	1	1	..	..	..	13
40	..	..	..	..	1	1	2	1	2	..	..	..	..	7
44	..	..	..	..	..	1	2	2	2	1	..	..	..	8
48	..	..	..	..	..	..	..	2	1	..	..	..	..	3
52	..	..	..	..	..	..	..	..	1	2	..	..	..	3
56	..	..	..	..	..	..	..	..	2	1	1	..	..	4
	1	28	24	19	17	7	10	9	9	5	1	..	..	130

The form, size, number, and distribution of the sample plots. When in 1908 I worked out my statistical method of examining formations my object was to obtain a means of getting an objective and exact determination of the composition of a plant community. Up till then investigators had to content themselves with a general floristic description and with a statement of the importance of the species in a given formation based upon individual judgement. One of the difficulties was the more or less diffuse boundaries between the formations. The first thing, however, to be done was to set aside the question of boundaries and to try to obtain a picture of the formation where it was typical. As one might often have to deal with comparatively large extents of ground covered with, at any rate apparently, very uniform vegetation, I first tried to obtain an expression for this by a thorough investigation of a sample plot of 100 sq. metres, or later of 10 sq. metres. A thorough investigation of such a sample plot usually occupies much time. It became, moreover, apparent that for a satisfactory result one plot was not enough, and by investigating more plots the work becomes much more laborious. The next consideration confronting me was this. Since it is much more difficult to investigate large sample plots than small ones, and since the reliability of the investigation is determined essentially by the number of the plots, the proper procedure would be, instead of using few large plots, to analyse the vegetation by means of numerous small ones. By this means it is possible to give exactly and objectively to each species a valency corresponding with its occurrence. This valency is a straightforward statement of the number of samples in which each species is encountered. It was in this way that the valency method arose. The valency numbers, the percentage of frequency (F per cent.), do not merely give the mutual relationship of the species occurring, in so far as it is dependent upon the numbers in which they occur, but it also gives us a means by transforming the floristic units, the species, into units of another kind, e.g. life-forms, and of giving to each species a value corresponding to its degree of frequency. A biological formation spectrum not based upon valency numbers but only on the species, making them all equivalent in value, seems to me misleading, and, however perfectly done, without value.

By means of a series of experiments I came to the conclusion that 0·1 sq. metre was a suitable size for ordinary investigation of formations. Other investigators favour large sample plots, attempting within these to give to the species a numerical value by making a rough estimate. But such material has not the objective character which alone can render it capable of being used as an exact basis for comparative investigations.

It is clear that in using my method it will usually happen that a varying number of species with low frequency are not included in the analyses. If a complete list of the flora is wanted the species in the analyses must be supplemented. In characterizing formations it is not essential that all the species of low frequency should be given frequency numbers. As already said, what is most important is to determine the constants of the formations, especially the species density and the formation dominants. To some extent the number of species is expressed by the species density; but to know which of the low frequency species are included in the analyses is not essential either for the species density or for the formation spectrum.

Since then the question of a complete list of the flora may be looked upon as something apart, which need not be settled by means of statistical analyses of the formations, our object is attained by limiting the number of the sample plots to that necessary for determining the species density and the dominants. We are then able to investigate a given vegetation by means of observations in several places, thus becoming certain to what extent one is dealing with the same formation.

The form of the sample plots. In the determinations of the F per cent. circular plots should always be used. Only with this shape can a uniform result be obtained. With all other shapes the result is not determined by the centre of the plot in the same way as it is with the circle, but is dependent also upon the position of the sides.

The size of the plots depends upon the degree of accuracy desired. The choice of the size is therefore a practical question. In ordinary investigations I have found it practical to use a plot of 0·1 sq. metre. If in using this size a species has an F per cent. of 100 it means that in the formation investigated the individuals of this species are arranged so densely that there is between them practically no interval large enough to contain a circle of 0·1 sq. metre in diameter. But beyond this the figure F per cent. = 100 tells us nothing of the density of the individuals. If we want more accurate information on this point the F per cent. must be determined by correspondingly smaller plots.

Unless it is necessary the size of the sample plot should not be altered. Analyses made with different sized plots cannot without further treatment be used in comparative investigations.

The number of sample plots. The necessary number of plots is attained when the particular to be determined has become constant. Since our most important concern is to determine the constants in the given formation, these constants being the species density and the dominants, the investigation can be limited to the number of samples necessary for this purpose (cf. pp. 527–8).

The distribution of the samples. If the vegetation be uniform, the distribution of the samples is of no consequence; but if we cannot at once see whether or not the vegetation is uniform then the samples must be taken in known order, so that it will be clear from the scheme of the observations whether the last samples are taken within the same formation as the first (cf. Chapter XI, pp. 384–90).

XVII

BOTANICAL STUDIES IN THE MEDITERRANEAN REGION

I. THE MEDITERRANEAN THEROPHYTE CLIMATE

DURING my journey in the Mediterranean region in 1909–10, on which my principal object was to study the life-forms of plants in a Therophyte climate, I also endeavoured, when I had the opportunity, to study the vegetation of the alluvial strand formations, so as to compare it with that of the corresponding formations in a Hemicryptophyte climate, especially in the regions around the North Sea and the Baltic.

Alluvial strand formations are found in many parts of the Mediterranean region, partly as clayey and partly as sandy deposits, corresponding in the main to the salt-marsh and dune formations of our coasts. The nature of the soil being essentially the same in both localities, an opportunity is afforded for studying climatic influence in the two areas as it is manifested in the life-forms of the species and the nature of the dominant formations.

As far as Italy is concerned I have only been able to examine sand formations, viz. in the following five parts of the west coast: (1) the western part of the Tuscan Maremma between the Arno and Leghorn; (2) the part round Stagno di Orbetello; (3) the part by Fiumicino north of the mouth of the Tiber; (4) the stretch along the southern edge of the Pontine Marshes west of Terracina; (5) the part by Cumae between Lago di Fusaro and the sea.

The method used is in the main the same as that adopted in my previous works. I wish however to say some further words about hydrotherm figures and their significance.

Hydrotherm figure and plant climate. The plant climates are characterized by the biological spectrum which is based on the adaptation of plants for surviving the unfavourable season. Since this season and all seasons are in the first place dependent on the temperature and precipitation, it is of special interest to express the annual course and interrelationship of these two factors in a single clear picture which will immediately indicate the character of the climate in so far as it is determined by the two said factors. This, as I have previously shown, can easily be done when the precipitation is denoted in centimetres, not, as is usually done, in millimetres. If the amount of the precipitation is given in centimetres, the same scale can be employed for the plotting of the temperature and precipitation curves, the numerals of the ordinate in the first case indicating degrees centigrade, in the second case centimetres. In order to avoid division of the particularly characteristic portions of the curves corresponding to the favourable and unfavourable seasons, the curves are made to begin with the spring, i.e. with April for the northern and October for the southern hemisphere; this will make all figures immediately comparable. In order to have a single term for these figures I have called them hydrotherm figures, and so as to show at first sight which is

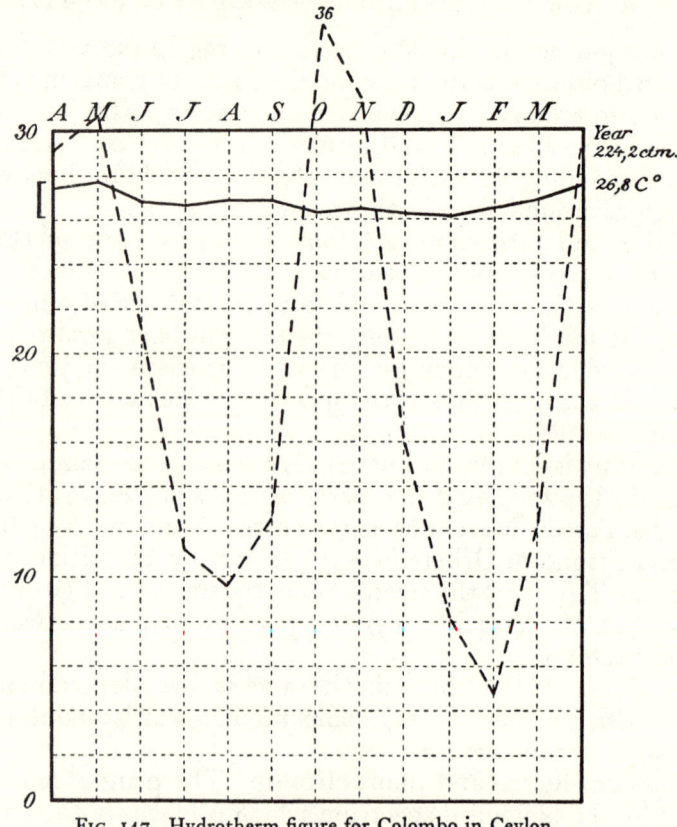

Fig. 147. Hydrotherm figure for Colombo in Ceylon.

the temperature and which the precipitation curve (to assist the memory), the latter is dotted, while the former is continuous.

Since the hydrotherm figure shows the interrelationship and annual course of the temperature and precipitation on which the relation between the favourable and unfavourable season depends, this figure becomes the chief expression of the physical climate as a condition of vegetation. And since the biological spectrum is based on a life-form system which again is based on the adaptation of plants for surviving the unfavourable season, we have good reason to expect that, at any rate in their main features, the hydrotherm figure and the biological spectrum will correspond; but how far this parallelism goes will only appear from a comparison between the biological spectra of the local floras and the hydrotherm figures of the areas in question. At present only few such investigations are available. The life-forms of the species in the various areas must first be determined, but the biological spectra which are already at hand show that, at any rate as far as the principal climates are concerned, there is close agreement between the hydrotherm figure and the biological spec-

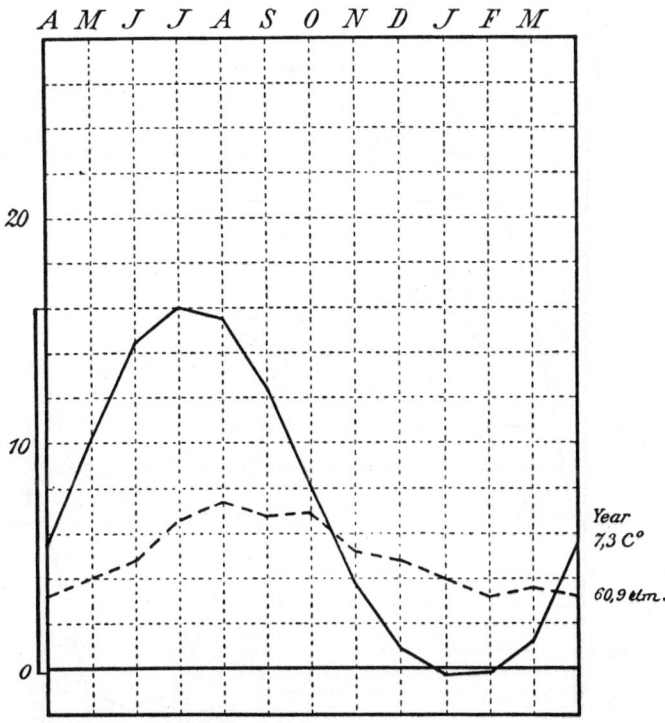

FIG. 148. Hydrotherm figure for Denmark. (Cf. p. 9. The divergences are due to new data.)

trum. Thus, for instance, there is no doubt that where we have hydrotherm figures of the type shown in Fig. 147 we shall find a preponderance of Phanerophytes in the biological spectrum of the local floras in any not too narrowly delimited area, no matter what is the systematic composition of the flora. Here we have a Phanerophyte climate which comprises at any rate all tropical regions without a marked dry season. Where we have hydrotherm figures of the type shown in Fig. 148 we shall probably always find that the Hemicryptophytes dominate in the biological spectrum; here we have the Hemicryptophyte climate prevailing in the temperate zones. And where the hydrotherm figure is of the type shown in Fig. 149 Hemicryptophytes and Chamaephytes, or in extreme cases Chamaephytes alone, dominate the biological spectrum. Here we have a Chamaephyte climate comprising the frigid zones towards the poles and the high mountain regions of the temperate zones. Finally we shall probably find that wherever we have hydrotherm figures like those shown in Figs. 150–3 the Therophytes will be dominant in the biological spectra. This at any rate is the case in the Mediterranean region and likewise in parts of the Californian and Mexican desert region. We do not as yet know whether south-western Australia and those areas of South Africa and western South

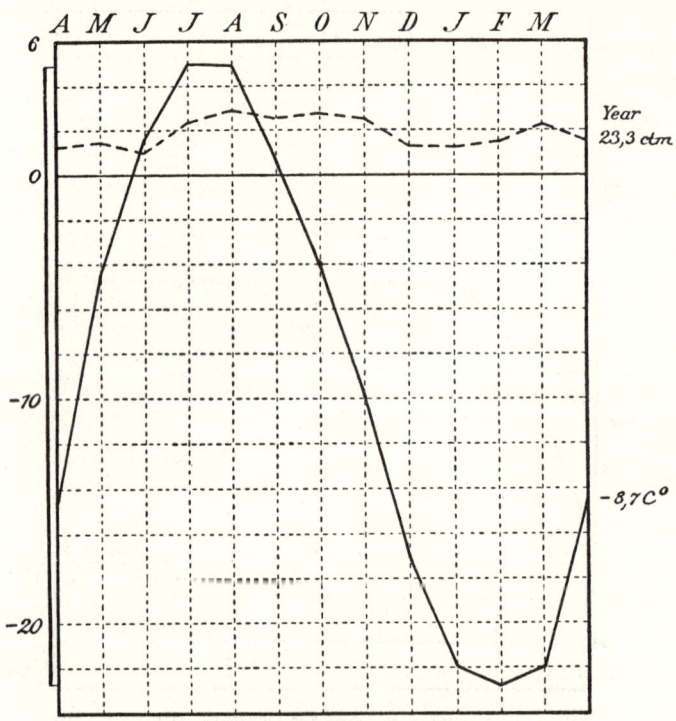

FIG. 149. Hydrotherm figure for Upernivik (Greenland).

America, where we have the same type of hydrotherm figure, also show a preponderance of Therophytes in the biological spectrum; it is quite possible that there may be regions where other circumstances, for instance difficulties of immigration, have prevented a life-form from becoming as numerous in a certain locality as might have been expected from a comparison of its hydrotherm figure with that of the corresponding areas in other countries.

Even though we can thus, if we take typical examples of plant climates, infer biological spectra from hydrotherm figures and conversely, we must not expect a complete parallelism. Because the hydrotherm figure is a product of two factors, it may very well happen that two somewhat different hydrotherm figures may in the main produce the same conditions of life.

It must not be expected, therefore, that the line between two plant climates will show the same hydrotherm figure everywhere. Moreover, it must be kept in mind that though the hydrotherm figure expresses the two most important factors in plant life, it is no exhaustive expression of the significance of these two factors. The hydrotherm figure is based on the mean temperature and the mean precipitation; future investigations must discover whether the use of other determinations, such as the monthly mean maxima and mean minima, would show more clearly the parallelism between the hydrotherm figure and the biological spectrum.

The Therophyte climate of the Mediterranean region. Throughout the Mediterranean region, where we have a hydrotherm figure like that of Figs. 150–3, Therophytes are dominant in the biological spectrum of every local flora, and in accordance with the fact that Therophytes have a capacity for taking root even in the smallest spot, the Therophytes as a rule dominate much more in the biological spectrum of the Therophyte climate than for instance the Phanerophytes and Hemicryptophytes in the spectrum of the Phanerophyte and Hemicryptophyte climates.

The hydrotherm figure typical of the Mediterranean region is characterized by its complete absence of parallelism between the temperature and precipitation curves. It thus denotes a climate hostile to vegetation or at any rate comparatively unfavourable to it. In the summer the temperature shows a high curve, the precipitation a low curve; inversely, the precipitation curve is fairly high in the winter but then the temperature is too low compared with that to which the plants have become adapted in the summer. Hence the epigeal perennials (Phanerophytes and Chamaephytes) must be adapted both to the physical drought of summer and the physiological drought of winter. This adaptation generally manifests itself in a xerophytic structure of various kinds. Only where the two curves intersect or at least approach each other at the transition from summer to winter and the reverse, in other words, in the spring and in the autumn, is there a favourable relation between temperature and precipitation. The Therophytes have the life-form best fitted to such a brief favourable space of time, since they can complete their life-cycle in the course of some few months. So they will have finished seeding, that is to say, their whole development, when the very warm and dry midsummer sets in.

In the cold-temperate zone the Hemicryptophytes, which are predominant there, have their perennating buds protected by the soil-surface in which they are situated. In the lowlands of the Mediterranean region, where the greatest danger is the excessive desiccation during the dry summer, the greatly heated soil-surface is hardly any protection to the perennating shoots with their rejuvenating buds. This probably accounts for the fact that several species which are pronounced Hemicryptophytes in other regions show a disposition, in the Mediterranean region, to 'get out of the earth', i.e. they show a tendency to form aerial perennating shoots

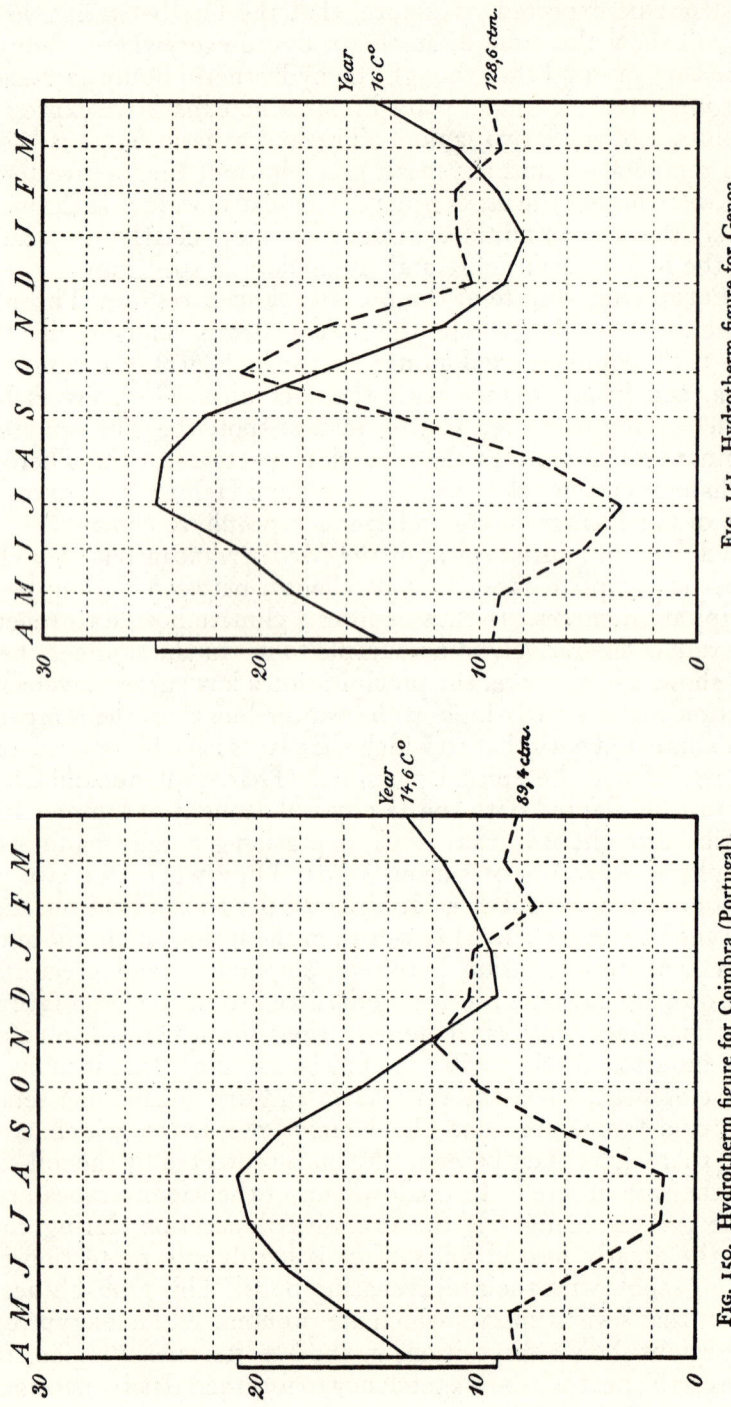

Fig. 151. Hydrotherm figure for Genoa.

Fig. 150. Hydrotherm figure for Coimbra (Portugal).

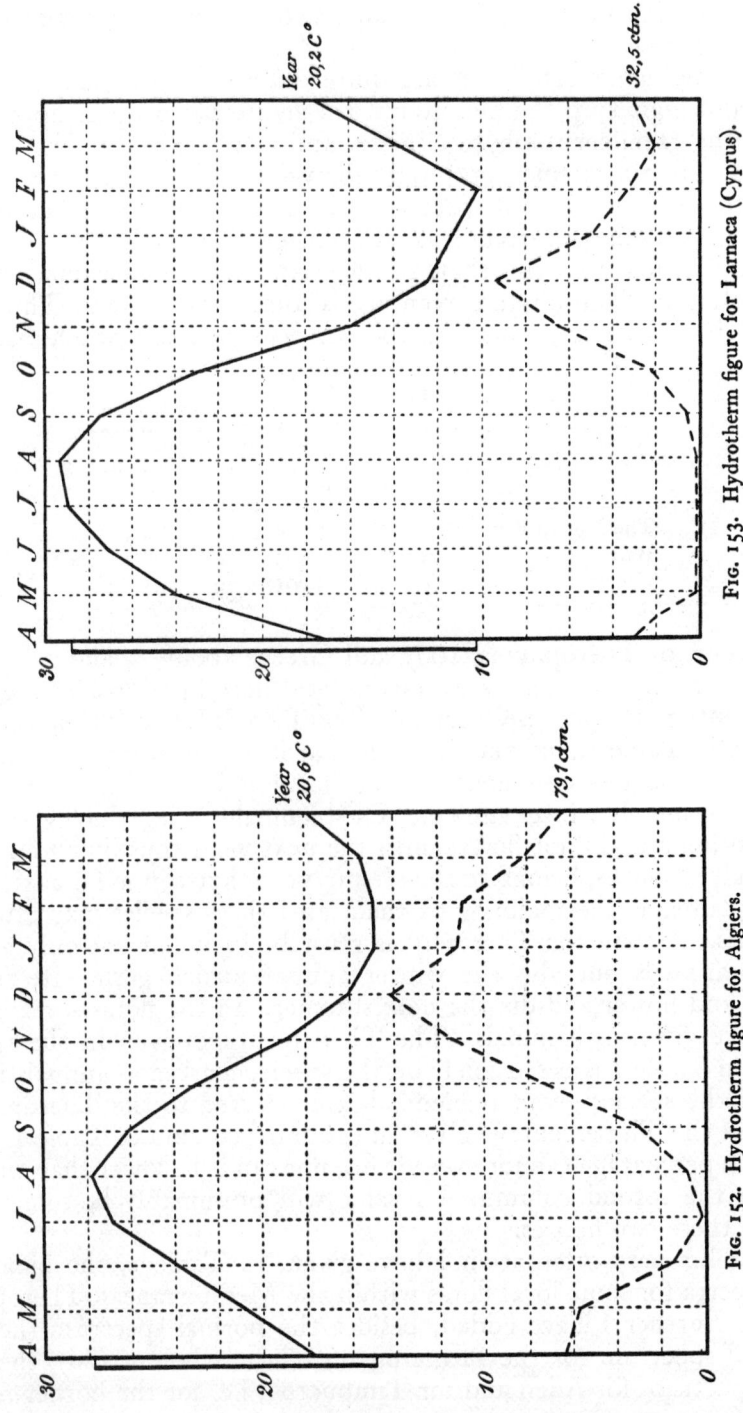

FIG. 152. Hydrotherm figure for Algiers.

FIG. 153. Hydrotherm figure for Larnaca (Cyprus).

bearing rejuvenating buds, thus forming a transition to the chamaephytic life-form.

In the highlands conditions are quite different. Gradually, as we go higher and higher up the mountains, the hydrotherm figure approaches that of the Hemicryptophyte climate, and as a result the Therophytes decrease while the Hemicryptophytes show a marked increase and become dominant in the biological spectrum. In other words, the Therophyte climate only reaches a certain level above the sea, varying in the different areas. It follows from this that, taking these large mountainous regions as a whole, the Therophyte percentage is comparatively low. Thus from Table 1 it will be seen that the Iberian peninsula as a whole has only

TABLE 1

	No. of species.	Th %	
		Min.	Max.
The Iberian Peninsula	5,589	27	32·5
Italy	4,963	24·5	28
Greece	3,691	29	34

27 per cent. of Therophytes, Italy and Greece about 25 and 29 respectively. As soon, however, as we take a local flora in a lowland area, the Th percentage at once rises very rapidly. Two different Th percentages are given in Table 1, and the reason is that it is doubtful in several cases whether the species designated in the floras as biennial are really true biennials or merely winter annuals. The biennial species which germinate in the spring and do not flower until the next year, thus living for more than twelve months, belong to the Hemicryptophytes in so far as they are rosette plants, as the majority of them are; these species may often be perennials. Among the Therophytes must be included not only all the summer annuals but also the winter annuals which germinate in the autumn and flower and die the next summer. In the floras several such species are given as biennials. The Th percentage given in the second column in Table 1 is based solely on the species marked as annuals in the floras; if the species given as biennials are referred to the Therophytes, we obtain the Th percentage given in the third column of Table 1. The actual Th percentage cannot exceed this, nor can it be lower than the one given in the second column. Its value will presumably be somewhere between these two figures.

For preliminary orientation I have given, in Table 2, the biological series spectra for some local floras within the Mediterranean Therophyte climate. Further I have added, besides the normal spectrum (i.e. the biological spectrum for the Phanerogams of the whole world), the biological spectrum for Aden and for Timbuctoo, i.e. for the border region

TABLE 2

The biological spectra (series-spectra) from areas within various plant climates

	No. of species.	Ph	Ch	H	Cr	Th
1. Aden	176	34	27	19	3	17
2. Timbuctoo	137	24	36	8	6	26
3. "	..	20	37	6	5	32
4. The Libyan Desert	194	12	21	20	5	42
5. Cyrenaica	375	9	14	19	8	50
6. Ghardaia	300	3	16	20	3	58
7. Transcaspian Lowlands	663	12	7	28	10	43
8. Argentario	866	12	6	29	11	42
9. Ferrara	657	8	3	39	15	35
10. Euganean Hills	966	7	3	43	14	33
11. Poschiavo, below 850 metres	447	10	5	55	9	21
12. Ekaterinoslav	1,046	5	3	55	13	24
Normal spectrum	..	46	9	26	6	13

between the Therophyte and Phanerophyte climates; also the biological spectra for two local floras—Poschiavo and Ekaterinoslav—within the Hemicryptophyte climate near the border region between this and the Therophyte climate. All the Mediterranean local floras show a very high Th percentage. The biological spectrum for Argentario on the west coast of Italy is a typical example of a spectrum in the European Mediterranean region. The spectral difference between this region and, for instance, the northern Sahara, is however very slightly pronounced in these series spectra, whereas, as we shall see later on, the class spectra, by which the Phanerophytes are subdivided, will show a marked difference.

In the biological spectrum of Aden we have an example of conditions in the border region between a sub-tropical and a tropical area with very dry climates. Where the winter is comparatively cold, the therophytic life-form will predominate, but where the temperature as a whole is so high that the winter temperature is not low enough to be an essential hindrance to the vital activities of the plants, there it will be an advantage to the plant, living in these dry regions, to benefit by the passing showers of the brief favourable periods.

Evergreen epigeal perennating species xerophytic enough to stand the climate will here be the most favoured. As the table shows, the Phanerophytes and Chamaephytes constitute 23 per cent. of the flora of Cyrenaica: in the Libyan Desert the number has risen to 33 per cent., but the Th percentage is still high, viz. 42; if, however, we go farther south, to Aden, the Th percentage falls to 17, while the Phanerophytes and Chamaephytes together constitute 61 per cent. Now this figure is perhaps somewhat too high, for I have here given for Aden the biological spectrum which I had previously based on Krause's *Aden Flora* from 1904 to 1905: later, however (1914–16), Blatter published a new *Aden Flora* comprising a considerably greater number of species; on a preliminary estimate it appeared that the Phanerophyte percentage remains the same as before, whereas when we make use of Blatter's material the Chamaephyte percentage has fallen to 21. But even if we reckon with this figure, the Phanerophytes and Chamaephytes together constitute 55 per cent., that is to say, the very figure which the normal spectrum shows for these two life-form series together. The Phanerophytes of Aden are principally (26 per cent. out of the 34 per cent.) Nano-phanerophytes, as was only to be expected, and thus we have here a plant climate characterized by the preponderance of Nano-phanerophytes and Chamaephytes.

In the southern Sahara we might expect to find the boundary line between the Mediterranean Therophyte climate and that of the tropical desert or semi-desert, with a spectrum similar to that of Aden. Recently O. Hagerup has published a paper (O. Hagerup, 1930), in which he goes very thoroughly into the life-forms of the species contained in the flora of Timbuctoo, and gives the biological spectrum of this flora. In accordance with the climate, which is hostile to vegetation (see the hydrotherm figure on p. 16 of O. Hagerup's paper), the flora of Timbuctoo is very poor in species (137 species). The author gives a detailed account of the life-form of each species, and particularly calls attention to the interesting fact that a comparatively large percentage of the species may occur in two or sometimes three life-forms. Thus many species which are usually Therophytes may appear as Chamaephytes where the environment is favourable. Further, all the Chamaephytes seed in the year of germination. This makes it possible for them to appear as Therophytes where the environment is so unfavourable that the individuals cannot survive through the most rigorous season. As a matter of fact, a particularly large number of Chamaephytes occur also as Therophytes. Thus we have here examples of the phenomenon also found in other regions, but apparently particularly conspicuous in these, that the life-form system employed is well suited to reflect the external conditions, since a change in the environment may show a corresponding change in the life-form of the species.

The author discusses the various methods of presenting the biological spectrum of such a flora. Finally he ascribes to each species the life-form prevalent for it in the flora in question. By this method we obtain for Timbuctoo the spectrum shown in Table 2, 2. By its comparatively low Th percentage, and especially by its high Chamaephyte percentage, it shows wide deviation from the Mediterranean spectrum, but approaches very near to the spectrum of Aden (Table 2, 1). Like the latter the spectrum of Timbuctoo is characterized by the excessive prevalence of epigeals, especially Chamaephytes and Nano-phanerophytes, combined with a rather high Th percentage as compared with the normal spectrum.

Such a life-form spectrum seems to be peculiar to very dry desert-like tropical regions.

It would, however, be of interest to ascertain what would be the effect on the biological spectrum if we were to take into account the fact emphasized by the author, that so many of the species occur in two (or three) life-forms. By recording the individual species not only under the life-form prevalent for the species in question but also under the other life-forms in which it *may* occur, we get 173 units (instead of 137 species), which gives the result shown in Table 2, 3. As will be seen, it is in the main the same as the preceding spectrum (Table 2, 2), the most notable difference being the rise in the Th percentage. But in both cases the chamaephytic life-form is predominant, in connexion with Nano- and Micro-phanerophytes. Together these three life-forms constitute 60 per cent. in the first spectrum and 55 per cent. in the second (while they only constitute 42 per cent. in the normal spectrum). A flora with such a spectrum does not belong to the Therophyte climate, but the rising Th percentage shows that we are here approaching the area of this climate.

As regards the northern limit of the Mediterranean Therophyte climate, it will be seen from Table 2 that the biological spectra for Ferrara and the Euganean Hills show approximation to the Hemicryptophyte climate. Argentario has 29 per cent. of Hemicryptophytes; in the flora of Ferrara the figure has risen to 39, and in the biological spectrum of the Euganean Hills the figure has reached 43; at the same time the Th percentage has decreased from 42 (Argentario) to 33 (Euganean Hills). A little farther north, just south of the Alps, we have the boundary between the Mediterranean Therophyte climate and the Central European Hemicryptophyte climate. The Poschiavo valley in the southern part of the Alps has thus, even in its lower part (below 850 metres), a pronounced hemicryptophytic climate spectrum, viz. 55 per cent. of Hemicryptophytes and only 21 per cent. of Therophytes.

In order to prove that the plant climate of the Mediterranean lowlands is really characterized by the prevalence of the therophytic life-form in the biological spectrum of all the local floras, it is not necessary to determine the life-form of all the species in the floras of the individual areas: in other words, it is not necessary to make a complete biological spectrum. It will suffice here to determine the Th percentage, and this is fairly easy to do, for this merely means that it is sufficient to determine for each species whether or not it is a Therophyte. A difficulty, though of no essential importance, arises from the fact that it is not always easy to decide whether the species given as biennials in the floras really are biennials and thus Hemicryptophytes, or whether they are winter annuals and thus Therophytes. In the following tables, therefore, two figures are generally given for the Th percentage, the first of which indicates the lowest, the second the highest Th percentage which can possibly occur in the local floras in question. The first figure is based solely upon the species given as annuals in the respective local floras; these are, if the lists are correct, in all cases Therophytes. The second figure results when we include among the Therophytes not only the species given as annuals but also the species designated as biennials in the respective floras. Some of these are very probably winter annuals and thus Therophytes, but in no case can the Th percentage exceed this figure, just as it cannot be lower than the first figure.

In thirty-seven local floras the average difference between the two Th percentages was about 4·5; judging by the cases in which I have tried to determine how many Therophytes are included among the species given as biennials, the actual Th percentage will as a rule hardly be more than 1–2 higher than the figure given in the tables as 'minimum Th %'.

TABLE 3

	No. of species.	Th %	
		Min.	Max.
Western France	1,625	32	..
Lower part of the Minho Valley	749	33	37
Odemira (South Portugal)	876	39	41·5
Lowlands of Madeira	213	51	..
Madeira as a whole	427	39	..
Highlands of Madeira	269	28	..

For the sake of clearness the floras examined have been arranged in groups in the following tables according to their longitudinal position, and within each table they are again arranged according to their latitudinal position. Beginning from the west, Table 3 shows the Th percentage in some floras along the coasts of the Atlantic. As far as western France is concerned, I have unfortunately only been able to use Lloyd's *Flora de l'ouest de la France* (Lloyd, 1863). It does not matter much that this book was written so long ago and that several species have no doubt been discovered since its publication. Experience has taught me that in statistical investigations of this kind it is of little consequence that a flora has not been exhaustively examined if only the investigation is not one-sided. But the drawback lies in the fact that Lloyd's *Flora* comprises an area which probably belongs in part to the Therophyte climate and in part to the Central European Hemicryptophyte climate. The comparatively high Th percentage for the flora as a whole shows that at any rate part of western France must come within the Therophyte climate. By determining the Th percentage in a series of minor local floras, we must try to arrive at the boundary line between the two plant climates.

On the west coast of the Iberian Peninsula I have determined the Th percentages in two of the local floras, a northerly one in the lower part of the Minho Valley (Merino, 1897) with 33 per cent. of Therophytes, and a southerly one in Odemira in Algarve (Sampaio, 1908–9) with 39 per cent. of Therophytes; the difference between the two figures shows good agreement with the difference between the hydrotherm figures of the two areas: the Minho Valley has a far less dry summer than Odemira. Finally Madeira affords an example of the south-western corner of the Mediterranean Therophyte climate area. On the basis of Vahl's flora lists (Vahl, 1904) I have determined the Th percentage partly for Madeira as a whole and partly for the lowland and highland tracts separately. It is in harmony with the fact that the Th percentage decreases with the altitude that Madeira's highlands have only 28 per cent. of Therophytes, while the lowlands have 51, and as the highland flora is comparatively rich in species, the Th percentage for the flora of the whole island is comparatively low, viz. only 39.

TABLE 4

	No. of species.	Th %	
		Min.	Max.
The Montpellier region	2,044	32	37·5
The Camargue (at the mouth of the Rhone)	233	37	40
Porquerolles	508	45	51
The Balearic Isles	661	35	38
Sierra Nevada, below *c.* 633 metres . .	1,060	51	..
Ghardaia (North Sahara) . . .	300	58	..
El Golea (North Sahara)	169	56	..

Table 4 shows the Th percentage in a somewhat more easterly series of local floras. In the most northerly, the Montpellier region, the Th percentage is lowest; here we are near the border-line of the Hemicryptophyte climate. I have used Loret and Barrandon's *Flore de Montpellier* (Loret et Barrandon, 1886) as my basis; the northern part of the area dealt with in this flora most probably comes within the Hemicryptophyte climate, while the southern part has a pronounced Therophyte climate spectrum. Thus the Camargue, the region round the mouth of the Rhone (Flahault et Combres, 1894), has at least 37 per cent. of Therophytes (Raunkiær, 1914). Porquerolles (Jahandiez, 1905), one of the islands off Hyères, has even as much as 45 per cent. of Therophytes.

On the other hand, the Balearic Isles, for which I have consulted Willkomm's plant list (Willkomm, 1876), have only about 35 per cent. of Therophytes, which may be due to the fact that these islands are mountainous with a preponderance of Hemicryptophytes in the higher regions, in which the altitudes in some places exceed 1,500 metres. For southern Spain I have used Boissier's plant list from the Sierra Nevada (Boissier, 1837); the lowlands (below *c.* 633 metres), as might be expected, show a very high Th percentage, viz. 51.

For the northern Sahara I have consulted Chevallier's plant lists for Ghardaia and El Golea (Chevallier, 1903), which shows respectively 58 and 56 per cent. of Therophytes.

Boissier has plant lists from various altitudes of the Sierra Nevada; hence we have here an opportunity of illustrating how the biological spectrum changes with the degree of elevation. The Th percentage decreases very rapidly as we reach higher levels, whereas the Hemicryptophyte percentage shows a corresponding rise. In the Alpine region we see, in addition, a rapid increase of the Chamaephyte percentage, and in the nival region, finally, the Chamaephytes are the dominant life-form, the Chamaephyte percentage being here considerably higher than in the normal spectrum, while the Hemicryptophyte percentage is only barely double that of the normal spectrum. These relations will appear clearly from a glance at Table 5, and also Table 6 showing the corresponding facts from the Pyrenees.

TABLE 5

Sierra Nevada (South Spain): biological spectra of zones of various altitudes; in *e* and *f* only the Therophyte percentage has been determined

	No. of species.	Ph	Ch	H	Cr	Th
a. Above 3,000 metres.	14	..	50	50	..	..
b. Above 2,900 metres.	20	..	40	55	..	5
c. Above 2,530 metres.	113	2	25	67	1	5
d. 1,500– *c*. 2,530 metres.	408	7	20	51	3	19
e. 633–1,500 metres	698	..	..	..	..	29
f. Below *c*. 633 metres.	1,060	..	..	..	..	51

TABLE 6

The Pyrenees: biological spectra of zones of various altitudes above 2,600 metres

	No. of species.	Ph	Ch	H	Cr	Th
Above 3,200 metres.	13	..	61	31	8	..
„ 3,000 „	23	..	48	44	4	4
„ 2,800 „	37	..	38	57	2·5	2·5
„ 2,600 „	183	2	32	60·5	1·5	4

Extensive floristic data are available for the determination of the Th percentage in the islands west and south of Italy. Sommier has given us plant lists for the Tuscan Islands (Sommier, 1902, 1903, 1909), Pantellaria, Linosa, and Lampedusa (Sommier, 1908), and in collaboration with Caruano also a flora of Malta (Sommier et Caruano, 1914–15). Beguinot has plant lists for the Ligurian Islands (Beguinot, 1907), the

Neapolitan and the Pontine Islands (Béguinot, 1905 b). The Maddalena Islands between Corsica and Sardinia have been investigated by Vaccari (Vaccari, 1894-9), and Lojacono has explored the Lipari Islands (Lojacono, 1878). Consulting this material I have determined the Th percentage for a series of island groups and islands. The results are seen in Tables 7-14.

TABLE 7

The Therophyte percentages of some Italian islands

	No. of species.	Th %	
		Min.	Max.
The Ligurian Islands	436	32	37
The Tuscan Islands	1,384	42	46·5
The Maddalena Islands	557	52	57
The Pontine Islands	576	54	58
The Neapolitan Islands	976	46	50·5
The Lipari Islands	467	56	59
Pantellaria	454	59	63·5
Malta	905	50	52
Lampedusa and Linosa	522	61	65

TABLE 8

The Therophyte percentages of the Ligurian Islands

	No. of species.	Th %	
		Min.	Max.
Gallinaria	200	28·5	33
Bergeggi	137	24	32
Palmaria	295	33	38
Tino	120	25	29
Tinetto	45	18	20

TABLE 9

The Therophyte percentages of the Tuscan Islands

	No. of species.	Th %
Palmaiola	121	49
Giannutri	187	59
Montecristo	388	57
Gorgona	456	52
Pianosa	469	56
Capraia	611	51
Giglio	682	50
Elba	1,054	44

TABLE 10
The Therophyte percentages of the Maddalena Islands

	No. of species.	Th % Min.	Th % Max.
Maddalena	458	43·5	48
Caprera	405	56·5	61
S. Stefano	145	55	59
Spargi	113	48·5	57

TABLE 11
The Therophyte percentages of the Pontine Islands

	No. of species.	Th % Min.	Th % Max.
S. Stefano	186	63	68
Ventotene	321	58·5	63
Zannone	253	56	59
Gavi	105	52	55·5
Ponza	404	54	58·5
Palmarola	247	60·5	65

TABLE 12
The Therophyte percentages of the Neapolitan Islands

	No. of species.	Th % Min.	Th % Max.
Capri	739	47	51·5
Nisida	255	51	56·5
Procida	299	47	52·5
Vivara	122	33·5	40
Ischia	757	48	53

TABLE 13
The Therophyte percentages of the Lipari Islands

	No. of species.	Th % Min.	Th % Max.
Stromboli	189	61	64
Basiluzzo	183	61·5	65
Panarea	270	62	65
Salina	282	58·5	61
Lipari	323	60	63
Vulcano	217	63·5	66

In Table 7 I have first shown the state of affairs in the individual groups of islands and in some isolated islands, the areas being arranged in succession from north to south. The most northerly area, the Ligurian Islands, has only 32 per cent. of Therophytes. Here we approach the boundary between the Therophyte and the Hemicryptophyte climates. All the other areas have a high or even a very high Th percentage, showing a pronounced Therophyte climate spectrum.

Table 8 shows the Th percentage for each island in the Ligurian group; the Th percentage is low, as it is for the group as a whole. Tables 9–13 give the Th percentages for the individual islands in the Tuscan group, the Maddalena Islands, the Pontine, the Neapolitan, and the Lipari Islands —29 islands in all. The number of species in each of these islands exceeds 100, but at the same time varies greatly, from 105 (Gavi) to 1,054 (Elba). Only Vivara has a Th percentage below 40, viz. 33·5; the Th percentage of all the rest of the islands ranges from 43·5 to 63·5.

TABLE 14

The Therophyte percentages of some Islets and Reefs of the Tuscan Group

	No. of species.	Th %
Topi	48	44
Praiola	27	44·5
Argentarola	11	36·5
Pan di Zucchero	85	35
Troia	15	33
Formiche di Grossete	14	35·5
Formica di Burano	9	33
Cappa	17	41
Scola di Pianosa	79	57
Scarpa di Pianosa	6	33
Cerboli	73	56

Finally, Table 14 shows the Th percentages for a number of islets and rocks in the Tuscan group. Though the number of species in each single case is quite low, viz. from 6 to 85, still the Th percentage is above 30 throughout the area, ranging from 33 to 57. Even though we cannot expect floras so poor in species to show a spectrum like that of the area in question as a whole, still it is interesting to note that none of the 11 minor floras forms an exception to the rule, since even Scarpa di Pianosa with no more than 6, Formica di Burano with 9, Argentarola, Troia, Formiche di Grossete, and Cappa with 11, 15, 14, and 17 species respectively, all have at least 33 per cent. of Therophytes. This illustrates what I have shown for other plant climates, viz. that the biological spectrum of even quite small floras as a rule shows the same characteristics that appear in the plant climate as a whole. Thus the Chamaephytes show a relative or even absolute preponderance over the other life-forms in the floras poor in species of the high arctic islands and on the nunataks which project above the inland ice of Greenland—corresponding to the great preponderance of Chamaephytes in the biological spectrum of the high arctic Chamaephyte climate.

Before I go on to deal with the Th percentages of the local floras on the peninsula of Italy I shall treat a series of local floras in the eastern part of the Mediterranean Therophyte climate-zone. The Th percentage of the Transcaspian lowlands has been determined by O. Paulsen (Paulsen, 1912); I have determined the Th percentage of Syria on the basis of Post's *Flora of Syria, Palestine*, &c.; for Cyprus I have used Holmboe's *Studies on the Vegetation of Cyprus* (Holmboe, 1914), and for the Libyan Desert Ascherson and Schweinfurth's *Illustration de la flore d'Égypte* (1889).

TABLE 15

The Therophyte percentages in some larger areas within the Eastern part of the Mediterranean Therophyte Climate

	No. of species.	Th %
The Transcaspian Lowlands	663	43
Syria	3,592	43
Cyprus	1,270	48
The Libyan Desert	194	42
Cyrenaica	711	53
Tripoli	578	52

As will be seen from Table 15, these four floras show a fairly high Th percentage, though it is lower than many of the Th percentages of the local floras mentioned above. It is lower, too, than the Th percentage found for Cyrenaica and Tripoli (also given in Table 15), based on the latest investigation of the floras of these countries (Durand et Baratte, 1910). We have here an opportunity of illustrating the fact that even rather incomplete floral lists may very well serve as a basis for the biological spectrum of a flora—provided the incompleteness of the list is not due to a one-sided investigation of the area under consideration. If completeness and comprehensiveness are aimed at, even a flora list which, for want of a full investigation, only contains a fraction of the species present will, in virtue of the statistical method employed, give almost the same biological spectrum as the whole number of species. Thus, in 1908, I determined the biological spectrum for Cyrenaica and Tripoli on the basis of some plant lists then at hand, viz. Cosson (Cosson, 1879, 1889), Daveau (Daveau, 1876), and Letourneux (Letourneux, 1889). These lists contained, for Cyrenaica and Tripoli, 375 and 369 species respectively, which gave a Th percentage of 50 and 51. I have now consulted the flora of Durand

and Barette published in 1910, according to which Cyrenaica has 711 and Tripoli 578 species, the Th percentages being 53 and 52 respectively. Though 336 species have here been added to the flora list of Cyrenaica and the number of species is thus almost doubled, the Th percentage is nearly the same. The flora list of Tripoli has been augmented by 208 species, that is by more than the half of its former number of species, but the Th percentage is practically the same.

TABLE 16

The Therophyte percentages of Attica and some islands round the Balkan Peninsula

	No. of species.	Th %	
		Min.	Max.
Arbe and surrounding islands	766	33	39
The Tremitic Islands and Pelagosa	448	50·5	55
Zakynthos	626	46	..
Thasos	476	48·5	52·5
The Northern Sporades	502	51·5	..
Attica	1,092	48	52·5
Aegina	566	58	62·5
Karpathos	375	50	..

Table 16 shows the Th percentage for Attica and some islands in the Aegean Sea; further for Zakynthos (Zante) in the Ionian Sea and for some islands in the Adriatic. For Thasos off the Macedonian coast I have consulted Haláczỳ's *Florula insulae Thasi* (Haláczỳ, 1893), and Bornmüller's supplement to it (Bornmüller, 1894); they have 476 species in all, 48·5 per cent. of which at least are Therophytes. For the northern Sporades I have used Haláczỳ's *Florula Sporadum* (Haláczỳ, 1897). For Attica I have consulted Heldreich's plant list in Mommsen's *Griechische Jahreszeiten* (Heldreich, 1877); some species have probably been added since Heldreich's time, but there is no reason to believe that these belong principally either to the Therophytes or to the other life-forms, so there is no need to suppose that a Th percentage based on a more recent list would differ appreciably from that I have based on Heldreich's list of 1877. In this connexion I should like to point out that I have as far as possible endeavoured to make the flora lists of the areas compared uniform as regards the delimitation of the species and the demarcation of the spontaneous flora (which is our starting-point here) from the cultivated plants and those weeds whose existence is dependent on cultivation. This explains why the number of species given by me for the local floras often deviates a little from the number found on counting the species given in the flora lists quoted.

For Aegina I have consulted Heldreich's *Flore de l'île d'Égine* (Heldreich, 1898), and for Karpathos (an island north-east of Crete) the plant list given by Stefani, Forsyth Major, and Borbey (Stefani, Forsyth Major, et Borbey, 1895). It will be seen from Table 16 that all the local floras have a high Th percentage, viz. between 48 and 58.

We will now examine the Th percentages for the islands in the Adriatic. For Zakynthos (Table 16) I have consulted Margot and Reuter's plant list dating from 1838 (Margot et Reuter, 1838), which comprises 626 species with a Th percentage of 46, thus a rather high one. This plant list is probably not very complete; a complete list for Zakynthos would perhaps comprise 8–900 species. But as I mentioned above when discussing the floras of Attica, Tripoli and Cyrenaica, there is no reason to suppose that there would be any essential difference between the Th percentage derived from the complete flora list and that found by me by means of Margot and Reuter's list; and the Th percentage thus found is in perfect agreement with the geographical situation of the island.

If we go farther north, we have, about midway off the east coast of Italy, north of the Gargano peninsula, the small Tremitic Islands. These, together with the island of Pelagosa situated in the middle of the Adriatic, have been investigated by Béguinot (Béguinot, 1910), whose plant list I have used for the determination of the Th percentage. Taken together these islands have 448 species, 226 of which are annuals, that is to say, at least 50 per cent. are Therophytes. In order to show how, in this case too, the biological spectra of quite small local floras agree with those of floras richer in species within the same plant climate, I have determined the Th percentage for each of the islands in the Tremitic group:

TABLE 17

The Therophyte percentages of the Tremitic Islands and Pelagosa

	Area in sq. km.	No. of species.	Th % Min.	Th % Max.
S. Domino	c. 2	298	51	57·5
S. Nicola	c. 0·5	236	55·5	59
Capperaia	c. 0·5	91	51·5	52·5
Cretaccio	c. 0·04	53	45	51
Pianosa	c. 0·02	35	37	43
Pelagosa	c. 0·3	96	40·5	41·5

the result will be seen in Table 17 where I have added the areas of the various islands (after Béguinot). Apart from S. Domino, which has an area of about 2 sq. km., the sizes of the islands range from 0·02 to 0·5 sq. km. The number of species on some of these very small, barren, calcareous islands is very low, varying from 35 (Pianosa) to 298 (S. Domino). In four of the islands the number of species is below 100; nevertheless they all show the Th percentage of the Therophyte climate, ranging from 37 (Pianosa) to 55·5 (S. Nicola).

The last example I shall mention here is Arbe and surrounding islands, a group off the coast of Dalmatia, the floras of which have recently been examined by Morton (Morton, 1915). From Table 16 it will be seen that the Th percentage there is at least 33, that is to say, approximately the same as that of the Ligurian Islands (see Table 8). This agrees well with the fact that, here as there, we are near the boundary line between the Therophyte climate and the Central European Hemicryptophyte climate.

Finally I shall give some examples of the Th percentage in some local floras on the Italian peninsula. As this country abounds in mountains which attain considerable heights in several places, there is rather a large difference in the hydrotherm figures for the various parts. The higher we rise above the level of the sea and the farther we go north, the more the hydrotherm figure will approach the relations characteristic of the Hemicryptophyte climate of Central Europe, and simultaneously a change will occur in the biological spectrum: the number of the Therophytes will decrease and that of the Hemicryptophytes will rise.

The floristic material for Italy is very large. It would require a great deal of work to utilize this material to the full in constructing the biological spectra for all areas; here we shall merely mention a few examples to illustrate the general facts for Italy. For a series of local floras the Th percentage has been determined, in some cases it has also been determined to what life-form class each species belongs. For my present purpose, however, the class spectrum is not necessary. Hence I have merely given the series spectrum, which admits of a more comprehensive view. It states the percentage proportion of Phanerophytes, Chamaephytes, Hemicryptophytes, Cryptophytes, and Therophytes (Table 18).

TABLE 18

Biological spectra from various parts of Italy

	No. of species.	Ph	Ch	H	Cr	Th
North-east Italy	1,800	9	7	50	13	21
Colline di Crea	996	..	..	..	..	33
The Orba Valley NW. of Genoa	327	..	..	..	..	33
Euganean Hills	966	7	3	43	14	33
Ferrara	657	8	3	39	15	35
Argentario	866	12	6	29	11	42
Campobasso	666	4	6	49	10	31
Vesuvius (c. 1,250 metres)	598	..	..	..	..	43
Monte Pollino (c. 2,271 metres)	1,260	..	..	..	..	33

In the lowlands the Th percentage is high throughout, though it is considerably lower in northern than in central and southern Italy; on the basis of Sommier's plant list (Sommier, 1902) I have determined the biological spectrum for Argentario (Table 18). This spectrum, which shows

42 per cent. of Therophytes and 29 per cent. of Hemicryptophytes, may probably be regarded as typical for the lowland flora of the central part of Italy. In southern Italy the Th percentage is no doubt somewhat higher. A comparison between the biological spectrum of Argentario and that of the Mediterranean region of north-eastern Italy (Table 22) will show that the Therophytes and the Hemicryptophytes have changed places as regards their absolute amounts; in the spectrum of Argentario there are 29 per cent. of Hemicryptophytes and 42 per cent. of Therophytes, while the Mediterranean region of north-eastern Italy shows 43 per cent. of Hemicryptophytes and only 30 per cent. of Therophytes.

For southern Italy I have determined the Th percentage for Monte Pollino (height 2,271 metres) on the basis of Terracciano's plant list (Terracciano, 1889-90). It is in harmony with the considerable height that the Th percentage is only 33. Mount Vesuvius which is only c. 1,250 metres, shows a correspondingly higher percentage, viz. 43 (Table 18).

TABLE 19

The biological spectra of the zones of various altitudes on Monte Pollino

	No. of species.	Ph	Ch	H	Cr	Th
Above 2,200 metres	42	..	29	71	..	..
2,000–2,200 ,,	38	5	29	66	..	..
1,800–2,000 ,,	52	8	25	65	..	2
Above 1,800 ,,	80	4·5	20	68	2·5	5
The flora of Monte Pollino as a whole	1,260	..	..	..	..	33

Table 19 shows for Monte Pollino how the biological spectrum changes with the level; above 1,800 m. there are in all 80 species, only 5 per cent. of which are Therophytes. Of the 42 species found above 2,200 metres none are Therophytes; the species which have been able to maintain themselves here have either the hemicryptophytic or the chamaephytic life-form, and of these two groups the Chamaephytes show comparatively the greatest increase. The biological spectrum for Argentario has only 6 per cent. of Chamaephytes, while that of Monte Pollino (above 2,200 metres) has 29 per cent. of Chamaephytes, that is to say, a Chamaephyte percentage five times as high, whereas the Hemicryptophyte percentage at the summit of Monte Pollino is only about two and a half times as high as Argentario's Hemicryptophyte percentage.

The biological spectrum for the flora of Campobasso (Table 18) is based on Villani's plant list (Villani, 1906-7). Here the Th percentage is only 31. This comparatively low Th percentage expresses the fact that we are here dealing with an area which consists largely of mountain land comprising considerable local heights. The species peculiar to the highland tracts and absent from the lowlands are principally Hemicryptophytes, while none or only a very few are Therophytes. Hence the biological spectrum for the flora as a whole shows a diminution of the Th percentage and a rise in the Hemicryptophyte percentage, so that the spectrum approximates to that which is peculiar to the Hemicryptophyte climate.

Jatta (1877), Matteuci (1894), and Grugnola (1894) have plant lists for Gran Sasso. By means of Grugnola's lists we may distinguish three groups of species belonging principally to different levels: (1) below 800 metres, (2) between 800 and 2,000 metres, and (3) above 2,000 metres. Of the 791 species, 214 belong to the first, 351 to the second, and 226 to the third group. Since only those species are given which, in the main, are peculiar to each zone, the figures are indeed lower than the actual number of species, but owing to the principle applied in the grouping of the species they will provide a useful foundation for a comparison between the Th percentages of the three zones. Table 20 shows the result. As usual the Th percentage decreases rapidly with the height: thus above 2,000 metres the biological spectrum shows only 3·5 per cent. of Therophytes. Determining the Th percentage by Matteuci's plant lists (Matteuci, 1894) we arrive at about 2 per cent. of Therophytes in the flora above 2,200 metres. This agrees very well with the figures found from Grugnola's list.

TABLE 20

The Therophyte percentages of Gran Sasso d'Italia

	No. of species.	Th %
Gran Sasso d'Italia: above 2,000 metres	226	3·5
„ between 800 and 2,000 metres	351	8
„ below 800 metres	214	31

Next we shall give the Th percentages of some local floras in northern Italy. Works consulted: for the province of Ferrara Revedin's *Contributo alla flora vascolare della provincia di Ferrara* (Revedin, 1909), for the Euganean Hills Béguinot's *Saggio sulla flora e sulla fitogeografie dei Colli Euganei* (Béguinot, 1904), and the *Prospetto delle piante vascolare finora indicata per i Colli Euganei e per la Pianura Padovana* (Béguinot, 1905); for the Orba valley on the northern side of the Ligurian Alps Morteo's *Florula alluvionale di un tratto del torrente Orba* (Morteo, 1906), and for the Crea hills east of Turin, Negri's *La vegetazione delle colline di Crea* (Negri, 1905–6). Finally I have consulted Gortani's *Flora friulana* (Gortani, 1905). As will appear from Table 18 the Th percentage for the local floras of northern Italy is relatively low compared with that of the more pronounced Therophyte climate, and there may be reason to discuss the question as to whether this region as a whole really belongs to the Therophyte climate zone. Here, where we are at any rate near the border-line of the Hemicryptophyte climate zone, we meet with the difficulty previously referred to, caused by the fact that Therophytes are specially favoured by cultivation. If, for the Po valley, we disregard the species principally occurring on cultivated land, the Th percentage will probably sink so much that the biological spectrum will become a Hemicryptophyte climate spectrum. Before man began to cultivate the Po valley, it was no doubt in the main covered by forests, that is to say, by phanerophytic formations, and there can hardly be any doubt that the Th percentage was then lower than now. This, however, is not the decisive fact. For if the species which have immigrated with the culture plants are able to maintain themselves without the aid of cultivation, they now belong to the flora just like those which have previously immigrated. It is not the manner or time of immigration that settles the question whether the species are to become natives of a flora, but their ability to prevail against competition. Whether they have been introduced by wind, water, animals or man is of no consequence, the main point is whether or not they can maintain a place for themselves without the aid of cultivation. If the biological spectrum is to be an expression of the climate, it must be based solely on the species able to maintain themselves even if all cultivation were discontinued and Nature left entirely to herself. Of course, in a pronounced Therophyte climate, we find a great many Therophytes on cultivated soil. But in most places it is easy to ascertain which of the therophytic species can hold their own on uncultivated soil (strand formations, forests, coppice, &c.), and which must thus be said to have become natives of the flora though originally introduced by man. In an intensively cultivated area like the Po valley it is much more difficult to arrive at the facts. At any rate I shall make no attempt to decide which of the species now in existence would probably disappear—sooner or later—if all cultivation ceased. In order to do this one must be familiar with the flora of localities more or less unaffected by cultivation which are scattered over the area. In constructing the spectra shown in Table 18 I have only omitted those species from the plant lists I have consulted which must a priori be assumed to occur exclusively on cultivated soil.

Judging by the biological spectrum for the Crea and the Euganean Hills, the boundary line between the Mediterranean Therophyte climate and the Central European Hemicryptophyte climate must lie somewhat farther to the north. For a more exact determination of this boundary line a knowledge of the biological spectrum for a series of local floras along the northern limit of the Po Valley would be necessary, but I have not the floristic material for such an investigation. I can only give a single example, viz. Bosco Montello in the province of Treviso north of Venice. Bosco Montello is a low range only up to 360 metres high, situated by the river Piave and comprising only about 6,000 hectares. According to Saccardo (Saccardo, 1895) it was a crown forest up to 1894; then it was partly cut up into lots for the poorer population of the surrounding district, and partly sold. In 1895, before any change had taken place, F. Saccardo gave a list, comprising in all 615 species, of the flora of this small area. Apart from a little cultivated soil the area was covered partly by forests, especially of *Quercus pedunculata*, partly by other formations, i.e. marsh plant formations along the river Piave, and mountain plants. Thus the area seems to show a flora of a relatively primitive character, where the original species have been able to maintain themselves. That afforestation, &c., has made changes in the quantitative distribution of the species is here of no consequence as it does not affect the biological spectrum. Now, a determination of the life-forms of the species of this flora shows that, even if we include all species, even those occurring on cultivated soil, yet the Th percentage is only about 17; and omitting the species of cultivated soil, the Th percentage will only be about 12; in both cases the spectrum shows a pronounced Hemicryptophyte climate. Table 21 shows the series spectrum for the whole flora.

TABLE 21

	No. of species.	F	Ch	H	Cr	Th
Bosco Montello	615	12	4	52	15	17

In a single area, viz. in the province of Udine, between the northern Adriatic and the Carnic Alps, so much floristic material is available that we can follow step by step the transition to the Hemicryptophyte climate. In his *Flora friulana* Gortani (Gortani, 1905) has divided this area into six regions, viz. (1) Regio mediterranea, the narrow strip of coast along the Adriatic; (2) Regio padana, a fairly broad lowland tract between the former and the foot of the mountains; (3) Regio submontana; upper limit, *c.* 400 (300–500) metres above the sea; (4) Regio montana: upper limit, *c.* 1,600–1,700 metres; (5) Regio subalpina: upper limit from 1,800 to 2,000 metres; (6) Regio alpina: from the preceding region to the highest point, the summit of Caglian, 2,782 metres high. For each of these regions I have determined the biological class spectrum, though for the sake of clearness I have merely given the series spectrum in Caglian, Table 22. A glance at the table will show that gradually as we proceed higher and higher up the mountains from the low coast, the predominance of the Therophytes in the biological spectrum will be followed by that of the Hemicryptophytes and finally by that of the Chamaephytes. On September 5th, 1902, 11 species were found on the summit of Caglian, 2,782 metres high, viz. 3 Hemicryptophytes and 8 Chamaephytes, that is to say, *c.* 73 per cent. of Chamaephytes and *c.* 27 per cent. of Hemicryptophytes.

We need not go farther north, for if we disregard the species whose presence in the local floras is due to cultivation, there is nowhere north of the Alps any local flora with a Th percentage approaching that which is characteristic of the Mediterranean Therophyte climate. In the

TABLE 22

North-eastern Italy: the biological spectra of the zones of various altitudes from the Po region to the summits of the Alps (cf. text)

	No. of species.	Ph	Ch	H	Cr	Th
The summit of Caglian (2,782 metres)	11	..	73	27	..	..
Regio alpina	395	3	17	71	5	4
„ subalpina	551	4.5	14	67	9	5.5
„ montana (c. 1,600–1,700 metres to c. 400 metres)	1,083	9	7	59	11	14
Regio submontana	1,161	9	5	52	13	21
„ padana	937	7	3	47	15	28
„ mediterranea	850	8	4	43	15	30
Normal spectrum	..	46	9	26	6	13

cold-temperate zone the Hemicryptophytes are dominant in the biological spectrum of all lowland floras, but if we ascend into the alpine region or pass towards the arctic region the Chamaephytes become more and more predominant in the spectrum; first they grow comparatively numerous, and finally even absolutely more numerous than the Hemicryptophytes, showing the same phenomenon as the local flora of the summit of Mount Caglian, which comprised only 11 species.

As previously mentioned, to draw the boundary line between the plant climates we must compare the biological spectra of the local floras with the normal spectrum. In the present case we must draw the line between the Therophyte and Hemicryptophyte climate at the point where we have:

$$\frac{\text{the H\% of the local spectrum}}{\text{the H\% of the normal spectrum}} = \frac{\text{the Th\% of the local spectrum}}{\text{the Th\% of the normal spectrum}}$$

Where the Therophytes have increased at a greater rate than the Hemicryptophytes we have a Therophyte climate; where the reverse is the case we have a Hemicryptophyte climate. If we apply this principle to the biological spectra shown in Table 22 it will at once be seen that the boundary line between the Therophyte and Hemicryptophyte climates here lies between Regio padana and Regio submontana. In Regio padana we have

$$\frac{47 \text{ (i.e. H\% of local spectrum)}}{26 \text{ (i.e. H\% of normal spectrum)}} < \frac{28 \text{ (i.e. Th\% of local spectrum)}}{13 \text{ (i.e. Th\% of normal spectrum)}}$$

Thus the Th percentage is comparatively higher here than the Hemicryptophyte percentage. In Regio submontana the reverse is the case. Here we have

$$\frac{52 \text{ (i.e. H\% of local spectrum)}}{26 \text{ (i.e. H\% of normal spectrum}} > \frac{21 \text{ (i.e. Th\% of local spectrum)}}{13 \text{ (i.e. Th\% of normal spectrum)}}$$

Thus the Hemicryptophyte percentage is comparatively higher than the Th percentage. In this way we can draw the boundary line wherever the floristic material is available and the life-forms of the species are determined.

For practical reasons the biological spectra by which I have attempted to characterize the Mediterranean Therophyte climate in the preceding pages are based on the Phanerogams alone. Formally there could be no objection to including the Cryptogams with the Phanerogams in the life-form system. Notably the Mosses as well as the Vascular Cryptogams could easily be placed in my life-form system. It would, however, be quite misleading to let the biological spectrum at the same time comprise Thallophytes and Phanerogams—quite apart from the fact that as a rule we do not know the number of Thallophytes contained in the local floras with anything like the accuracy with which we know the number of phanerogamic species. Phanerogams constitute the principal element in the plant covering of the earth: compared with these the areas covered with algae and fungi are quite negligible though these groups comprise a great number of species.

As regards the Lichens and Mosses, their soil cover is, at any rate locally, very considerable; but compared with the majority of Phanerogams the individuals are small while the number of species is large. If then, they were included with the Phanerogams as a basis for the biological spectrum, this would mean that the spectrum would entirely lose its character as a clear picture of the vegetation. From a practical point of view it is besides impossible to include them with the Phanerogams, since the local plant lists do not as a rule include them, and it is a matter of course that, in a comparative investigation, it is an indispensable requirement that the spectra to be compared should comprise the same groups of plants; biological spectra which only include Phanerogams are not comparable with those also comprising other groups, e.g. Lichens and Mosses.

For the Pteridophytes the case is different. Both morphologically and physiognomically the Pteridophytes are so nearly allied to the Phanerogams that it would be quite natural to treat these two groups together. At first, therefore, I tentatively included them in the spectrum, but I soon had to give it up, for in too many cases the Pteridophytes were not included in the flora lists. Hence I have subsequently built up the biological spectrum on the Phanerogams alone. This is not only practically a necessity because the flora lists are often confined to the Phanerogams, but in my opinion it is also an advantage because the great role which the Pteridophytes as well as the Lichens and Mosses play or may play in the vegetation will appear more clearly when these groups are treated separately. In this way further means of characterizing areas defined by the Phanerogam spectrum are obtained, and in several cases subdivisions may be made by means of the Cryptogams.

Separate treatment of the individual groups of Cryptogams—especially Lichens, Hepaticae, Mosses and Pteridophytes—involves the further advantage that we need not be confined to the life-form system used for the Phanerogams, but can use the system most appropriate for the individual group in each case. The relationship to each other of these groups, and their relation to the Phanerogams in regard to geographical distribution, may then be subjected to a comparative investigation, which is all the more needed, since the conditions of dispersal are not the same for the Cryptogams as for the Phanerogams.

The first question to be considered when dealing with the geographical distribution of the Cryptogams is the number of species of this group contained in the individual floras; then the various floras are made comparable by taking the flora of the whole earth as the standard. If, for instance, we take the Pteridophytes we must first determine the proportion of the pteridophytic and phanerogamic species in the flora under consideration, and the ratio so obtained is then divided by the same ratio in the earth's flora as determined once for all. I call the resulting figure the Pteridophyte quotient (Pt.Q.) of the flora. This quotient expresses how many times more (or less) Pteridophytes a flora contains than it would contain if the ratio were the same as in the earth's flora as a whole.

In determining the number of the earth's Phanerogams and Pteridophytes, which forms the common standard with which all floras are to be compared, it is of course important to approximate as much as possible to the true figures, but on the other hand the importance and applicability of the standard do not stand or fall by its absolute correctness. For since the same standard is applied throughout, the proportion of the quantities found by measurement will remain the same whether or not the standard is absolutely correct.

As I explained when I was dealing with the construction of the normal spectrum (pp. 432–3), I have put the number of the earth's Phanerogams at c. 140,000 species. Next I have tried to determine the number of cryptogamic species: Pteridophyta, Musci, Hepaticae, and Lichenes. In order to facilitate the computation I have slightly modified the resulting figures so as to obtain an integer when the figure for the Phanerogams of the earth is divided by the number of species for the individual groups of Cryptogams. As a result I assume the following figures for the whole earth:

Phanerogamae . . .	140,000 species	
Pteridophyta . . .	5,600 ,,	(25)
Musci	12,700 ,,	(11)
Hepaticae . . .	4,000 ,,	(35)
Lichenes . . .	6,350 ,,	(22)

I have added in brackets the figures indicating how many times the figure for the earth's Phanerogams exceeds that of the individual groups of Cryptogams—respectively 25, 11, 35, and 22.

Let us take as an example the flora of Iceland, which comprises 1,026 Phanerogams and 50 Pteridophytes. From the equation

$$\frac{\text{Earth's Phanerogams}}{\text{Earth's Pteridophytes}} \frac{140,000}{5,600} = \frac{25}{1} = \frac{\text{Iceland's Phanerogams}}{x} \frac{1,026}{}$$

we get $x = 41$, that is to say, that Iceland should have 41 species of Pteridophytes if the ratio borne by its Pteridophytes to its Phanerogams were the same as this ratio for the whole earth. Since the flora of Iceland comprises 50 Pteridophytes, the Pt.Q. will be 50/41, i.e. c. 1·2, which means that the number of Pteridophytes in Iceland is 1·2 times what it would be if the numerical relationship between Phanerogams and Pteridophytes were the same in Iceland as in the earth's flora as a whole.

If we assume, as I have done here, that the earth's flora as a whole contains 25 times as many Phanerogams as Pteridophytes, we can find the Pt.Q. of a local flora by multiplying the number of the Pteridophytes in the flora in question by 25 and dividing the product by the number of Phanerogams in the local flora.

With Iceland we may compare Formosa, which has 1,613 Phanerogams and 261 Pteridophytes. Provisionally assuming these figures to be correct, Formosa's Pt.Q. will be $\frac{25 \times 261}{1,613} = c.\ 4$, showing that Formosa's flora has 4 times as many Pteridophytes as it would have if the proportion of Phanerogams and Pteridophytes were the same as in the earth's flora as a whole.

As a third example we may take Argentario off the west coast of Italy, with 866 Phanerogams and 27 Pteridophytes. This gives Pt.Q. $= \frac{25 \times 27}{866} = c.\ 0·8$ which shows that the environment, probably the climate, is unfavourable to Pteridophytes.

As a fourth and last example I choose the proportion of the Mosses in the flora of Sardinia. According to Barbey (Barbey, 1885) Sardinia has 1,793 Phanerogams and 176 Mosses. This gives us a Moss quotient $\frac{11 \times 176}{1,793} = c\ 1·1.$

I shall now show by some examples how the facts appear if, by means of the method indicated above, we use Pteridophytes, Mosses, and Lichens in a comparison between the Mediterranean Therophyte climate and other plant climates, e.g. the Hemicryptophyte climate.

In Table 23 the Pt.Q. for a number of larger or smaller areas in the Mediterranean region will be found. It appears that in all the 20 floras the Pt.Q. is below 1, often very much below. This means that in the

Table 23

The Pteridophyte quotient of some floras within the Mediterranean Therophyte climate

	Pt. Q.
Spain	0.4
Italy	0.5
Greece	0.4
Syria	0.2
Tunis and Algeria	0.4
Tripoli	0.2
Galicia	0.8
Sardinia	0.7
Ligurian Islands	0.6
Argentario	0.9
Central Italy	0.6
Vultur	0.4
Linosa	0.5
Lampedusa	0.2
Malta	0.3
Friaul	0.7
Venice	0.5
Samos	0.6
Karpathos	0.3
Cyprus	0.4

individual floras within the Mediterranean Therophyte climate there is a smaller proportion of Pteridophytes than there would be if these and the Phanerophytes were equally favoured. The Therophyte climate is antagonistic to Pteridophytes; and the more pronounced the drought in summer is, the lower the Pt.Q. seems to be. Thus in Tripoli it is only 0.16, in Syria 0.23, on Karpathos 0.33, and on Lampedusa 0.24. Towards the limit of the Hemicryptophyte climate the Pt.Q. increases, thus in northern Spain (Galicia) it is 0.79 and in northern Italy (Friaul) 0.72. But of course in the individual cases the magnitude of the Pt.Q. is dependent on various factors. If, for example, central Italy has a lower Pt.Q. than Argentario, this is no doubt due to the fact that the average areas of the pteridophytic species are larger than that of the phanerogamic species; other conditions being equal, large areas have therefore a lower Pt.Q. than small areas.

In Table 24 I have given for comparison the Pt.Q. for some local floras in the Hemicryptophyte climate of the northern hemisphere. Here the Pt.Q. is somewhat higher, as a rule about 1.

As examples of very high quotients I have given the Pt.Q. for some of the South Sea Islands in the Phanerophyte climate (Table 25). The very high quotient here, besides being due to the climate, which is favourable

TABLE 24

The Pteridophyte quotient of some floras within the Hemicryptophyte climate

	Pt.Q.
Denmark	1·1
The surroundings of Stuttgart	1·0
The North German Plain	1·0
West Lancashire	1·4
Racine (Wisconsin)	0·9
Alabama	0·9

TABLE 25

The Pteridophyte quotient of some South Sea Islands within the Phanerophyte climate

	Pt.Q.
Hawaii	5·1
Fiji	6·1
Norfolk Island	8·3
The Society Islands	11·3

to Pteridophytes, is especially caused by the fact that as a rule Pteridophytes spread more easily than Phanerogams. In accordance with this the most isolated of these islands, the Society Islands, have the highest Pt.Q.

For the Mosses, too, the Mediterranean Therophyte climate is less favourable than the Hemicryptophyte climate. Table 26, 1–5, shows that in these five local floras from the Mediterranean region the moss quotient was about one, while as seen in Table 26, 6, the Hemicryptophyte climate has a considerably higher Musc.Q.

TABLE 26

The Moss quotient of some floras within the Therophyte climate (1–5) and the Hemicryptophyte climate (6)

	Musc.Q.
1. The Tuscan Islands as a whole	1·2
2. Montecristo	1·1
3. Capraia	1·4
4. Sardinia	1·2
5. Malta	1·0
6. West Lancashire	4·6

The Therophyte climate seems to be more favourable to Lichens than to Mosses; still, as will be seen from Table 27, the Therophyte climate favours Lichens less than the Hemicryptophyte climate; this appears too from the fact that Sardinia and Malta have a lower Lichen quotient than Italy, Italy's comparatively high Lichen quotient being especially due to the higher areas situated above the Therophyte climatic zone of the lowlands.

TABLE 27

The Lichen quotient of some floras within the Therophyte climate (1–3) and the Hemicryptophyte climate (4)

	Lich.Q.
1. Italy	3·5
2. Sardinia	2·1
3. Malta	2·4
4. Denmark	4·3

II. THE VEGETATION OF ALLUVIAL STRAND FORMATIONS OF THE WESTERN COAST OF ITALY

An exact objective delimitation and analysis of the formations must be the first aim in a thorough study of their ecology, but as yet only few objective analyses have been made. Hence there is everywhere an abundance of formations which it would be desirable to study. Under such conditions it will often, at least in part, be a matter of personal predilection what kind of formations a botanist will examine first. Even if he takes an interest in all kinds of vegetation, still each person will feel specially attracted by certain types.

Our northern heaths and flowery meadows, the flora of the beechwood in spring, the flowery Alpine carpet, and the maquis of the Mediterranean regions are among the types of vegetation nearest to my heart, and so it was especially the maquis that I wished to study on my journey in the Mediterranean countries in 1909–10. If, in the following, I write most of the sand-dune vegetation, especially the dune maquis, it is not because the latter attracts me most; the rich maquis of the mountain slopes, steeped in fragrance and sunlight and with the view of the blue sea, seems even more glorious to me. But from my earliest youth the sand dunes of western Jutland along the coast of the North Sea have captured my interest, and so it seemed natural to me to examine especially the sand-dune vegetation of the Mediterranean countries, particularly because it afforded an excellent opportunity of studying that difference in the vegetation which results from the difference in climate, since the soil of our northern and Mediterranean dunes is essentially the same.

In height and extent the Italian sand dunes that I have seen cannot compare with the dunes of the west coast of Jutland. Only on Tombolo di Feniglia near Argentario did I see extensive shifting dunes reminding me of the great white dunes of Jutland, and there as here, the sand is mainly quartz sand, only occasionally, e.g. at Capalbio and Fiumicino, has it an admixture of greater or smaller quantities of magnetic iron sand.

In Table 28 I have recorded the size of the grains of sand in the different

TABLE 28

The size of the sand grains of the Italian dunes

1^1 = Fragments of shells.

	Percentage.				
mm.	0·25	0·5	1	2
1. The Lido of Venice: the beach	24	75·9	0·1[1]	..	..
2. The Lido of Venice: low young dunes	18	81·8	0·2[1]	..	..
3. South of Marina di Pisa: the beach	1	46·4	52·5	0·1[1]	..
4. The summit of young dunes, 1–2 metres high	1	92	6·9	0·1[1]	..
5. The summit of young dunes, 1–2 metres high	1	89	9·9	0·1[1]	..
6. *Circa* 4 km. south of Marina di Pisa: the summit of older dunes, 3–4 metres high	1	80·9	18	0·1[1]	..
7. Tombolo di Feniglia: between the low- and high-water marks	4	93	3	..	..
8. Tombolo di Feniglia: the summit of the large shifting dune	4	95·1	0·9	..	..
9. Tombolo di Feniglia: the summit of the large shifting dune	4	95·5	0·5	..	..
10. Tombolo di Feniglia: the summit of old dunes	2	97·5	0·5	..	..
11. Between Capalbio and Ansedonia: the outermost sand-wall, just reached and even partly washed by the high tide	0·5	14	84·8	0·5	0·2
12. Between Capalbio and Ansedonia: the second sand wall	0·5	14	85	0·5	..
13. Between Capalbio and Ansedonia: denuded area	5	83	12	..	..
14. Fiumicino: between the low- and high-water mark	4	94	2	..	..
15. Fiumicino: the outermost sand wall, now and then washed	2	21	72	5	..
16. Fiumicino: the summit of a high dune, *c.* 2 metres	2	75	23	..	..
17. Fiumicino: low, partly denuded flats near the sea	0·5	11·5	46·5	41	0·5
18. Fiumicino: low, partly denuded flats near the sea	1	37	45	15	2
19. Antium: sand washed into the cellars of the ruins	0·5	25·5	69	5	..
20. South coast of Pontine swamps: beach	4	88	8	..	..
21. South coast of Pontine swamps: shifting dunes	18	82	..	..	..
22. Selva vetere: seaside dunes	34·6	65·4	..	..	..
23. Between Torre Gaveta and Cumae: shifting dune between Juniper dunes	20	80	..	..	..
Average of 1–23	6·6	66·9	23·5	2·9	0·1
Rounded off	7	67	23	2	..

parts of the regions examined by me.[1] The greater part of the sand consists of grains whose size ranges from 0·25 to 0·5 mm. in diameter. Where the wind deposits sand and the dunes are growing, it will probably as a rule be only a very small percentage weight of the sand which consists of grains above 0·5 mm. in diameter. In Table 28, Nos. 2, 4, 5, 8, 9, 21, and 22 belong to this category, and the analyses of these 7 samples show on the average 97 per cent. with grains below 0·5 mm. (Table 29, *b*). In

[1] I owe thanks to Dr. Carsten Olsen, who has kindly made these determinations.

older dunes, that is where no sand is at present being deposited, the proportion of grains of the different sizes may vary much according as sand has been carried away from the surface or not. Where, as in No. 10, Table 28, the relationship between the classes of grain size is the same as that in the recent dunes, the wind has probably not carried away sand to any great extent; if, however, as in Nos. 6 and 16 (Table 28) the percentage of grains of 0·5–1 mm. is comparatively high (Table 29, *c*), the wind has no doubt carried away more or less sand, and principally the smaller and lighter grains. This phenomenon is of course specially marked in severely wind-blown dune areas. Here, as seen in Nos. 13, 17, and 18 (Table 28), the proportion of the different classes of grain size will be quite different, since it is chiefly the large grains that have been left (Table 29, *f*). Here we have not only a large percentage weight of grains of 0·5–1 mm. in size, but also a fairly high percentage of grains 1–2 mm. in diameter, and even some grains exceeding 2 mm. in diameter.

On the flat sandy beach we see in the main the same as we see in the old dunes; sometimes the surface consists of copious amounts of blown sand with a size of grain as in the more recent dunes (Table 28, No. 1), sometimes the larger grains are comparatively predominant, the smaller ones having been carried away by the wind (Table 28, Nos. 3 and 20).

In the line of low banks of sand thrown up during storms, and more or less within reach of and partially washed by the waves, the state of things is mainly as in the wind-blown dune area; only here it is not the wind but the water that sifts the sand, since the retreating waves still more easily carry the lighter grains with them, while the larger and heavier will be left. Hence, as will be seen from Nos. 11, 12, and 15 (Table 28) we have here a high percentage of grains with a diameter of 0·5–1 mm. (Table 29, *e*). Outside these low banks, in the zone between the high- and low-water mark, the relations of the sizes of grain are in the main the same as in the recent dunes—judging by two samples examined by me, viz. No. 7 (Table 28) from Tombola di Feniglia, and No. 14 (Table 28) from Fiumicino, which quite agree.

It is a well-known fact that in different parts of the same dune area, nay in the same dune, the sand may vary considerably as regards the proportion between the different classes of grain size, and as mentioned above, I think that this is due to the sifting of the wind. Hence, if we wish to compare the sand dunes of two different countries as to the size of grain, the samples, in order to be comparable, must be taken under the same conditions, preferably from the youngest layers of the young growing dunes. If the samples are taken from the deeper layers or from old dunes, it may happen that they contain sand which has been more or less exposed to sifting by the wind. Unfortunately, however, in comparing the Italian and the Jutland dunes as to the size of grain, I cannot base my comparison on a material which complies with these requirements, since the **samples**

BOTANICAL STUDIES IN THE MEDITERRANEAN REGION

which I have collected for this purpose from the sand dunes of Jutland have not yet been examined. Hence I must here merely base my comparison of the sand dunes of the two countries on the average of samples taken in various parts of the two areas, from old as well as young dunes, and from the sandy beach.

Warming (1909) has a determination of the size of grain in 16 samples of sand from the west coast of Jutland. One of the samples No. 9 (Søndervig) I have to omit on account of an error in the percentage figures which, when added, make only 90 instead of 100. The remaining samples fall into two groups, viz. from northern Jutland (samples 1–8) and from Fanø (S. Jutland) (samples 10–16). In Table 29, h–k, I have given the

TABLE 29
Comparison between the size of the sand grains from Italian and Danish dunes

	mm.	Percentage.			
		0·25	0·5	1	2
a. Italian dunes: between the low- and high-water mark (Table 28, Nos. 7 and 14)	4	93·5	2·5	..	..
b. Italian dunes: young dunes (Table 28, Nos. 2, 4, 5, 8, 9, 21, and 22)	11	86	3	..	..
c. Italian dunes: old dunes (Table 28, Nos. 6 and 16)	1·5	78	20·5	..	..
d. Italian dunes: the beach (Table 28, Nos. 1, 3, and 20)	10	70	20	..	..
e. Italian dunes: sand-walls (Table 28, Nos. 11, 12 and 15)	1	16	81	2	..
f. Italian dunes: denuded areas (Table 28, Nos. 13, 17, and 18)	2	44	34	19	1
g. Italian dunes: Table 28, Nos. 1–23 in all	7	67	23	3	..
h. Northern Jutland (Nos. 1–8 by Warming)	44	53	2	1	..
i. Fanø (Nos. 10–16 by Warming)	51	48	1	..	..
k. Northern Jutland and Fanø together	47	50	2	1	..

average for northern Jutland, for Fanø and for both. In Fanø 51 per cent. of the sand consists of grains with a diameter below 0·25 mm.; the corresponding figure for northern Jutland is only 44. If we compare with this the Italian dunes, it proves that here we have a much lower percentage of sand with a size of grain below 0·25 mm.; the average of all the 23 samples from the Italian dunes shows only 7 per cent. of sand with a size of grain below 0·25 mm. (Table 29, g), and even the recent dunes taken separately only show 11 per cent. of sand with this size of grain (Table 29, b). As regards the amount of sand with a size of grain below 0·25 mm. none of the Italian samples approximate to the average of the Jutland samples. The sample approaching most nearly to it is the sand from Selva vetere (Table 28, No. 22), the dunes of which show 34·6 per cent. with grains below 0·25 mm. Out of the 23 Italian samples 17 have only 0·5–4 per cent. of sand with a size of grain below 0·25 mm., while the samples from Jutland all show higher figures.

In addition the percentage amount of sand with a size of grain of 0·5–1 mm. is much higher in the Italian than in the Jutland sand dunes, viz. 23 as against 2.

On the whole there can be no doubt that as a rule the sand is more coarse-grained in the Italian than in the Jutland sand dunes. The difference, however, is hardly so great as to be of any importance with respect to the power of the sand to retain the water in the two areas.

I. THE TUSCAN MAREMMA FROM THE LOWER COURSE OF THE ARNO TO LEGHORN

On both sides of the lower course of the Arno there occur extensive alluvial formations. South of the Arno the alluvial land only extends to Leghorn, but stretches far inland; on the north it becomes quickly narrower, but, according to the map, extends nearly as far as Spezia. I myself have seen only the stretch between the mouth of the Arno and Leghorn.

Between Pisa and Leghorn the land is quite flat; near Pisa the soil is cultivated, vegetables especially being grown, but soon large stretches of uncultivated land extend on both sides of the railway, with a close vegetation of herbaceous plants, especially Glumiflorae. In many places the ground is swampy and covered with water in the wet season: Palude Maggiore east of the railway is such an extensive swampy stretch. Here and there a vegetation of *Phragmites communis* and *Typha* is seen. Towards the station of Tombolo, about half-way between Pisa and Leghorn, the land is slightly higher in spots, with small woods of Conifers (*Pinus Pinea, P. Pinaster*) with evergreen and deciduous broad-leaved trees; on the west is seen the wood of the Maremma on the belt of old dunes nearer the sea. Past Tombolo, east of the railway, large woodless partly swampy stretches, Padula di Stagno, are again seen. A little before Leghorn, between the town of Stagno and the sea, the railway passes the canals leading from the Maremma swamps to the sea.

I had occasion to study the partly wooded country between the mouth of the Arno and Leghorn on some excursions at the close of December 1909 and the beginning of January 1910. On one of these excursions my son and myself followed the coast southward from Marina di Pisa at the mouth of the Arno to a point about half-way between the Arno and Leghorn. Here notices warning us of military gun-practice stopped our further progress along the shore, so we turned inland, traversing the Maremma wood and with some difficulty passing the swampy lower tract, largely covered with water, in which Fosso il Lamone has its course. Then we wandered southward through the old sand dune area east of Fosso il Lamone, covered with wood and scrub, till we were stopped in the evening by the canals that come from the Maremma swamps and fall into the sea at Foce Calombrone. We then followed the canals eastward until late in the evening we reached the bridge carrying the Pisa–Leghorn Railway over the canals; thence, under cover of the night, we followed the rails to Leghorn.

On another excursion we examined the sand-dune area south of Marina di Pisa. A third excursion started from the station Tombolo on the Pisa–Leghorn Railway and had for its goal the eastern part of the Maremma wood, east of the place where, on our first excursion, we had turned southward after passing Fosso il Lamone.

Like our sand-dune area, for instance along the west coast of Jutland, the land between the mouth of the Arno and Leghorn consists of a system of dunes and valleys running parallel to the coast, but the dunes are lower and less rugged, and the valleys or plains between them broader; notably there are three broad longitudinal valleys, each with a primitive drainage canal. Moreover there occur among the dunes partially swampy depressions and shorter valleys; thus in the terrain between Fosso il Lamone and the sea there is, at any rate in some places, a narrow valley covered with herbs and scattered shrubs, while the dune is covered with coppice and low wood. Only on a quite narrow line of dunes, up to 100 metres broad, stretching along the sea, the vegetation has a similar appearance to that of our dunes, consisting of a more or less open growth of herbs and some dwarfy shrubs; then follows a nanophanerophytic vegetation, on the seaward side of scattered *Juniperus macrocarpa*, then a denser growth of this species and especially *Pinus Pinaster*; inland, on the older dunes, the vegetation grows taller and more mixed and passes gradually into the micro-mesophanerophytic Maremma wood of evergreen and deciduous species.

From the sea inward the following belts can be distinguished:

(1) the sandy beach;
(2) the recent dunes, with a *Psamma arenaria*-Formation;
(3) slightly older dunes, with a richer and more varied vegetation;
(4) older dunes with a *Juniperus–Pinus*-Formation, transitional to
(5) a *Pinus Pinaster*-Formation.
(6) Maremma wood on old sand-dune area.
(7) Low woodless, partly swampy, tracts.

The sandy beach. Immediately south of the seaside resort Marina di Pisa at the mouth of the Arno, the flat barren beach is quite narrow, only about 5 metres wide; farther south it grows quickly broader and probably averages 30–50 metres; in two places where I measured it, it was 55 and 40 metres. The sand is greyish brown with a number of small mussels and other shells, e.g. of *Solen*: in addition there occur some of the peculiar spherical *Posidonia* balls met with in so many places on the coasts of the

Mediterranean. The sand is loose and the going is heavy except near the water; at a certain but varying depth, however, the ground suddenly gets quite firm. In most places the beach passes smoothly into the sea, but in some places there are short stretches of low sand-bank, the sea washing away portions of the sand here and there.

The most recent dune area forms no continuous belt but consists of a quite irregular system of small dunes, which as a rule do not rise more than 3-4 metres above the beach. This area, which is covered with a *Psamma arenaria*-Formation, was in three places 35, 40, and 45 metres broad. Fig. 154 shows the seaward dune area looking northward along the coast; the photograph was taken about 3 km. south of Marina di Pisa, which is seen close by the sea on the horizon. The sea is seen on the extreme left, then follows the dune area characterized by its vegetation of *Psamma*, which on the extreme right gradually passes into the nano-phanerophytic *Juniperus macrocarpa*-Formation and thence into a taller vegetation of *Pinus Pinaster* with *Juniperus macrocarpa*. Fig. 155 shows the extreme seaward edge of the vegetation: in the foreground is seen washed-up material of various kinds, amongst other things *Posidonia* balls; then a narrow strip characterized by a scattered growth of *Euphorbia paralias*, then low *Psamma* dunes.

About 10 metres inside the extreme edge of the vegetation I made an investigation of the *Psamma*-Formation by means of my formation-statistical method, investigating at definite intervals 50 circular samples of 1/10 sq. metre in size. For each sample the species occurring in the sample were noted. Only 3 species occurred in the 50 samples, viz. *Psamma arenaria* in 41, *Euphorbia paralias* in 15, and *Eryngium maritimum* in 6. To arrive at the percentage frequencies of the samples these figures must be multiplied by 2, since only 50 samples were examined. Consequently the frequency percentages of the species found here are respectively 82, 30, and 12 (cf. Table 30, A): thus this formation is very poor in species. It is likewise, as will be seen from Fig. 156, very open. This fact appeared also from the formation statistics, 8 per cent. of the samples being devoid of vegetation.

When leaving the beach and entering the older sand-dune terrain, the vegetation gradually grows richer in species. Table 30, B, shows the result of the statistical analysis about 35 metres inside the extreme margin of vegetation. In these samples there occurred in all 8 species, but the vegetation as a whole was fully as open as in A, as much as 20 per cent. of the samples being entirely without vegetation. Somewhat farther inland *Juniperus macrocarpa* has begun to invade the *Psamma* dune, and another few metres eastward *Juniperus* and *Pinus* are entirely dominant.

Table 30, C, shows the state of things about 50 metres inside the beach but still on the same line from which the samples of Table 30, A and B, were taken. Here the vegetation was denser with plants in all samples.

FIG. 154. The dune area *c.* 3 km. south of Marina di Pisa which is seen in the background (on the sea). To the right, wood of *Juniperus* and *Pinus*.

FIG. 155. Dune area *c.* 2 km. south of Marina di Pisa; looking northward along the seaside of the outermost dunes with *Euphorbia Paralias*-Formation; inside this the *Psamma* dune.

FIG. 156. Dune area *c*. 3 km. south of Marina di Pisa, seen from the seaside dune towards south-east. In front: seaside dune with *Psamma*; then old dunes with low wood of conifers, first *Juniperus*, then *Pinus*.

FIG. 157. Tenuto del Tombolo west of Tombolo railway station. The easternmost of the three large longitudinal depressions going north–south through the Maremmas of Pisa-Leghorn. In front water with tufts of *Juncus acutus* and *J. maritimus*; to the right higher ground with maquis: here *Quercus Ilex*. To the left on a higher level wood of *Pinus Pinea* with maquis.

Juniperus has begun to immigrate and has already a frequency percentage of 38; *Psamma arenaria* is decreasing, while a couple of chamaephytic species, *Helichrysum italicum* and *Dorycnium hirsutum*, are very conspicuous.

TABLE 30

Frequency of the species in formations on the seaside dunes south of Marina di Pisa. A, 10 metres; B, *c.* 35 metres; and C, *c.* 50 metres within the outermost limit of vegetation. 50×0·1 sq. metres.
A with 8 per cent. of the samples, B with 20 per cent. of the samples, C with no samples, without plants

	Life-form.	F%		
		A	B	C
Juniperus macrocarpa S. et S.	N	..	..	38
Sorghum hirsutum Ser.	Ch	..	28	72
Euphorbia Paralias L.	Ch	30	16	10
Helichrysum italicum (Roth) G. Don.	Ch	..	..	42
Polygonum maritimum L.	Ch	..	2	..
Eryngium maritimum L.	H	12	10	2
?Hypochoeris radicata L.	H	..	..	12
Silene Otites Sm.	H	..	2	12
Solidago Virga aurea L.	H	..	20	..
Psamma arenaria R. et S.	G	82	66	60
Echinophora spinosa L.	Th	..	4	4
Density of species	..	1·2	1·5	2·5
No. of species	..	3	8	9

In addition to the species recorded in Table 30 the following were observed in the *Psamma* dune: *Cynodon Dactylon, Seseli tortuosum, Medicago marina, Cakile maritima, Odontites lutea, Carthamus lanatus, Medicago litoralis,* and in one or two places *Tamarix gallica.*

The Maremma wood on the comparatively low belts of old dunes running parallel with the coast is composed of a mixture of evergreen and deciduous Mesophanerophytes. The wood is now fairly dense, now more open, with a rich undergrowth of Micro and Nanophanerophytes. The deciduous Mesophanerophytes are *Alnus glutinosa,* and *Quercus pedunculata,* further *Ulmus campestris.* The evergreen Mesophanerophytes include especially *Pinus Pinaster, Pinus Pinea,* and *Quercus Ilex* (up to *c.* 20 metres high); further *Smilax aspera* and *Hedera Helix,* which occur very often as Mesophanerophytes and grow as high up as the crowns of the taller trees. They contribute essentially to the peculiar character of the wood.

In the undergrowth various forms of *Phillyrea variabilis* and *Myrtus communis* play an important part; further *Juniperus macrocarpa, Cotoneaster pyracantha, Erica scoparia, Daphne Gnidium, Lygeum Spartum, Lonicera implexa, Cistus salvifolius, Helianthemum Chamaecistus,* and

Rubus thyrsoideus. Of herbaceous plants the following were noted: *Rubia peregrina, Lactuca muralis, Brachypodium silvaticum, Dianthus carthusianorum,* and *Sedum sexangulare.*

The woodless depressions and the Maremma swamps. The lowest areas, which occur especially in the southern part, were, in January, more or less covered with water, from which projected *Juncus acutus, Juncus maritimus, Saccharum Ravennae, Scirpus Holoschoenus,* and other species. On the parts not inundated the following species were noted: *Agrostis alba, Althaea officinalis, Brunella vulgaris; Centaurea Jacea, Chlora perfoliata, Cynodon Dactylon, Euphorbia pubescens, Inula viscosa, I. crithmoides, Limonium vulgare, Linum maritimum, Medicago lupulina, Mentha aquatica, Odontites lutea, Plantago Coronopus, Samolus Valerandi* (inundated soil), *Salicornia herbacea, Suaeda maritima, Schoenus nigricans,* and *Tamarix gallica.* Here and there, on higher soil, grew a coppice of *Phillyrea* and *Myrtus* with *Daphne Gnidium, Cistus salvifolius, Erica scoparia,* and other species.

I have made no statistical analysis of the formations here and shall not attempt any general description of the vegetation. A glance at the photographs (Figs. 157–9) will give the reader a better impression of the general character of the Maremma than any lengthy descriptions. The pictures are all taken from the part west of Tombolo, about half-way between Pisa and Leghorn.

Fig. 157 shows the view looking south over the most easterly of the three large longitudinal depressions stretching from north to south through the Pisa–Leghorn Maremma; in the foreground inundated soil with tussocks of *Juncus acutus* in the middle, and a *Juncus maritimus* vegetation on the left; on slightly higher soil in the right background is seen maquis with *Quercus Ilex*; on the left, on the higher ground, the western edge of the eastern strip of wood is seen, consisting in great part of *Pinus Pinea* with *Quercus Ilex.*

Fig. 158 is from the same depression as the preceding picture. In the foreground are *Juncus maritimus,* &c., then small groups of *Phillyrea, Myrtus,* and other maquis shrubs; in the background wood consisting of *Quercus Ilex* and *Pinus Pinea.* Fig. 159 is a partly inundated depression farther westward. In the foreground are water, coppice of *Phillyrea and Erica,* then *Quercus Ilex*; on the left *Juncus maritimus*; farther back *Phillyrea* coppice, and then wood consisting of *Pinus Pinea, P. Pinaster, Quercus Ilex* and, especially in the lower parts, *Alnus glutinosa.*

Whether the Cascine, the well-known pleasure-ground or park along the Arno, west of Florence, is entirely planted or consists partly of remains of the original phanerophytic vegetation, I do not know; but it is highly reminiscent of certain parts of the Maremma wood, and I do not doubt that the lowland tracts on either side of the lower course of the Arno were

FIG. 158. Near Fig. 157. One of the depressions partly inundated during the winter. In front water with *Juncus maritimus*, then scrub, 1·5–4 metres high. of *Erica, Phillyrea, Lygeum, Smilax aspera, Juniperus macrocarpa*, &c.

FIG. 159. Near Figs. 157 and 158. In front water and scrub of *Phillyrea* and *Erica*; to the left *Juncus maritimus*. In the background wood and scrub of *Quercus Ilex, Phillyrea, Pinus Pinea, P. Pinsapo, Alnus*, &c.

once covered by similar woods, which have been cut down to give place to extensive horticulture around the towns, and fields and pasture farther out. It is chiefly due to the malaria mosquito *Anopheles*, to which the Maremma swamps are a regular Eldorado, that the Maremma wood has not long since been consumed by fire like the other woods. Still, in the winter months *Anopheles* cannot protect the wood, and its disappearance is probably only a question of time, if more interest is not taken in silviculture in Italy in the future.

II. STRAND FORMATIONS AT ARGENTARIO

The strand formations we are here going to consider consist especially of the two narrow spits of land connecting Argentario with the mainland. Argentario, it is true, is not nowadays an island, but there is no doubt that it has once been an island which has only comparatively recently become connected with the mainland by the two low spits of sand, Tombolo della Gianella in the north and Tombolo di Feniglia in the south. Together with the coast of Italy and Argentario these sand spits enclose a lake or lagoon called Stagno di Orbetello. A little south of the middle of the eastern side of this lake a narrow peninsula extends from the coast of Italy about 3 km. into the lagoon, and on the extreme point of this peninsula stands the small town of Orbetello, from which a dike with a bridge has recently been carried south-westward to Argentario, which the pedestrian can now reach in three ways. Orbetello is very conveniently situated as a starting-point for excursions to Argentario and the above-mentioned spits of sand, so during my visit here for some days in January 1910, I chose this town as my place of sojourn.

Orbetello is a small town with old primitive ramparts, running right down to the water for long stretches. It is a quiet town outside the usual tourist routes, and an ordinary tripper will hardly find much there to interest him. But any one who has a love of Nature, and who has spent a long series of happy days studying the comparatively undisturbed vegetation to which one can gain access in this part of the country, will not easily forget Orbetello and the country around its lagoons.

Looking westward from Orbetello, there rises straight in front of the observer Argentario's beautiful island-like group of mountains, on the slopes of which great stretches are covered with maquis. Looking northwestward and southward, the still water of the lagoon is bounded in the distance by the low narrow strips of land whose vegetation it was my special object to study. When there was a wind, it was bitterly cold, but most of the time the weather was quiet and fine. On calm dark nights a mass of lights were seen on the lagoon, the flares on the fishing-boats.

In addition to the sand-dune area of Tombolo della Gianella and Tombolo di Feniglia I studied the sand dunes east of the mountain group Ansedonia, as far as Lago di Burano.

Tombolo della Gianella. From Torre S. Liberata on Argentario the coast-line forms a faintly concave curve north-north-eastward to the mouth of the Albegna. The dune area is low and narrow, narrowest at Argentario, becoming slightly wider towards the Albegna. While on Tombolo di Feniglia there are large tracts devoid of vegetation, with travelling dunes, the dunes of Tombolo della Gianella are all covered with a low maquis. Judging by the direction in which the dunes travel on Tombolo di Feniglia the south-west wind must be assumed to be the prevailing wind here, the direction of the dunes being mostly from the south-west to the north-east. Hence the reason why the dunes of Tombolo della Gianella are low and covered with vegetation must probably be that Argentario's mountain protects them against the south-west and partly also against the west winds.

Fig. 160 shows the view of Tombolo della Gianella from the heights at Torre S. Liberata; on the left the sea, the coast, and the dunes covered with shrubs; on the right Stagno di Orbetello and in the background the mountains to the east.

A little east of Torre S. Liberata, proceeding inland towards Stagno from the sea, the succession is as follows:

1. A sandy beach, 12–15 metres wide, whose inner part is formed by the highest tide, which is somewhat more than 0·5 metre higher than the daily high tide. This part is covered with washed-up portions of plants, among which is a large amount of *Posidonia* balls and fragments of *Posidonia* shoots. Among these washed-up fragments of plants on the highest part of the beach we meet with the first vegetation, consisting of a little *Psamma arenaria*, *Agropyrum junceum*, and some few individuals of *Inula crithmoides*, *Anthemis maritima*, and *Eryngium maritimum*.

2. A dune area, 10–20 metres wide; on the seaward side *c*. 0·5 metre higher than the sandy beach, sloping up inland more or less regularly towards the crest of the dune. Inside this follows flat cultivated land, probably old levelled dune nearest the wall of dunes. The dune as a whole is only *c*. 2 metres higher than the highest tide-mark, though here and there it has been blown up into somewhat higher accumulations. The vegetation is nanophanerophytic maquis, a *Juniperus*-Formation (Fig. 161) on the most seaward, lower part, consisting of the Phanerophytes and Chamaephytes enumerated in Table 31, A. Here are a number of grasses, such as *Psamma arenaria* and *Dactylis glomerata*, and some rosettes of dicotyledonous plants. In addition to the Chamaephytes recorded in Table 31, A, there occurred some few specimens of *Crithmum maritimum* and *Inula crithmoides*.

A little farther inland, on older and somewhat higher ground, the maquis consisted of a *Rosmarinus officinalis–Erica multiflora*-Formation as recorded in Table 31, B; besides the species given there, *Asparagus acutifolius* was found, and some few herbs.

FIG. 160. Tombolo della Gianella, seen eastward from the heights at the western end of the spit. From left to right are seen: (1) the sea, (2) beach, (3) dune area with nanophanerophytic scrub, (4) level cultivated area, (5) Stagno di Orbetello. In front *Olea* scrub, *c.* 1 metre high.

FIG. 161. Tombolo della Gianella. From left to right are seen: (1) sea, (2) beach, (3) somewhat higher sandy ground with maquis of *Juniperus, Erica multiflora, Rosmarinus, Phillyrea, Smilax aspera, Psamma,* &c., (4) narrow dune wall with maquis: the land inside this is level and cultivated. In the background are seen some few Stone Pines, remnants of the Maremma wood now destroyed by cultivation.

The scrub about 2 metres high, covering the landward side of the dunes, consisted especially of *Quercus Ilex* and *Phillyrea variabilis*, up which *Smilax aspera* climbed. Fig. 161 shows the oblique contour of the top of this copse formed by the wind.

TABLE 31

The frequency of Phanerophytes and Chamaephytes in three analyses of the dune maquis on Tombolo della Gianella (cf. text) 50×0·1 sq. metres

	Life-form.	F%		
		A	B	C
Asparagus acutifolius L.	N	10	..	..
Cistus monspeliensis L.	N	..	..	10
„ salvifolius L.	N	6	22	70
Daphne Gnidium L.	N	12	4	4
Erica multiflora L.	N	38	38	46
„ scoparia L.	N	..	2	..
Juniperus macrocarpa S. et S.	N	96	16	..
Lonicera implexa Ait.	N	28	6	..
Myrtus communis L.	N	..	12	2
Phillyrea variabilis Timb.	N	2	18	10
Pistacia Lentiscus L.	N	..	10	..
Quercus Ilex L. (here N.)	N	6	2	..
Rosmarinus officinalis L.	N	14	42	60
Smilax aspera L. (here N)	N	6	22	2
Spartium junceum L.	N	4	2	..
Dorychnium hirsutum Ser.	Ch	2	2	..
Helichrysum italicum (Roth) G. Don.	Ch	12	28	..
Medicago marina L.	Ch	2	..	..
Ruscus aculeatus L.	Ch	..	2	..
Sedum sp.	Ch	..	8	..
Density of species		..	2·4	2·4	2
No. of species		..	14	17	8

The *Psamma arenaria*-Formation of which there is only a hint near Argentario, becomes more marked as one proceeds towards the Albegna, so that a *Psamma* dune is developed, low it is true, but distinct. Inside this comes a small valley, and next, the older dune area covered with the maquis.

About half-way between Argentario and the Albegno there occurred:

1. A sandy beach, 10–12 metres wide, with a high-water mark of washed-up parts of plants in its highest part.
2. *Psamma* dune, 7–8 metres broad and *c.* 1 metre high, with *Psamma arenaria*, *Sporobolus pungens*, *Euphorbia Paralias*, and some few young specimens of *Juniperus*.
3. Dune valley, *c.* 8 metres broad, with scattered young *Juniperus*.
4. A *Juniperus*-Formation, 3–4 metres broad, on older dunes.

5. Nanophanerophytic maquis on old dunes, 100–150 metres broad.

6. Cultivated land with a few Stone Pines.

The maquis (No. 5) was more meagre and poorer in species than nearer Argentario, but here too *Rosmarinus officinalis* and *Erica multiflora* were frequent, to which was added *Cistus salvifolius*, very much the most frequent species over large stretches. In a spot like this I determined the frequency percentage of the Phanerophytes and Chamaephytes. The result is recorded in Table 31, C, from which it is seen that it is a *Cistus salvifolius–Rosmarinus officinalis*-Formation with *Erica multiflora*.

Farther on, towards the mouth of the Albegno, there is no vegetation of shrubs on the seaward side of the low dunes, nor does there ever seem to have been any in this place. Farther inland, however, around the northern part of Stagno di Orbetello, there has probably been a Maremma wood and Maremma maquis, remains of which are still found. The fate of this phanerophytic vegetation may be gathered from Fig. 162 taken near the Albegno and showing large heaps of charcoal. At the landing stage on the left are seen the small vessels which will carry it away, and behind the house the remains of the wood.

Around the Albegno, where the wood has long since been cut down, the higher levels of the more or less clayey soil had once been cultivated, but were now laid out as pasture, partially covered with *Inula crithmoides*, *I. viscosa*, *Salicornia macrostachya*, and other species of the kind growing on uncultivated soil. The lower areas, which showed no sign of having ever been cultivated, were covered entirely with a succulent vegetation consisting especially of *Chenopodiaceae*, viz. *Obione portulacoides* (frequency percentage 80–100), *Salicornia macrostachya* and *S. fruticosa*; further *Artemisia coerulescens* (?), *Inula crithmoides*, *Statice Limonium*, some few tufts of *Juncus acutus*, and some few *Tamarix*; here and there a grass carpet.

Tombolo di Feniglia. While the sand-dune area of Tombolo della Gianella is quite narrow and low and entirely covered with maquis, conditions are quite different on Tombolo di Feniglia. Here there is a comparatively broad irregular belt of dunes, consisting partly of more or less wind-blown stretches with scattered sand-hills with maquis, and for the rest entirely without plants or with a very open vegetation; and partly of moving dunes devoid of vegetation, which advance from the south-west towards the north-east and are gradually burying the Maremma maquis and Maremma wood growing on the more or less clayey soil along the southern side of Stagno di Orbetello. Fig. 163 will perhaps give the state of things more clearly. It shows Tombolo di Feniglia viewed from a group of mountains on Argentario. From right to left will be seen:

A, the sandy beach of the sea with low dunes; B, a broad partly wind-

FIG. 162. The north-eastern end of Tombolo della Gianella. Charcoal loading just south-west of the mouth of Albegno. Round the house large heaps of charcoal, i.e. the remains of the wood.

FIG. 163. Tombolo di Feniglia seen from the abutting heights on Argentario (to the east of Port Ercole). To the right the sea, to the left Stagno di Orbetello.

FIG. 164. South coast of Tombolo di Feniglia looking westward (towards Argentario). From left to right: (1) the sea, (2) beach with *Posidonia* balls along the high-tide line, (3) storm breakwaters with wreckage and many *Posidonia* balls; then (4) a quite low wall of sand c. 1 metre higher than the high-tide line, with *Euphorbia Paralias*-Formation (see Fig. 165).

FIG. 165. South coast of Tombolo di Feniglia seen eastward; from the right to the left: (1) the sea, (2) beach with a dark zone of *Posidonia* balls on storm breakwaters, (3) *Euphorbia Paralias*-Formation on a low dune area, (4) flat depression with *Euphorbia Paralias* and a little *Psamma*, (5) somewhat higher dune area with *Juniperus*, &c.

blown area consisting of larger or smaller sand-hills or groups of hills separated by sand-levels denuded by the wind; C, a line of almost bare shifting dunes, moving from the south-west to the north-east towards Stagno di Orbetello and burying the Maremma wood and Maremma maquis growing along this; the picture shows clearly how the phanerophytic vegetation intrudes into the sand-hills; D, low, partly more or less clayey, partly sandy soil with Maremma maquis and, especially towards the east, Maremma wood; E, the shore of the lagoon.

I shall now discuss the different sections.

A. The Beach and Dunes (see Fig. 164). On the western part of the south coast and in the main also farther eastward, the following zones were observed:

1. Within the daily high-water mark a slightly rising sandy beach, c. 18 metres wide; to seaward with thrown-up fragments of plants, especially of *Posidonia*, to landward *Posidonia* balls; some blown sand; here and there wind-ripple marks.

2. Spring-tide belt, c. 8 metres wide, rising about 1 metre higher than the preceding terrain, with thrown-up sticks, wreckage, and many *Posidonia* balls more or less buried in blown sand; some few young individuals of *Euphorbia Paralias* were seen in this belt. Nowhere else have I ever seen such large quantities of *Posidonia* balls. They occur in greatly differing sizes and stages of development. The shore all along the bay between Argentario and Ansedonia, especially towards Ansedonia, was covered with fragments of *Posidonia*, making the sea in some places look very much like barley soup, or a pulpy mass consisting partly of recently broken-off *Posidonia* shoots and partly of shoots more or less worn or torn by continual friction against the sand or each other, till at last there was only a very small fragment, prickly from the free ends of the strands of strengthening tissue. Besides these fragments the water contained immense quantities of shorter or longer needle-shaped fragments of these strands of hard tissue. The formation of the balls presumably takes place in the following way. The small fragments of the shoots, with their spikes of the free ends of strengthening tissue, being rolled backwards and forwards with the mass of free needle-shaped fragments of the resistant tissue, an increasing number of these are caught among the needles of the rolling fragments, which at last become quite covered with the entangled and thrust-in needle-shaped fragments of strengthening tissue. To begin with, the balls are generally lenticular-oval; they are washed backwards and forwards at the water's edge, with now one side, now the other, turned downwards, and so take up fresh needles, principally on the flat sides, which thus become more and more rounded until an oblong rounded body, oval in longitudinal section, results. Owing to its shape this body will mostly roll at right angles to its longitudinal axis and so chiefly take up needles in a belt round its middle, thus approaching more and more to the spherical

shape, which it finally attains if it is allowed to lie rolling at the water's edge for a sufficiently long time. Most of the balls thrown up on to the beach are spherical or sub-spherical, but fine large balls are met with which are oblong with an oval longitudinal section, and of course the thrown-up masses contain forms representing the various stages of development.

3. A low range of dunes or wall of sand, *c.* 20–25 metres wide, generally only *c.* 1 metre higher than the spring-tide belt, a little lower on the landward side, so that a slight depression, a valley, occurs between the low dune and the older terrain.

The vegetation of the dunes is an open *Euphorbia Paralias*-Formation (Fig. 165), corresponding to the front line of vegetation on the dunes between the Arno and Leghorn. In addition to *Euphorbia Paralias* the vegetation consisted chiefly of *Sporobolus pungens* with further scattered individuals of *Diotis maritima, Agropyrum junceum, Psamma arenaria, Echinophora spinosa, Pancratium maritinum, Medicago marina, Eryngium maritimum,* and *Crucianella maritima*. The statistical results for two localities will be seen in Table 32, showing that the Chamaephytes and Geophytes preponderate—if *Sporobolus* can be regarded as a Geophyte: I am not quite sure that it does not survive the unfavourable season as a Chamaephyte.

TABLE 32

Two analyses of *Euphorbia Paralias*-Formation on Tombolo di Feniglia (cf. text). 50×0·1 sq. metres

	Life-form.	F%	
		A	B
Diotis maritima Sm.	Ch	4	..
Euphorbia Paralias L.	Ch	**48**	**48**
Medicago marina L.	Ch	4	2
Eryngium maritimum L.	H	..	2
Agropyrum junceum (L.) Beauv.	G	..	6
Pancratium maritimum L.	G	4	4
Psamma arenaria (L.) R. et S.	G	2	..
Sporobolus pungens Kth.	G(–Ch)	**58**	**52**
Echinophora spinosa L.	Th	2	..
Density of species	..	1·2	1·1
No. of species	..	7	6

As will be seen, *Psamma* is only just represented. In any case there is no real *Psamma* dune; only in a single place, about the middle of Fig. 165, did there occur a single *Psamma* dune which, however, was considerably higher than the part covered with the *Euphorbia Paralias*-Formation. This will illustrate the remarkable power of the *Psamma* to accumulate sand and form dunes as compared with the other species occurring here.

It is not easy to say what is the cause of the scarcity of the *Psamma* here. But since there is no *Psamma*-Formation or similar agent to accumulate and fix the sand along the sea, the wind carries the sand landwards and 'shifting dunes' are formed.

B. The rather wide, partly denuded dune area lying inside the low range along the shore consists of very irregular and heterogeneous ground: partly of scattered sand-hills covered with scrub; partly of more or less deeply denuded stretches with a very open vegetation principally of herbs, or destitute of vegetation; partly of terrain where sand is now more or less abundantly deposited and accordingly either devoid of vegetation or with an open vegetation especially of herbs.

1. The shrub-clad sand-hills are hardly the remains of a previously continuous shrub-clad dune area. They are more probably a further development of such isolated *Psamma* dunes as were mentioned above (Fig. 165). Owing to the great power of the *Psamma* to fix the sand, these dunes have not been subject to such violent and sudden changes as the rest of the area, hence shrubs have been able to take root here and prevent the wind from breaking up the dunes. The shifting dunes have travelled past them and have even contributed to their growth. Here and there, however, the wind may sometimes have got the better of such a shrub-clad sand-hill and razed it. At any rate sand-hills are seen here and there with only one bush at the summit while the sides, more or less steeply cut away by the wind, show a tangle of denuded roots protruding in all directions for a length of several metres.

The shrub-clad sand-hills only rise about 1–4 metres above the surrounding terrain, in most cases *c*. 2 metres. Their circumference is as a rule 15–50 metres. The scrub is formed of *Juniperus* (*macrocarpa*?), *Pistacia Lentiscus, Myrtus communis, Phillyrea variabilis,* and *Rosmarinus officinalis,* here and there with *Clematis* (*flammula*?), *Erica multiflora, Daphne Gnidium,* and *Smilax aspera.* None of them grow higher than Nanophanerophytes. *Juniperus* occurs farthest seawards and is often the only shrub on the front line of the sand-hills. On such a *Juniperus* dune the summit was in one case covered with *Psamma*, which corroborates the conjecture that these sand-hills have begun as *Psamma* dunes. Farther landward, too, some few dunes may be covered with a single species: *Myrtus, Pistacia, Phillyrea,* or *Rosmarinus.*

On 16 sand-hills more closely studied the following conditions were observed. Three were covered with *Juniperus* alone, 3 with *Pistacia*, 1 with *Phillyrea*, 1 with *Myrtus*, 1 with *Rosmarinus*, and 7 with 2–5 of the species mentioned. *Pistacia* was found on 9 in all, *Juniperus* and *Myrtus* on 7, *Rosmarinus* on 5, *Phillyrea* on 4, and *Clematis, Erica, Daphne,* and *Smilax* each on 1. The larger species, especially *Pistacia* and *Myrtus*, can work their way up through the sand if the accumulation of sand is not too overwhelming. Among the low species the same is the case, at any rate with *Rosmarinus* and *Erica*, though like *Daphne Gnidium* they principally occur where there is but little accumulation.

2. The more or less deeply denuded parts have their greatest extension farther inland between the north-western extremities of the large shifting dunes, but in some places they extend between the sand-hills right down to the range of lower dunes (see above, A. 3). Rather long stretches have been denuded by the wind right down to a firm, more or less dark base, almost on a level with the upper part of the spring-tide belt. Just behind the lower range these areas have a very scattered vegetation of species from the lower dunes: *Euphorbia Paralias*, *Crucianella maritima*, *Sporobolus pungens*, and *Diotis maritima*, all being able to form small dunes. Farther inland other species are added, e.g. *Eryngium maritimum*, *Agropyrum junceum*, and *Polygonum maritimum* (forming dunes). In addition a quantity of seedlings of Therophytes occurred here. A large quantity of withered shoots of *Echinophora spinosa*, as well as the withered shoots of *Eryngium maritimum*, might form small dunes.

Still nearer Stagno di Orbetello, especially between the easternmost and next easternmost of the large dunes, the vegetation of the extremely denuded area passed into a vegetation of shrubs forming the maquis of the level terrain along Stagno di Orbetello. In addition to the dune plants mentioned above and *Juniperus*, there occurred here a more or less scattered vegetation of *Pistacia*, *Myrtus*, *Phillyrea*, *Rosmarinus*, and *Erica multiflora*. A similar vegetation, but in an earlier stage, was seen to the west of the north-easternmost large dune. In this connexion I may mention a corresponding area at the western extremity of Tombolo di Feniglia in the shelter of Argentario. Along the base of the mountain group of Argentario there is a narrow tract of cultivated ground; then follows a low dry and sandy area, probably a dune area long since denuded by the wind, and in the course of time invaded by a meagre maquis, an *Erica multiflora*-Formation, the composition of which may be seen in Table 33. This vegetation was quite low, not more than 15–50 cm. In other more shel-

TABLE 33

Analysis of a very poor maquis at the west end of Tombolo di Feniglia (the life-form given to the species are those here presented by them)

	Life-form.	F%
Cistus monspeliensis L.	N	8
„ salvifolius L.	Ch	28
Erica multiflora L.	Ch–N	96
Helianthemum Fumana Mill.	Ch	48
Rosmarinus officinalis L.	Ch–N	88
Teucrium montanum L.	Ch	4
Undetermined Chamaephyte (Polygala sp. ?)	Ch	2

FIG. 166. Tombolo di Feniglia. The partly denuded shifting dune with two elms 8–10 metres high, having their roots laid bare. In the background the wood on the border of the dune.

FIG. 167. South side of Tombolo di Feniglia: dunes kept together by *Phillyrea*. *Euphorbia Paralias*, *Eryngium*, &c., on the sand between the dunes.

tered spots the maquis was higher, 1–1·5 metres and some other species were added, especially *Myrtus, Phillyrea,* and *Pistacia.*

In other places the denudation has not gone so deep. The wind alternately accumulates or blows away the sand at shorter or longer intervals. Certain circumstances show that in places at any rate comparative calm must have prevailed for a long time so that Phanerophytes were able to colonize the dunes and grow up before the sand was again blown away. Thus Fig. 166 shows two Elms, 8–10 metres high, which have developed on an old sand dune but from the roots of which the wind has partly carried away the sand. If the denudation is continued these elms will no doubt die. The irregular crowns of the two elms extend from the south-west to the north-east in the same direction as the dunes are moving, and show the prevailing direction of the wind.

C. The shifting dunes. The shrub-clad dunes and the denuded stretches destitute of any covering of fresh sand only constitute a small part of the whole area, the greater part being covered by blown sand. Where there is but a slight accumulation of sand there is a more or less open vegetation of herbs and some few shrubs, where more sand is accumulated, sand-hills are formed that bury the existing vegetation and are themselves almost or entirely devoid of vegetation.

In a strong south-westerly gale the blown sand rises like a mist above the isthmus and is carried towards Stagno di Orbetello until it is arrested by the maquis and the wood along the southern side of the lake. Here, along the margin of the phanerophytic formation, the sand gradually forms a dune whose windward face slopes gently, whereas the leeward face shows a steep slope down which the sand rolls, thus causing the dune to travel slowly but surely inland, burying maquis and wood on its way.

It is difficult to give, in a few words, a clear description of the very different forms of ground and other conditions here, hence I shall for the most part refer the reader to the photographs, Figs. 167–78, which will give better than many words an impression of the conditions.

Beginning at the sea I follow the dunes inland to Stagno di Orbetello.

Fig. 167 shows a dune area near the sea looking east; there was but little blown sand at the time and the level or gently sloping stretches are, therefore, partially covered with a very scattered vegetation, especially of *Euphorbia Paralias* and *Eryngium maritimum.* On the right are seen a couple of the old dunes with *Juniperus* scrub referred to above; fresh dunes have been deposited round them; on the left a partly destroyed old dune projects above the new dune. The remaining part of the old dune is kept together by the huge root system of the *Phillyrea* growing on the summit: its ragged vane-like crown shows the prevailing direction of the wind to be south-west.

Fig. 168 shows another tract of the dune farther inland with two old dunes. As long as these are not entirely buried in the advancing dune they are partly or entirely separated from it by a depression. The dune to the right is covered with a dense growth of *Clematis*, that to the left is not clad with scrub, but on its windward southern end it has a comparatively large *Quercus* (*Suber*?) with a wry crown. This oak is no survival of the original Maremma wood, for it is standing on old dune ground: the dunes here have probably once been covered at least partially with wood.

Fig. 169 shows conditions still farther inland, along the windward side of one of the large dunes; in the foreground there is comparatively low soil with no great accumulation of sand and hence partly with a vegetation of herbs, especially *Inula viscosa*, further *Euphorbia Paralias* and small individuals of *Juniperus*. From here the ground rises towards the background to a fresh dune covering an old dune. The latter once bore a vegetation which must at any rate partially have consisted of Phanerophytes the crowns of which project above the sand and are still alive. In front, where the covering of sand is thinnest, there occur amongst others *Rubus* (*discolor*?) and *Daphne Gnidium*, which have both escaped choking. Most of the Phanerophytes in the background are *Pistacia Lentiscus* and *Quercus Ilex*.

Fig. 170 shows conditions along the landward side of the north-west end of one of the large dunes, which has here come right out to Stagno di Orbetello seen in the background; the town of Orbetello is also seen in the background and, on the left, a part of Argentario. The dune rises gradually on the right: on the left is seen a large denuded stretch with some few trees (*Quercus Ilex*); in the foreground a low dense growth of *Myrtus communis*, standing on the old dune ground. It has kept pace with the growth of the dune similarly to *Salix repens* on the dunes along the North Sea; a little farther back to the left, a dense growth of *Pistacia* and *Phillyrea*, likewise on the old dune ground, but in the middle of the sand of the new dune, surrounded by a small depression; to the right of it a few bushes of *Phillyrea*, of which only the tops project above the sand, but which are still alive; on the extreme right the Maremma wood, which is gradually being buried by the dune.

Fig. 171. Still nearer Stagno di Orbetello. Here and there the remains of killed shrubs project above the sand. The small bush in the left foreground is a *Rosmarinus officinalis* which is still alive, probably it is growing on the summit of an old dune buried in the shifting dune. In other places, on old, now nearly buried, dunes, is likewise seen living low scrub of *Juniperus*, *Pistacia*, *Clematis*, *Phillyrea*, and *Smilax aspera*. In the far background several *Quercus Ilex*, partly killed by the dune overwhelming the wood.

Fig. 172 shows the north-western end of a dune, here projecting right into Stagno di Orbetello. In the foreground low dunes with *Euphorbia Paralias*, then the edge of the shifting dune, 2–3 metres high; the shrub

FIG. 168. Tombolo di Feniglia. In front a dune covered with *Clematis*.

FIG. 169. Tombolo di Feniglia. In front the old partly denuded shifting dune with herbs, especially Therophytes; towards the background the new bare dune with projecting surviving specimens of *Rubus, Daphne Gnidium*; and, in the background, *Pistacia, Quercus Ilex*, &c., standing on an old dune and now partly buried in the sand of the new one.

Fig. 170. Tombolo di Feniglia. View of Stagno di Orbetello. The town of Orbetello is seen in the background to the right. In front is seen *Myrtus* dune; farther back *Pistacia, Phillyrea, Quercus Ilex*, &c., on an old dune and now partly buried in the new one.

Fig. 171. Tombolo di Feniglia, looking towards north-west and towards the inner side of the copse between the dune area and Stagno di Orbetello. In front the moving sand with partly dead plants; then in depressions and on old low dunes *Juniperus, Pistacia, Clematis, Phillyrea*, and *Smilax aspera*; to the left *Quercus Ilex*, partly killed by the dune advancing and enveloping the wood.

FIG. 172. Tombolo di Feniglia. In the front low dunes with *Euphorbia Paralias*, then a shifting dune 2–3 metres high which has reached Stagno di Orbetello. *Pistacia Lentiscus*, *Rubus*, and other bushes project over the dune. Along the shore of the lake: *Juncus acutus*-Formation, and inside this *Tamarix*.

FIG. 173. Tombolo di Feniglia. A big dune burying the Maremma wood. The vegetation consists of *Phillyrea*, *Pistacia*, *Myrtus*, *Rubus*, *Pteridium*, *Paliurus*, &c.; many detached herbs brought by the wind are seen.

FIG. 174. Tombolo di Feniglia. The maquis being choked by the shifting dune.

FIG. 175. Tombolo di Feniglia. A shifting dune 5–6 metres high burying the maquis. In the centre *Quercus Ilex*.

FIG. 176. Tombolo di Feniglia. *Populus-Ulmus* wood partly buried in a dune 4–5 metres high; to the right an *Ulmus* with large galls.

FIG. 177. Tombolo di Feniglia. Northern border of a large shifting dune with partly buried *Quercus Ilex*, some of which are already killed.

FIG. 178. Tombolo di Feniglia. The border of the easternmost shifting dune 2–3 metres high covering the trunk of a *Pinus Pinea*.

FIG. 179. The coast between Ansedonia and Capalbio looking southwards from the mountain knot of Ansedonia. Inside the beach a low dune wall with *Juniperus* scrub, than a level grass-clad depression. See p. 600.

in front of this is *Pistacia Lentiscus*. Along the shore behind the dune is seen a border of *Juncus acutus*, then *Tamarix*, and then maquis.

The preceding pictures showed the windward face of the dunes, the slope of which is generally gradual, so that the rise as one ascends is not very noticeable. The succeeding pictures show the steep landward face of the dunes where the sand is pouring down between the trees and shrubs of the Maremma wood.

Fig. 173 shows the leeward side of the easternmost of the large dunes. On the right and left the sand has rolled slowly down and filled out the space between the stems of the Phanerophytes, which stand surrounded by sand but still alive. Here the vegetation consisted of *Phillyrea, Pistacia, Myrtus, Rubus, Pteridium aquilinum*, and the deciduous *Paliurus Spina-Christi*, seen in the centre of the picture with fruits on the leafless branches. In one place, in the centre of the picture, a great accumulation of sand has taken place; the dune here projects in a tongue and has quite buried the shrubs; on the surface, projecting above the sand, are seen quantities of withered shoots of dune plants: *Eryngium, Echinophora*, and others, which the wind has carried along together with the sand.

Fig. 174 shows a dune which has entirely buried a long stretch of the maquis. The plants in the right foreground are *Phillyrea, Rubus, Clematis*, and *Pteridium*; in the left background *Phillyrea* scrub partly buried in sand.

Fig. 175 shows the upper edge of the leeward side of a dune 5–6 metres high. The maquis has been partly buried. At the top the crown of a still living *Quercus Ilex* projects above the sand. Shoots of *Clematis* and *Rubus* are growing up in great quantity among the branches of the oak.

Fig. 176 shows part of a Maremma wood, the stems of which have been buried in the dune which is 4–5 metres high. The wood here consisted chiefly of *Populus* and *Ulmus campestris*. On the extreme right *Ulmus* is seen covered with large *Schizoneura lanuginosa* galls, up to 8 cm. long.

Fig. 177 shows the dunes in another place where the partially buried wood consisted chiefly of *Quercus Ilex*, some of which were already dead. The plant on the extreme right is the crown of a *Pistacia* projecting above the sand.

Fig. 178 shows a still living Stone Pine farther landward, buried right up to the crown in a layer of sand 2–3 metres deep. Behind and to the left of the pine is seen scrub of *Myrtus* and *Pistacia* which has kept pace with the growing dune. In the far background is seen a glimpse of the *Populus–Ulmus* wood shown in Fig. 176.

D. The low, partly more or less clayey, partly sandy terrain between the belt of dunes and Stagno di Orbetello is quite narrow towards Argentario, but widens towards the east. Apart from small very damp areas along the lake it is covered with a vegetation of Phanerophytes, partly maquis and partly wood. Though this is an out-of-the-way place and so has not perhaps been so much affected by human activities as parts nearer inhabited places, it cannot by any means be taken for granted that the vegetation is unaffected by cultivation. To the westward the vegetation consists of a maquis 1–5 metres, most often 2–4 metres, high, with a few low trees. This part has evidently once been covered with a taller wood.

Table 34

Two analyses of the maquis along the south coast of Stagno di Orbetello. (Life-form as in Table 33)

	Life-form.	F% A	F% B
Clematis sp.	N	2	4
Daphne Gnidium L.	N	10	14
Lonicera implexa Ait.	N	2	..
Myrtus communis L.	N–M	40	32
Paliurus Spina-Christi Mill.	N	14	2
Phillyrea variabilis Timb.	N–M	60	72
Pistacia Lentiscus L.	N–M	46	40
Quercus Ilex L.	M	8	6
Rosa sempervirens L.	N	2	10
Spartium junceum L.	N	..	2
Smilax aspera L.	N–M	32	52
Density of species	..	2·2	2·3
No. of species	..	10	10

The composition of the maquis will appear from two formation-statistical analyses, the results of which are seen in Table 34; if the formation is to be named after a single dominant species, it must be called a *Phillyrea*-Formation, *Phillyrea variabilis* occurring with the highest frequency percentages, 60 and 72, in the two analyses. Besides *Phillyrea, Myrtus, Pistacia,* and *Smilax* formed the bulk of the plant covering; in some places *Pistacia* was dominant; in many places *Smilax aspera* had become so interwoven with the shrubs that the vegetation was quite impassable. Besides the species recorded in Table 34 other plants occurring here are especially *Pteridium aquilinum, Rubus (discolor?), Cistus salvifolius, C. monspeliensis, Erica arborea, E. multiflora, E. scoparia, Rosmarinus officinalis, Juniperus, Hedera Helix,* and some few small individuals of *Quercus Robur.*

In the eastern part, where the phanerophytic formation is broader, there is found some low wood, formed partly of *Quercus Ilex* partly of *Populus tremula* and *Ulmus campestris* projecting above the maquis, in this place consisting especially of *Pistacia*; in some places there is *Paliurus* scrub. Here and there small open spots appear with a vegetation rich in species and consisting of Hemicryptophytes, Therophytes, and some few Geophytes, e.g. *Agrostis alba, Asphodelus ramosus, Bellis perennis, Brunella vulgaris, Cirsium arvense, Crocus biflorus, Cynodon Dactylon, Daucus Carota,*

Erythraea sp., *Lotus angustissimus*, *L. tenuis*, *Plantago Coronopus*, *P. lanceolata*, *Ranunculus bulbosus*, *Taraxacum* sp.

E. **Formations of herbs and of halophilous shrubs growing beside the lake.** On Tombolo di Feniglia the maquis generally extends almost to the water's edge, so that, apart from the helophytic formations, there is as a rule no ground covered with herbs. Where smaller areas are partly covered with herbs, as at the south-western corner of the lake and in some parts along its western side, this is no doubt due to the fact that the original vegetation of shrubs has been cut down to give place to pasturage, and as a matter of fact some few remnants of the maquis are seen among the herbs. Where a Helophyte belt occurs, a *Phragmites*-Formation is generally formed, with an occasional admixture of *Typha*. In some places there is a *Scirpus maritimus*-Formation. On the more or less swampy soil behind the *Phragmites*-Formation there is often a *Juncus maritimus*-Formation, and then a belt rich in species, principally herbaceous, especially *Juncus acutus*, *Scirpus Holoschoenus*, *Inula viscosa*, and *Inula crithmoides*. In some parts there was nothing outside the *Juncus maritimus*-Formation, in others, where there was shallow water, an *Arthrocnemum glaucum*-Formation occurred.

At the south-western corner of the lake there is a large level tract without maquis, *c.* 120 metres wide. Here the sea is very shallow. The bottom near the shore is sandy, stony, and quite destitute of vegetation. The front line of vegetation on the more or less sandy clay soil is a *Juncus maritimus*-Formation; it forms a belt only a few metres wide, here and there sending out a tongue into the sea. There are also some few isolated insular areas. The plants are, besides *Juncus maritimus*, especially *Inula crithmoides*, *Aster Tripolium*, *Arthrocnemum glaucum*, *Artemisia coerulescens* (?), and some few other species. Then, on somewhat higher soil, a *Juncus maritimus–Artemisia*-Formation with *Statice Limonium*, *Inula crithmoides*, *Carex extensa*, *Arthrocnemum*, *Aster Tripolium*, and many other herbs; then maquis, first scattered bushes, among others *Tamarix*, and next dense scrub with *Pistacia* as the dominant species.

On a recent alluvium near the north-western corner of Stagno di Orbetello, passing landward from the sea, the order of succession was as follows:

1. *Salicornia fruticosa* at the water's edge, landward mixed with *Obione portulacoides*.

2. *Arthrocnemum glaucum* at a slightly higher level with *Triglochin paluster* especially on the landward side. Gradual elevation of the bottom by the deposition of ooze at high tide.

3. **Herbaceous carpet rich in species**, at a slightly higher level, often with great quantities of *Juncus acutus*; further, *Inula crithmoides*, *I. viscosa*, *Statice Limonium*, various *Gramineae*, *Cyperaceae*, &c. At the highest levels a few small plants of *Pistacia* forming together with a *Rubus*

the remnants of the original maquis which has probably been removed to give place to pasture.

The sand dune area between Ansedonia and Lago di Burano shows very uniform conditions, being occupied all the way by a continuous range of dunes with *Juniperus* scrub (Figs. 179–80), now *Juniperus macrocarpa* and now *J. phoenicea*. About half-way between Capalbio and Ansedonia the order of succession from the sea landward was as follows:

1. In front a low wall with a sharp crest, here and there immersed at spring tides.
2. A flat valley.
3. A higher sand-bank with a rounded summit.
4. A flat valley with wreckage, branches, &c., and with enormous masses of *Posidonia* balls. Spring-tide line.
5. Sand dune, 4–6 metres high, with *Juniperus phoenicea* scrub moulded by the wind, often ragged and torn (Fig. 180); the landward face covered with scrub, especially of *Pistacia* and *Phillyrea*; the scrub is 2–4 metres high.
6. A comparatively broad and deep valley with a cropped vegetation of herbs. This part was probably once covered with maquis and wood, for which pasture has now been substituted.
7. Broad range of older dunes, covered especially by *Pistacia* and *Phillyrea*.
8. Broad depression, a continuation of Lago di Burano. The higher levels have no doubt formerly been covered by wood which has been cut down to obtain pasturage. Now only a few trees (*Populus alba*) and some *Tamarix* bushes are left. The bulk of the vegetation consisted of *Juncus acutus* and *J. maritimus*.
9. Higher, cultivated land and pasture. Here were quantities of tall dead stems of various herbs, especially *Inula viscosa* and, in spots, a liliaceous plant (*Ornithogalum*?). Towards the east, near to Capalbio, the vegetation of the dunes consists especially of *Juniperus macrocarpa*, while *J. phoenicea* is the dominant nearer to Ansedonia. The herbaceous vegetation of the dunes consists especially of scattered *Medicago marina, Polygonum maritimum, Xanthium spinosum, Echinophora spinosa, Sporobolus pungens, Eryngium maritimum, Agropyrum* sp., and *Cyperus aegyptiacus*.

A dune area similar to that between Ansedonia and Capalbio forms a continuation towards the south-east. Near Chiarone the depression in which Lago di Burano is situated is filled with huge masses of reeds. Past Chiarone the reed vegetation is replaced by areas of pasture with tufts of *Juncus acutus*; still farther on, the depression passes into higher, cultivated land.

III. THE STRAND FORMATIONS AT THE MOUTH OF THE TIBER

North of Fiumicino I was able to observe the conditions on the extensive new formations at the mouth of the Tiber. Here the strand forma-

FIG. 180. Between Ansedonia and Capalbio: seaside dune with *Juniperus* scrub, seen from the beach.

FIG. 181. A little north of Fiumicino at the mouth of the Tiber. Beach with wreckage, then the colonizing vegetation.

Fig. 182. A little north of Fiumicino looking northward: the limit between the *Euphorbia Paralias*-Formation and the *Psamma*-Formation (to the right).

Fig. 183. A little north of Fiumicino, looking northward: the seaward limit of the *Juniperus macrocarpa*-Formation, with a few *Tamarix*.

tions consist of low level stretches of sand, as a rule hardly rising more than about a metre above high-water mark. There is some blown sand here and the ground is uneven because sand accumulates here and there around some few plants or groups of plants; but regular dunes do not occur, the highest heaps, formed round scattered groups of *Psamma*, are generally less than 1 metre high. Only in one place did I see a *Psamma* dune about 2 metres high. Figs. 181–4 will give the reader the best idea of the condition of the ground. Nearest the sea there is a trace of low line ridges with intervening valleys or depressions, but the difference in height between ridge and valley is very slight. A little to the north of Fiumicino, passing inland from the sea, conditions were as follows:

1. From high-water mark over the front ridge to the landward side of the following valley, *c*. 40 metres (Fig. 181); the seaward face of the flat, broad ridge was covered with wreckage and other washed-up material. In some parts, at any rate, the ridge is overflowed at spring-tides: the black, heavy, coarse-grained sand is deposited on the gently sloping landward face, while the water with all the mud remains in the valley inside. Gradually the water sinks into the soil, while the mud is deposited as a greasy layer which makes the floor of the valley more or less conspicuously like a sandy marsh. The ridge is destitute of vegetation except for a very few plants of *Euphorbia Paralias, Sporobolus pungens,* and *Echinophora spinosa*. On the flat eastern (landward) face of the ridge there are the beginnings of a very open *Euphorbia Paralias*-Formation which also occurs at the higher levels of the valley, whereas its lowest levels are quite destitute of vegetation. In addition to *Euphorbia Paralias* there occur *Sporobolus pungens, Polygonum maritimum,* and *Salsola Kali*; the latter forms tiny dunes.

2. From ridge No. 1 across the next flat ridge and the following valley to its eastern side, *c*. 40 metres. Ridge No. 2 and the higher levels of valley No. 2 have a similar *Euphorbia Paralias*-Formation, but more abundant. In addition to the above-mentioned species the following were observed: *Senecio leucanthemifolius, Eryngium maritimum, Anthemis maritima, Echinophora spinosa, Raphanus Raphanistrum, Erodium cicutarium, Ononis variegata, Medicago litoralis, Silene italica,* and *Tyrimnus leucographis*. There occurred, too, some few *Psamma arenaria*, which, however, become commoner on the succeeding terrain. Many places, especially down the valley, are almost or entirely devoid of vegetation. A few tufts of *Juncus acutus* were seen in some parts of the valley.

Within 1–2 there is no regular alternation of longitudinal ridges with longitudinal valleys; true, there occur depressions with a different vegetation from that of the higher levels, but they do not form regular longitudinal valleys.

3. After 2 comes a somewhat higher but still rather flat area, 40–50 metres wide (Fig. 182). Here are some scattered *Psamma* tufts and low

Psamma dunes, a meagre *Psamma arenaria*-Formation. In addition to *Psamma arenaria*, the same species that occurred in 1–2 are to be found here, especially *Sporobolus*, *Anthemis*, and *Senecio*, besides a few other species, notably *Plantago Coronopus* in quantity.

4. Transitional area with scattered low cushions of *Juniperus macrocarpa* (0·25–0·5 metre high) (Fig. 183) and groups of *Tamarix*, 1–1·5 metre high; on the grey soil, especially *Sporobolus*, *Plantago Coronopus*, *Anthemis*, and various other species. Here and there lower areas with a little water and partly clayey soil; probably this part is at times washed by the sea. In such places were to be found in the middle *Typha angustifolia* and at the edge *Salicornia europaea*, *Juncus acutus*, and *Tamarix*. In the inner part of 4 a fairly long longitudinal valley was seen with *Juncus maritimus*, *J. acutus*, and at the edge *Plantago Coronopus*.

5. The *Juniperus macrocarpa*-Formation. The farther we pass landward, the more frequent is the occurrence of *Juniperus macrocarpa*. As a rule it occurs in small belts, 0·1–1·5 metre high and 0·5 to several metres broad, between which the grey sand shows a scattered vegetation of *Plantago Coronopus* and other of the plants mentioned above. In addition there occur some species not seen in the area outside; in the lowest parts, mainly longitudinal depressions, especially *Juncus acutus*, further *Juncus maritimus*, *Scirpus Holoschoenus*, and *Schoenus nigricans*. In the western part of 5 *Juniperus macrocarpa* is the only shrub; passing eastward a few *Phillyrea* are added; soon *Pistacia*, too, occurs and *Cotoneaster Pyracantha*, *Juniperus phoenicea*, and *Daphne Gnidium*; finally *Juniperus* disappears entirely or almost entirely and we get

6. The Maremma Maquis (Fig. 184) alternating with flat, partly swampy valleys with *Juncus acutus*, *J. maritimus*, *Schoenus nigricans*, *Scirpus Holoschoenus*, *Inula viscosa*, *Typha*, *Cladium Mariscus*, *Saccharum Ravennae*, *Mentha aquatica*, and other species. The maquis, which is 2–4 metres high, consists especially of *Phillyrea*, *Pistacia*, *Spartium junceum*, *Smilax aspera*, and others. It extends eastwards a little way past the station Fiumicino and is succeeded by more or less swampy pasture with some few scattered low shrubs. This area, which is quite narrow, is succeeded by cultivated soil, now mainly large pastures without trees or shrubs, or only a very few trees at field boundaries, or solitary houses. Over very long stretches not a single house is seen.

The coastal area at Fiumicino being low and flat but at the same time broad, at the first glance it leaves a very different impression to that given by, e.g. the Arno–Leghorn coast or the part round Argentario. But a study of the vegetation from the sea landwards soon shows that the order of succession at Fiumicino is in reality quite the same as that observed on the other parts of the Italian west coast dealt with, the main stages of which are as follows: *Euphorbia Paralias*-Formation—*Psamma arenaria*-Formation—*Juniperus*-Formation—maquis; the only difference being

FIG. 184. A little north of Fiumicino: Maremma maquis 2–4 metres high, the greater part formed of *Phillyrea*. In front swamp.

FIG. 185. *Circa* 2 km. west of Terracina, looking westward along the south side of the Pontine Swamps; in the background M. Circeo. From the sea landwards: (1) rather sandy ground with *Psamma*-Formation, (2) flat valley with *Crucianella maritima, Medicago marina, Teucrium Polium*, &c., (3) from the bottom of this valley and up the south slope of the next following dune area a prominent *Helianthemum halimifolium*-Formation (Table 35), (4) dune wall with maquis of *Phillyrea, Erica multiflora, Rosmarinus, Calycotome, Pistacia, Daphne Gnidium, Helianthemum halimifolium, Smilax aspera, Quercus Ilex*. Inside the dune wall level sandy ground with vineyards.

that while the first three extend over more than 100 metres at Fiumicino, they are, in other places, as for instance on Tombola della Gianella and at Capalbio, confined to an area of some few metres.

IV. THE SOUTHERN COAST OF THE PONTINE SWAMPS

I do not know how far the new formations along the lower course of the Tiber extend, but at any rate the alluvial coastal formations are not to be found at Anzio and Nettuno which I visited in the hope of being able to study the sand-dune formation on this part of the western coast of Italy also. Here, however, there are cliffs subject to erosion by the sea, which floods the cellars in old Antium partly excavated in the loose sandstone of the soil.

The whole of the long stretch along the south-western side of the Pontine swamps—Lido di Fogliano, Lido di Paola, &c.—probably consists of sand dunes right down to M. Circeo. This tract I have not seen, however, whereas I was able to examine the southern coast of the Pontine swamps between Terracino and M. Circeo. Here there is a sand-dune coast highly reminiscent of Tombola di Feniglia at Argentario. On the whole conditions around the Pontine swamps somewhat resemble those at Stagno di Orbetello: M. Circeo corresponds to Argentario; Lido di Paola and Lido di Fogliano to Tombola della Gianella, and the stretch Terracina–M. Circeo to Tombola di Feniglia, while the Pontine swamps themselves correspond to Stagno di Orbetello, but are in a more advanced stage of development.

Nearest Terracina the sand-dune area is quite narrow, but becomes broader as we pass westward. A little way to the west of Terracina, passing inland from the sea, the order of succession was as follows:

1. A bare beach, *c.* 20 metres wide, rising towards the landward side and bounded there by a verge of sand 0·5–0·75 metre high, which at any rate in part is due to erosion during a recent storm. Then

2. Older and slightly higher ground with an open *Psamma*-Formation; here, in addition to *Psamma arenaria*, especially *Sporobolus pungens*, *Agropyrum junceum*, *Cyperus aegypticus*, and *Pancratium maritimum*. On the landward side this quite low *Psamma* dune then passes into

3. An often slightly lower older terrain with a scattered growth of *Teucrium Polium* and *Crucianella maritima* together with *Senecio leucanthemifolius* and other species also observed on the corresponding low flats at Fiumicino. Rising landward to

4. A range of dunes, 1–2 metres higher than 3, with a maquis 0·25–1 metre high, consisting especially of *Pistacio Lentiscus*, besides *Phillyrea, Erica multiflora, Teucrium Polium, Helianthemum halimifolium, Cistus salvifolius, Rosmarinus officinalis, Calycotome spinosa, Daphne Gnidium,* and *Smilax aspera*.

5. Lower ground with vineyards on sandy soil.

Somewhat farther westward similar conditions prevail but in a more advanced stage of development; thus there is an incipient vegetation of

Psamma and a few other species in No. 1; No. 3 gets broader, forming in places rather low areas with a vegetation different from the usual type; finally in No. 4 two belts or formations may be distinguished: in front a *Helianthemum halimifolium*-Formation and then the usual maquis.

In the immediate vicinity of the westernmost of the two canals which drain the Pontine swamps the order of succession from the sea landward was as follows:

1. Low belt, *c.* 20 metres wide, with young *Psamma arenaria*.
2. Somewhat higher range of dunes, 5–7 metres broad, with an older *Psamma arenaria*-Formation, mainly of the same composition as that of No. 2 above.
3. Low flat valley-like belt, 20–30 metres wide, of grey sandy soil, with especially *Crucianella maritima, Teucrium Polium, Eryngium maritimum, Medicago marina,* and *Senecio leucanthemifolius;* here, too, some few young *Juniperus (macrocarpa?)*.
4. A sand-dune area 2–3 metres higher, with a vegetation of shrubs:
   a. Up the southern slope a grey *Helianthemum halimifolium*-Formation (Fig. 185) with especially *Cistus salvifolius* and *Teucrium Polium,* a little *Pistacia* and a few other species (Table 35).

TABLE 35

Two analyses of a *Helianthemum halimifolium*-Formation on the south coast of the Pontine Swamps. 50×0·1 sq. metre

	Life-form.	F%	
		A	B
Calycotome spinosa (L.) Lk.	N	2	10
Cistus salvifolius L.	N–Ch	..	32
Daphne Gnidium L.	N	2	2
Helianthemum halimifolium Willd.	N	84	74
Rosmarinus officinalis L.	N	10	2
Thymelaea hirsuta Endl.	N	..	2
Dianthus Carthusianorum L.	Ch	..	2
Teucrium Polium L.	Ch	38	44
Undetermined Crucifer	Ch	2	4
Species density	...	1·4	1·7
No. of species	..	6	9

b. Crest of dune with *Erica multiflora*, &c., as in 4 (p. 603); also some few *Quercus Ilex.*

5. Vineyards on low sandy soil.

Parallel with the west coast the Sisto, a river or canal, runs through the Pontine swamps; having reached the sand-dune area of the south coast, it bends towards the east and falls into the sea through two canals. To the west of these canals the sand-dune area is broad, and seaward more or

FIG. 186. The south coast of the Pontine Swamps, about 1 km. west of Porto Badino, looking eastward. Broken dunes, of which only some knolls remain, kept together by the roots of various shrubs of the maquis.

FIG. 187. Near Fig. 186: shifting dunes burying the maquis.

FIG. 188. Near Figs. 186 and 187. The wood and the maquis are cleared away and only the cork-oaks are left. Behind the cottage is seen a deciduous Maremma wood on lower, partly swampy ground.

FIG. 189. Seaside dune west of Lago di Fusaro, nearly half-way between Torre gaveta and Acropon. Along the foot of the dune a *Psamma*-Formation; on the dune a *Juniperus macrocarpa*-Formation.

less irregular and broken, here and there with shifting dunes invading the lower soil covered by maquis and wood. Passing inland from the sea the following conditions were observed:

1. Beach entirely or almost entirely devoid of vegetation.
2. Low, more or less pronounced *Psamma* dune.
3. Here and there remnants of grey flat sandy soil with a scattered vegetation of herbs and Chamaephytes.
4. Ragged higher sand dunes kept together by the roots now of one now of another kind of maquis plant: *Phillyrea, Pistacia, Rosmarinus, Myrtus, Juniperus macrocarpa* (Fig. 186), as was the case on Tombola di Feniglia; here and there moving dunes overwhelming the next formation (Fig. 187).
5. Wide stretch with a maquis, c. 1 metre high, on the lower stretches as much as 2 metres high, of *Erica arborea, Phillyrea, Daphne Gnidium, Helianthemum halimifolium, Pistacia,* and *Rosmarinus* (Fig. 187).
6. Maquis similar to that of 5, but with scattered *Quercus Suber* and *Q. Ilex* (Fig. 187, in the background). A statistical investigation of the frequency of Chamaephytes and Phanerophytes gave the result recorded in Table 36. Here and there the land has been cleared and brought under cultivation. The cork trees are left (Fig. 188).

TABLE 36

Analyses of a maquis on the south coast of the Pontine swamps. 50×0·1 sq. metre

	Life-form.	F%
Cistus salvifolius L.	N–Ch	78
Daphne Gnidium L.	N	2
Erica arborea L.	M–N	70
Helianthemum halimifolium Willd.	N.	66
Juniperus macrocarpa S. et S.	N	2
Myrtus communis L.	N	6
Phillyrea variabilis Timb.	N	6
Quercus Ilex L.	MM	4
,, Suber L.	MM	4
Species density	..	2·4
No. of species	..	9

7. River (the Sisto).
8. Deciduous phanerophytic formation of *Alnus glutinosa, Quercus Robur, Ulmus campestris, Populus alba, Salix* sp., *Paliurus Spina-Christi,* and *Crataegus*. How far this wood extended I could not see; the fate awaiting it was evident from large stores of charcoal.

The coast between Terracina and Gaeta, as far as it can be seen from the Terracina mountains, is a low coast of dunes, from a distance exactly resembling the coast at Capalbio, &c., viewed from Ansedonia. Apart from the extreme northern tract where special conditions prevail because a small river runs out there, I have had no opportunity of examining this coast, nor the extensive, low, partly swampv ground, 'Selva vetere', lying inside it.

At the north-western corner, north of the canal, conditions from the sea inland were as follows:

1. Beach.
2. Low recent dune formation, 0·5–1 metre high, with *Psamma*-Formation which, in addition to *Psamma arenaria*, consisted of *Euphorbia Paralias, Teucrium Polium, Crucianella maritima, Pancratium maritimum, Anthemis maritima*, and a few others.
3. Somewhat older and higher dunes with *Pistacia, Pteridium, Tamarix, Teucrium marum, Daphne Gnidium, Myrtus*, and *Phillyrea*.

In the same place I had an opportunity of examining a part of the reed swamp, which here consisted of *Cladium Mariscus, Typha, Phragmites, Saccharum Ravennae, Juncus maritimus*, and *J. acutus*.

V. The Strand Formations to the West of Lago del Fusaro

From Gaeta southward to Torre Gaveta off the south end of Lago del Fusaro there is a narrower or broader margin of sandy ground, the vegetation of which I have, however, only been able to examine on the stretch between Lago del Fusaro and the sea.

Inside a quite narrow sandy beach only a few metres wide there occurred a more or less ragged sand dune up to 4 metres high, with *Juniperus*. On the landward side it passed imperceptibly into a lower area of dunes covered by maquis and low wood. About half-way between Torre Gaveta and Acropon, from the beach to Lago del Fusaro, conditions were as follows:

1. A bare sandy beach, 3–5 metres wide, with shingle, thrown up branches, seaweed, &c., either passing imperceptibly into the sand dune, or bounded in places by a low verge formed at spring-tides.
2. A sand dune, up to about 4 metres high, the base and part of the seaward face of which was covered by a pronounced *Psamma*-Formation in a belt 6–10 metres wide, while the more or less ragged crest was covered with a *Juniperus macrocarpa*-Formation (Fig. 189), passing inland into a dune maquis on the lower terrain inside the dune.

The *Psamma*-Formation consisted chiefly of *Psamma arenaria*, which, however, is neither very vigorous nor dense; a statistical analysis gave a frequency percentage of 72. In addition to *Psamma* there occurred, in very scattered growth, some few *Euphorbia Paralias, Crucianella maritima, Anthemis maritima, Ononis variegata,* a grass, and some few quite young plants of *Juniperus macrocarpa*.

The *Juniperus macrocarpa*-Formation along the crest of the dune consists mainly of *Juniperus macrocarpa* 0·35–2 metres high; further *Phillyrea* (rather broad-leaved), *Rosmarinus, Cistus salvifolius*; and *Smilax aspera, Medicago marina, Sporobolus pungens,* and some other species. Table 37 shows the results of a statistical analysis of the Phanerophytes and Chamaephytes of the *Juniperus macrocarpa*-Formation on the seaward face of the dune. *Juniperus macrocarpa* is the dominant.

TABLE 37

Analysis of *Juniperus macrocarpa*-Formation on the seaside dune west of Lago del Fusaro. (Only Phanerophytes and Chamaephytes are dealt with.) 50×0·1 sq. metre

	Life-form.	F%
Juniperus macrocarpa S. et S.	N	**100**
Cistus salvifolius L.	N–Ch	4
Phillyrea variabilis Timb.	N	2
Rosmarinus officinalis L.	N	2
Smilax aspera L.	N	2
Medicago marina L.	Ch	2
Species density	..	1·1
No. of species	..	6

Down the landward face of the dune the vegetation becomes higher and more luxuriant. *Juniperus* is here more abundantly interspersed with *Quercus Suber, Q. Ilex, Phillyrea, Rosmarinus,* and *Cistus salvifolius*; further with *Daphne Gnidium, Pistacia Lentiscus,* and *Teucrium Polium.*

TABLE 38

The wood west of Lago del Fusaro. (Only the Phanerophytes and the Chamaephytes are dealt with.) 50×0·1 sq. metres

	Life-form.	F%
Quercus Ilex L.	MM	24
„ Suber L.	MM	4
Cistus salvifolius L.	N–Ch	64
Daphne Gnidium L.	N	4
Juniperus macrocarpa S. et S.	N	36
Myrtus communis L.	N	8
Phillyrea variabilis Timb.	N	10
Pistacia Lentiscus L.	N–M	8
Rosmarinus officinalis L.	N	78
Smilax aspera L.	N–M	4
Teucrium Polium L.	Ch	2
Species density	..	2·4
No. of species	..	20

On the lower area, between the seaward range of dunes and Lago del Fusaro, conditions were as follows: nearest the seaward range the vegetation is low, 0·5–2 metres high; its composition may be gathered from Table 38, which shows that here we have a *Rosmarinus–Cistus salvifolius*-Formation. A good deal of *Juniperus macrocarpa* was still to be found, but passing inland on to lower and damper soil, it decreased considerably towards Lago del Fusaro; simultaneously *Quercus* increased in number and height, and a meso-phanerophytic wood appeared, especially of *Quercus Ilex*, with a nano-micro-phanerophytic undergrowth of the other species recorded in Table 38. Besides, there occurred *Ruscus aculeatus, Rubus, Asparagus acutifolius, Juncus acutus, Inula viscosa, Saccharum Ravennae*, and others.

Finally, along Lago del Fusaro there is a border of a *Phragmites communis*-Formation with *Saccharum Ravennae, Juncus acutus*, and *Inula viscosa*.

A COMPARISON BETWEEN THE VEGETATION OF THE MEDITERRANEAN SAND DUNES AND THAT OF THE DUNES IN THE HEMICRYPTOPHYTE CLIMATE AREA

Finally, in making a comparison between the vegetation of the Mediterranean, especially the Italian, sand dunes, and that of the dunes in a Hemicryptophyte climate area, viz. the dunes along the west coast of Jutland, it will be necessary to make a few explanatory statements about the concepts of succession and climax formation.

In sand-dune areas where 'new soil' is constantly being formed there is an excellent opportunity of making studies on succession, that is, of observing how the composition and environment of the vegetation are altered in the course of time in a certain topographically defined spot.

Much has been written on succession, especially by English and American authors, and a number of systems have been put forward, e.g. by Clements, Cowles, Crampton, and others. I do not propose to swell their number, but for the sake of clearness it is necessary, on examining the order in which the formations succeed one another in the same spot, to assemble in groups those environmental changes which determine the succession. These changes, it appears to me, fall naturally into two groups, viz. (I) changes in climate, and (II) changes due to other conditions.

Remembering the numerous testimonies to climatic changes in the past which we possess, there is no reason to doubt that at the present day, too, such changes occur, though at so slow a rate that we have no sure record of them in the short span of years in which exact measurements have been made. Hence, even if climatic changes do possibly occur within the areas with which we are here concerned, the slowness with which they take place makes it possible to disregard them as a cause of the often rather quickly occurring changes in the vegetation observed in the sand-dune area.

The changes of environment due to circumstances other than climatic changes are numerous and still active. They may be divided into two groups: (A) the chemico-physical, i.e. those directly due to the action on the soil of chemical and physical factors, and (B) the biotic, i.e. those due to organisms.

Even in a soil quite destitute of organisms changes constantly occur, e.g. weathering, washing out, &c.: the same spot is not of the same character, not subject to the same conditions, in one year as in the preceding year. In a soil covered with vegetation the changes may be even greater and more complicated, since the vegetation itself, both the higher plants and the micro-organisms living in the soil, cause changes in the soil, often of a very radical nature, e.g. with respect to reaction (hydrogen-ion concentration), nitrification, &c., &c. This multitudinously graded interaction between the physico-chemical conditions, the micro-organisms, and the covering of higher plants is among the questions which it is difficult to explain in detail, but there is no doubt that changes constantly occur here which more or less influence the composition and character of the vegetation even though in many cases they only take place very slowly. Let us imagine that a heath has been untouched by cultivation for centuries. The physiognomy of the species and their proportion will hardly be the same to-day as it was hundreds of years ago. Everything would seem to show that even where we are concerned with stabilized formations, this does not mean that the formation is unchangeable even though the climate does not change, it simply means that those changes of environment that take place occur so slowly that the corresponding changes in the vegetation can keep pace with them. A complete stand-still will hardly be found anywhere. If we take the term 'formation' in a limited sense as I do, it is thus rather useless to talk of a climax formation; either every stabilized formation must be called a climax formation, or there are no climax formations at all. Even if we took it for granted that the climate remained for ever unchanged, there would, with our knowledge of nature, be no reason to assume that any one of the now existing countless formations would be quite the same in a thousand years as now.

The fact is that when botanists talk of climax formations, they take the term 'formation' in a very wide sense, in which it almost coincides with my life-form classes and formation classes. In this sense we may perhaps speak of climax formations. Thus there can hardly be any doubt that, if the climate and the relation between land and sea, &c., remain unchanged, the perpetually warm and moist tropical regions, when left to themselves, will remain covered with evergreen woods, evergreen phanerophytic formations. If, therefore, we regard the evergreen tropical woods as one single formation, it will be the climax formation of the moist and warm tropical regions. This, however, may be much more simply expressed, it appears to me, by saying that the evergreen Phanerophytes are the dominant life-form of these regions—in short—the term 'Phanerophyte climate' will sufficiently cover the facts.

It is not, however, the formation in its widest sense which is the actual subject of the doctrine of formations. These 'formations' or types of vegetation belong partly to physiognomical plant geography partly to the doctrine of plant climates as based on life-form statistics.

Such a varied and composite thing as for instance the evergreen tropical wood or the deciduous wood of the Hemicryptophyte climate area cannot be made the direct subject of a scientific investigation, it must first be resolved into its individual components, the more narrowly delimited formations, and these latter are the real subject of the doctrine of formations.

But there is no reason to believe that a tropical wood in a given spot would remain the same formation, in the narrow sense, even if the climate did not change. From day to day, from year to year, changes occur in the soil owing to the influence of the atmospherical agencies and the vegetation, changes tending to produce soil conditions already existing in other places with a vegetation adapted exactly to such conditions. And there is no reason to believe that the vegetation in the first-mentioned place will not gradually undergo corresponding changes.

Or if, for instance, we take a wooded range of hills in Zealand (Denmark), we may presumably venture to say that, provided the climate remains unchanged, the climax formation of this spot will be a phanerophytic formation, more especially a formation of deciduous Phanerophytes, but it is extremely doubtful whether it will be a *Fagus silvatica*-Formation, as it was before the interference of man.

Thus, if we take the term 'formation' in its narrow sense, this means that either there exists no climax formation, or the term must be used to designate any stabilized formation, i.e. relatively stabilized, for absolute stabilization does not exist. If, however, the term is used more comprehensively, a climax formation will be a very vague concept without any concrete sense, for as far as we have hitherto seen, the concrete, narrowly delimited formation with its definite composition and physiognomy, is never the climax but simply a link in an endless chain.

Wherever a change in the environment takes place, whether due to cultivation or other circumstances, we observe that corresponding changes take place in the composition, or at any rate in the physiognomy, of the vegetation. A small change in the degree of moisture, or the addition of even quite minimal amounts of certain substances, will often result in very conspicuous changes in the composition and character of the vegetation. Generally the change of environment takes place so slowly that the corresponding changes in the vegetation keep pace with them, at any rate apparently, and the formations with which we are concerned are thus, at least apparently, stabilized. If, however, the change of environment takes place suddenly, for instance if water is dammed up by barrage, or changes in illumination occur owing to the felling of trees, &c., some time will elapse before the equilibrium is restored, for the immigration of species suited for the new environment will take a certain time. Under such conditions vegetations may very well be found which at first sight look like ordinary stabilized formations, but if observed for a number of years, they will be found to change from year to year. Thus in woods where trees are cut down now and then, one must be careful not to regard the flora of the ground as a perfectly valid stabilized expression of the momentary environment.

The climatic changes of environment from month to month in the course of the year give the vegetation a different aspect at the different seasons. And even where the climate does not change, one year is not like the next. Hence a formation may differ somewhat in appearance from one year to another, but in the long run the formation and its various aspects will remain essentially unchanged if the soil remains unchanged.

Now, however, the fact is that, even if we assume the climate to be unchanged, changes in the soil will continually occur from year to year. And as regards the sand-dune area, it is natural to suppose that the changes here are specially rapid, since not only weathering and the influence of the vegetation but also the wind plays a very important role. It might therefore be supposed that here there could be no climax formation, especially so on the mobile sand dune. And yet the truth is that the sand dune is the very place where we can find conditions that somewhat justify the use of the term 'climax formation'. For in spite of the great mobility of the sand dunes the special nature and cyclical course of the changes will make one year like another, so that in the long run there will be no change.

I shall begin by describing conditions in the Jutland sand dunes and subsequently compare them with the Italian dunes. Apart from the beach the area in both regions consists of two tracts, the more or less mobile seaward dune and the older fixed dune.

As soon as there is a wind, changes occur in the sand of the front row of dunes. The wind carries away sand from the surface and deposits it farther inland where there is more shelter, and fresh supplies of sand from the beach are deposited on the dunes. This may under certain circumstances lead to a neutralization of those progressive changes in the soil which in other localities are the cause of the steady though slow change in the vegetation. True, the dune, which for the time being is mobile, will as a rule not long remain so; it will become part of the fixed dune area, a fresh range of dunes being formed in front of the old one. But the same dune may not rarely remain in the front for years, almost as much sand being carried away annually as new sand is deposited. In that case, the surface will in the main be the same from year to year, it will consist of fresh sand from the beach. Since this is a radical change in the soil, whose upper layer is thus constantly renewed, the progressive changes which in other places cause changes in the vegetation are here eliminated. The influence on the soil of the physico-chemical and biological factors must start afresh every year. In other words, the change will become a rhythmical repetition, and will cease to cause a progressive development. As long as this state of things lasts and the climate does not change, the environment in the same spot will be the same year after year, and consequently there will be no essential change in the vegetation. Such conditions may be found in places where no erosion of the coast takes place for a number of years and new dunes are not formed outside those already in existence. In several parts of the west coast of Jutland I know of such topographically determined dunes bearing to this day a *Psamma arenaria*-Formation quite like that found there more than thirty years ago.

But of course the above-described state of equilibrium will cease sooner or later. Sooner or later conditions will alter in the front line of dunes. Either the wind or the wind and sea together will break it down, and a fresh dune will be formed inside the site of the old one; or a fresh seaward dune will be deposited outside the old one, which will then be sheltered and become part of the fixed dune area, and subject to the more or less slow changes to which the latter is constantly exposed by weathering, washing out and various biological conditions.

I shall not here go more closely into the details of the floristic composition of the vegetation in the various parts of the dune area of Jutland; these have been described at length by Warming (Warming, 1909). Nor shall I make any floristic comparison between the dunes of Jutland and those of Italy. Since we are here concerned with two very different climates, they have only a small percentage of species in common. In order to be able to compare the dune vegetation of the various climates as well as that of the various different zones of the dune area in the same country, a common measure or standard independent of the components of the flora is needed.

It will be difficult to find an area in which there occurs, within such narrow limits as the dune area of Jutland, such a multitude of formations, distributed apparently in such a motley mosaic, but on closer inspection the distribution shows an absolute conformity to law. If, on the 'grey dune' of the fixed dune area, we pass from the summit to the floor of the dune valley, during this passage of only a few metres we shall encounter a series of formations, chiefly dependent on the distance from the ground-water and therefore arranged in zones, some of them, e.g. the *Nardus*- and *Sieglingia*-Formations, being only 1 metre wide or less. And the nature and succession of the formations are repeated from dune to dune within the same area. It is a common feature of all these formations on the 'grey dune' that the Hemicryptophytes are the dominant life-form, that is to say the vegetation is a Hemicryptophytium.

Things are otherwise in the front row of dunes. Here the cryptophytic (geophytic) life-form is dominant. In addition there occur a number of

Therophytes, but they are not physiognomically dominant. The vegetation of the front row of dunes is a Cryptophytium.

If, however, we pass farther away from the beach to the oldest part of the dune which has long been exposed to washing out, &c., and hence is probably poorer in nourishment and more acid, with a greater hydrogen-ion concentration, the soil is covered with a heath vegetation in which the chamaephytic life-form is absolutely dominant. The vegetation here is a Chamaephytium.

In the dune area of Jutland if we thus wander from the beach landwards to older and older dunes we meet with three regions or facies of vegetation, differing in the dominating life-form, this being respectively that of the Cryptophytes (Geophytes), the Hemicryptophytes and the Chamaephytes. It must be stated that the word facies is not here used as a narrow limited term, but only as an indication of a subdivision within a larger unit, viz. the vegetation of the dune.

When the dune heath is as a rule only met with at some distance from the sea, this is not because it cannot thrive in the immediate vicinity of the sea: on the contrary, where the sea has recently eroded the land, for instance in certain localities in the north of Jutland and on Fanø, heath may be met with quite near the shore.

In the hemicryptophytic facies of the sand-dune area smaller or larger tracts of other classes of formations are met with in several places, especially of course the water and swamp plant formations of water-covered depressions, but occasionally also tracts of deciduous nano-phanerophytic formations of *Salix*, *Hippophaë*, and *Rosa*. *Salix repens* especially often forms fairly extensive chamae-nano-phanerophytic growths, and in northern Jutland *Hippophaë* forms rather large copses in several places, especially in dune valleys near the sea. But in its broad features the ground between the front line of dunes and the heath is a Hemicryptophytium.

If, now, we compare with Jutland conditions in Mediterranean regions, especially on the west coast of Italy, here as there we have a Cryptophytium on the front line of dunes, though it is often but slightly developed. On the other hand, the hemicryptophytic facies which is so extensive on the dunes of Jutland, is entirely absent on the Italian dunes. On the Italian dunes the zone corresponding to the dune heath and 'grey dune' of Jutland is covered by evergreen, xerophytic nano-phanerophytic formations, the dune maquis, an evergreen Nanophanerophytium.

In the Hemicryptophyte climate area, new soil, apart from swampy places, is quickly covered with Hemicryptophytes which survive the winter, or unfavourable season, because the perennating buds are placed in the soil surface and protected by it against the desiccating influence of the cold. Where one or several of the comparatively few gregarious Chamaephytes and Phanerophytes able to maintain themselves in the Hemicryptophyte climate can grow in the locality, a phanerophytic or chamaephytic vegetation will gradually become dominant, and we get a wood or heath which thus, for the time being, is—not a climax formation, for the formation will be perpetually changing—but a climax type of vegetation; as far as Jutland is concerned partly a deciduous Phanerophytium, partly an evergreen, xerophytic Chamaephytium.

In the Mediterranean Therophyte climate conditions are otherwise. There we have two more or less unfavourable seasons, the cold winter and the very dry and warm midsummer. In neither of these seasons is it of any use to have the shoot-apices with the embryonic tissue placed in the soil-surface. Conditions in the winter render it unnecessary, and in midsummer the strongly heated and dry soil is hardly the best place for the delicate shoot-apices. It would seem natural to suppose that a position a little above the surface would be better for the shoot-apices enclosed by the older parts of leaves. Hence we sometimes see that the rhizomes of Mediterranean Hemicryptophytes tend to grow up above the surface and thus form a transition to the Chamaephytes. True, the same phenomenon may occasionally be observed in the Hemicryptophyte climate, where, for instance, *Tussilago Farfarus* occurs in dense growths, or where the plant is strongly shaded by other species. There the rhizome will in the summer season grow up above the surface as if the plant were a Chamaephyte. But it does not long remain a Chamaephyte, for these aerial rhizones die away in the autumn. In the Mediterranean regions this need not happen in such cases, and if the aerial parts of the rhizomes do not die away in the dry summer time either, we have a true Chamaephyte. As a matter of fact it is very probable that some few of the Mediterranean species which I have recorded as Chamaephytes, are in reality Hemicryptophytes which have become Chamaephytes in this way. They are species that behave like Chamaephytes in a Therophyte climate but like Hemicryptophytes in the Hemicryptophyte climate.

The Chamaephytes play a greater part in the youngest portions of the Mediterranean dunes than in the corresponding parts of the dunes in Jutland. Along the base of the front line of dunes, at and immediately above the spring-tide line, the Jutland dunes have a vegetation in which *Agropyrum junceum* as a rule dominates, and where a number of Therophytes are likewise to be met with, e.g. *Cakile maritima, Salsole Kali,* and others; and also frequently, *Honckenya peploides,* a Hemicryptophyte which is fairly well able to keep pace in its growth with the dune and avoid being choked by the sand. In the Italian sand-dune area a number of Therophytes are of course met with in this zone, but also some Chamaephytes, especially *Euphorbia Paralias* (Figs. 155, 165, and 182), likewise *Diotis maritima, Crucianella maritima,* and *Medicago marina.* But even in the transitional area between the Cryptophytium of the front row of dunes and the Nanophanerophytium of the old dunes, a facies more or less characterized by Chamaephytes may be found. On this point, however, a change seems to take place as we proceed farther south from the northern limit of the Therophyte climate. Conditions in the Tuscan sand-dune area between the Arno and Leghorn are recorded in Table 30, p. 585, which shows the result of the statistical analysis of the vegetation at three different distances from the coast, viz. (A) 10 metres, (B) 35 metres, and (C) c. 50 metres from the front line of vegetation. The column succeeding the list of species records the life-form of the individual species. Now, as previously described in more detail, by adding up, for each formation in this table, the frequency numbers for the species with the same life-form, the figures recorded in the first five columns of Table 39 are arrived at; the percentage distribution of these figures is given in the five last columns, and we have here the biological spectrum of the individual formations. This shows the first formation characterized by the fact that Cryptophytes (Geophytes) are dominant (A in Table 39), and this forma-

TABLE 39

The biological formation spectra of the formations of Table 30 (p. 585)

	Points.					Percentages.				
	Ph	Ch	H	Cr	Th	Ph	Ch	H	Cr	Th
A	..	30	12	82	..	..	24	10	66	..
B	..	46	32	66	4	..	31	22	44	3
C	38	124	26	60	4	15	49	10	24	2

tion is succeeded by one in which Chamaephytes are dominant (C in Table 39); B in Table 39 shows the transition between these two formations. In the sand-dune area of southern France similar conditions prevail. Table 40, which shows the biological spectrum of four different formations at Palavas, is directly comparable with Table 39. Table 40, A, corresponds to Table 39, A; Table 40, B, to Table 39, B; Table 40, C and D, correspond to Table 39, C. In Table 40, C, *Anthemis maritima* was dominant, and in Table 40, D, *Helichrysum Stoechas*, just as in Table 39, C, from the Tuscan dunes.

TABLE 40

The biological formation spectra from the dunes at Palavas in South France. A, B, and C and D correspond to A, B, and C in Table 39 respectively

	Points.					Percentages.				
	Ph	Ch	H	Cr	Th	Ph	Ch	H	Cr	Th
A	..	10	2	100	8	..	8	2	83	7
B	..	118	4	132	2	..	46	1·5	51·5	1
C	..	226	18	34	24	..	75	6	11	8
D	..	186	0	52	6	..	76	0	21	3

Farther south, for instance on Tombola della Gianella, at Capalbio and near Lago del Fusaro, the dune maquis extends to the crest of the front row of dunes. The Cryptophytium of these dunes is only slightly developed here and is not separated from the Nano-phanerophytium of the older dune area by a transitional zone characterized by Chamaephytes. It is doubtful, however, whether this is due to climatic causes or to the slow rate of growth of the dunes in this place. Even the front line of dunes is pretty old and so has had time to become clothed with dune maquis right down to the beach.

The xerophytic aspect of the dune maquis and the dune heaths. To enable a plant to exist, there must be a certain life-promoting harmony between its structure and its environment, it must be adapted to its environment. It is the problem of obtaining water which most profoundly affects the vegetation; hence, where conditions cause dryness, this at once sets its mark on the plants. It does not follow, however, that the structural features in plants which must be regarded as an adaptation to aridity are a proof that they live in a dry environment. Structural characters which are essential to the existence of the plant under certain conditions may very well be present in an environment where they are of no use, or even do a great deal of harm, provided that they are not sufficiently harmful to prevent the plant from holding its own in the competition with other species. If plants with a xerophytic aspect are present in a moist environment, this does not mean that certain forms of dryness which will explain their presence on moist soil have been overlooked, it only means, perhaps, that certain conditions are present which enable the plants, in spite of their structure, to maintain themselves even on moist soil.

Adaptation to drought is of course most conspicuous in the structure of the leaf, but there exist other adaptations to dryness than a xerophytic leaf structure, hence we may find species on very dry soil whose leaf structure is not markedly xerophytic. As regards the question of water, the chief requirement is that there is a state of equilibrium between intake and output. Loss of water by transpiration must be made good with sufficient rapidity by water extracted from the soil through the activity of the root. This state of equilibrium may be disturbed either by a change in environment tending to produce a more rapid transpiration, or by a diminution of the available water in the soil. There are two ways in which the plant may restore the disturbed equilibrium; either by structural features which reduce transpiration, or by an increased power of absorbing water from the soil. In most cases the plant has probably both adaptations, but there are also plants which, even on very dry soil, do not show a corresponding xerophytic leaf structure. Thus the leaf of *Peganum Harmala* is not xeromorphic, and cut-off shoots quickly become flaccid; nevertheless it is able to grow on very dry soil where most other species have pronouncedly xeromorphic leaves. In order to understand the appearance of the vegetation as an expression of environment, it must be remembered that the ability of the species to absorb water from the soil may vary greatly (Fitting, 1911). Though the leaves of *Peganum Harmala* are not xeromorphic, the species may be no less xerophytic on that account, that is, it may be no less adapted to dry conditions than even the most xeromorphic desert plants.

If we wish to understand the dependence of the physiognomy of a vegetation on its relation to the water question, and the often very great differences which the species in the same locality may present in this respect, one fact, which is also decisive for the characterization of the life-forms as a whole, must be primarily kept in mind, viz. that every plant must in the first place be adapted to survive the unfavourable season. It follows from this that, in so far as adaptation to drought is expressed in the leaf structure, the latter must be adapted to the conditions prevailing in the unfavourable part of the period in which the leaves live. And as a matter of fact it is primarily evergreen leaves which are xeromorphic. In the Mediterranean dune maquis as well as in the Jutland dune heath the summer-green leaves of most species show hardly any or no xeromorphic characters. In the Therophytes the whole plant, and in the perennating species the individual shoots, are adapted to complete their development in a comparatively short, favourable period. As regards the unfavourable season the leaves behave in the same way as the Therophyte life-form, i.e. they entirely avoid it; they die away without on that account endangering the existence of the species.

In the dune maquis as well as the Jutland dune heaths the evergreen species dominate and the great majority have pronouncedly xeromorphic leaves. But xeromorphy is hardly caused by the same factor in the two localities. In the dune maquis leaf xeromorphy is primarily an adaptation to the drought of midsummer, whereas, in the Chamaephytes of the Jutland dune heath, it is an adaptation to the physiological drought of winter. It is true that some botanists have held that the xeromorphic character of the Jutland heath must be regarded as an adaptation to the dryness of the soil in summer. But this is hardly the case, for among the xeromorphic evergreen species such as *Calluna vulgaris*, *Erica tetralix*, *Empetrum nigrum*, *Arctostaphylos Uva-Ursi*, and *Vaccinium Vitis-Idaea* there occur a number of species which have no xeromorphic characters, e.g. *Vaccinium uliginosum*, *Lotus corniculatus*, *Arnica montana*, *Trientalis europaea*, and many others, and these are all summer-green, their leaves withering in the winter. True, some species have xeromorphic leaves, e.g. *Festuca ovina*, *F. rubra*, *Deschampsia flexuosa*, and *Nardus stricta*; but these are just the species whose leaves are partly winter-green, since they only wither more or less at the tips. On the whole, the xeromorphy of the evergreen species of the Jutland heaths is sufficiently explained by the fact that the leaves are evergreen and consequently must be adapted to the physiological drought of winter. On the other hand, the soil in summer is not sufficiently dry to necessitate specially xeromorphic characters in the summer-green species.

In the Mediterranean sand-dune area conditions are otherwise. The very dry and warm midsummer is the critical season here, and so about half of the species of the various areas, viz. the Therophytes, are adapted to complete their development from seed to seed before the dry season sets in. In the non-xerophytic perennating species the leaves wither in the dry season, but with the autumn rains a more or less favourable time commences again in which many species produce new leaves which survive the winter though they have no xerophytic characters. The species whose leaves are so xerophytic that they can withstand the physical drought of midsummer, will probably as a rule easily survive the moist winter even if this is in some degree a season of physiological drought. The leaf xeromorphy of the evergreen plants of the dune maquis is, therefore, primarily an adaptation to the physical drought of midsummer, whereas the leaf xeromorphy of the evergreen species of the Jutland dune heaths is necessitated by the physiological drought of winter.

In order fully to understand the occurrence of the plants in relation to environment it is necessary to know all the adaptations in the individual species and the value of the various adaptations. There can hardly be any doubt, however, that there exist adaptations, e.g. intracellular, of which we are as yet ignorant, and we are still very far from being able to determine and weigh the value to the plant of the various adaptations we know.

LITERATURE

ARCANGELI, G. (1882), Compendio della flora italiana. 1882.
ASCHERSON ET SCHWEINFURTH (1889), Illustration de la flore d'Égypte. Mém. Inst. Égypt. II.
BARBEY, W. (1885), Florae Sardoae Compendium. Lausanne.
BARBEY, W.: see STEFANI.
BARRANDON, A.: see LORET, H.
BARRATTE, G.: see DURAND, E.
BÉGUINOT, A. (1904), Saggio sulla flora e sulla fitogeografica dei Colli Euganei. Memorie della Società geografica Italiana, **11**, 1904.
—— (1905 a), Prospetto delle piante vascolari finora indicate per i Colli Euganei e per la Pianura Padovana. 1905.
—— (1905 b), La vegetazione delle Isole Ponziane e Neapolitane. Annali di Botanica, vol. iii, fasc. 3. Roma 1905.
—— (1907), La vegetazione delle Isole Liguri di Gallinaria, Bergeggi, Palmaria, Tino e Tinetto. Res Ligustica. **39**, 1907.
—— (1910), La vegetazione delle isole Tremiti e dell'isola di Pelagosa. Memorie della Società Italiana delle Scienze, Serie 3ª, 16, 1910.
BORNMÜLLER, J. (1894), Nachtrag zu 'Florula insulae Thasos'. Oesterr. Bot. Zeitschr., 1894, pp. 124, &c., 173, &c., 212, &c.
CARUANO, A.: see SOMMIER, S.
CHEVALLIER, L. (1903), Deuxième note sur la flore du Sahara. Bull. de l'herbier Boissier, 1903, pp. 669–84, 757–79.

Combres, P.: see Flahault, Ch.
Cosson, E. (1875), Plantae in Cyrenaica et Agro Tripolitano natae. Bull. Soc. Bot. France, **22**, 1875, p. 45, &c.
—— (1889), Plantae in Cyrenaica et Agro Tripolitano, anno 1875, a cl. J. Daveau lectae. Bull. Soc. Bot. France, **36**, 1889, p. 100, &c.
Daveau, J. (1876), Excursion à Malte et en Cyrénaïque. Bull. Soc. Bot. France, **23**, 1876, p. 17, &c.
Durand, E., et Barratte, G. (1910), Florae Libyae Prodromus ou catalogue raisonné des plantes de Tripolitaine. 1910.
Fitting, H. (1911), Die Wasserversorgung und die osmotischen Druckverhältnisse der Wüstenpflanzen. Zeitschr. für Botan., **3**, p. 209, &c.
Flahault, Ch., et Combres, P. (1894), Sur la flore de la Camargue et des alluvions du Rhône. Bull. Soc. Bot. France, **44**, 1894, pp. 37–58.
Forsyth Major, C. I.: see Stefani.
Gortani, Luigi e Michele (1905), Flora friulana. 1905.
Grecescu, D. (1898), Conspectus Florei Romaniei. 1898.
Grugnola, G. (1894), La vegetazione al Gran Sasso d'Italia. 1894.
Guadagno, M. (1911), Prime notizie sulla vegetazione delle Isole Sirenuce. Bulletino dell'Orto Botanico della R. Università di Napoli, tom. iii, 1911.
Gussone, J. (1854), Enumeratio plantarum vascularium in insula Inarime, &c. 1854.
Hagerup, O. (1930), Études des types biologiques de Raunkiær dans la flore autour de Tombouctou. (Det Kgl. Danske Videnskabernes Selskab. Biologiske Meddelelser, ix. 4.)
Halácsy, E. v. (1897), Florula Sporadum. Oesterr. bot. Zeitschr., 1897, pp. 60–2, 92–9.
—— (1892, 1893), Florula insulae Thasos. Oesterr. Bot. Zeitschr., 1892, p. 412, &c.; 1893, p. 22, &c.
—— (1904), Conspectus Florae Graecae. 1901–4.
Heldreich, Th. de (1877), Die Pflanzen der attischen Ebene. A. Mommsen, Griechische Jahreszeiten. Heft V, 1877.
—— (1898), Flore de l'île d'Égine. Bull. de l'herbier Boissier, **6**, 1898, p. 221, &c., p. 289, &c., p. 379, &c.
Holmboe, J. (1914), Studies on the Vegetation of Cyprus. Bergens Museums Skrifter, Ny Række, **1**, 1914.
Jahandiez, Emile (1905), Les Îles d'Hyères. 1905.
Jatta, A. (1877), Ricordo botanico del Gran Sasso d'Italia. Nuovo Giornale Bot. Ital., **9**, 1877, pp. 197–218.
Knoche, Herman (1921–3), Flora balearica, i–iv, 1921–3.
Letourneux, A. (1889), Note sur un voyage botanique à Tripoli de Barbarie. Bull. Soc. Bot. France, **36**, 1889, p. 91, &c.
Lloyd, J. (1862), Flore de l'ouest de la France. 2e édit., 1862.
Lojacono, M. (1878), Le isole Eolie e la loro vegetazione con enumerazione delle piante spontanee vascolari. Palermo 1878.
Loret, H., et Barrandon, A. (1886), Flore de Montpellier. 1886.
Margot, H., et Reuter, F. G. (1838), Essai d'une flore de l'île de Zante. 1838.
Martelli, U.: see Matteuci, D.
Matteuci, D., et Martelli, U. (1894), Da Perugia al Gran Sasso d'Italia. Nuovo Giornale Bot. Ital., Nuova serie, **1**, 1894, pp. 34–52.
Merino, R. P. Baltasar (1897), La vegetación espontanea y la temperatura en la cuenca del Miño. Tuy 1897.
Morteo, E. (1906), Florula alluvionale di un tratto del terreno Orba. Malpighia, **20**, 1906, p. 487, &c.
Morton, Fr. (1915), Pflanzengeographische Monographie der Inselgruppe Arbe, umfassend die Inseln Arbe, Dolin, S. Gregorio, Gali und Pervicchio samt den umliegenden Scoglien. Engl. Jahrb. **53**, 1915, Beiblatt, pp. 69–273.

NEGRI, G. (1905–6), La vegetazione delle colline di Crea. Accademia Reale delle Scienze di Torino, 1905–6.

PASQUALE, G. A. (1869), Flora Vesuviana o catalogo ragionato delle piante del Vesuvio confrontate con quelle dell'isola di Capri. Atti della R. Accademia delle scienze fisiche e matematiche, vol. iv. Napoli 1869.

PAULSEN, OVE (1912), Studies in the Vegetation of the Transcaspian Lowlands. Copenhagen, 1912.

POST, G. E. (1896), Flora of Syria, Palestine, and Sinai from the Taurus to Ras Muhammad, and from the Mediterranean Sea to the Syrian Desert.

REUTER, F. G.: see MARGOT, H.

REVEDIN, P. (1909), Contributo alla flora vascolare della provinzia di Ferrara. Nuovo Giornale Bot. Ital., Nuova serie, **16**, 1909, pp. 269–334.

RIKLI, M., u. RÜBEL, E. (1928), Zur Kenntnis von Flora und Vegetationsverhältnissen der Libyschen Wüste. Festschrift Hans Schinz. 1928.

RÜBEL, E.: see RIKLI, M.

SACCARDO, F. (1895), Florula del Montello. Bulletino della Società veneto-trentina di scienze naturali, tom. vi, no. 1, 1895 (Padova), pp. 5–18.

SAMPAIO, GONÇALO (1908–9), Flora vascular de Odemira. Boletim da Sociedade Broteriana, **24**, 1908–9, pp. 7–132.

SCHWEINFURTH: see ASCHERSON.

SOMMIER, S. (1902, 1903), La flora dell'Arcipelago Toscano. Nuovo Giornale Bot. Ital., Nuova serie, **9**, 1902, p. 319, &c.; **10**, 1903, p. 131, &c.

—— (1908), Le isole pelagie Lampedusa, Lionosa, Lampione e la loro flora con un elenco completo delle piante di Pantellaria. Firenze 1908.

—— (1909), La flora dell'isola di Pianosa nel mar Tirreno. Nuovo Giornale Bot. Ital., Nuova serie, **16**, 1909.

SOMMIER, S., et CARUANO, A. (1914–15), Flora melitensis nova. Bolletino del R. Orto botanico di Palermo, Nuova serie, **1**, 1914–15.

STEFANI, C. DE, FORSYTH MAJOR, C. I., et WILLIAM BARBEY (1892), Samos. Lausanne 1892.

—— (1895), Karpathos. Lausanne 1895.

TERRACINO, N. (1889–90), Synopsis plantarum vascularium Montis Pollini. Annuario del R. Instituto Bot. di Roma, Ann. IV, 1889–90.

VACCARI, A. (1894), Flora dell'Arcipelago di Maddalena. Malpighia **8**, 1894, p. 227, &c.

—— (1896), Supplemento alla flora dell'Arcipelago di Maddalena. Malpighia **10**, 1896, pp. 521–34.

—— (1899), Secondo supplemento alla flora dell'Arcipelago di Maddalena e indice alfabetico generale. Malpighia, **13**, 1899, pp. 200–10.

VAHL, M. (1904), Madeiras Vegetation. 1904.

VALLOT, J. (1885), Flore glaciale des Hautes-Pyrénées. Bull. Soc. Bot. France, **32**, 1885, pp. 133–42.

VILLANI, A. (1906, 1907), Primo contributo allo studio della flora Campobassana. Malpighia, **20**, 1906, p. 49, &c., p. 333, &c.; **21**, 1907, p. 3, &c.

WARMING, EUG. (1909), Dansk Plantevækst. 2. Klitterne.

WILLKOMM, M. (1876), Index plantarum vascularium quas in itinere vere 1873 suscepto in insulis Balearibus legit et observavit Mauritius Willkomm. Linnaea, 1876, p. 1, &c.

—— (1896), Grundzüge der Pflanzenverbreitung auf der ibirischen Halbinsel. (Engler u. Drude, Die Vegetation der Erde, i), 1896.

INDEX

A number of the less important place-names have been omitted, except where a considerable description of their vegetation occurs in the text.

Abo Skerries (Norway), Chamaephyte percentage on, 132.
Abundance, degrees of, 401.
Actinometer, 484–5.
Adaptation, *see* Life-form, Unfavourable Season.
Aden, biological spectrum of, 116, 122, 555.
Alabama (U.S.A.), 440, 442, 577.
Alaska, biological spectra of, 121, 126.
Alder swamps, 255–60.
Alder wood, 229, 251, 253, 255, 257.
— ground flora of, 256–60, 506.
Aleutian Islands (Arctic), 125, 130.
Algeria, flora of, 440, 576.
Allindelille Fredskov (Zealand), 483, 495, 527, 530.
Alluvia, Mediterranean coastal, 192, 343–67. *And see under* Coast vegetation.
Alluvial beaches, 148–50, 168, 171.
— — in Denmark, 168–72, 187.
— formations on Italian coast, 578–608.
— — on West Indian coasts, 148–94.
Alpine region, life-forms of, 36–7, 39, 135–8, 288, 302, 571.
Alps, biological spectra of, 135–8.
Altitude and proportion of Therophytes, 347.
Altitudes, biological spectra at different, 135–42, 560, 568–9, 572.
America, arctic, biological spectra of, **123–7**. *And see* Baffin Land, Boothia, Melville Island, &c.
Amgun Bureja (Amur Province), biological spectrum of, 187.
Anemone nemorosa, 206–17.
— — nitrate content of, 488–516.
Annual plants, 1, 97, 196, 344, 554, 557–8.
Antarctic Chamaephyte climate, 283, **298–302**.
— islands, biological spectra of, **298–300**.
Aosta Valley (Switzerland), biological spectrum of, 138.
Arabian Sea, Pteridophyte quotients of islands in, 440.
Arctic Chamaephyte climate, **123–34**, 283–98, 406, 432.
— zone, 118, 123–34, 137.

Arctic-nival region, 113, 134, 136.
Arenas Gordas (Spain), 360.
Argentario Island (Italy), 345, 438, 575–6.
— — biological spectrum of, 406, 428, 555–7, 567–8.
— strand formations of, 587–600.
Arran, Island of (Scotland), 442.
Arrhenius, O., 448.
Ashwood swamp, ground flora of, 539–43.
Asia, arctic coast, biological spectra of, 125, 287. *And see* Chukotski, Taimyr, &c.
Association, use of term, 308, **380**.
Atmometer, 481–3, 486.
Attica, Therophyte percentage of, 565.
Aubert, S., 401–2.
Australian alps, 442.

Baffin Land, biological spectra of, 119, 127, 406.
Balearic Isles, Therophyte percentages in, 559.
Balkan Peninsula, biological spectra of islands off, 565.
Banks Island, biological spectrum of, 126–7.
Baranetzky, 21.
Barrandon, A., 343.
Beaches, *see under* Alluvial, Sandy.
Bear Island (Arctic Ocean), biological spectrum of, 124, 128, 295.
Beechey Island (Arctic), biological spectrum of, 127–8.
Beechwood, 201, 206–7, 210–11, 213, 216–17, 220, 229, 255, 257–9, 495, 498–504, 506.
— ground flora of, 232–40, 394, 413, 460, 474–5, 480, 495–6, 498–504, 523.
— pH in, 455–6, 459–61, 463, 465, 468–70, 472–7, 480–1, 523.
Beetles, tunnelling, 157.
Biennials, 48, 97, 344, 554, 557, 558.
Biochore, Chamaephyte, 10 %, 123, 130–5, 140.
— — 20 %, 122–30, 137, 139, 140, 142.
— — 30 %, 123, 139, 140.
Biochores, **117**, 123–43.
Biological formation spectra, 237, 243, 250, 259, 263, 271–2, 280, 282, 296, 322, 545, 615.

INDEX

Biological spectrum, **115,** 117–18, 121–2, 127, 133–43, 148, 171, 180–201, 216–17, 223, 232, 234–9, 263, **283,** 302, 308–9, 328, 331–3, 335–9, 342–5, 347–9, 352–3, 357–8, 360, 362, 367, 405–7, 409, 421, 424–35, 547–50, 553–60, 563–73, 615.
— — change in, 133–4, 180–3, 186, 192–3.
— — normal, *see* Normal.
— — statistics, *see* Statistics.
— type, *see* Life-form.
Biotic changes, 609, 612.
Birchwood, 255.
Birger, S., 133, 194.
Blytt, A., 140.
Bog ('Mose'), 264–74, **322–4.**
— flora of, 231, 267–8.
Bogs, sphagnum, 244–55.
Boothia Felix, biological spectra of, 127–8.
Boreal zone, 134, 181, 184, 285.
Børgesen, F., 150, 158, 169, 176.
Bornebusch, C. H., 489, 510.
Boysen-Jensen, P., 484.
Braun, A., 283.
Brazilian Campos, life-forms in, 189–91.
Breathing roots, aerial, 155, 157–8.
Briquet, J., 399.
Brockmann-Jerosch, H., 135.
Bud-covering in Chamaephytes, 34.
— — in Hemicryptophytes, 42, 46, 53.
— — in Phanerophytes, 17, 24, 26–30.
Bud-scales, 24, 26–9, 101.
Buds, importance of protection of, 113.
— perennating, 16, 17, 19, 20, 23–4, 33–5, 39–41, 57, 61, 64, 66, 69, 74, 79, 81, 87–8, 90, 96–7, 107–10, 142, 410–11, 426–7, 553, 613.
Burkill, I. H., 140, 147.

Callmé, A., 194.
Camargue, Île de la (France) biological spectrum of, 188, 192, 194, 348, 359–62, 559.
Campos, Brazilian, life-forms in, 189–91.
Candolle, A. P. de, 283.
Carpathians, biological spectra of, 139.
Caucasus, biological spectra of, 138–9.
Cette (France), 343, 358–9.
Chamae-nanophanerophytic formations, 613.
Chamaephyte biochore, *see* Biochore.
— climate, 120, 123–34, 142–3, 283–302, 367, 406, 432.

Chamaephytes **1, 17**–19, 22, **33–9,** 42, 64, 100, 102–3, 106, 110, 114–16, 120–37, 139, 140, 142–3, 166–7, 169, 187, 194–5, 201, 221–2, 250, 265–6, 271, 279, 283–302, 306, 312–13, 317, 321, 323, 326–7, 331, 335, 337, 346, 349, 350–2, 357, 359, 360, 362, 367, 370, 373–4, 406–8, 415, 418, 422, **426–7,** 430–1, 495, 549, 555–7, 563, 567–8, 571–2, 585, 588–90, 592, 605, 607, 613–14, 617.
— active, **1, 35, 37,** 38.
— deciduous, 322, 327, 408, 409, 422.
— evergreen, 312, 322–3, 327, 362, 408–9, 415, 422.
— herbaceous, 100.
— passive, **1, 35–8.**
— suffruticose, **1,** 33, **35–6,** 39, 40, 42, 106.
Chamaephytium, 613.
Chemico-physical changes, 609, 612.
Chidley Peninsula (Labrador), biological spectrum of, 119, 127.
Chilkat-Land (Alaska), biological spectrum of, 121, 126.
Choripetalae, biological spectrum of, 434.
Chukotski Peninsula (E. Siberia), biological spectrum of, 125, 287.
Classification, system of, 305–8, 351, 371.
Clay, formations on, 218–44, 274–8.
— in alluvia, 154–6, 160–2, 170–1, 354, 362–5, 367.
Climate and life-form, 8, 12, 15, 16, 19, 22, 30, 33–6, 39, 42, 47, 50, 55, 64–6, 88, 97–106, 110–43, 148, 168, 180–94, 406–7, 550, 553–5, 576.
— and plant protection, 16, 17, 19, 22–4, 26, 29, 30, 33, 113.
— and vegetation, 8–12, 14, 111–18, 324, 439, 556.
Climates, method of comparing, 372.
Climatic changes, 608, 611.
— formations, 191.
Climax formation, 609–10.
Clova (Scotland), biological spectrum of, 122, 140–2, 187.
Coast vegetation, 148–94, 310–16, 343, 345, 348, 353–67, 547–620.
Co-dominants, **519,** 520, 523–6.
Cold temperate regions, life-forms in, 119, 133–42, 184, 553.
Cold zone, life forms in, 122–42.
Colonization, 240–1.
Colorimetric method of determining pH, 444–5, 449.
Coltsfoot, life-form of, 105–10.

INDEX

Commander Islands (Arctic), 125–6.
Comparimeter, use of, 452–3.
Comparison of Danish and Mediterranean floras, 346, 353, 360–3, 367.
—— Danish and West Indian coast floras, 166–72, 188, 191–2.
—— dunes in Mediterranean and Hemicryptophyte climates, 608–18.
—— floras on new and old ground, 180, 185–9, 189, 192.
—— Hemicryptophyte and Therophyte climates, 547.
—— Italian and Danish dunes, 578, 580–2, 608–18.
Competition, 4, 167, 181, 183, 193, 202, 379–80, 397, 408, 443, 495–6, 514, 527, 541, 570.
— effect of lack of, 162, 168, 193.
Compositae and normal spectrum, 116.
Coniferous Woods, *see* Spruce woods.
Connecticut (U.S.A.), 440, 442.
Coppermine River (Canada), biological spectrum of country near, 126, 287.
Corals on alluvial beach, 151.
Cotta, H., 199, 200.
Crozet Island (Indian Ocean), biological spectrum of, 298–9.
Cryptogams and biological spectra, 573–4.
— on dunes, 332–5.
— significance for characterizing plant climates, 435–42. *And see* Pteridophytes, Lichens, Mosses.
Cryptophytes, **I, 17–19,** 39, **64–97,** 105, 107, 110, 133, 187, 195, 251, 284–6, 288, 296, 306, 406, 430–1, 434, 567, 612–14.
Cryptophytium, 612–14.
Cultivated ground, life-forms on, 121, 180–1, 193, 570.
Cultivation, effect on flora, 4, 5, 180, 183–4, 196, 198–200, 219, 221, 246, 319, 331, 349, 570.
Cushion plants, 1, **17,** 35, 38–9.
Cyprus, 440, 564, 576.
Cyrenaica, biological spectrum of, 122, 555, 564–5.
Czapek, 37–8, 58–9.

Dalmatia, islands off, 567.
Danish West Indies, *see* West Indies.
Darwin, 4, 283.
Darwinian theory, 3, 62–3.
Death Valley (California) biological spectrum of, 120–1, 428.

Denmark, biological spectrum of, 122, 182, 187, 195, 309, 405–6, 428.
— coast of, 157, 159, 166–72, 309–16.
— flora of, 99–102, 105–7, 109, 113–14, 121, 166–72, 182–4, 188, 191–2, 199, 217–79, 346, 353, 360–3, 367, 399, 407, 438, 440, 577–82, 608–18.
— plant communities of, **217–79,** 303–42.
Depth of geophytes in soil, regulation of, 69–71, 76–9, 86–7.
— of hemicryptophytes in soil, regulation of, 40, 56–7.
Deserts, 13, 19, 97, 394.
— Sub-tropical, 13.
— Tropical, 13, 556–7.
Desiccation, protection against, 17, 26, 35, 64, 67, 130, 613.
Disko Island (Greenland), chamaephyte percentage in, 132.
Djurdjura (Algeria), biological spectrum of, 299, 302.
Dolgi Island (Russia), biological spectrum of, 295.
Dominance, area of, **517–27,** 536–7.
— determination of, 204–5.
Dominant percentage, **527,** 528–32, 534–5, 537.
— species, 204, 207–10, 213, 215, 227–8, 233, 248–9, 254, 256, 258–9, 261, 263–70, 272, 274–6, 319, 324–6, 332, 334, 379, 397, 416, 493–4, 584, 599, 615, 617.
Dominants, number of, 531–43.
— physiognomical, *see* Physiognomical.
Drought, protection against, 17, 24, 26, 32, 67, 97, 616.
Dune-heath, 328, 362, 613, 616.
Dunes, inland, 331–41.
— sand, *see* Sand dunes.
Dyrehave, the (= Dyrehaven) (Copenhagen), 384, 413, 447, 454, 470, 491, 495, 498, 500, 508, 512–13.

Ecology, **202–3,** 283, **368, 423.**
Edaphic formations, 191.
Egebjergene spruce wood (Copenhagen), 240–3.
Egesaaten (Copenhagen), soil acidity in, 463–5, 475–6.
Eggers, H. F. A., 149, 158.
Ekaterinoslav (Russia), biological spectrum of, 555.
Ellesmereland (Arctic), biological spectrum of, 119, 127.

Environment, importance of, 3–8, 202, 306–8, 372, 616.
Epigeal life-forms, 431, 553, 556.
Epiphytes, 102, 114, 430.
— phanerophytic, 32–4, 301–2.
Equator, 8, 118, 134, 143.
Equiconditional regions, **5–10,** 12, 14–16, 98–104, 201–2.
Equilibrium of flora, disturbance of, 198, 379, 611–12.
Eremitage, Plain of the (Copenhagen), 447, 454–63, 465.
Ernst, A., 197.
Euganean Hills (Italy), biological spectrum of, 344, 555, 557, 567, 570.
Europe, Central, 39, 47, 102–3, 106, 110, 192, 343, 351.
— Northern, 102, 185, 288.
— Southern, 106, 110.
Evaluation of formations, methods of, 410–13.
Evaporation, adaptations to prevent excess, 368–9.
— in woods, 481–7.
Experiments on woodland and pasture soils, 477–87.
Exposure, effect of, 235, 502–3.

Falkland Islands, biological spectrum of, 298, 300.
Fanø (Denmark), 157, 168–9, 171–2, 188, 192, 264, 271–9, 309, 327, 353, 361, 367, 525–8, 530, 532, 543, 581.
Faroes, flora of, 131, 133, 140–2, 442.
Felled beechwood, flora of, 238–40.
— wood, alteration in pH of soil, 471–3.
Ferdinandsen, C., 530.
Ferrara (Italy), biological spectrum of, 555–7, 567, 569.
Feilden, H. W., 124.
Fields, flora of, 221, 223.
Finland, Chamaephyte percentage in, 131–2.
— frequency figures for, 399, 401.
Fire, effect on heath, 388.
Fjeldmark, 296–7.
Flahault, C. and Combres, P. (Flora of Camargue), 348, 359, 364.
Floristic composition of vegetation, 202–3, 205–17, 309, 379–90 408–9.
— plant geography, 111–12.
Formation, concept of, 539.
— dominants, **527–31.**
— use of term, **308, 380, 416,** 609–10.

Formations, corresponding, 166–9, 171.
— floristic characterization of, 381–90.
— investigations of by means of leaf size, 368–78.
— statistics of, 201–82, 303–42, 355–62, 364–7, 379, 424, 584.
— study of, 425, 518.
— theory of, **202,** 203–5, 306, 407, 518.
Formosa, 437, 440, 575.
Fortun-Indelukke (Copenhagen), 448, 455–59, 461, 468–9, 471–2, 476, 495, 498, 504.
France, Southern, 192, 373, 376, 615. *See also* Mediterranean.
France, Western, Therophyte percentage in, 558.
Frank, A. B., 488, 490.
Franz Josef Land, biological spectrum of, 122, 124, 128, 295.
Frederiksvaerk (Denmark), 520, 522.
Frequencies, law of distribution of, 390–404.
Frequency, 206–7, 216, 223–4, 238–9, 253, 259, 276, 279, 318–19, 324, 332, 334, 337, 349, 355–7, 361, 365–6, 372, 381–6, 389–411, 413, 415–21, 443, 446, 454, 494, 517–43, 545–6, 589, 605, 615.
— curve, 228, 392–3, 396, 400–1, 403, 524, 531.
— degree of, as biological characteristic of plant formation, 404–9.
— density, 535.
— dominants, 493–4, 498, 500, **519–20,** 522–7, 531–2.
— percentages, 390–404, **518–**19, 524, 526–7, 532–4, 537–8, 545–6, 584–5, 590, 592, 598, 604, 606.
— relative, 103.
— sum of, **531–5,** 537–8, 542.

Gamopetalae, biological spectrum of, 434.
Ganges Delta, 440.
Garigue, 373.
Geophytes, 1, **64–90,** 100, 102, 114, 119–21, 216–17, 222–3, 226, 232, 235, 237, 239, 259, 277, 279, 311, 313, 317–18, 323, 332–3, 337–8, 357–60, 362, 408, 421–2, **426,** 430, 592, 598, 612, 615.
— bulb, 1, **66–8, 82, 84–8.**
— rhizome, 1, 66, **68–74,** 91, 119, 288, 297.
— root, 1, 66, **88.**
— root tuber, 1, 66, 78, **80–3.**
— stem tuber, 1, 66–7, **74, 76–8.**
Georgia (U.S.A.), biological spectrum of, 119, 182, 428.

INDEX

Geotropic sensitiveness, conditions affecting, 71–4, 92.
Geotropism, negative, 19–21, 34–8, 52, 55, 58–62, 68.
— positive, 92, 96.
— transverse, 35, 37–8, 68, 71.
Germany, 199, 200, 440, 577.
Glacial clay, formations on, 218–63.
— sand region, 263–74.
Glades, pH in, 474–5.
Gran Sasso (Italy), biological spectra of, 569.
Grassfields, annual flora of, 223–8.
— 8-year-old, flora of, 223–8.
Gravity, effect on plant habit, 35, 37–8, 52, 55, 58–9.
— insensitiveness to, 38, 60.
Great Bear Lake (Canada), biological spectrum of country near, 126, 287.
Great Britain, frequency figures in, 399.
Greece, flora of, 554, 576.
Greenland, biological spectra of, 122, 127–9, 287, 289–93, 297, 564.
— East, biological spectrum of, 122, 127–8, 287.
— islands near, biological spectra of, 290, 292–4, 297.
— North, biological spectrum of, 289–93.
— N.E., biological spectra of, 289–92, 297.
— N.W., biological spectra of, 289–91, 297.
— West, biological spectra of, 128, 287.
Grevillius, A. Y., 194.
'Grey Dune', 311, 325–8, 331, 360, 362, 525, 612–13.
Grindsted (Jutland), 388–9.
Gymnosperms, biological spectrum of, 434.

Haberlandt, G., 62.
Habit in trees, cause of, 19–22.
Habitat and life-form, 189–90, 443.
Hagerup, O., 556.
Halophilous shrubs, 599.
Hare Island (W. Greenland), biological spectrum of, 289–90.
Heath, 199, 201, 263–74, 303–6, 310, 312, 316–31, 352, 373–4, 385–9, 393–4, 444, 497–8, 613.
Heath-bog, **323–4.**
Heath soil, pH of, 478–9.
Hecistotherms, **6.**
Heer, O., 135, 137.
Height and mass of formation, 418–21.
Heliotropism, positive, 38, 59.
— negative, 37, 59, 60, 92, 96.

Helophytes, 1, 64, **90–6,** 102, 114, 119, 201, 279, 288, 296, 317, 321, 362, 422, **426,** 430, 492, 599.
Hemicryptophyte and Chamaephyte climate, 120, 123, **285.**
Hemicryptophyte climate, 119–21, 123, 134, 143, 181–4, 186–7, 192–5, 197, 200, **284,** 343–4, 347, 352–3, 360, 367, 406, 428, 432, 439, 549, 553–5, 557–8, 567, 569–72, 575, 577, 608–18.
Hemicryptophytes, **1,** 15, 17–19, 35, 37–8, **39–64,** 68, 70, 88, 90–1, 97, 100, 102–7, 110, 114, 119–21, 123, 130, 133, 137, 181–3, 186–90, 192–3, 195–6, 199–201, 217, 220–2, 226, 232, 237, 239, 243–4, 259, 263, 277, 279, 284, 286–8, 293–4, 296–7, 306, 311–13, 317–18, 323, 331, 333, 335, 337–8, 343–5, 347–9, 352–3, 360, 362, 364, 406–8, 421–4, 426–7, 430–1, 434, 549, 553, 557, 559–60, 567–8, 571–2, 598, 612–14.
— mesomorphic, 317.
— partial rosette, **1, 46–8,** 107.
— rosette, **1, 48–54,** 105, 107, 110, 480, 486.
— with stolons, **45–6.**
— without stolons, **42–5.**
— xeromorphic, 313, 317, 332.
Hemicryptophytium, 612–13.
Hérault (France), flora of, 343–7, 353.
Herb field, 296.
Hesselman, H., 489, 491, 510.
'Hilling', 40, 46, 57, 76, 79, 86.
Hjälmar Lake (Sweden), islands in, 194–7.
Hjelt, H., 399.
Holland, 194, 353.
Hope Island (Spitsbergen), biological spectrum of, 122, 124, 128.
Horn, Cape, 301, 442.
Humboldt, 283.
Humboldt Glacier (N. Greenland), biological spectrum of, 290, 295.
Humus, formations on, 244–63.
Hydrogen-ion concentration and depth of soil, 448–9.
— — and species density, 497.
— — influence of vegetation on, 443–87.
— — methods of determining, 449–53.
— — variation in different localities, 445–8.
Hydrophytes, **1,** 64, **96–7,** 102, 114, 119, 201, 279, 296, 317, 362, 422, 426, 430, 492.
Hydrotherm figures, **9–15,** 116–18, 134, 190, 301, 439, 547–53, 559.

INDEX

Hydrotherm figure for—
  Algiers, 552.
  Batavia (Java), 12.
  Cape Horn, 301.
  Coimbra (Portugal), 551.
  Colombo (Ceylon), 548.
  Denmark, 9, 549.
  Genoa, 551.
  Jakobshavn (Greenland), 301.
  Larnaca (Cyprus), 552.
  Mediterranean, 553.
  Nagasaki (Japan), 15.
  Naples, 14.
  South Georgia (S. Atlantic), 301.
  Sumatra, 11.
  Sweers Island (Gulf of Carpentaria), 13.
  Upernivik (Greenland), 550.
Hygrophilous communities, 321.
Hypogeophytes, **495** n., 496, 500, 504.

Iberian Peninsula, life-form percentages in, 554, 558–9.
Iceland, biological spectrum of, 122, 124, 128.
— flora of, 575.
Illumination, effect of, on soil, 463, 465–6, 468, 471, 473, 480, 487. *And see* Light.
Indian Ocean, biological spectra of islands in, 298–9.
— — Pteridophyte quotients of islands in, 440–1.
Inflorescence buds, production of, 106–10.
— — survival of, 108–9.
Inland plains, vegetation of, 316–31.
Invasion by new species, effect on biological spectrum, 180–2, 185–7, 192–4, 198–9.
— of America by European plants, 181–2.
— of dunes by heath, 328–9.
Ireland, flora of, 437–8, 440.
— frequency figures for, 399–401.
Irmisch, Th., 105–6, 108–10, 283.
Islands, newly formed, biological spectra of, 194–5.
— Pteridophyte-quotient of, 439–40. *And see under* Arctic, Antarctic, Indian Ocean, &c.
Isobiochore, **406,** 432.
Isochamaephyte lines, 130.
Isotherms, 117, 123, 130, 132, 134, 284–5, 352, 406, 428.
Italy, biological spectra of, 122, 351, 567–8, 572.
— flora of, 440, 554, 576–8.
— sand formations in, 547, 578–608.

Jaegersborg Hegn (Copenhagen), 458, 461, 468–9, 473, 491, 498, 500, 502, 508, 512.
James Bay (Labrador), biological spectrum of, 127, 187.
Jan Meyen (Arctic), biological spectrum of, 124, 128, 295.
Jensen, C., 141.
Jonstrup Vang (Copenhagen), 206, 210–17, 220, 223, 229–39, 241, 243–5, 247, 250–63.
Joux, Vallée de (Jura), 401–3.
Jungersen, H., 156.
Junkersdal (Norway), Chamaephyte percentage in, 131.
Jutland, 186, 189, 194, 199, 263, 271, 303–42, 353, 360–2, 373–5, 444, 578, 580–1, 583, 608, 611–14, 617–18.

Karelia keretina, Chamaephyte percentage in, 132.
Kerguelen Island, biological spectrum of, 298–9.
Kihlman, A. O., 399.
King Karl's Land, biological spectrum of, 122, 124.
King William's Land, biological spectrum of, 289.
Kjellman, F. R., 125.
Klein, R., 489.
Klincksieck, P., 181–2.
Kolguev Island (Russia), biological spectrum of, 124, 132.
Köppen, 117.
Kotula, B., 139.
Krakatoa Island, 194, 197–8.
Kuriles, Southern (Japan), biological spectrum of, 187.

Labrador, biological spectra of, 119, 127–8, 182.
Lagerberg, T., 397–8, 410–11, 419.
Lagoa Santa (Brazil), 190–1, 440, 442.
Lagoons, formation of, 155–8.
Lake vegetation, 599.
Lamarck, 62–4.
Lambert's Land (Arctic), 292.
Lampedusa (Mediterranean), life-form percentage of, 561.
Lancashire (England), 440, 442, 577.
Langlig (Denmark), biological spectrum of, 188–9.
Lapponia enontekiensis, 131–3.
— imatrensis, 132.
— inarensis, 132.

INDEX

Lapponia murmanica, 132.
— ponojensis, 132.
— tulomensis, 132.
Leaf size, use in biological plant geography, 368–78.
Leaves, classification of, 369–378.
— protective adaptation of, 26, 41, 616.
Leghorn (Italy), 582–3, 586.
Leptophyll, **371**, 374–5, 409.
Libyan Desert biological spectrum of, 122, 555, 564.
Lichen quotient, 437, 441–2, 577–8.
Lichens and biological spectra, 435–6, 573.
— frequency figures for, 395.
— on dunes, 332.
Lidforss, 61–2.
Life-form, determination of, 113–15, 142, 171.
— dominance, change in, 193–6, 349, 360.
— percentages, 99, 103, 105, 114–17, 119–30, 133, 137–42, 181–4, 187–90, 192, 195–8, 200, 283, 286–302, 321, 327, 331, 344–8, 357, 406, 408, 426, 431–2, 554–73.
— plant's power of changing, 34, 55, 105, 189, 553, 556.
— system of, **16-19, 112, 283, 380-1**.
Life-forms, characteristic, 14, 33, 100–5, 114–17, 119, 120, 129, 284, 317–18, 349, 407, 432, 614.
— dominant, 12, 39, 65, 100–6, 188, 280, 288, 296, 308, 311, 323, 328, 343, 353, 358, 373–4, 407, 426, 549, 553, 557, 571–2, 609, 612–13, 615.
— of plants on new soil, 148–200.
— on West Indian coasts, 178–9.
— poorly protected, 16, 32, 106, 118, 120, 167.
— protected, 17, 19, 30, 34, 64, 106, 120, 167, 177, 553.
— recessive, 192–3.
— relationship between, 11, 12, 16, 55, 189.
Light, effect on plant growth, 5, 6, 55–6, 58–60, 71–4, 92, 94, 110, 159, 235, 495, 542. *And see* Shade.
Light intensity, determination of, 484–5.
Ligurian Islands, flora of, 576.
— — life-form percentages in, 344, 560–1, 563, 567.
Linosa (Mediterranean), 561.
Lipari Islands (Italy), life-form percentages in, 561–2.
Liverworts and biological spectra, 435–6.
Liverwort quotient, 437, 441–2.

Livingston, B. E., 481.
Loret, H., 343, 345, 347.
Lundager, A., 291.
Lutken, Ch., 221.
Lyngby Bog (near Copenhagen), 248–9, 253–5.

Maaløv-Krat (Copenhagen), 221–9, 232, 244, 246–50.
Maasø (Finmark), Chamaephyte percentage in, 131.
Macrophyll, **371**.
Maddalena Islands (Sardinia), life-form percentage in, 561–2.
Madeira, biological spectrum of, 122, 558–9.
Magerø (Finmark), Chamaephyte percentage, 131.
Maige, 58–9.
Malta, flora of, 440, 560–1, 576–8.
Mangrove islands, 159–60.
— vegetation, 149–50, 153–67, 169, 171–2.
Maquis, 349–52, 359, 362, 369, 373, 378, 384.
— dune, 578, 587–91, 594, 595, 597–602, 604–5, 613, 616–18.
Maremma, 582–7, 590–1, 596–7.
Marine alluvia, vegetation of, 353–67. *And see* Coast vegetation.
Meadow, 201, 317–24, 447, 492, 506, 508.
— Grass, 260.
— Sedge, 260–1, 318–20, 322.
— Woodland, 260–3.
Mediterranean, 36, 55, 103, 192, 280, 343–67, 369, 378, 547–620.
Megaphanerophytes, 16, **32**-3, 102, 114, 117, 193, 201, 421, 430.
— deciduous, 34, 279, 421.
— evergreen, 34, 279, 421.
— — with bud-covering, 34.
— — without bud-covering, 33, 97.
Megaphyll, **371-2**.
Megatherms, **6**.
Melville Island (Arctic), biological spectrum of, 126–7.
Mentz, A., 323.
Mesophanerophytes, 16, **32**-3, 102, 106, 114, 117, 189, 193, 201, 300–2, 349, 362, 430, 608.
— deciduous, 34, 279, 370, 422, 585.
— evergreen, 34, 279, 370, 422, 585.
— — with bud-covering, 34.
— — without bud-covering, 33.

Mesophilous communities, 321.
Mesophyll, **371–2**.
Mesotherms, **6**.
Methods of investigation of flora, 280–2, 355, 382–4, 397, 399, 401, 409–13, 417–20, 429–30, 444, 517–18.
Micro-mesophylls, **372**.
Microphanerophytes, 16, **32–4**, 102, 105–6, 155, 157–8, 189, 300, 307, 349, 351, 362, 422, 495, 557, 585.
— deciduous, 34, 279, 422.
— evergreen, 34, 422.
— — with bud-covering, 34.
— — without bud-covering, 34.
— — xeromorphic, 351, 359.
Microphyll, **371–2**, 375.
Microtherms, **6**.
Minho Valley (Portugal), Therophyte percentage in, 558–9.
Mitscherlich, 481.
Molisch, H., 488–9.
Møller, A., 199.
Møller-Holst, 106.
Monocotyledons, biological spectrum of, 434.
Montello, Bosco (Italy), biological spectrum of, 570–1.
Montpellier (France), life-form percentage of, 559.
Moor, 264–5, 267, 270.
More, A. G., 399, 400, 438.
Mortensen, H., 220, 229.
Moss quotient, 437, 441–2, 575, 577.
Mosses and biological spectra, 435–6, 573.
— frequency figures for, 395.
— on dunes, 335, 337–40.
Mountain regions, vegetation of, 245, 347, 554, 559.
Muddy beaches, 153–5, 160–3.
Mutation, 3, 62–4, 104.

Nanism, 30, 32.
Nano-microphylls, **372**.
Nanophanerophytes, 1, 16, 27, 30, **32–6**, 42, 102–3, 105–6, 114–15, 120, 159, 189, 193, 271, 279, 307, 313, 321–3, 327, 349–51, 362, 367, 373, 384, 422, 430–1, 495, 556–7, 584–5, 588–9, 613.
— deciduous, 34, 422, 613.
— evergreen, 34, 351–2, 359, 422, 613.
— — with bud-covering, 34.
— — without bud-covering, 34.
Nanophanerophytic-chamaephytic formation, 367.

Nanophanerophytium, 613–14.
Nanophylls, **371–2**, 374–5, 409.
Naturalized plants and biological spectrum, 180–3.
Neapolitan Islands, life-form percentage in, 561–2.
Nemec, B., 61–2.
New Mexico, 440.
New Zealand, 101.
New Zealand, biological spectra of islands to south of, 298–9.
Nielsen, P., 106–8, 110.
Nitrate content of *Anemone nemorosa*, 488–516.
Nitrates in plants, methods of determining, 488–91, 511, 513.
Nordby salt marsh (Fanø), 168–9, 188, 192, 274, 276.
Nord-Reisen (Norway), Chamaephyte percentage in, 131.
Normal (biological) spectrum, **115–17**, 119–23, 128, 133–4, 187–9, 192, 284–7, 343–4, 348, 352, 357, 406, **425–34**, 556, 572.
North Africa, 350.
North America, biological spectra in, 118, 120, 122, 125, 181–2, 184.
Northern Hemisphere, life-forms of, 284.
Norway, biological spectra in, 131, 140.
Novaya Zemlya, biological spectrum of, 122–4, 128, 287.
Nugsuak Peninsula (W. Greenland), biological spectrum of, 290.
Nunataks, biological spectra of, 129, 292, 564.

Oakwood, ground flora of, 220–32, 462, 464–8, 480, 486, 495–6, 504–6.
— pH in, 455–6, 461–70, 480.
Obi Valley (Siberia), biological spectrum of, 187.
Oceanic Islands, Pteridophyte quotients of, 440.
Odemira (Portugal), Therophyte percentage in, 558–9.
Olsen, C., 391, 393, 395–6, 444–5, 447–8, 450, 452–4, 489, 529–30, 543.
Oltmanns, F., 37.
Orbetello (Italy), 587–8, 590–1, 594–7, 599.
Orkneys, Chamaephyte percentage in, 131.
Ørsted, A. S., 149.
Ostenfeld, C. H., 141–2, 291.
Ostrobottnia borealis, Chamaephyte percentage in, **132**.

INDEX

Pacific Ocean, biological spectra of islands in, 298–9.
— — Pteridophyte quotient of islands in, 440–1.
Palavas (S. France), 343, 354–8, 615.
Pantellaria (Mediterranean), life-form percentage in, 561.
Pasture, vegetation of, 330, 492, 506, 508.
— pH in, 455–80, 492.
Peary Land (Greenland), 292.
Pelagosa (Mediterranean), 565–6.
Penzig, O., 197.
Percentage area, see Soil cover.
— proportion between life-forms, 100, 217, 374.
— — between species, 208–13, 215, 379, 410.
Phanerophyte climate, **116–17**, 120, 143, 186–7, 189, 192, 194, 198, 200, 406, 428, 431, 549, 553, 555, 576, 609.
Phanerophytes, **1, 16–35,** 39, 41–2, 55, 64, 100–1, 113, 115–21, 133, 160, 162–4, 166–9, 172, 175, 177, 180, 184, 186–97, 199–201, 220–1, 251, 266, 284, 286–7, 296, 300–2, 306, 321, 349–52, 359, 362, 370, 373, 407, 418, 426, 427, 431, 434, 495–6, 506, 549, 553, 555–6, 567, 570, 576, 588–90, 595–7, 605, 607, 613.
— deciduous, 1, **16,** 34, 101, 317, 320, 362, 369–70, 427, 605, 610.
— evergreen, 1, **16,** 27, 30, 34, 101–2, 369–70, 373–4, 427, 609.
— herbaceous (phanerophytic herbs), **32–5,** 42, 102.
— with bud-covering, 1, 26–8, 41.
— without bud-covering, 1, 26.
— with real bud-scales, 28, 102.
Phanerophytic herbs, see Phanerophytes, herbaceous.
— epiphytes, 32–4.
— parasites, 33.
— spectrum, 199.
— stem-succulents, 32.
Phanerophytium, 614.
Phenotypes, 493.
Physiognomical dominants, 494, 496, 522–3, 525, 539.
Physiognomy of vegetation, 180, 204, 310, 349–52, 380, 409–16, 418, 421, 494, 610.
Phytobiological spectrum, 405.
Phytoclimatic regions, 101, 117, 123, 201, 216, 280, 407.
Pisa (Italy), 582–3, 586.
Plant climates, 7, 8, 99, 102–3, 105, 116–23, 134–43, 181–7, 190–2, 195, 198, 200–1, 204, 216, 283–4, 349, 375, 404–7, 415, 421–2, 425–9, 431–2, 547–53, 556, 563, 575, 610. See also under names of life-forms.
— — characterization by cryptogams, 435–42.
Plant climatology, **202–3,** 425.
Plant Geography, life-forms as a basis of, 111–47, 283, 306–7, 425–6.
Plummer, 448–9.
'Points' = degree of Valency, q.v.
Poles, the, 8, 30, 98, 118, 134–5, 143, 550.
Pollino, Monte (Italy), biological spectra on, 567, 568.
Pontine Islands (Italy), life-form percentages in, 561–2.
Pontine swamps, 603–6.
Porquerolles, island of (S. France), 348, 559.
Porsild, M., 289.
Poschiavo (Switzerland), biological spectrum of, 121–2, 135–6, 555, 557.
Precipitation curves, 9–15, 117, 548, 553.
— effect on nitrate content, 502.
Pribilov Islands (Alaska), 125–6.
Prop roots, 153, 156.
Prosperity, degree of, 418.
Protohemicryptophytes, **1,** 33, **41, 42–6,** 51, 55–6, 480.
Pteridophyte quotient, 436–41, 574–6.
Pteridophytes and biological spectra, 435–6, 438, 573–4, 576.
Pyrenees, life-form percentages in, 560.

Quadrats, 206–9, 213, 371.

Racine (Wisconsin), 438, 440, 577.
Radde, G., 138.
Rain forest, tropical, 185, 369, 393.
Ravn, F. Kolpin, 308.
Resvoll-Holmsen, H., 391, 393, 395–6.
Rhizome Geophytes, see under Geophytes.
Rhizomes, power of travelling, 69, 70.
Rhone delta, 188, 192, 194, 359–67.
— valley, 351–2.
Ringkjøbing Fjord (Denmark), 264, 353, 365.
Rock vegetation on sea-shore, 151, 153.
Rosette Hemicryptophytes, see under Hemicryptophytes.
— plants, 1, **41**–2, 46–56, 105–8.
— — partial, 1, **46**–8, 55–6, 107.
Russia, Arctic, biological spectra of, 124–5.
And see Siberia.

Saccardo, F., 570.
Saelan, T., 399.
Sahara, biological spectra of, 122, 555–6, 560.
St. Croix (West Indies), 153–80, 186–9, 191–2, 194.
St. Domingo, 148.
St. Jan (West Indies), 101–2, 113–17, 119, 121, 148–57, 187–9, 428.
St. Lawrence Island (Alaska), biological spectrum of, 121, 125–6, 132–3, 428.
St. Thomas (West Indies), 101–2, 113–17, 119, 121, 148–50, 153, 155–7, 187–9, 428.
Saintes-Maries (S. France), 343, 354, 359, 363–4.
Saline formation, 274–9.
Salt marsh, biological spectrum of, 278.
Salt marshes, 167–8, 192, 274–9, 444.
Salt pond (panne), **149**, 155–6.
Samos (Greece), biological spectrum of, 122.
— Pteridophyte quotient of, 440.
Sample plots, 223, 235, 237–8, 243, 256, 261, 263, 297, 355, 383–4, 397, 544–6. *And see* Unit areas.
Sand dunes, 157, 189, 365, 578–82, 592–3, 595–7, 601–7.
— — vegetation of, 174, 193–4, 271–2, 309–16, 354–62, 294, 525–6, 583–5, 588–92, 595–7, 600, 602–7.
— grains, size of, 579–80.
Sandhills, 593–4.
Sandy alluvia, 348, 353–4, 364, 367.
— beach, 150, 153–7, 159, 172–7, 359, 583, 588, 606.
Sardinia, flora of, 575–8.
Savanna, 12, 189–91.
Schimper, A. F. W., 7, 191, 488–9.
Schröter, C., 137.
Schwendener, S., 283.
Scilly Isles, Chamaephyte percentage in, 131.
Scrub, deciduous, 320–2, 331.
— evergreen, 372–3.
— flora of, 221–3.
— nano-microphanerophytic, 362, 608.
— nanophanerophytic, 321, 332, 337–41, 590.
— nanophanerophytic-chamaephytic, 321.
— on sea-shore, 158, 160, 162, 165, 168, 174–5, 177, 313–15, 589, 593, 599.
Seedlings, observations on, 106–10.
Seychelle Islands (Indian Ocean), biological spectrum of, 114–17, 122, 406, 428, 440–1.
Shade in woods, effect on flora, 231, 233, 235, 238, 240–1, 243, 260, 296–7, 513, 541–2.
Shetlands, Chamaephyte percentage in, 131.

Skagens Odde (Denmark), plant formations of, 303–42, 361.
Skallingen (Denmark), 171–2, 186–9, 194, 309, 353.
Skaw, The, *see* Skagens Odde.
Siberia, 124, 131–3, 185–6.
— New, biological spectrum of, 125.
— West, biological spectrum of, 124, 132.
Sierra Nevada (Spain), biological spectrum of, 299, 302, 559–60.
Simmons, H. G., 146, 289–91.
Sitka (Alaska), biological spectrum of, 121, 126, 187.
Society Islands, 441, 577.
Soil acidity, *see* Hydrogen-ion concentration.
— changes in, 609, 611.
— cover, use of degree of, **409–16**, 419–20.
— effect on life-form relationships, 189–90, 192.
— effect of vegetation types on pH of, **443–87**.
— humidity and distribution of vegetation, 159, 218, 272–3, 309, 324, 327, 444.
— — and species density, 497–8.
— plants on new, 148–200, 613.
— reaction, *see* Hydrogen-ion concentration.
Sørensen, S. P. L., 444.
South America, biological spectra of islands near, 298, 300.
South Georgia (S. Atlantic) biological spectrum of, 298–9.
Spain, 343, 353–4, 360, 362, 373, 400, 576.
Species constants, 494.
Species density, 239, **261–3**, 279, 328, 331, 334–5, 337, **494–8**, 504, 506, 517–19, 522–3, 525, 527, 531–44.
Species, discontinuity of, 355.
Spectrum, biological, *see* Biological spectrum.
— normal, *see* Normal spectrum.
— transitional, 428.
Spitsbergen, biological spectrum of, 122, 124, 287, 428.
Spruce woods, 220, 229, 235, 240–5, 250–3, 448–9, 455.
— — ground flora of, 457–8.
— — pH in, 455–8, 461, 468–9, 471–2, 480–1.
Stahl, E., 488.
Stampeskov (Copenhagen), 460, 465–9, 470, 473, 475, 486, 498, 503–4.
Starch grains, influence on root's reaction to gravity, 61–2.
Staten Island (Tierra del Fuego), biological spectrum of, 298, 300.

Statistics of leaf size, 374–7.
— of life-forms, 99–104, 111–43, 148–200, 287–302, 309, 344–8, 428, 554–73, 615.
— of plant formations, 201–82, 303–42, 356, 358–62, 364–7, 379–423, 444, 527–46, 585, 589, 594, 598, 604–5, 607.
Stem-succulents, 33–4, 102, 114–15, 120, 430.
Steppe, herbaceous, 318.
Steppes, 14, 17, 19, 65, 97, 327, 331.
Stirling, J., 442.
Stolons, Cryptophytes with, 78, 86–7, 91, 96, 106.
— Hemicryptophytes with, 45, 50, 52–3, 57–9, 106.
— possible origin of, 60–2.
— rosette plants with, 50, 52–3, 106.
Stony plains, vegetation of, 341–2.
Storage organs in Geophytes, 65–7, 74, 77, 79, 80, 84, 87–8.
Stuttgart (Germany), biological spectrum of, 122, 187, 345.
— Pteridophyte quotient of, 440, 577.
Sub-tropical regions, life-forms of, 13, 14, 30, 120, 143.
Succession, 608.
Swamp vegetation, 582, 586–7, 592–6.
Sweden, Central, 197.
Switzerland, nival regions of, 135–7.
Syria, life-form percentage in, 564.
— Pteridophyte quotient of, 576.

Taimyr Peninsula (Siberia), biological spectrum of, 125.
Temperate regions, life-forms of, 23, 36, 47.
Temperature and life-forms, 6, 7, 16, 22–3, 191.
— curves, 9–15, 117, 300–2, 548, 553.
— effect on plant's reaction to gravity, 61–2. 343–5, 347–9, 352, 406, 439, 547–78, 614.
Therophyte climate, **97,** 120–1, 143, 194, 343–5, 347–9, 352, 406, 439, 547–78, 614.
Therophytes, **1, 19,** 39, **97–8,** 100, 106, 114, 119–21, 133, 167, 180–2, 187, 192–3, 195–6, 222–3, 226, 277, 279, 284, 286–7, 296, 300, 302, 306, 331, 343–9, 352, 360, 362, 364, 367, 375, 421–2, 427, 430–1, 550, 553–78, 594, 598, 612, 614, 617.
Therophytic-hemicryptophytic formation, 364.
Tidal belt in W. Indies, 166–8, 171.
Timbuctoo, biological spectrum of, 555–6.
Tombolo (N. Italy), 582–3, 586.
— di Feniglia, 353, 578–9, 587–8, 590, 592, 594, 599.

Tombolo della Gianella, 353, 587–9.
Transcaspian Lowlands (Russia), biological spectrum of, 555, 564.
Transition between Chamaephytes and Hemicryptophytes, 35, 37–8, 130, 614.
— — Chamaephytes and Phanerophytes, 17, 22, 33, 35.
— — Chamaephytes and Therophytes, 553, 556.
— — Nanophanerophytes and Suffruticose Chamaephytes, 35.
— — passive and active Chamaephytes, 38.
— — Therophyte and Hemicryptophyte climates, 571.
Transitional zone between heath and meadow, 319–22.
Tremiti Islands (Italy), life-form percentage in, 565–6.
Treub, M., 197.
Tripoli, biological spectrum of, 122, 564–5.
— Pteridophyte quotient of, 440, 576.
Tristan da Cunha, biological spectrum of, 298–9.
Tromsø (Norway), 131.
Tropical regions, life-forms of, 6, 12, 15, 16, 30, 32–3, 100, 115–18, 143, 148, 609.
Tundra, 296–7.
Tunis, 354, 362, 440, 576.
Tuscan Islands, biological spectrum of, 121–2, 560, 562, 577.
Tuscan Maremma, 582–7.
Tussilago farfarus, life-form of, 105–10.

Underground stems, 39, 40, 64–97.
Unfavourable season, adaptation to survive, 8, 9, 15, 17, 22, 24, 29, 30, 32–4, 51, 61, 96–7, 112–14, 142, 148, 204, 283, 416, 421, 425–6, 547–8, 553, 556, 617–18.
— — buds surviving, 17, 19, 33–5, 40, 42, 45, 57, 64, 112–13, 426.
— — shoots surviving, 17, 19, 33–5, 40, 42, 45, 57, 64, 74, 79, 88, 96–7, 106–10, 112, 614.
— soil conditions, effect on life-form, 189, 192–3, 367.
Units of area for investigation of vegetation, **206–17,** 220, 223, 235, 297–8, 355, 382–4, 397–9, 410–13, 417–18, 518, 544–6.
— — apparatus for delimiting, 206, 209, 213, 355, 384, 411–12, 419.

Vaccari, L., 138.
Vahl, M., 391, 393.

Valders (Norway), biological spectra of, 140.
Valencies of particular species, 224–6.
Valency, 204, **206**–13, 215, 217, 235, 280, 296, 373–4, 382, 385, 389–90, 407–10, 417–18, 517–18, 545.
— method, **206**, 208, 217, 296, 307, 372–3, 382, 397, 410, 417, 517.
Varde (Jutland), 385–7.
Vardø (Norway), life form percentage in, 122, 124, 131.
Vaygach (Arctic Russia), biological spectrum of, 123–4, 132, 285, 287.
Vegetation types, influence on pH of soil, 443–87.
— uniformity of, 8, 10, 201.
Vegetative reproduction, 37, 39, 40, 46, 57, 78, 87.
Virgin Islands (W. Indies), 105–6.
Vöchting, 61–2.
Vries, H. de, 3, 59, 63.

Warming, E., 108–9, 150, 184–6, 188–91, 199, 283, 296, 308, 311–12, 337, 360, 434, 581, 612.
— criticism of life-form theory, 184–6, 188–91, 199.
Water and plant life, 7, 368, 616–17.
Water-table, 260–1, 272–3, 309, 310, 316, 444.
Watson, H. C., 147, 399.
Weis, Fr., 511–12.
West Indies, Danish, 101, 105, 113, 148–80, 187–8, 191, 194, 199, 369. *And see* St. Croix, St. Jan, St. Thomas.
Wiesner, J., 59.
Willis, J. C., 140, 147.
Winter, *see* Unfavourable Season.
Woodland, 184–5, 189–91, 197–9, 306, 318, 321, 397, 447, 454, 543, 598, 607, 610. *And see* Alder-, Beech-, Oak-, Spruce-wood, &c.
— soil, 454–81, 502, 511–13.
— — pH of, 454–81, 523.
Wood, Maremma, 585–7, 590–1, 596–7.
Wood margin flora, 236.

Xeromorphy, 375, 422, 616–18.
Xerophilous vegetation in W. Indies, 155, 162.
Xerophily, 4, 24, 32, 368.

Yakutat Bay (Alaska), biological spectrum of, 126.
Yalmal Peninsula (Siberia), 125.
Yenisei, R. (Siberia), 125, 186–7.
York, Cape (N. Greenland), biological spectrum of, 290, 295.
Yugor (Arctic Russia), biological spectrum of, 124, 132.

Zealand (Denmark), island of, 524.
Zonation of vegetation, 229, 272–3, 312–19, 354–5, 358, 360, 365, 444
— — on seashore, 155, 160–4, 172, 176, 583–4, 588–9, 591, 599–601, 605.

HISTORY OF ECOLOGY
An Arno Press Collection

Abbe, Cleveland. **A First Report on the Relations Between Climates and Crops.** 1905

Adams, Charles C. **Guide to the Study of Animal Ecology.** 1913

American Plant Ecology, 1897-1917. 1977

Browne, Charles A[lbert]. **A Source Book of Agricultural Chemistry.** 1944

Buffon, [Georges-Louis Leclerc]. **Selections from Natural History, General and Particular, 1780-1785.** Two volumes. 1977

Chapman, Royal N. **Animal Ecology.** 1931

Clements, Frederic E[dward], John E. Weaver and Herbert C. Hanson. **Plant Competition.** 1929

Clements, Frederic Edward. **Research Methods in Ecology.** 1905

Conard, Henry S. **The Background of Plant Ecology.** 1951

Derham, W[illiam]. **Physico-Theology.** 1716

Drude, Oscar. **Handbuch der Pflanzengeographie.** 1890

Early Marine Ecology. 1977

Ecological Investigations of Stephen Alfred Forbes. 1977

Ecological Phytogeography in the Nineteenth Century. 1977

Ecological Studies on Insect Parasitism. 1977

Espinas, Alfred [Victor]. **Des Sociétés Animales.** 1878

Fernow, B[ernhard] E., M. W. Harrington, Cleveland Abbe and George E. Curtis. **Forest Influences.** 1893

Forbes, Edw[ard] and Robert Godwin-Austen. **The Natural History of the European Seas.** 1859

Forbush, Edward H[owe] and Charles H. Fernald. **The Gypsy Moth.** 1896

Forel, F[rançois] A[lphonse]. **La Faune Profonde Des Lacs Suisses.** 1884

Forel, F[rançois] A[lphonse]. **Handbuch der Seenkunde.** 1901

Henfrey, Arthur. **The Vegetation of Europe, Its Conditions and Causes.** 1852

Herrick, Francis Hobart. **Natural History of the American Lobster.** 1911

History of American Ecology. 1977

Howard, L[eland] O[ssian] and W[illiam] F. Fiske. **The Importation into the United States of the Parasites of the Gipsy Moth and the Brown-Tail Moth.** 1911

Humboldt, Al[exander von] and A[imé] Bonpland. **Essai sur la Géographie des Plantes.** 1807

Johnstone, James. **Conditions of Life in the Sea.** 1908

Judd, Sylvester D. **Birds of a Maryland Farm.** 1902

Kofoid, C[harles] A. **The Plankton of the Illinois River, 1894-1899.** 1903

Leeuwenhoek, Antony van. **The Select Works of Antony van Leeuwenhoek.** 1798-99/1807

Limnology in Wisconsin. 1977

Linnaeus, Carl. **Miscellaneous Tracts Relating to Natural History, Husbandry and Physick.** 1762

Linnaeus, Carl. **Select Dissertations from the Amoenitates Academicae.** 1781

Meyen, F[ranz] J[ulius] F. **Outlines of the Geography of Plants.** 1846

Mills, Harlow B. **A Century of Biological Research.** 1958

Müller, Hermann. **The Fertilisation of Flowers.** 1883

Murray, John. Selections from *Report on the Scientific Results of the Voyage of H.M.S. Challenger During the Years 1872-76.* 1895

Murray, John and Laurence Pullar. **Bathymetrical Survey of the Scottish Fresh-Water Lochs.** Volume one. 1910

Packard, A[lpheus] S. **The Cave Fauna of North America.** 1888

Pearl, Raymond. **The Biology of Population Growth.** 1925

Phytopathological Classics of the Eighteenth Century. 1977

Phytopathological Classics of the Nineteenth Century. 1977

Pound, Roscoe and Frederic E. Clements. **The Phytogeography of Nebraska.** 1900

Raunkiaer, Christen. **The Life Forms of Plants and Statistical Plant Geography.** 1934

Ray, John. **The Wisdom of God Manifested in the Works of the Creation.** 1717

Réaumur, René Antoine Ferchault de. **The Natural History of Ants.** 1926

Semper, Karl. **Animal Life As Affected by the Natural Conditions of Existence.** 1881

Shelford, Victor E. **Animal Communities in Temperate America.** 1937

Warming Eug[enius]. **Oecology of Plants.** 1909

Watson, Hewett Cottrell. **Selections from *Cybele Britannica.*** 1847/1859

Whetzel, Herbert Hice. **An Outline of the History of Phytopathology.** 1918

Whittaker, Robert H. **Classification of Natural Communities.** 1962

DATE DUE

DEMCO 38-297